D0078848

Introduction to Population Genetics

Richard Halliburton
Western Connecticut State University

PEARSON

Prentice Hall

Upper Saddle River, NJ 07458

Library of Congress Cataloging-in-Publication Data

Halliburton, Richard
    Introduction to population genetics / Richard Halliburton.
        p. cm.
    Includes bibliographical references (p.).
    ISBN 0-13-016380-5
        1. Population genetics. I. Title.

QH455.H34 2004
576.58—dc22   2003060113

Executive Editor: *Teresa Ryu Chung*
Assistant Editor: *Colleen Lee*
Editor in Chief: *John Challice*
Assistant Managing Editor, Science: *Beth Sweeten*
Production Editor: *Donna King*
Buyer: *Alan Fischer*
Manufacturing Manager: *Trudy Pisciotti*
Senior Marketing Manager: *Shari Meffert*
Art Editor: *Jess Einsig*
Managing Editor, Audio/Visual Assets: *Patty Burns*
Art Studio: *Progressive Publishing Alternatives*
Art Director: *Jayne Conte*
Cover Designer: *Suzanne Behnke*
Cover Image: *Photo Researchers/Science Photo*
Editorial Assistant: *Mary Burket*

© 2004 by Pearson Education, Inc.
Pearson Prentice Hall
Pearson Education, Inc.
Upper Saddle River, NJ 07458

All rights reserved. No part of this book may be reproduced, in any form or by any means, without permission in writing from the publisher.

Pearson Prentice Hall® is a trademark of Pearson Education, Inc.

Printed in the United States of America

10 9 8 7 6 5 4 3 2 1

ISBN 0-13-016380-5

Pearson Education Ltd., *London*
Pearson Education Australia Pty. Ltd., *Sydney*
Pearson Education Singapore, Pte. Ltd.
Pearson Education North Asia Ltd., *Hong Kong*
Pearson Education Canada, Inc., *Toronto*
Pearson Educación de Mexico, S.A. de C.V.
Pearson Education—Japan, *Tokyo*
Pearson Education Malaysia, Pte. Ltd.

Brief Contents

Contents

Preface

Population genetics is increasingly relevant to real-world problems such as mapping of genes associated with human diseases, conservation of endangered species, and antibiotic and drug resistance. This book is an attempt to explain the principles of population genetics to biology students, most of whom will not become population geneticists. I have tried to show the applications of population genetics by using real-world examples from a variety of disciplines, including ecology, evolutionary biology, conservation biology, molecular biology, medicine, human genetics, and epidemiology.

Students bring a wide range of interests and backgrounds to a population genetics course. Some of the existing texts are elementary, designed to provide only a basic background; others are advanced, intended primarily for graduate students and specialists. This book tries to steer a middle ground. It is intended for advanced undergraduates and beginning graduate students who have taken a course in genetics and have some background in probability and statistics. The mathematics is not particularly advanced. A good working knowledge of algebra and familiarity with logarithmic and exponential functions are assumed. Calculus and linear algebra are used only occasionally. Basic probability theory is used extensively; Appendix A provides a review of the probability theory used in the book.

Mathematical models are an essential part of population genetics, and students often have trouble with them. Three features of the book are intended to help students understand and appreciate mathematical models in population genetics. First, they are presented as "what-if" exercises—as hypotheses that can be tested and modified by looking at real populations, as opposed to unrealistic mathematical descriptions of some hypothetical population. Second, I have emphasized the biological rationale and assumptions of the models and have tried to explain their biological implications in nonmathematical language. Third, I have tried to explain the equations and the steps in the derivations carefully, without sacrificing (much) mathematical rigor. Secondary mathematical details are often segregated into boxes, which can be read or skipped, as desired. My experience has been that students can do most of the math in this book if they are led through the steps slowly and carefully.

An important feature of this book is the emphasis on using a spreadsheet as a learning tool. I believe that one of the best ways to learn population genetics is to simulate models and analyze data with a spreadsheet. Boxes throughout the text provide detailed examples of data analysis, which students can replicate on their own computers and use as guides to working the end of chapter problems. Many problems require students to analyze models with spreadsheet simulations.

The organization of the book is mostly traditional. Chapter 1 is an overview of population genetics, and a gentle introduction to mathematical modeling. Chapter 2 introduces the ideas of genotype frequency, allele frequency, and observed heterozygosity. It then provides an extensive overview of the different kinds of genetic variation found in natural populations, including various kinds of molecular markers. These sections can be read at the beginning of a course, or referred to as needed.

The heart of the book is Chapters 3 through 10. Chapter 3 is an introduction to the Hardy-Weinberg principle and the idea of expected heterozygosity. F is introduced in its most general meaning, as a measure of the difference between the expected and observed heterozygosity. Chapter 4 introduces linkage and gametic disequilibrium and the approach to equilibrium, along with some applications. Sections 4.4 and 4.5 introduce the principles of human gene mapping and disease diagnosis based on linkage disequilibrium. I cover these topics because my students are interested in them; however, they can be skipped without loss of continuity. Chapters 5 through 9 cover the basic evolutionary processes of population genetics: natural selection, mutation, genetic drift, nonrandom mating, and gene flow. Each is introduced as a process acting alone, then interactions with other processes are discussed. Many examples, both conventional and molecular, are described.

Molecular population genetics and the neutral theory are discussed throughout the text. Chapter 10 is a detailed discussion of the neutral theory and its predictions, relevant observations from experimental and natural populations, and ways to detect natural selection on protein and DNA sequences.

The last three chapters contain material not usually covered in a first course on population genetics. They are included here as introductions to advanced topics for students and instructors who wish to pursue them. Chapter 11 is an introduction to methods of molecular phylogenetics. Chapter 12 reviews various kinds of natural selection beyond the simple one-locus, two-allele model. Chapter 13 is an introduction to quantitative genetics, with some emphasis on evolutionary quantitative genetics. Only a few sections in these chapters should be considered essential; for example, most instructors will want to cover Section 11.3 (Gene genealogies and coalescence) and 12.9 (Sexual selection). Students with some prior knowledge of quantitative genetics will want to read Section 13.6 (Evolutionary quantitative genetics). Instructors may wish to cover other sections depending on the interests of their class.

The problems at the end of each chapter include simple derivations, applications of formulas, data analysis, and simulation exercises. Many of them require students to investigate models or analyze data with a spreadsheet. Several require written analysis or short essays.

Acknowledgments

Much of the first draft of this book was written during a sabbatical leave at the University of California at Davis. I thank the Section on Evolution and Ecology for hosting me, and Western Connecticut State University for granting me the leave.

Several professors at UC Davis patiently answered my many questions, probably wondering how someone who was supposedly writing a textbook could ask such naïve questions. I especially thank John Gillespie, Timothy Prout, and Michael Turelli for their help. Tim Prout initially suggested that I write the book. In addition to answering many questions and providing much encouragement, he read several chapters and offered his usual insightful comments.

In addition to the reviewers listed below, several people have read one or more chapters of the manuscript and offered useful suggestions. They include John Briggs, John Gillespie, Josephine Hamer, Tom Philbrick, Tim Prout, Howard Shellhammer, and Michael Turelli. I have probably forgotten someone; if so, I apologize. Ansel Halliburton saved me hours of tedium by programming the Literature Cited database for the correct format.

I have heard horror stories about authors' relationships with publishers. I am pleased to say that I have had none of those experiences with Prentice Hall. Teresa Chung was patient beyond reason with my slow writing, and applied just the right amount of pressure to keep me on track. Mary Burket obtained the necessary permissions to reproduce previously published material and saved me many hours. Donna King of Progressive Publishing Alternatives answered my many questions about the publishing process. They have all been unfailingly pleasant to work with.

Finally, I thank my students, for putting up with early drafts of the manuscript instead of a legitimate textbook, and for continuing to ask hard questions.

No book of this size is likely to be without errors. I would appreciate readers calling my attention to any they find.

Richard Halliburton
Western Connecticut State University
halliburtonr@wcsu.edu

Reviewers

Prentice Hall has commissioned numerous reviewers to evaluate the manuscript. These reviewers have been extraordinarily helpful; I am embarrassed to think what the book would have been like without their help. Many are listed below; others have asked to remain anonymous. I am very grateful to all of them.

Mary Ashley, *University of Illinois–Chicago*

Michael Blouin, *Oregon State University*

Butch Brodie, *Indiana University*

Ashley Carter, *Yale University*

Deborah Charlesworth, *University of Edinburgh*

Keith Crandall, *Brigham Young University*

Cheryld Emmons, *Alfred University*

John A. Endler, *University of California–Santa Barbara*

Bryan Epperson, *Michigan State University*

Michael Gilpin, *University of California–San Diego*

Dan Graur, *Tel Aviv University*

Wade Hazel, *DePauw University*

Donal Hickey, *University of Ottawa*

Guy Hoelzer, *University of Nevada–Reno*

David Houle, *Florida State University*

Norman Johnson, *University of Massachusetts–Amherst*

J. Spencer Johnston, *Texas A&M University*

Thomas Kocher, *University of New Hampshire*

Robert Krebs, *Cleveland State University*

Kate Lajtha, *Oregon State University*

John H. McDonald, *University of Delaware*

Howard Meyer, *Oregon State University*

William Morris, *Duke University*

Michael Nachman, *University of Arizona*

David Notter, *Virginia Polytechnic Institute and State University*

Sergey Nuzhdin, *University of California–Davis*

Dennis Nyberg, *University of Illinois–Chicago*

Maria Orive, *University of Kansas*

Andy Peters, *University of British Columbia*

Timothy Prout, *University of California–Davis*

Joseph Quattro, *University of South Carolina*

Leslie Real, *Emory University*

Bonnie Ripley, *University of San Diego*

Malcolm Schug, *University of North Carolina–Greensboro*

Eric Scully, *Towson University*

Wendy Sera, *Baylor University*

Montgomery Slatkin, *University of California–Berkeley*

Ken Spitze, *University of Miami*

Lori Stevens, *University of Vermont*

Alan Stiven, *University of North Carolina–Chapel Hill*

Daniel Thompson, *University of Nevada–Las Vegas*

David Tonkyn, *Clemson University*

Marcy Uyenoyama, *Duke University*

Ronald Van Den Bussche, *Oklahoma State University*

Stuart Wagenius, *University of Minnesota*

John Wakeley, *Harvard University*

Bruce Walsh, *University of Arizona*

Michael Whitlock, *University of British Columbia*

Claire Williams, *Texas A&M University*

Jason Wolf, *University of Tennessee–Knoxville*

Paul Wolf, *Utah State University*

Clifford Zeyl, *Wake Forest University*

Jianzhi Zhang, *University of Michigan*

Introduction

*Population biology, at that time [the 1960s], was still a new label
for a new combination of scientific concerns and perspectives. It had
subsumed a number of disciplines that, throughout earlier decades,
earlier centuries, were generally treated as distinct: evolutionary biology,
taxonomy, biogeography, ecology, the demography of plant and
animal species, genetics.*

—*David Quammen (1996)*

Population genetics is frequently considered an abstract and theoretical subject, of no relevance to the real world. The truth is that population genetics is
an increasingly important component of many areas of mainstream biology.
Some examples include:

Plant and animal breeding

Pesticide and herbicide resistance in agriculture

Infectious disease and drug resistance in bacteria and other human pathogens

DNA fingerprinting and forensic genetics

Habitat fragmentation and preservation of endangered species

Mechanisms of genetic differentiation and speciation

Applications and implications of the Human Genome Project

Identification of genes causing susceptibility to complex human diseases

Genetic counseling

Human evolution and origins of modern humans

Effects of releasing genetically modified organisms into the environment

Population genetics has much to contribute to these and many other issues in
biology. We will touch on these topics, and others, as we study the conceptual
and empirical aspects of population genetics. We will see that population genetics is indeed theoretical (although not entirely so), but that modern molecular

and statistical techniques provide unprecedented opportunities to test these theories and apply them to real-world problems. These theories can help us minimize the spread of some infectious diseases, convict criminals or free innocent persons, identify individuals who may be susceptible to certain diseases, and help to save endangered species on the brink of extinction. This is an exciting time to be studying population genetics!

1.1 Population Biology

Before we begin our study of population genetics, we should first consider the nature of a population, and how population genetics fits into the broader discipline of population biology. Population biology consists of two artificially separate disciplines, population ecology and population genetics. **Population ecology** considers the factors that determine population size, such as birth rates, death rates, intraspecific and interspecific competition, predation, and parasitism. **Population genetics** considers the factors that determine the evolution of a population, such as natural selection, genetic drift, mutation, recombination, and gene flow. The two disciplines are usually studied separately, at least initially, but are really very closely related. For example, the genetic composition of individuals in a population determines that population's growth rate; the effect of natural selection sometimes depends on population size. Even though the subject of this book is population genetics, we shall often see how ecological factors affect the genetics of populations.

Our first job is to define what we mean by a population. E. O. Wilson (1992) in *The Diversity of Life* defined a population as any group of organisms belonging to the same species at the same time and place. Roughgarden (1979) defined it as "a collection of organisms that we have lumped together because we believe they function together as a unit." Ayala (1982) wrote, "A population is a community of individuals linked by bonds of mating and parenthood; in other words, a population is a community of individuals of the same species."

Many other biologists have defined populations in slightly different ways. Without belaboring the point, we can say that certain ideas are common to all of these definitions:

- A population consists of members of the same species.
- Populations are geographically continuous (localized).
- Any individual can potentially mate with any other individual in the population.
- A population can be studied as a *unit*; for example, you can study the size of a population, or the frequency of a particular allele in that population.

Given these attributes of biological populations, we can make several statements about the general characteristics of most populations. These are based on many decades of research:

- Individuals within a population are (usually) genetically different from one another.
- Different populations of the same species can differ from one another. For example, house sparrows in New England differ slightly in size and coloration from house sparrows in England or California.

- Populations can change through time. Living armadillos are different from fossil armadillos. Peppered moths today are different from peppered moths in 1850, or even 1950.

The goal of population biology is to explain these, and many other, observations. It uses the tools of genetics, ecology, paleontology, biochemistry, molecular biology, mathematics, probability theory, computer simulation, and others. To provide an idea of the diversity of subjects that population biology addresses, we list a few questions and examples:

- How is the number of individuals in a population regulated? Why are some populations relatively stable year after year, while others experience erratic outbreaks and crashes?
- Why do some animal populations cycle more or less regularly? The Canada lynx and snowshoe hare are known to have undergone cyclic fluctuations in population size for many decades.
- Why are some species rare and others common? The passenger pigeon was once the most abundant bird on the planet, but the ivory billed woodpecker was always rare.
- What are the causes of extinction? Both of the birds mentioned above are extinct. Why did the most abundant bird in the world go extinct in a period of 20 years or so?
- How do populations change over time, and what causes the change? Why are modern peppered moths and armadillos different from their ancestors? Why do certain *Drosophila pseudoobscura* populations undergo cyclic changes in the frequency of specific chromosomal inversions.
- Why do some species appear *not* to change much over long periods of time? The hermit crab has existed, virtually unchanged in external anatomy for 200 million years.
- Why are some species geographically widespread, and others very localized? The house sparrow is found essentially worldwide; Kirtland's warbler is found only in one small forest in Michigan.
- How does competition among different species affect their population sizes and evolution? How can several different species of warblers inhabit the same forest?
- How do predation and parasitism affect the size and evolution of a population?
- How much genetic variation is there in natural populations, and how does the amount of genetic variation affect the rate of evolution? Cheetahs have very little genetic variation, and are endangered. Are these two facts related?
- What is the best way to preserve an endangered species, or subspecies? Consider the controversies over the spotted owl and the Mt. Graham red squirrel. How can we save certain Galapagos tortoise populations from extinction?
- Why do some introduced species cause no problems in their new habitats, while others become serious pests and endanger native species. Cheatgrass is an introduced grass that grows rapidly, reproduces prolifically, and chokes out native grasses; thousands of acres in the western United States consist of nothing but cheatgrass. Can population biology help us understand and prevent such invasions?

- How can we increase the quantity and quality of products from agriculturally important species such as dairy cattle, wheat, rice, and so forth?
- Why do some infectious diseases spread to epidemic proportions, while others do not? Compare the effects of AIDS, Ebola, herpes, influenza, and common cold viruses. What is the best strategy for controlling these diseases?
- Why have so many strains of bacteria become resistant to antibiotics, and what can we do to prevent or minimize this serious health problem?
- Why are some species clearly distinct from their close relatives, while others are almost imperceptibly different? Certain species of *Drosophila* are indistinguishable to humans, but do not interbreed, while others differ dramatically in morphology.
- What are the causes of speciation, the splitting of a single species into two?
- How do genetic and environmental factors interact to control complex characteristics such as survival, body size, growth rate, or human intelligence?
- How genetically different are human racial or ethnic groups, and why does it matter, or not matter?

Most of these questions involve both genetic and ecological issues. We might summarize the main question of population biology as follows: How do genetics and ecology interact to determine the size of a population and its distribution and evolution? In this book we will concentrate on the genetic aspects, but we shall often see how ecological factors affect the genetics and evolution of populations.

Nearly all biologists agree that the theory of evolution is the unifying theory of biology, even if they do not study evolution directly. Theodosius Dobzhansky, one of the great evolutionary biologists of the twentieth century, expressed this idea in his famous statement, "Nothing in biology makes sense except in the light of evolution" (Dobzhansky 1973). Population biology is nothing less than an attempt to understand the mechanisms of evolution. Population biology aims to go beyond description, beyond the cataloging of facts, to understanding and explanation. The changes in frequencies of melanic moths in Great Britain have been well documented. Population biology attempts to explain *why* these changes have occurred. And with understanding comes the possibility of *prediction*. If we know the genetic and ecological basis of these changes, we can predict how the frequency of melanic moths will change if air pollution controls are initiated. Or perhaps we can predict that we will never find a cure for the common cold, because the virus will inevitably evolve resistance to any medical treatment.

1.2 An Overview of Population Genetics

Population genetics is the study of the factors that determine the genetic composition of a population, and how they act. We can imagine four fundamental questions of population genetics:

1. What are the processes that cause populations to change over time, and how do they act on natural populations?
2. What are the processes that cause populations to diverge from one another, and under what circumstances can they produce new species?

3. How much and what kind of genetic variation is present in natural populations?

4. What processes are responsible for the preservation of genetic variation, as opposed to its elimination?

One of the problems in studying population genetics is that everything depends on everything else. You cannot understand genetic drift without knowing something about mutation. You cannot understand the consequences of gene flow without knowing something about genetic drift and natural selection. The purpose of this section is to provide a brief overview of the main ideas of population genetics so that you will have a basic understanding of them all before beginning a detailed study of any one. The devil is in the details, and the rest of this book is the devil.

We can consider several basic evolutionary processes that act on a population. The first and most fundamental is **mutation**. Mutation is defined as any heritable change in the genetic material. It is an error that occurs at random with respect to its benefit or detriment to the organism (although most mutations are deleterious to some extent). Mutation is the ultimate source of all genetic variation. It provides the raw material on which other evolutionary processes can act; without mutation, no evolution could occur.

Recombination is a secondary source of genetic variation. It creates new combinations of alleles, but not new alleles. This is important because natural selection acts on phenotypes, which are, at least in part, the result of many interacting genes. New combinations of alleles can lead to new phenotypes upon which natural selection can act.

The process that beginning students are most familiar with is probably **natural selection**. This is Darwin's idea that individuals with heritable favorable variations survive and reproduce at a higher rate than other individuals in the population. They leave more offspring, and pass these variations to their offspring; therefore, the frequency of these variations increases in the next generation. Natural selection is well documented, but it is difficult to study quantitatively because many variables are involved.

An important idea related to natural selection is the concept of fitness. Fitness is an individual's lifetime reproductive output—the number of offspring that it leaves during its lifetime. It can be expressed as either an absolute value or a relative value, compared to other individuals in the population. The average fitness of all individuals in the population is called the population fitness, or mean fitness. One might assume that natural selection would act to increase the mean fitness of a population, but as we shall see, that is not always true.

Another important evolutionary process is **genetic drift**. Genetic drift is random fluctuation of allele frequencies due to random sampling of gametes and other chance events that occur each generation. It is completely random, and can in fact cause beneficial variations to be eliminated or deleterious variations to become common in a population. Its long-term effect is to decrease the amount of genetic variation within a population. Genetic drift will cause isolated populations to diverge from one another, as it acts randomly and independently in each population. We shall see that the combination of genetic drift and mutations that are neutral with respect to natural selection can explain much, but not all, of the molecular variation present in natural populations.

Gene flow may occur when individuals move from one area to another, sometimes into a new population. If they survive and reproduce in that new population, then they bring their genes into that new population and gene flow has occurred. The main effect of gene flow is to make populations more similar to one another than they would be without it. The long-term effect is the opposite of genetic drift.

Another factor affecting the genetics of populations is **nonrandom mating**. This occurs when mating individuals are genetically related to one another, or are phenotypically more (or less) similar to each other than two individuals chosen at random. It is common in both plants and animals, and can have important effects on the amount and nature of genetic variation in a population. It can lead to divergence among populations, and ultimately to speciation. Nonrandom mating is closely related to natural selection—for example, females sometimes choose males nonrandomly based on some phenotypic characteristic, a process known as sexual selection.

The fundamental quantities in population genetics are allele frequencies and genotype frequencies. These are easily calculated from a sample of individuals from a population, and provide information about how much genetic variation exists in a population, and how similar or different two populations are. Much of theoretical population genetics consists of studying how allele frequencies and genotype frequencies are expected to change under various combinations of genetic and ecological processes.

A fundamental principle of population genetics is the **Hardy-Weinberg principle**. It says that in the absence of the processes described above, allele frequencies and genotype frequencies will remain constant in a population. The Hardy-Weinberg principle is valid only if a long list of assumptions is met, and much of early theoretical population genetics was directed toward predicting what will happen when these assumptions are violated.

One of the most important questions in population genetics is how much genetic variation is present in natural populations. The short answer is that, when examined at the protein or DNA level, most populations have enormous amounts of genetic variation. Much of this variation is not visible at the morphological level.

This raises the question of why this extensive variation exists. One hypothesis asserts that it is due to varying kinds of natural selection that tend to favor one allele, then another, so that no single allele has a permanent advantage. This is known as balancing selection. An alternative hypothesis asserts that most of this molecular variation is a result of the opposing effects of mutation, which generates variation, and genetic drift, which eliminates it. This is the essence of the neutral theory of molecular evolution. There is much evidence supporting both of these hypotheses, so the question is not which is true, but which is more important in any given situation.

Mutation, genetic drift, and natural selection can cause isolated populations to diverge from one another. Gene flow retards this divergence. The balance between these processes determines how different populations are from one another. As populations diverge, reproductive isolation may develop, so that members of different populations may be unable to breed successfully with one another. This is an important part of the process of speciation.

There are three main areas of population genetics: Theoretical population genetics uses mathematical and computer modeling to understand the effects of different evolutionary processes under various conditions. Experimental population genetics attempts to test these theoretical predictions under controlled laboratory and field conditions. Empirical population genetics studies genetic variation in natural populations, using the findings of theoretical and experimental population genetics as guides. All three areas are necessary and there is extensive feedback among them. The best studies integrate all three. We will move freely back and forth among them in this book.

This has been a very brief overview of the main ideas of population genetics. The main purpose is to help you to understand what is meant when, for example, gene flow is mentioned in the chapter on natural selection, before gene flow has been studied in detail.

1.3 A Short History of Population Genetics

Much (but not all) of modern population genetics is concerned with analysis of DNA and protein sequences. However, the theoretical and empirical framework of population genetics can be traced back to the early 1990s, and even earlier. We shall see how the questions asked by late-nineteenth- and early-twentieth-century naturalists about the nature of variation and selection have led directly to the questions being asked by today's molecular population geneticists.

Darwin and Mendel established the foundations of population genetics, although neither of them realized it. Darwin emphasized the importance of variation and heredity in the evolutionary process, and Mendel provided a mechanism for heredity. Unfortunately, Darwin had no knowledge of Mendel's work, and his understanding of heredity was completely wrong. One might think that, when Mendel's work was rediscovered in 1900, its importance to Darwin's theory of natural selection would be obvious. As we shall see, that was not the case at all.

Variation and Heredity in the Nineteenth Century

Darwin published the first edition of *The Origin of Species*[1] in 1859. In it, he described and defended his revolutionary idea of natural selection. The essence of Darwin's argument can be succinctly summarized: (1) More individuals are born than can survive under conditions of limited food and resources. Therefore, there must be, in Darwin's words, a "struggle for existence." (2) Individuals in a population vary with respect to many characteristics. (3) Some of these variations are heritable. (4) Some of these heritable variations give their bearer an advantage with respect to survival or reproduction. (5) Individuals with these heritable advantageous variations will survive and reproduce at a higher rate than other individuals in the population. (6) Therefore, the frequency of these advantageous variations will increase in the offspring generation. (7) In this way, advantageous variations can accumulate in a population. (8) This slow accumulation of advantageous variations allows a population to diverge more or less without limit from its original state.

---

[1]The full title is *On the Origin of Species by Means of Natural Selection, or the Preservation of Favoured Races in the Struggle for Life*, but it is usually referred to as simply *The Origin of Species*.

Although Darwin's documentation and defense of evolution was accepted by most nineteenth-century biologists, there was controversy over the mechanism of natural selection. The dispute revolved around the nature of variation and its inheritance. Darwin and other biologists of his time recognized two kinds of variation. The most obvious kind was the discrete, discontinuous variation, which Darwin called "sports." Darwin cited the example of sheep with short, stubby legs. These discrete variations were frequently used by plant and animal breeders to breed unique strains. The second kind of variation consisted of small, continuous variations commonly seen in every species, such as variations in body weight. Darwin called these individual differences.

Darwin believed that sports were very rare, often infertile and sickly. For that reason, he believed that they were not important in evolution, and that natural selection acted primarily on small, individual variations. Therefore, evolution must be a slow, gradual process.

Francis Galton (1822–1911)

This idea was criticized by some of Darwin's contemporaries, including his cousin Francis Galton. Galton was an important figure in nineteenth-century biology, and was one of the earliest biologists to use mathematical techniques to study variation. He was very interested in the idea of regression and attempted to document and quantify it. According to Galton, the offspring of a selected group would be closer to the population mean than their selected parents. For example, Galton showed that the children of tall parents were usually closer to the population mean than their parents. Galton concluded that small advantageous variations would be lost by regression toward the mean, and that natural selection would be ineffective on small, individual differences. Therefore, argued Galton, evolution must act on discontinuous traits (sports) and must progress by discontinuous jumps, and not by slow gradual change as Darwin proposed.

Karl Pearson was one of the founders of statistics. He developed basic methods for the statistical analysis of variation in populations. He argued that Galton was wrong, and that regression toward the mean would occur only if the selected group mated with the population as a whole. If the selected group bred within itself, regression would not occur. Thus, Pearson supported Darwin in the belief that natural selection acts on continuous variation and that evolution proceeds by slow, gradual change.

Others sided with either Darwin or Galton. The debate was due to a misunderstanding of the hereditary process. Recall that Mendel's work was unknown until 1900. The prevailing ideas of heredity in the late nineteenth century were the inheritance of acquired characteristics and the idea of blending inheritance. It was commonly observed that offspring were often intermediate between their parents for some characteristic. For example, marriages between a tall person and a short person usually produce children of intermediate height. These observations led to the erroneous conclusion that the hereditary material blended, and that the offspring would always be intermediate between the two parents. Under this mechanism of heredity, Galton's criticism was valid, and advantageous variations would indeed be swamped by regression toward the mean. Darwin was never able to satisfactorily overcome this criticism. Resolution of the problem had to await the rediscovery of Mendel's work and its application to natural selection. Even then, the connection was not obvious.

The Conflict Between Mendelians and Biometricians

At the beginning of the twentieth century, there were two views about the evolutionary process. One group, the gradualists, believed that natural selection acted on small individual variations and that evolution was a slow, gradual process. The other group, the saltationists, believed that natural selection was ineffective on individual variations, and that evolution acted on discontinuous variations or sports, and consequently proceeded by discontinuous jumps. Neither group had any clear understanding of the mechanism of inheritance, upon which any theory of evolution must depend.

When Mendel's paper was rediscovered in 1900, the two groups responded to it very differently. The saltationists supported Mendelian inheritance because they thought it provided a mechanism for discontinuous evolution. They assumed that the discreteness of the hereditary units, as hypothesized by Mendel, inexorably led to evolution by discrete steps. This group came to be known as the Mendelians. The gradualists argued that experiments with continuously varying characters suggested some kind of blending inheritance. These experiments appeared to be inconsistent with the idea of Mendelian inheritance of discrete units. This group came to be called the biometricians, after one of their leaders, Karl Pearson, who founded the field of biometry. Both groups assumed, erroneously, that Mendelian inheritance was incompatible with the process of gradual evolution.

The validity of Mendelism was vigorously debated at the beginning of the twentieth century, but accumulation of much experimental evidence eventually convinced most biologists that discrete characters are inherited in Mendelian fashion. Much of this work was done by T. H. Morgan and his students with *Drosophila melanogaster*, and culminated in the first book on Mendelian genetics, *The Mechanism of Mendelian Heredity* (Morgan et al. 1915). A second major influence was the work of the Swedish geneticist Nilsson-Ehle who, in 1909, showed how the inheritance of a nearly continuous character, kernel color in wheat, could be explained by the action of several genes, each inherited in Mendelian fashion. A rigorous mathematical demonstration of this was published by R. A. Fisher in 1918.

The problem thus became how to reconcile Mendelism with gradual evolution by natural selection. The British mathematician G. Udny Yule first pointed out in 1902 that Mendelian inheritance and gradual evolution were not necessarily incompatible. Pearson responded in 1904 with a mathematical analysis and concluded that the theoretical consequences of Mendelian inheritance were inconsistent with his researches on correlations among relatives. Yule (1907) countered with an analysis showing that Pearson's theoretical objections were based on an assumption of complete dominance, and if one assumed partial dominance, those objections disappeared.

These mathematical analyses had little effect on mainstream biologists. It was several years before the experimental results described above finally convinced most biologists that Mendelian inheritance was compatible with gradual evolution. Thus, it took more than a decade to integrate Mendel's work into the framework of evolutionary biology. This integration was delayed by the conflict between Mendelians and biometricians, which resulted as much from ego and personality conflicts as it did from scientific debate. Provine (1971) provides an interesting and scholarly discussion of the conflict.

Theoretical Population Genetics

Perhaps the first paper in theoretical population genetics was that of Yule in 1902. Today, every genetics student knows that when a homozygous dominant strain is crossed to a homozygous recessive strain, the F_2 will show a 3:1 ratio of dominant to recessive phenotypes. Yule showed that if the F_2 individuals were bred to one another, this 3:1 ratio would continue through future generations without change. This was the first hint of what came to be known as the Hardy-Weinberg principle. However, Yule erroneously thought that the 3:1 ratio would be approached irrespective of initial frequencies in a population. William Castle and Karl Pearson reached similar conclusions at about the same time, but these ideas received little attention because of the conflict between the Mendelians and the biometricians.

In the early years of the twentieth century, it was commonly believed that a dominant allele would increase in frequency and take over the population, or, as Yule believed, approach a 3:1 phenotypic ratio. William Castle in 1903 first suggested that this is not true, claiming that if selection stopped, "the race remains stable at the degree of purity then attained." However, it was not until 1908 that British mathematician G. H. Hardy and German physician Wilhelm Weinberg independently and clearly demonstrated that, in the absence of disturbing forces, allele frequencies and genotype frequencies will remain constant. This statement has become known as the Hardy-Weinberg principle, and is the cornerstone of theoretical population genetics. Most of the theoretical work for the next 20 years or so involved working out the consequences when the assumptions of the Hardy-Weinberg principle are violated.

Most of early population genetics was theoretical. This was mainly because it was difficult to obtain data on variation and change in natural populations. It was also recognized that the assumptions of the Hardy-Weinberg principle are unrealistic, and it was considered important to understand what would happen when these assumptions are violated. The three main figures in this quest were R. A. Fisher, J. B. S. Haldane, and Sewall Wright. Between 1918 and 1932, they established the theoretical foundations of population genetics.

R. A. Fisher (1890–1962)

R. A. Fisher (1890–1962) was a British statistician. He published his first paper on population genetics in 1918. In that paper he demonstrated that variation in continuous characters could be explained by Mendelian inheritance of several genes. He also described the very important statistical technique of analysis of variance. In later papers, Fisher made the first serious attempt to synthesize Mendelism and gradual evolution (although Yule had made a beginning years earlier). He wrote a variety of papers on genetic correlations, mutation, natural selection, quantitative genetics, dominance, and related topics. In 1930, he published *The Genetical Theory of Natural Selection*, an extended review of his ideas. This book has intrigued and infuriated geneticists for more than 70 years. Fisher's writing was dense and often cryptic. He sometimes made assumptions or approximations without mentioning them, which made his reasoning difficult to follow. In the book, Fisher demonstrated mathematically that blending inheritance is incompatible with natural selection because it leads to a decrease in variation each generation, and that Mendelian inheritance preserves variation and thus is consistent with Darwin's theory. He introduced what he called the "fundamental theorem of natural selection," which says, in Fisher's words, "The rate of increase in fitness of any organism at any time is equal to its genetic variance at that time." If you find that sentence difficult to understand, you are not alone. Population geneticists have debated its meaning and relevance since it was first published. Fisher also discussed mutation, natural selection, genetic variation, the evolution of dominance, mimicry, and a number of topics related to human evolution.

Fisher's book was very influential; it represented the first extensive analysis of Mendelian heredity and natural selection. Population geneticists and evolutionary biologists are still mining it for insights into modern problems. Crow (1990) provides an interesting overview of Fisher's life and work.

J. B. S. Haldane (1892–1964) was the son of a British physiologist. It is difficult to summarize Haldane with a label. He was, among other things, a classicist, linguist, essayist, biochemist, physiologist, geneticist, and a one-time communist. He is known for such diverse contributions to biology as the Oparin-Haldane theory of the origin of life, the Briggs-Haldane equations of enzyme kinetics, and the mathematical theory of natural selection—all this, in spite of the fact that Haldane never earned a scientific degree. His undergraduate degree was in classics, and he never obtained a graduate degree, in science or any other field. As Crow (1992) put it, "if you know as much as Haldane, you don't need one."

Haldane's contributions to population genetics were many and varied. Between 1924 and 1932 he published a series of papers with the general title, "A Mathematical Theory of Natural and Artificial Selection." In these papers he worked out the basic theory of natural selection, and its interactions with gene flow, inbreeding, dominance, mutation, epistasis, and other processes. He did all of this without the aid of a computer!

J.B.S. Haldane (1892–1964)

In 1932 Haldane summarized this work in a book entitled *The Causes of Evolution*. Unlike Fisher, Haldane wrote clearly and cared whether the public could understand him. Most of the book, with the exception of the mathematical appendix, can be understood by nonspecialists.

Haldane died in 1964 from cancer, of which he wrote

> *I wish I had the voice of Homer*
> *To sing of rectal carcinoma*
> *Which kills a lot more chaps, in fact,*
> *Than were bumped off when Troy was sacked.*

For more on the fascinating life of Haldane, see the biography by Clark (1969); see Crow (1992) for a brief perspective on Haldane's life and work. Many of Haldane's most important scientific papers are reprinted in the collection by Dronamraju (1990).

Sewall Wright (1889–1988) was born in Illinois, USA. Much of his early work was related to agriculture. His earliest work focused on coat color in guinea pigs. From these studies Wright recognized the importance of gene interaction, a conclusion that influenced much of his later work. In 1921, Wright published a series of five papers, entitled "Systems of Mating." These were his first important contributions to population genetics. In them he analyzed the effects of inbreeding, nonrandom mating, and selection on the genetic composition of a population. He

Sewall Wright (1889–1988)

introduced the important concept of the inbreeding coefficient, which we will meet many times in this book.

Wright's work with guinea pigs and inbreeding led him to consider methods of artificial selection. From this, he began to think about selection and evolution in natural populations. In 1931 he published a long paper entitled "Evolution in Mendelian Populations," in which he laid out his view of evolutionary genetics. His view differed significantly from Fisher or Haldane. Wright placed more emphasis on gene interaction, genetic drift, and population subdivision. He believed that a series of relatively small, partially isolated populations with limited gene flow among them provided the situation most conductive to evolution. In such populations, favorable gene complexes could become established by genetic drift, and then increase in frequency by natural selection. These ideas led to his shifting balance theory of evolution, an idea that is still controversial.

Two of Wright's most important contributions are the concepts of the inbreeding coefficient and of effective population size. Almost everything in population genetics today depends on one or both of these concepts. Wright summarized his life's work in a four-volume treatise (Wright 1968–1978), although he continued to publish until he died in 1988. Provine (1986) has written a biography of Wright; he has also edited a collection of Wright's important papers (Wright 1986).

To summarize, the cornerstone of modern population genetics was laid in 1908 by Hardy and Weinberg. Over the next 20 years or so, Fisher, Wright, and Haldane examined the theoretical consequences of Mendelian inheritance in natural populations, using the Hardy–Weinberg principle as their starting point. By 1932, each of them had published a major summary of their work. These three publications remain the foundation of theoretical population genetics. The best account of these years is Provine's (1971) history.

Genetics of Natural Populations

As the theoreticians were working out the population implications of Mendelian inheritance, field biologists were beginning to think about genetic variation in natural populations. The earliest work with natural populations was initiated by the Russian biologist Sergei Chetverikov.[2] In 1926, Chetverikov wrote a remarkable paper in which he reached, without mathematics, many of the same conclusions derived by Fisher, Wright, and Haldane. He argued, among other things, that natural populations should harbor large amounts of concealed genetic variability. A year later, Chetverikov published a short abstract summarizing his initial studies of natural populations of *Drosophila*. He captured wild flies and performed full sib matings with their offspring. Analysis of progeny of the sib matings revealed hidden genetic variability, as Chetverikov had predicted.

Chetverikov's methods were rather primitive by today's standards, but they were the best available at the time. He was the first to actually attempt to measure the amount of genetic variation present in a natural population. Even his small scale and primitive attempts revealed a surprising amount of hidden variation. This work was continued and expanded by Chetverikov's Russian students

---

[2]Sometimes transliterated as Tshetverikov or Tschetwerikoff.

and colleagues N. V. Timofeeff-Ressovsky, N. P. Dubinin, and others, but especially by Theodosius Dobzhansky.

Theodosius Dobzhansky (1900–1975) was born in Russia and moved to the United States in 1927 to work with T. H. Morgan. His early work in Russia concerned morphological variation in ladybird beetles, but he is best known for his work with natural and experimental populations of *Drosophila*. Dobzhansky's work can perhaps be classified into three overlapping categories. First was his investigation into the levels and kinds of genetic variation in natural populations. This started with his studies of ladybird beetles, and continued with studies of morphological and chromosomal variation in *Drosophila pseudoobscura*, and hidden variation in various *Drosophila* species. Second was his work with inversion polymorphisms in *D. pseudoobscura*, begun in 1936. Dobzhansky initially believed these inversion polymorphisms were neutral with respect to natural selection, but with extensive observation and experimentation, he demonstrated that they were subject to natural selection. Dobzhansky's third major area of research was speciation. He suggested that reproductive isolation was the essential characteristic of different species and pioneered the technique of genetic analysis of species differences by analyzing the hybrid and backcross progeny of species that could be hybridized. For a summary of Dobzhansky's life and major scientific accomplishments, see Ayala (1977).

Theodosius Dobzhansky
(1900–1975)

Many of Dobzhansky's papers were published in a series with the general title "Genetics of Natural Populations." These have been reprinted, along with extensive commentary and historical information, in Lewontin et al. (1981). These papers, important as they are, comprise but a small part of Dobzhansky's scientific output. The series consists of 43 papers, but Dobzhansky published hundreds more.

Many believe that Dobzhansky's most important contribution was his 1937 book, *Genetics and the Origin of Species*. In it, Dobzhansky argued that mutation is the ultimate source of genetic variation, but natural selection is the main process driving adaptive evolution. Like Chetverikov, he argued that hidden variation is common. Furthermore, he argued that the variation seen within local populations is the same kind as the variation among different populations, or even among different species. He introduced the concept of reproductive isolation as the criterion for distinguishing different species, and claimed that it is caused by the same kinds of genetic differences that characterize individuals within a population, or among different populations.

The effect of the book was to initiate what has been called "the modern synthesis" of evolutionary biology. Dobzhansky united systematics and genetics by showing that species differences could be studied using genetic techniques. He showed that natural selection not only *could* act on small genetic differences, but that it *did*. He provided data supporting the theoretical analyses of Fisher, Wright, and Haldane. Most importantly, he wrote in a language that nonmathematically inclined biologists could understand, explaining how the mathematical theories applied to natural populations. The book went through three editions, and a fourth under a different title (Dobzhansky 1970). Its impact has been immeasurable. See Powell (1987) and Lewontin (1997) for discussions.

Dobzhansky was also a great teacher. A list of his students and "grandstudents" reads like a list of Who's Who among important players in modern population and evolutionary genetics.

While Dobzhansky and his students were working in the United States, E. B. Ford, P. M. Sheppard, H. B. D. Kettlewell, and others founded the British school of ecological genetics. They pioneered the experimental manipulation of natural populations, including movement of individuals from one population to another, and the establishment of artificial populations. Ford was particularly interested in polymorphisms of discrete characters. He and his colleagues studied genetic polymorphisms in natural populations of butterflies, moths, and snails, and attempted to understand these polymorphisms in terms of natural selection, genetic drift, and gene flow. One of their important conclusions was that ecological differences often work to maintain genetic polymorphisms within and among populations. For example, in some species of snails, different shell coloration and banding patterns are favored in different habitats, and at different times of the year. For a review of these important studies, see Ford (1975).

Molecular Population Genetics

The molecular era in population genetics began in 1966 when Lewontin and Hubby (1966) and Harris (1966) introduced the technique of protein electrophoresis to population genetics. The two great advantages of electrophoresis were that it could be applied to essentially any organism, and that it allowed us to estimate variability of a gene without any a prior idea of whether it was variable or not.

These two studies motivated many more, on a variety of organisms. It soon became obvious that genetic variability, as estimated from protein variability, was more extensive than anyone had previously thought. It became necessary to explain why all of this variation exists.

The initial assumption was that the observed variation is maintained by some form of natural selection. However, even in the first study, Lewontin and Hubby (1966) pointed out that this might lead to an unreasonably low survival rate in a population. An alternative was independently proposed by Kimura (1968) and by King and Jukes (1969). They proposed that most of the variation seen at the protein level is, in fact, neutral with respect to natural selection, and is maintained by a balance between the opposing processes of mutation and genetic drift. This idea has become known as the neutral theory of molecular evolution, or, more commonly, the neutral theory.

The neutral theory was immediately controversial. The "neutralists" argued that the patterns seen in electrophoretic studies are inconsistent with the action of natural selection, and that the neutral theory best explains the data. The "selectionists" argued that genetic drift is a minor factor, and that most of the observed variation is maintained by some form of natural selection. During the 1970s, much time and effort was spent on the sometimes acrimonious controversy. By the end of the 1970s many participants had begun to realize that the question could not be answered with the tools then available.

In the 1980s, it became possible to manipulate and analyze DNA and to apply these techniques to population genetics. Initially, restriction enzymes were used to analyze easily isolated pieces of DNA such as mitochondrial DNA. Later, DNA sequences were analyzed directly. Kreitman (1983) was the first to apply DNA sequencing techniques to population genetics. He sequenced 11 copies of the alcohol dehydrogenase locus of *Drosophila melanogaster*, and found that all were different! This was a revolutionary study and, like the first electrophoresis studies, has led to a flood of similar surveys in various organisms. The general result is that genetic variation is even more pervasive at the DNA level than at the protein level. Again, we must ask: Why so much variation?

The molecular and statistical tools for studying DNA variation are much more powerful than for protein variation. Since the 1980s, advances in molecular techniques, statistical methods, and computer technology have combined to revolutionize the way population geneticists think and work. Modern molecular techniques allow us to routinely compare the DNA sequences of dozens, or even hundreds, of individuals, and to compare coding sequences (sequences that code for proteins) to noncoding sequences. The neutral theory has become the most sophisticated mathematical theory in biology, and has produced precise and detailed predictions about the amounts and kinds of DNA variation to expect under different circumstances. Sophisticated statistical techniques have been developed to test these predictions, taking advantage of computer power and availability undreamed of in the 1960s and 1970s.

In the last few years, a number of genome projects have been completed, including a first draft of the human genome. This has made possible the study of comparative genomics. We can now compare genes and gene organization in different species and ask questions about comparative population and evolutionary genetics that were not even imaginable a few years ago (e.g., Charlesworth et al. 2001; Eisen 2000; Koonin et al. 2000).

1.4 Mathematical Models

The theories of population genetics explain the process of evolution just as the theory of gravitation explains the motions of the planets. Both biology and physics are immensely richer as a result. And just as the theories of physics require mathematics, so do the theories of population genetics.

A glance through this book will show that it is full of equations and mathematical models. Population genetics, and its sibling, population ecology, are probably the most mathematical of the biological sciences. Every textbook (including this one) attempts to justify its extensive use of mathematics. Most (including this one) are unconvincing, at least initially. Students are frequently skeptical of mathematical models, and resist them. One of the goals of this book is to convince you that mathematical models are necessary tools in population genetics, just as a microscope is a necessary tool in cell biology. And that it does not take a mathematical genius to understand them!

The main problem encountered by most beginning students of population genetics is not the mathematics itself, but the difficulty in thinking quantitatively—a new experience for many biology students. If you have equations that you cannot solve, you can usually find a mathematician to help you, or you can simulate them on a computer. The challenging part for most biologists is formulating the equations. This book will give you guidance and practice. As you become more experienced, you may even come to enjoy working with mathematical models!

A mathematical model is simply a mathematical representation of some biological (or other) process. One function of a mathematical model is to *describe* laboratory or field observations in a quantitative way. Think of a model as an equation that gives a set of "expected values" for a series of observations. We will first consider an example, and then discuss why such models are useful and necessary.

An Example

Figure 1.1 shows how population size changes through time for several species, including bacteria, yeast, protozoa, and sheep. For now, look only at the data points and ignore the lines. We see that all of these populations have the same basic growth form: The population initially grows slowly, then reaches a point of rapid increase; growth then slows down and eventually stops, and the population size remains more or less constant.

It might be surprising that the basic form of population growth is so similar in these very different species. It is natural to wonder whether this similarity is due to biological principles that are common to all of these organisms. Are there fundamental biological principles that can explain why all of these populations (and many others) grow as they do? Mathematical models can help us answer this question.

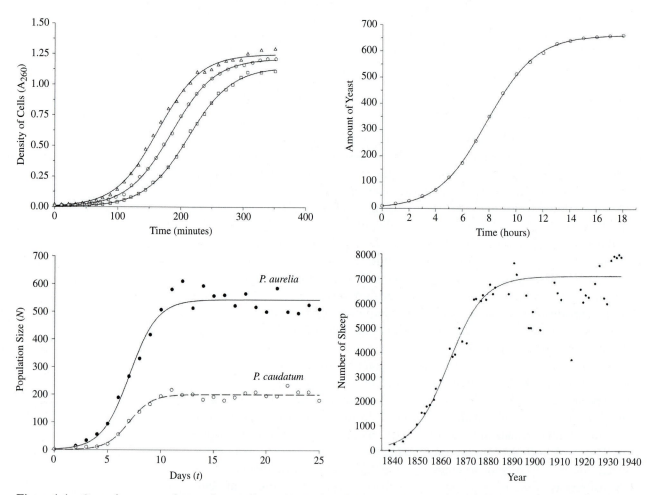

Figure 1.1. **Growth curves of several organisms**   (a) *E. coli* (Lebda 1999); (b) Yeast (Pearl 1927); (c) *Paramecium* (Gauss 1934); (d) Sheep; modified from Renshaw (1991); data from Davidson (1938a). Solid lines represent best fit of equation (1.1) to the data points.

It turns out that all of these examples can be described by the same equation:

$$\frac{dN}{dt} = rN\left(\frac{K - N}{K}\right) \qquad \textbf{(1.1)}$$

where N represents the population size, t represents time, and r and K are "parameters" of the equation, to be discussed below. The derivative dN/dt is the rate of change of the population size over time; i.e., its growth rate. Equation (1.1) says that the growth rate of a population at any given time depends on the population size at that time, and on these things we have called r and K. This equation can be derived from basic biological principles; we will not do so, but see Gotelli (2001) for a particularly clear derivation. For our purposes, you should just accept it as a description of the kind of population growth seen in Figure 1.1.

A **parameter** is simply a constant in an equation, whose value we specify or estimate. The parameter r in equation (1.1) is the maximum reproductive rate of an individual. It varies among species, of course—bacteria reproduce faster than sheep. But this model assumes that all bacteria in the population have the same r, and that all sheep have the same r. The other parameter, K, is called the carrying capacity, and is a measure of how well individuals in the population can survive and reproduce under crowded conditions. It too varies for different species. We can estimate r and K from laboratory or field studies. See any ecology text for more on r and K.

Equation (1.1) is a differential equation. It can be solved to give the population size as a function of time, and of r and K. We can then plot population size $vs.$ time for each species. The results are the solid lines in Figure 1.1. Note that, especially for the microorganisms, the model (equation) describes the actual population growth quite well.

The point of this example is that we have a model (equation) which *describes* the growth of populations, and this model is derived from basic biological principles, symbolized in the equation by r and K. Note that the model is useless without some estimate of r and K. But once we have these estimates, we can *predict* future population growth.

How to Work with Mathematical Models

> *First you guess. Don't laugh, this is the most important step. Then you compute the consequences. Compare the consequences to experience. If it disagrees with experience, the guess is wrong. In that simple statement is the key to science. It doesn't matter how beautiful your guess is or how smart you are or what your name is. If it disagrees with experience it's wrong. That's all there is to it.*
>
> —*Richard Feynman*

There are several steps to working with a mathematical model. If you keep each step separate in your mind, you will have much less trouble than if you try to combine them.

1. State what it is you want to know about the population.
2. State the biological assumptions of the problem.
3. Express the assumptions and questions in the form of one or more equations. Explicitly define all symbols in the equations.

4. Solve the equations. This can mean several things, depending on the equations and the goals of the model. You will learn more about this as we study specific models.

5. Analyze what the solutions to the equation(s) mean in terms of the biological problem. What does the model predict? What is the long-term outcome of the process being studied? Is there a point at which the population will remain unchanged? Is there more than one possible outcome?

6. Compare the predictions of the model with observations and biological knowledge of the population.

7. If the observations do not match the predictions of the model, return to step 2, modify the assumptions, and start over.

8. Repeat until the model describes what you see in the biological world. You should remain skeptical even if you are lucky enough to get a model that seems to "work." We will see that superficial agreement between a model and observations of natural populations does not necessarily mean that we understand the whole story.

Clearly, this is an ongoing process and you will often go through several models before you get one that is a reasonable approximation of reality. We will illustrate the process with an example.

A Simple Model of Recurrent Mutation

We illustrate the process of model building and analysis by working through a simple model of mutation. Consider a single locus with a wild type allele, designated A, and a mutant allele, designated a. Let p and q $(= 1 - p)$ be the frequencies of A and a. Every generation a certain proportion, u, of the wild type alleles mutates to a. The parameter u is the mutation rate. There is no back mutation; that is, a does not mutate to A. We will ignore natural selection and all other evolutionary processes. We wish to know how the allele frequency changes over time. The answer should be obvious if you think about it, but we will derive it just for practice.

1. Statement of the problem:
We wish to know how the allele frequencies change with time. Note that since $q = 1 - p$, we need to consider only the change in either p or q.

2. Assumptions:
Mutation acts only in one direction, A to a.
The mutation rate from A to a is u per generation; in other words, a fraction u of A alleles mutate to a each generation.
Natural selection and genetic drift are negligible.

3. Formulation of the equation:
It is easy to write the equation for p:

$$\begin{pmatrix} \text{frequency of } A \\ \text{next generation} \end{pmatrix} = \begin{pmatrix} \text{frequency} \\ \text{now} \end{pmatrix} - \begin{pmatrix} \text{proportion of } A \\ \text{that mutates to } a \end{pmatrix}$$

In symbols this becomes

$$p_{t+1} = p_t - up_t$$

where p_t is the frequency of the A allele in any generation, and p_{t+1} is its frequency one generation later. The second term on the right is a proportion of a proportion. The proportion of A alleles in the population is p_t, and the proportion of those that mutate is u, so the total proportion of A alleles lost is up_t. Imagine cutting a pie into halves; then cut each half into thirds. Each piece is then $1/2 \times 1/3 = 1/6$ of the pie.

The above equation can be rewritten as

$$p_{t+1} = p_t(1 - u) \tag{1.2}$$

This is our basic equation. This kind of equation is called a **recursion equation**. In general, a recursion equation describes how some quantity (here, the frequency of A) changes from one time period (usually a generation) to the next.

4. Solve the equation:

To solve a recursion equation means to find an equation that gives the value of the variable at any future time, given its current (or initial) value. Here, we want to find an equation for p_t in terms of p_o, where p_o is the initial allele frequency, and p_t is the frequency t generations later. It is easy to solve equation (1.2). If p_o is the initial allele frequency, then one generation later the allele frequency is

$$p_1 = p_o(1 - u)$$

In the second generation, it is

$$p_2 = p_1(1 - u) = [p_o(1 - u)](1 - u) = p_o(1 - u)^2$$

In the third generation, it is

$$p_3 = p_2(1 - u) = [p_o(1 - u)^2](1 - u) = p_o(1 - u)^3$$

It should be obvious that the pattern (solution) is

$$p_t = p_o(1 - u)^t \tag{1.3}$$

Most recursion equations are more difficult to solve than this one, and special techniques are sometimes needed.

5. Analyze the solution:

Now, u is a small fraction; therefore, $(1 - u)$ is less than one, and $(1 - u)^t$ approaches zero as t gets large. So, p approaches zero with time. This makes sense, since every generation the population loses a few A alleles by mutation, and never regains them. Our model predicts that eventually all A alleles will mutate to a.

We can get some idea of how fast the allele frequency changes. Mutation rates are low, on the order of 10^{-5} or less. If we let $u = 10^{-5}$ and $p_o = 0.95$, we can plot p_t versus t (Figure 1.2). Note the log scale in the figure. We see that the allele frequency changes imperceptibly for thousands of generations. It takes about a thousand generations for p to decrease from 0.95 to 0.94, and 10 thousand to decrease to 0.85. p does indeed approach zero, but at a very slow rate. We conclude that recurrent mutation is an insignificant force in changing allele frequencies, except over a very long time.

Figure 1.2 **Change in frequency of the *A* allele due to one-way mutation of *A* to *a*, as predicted by equation (1.3).** Mutation rate is $u = 10^{-5}$. Note log scale for generations.

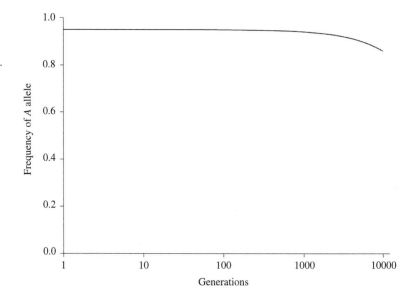

6. Compare the prediction with observations:

Our model predicts that after a few thousand generations, more or less, the mutant allele should be common. Our biological experience tells us that this is not the case; alleles produced by recurrent mutation are rare in any population. So, the prediction of the model is inconsistent with observations.

7. Reexamine the assumptions and revise the model:

One thing we have left out of our model is back mutation. It is possible that *a* alleles can mutate back to *A*. Will this make the model closer to reality? It is easy to modify the recursion equation to allow for back mutations. Let *v* be the mutation rate from *a* to *A*. Then the recursion equation becomes

$$\begin{pmatrix} \text{frequency of } A \\ \text{next generation} \end{pmatrix} = \begin{pmatrix} \text{frequency} \\ \text{now} \end{pmatrix} - \begin{pmatrix} \text{proportion of } A \\ \text{which mutate to } a \end{pmatrix}$$

$$+ \begin{pmatrix} \text{proportion of } a \\ \text{which mutate to } A \end{pmatrix}$$

In symbols, this is

$$p_{t+1} = p_t - up_t + vq_t \tag{1.4}$$

This equation still predicts that mutant alleles should be more common than wild type alleles (Problems 1.3 and 1.4). Our revised model is still inconsistent with biological observations.

8. Revise the model again:

Look again at the assumptions in step 2. The assumption of no natural selection should bother you. We know that mutant alleles are usually deleterious in some way, and therefore natural selection will work to eliminate them from the population. The next step in our model building process would be to consider the effects of natural selection against the *a* allele. We will defer that until Section 6.5, when we study the interactions between mutation and natural selection.

There is another, semantic assumption hidden in this entire process: We have assumed that there is a single wild type allele and a single mutant allele. When we examine the DNA sequence of a gene, we shall see that this is an unrealistic assumption. When we consider molecular models of mutation, we must revise our entire approach. It turns out that, in addition to mutation rates and natural selection, we must include genetic drift in any realistic model of mutation. We will consider these ideas in future chapters.

Why Mathematical Models?

The biological world is unimaginably diverse and complex. To make sense of it, we must begin by simplifying. This is true for biological models (experiments), as well as for mathematical models. The goal of both is to identify those processes that are most important in cause and effect. We do this by trying to control other variables.

The two basic functions of a mathematical model are description and prediction. We can describe laboratory or field observations without mathematical models, of course, but if we wish to predict, we need some theoretical basis from which to work. Mathematical models provide this theoretical basis: They suggest hypotheses to be tested with laboratory and field data. The recurrent mutation example suggests that mutation alone is ineffective in changing allele frequencies over the short term, and that natural selection is probably more important in determining allele frequencies in a population. These are testable hypotheses for field or laboratory experiments.

One of the greatest benefits of working with mathematical models is that they focus our thinking. When you write down an equation in a model, you are making explicit decisions about what variables you think are important and how they interact. This does not necessarily mean that you believe the model is an accurate description of reality. You are simply asking the question, What would happen if these assumptions were true? You can then compare these predictions with biological observations and make changes if necessary, as we did with the recurrent mutation model.

Population processes often occur over time scales and geographic regions that cannot be observed directly. Mathematical models allow us to look into the future, or to take a bird's eye view of the population as a whole. We do not know how resistance to a new insecticide might evolve over space and time, but we can make predictions based on mathematical models. The accuracy of these predictions depends, of course, on the assumptions of the models.

We should not expect a model to make precise predictions, because all models (like all experiments) are simplifications of reality. Rather, the usual function of a model is to suggest which possibilities are plausible, which variables are more important, and what values of these variables lead to biologically realistic outcomes. These hypotheses can then be tested in the laboratory or in the field.

We see that population genetics requires interaction between three different approaches: field studies of natural populations, laboratory or field experiments, and mathematical modeling (theory). Field studies tell us what is actually going on in nature. The goal of experimental or theoretical studies is to help us understand these observations. From observations of natural populations, we make hypotheses about cause and effect. Only by experimental manipulation (in the field or in the laboratory) of the variables under study can we test these hypotheses. For example, we can change the mutation rate in a population of bacteria or fruit flies by irradiation or chemical mutagenesis and see how this affects the growth rate of the population. This may help us to understand whether increased mutation rates (due to, for example, the hole in the ozone layer) are a significant danger to natural populations. The role of theory is to guide us in our search for cause and effect. For example, our analysis of mutation suggests that increased mutation rates *alone* are not likely to be a significant factor in the short-term evolution of populations. But theory can tell us whether increased mutation rates, in combination with natural selection, might have significant long-term consequences.

Theoretical models must be based on our knowledge of natural populations, and should predict biologically realistic outcomes from biologically realistic estimates of the relevant parameters. If they do not, either we have overlooked an important variable, or we have bad estimates of the ones we have included. This sends us back to the laboratory or the field. The best models are motivated by puzzling observations or experimental results. They stimulate empirical work and further our knowledge, even if they are ultimately shown to be wrong. A good example of a theoretical model that has stimulated much laboratory and field study is the neutral theory of molecular evolution, to be discussed in Chapter 10.

As illustrated by the recurrent mutation example, modeling is a process. We often begin with a simple unrealistic model with many assumptions, and proceed by making more complex and realistic models with fewer assumptions. The goal is to find a model that adequately describes and predicts the behavior of natural populations, at least under certain conditions.

When working with mathematical models, you should keep in mind three questions: (1) What are the assumptions of the model? (2) What are its predictions? (3) Are these predictions consistent with experimental and field observations? The last is most important. If there is an inconsistency, the model is wrong, not nature.

Kinds of Mathematical Models

Mathematical models can be classified based on several criteria. One is whether they consider the population in discrete time or in continuous time.

Discrete time models assume that the processes being studied occur at discrete time intervals, for example, every generation or every year. Discrete time models are usually formulated as recursion equations. They are sometimes difficult or impossible to solve explicitly and sometimes yield much more complicated results than their continuous time counterparts. The mutation model in this chapter is a discrete time model.

Continuous time models assume the processes being studied occur continuously. The population growth model at the beginning of this section is a continuous time model. Continuous time models are usually formulated as differential equations, and methods for solving them have been extensively developed by mathematicians. We can sometimes approximate a discrete time model with a continuous time model that is easier to solve.

Another way to classify mathematical models is whether or not they incorporate random effects. These are called deterministic models or stochastic models.

Deterministic models ignore random effects. They are called deterministic because once the initial conditions and variables are specified, the outcome is determined uniquely. Deterministic models are usually easier to work with than stochastic models. Both examples in this section are deterministic models.

Stochastic models incorporate random effects. Stochastic is just a fancy synonym for random. Stochastic models use the tools of advanced probability theory and are often difficult to work with. The solution to a stochastic model is not unique, but is expressed in terms of probability. In this book we will introduce some stochastic models, but will not work through the mathematics necessary to analyze them completely. The most familiar example of a stochastic model is the theory of genetic drift (Chapter 7). The neutral theory of molecular evolution (Chapter 10) is another.

Stochastic models sometimes make different predictions than do their deterministic counterparts. For example, in very small populations, stochastic effects can overpower deterministic effects, and cannot be ignored. We must decide which kind of model is appropriate, based on the biology and ecology of the population we are studying.

Thus, a mathematical model can be formulated in either discrete time or continuous time, and is either deterministic or stochastic. We will encounter all of these variations.

How Much Mathematics Do You Need to Know?

The mathematics in this book is not particularly difficult. A willingness to think quantitatively is far more important than the ability to differentiate and integrate complicated functions. To get the most out of this book, you should have the following mathematical skills and knowledge. You must

- Have a thorough working knowledge of algebra.
- Be familiar with logarithmic and exponential functions.
- Understand the concepts of differentiation and integration, and be able to differentiate and integrate algebraic, logarithmic, and exponential functions.
- Understand the basic concepts of probability: random variables, distribution, expectation, and variance.
- Understand the basic concepts of statistical testing, such as the chi-square test, the t-test, and confidence intervals.

Adequate preparation would consist of one or two semesters of calculus, and a course in probability and statistics. A basic understanding of probability is more important than calculus, at least initially. It is assumed that you have had courses in general biology, genetics, and ecology.

One of your most important tools for learning population genetics is a good spreadsheet program. You will often want to iterate or to graph an equation, and this can usually be done quickly and easily with a spreadsheet. Many of the exercises throughout this book assume basic familiarity with the calculating and graphing capabilities of a spreadsheet.

If you are serious about continuing your study of population genetics beyond this book, you should consider courses in differential equations, linear algebra, probability theory, and stochastic processes. You can do good population genetics research without knowing advanced mathematics, but the more you understand of the theory, the better you will be able to interpret empirical results in terms of theoretical predictions. Familiarity with software packages such as Mathematica or MathCad is also useful, as is some knowledge of computer programming.

Summary

1. Population genetics is relevant to many areas of mainstream biology.

2. A population is defined as a geographically continuous group of individuals of the same species.

3. Population biology consists of two closely related disciplines, population ecology and population genetics.

4. Population geneticists are concerned with four primary questions (see pp. 4–5).

5. The basic evolutionary processes are mutation, recombination, natural selection, genetic drift, gene flow, and nonrandom mating. Most of this text is concerned with the ways in which they interact and affect natural populations.

6. The fundamental quantities of population genetics are allele frequencies and genotype frequencies.

7. The Hardy-Weinberg principle says that, in the absence of disturbing forces, allele frequencies and genotype frequencies will remain constant.

8. Most natural populations have large amounts of hidden genetic variability.

9. The early history of population genetics was marked by a flawed theory of heredity. Even after the rediscovery of Mendel's paper, it was many years before Mendelian inheritance and Darwinian natural selection were integrated.

10. The foundations of theoretical population genetics were laid by Fisher, Haldane, and Wright between 1918 and 1932.

11. The study of the genetics of natural populations began with Chetverikov, and continued with Dobzhansky, and others.

12. Most natural populations have much variation at the protein and DNA level. Two hypotheses to account for this variability are balancing selection and the neutral theory of molecular evolution.

13. Mathematical models are essential tools of population genetics. A procedure is suggested for working with mathematical models (pp. 17–18).

14. Mathematical models can be formulated in either discrete time or continuous time, and can be either deterministic or stochastic.

15. Theoretical, empirical, and experimental population genetics all interact with one another in ways that ultimately help us to understand the genetics and evolution of natural populations.

Problems

1.1. Theoretical models are frequently criticized as being irrelevant, because their assumptions are unreasonable. Discuss this claim; write about one page, double-spaced.

1.2. Page 5 contains the sentence, "Natural selection is well documented, but it is difficult to study quantitatively because many variables are involved." What are some of the problems that might make the quantitative study of natural selection difficult?

1.3. An equilibrium allele frequency is a frequency that does not change from one generation to the next. Find an equilibrium allele frequency for equation (1.4) by setting $p_{t+1} = p_t = \tilde{p}$, where $\tilde{p}$ is the equilibrium allele frequency. Remember that $q = 1 - p$ for any p, and solve for $\tilde{p}$ in terms of u and v.

1.4. Back mutation ($a \rightarrow A$) is usually less frequent than forward mutation ($A \rightarrow a$). Assume the forward mutation rate is $u = 10^{-5}$ and the back mutation rate is $v = 10^{-6}$. Use your answer to Problem 1.3 to find the equilibrium frequency of the A allele. Does this make biological sense? Discuss. What factors might be involved that this model has not considered?

Genetic Variation

*If we could examine the diploid genotype of a typical individual
chosen from a sexually reproducing population, at what proportion
of its loci would it be heterozygous?*

—R. C. Lewontin (1974)

In 1859 Charles Darwin published his theory of natural selection. His theory
depended on the existence of hereditary variation among individuals. Darwin
spent much of his book documenting the existence of such variation in both domesticated and wild species of plants and animals.

Natural selection acts on those hereditary variants that increase an individual's
fitness. For now, we can define fitness as an individual's ability to survive and reproduce. Individuals with favorable variations will survive and reproduce more successfully than other individuals, and, consequently, the frequency of those favorable
variations will increase in a population. Therefore, it is clear that evolution by natural selection depends on the existence of genetic variation, and that the rate of evolution depends, at least in part, on the amount of genetic variation in a population.
This latter point was formalized by R. A. Fisher in 1930 in his Fundamental Theorem of Natural Selection, which says, roughly, that the rate of increase of fitness is
equal to the genetic variation in fitness. Therefore, if we are to understand evolution by natural selection, we must have estimates of how much genetic variation
exists in natural populations, and how much of this variation affects fitness.

Genetic variation is important even if it is not subject to natural selection.
Variation that is neutral today may be advantageous (or deleterious) at a different
time or in a different place. Patterns of neutral variation can indicate much about
the history of a population, gene flow among populations, and the past and present effects of natural selection. The usefulness of DNA fingerprinting in forensics
and individual identification depends on the existence of substantial neutral genetic variation among individuals.

We shall see that most populations contain enormous amounts of genetic
variation. This raises two (related) questions: (1) Why is there so much variation?

and (2) How much of this variation is subject to natural selection? These are two of the most important questions of population genetics, and we will return to them frequently throughout this book.

We have two goals in this chapter: First, we will see how to quantify the amount of genetic variation, in the form of genotype frequencies and allele frequencies. Second, we will survey the kinds of genetic variation seen in natural populations. In future chapters we will study patterns of genetic variation within and among populations, and seek explanations for these patterns.

2.1   Quantifying Genetic Variation

We first need to define a few terms. A **locus** is a position on a chromosome; it is a sequence of DNA that may or may not code for a protein. We define a **gene** as a sequence of DNA that codes for a protein. The term gene is sometimes interpreted more broadly, and the two terms are frequently used interchangeably. We will use the more general term locus when we are not concerned with the function of a particular DNA sequence. An **allele** is an alternative form (alternative DNA sequence) of a locus. Many loci have more than one allele. A locus is **monomorphic** if there is only one allele in the population. A locus is **polymorphic** if there are two or more alleles in the population at "appreciable" frequencies (Cavalli-Sforza and Bodmer 1971). The definition of appreciable is arbitrary. Two commonly used criteria for a polymorphic locus are that the most frequent allele has a frequency of less than 0.99 (99 percent criterion) or less than 0.95 (95 percent criterion). Sometimes we ignore frequency criteria and consider a locus to be polymorphic if it has two or more alleles, irrespective of frequency.

Genotype Frequencies and Allele Frequencies

For polymorphic loci, the fundamental quantities of interest are the genotype frequencies and allele frequencies. We begin with a locus with two alleles, designated A_1 and A_2. Consider the following example population:

Genotype	Number of Individuals
A_1A_1	50
A_1A_2	30
A_2A_2	20

We will designate the population size by N, in this case, 100.

The genotype frequencies are calculated in the intuitive way. Let P_{11}, P_{12}, and P_{22} represent the frequencies of A_1A_1, A_1A_2, and A_2A_2, respectively. Then,

$$P_{11} = \frac{50}{100} = 0.50$$

$$P_{12} = \frac{30}{100} = 0.30$$

$$P_{22} = \frac{20}{100} = 0.20$$

To generalize, if we let N_{11}, N_{12}, and N_{22} represent the numbers of the three genotypes, and let N be the population size $(= N_{11} + N_{12} + N_{22})$, then the genotype frequencies are

$$P_{11} = \frac{N_{11}}{N} \qquad (2.1)$$

$$P_{12} = \frac{N_{12}}{N} \qquad (2.2)$$

$$P_{22} = \frac{N_{22}}{N} \qquad (2.3)$$

Note that $P_{11} + P_{12} + P_{22}$ must equal 1.

For loci with only two alleles, we will sometimes designate the alleles by A and a (with no implication of dominance), and the genotype frequencies of AA, Aa, and aa by P, H, and Q, respectively. We will use this notation when we need to minimize the use of subscripts.

Calculating allele frequencies is somewhat more complicated than calculating genotype frequencies. First note that every diploid individual carries two copies of each locus; therefore, there are $2N$ copies of the locus in the population. Now count the number of copies of each allele. Every A_1A_1 individual carries two A_1 alleles and every A_1A_2 individual carries one. The A_2A_2 individuals have zero A_1 alleles. In our example, there are $(50 \times 2) + (30 \times 1) = 130$ copies of the A_1 allele. Therefore, the frequency of the A_1 allele, which we will designate by p (lowercase), is

$$p = \frac{2(50) + 1(30)}{2(100)} = 0.65$$

Similarly, the frequency of the A_2 allele, designated q, is

$$q = \frac{2(20) + 1(30)}{2(100)} = 0.35$$

The general formulas are

$$p = \frac{2N_{11} + N_{12}}{2N} \qquad (2.4)$$

$$q = \frac{2N_{22} + N_{12}}{2N} \qquad (2.5)$$

Note that $p + q = 1$ if there are only two alleles. It is easy to show (Problem 2.2) that equivalent formulas are

$$p = P_{11} + \frac{1}{2}P_{12} \qquad (2.6)$$

$$q = P_{22} + \frac{1}{2}P_{12} \qquad (2.7)$$

Note that we usually use P (uppercase) to designate genotype frequencies and p (lowercase) to designate allele frequencies. You must keep this difference in mind. Unfortunately, both P and p are also used to symbolize other quantities. You must pay close attention to the meanings of symbols.

Allele frequencies can be thought of in two ways. They are the *frequencies* of the alleles in a population, as we have just described. They can also be thought of as *probabilities*. Imagine picking an allele at random from a population. The probability that you pick an A_1 allele is p, and the probability that you pick an A_2 allele is q. We will switch back and forth between these interpretations throughout this book, and you should make sure you understand them before continuing.

Many loci have more than two alleles. The calculations are a straightforward extension of the two allele case. We usually designate the alleles by subscripts, that is, $A_1, A_2, \ldots, A_k$, where k is the number of alleles at the locus. Let N_{ii} and P_{ii} be the number and frequency of the A_iA_i homozygote, and N_{ij} and P_{ij} be the number and frequency of the A_iA_j heterozygote. For example, P_{33} is the frequency of A_3A_3, and P_{34} is the frequency of A_3A_4. Then the genotype frequencies are

$$P_{ii} = \frac{N_{ii}}{N} \tag{2.8}$$

for homozygotes, and

$$P_{ij} = \frac{N_{ij}}{N} \tag{2.9}$$

for heterozygotes.

We designate the frequency of the A_i allele by p_i. To calculate p_i, note that the A_iA_i genotype contains two copies of the A_i allele, and all heterozygotes of the form A_iA_j contain one copy. The frequency of the A_i allele is then

$$p_i = \frac{2N_{ii} + \sum_{j \neq i} N_{ij}}{2N} \tag{2.10}$$

where the summation means add up the numbers of all heterozygotes that contain an A_i allele. Often the equivalent formula is more convenient:

$$p_i = P_{ii} + \frac{1}{2}\sum_{j \neq i} P_{ij} \tag{2.11}$$

Note that in equations (2.10) and (2.11), N_{ij} and P_{ij} refer to the number and frequency of genotype A_iA_j, so that $N_{ij} = N_{ji}$ and $P_{ij} = P_{ji}$. In other words, the order of alleles is irrelevant. When applying these equations, each heterozygote is counted only once in the summation. Box 2.1 shows an example of how to use these formulas. For further practice, see the Problems.

Using equations (2.10) or (2.11) to calculate allele frequencies requires that heterozygotes be distinguishable from homozygotes. With complete dominance, allele frequencies cannot be calculated without making special assumptions. We address this issue in Chapter 3.

Estimates of Genotype and Allele Frequencies

The preceding discussion assumes that we know the genotypes of all individuals in the population of interest. This is almost never true. We usually take a sample from

---

Box 2.1   Calculating allele frequencies for multiple alleles

Consider a gene with four alleles, designated $A_1, \ldots, A_4$. From a population of 1000 individuals, you observe the following numbers:

Genotype	Number of Individuals	Frequency
A_1A_1	135	0.135
A_2A_2	95	0.095
A_3A_3	5	0.005
A_4A_4	50	0.050
A_1A_2	250	0.250
A_1A_3	90	0.090
A_1A_4	175	0.175
A_2A_3	35	0.035
A_2A_4	135	0.135
A_3A_4	30	0.030
Total	1000	1.000

The total population size is $N = 1000$. The genotype frequencies in the third column are the genotype numbers in the second column divided by the total (1000).

The allele frequencies are calculated from equation (2.10) as follows:

$$p_1 = \frac{2N_{11} + \sum_{j \neq 1} N_{1j}}{2N} = \frac{2(135) + 250 + 90 + 175}{2(1000)}$$
$$= 0.3925$$

$$p_2 = \frac{2N_{22} + \sum_{j \neq 2} N_{2j}}{2N} = \frac{2(95) + 250 + 35 + 135}{2(1000)}$$
$$= 0.3050$$

$$p_3 = \frac{2N_{33} + \sum_{j \neq 3} N_{3j}}{2N} = \frac{2(5) + 90 + 35 + 30}{2(1000)}$$
$$= 0.0825$$

$$p_4 = \frac{2N_{44} + \sum_{j \neq 4} N_{4j}}{2N} = \frac{2(50) + 175 + 135 + 30}{2(1000)}$$
$$= 0.2200$$

---

a population and wish to make inferences about the population from that sample. (See Box 2.2 for an essential discussion of populations and samples.) We can calculate the genotype frequencies and allele frequencies of the sample using the formulas given above, but we must remember that these sample frequencies are only *estimates* of the population frequencies. We usually put a $\wedge$ (circumflex; nickname, "hat") over a symbol to indicate that it is an estimate of the population value.

Consider allele frequencies. Let p_i represent the (unknown) frequency of the A_i allele in the population, and let $\hat{p}_i$ be the frequency of the A_i allele in the sample, as calculated from equation (2.10) or (2.11). Then $\hat{p}_i$ is an estimate of p_i, and is a random variable. (See Appendix A if you are unfamiliar with random variables.) Any random variable has an expectation (mean) and a variance. It can be shown from probability theory that the expectation of $\hat{p}_i$ is p_i, and that its variance is approximately

$$Var(\hat{p}_i) = \frac{\hat{p}_i(1 - \hat{p}_i)}{2N} \qquad \textbf{(2.12)}$$

Box 2.2 Populations and samples

We need to carefully define several statistical terms: A **population** consists of every possible object in the study, or every possible outcome of an experiment. These may be individuals in a biological population, genes in a genome, automobiles in a city, or any other group. The essential idea is that a population is a *complete collection* of objects, or outcomes. A **sample**, on the other hand, is a subset of the population. Thus, a sample is a small part, which we can study, of a larger population, which we usually cannot study in its entirety. We make inferences about the population based on the sample. We usually want the sample to be **random**, in which every individual in the population has an equal chance of being included in the sample.

A population can be described by one or more parameters. A **parameter**, for our current purposes, is a characteristic of a population. Examples are means, variances, allele frequencies, etc. We usually do not know the true values of these parameters, and we use the sample to estimate them. For example, a sample allele frequency is an estimate of the population allele frequency.

These sample estimates are random variables (see Appendix A). If you study a different sample from the same population, you will probably get a different value for the estimate. Like all random variables, estimates have an expectation (mean) and a variance (see Appendix A). It is desirable that the expectation of the estimate be equal to the population parameter it is estimating. If this is so, the estimate is **unbiased**. Otherwise, it is biased. It is also desirable that the variance of the estimate be as small as possible. One branch of statistics is concerned with

techniques of finding unbiased, minimum variance estimates of population parameters.

It may seem natural to create an estimate by replacing the population values by their sample values. But these so called "natural estimators" are frequently biased. For example, the population variance can be estimated by the quantity

$$\frac{1}{n}\sum_{i=1}^{n}(x_i - \bar{x})^2$$

where n is the sample size. But this is a biased estimate. The unbiased estimate of the population variance is

$$s^2 = \frac{1}{n-1}\sum_{i=1}^{n}(x_i - \bar{x})^2$$

Parameters are frequently designated by Greek letters; for example, μ and σ^2 are often used for the population mean and variance, respectively. Their estimates are often indicated by $\bar{x}$ and s^2. In population genetics we often use the same symbol for both the parameter and the estimate. In this case, we usually put a circumflex ("hat") over the symbol to indicate the estimate. For example, if p is the allele frequency in a population, then $\hat{p}$ is its estimate.

The distinctions between a population and a sample, and between a parameter and an estimate are fundamental. Statistics is the art and science of studying a sample, and using that sample to make inferences about the population it came from. A common problem in population genetics is to find an estimator of some population parameter of interest, and to determine the characteristics (e.g., the mean and variance) of that estimator.

The square root of the variance is called the standard deviation. We can use the standard deviation of $\hat{p}_i$ to obtain some idea of the reliability of our estimate. We can construct an approximate 95 percent confidence interval for p_i (the unknown true value) as follows:

$$\hat{p}_i - 2\sqrt{\frac{\hat{p}_i(1 - \hat{p}_i)}{2N}} \leq p_i \leq \hat{p}_i + 2\sqrt{\frac{\hat{p}_i(1 - \hat{p}_i)}{2N}} \tag{2.13}$$

This means that the true population allele frequency, p_i, will be between the upper and lower limits about 95 percent of the time. This can be a disturbingly wide range.

For example, if our sample allele frequency is 0.3 and the sample size is 20, the lower and upper limits are 0.16 and 0.44. The range is even larger for smaller samples. The lesson is that good estimates of allele frequencies require large sample sizes.

Polymorphism and Heterozygosity

There are two common ways to quantify the amount of genetic variation in a population: the proportion of polymorphic loci (P) and the average heterozygosity (H). We will illustrate these ideas with an example. Table 2.1 shows results of a hypothetical sample of ten individuals from a population. Six loci were studied in each individual. The loci are designated A through F and alleles are designated 1 or 2 at each locus. The three columns under the heading "Genotypes" are the numbers of individuals of each genotype. The column labeled "p_1" is the frequency of the 1 allele at each locus, as calculated from equation (2.10). (You should confirm these numbers for practice.)

The proportion of polymorphic loci in a population is defined as

$$P = \frac{\text{number of loci that are polymorphic}}{\text{total number of loci studied}} \tag{2.14}$$

In Table 2.1, the column labeled Polymorphic indicates whether each locus is polymorphic or monomorphic (95 percent criterion). We see that loci B, C, and D are polymorphic. Therefore, the polymorphism in this sample is $P = 3/6 = 0.50$. This is, of course, an *estimate* of the polymorphism in the population, since not all individuals in the population, nor all loci, were examined.

The proportion of polymorphic loci is not a particularly sensitive indicator of how much genetic variation there is in a population. One problem is that the definition of polymorphism is arbitrary, as previously mentioned. Second, one locus may have two alleles and a second locus may have 20 alleles. Both loci are considered polymorphic and P makes no distinction between them. In fact, we would consider the second locus much more variable than the first. For these reasons, the proportion of heterozygotes at a locus is considered a more sensitive indicator of variability.

TABLE 2.1 A hypothetical population showing genotypes at six loci.
The column labeled p_1 is the frequency of the 1 allele at each locus. The column labeled Polymorphism contains an M if the locus is monomorphic or a P if it is polymorphic. Heterozygosity at each locus is the observed proportion of heterozygous individuals.

Locus	Genotypes			p_1	Polymorphism	Heterozygosity
	11	**12**	**22**			
A	10	0	0	1.0	M	0
B	3	4	3	0.50	P	0.40
C	6	3	1	0.75	P	0.30
D	1	1	8	0.15	P	0.10
E	0	0	10	0	M	0
F	10	0	0	1.0	M	0
						Average = 0.133

Average observed heterozygosity is defined as the proportion of heterozygotes, averaged over all loci. If there are n loci, then

$$H_{obs} = \frac{1}{n} \sum_{i=1}^{n} H_i \tag{2.15}$$

where H_i is the observed frequency of heterozygotes at the ith locus, and H_{obs} is the average of the H_i over all loci studied. In Table 2.1, the column labeled "Heterozygosity" is the observed frequency of heterozygotes for each locus. The average of these values is $H_{obs} = 0.133$. Again, this is an *estimate* of the population heterozygosity. For more than two alleles, H_i represents the frequency of all heterozygotes at the ith locus.

Note that H_{obs} is the average *observed* heterozygosity. We shall see in Chapter 3 that it is also possible to calculate an *expected* heterozygosity, based on certain assumptions.

2.2 The Classical and Balance Hypotheses

Heterozygosity is the main means of quantifying the amount of genetic variation within a population. How much heterozygosity do we expect to find in a typical population? Dobzhansky (1955) characterized the two schools of thought regarding this question, and called them the classical hypothesis and the balance hypothesis.

The classical hypothesis claims that there is a single best allele at any locus, and that most individuals are homozygous for this "wild type" allele. Heterozygosity is due primarily to the presence of rare mutations, most of which are deleterious and will be eliminated by natural selection. Very rarely, a beneficial mutation will occur and will quickly take over the population and become the new wild type. Thus, the classical hypothesis predicts that heterozygosity should be rare. One advocate of the classical hypothesis, Muller (1950), suggested that it would be on the order of 0.1 percent in humans.

The balance hypothesis claims that most individuals are heterozygous at many loci. There is usually no single best allele at a locus, and many alleles are maintained by various forms of balancing natural selection, such as heterosis, frequency dependent selection, environmental variation, and so on. We will discuss balancing selection in future chapters; for now it is enough to understand that balancing selection is a general term for kinds of natural selection that act to maintain two or more alleles in a population. Thus, the balance hypothesis predicts that heterozygosity should be common. Dobzhansky was a proponent of the balance hypothesis, based on his work with *Drosophila pseudoobscura* (Section 2.6). Wallace stated the most extreme form of the balance hypothesis when he claimed that "the proportion of heterozygosis among gene loci of representative individuals of a population tends toward 100 per cent" (Wallace 1958a, p. 418).

Because evolution depends on the existence of genetic variation, the classical and balance hypotheses have important, and very different, implications about the evolution of natural populations. Therefore, one of the earliest and most important problems in population genetics has been to estimate how much genetic variation actually exists in nature. As we shall see, it has been a difficult problem, but the answer is now clear: Most natural populations contain large amounts of genetic variation. However, it is still not clear how much of this variation is due to balancing selection, and how much is due to other causes. For an excellent historical overview of this subject, see Lewontin (1974).

2.3 Genetic and Environmental Variation

Variation among individuals is ubiquitous. Humans vary in many obvious characteristics, for example hair color, eye color, skin pigmentation, height, and weight. It may not be obvious to untrained eyes, but careful examination of almost any other species will reveal similar variation. Other characteristics such as blood type and disease susceptibility are less visible, but are also highly variable.

It is not immediately clear whether this variation is due to genetic or environmental differences, or to a combination of both. For example, it is well known that both genetic and environmental factors affect body size and developmental rate in many organisms.

A classic study documenting the existence of both environmental and genetic variation was conducted by Clausen et al. (1941). *Potentilla glandulosa* is a widespread member of the rose family. In California it grows from near sea level to about 11,000 feet in elevation. Clausen et al. collected samples from three different locations, Stanford (100 ft. elevation), Mather (4600 ft. elevation) and Timberline (10,000 ft. elevation). They took cuttings of each sample and grew them in experimental plots at all three locations. Thus, genetically identical clones from each location were grown in all three locations. The results are illustrated in Figure 2.1. Each row represents a plant from one of the three locations; the three photographs in a row represent clones from that plant grown at three different locations. Because the three plants in a row are genetically identical clones, differences in a row represent differences due to different environmental conditions.

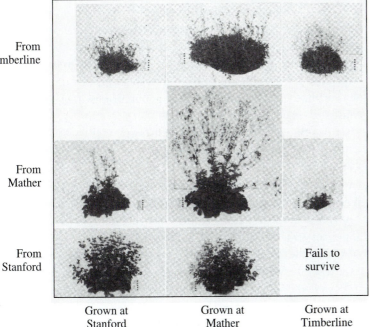

From
Timberline

From
Mather

From
Stanford

Fails to
survive

Grown at
Stanford

Grown at
Mather

Grown at
Timberline

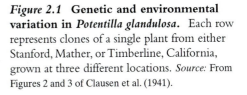

Figure 2.1 **Genetic and environmental variation in *Potentilla glandulosa*.** Each row represents clones of a single plant from either Stanford, Mather, or Timberline, California, grown at three different locations. *Source:* From Figures 2 and 3 of Clausen et al. (1941).

Similarly, the three plants in a column are genetically different (different origins) but grown in identical environments. Therefore, differences in a column are due to genetic differences among plants. Even a cursory glance at the figure shows that a plant's morphology and vigor are determined by both its genetic composition and its environment.

This study serves as a caution: Not all variation is genetic, and environmental differences can have important effects on morphology, survival, and reproduction. We must not forget this as we study genetic variation and its effect on evolution.

2.4 Variation for Discrete Morphological Characters

Some of the earliest studies of genetic variation in natural populations examined discrete morphological variants in *Drosophila* species (e.g., Dubinin et al. 1937; Spencer 1947, 1957). Typical results suggested that about 1 to 2 percent of wild caught flies showed a detectable morphological variation from the "wild type." These variants were frequently similar to the mutations in eye color, wing shape, and bristle morphology studied by *Drosophila* geneticists in the laboratory. However, not all of these variants were genetic, and some that were had a complex genetic basis. Lewontin (1974) reviewed these studies and suggested that the frequency of flies that showed morphological variation due to a single gene varied between 0.08 and 1.27 percent, depending on locality and time of collection.

This technique, laborious as it is (Dubinin et al. looked at more than 129,000 flies!), has two serious flaws. First, it depends on the investigator's ability to identify a morphological variant. This depends on the investigator's experience, visual acuity, attention span, and many other variables. Second, it can only detect variation that is actually expressed. Any recessive allele present in a heterozygous state will be undetected.

It is possible to detect hidden recessive alleles by breeding the progeny of wild caught females. A single wild caught female (assumed to have mated with a wild male) is placed in a vial. If either she or her mate was heterozygous for a recessive allele, half of the F_1 will be heterozygous for that allele. If the F_1 are allowed to mate with each other, some of the F_2 generation will be homozygous for that allele, and will express it phenotypically. Thus, the proportion of wild caught females that show morphological variants in the F_2 generation is an estimate of the frequency of hidden recessive alleles.

Spencer (1957) examined 736 families derived from wild caught *Drosophila mulleri* females. Of those, 224 families produced a total of 263 morphological variants in the F_2 (about 0.36 variants per family, or 0.18 per fly). Spencer estimated that he could detect only a small proportion of the variants, and concluded that an individual fly is heterozygous for about one recessive allele which can cause discrete morphological variation. Studies of other *Drosophila* species give similar results (Spencer 1957; Lewontin 1974; Powell 1997).

These *Drosophila* studies suggest that single gene variation affecting visible morphological traits is rare. Whether this is typical of other kinds of genetic variation is unclear. Powell (1997) estimates that only about 25 percent of the genes on the X chromosome of *Drosophila melanogaster* can give rise to visible variants, and there is no reason to think this is atypical. As Powell says, "it is clear that the study of frequencies of viable visible mutants does not examine the majority of genes" (Powell 1997, p. 22).

Other early studies of morphological variation in natural populations examined readily observable characteristics such as shell color and banding patterns in snails, wing patterns in butterflies, and flower color in plants. These studies documented the existence of genetic variation, estimated allele and genotype frequencies, and described patterns of variation. In one extreme example, Tan (1946) reported 15 alleles at a gene affecting wing pattern in the ladybird beetle, *Harmonia axyridis* (Figure 2.2).

Many early studies also provided evidence that visible polymorphisms can be subject to natural selection. For example, snails (*Cepaea nemoralis* and *C. hortensis*) with different banding patterns differ in their temperature and humidity tolerances (Lamotte 1959). The different banding patterns also provide cryptic coloration in different habitats (Cain and Sheppard 1950, 1954a).

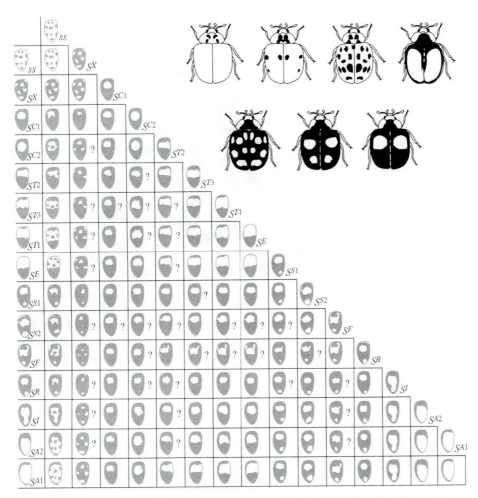

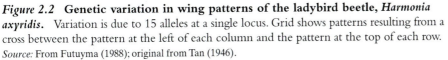

Figure 2.2 **Genetic variation in wing patterns of the ladybird beetle, *Harmonia axyridis*.** Variation is due to 15 alleles at a single locus. Grid shows patterns resulting from a cross between the pattern at the left of each column and the pattern at the top of each row. *Source:* From Futuyma (1988); original from Tan (1946).

The genes controlling these visible polymorphisms are probably atypical. They were studied because they produce obvious variants that can be studied genetically and ecologically. In fact, most genes do not produce such obvious effects. These studies provide interesting and important examples of genetic variation, but do not give us any information about polymorphism in typical genes. For reviews of these early studies of polymorphism, see Mayr (1963) and Ford (1975).

2.5 Variation in Quantitative Characters

We concluded in the previous section, that *single gene* variation affecting visible morphological traits is rare. This does not mean that variation for morphological traits in general is rare.

Quantitative traits are characters that are either measured on a continuous scale—for example, height or weight—or as a discrete count—for example, the number of abdominal bristles on a fruit fly or the number of scales along the lateral line of a fish. Phenotypic values of quantitative traits are almost always jointly determined by the effects of many genes and environmental effects. Important morphological characteristics such as body size are quantitative traits. Moreover, many important life history characteristics, such as development time, fecundity, and so on, are quantitative traits.

It has long been known that artificial selection can change the mean value of almost any quantitative trait in almost any population of almost any species. Three examples of artificial selection on quantitative traits are shown in Figure 2.3. The success of many such laboratory experiments indicates the existence of extensive genetic variation for quantitative traits, although we seldom know the precise genetic basis for any trait.

Natural populations also show substantial variation for quantitative traits. Some examples are body size and beak size and shape in Darwin's finches (Grant and Grant 1995) and song sparrows (Smith and Zach 1979); body size in fruit flies (Prout and Barker 1989); abdominal bristle number in fruit flies (Mackay 1995, 1996); floral morphology in monkeyflowers (Bradshaw et al. 1998); and several life history traits in *Daphnia* (Lynch et al. 1999). Falconer and Mackay (1996) and Roff (1997) discuss many other examples. The conclusion is that there is extensive genetic variation for most quantitative traits in most natural populations.

We examine quantitative traits and their evolution in Chapter 13. For now, we simply note that genetic variation for quantitative characters is widespread, and that this variation is, at least sometimes, subject to natural selection.

2.6 Chromosomal Variation

Some species exhibit variations in chromosome structure among individuals, for example inversions in several species of *Drosophila*, and translocations in several species of the plant genus *Clarkia*.

In some species of the order Diptera (flies, mosquitoes, etc.) the chromosomes in the salivary glands of the larvae replicate many times without separating. The result is large, easily observed chromosomes, with regular and repeatable patterns of light and dark bands (Figure 2.4a). These chromosomes are called polytene chromosomes. In several species of *Drosophila*, these chromosomes and their

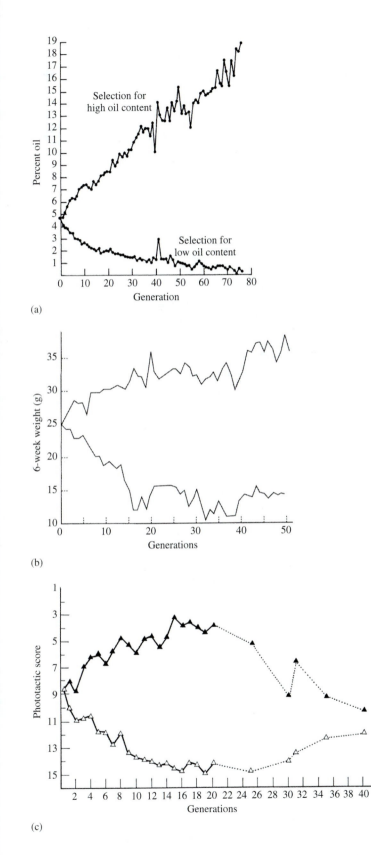

(a)

(b)

(c)

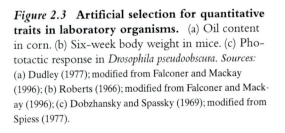

Figure 2.3 **Artificial selection for quantitative traits in laboratory organisms.** (a) Oil content in corn. (b) Six-week body weight in mice. (c) Phototactic response in *Drosophila pseudoobscura. Sources:* (a) Dudley (1977); modified from Falconer and Mackay (1996); (b) Roberts (1966); modified from Falconer and Mackay (1996); (c) Dobzhansky and Spassky (1969); modified from Spiess (1977).

Figure 2.4 **Polytene chromosomes of *Drosophila*, showing banding patterns.** (a) Two synapsed chromosomes homozygous for the same inversion pattern. (b) Two synapsed chromosomes differing by two inversions. *Source:* From Mettler et al. (1988).

(a)

(b)

banding patterns have been studied in detail, and numerous genes have been mapped to specific bands. Polytene chromosomes make it relatively easy to study chromosomal variation within species because the chromosomes are so large and clear. For example, if an individual contains two homologous chromosomes that differ by an inversion (i.e., one chromosome contains an inversion and the other does not), the chromosomes will form an inversion loop when they synapse. (For a discussion of inversion loops, see any genetics textbook.) Since homologous chromosomes remain synapsed in the larval salivary glands, the inversion loops are easy to observe (Figure 2.4b).

The best-known example of chromosomal variation within a species is the case of inversion polymorphisms of the third chromosome of *Drosophila pseudoobscura*. There are more than 20 chromosome types, differing by one or more inversions on the third chromosome. Many populations are polymorphic for several of them. Dobzhansky and his colleagues have studied them in great detail, and documented their geographic distributions and seasonal and long-term changes in frequencies. Figure 2.5 illustrates one example, summarizing many years of work by many scientists (Anderson et al. 1991). The figure follows the frequencies of four chromosome arrangements from about 1940 to 1981 in four geographically widespread locations. Note that both the frequencies and the long-term patterns vary dramatically in the different locations.

Dobzhansky believed that these inversion polymorphisms are maintained by natural selection, with chromosomal heterozygotes (heterokaryotes) usually

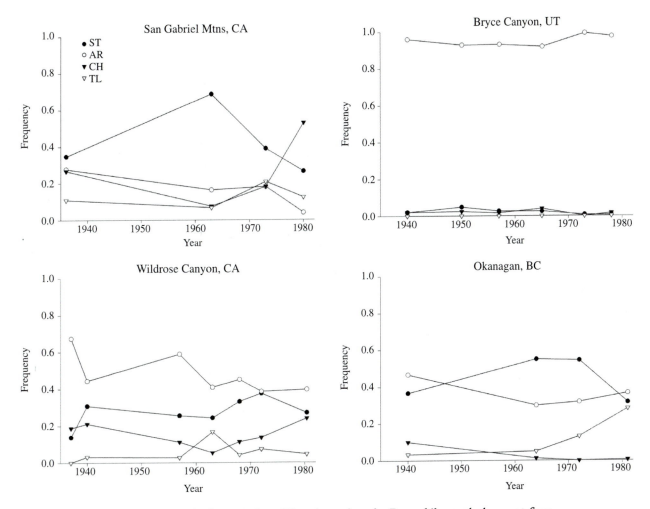

Figure 2.5 **Long-term changes in frequencies of four inversions in *Drosophila pseudoobscura* at four locations.** Abbreviations indicate names of inversions: ST = Standard; AR = Arrowhead; CH = Chiricahua; TL = Treeline. Note the long-term increase of TL in Okanagan, B.C. *Source:* Data from Anderson et al. (1991).

having higher fitnesses than chromosomal homozygotes (homokaryotes). He also believed that different inversions carry different alleles for the genes within them, and that different inversions are favored by natural selection in different circumstances.

These studies have had great influence on a generation of population geneticists and evolutionary biologists. Before the availability of molecular techniques and data, they were the most sophisticated studies in population genetics. Most of the papers were published in Dobzhansky's *Genetics of Natural Populations* series (see Lewontin et al. 1981). These studies have been continued by some of Dobzhansky's colleagues. See Anderson et al. (1991) for the most recent summary of more than 40 years of research.

The presence of inversion polymorphisms varies widely in *Drosophila* species. For example, all five arms of the chromosomes of *D. willistoni* carry inversions; there are more than 50 different inversions known in the species (Sorsa 1988). On the other hand, *D. simulans* is essentially lacking in inversions (Powell 1997).

Some species of mosquitoes have inversions, and they have been useful in distinguishing between cryptic species of *Anopheles*.

2.7 Hidden Variation for Fitness

In some organisms, most notably certain *Drosophila* species, it is possible to make individuals homozygous for every gene on a chromosome. This technique depends on the existence of so-called balancer stocks. These stocks have two main characteristics. First, two homologous chromosomes have different dominant marker alleles. Each of these alleles is phenotypically dominant, but is lethal when homozygous. The second feature is that each of the homologous chromosomes contains a different set of overlapping inversions. The result is that there is no recombination between the chromosomes. An example of a balancer stock is the *Cy/Pm* stock of *Drosophila melanogaster*. Curly (*Cy*) is a dominant allele producing abnormal wings. *Plum* (*Pm*) is a dominant allele producing an abnormal eye color. Both are on the second chromosome and both are lethal when homozygous. *Curly/Plum* flies have one second chromosome containing *Cy* and one containing *Pm*. There in no recombination between these chromosomes because of the multiple inversions. The maintenance of this stock is shown in Figure 2.6.

Balancer stocks can be used to make individuals that are homozygous for entire chromosomes. The experimental design is illustrated in Figure 2.7. A single wild caught male, with chromosomes designated $+_1$ and $+_2$, is bred to a *Cy/Pm* female. The F_1 contain one wild type chromosome, either $+_1$ or $+_2$ (intact, because there is no crossing over in *Drosophila* males), and one marker chromosome, either *Cy* or *Pm*. A single F_1 male ($+_1/Cy$ in the example) is then backcrossed to the marker stock. Because there is no crossing over, the $+_1$

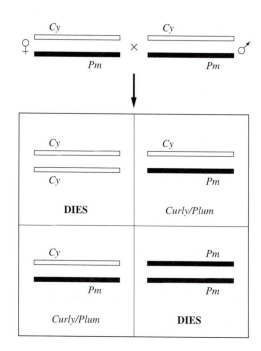

Figure 2.6 **Maintenance of a
Curly/Plum (Cy/Pm) balancer stock
of Drosophila melanogaster.**

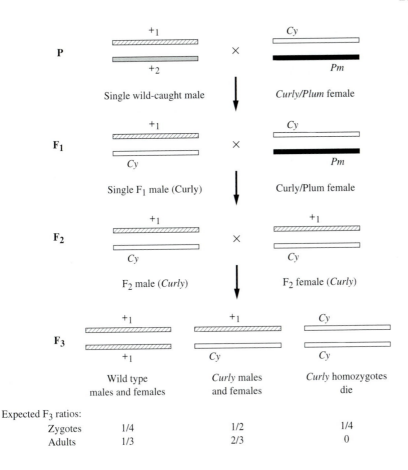

Figure 2.7 **Breeding scheme for using a balancer stock (*Curly/Plum*) to make individuals homozygous for an entire chromosome.**

chromosome is transmitted intact to the F_2. The F_2 flies that survive (*Cy/Cy* flies die) have the $+_1$ chromosome and either the *Cy* or the *Pm* chromosome. F_2 males with the *Cy* marker are mated to F_2 females with the *Cy* marker. The surviving F_3 flies (again, *Cy/Cy* flies die) are phenotypically curly (*Cy/*$+_1$) or wild type ($+_1/+_1$), in an expected ratio of 2 to 1. The wild type flies are homozygous for the entire $+_1$ chromosome. This procedure is sometimes called **chromosome extraction**, because it extracts a chromosome from nature and preserves it intact.

Now, if the $+_1$ chromosome contains a recessive mutation that is lethal when homozygous, then it will not survive in the F_3, and all F_3 individuals will be *Curly*. By performing this series of crosses with many wild caught males, it is possible to estimate the proportion of those males that carry a recessive lethal mutation. This has been done for several species of *Drosophila*. Table 2.2 summarizes some of the main studies. The proportion of chromosomes carrying recessive lethal mutations is surprisingly high, ranging from 9.5 to 61.3 percent. Note, this does not tell us *how many* lethal mutations are on a chromosome, as one or several will produce the same result. However, it is possible, by making certain assumptions, to estimate the number of lethal mutations per chromosome. Lewontin (1974) estimated from these data that there is about one recessive lethal mutation per genome (or two recessive lethals in a diploid fly), and that about one locus in a thousand typically carries a recessive lethal.

TABLE 2.2 Summary of *Drosophila* studies estimating the frequency of chromosomes that are lethal or semilethal when homozygous.

Semilethal is defined as having a relative viability of 0 to 50 percent of normal. *Source*: From Dobzhansky and Spassky (1954).

Species	Chromosome	Population	Chromosomes Tested	Percent Lethals and Semilethals
D. prosaltans	II	Brazil	304	32.6
	III	Brazil	284	9.5
D. willistoni	II	Brazil	2004	41.2
	II	Rio Grande Sul	645	28.4
	II	Florida	109	31.1
	II	Cuba	25	36.0
	III	Brazil	1166	32.1
	III	Florida	122	32.8
	III	Cuba	39	25.6
D. melanogaster	II	Nebraska	133	25.6
	II	Pennsylvania	117	28.2
	II	New York	449	31.8
	II	New York	78	34.6
	II	Massachusetts	3549	36.3
	II	Washington	138	39.1
	II	Texas	98	41.8
	II	Ohio	343	43.1
	II	Virginia	805	43.0
	II	Florida	468	61.3
	II	Israel	243	38.7
	II	Crimea	1630	24.8
	II	Ukraine	2700	24.3
	II	South Caucasus	2738	18.7
	II	North Caucasus	795	12.3
D. pseudoobscura	III	California	109	33.0
	III	California	326	21.3
D. persimilis	III	California	106	25.5

Recessive lethals, like recessive visible mutations, seem to be rare, and of the same order of magnitude (about one or two per individual), at least in *Drosophila*.

It is possible to use variations of this technique to estimate the proportion of chromosomes for which homozygosity has some effect on viability but is not lethal. From Figure 2.7, we expect about one-third of the F_3 flies to be wild type ($+_1+_1$). If homozygosity for the $+_1$ chromosome decreases viability, the observed frequency of wild type flies will be less than one-third. For example, if in a particular cross the frequency of wild type flies in the F_3 is 0.30, then the relative viability of the $+_1+_1$ homozygotes is 0.30/0.33 (= observed proportion/ expected proportion), or about 0.91 compared to heterozygotes. (The experimental design has to be modified slightly to be able to compare homozygotes to heterozygotes without the marker; the details need not concern us here.) In a typical experiment, Dobzhansky et al. (1963) estimated relative viabilities of 208 second chromosomes and 252 third chromosomes from an isolated population of *Drosophila pseudoobscura*. The average frequency of wild type flies in the F_3 generation was about 0.25 in each case, corresponding to relative viabilities of about 0.75 compared to heterozygotes. The *distribution* of viabilities is shown in Figure 2.8. We see that the distribution is bimodal. One group has viability at or near zero. These are the recessive lethals and near lethals discussed above. The second group has average viability slightly lower than the heterozygotes. Most chromosomes show a lower viability when homozygous than when heterozygous, but a few actually show a higher viability when homozygous. These results are typical of many other studies with several species of *Drosophila*.

Clearly, there is a great deal of genetic variation for viability, at least in *Drosophila*. This may be surprising; we might expect natural selection to favor the most viable chromosomes and eliminate all the others. Why has this not happened? One answer has to do with viabilities of chromosomal heterozygotes. It appears that most chromosomes carry recessive viability-reducing alleles at one or more loci. Chromosomal homozygosity produces genic homozygosity at these loci, which lowers viability. Conversely, chromosomal heterozygosity keeps recessive viability reducing alleles in heterozygous state, where they are not expressed. Therefore, most chromosomes have a higher fitness when heterozygous than when homozygous, a condition known as **chromosomal heterosis**. This prevents any single chromosome from becoming fixed in the population. But what about the few chromosomes that show higher viability when homozygous? Why have they not become fixed? The answer probably involves complex forms of balancing selection having to do with gene interaction, pleiotropy, different forms of natural selection, and so forth. These topics are discussed in Chapters 12 and 13.

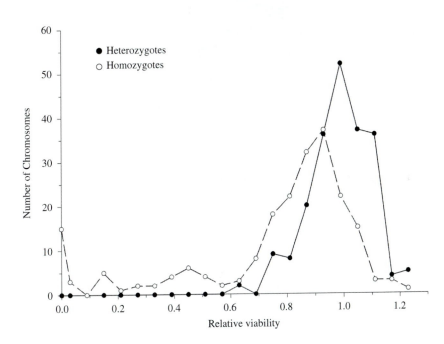

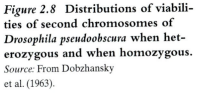

Figure 2.8 **Distributions of viabilities of second chromosomes of *Drosophila pseudoobscura* when heterozygous and when homozygous.**
Source: From Dobzhansky et al. (1963).

2.8 Human Blood Groups

It has long been recognized that certain kinds of blood transfusions are incompatible. This led to the discovery of several blood group genes. These genes produce proteins found on the surface of red blood cells. Because of their medical importance, these genes have been studied extensively. Many exhibit extensive polymorphism. Hedrick and Murray (1978) list 17 polymorphic loci (0.99 criterion), with heterozygosities ranging from 0.016 to 0.700, with an average of 0.368. The average heterozygosity for 60 blood group loci (including monomorphic loci) was 0.105.

Blood group loci sometimes show great geographic and racial differences. Figure 2.9 shows the worldwide distribution of the *B* allele of the ABO blood group system. The *B* allele occurs in relatively high frequency in Asia, but is almost entirely absent in the Americas. Table 2.3 gives several other examples of variation among three racial groups. We see that some loci are very similar among the three groups, while others vary dramatically. In an extreme case, the *Fy* allele at the Duffy locus has a frequency of 0.03 in Caucasian populations and 0.939 in African populations. Cavalli-Sforza and Bodmer (1971) and Mourant et al. (1976) describe many other examples of geographic and ethnic variation of blood group loci.

Are blood group genes typical of most genes? Most blood group polymorphisms have been discovered because of their medical importance, and thus are probably not representative of polymorphism in the human genome in general. There is also a bias because blood groups are recognized only when a variant is

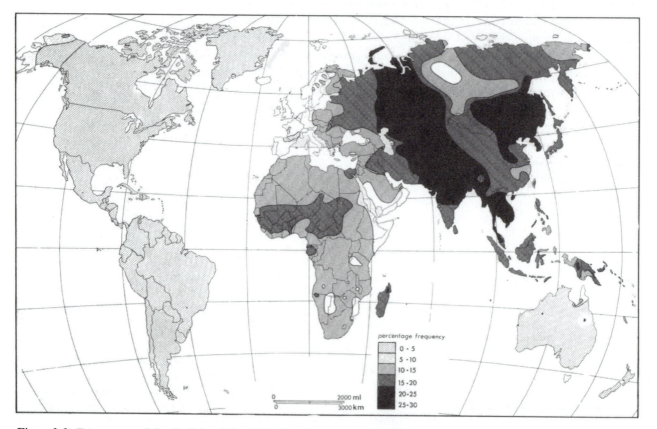

Figure 2.9 **Frequency of the *B* allele of the ABO blood group in aboriginal populations of the world.** *Source:* Modified from Mourant (1983).

TABLE 2.3 Allele frequencies at various blood loci in three racial groups.

Caucasian is northern European; African is central or southern Africa; Asian is China. *Source*: Simplified from Cavalli-Sforza and Bodmer (1971).

Locus	Alleles	Caucasian	African	Asian
ABO	A	0.2876	0.1780	0.1864
	B	0.0612	0.1143	0.1700
	O	0.6612	0.7077	0.6436
Lutheran	Lu^a	0.0354	0.0272	0
	Lu^b	0.9646	0.9728	1.0
Kell	K	0.0462	0.0029	0
	k	0.9538	0.9971	1.0
Secreter	Se	0.5233	0.5727	—
	se	0.4767	0.4273	—
Lewis	Le	0.8156	0.3188	0.7575
	le	0.1844	0.6812	0.2425
Duffy	Fy	0.0300	0.9393	0.0985
	Fy^a	0.4200	0.0607	0.9015
	Fy^b	0.5492	0	0
Kidd	Jk^a	0.4358	0.7818	0.3103
	Jk^b	0.5142	0.2182	0.6897
Auberger	Au^a	0.6213	0.6419	—
	Au^b	0.3787	0.3581	—
XG	Xg^a	0.675	0.55	0.54
	Xg^b	0.325	0.45	0.46

discovered. Again, we have interesting examples of polymorphism at specific kinds of loci, but little information about whether polymorphism is rare or common in the genome as a whole.

2.9 Allozyme Variation

The studies described above confirm that there is genetic variation in many populations. However, we still cannot answer the question, What proportion of loci in a typical individual are heterozygous? Or, similarly, What proportion of loci in the genome of a species are polymorphic? In fact, much of the variation we have discussed cannot be attributed to specific loci, and loci that are variable were detected *because* they are variable. It may be much easier to detect polymorphic loci than monomorphic ones, so that the polymorphisms we have discussed so far may represent a biased and misleading sample of all loci. Therefore, knowing that *some* loci are polymorphic gives us no information about *what proportion* of loci are polymorphic.

A way out of this dilemma was discovered in 1966. The key was to borrow techniques from the biochemists, and look at proteins. Proteins are the products of genes; therefore, variation in proteins indicates genetic variation. Proteins consist of sequences of amino acids. Some amino acids carry a positive or negative charge; others are neutral. Thus, a protein has a net charge, which is the sum of the charges of its constituent amino acids. If a variant protein (caused by a variant DNA sequence) has an amino acid substitution that changes the net charge of the protein, that variant protein will behave differently in an electrical field. This is the basis of the technique of **protein electrophoresis**, and protein variants revealed in this way are called **allozymes**, short for alloenzymes (Prakash et al. 1969). The laboratory procedures are summarized in Box 2.3, and Box 2.4 shows how allele frequencies, polymorphism, and heterozygosity can be estimated from allozyme data.

BOX 2.3   Protein electrophoresis

The general principle of protein electrophoresis is that proteins with different net charge will move through an electrical field at different rates, and can therefore be separated. There are two basic steps: (1) separation of the proteins in a starch or polyacrylamide gel; (2) visualization of the desired proteins.

In the first step, tissue samples (or whole organisms, in the case of small insects) are homogenized and loaded into sample wells of a gel. The gel is then subjected to an electrical current. The proteins will move through the gel (migrate) at rates that depend primarily on their net charge. After a certain amount of time (usually several hours) the gel is removed and soaked in a stain that is specific for the protein of interest. A dark band will appear on the gel at the location of the protein. Proteins with different charges will reveal bands at different places on the gel (Figure 2.10). Since the protein is the product of a specific gene, different bands for the same protein represent different alleles of the gene producing that protein. Therefore, if an individual produces two bands on a gel, that individual is heterozygous for that protein. Multimeric proteins are more complicated to interpret, but the principle is the same.

Gels typically contain ten to twenty sample wells, so many individuals can be compared on one gel. Different gels can be stained for different proteins, allowing several loci to be surveyed in each individual, as long as the experimenter has enough tissue for several gels. Box 2.4 shows how allele frequencies, polymorphism, and heterozygosity can be estimated from electrophoretic data.

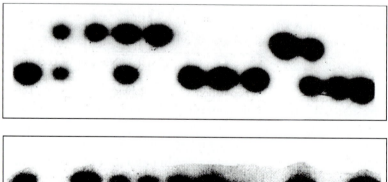

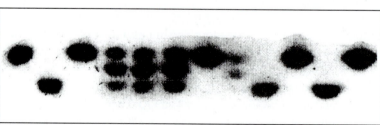

Figure 2.10 **Typical patterns after staining electrophoretic gels.**
The 12 vertical columns (lanes) represent tissue samples from 12 individuals. (a) A monomeric enzyme with two alleles. Homozygous individuals show one band and heterozygotes show two. (b) A dimeric enzyme with two alleles. Homozygous individuals show one band and heterozygotes show three. *Source:* From Ayala (1982).

Box 2.4 Estimating allele frequencies and heterozygosity from electrophoretic gels

We will use the gels in Figure 2.10 to illustrate the principle. Call the proteins in Figure 2.10 locus A and locus B, respectively. Assume there is a third locus, C, at which all individuals showed identical patterns on the gel.

At locus A, individuals showing one band are homozygous, and individuals showing two bands are heterozygous. Call the alleles Fast (F) and Slow (S) for their relative migration rates. If the sample well is assumed to be at the bottom of the gel, then individuals 1, 6, 7, 8, 11, and 12 are homozygous for the slow allele and individuals 3, 5, and 9 are homozygous for the fast allele. The remaining individuals are heterozygous.

Protein B is a dimeric protein, so heterozygotes show three bands. Therefore, individuals 4, 5, 6, and 8 are heterozygous. The remaining individuals are homozygous for either the Fast or Slow allele.

At locus C, all individuals are homozygous for the same allele. Call it slow just to give it a name.

Now we can calculate genotype frequencies and allele frequencies at each locus as follows:

Locus	Genotype	Number of Individuals	Genotype Frequency	Frequency of Slow Allele	Frequency of Heterozygotes
A	$A_S A_S$	6	0.50	0.625	
	$A_S A_F$	3	0.25		0.25
	$A_F A_F$	3	0.25		
B	$B_S B_S$	3	0.250	0.417	
	$B_S B_F$	4	0.333		0.333
	$B_F B_F$	5	0.417		
C	$C_S C_S$	12	1.000	1.000	
	$C_S C_F$	0	0		0
	$C_F C_F$	0	0		

Genotype frequencies are calculated using equations (2.1), (2.2), and (2.3) and allele frequencies are calculated using equation (2.4) or (2.5).

Two of three loci are polymorphic, so, from equation (2.14) $\hat{P} = 0.667$.

Heterozygosity is the average of the last column; therefore, from equation (2.15) $\hat{H}_{obs} = 0.193$

One great advantage of protein electrophoresis over other techniques is that we can examine loci without knowing ahead of time whether they are polymorphic or not. The only requirement is that there be some way to detect the gene product on a gel. In other words, we can consider allozyme loci to be something approaching a random sample (with respect to variability) of genes in the organism. For the first time, we have a way to estimate what proportion of loci are polymorphic.

These techniques were first used in 1966, on *Drosophila pseudoobscura* (Hubby and Lewontin 1966; Lewontin and Hubby 1966) and humans (Harris 1966). The results are summarized in Table 2.4. The main conclusion is that there appears to be much genetic variation in both species. About a third of the loci examined are polymorphic, and a typical individual is heterozygous at about 10 percent of its

TABLE 2.4 Polymorphism and heterozygosity as estimated by the first electrophoretic studies of *Drosophila* and humans.

Source: *Drosophila* data are from Hubby and Lewontin (1966) and Lewontin and Hubby (1966); human data are from Harris (1966).

Species	Sample Size	Number of Loci	Polymorphism	Heterozygosity
D. pseudoobscura	43	18	0.30	0.115
Humans	varied	10	0.30	0.099

loci. This much variability was completely unexpected. A second thing to note is that estimates of variation are almost identical in *Drosophila* and humans, perhaps a surprising result.

These two studies were the first of many that examined electrophoretic variation in hundreds of different organisms in essentially all taxonomic groups. A general summary of these results is shown in Table 2.5. There are differences among taxonomic groups, but the pattern is similar to the initial studies. Some populations show little or no allozyme variation, but protein electrophoresis has revealed large amounts of genetic variation in most organisms. Typically, an individual is heterozygous at about 5 to 15 percent of its loci. This amounts to hundreds or thousands of loci.

TABLE 2.5 Summary of allozyme studies that have estimated polymorphism and heterozygosity in various taxonomic groups.

N represents the number of species studied in each group. $\overline{P}$ and $\overline{H}$ are the average polymorphism and heterozygosity of all species in the group.
Source: Slightly modified from Nevo et al. (1984), except human data, which are from Harris and Hopkinson (1976).

	Polymorphism		*Heterozygosity*	
Group	N	$\overline{P}$	N	$\overline{H}$
Invertebrates, except insects	200	0.407	203	0.112
Insects, except *Drosophila*	130	0.351	122	0.089
Drosophila	39	0.480	34	0.123
Fish	200	0.209	183	0.051
Amphibians	73	0.254	61	0.067
Reptiles	84	0.256	75	0.083
Birds	56	0.302	46	0.051
Mammals	181	0.191	184	0.041
Humans	1	0.231	1	0.063
Monocots	12	0.378	7	0.116
Dicots	56	0.235	40	0.052
Gymnosperms	5	0.734	7	0.146
Summary				
Invertebrates	371	0.375	361	0.100
Vertebrates	596	0.226	551	0.054
Plants	75	0.295	56	0.075
Overall average	1042	0.284	968	0.073

Protein electrophoresis reveals only genetic variation that is expressed at the protein level. It is generally believed that these estimates of protein variation are underestimates. Only about one-third of all amino acid substitutions will result in a change in the charge of a protein, so the estimates in Table 2.5 may reveal only a third or so of the actual protein variation.[1] There are ways to get around this problem. See, for example, Singh et al. (1976), Coyne (1976), Coyne et al. (1978), and Ramshaw et al. (1979). A more serious problem is that, due to the redundancy of the genetic code, many nucleotide changes will not change the amino acid sequence of the protein. Thus, there may be variation at the DNA level that is unexpressed and undetectable at the protein level.

2.10 DNA Sequence Variation

Just as electrophoresis has allowed us to look at the products of genes without regard to phenotype in the traditional sense, newer techniques now allow us to look at variation directly at the DNA level. These techniques fall into two main kinds: direct methods using DNA sequencing technology, and indirect methods using restriction enzyme analysis and other molecular techniques to infer variation in DNA sequence. We will consider indirect methods in Section 2.11. The essential point is that all of these techniques allow us to compare a homologous piece of DNA from two individuals, and to estimate the number of nucleotides that differ between them. The DNA examined need not necessarily code for a protein. This makes possible comparisons among DNA sequences with different functions, for example, protein coding versus noncoding regions.

Kreitman (1983) was the first to apply DNA sequencing techniques to population genetics. He sequenced 11 copies of the alcohol dehydrogenase (*Adh*) gene of *Drosophila melanogaster*. He sequenced a total of 2721 nucleotides, which included the coding sequence itself, three introns, and portions of the untranscribed regions upstream and downstream from the gene. His results are summarized in Table 2.6. Of the 2721 nucleotides, 43 (1.6 percent) were polymorphic, that is, they were not the same in all 11 copies. Disregarding insertions and deletions, there were nine unique sequences, or **haplotypes**. A haplotype is a unique sequence of linked genetic markers, in this case, nucleotides. If insertions and deletions are considered, no two copies were identical; *all 11 copies were unique!* Of the coding sequence only, 8 of the 11 copies were unique (8 haplotypes), and 14 of 768 sites were polymorphic.

These results are astonishing. If *Adh* is a typical gene, then virtually all individuals carry unique alleles (haplotypes). Heterozygosity is essentially 100 percent! We can calculate that any two copies of the *Adh* gene differ at about 0.6 percent of sites, on the average (We will see how to make this kind of calculation shortly.) If this is true, and a "typical" gene has about 1000 nucleotides, then any two copies should differ at about six sites on the average.

---

[1]Some amino acid substitutions that do not change the net charge may change the three-dimensional structure of the protein in ways that alter its electrophoretic mobility, so electrophoresis probably detects somewhat more than one-third of amino acid substitutions.

TABLE 2.6 Summary of Kreitman's (1983) study of the DNA sequence of the region containing the alcohol dehydrogenase gene (*Adh*) in *Drosophila melanogaster*.

Eleven copies of the gene were sequenced.

Region	Number of Base Pairs	Number of Polymorphic Sites	Proportion of Polymorphic Sites
Coding sequence	768	14	0.0182
Introns	789	18	0.0228
Transcribed, not translated	335	3	0.0090
Upstream (5') untranscribed	63	3	0.0476
Downstream (3') untranscribed	767	5	0.0065
Entire sequence	2722	43	0.0158

We can compare Kreitman's sequencing data to allozyme data. Protein electrophoresis of hundreds of individuals had previously revealed only two alleles, designated *Fast* and *Slow* for their migration rates in starch or acrylamide gels. But DNA samples (coding region only) revealed eight alleles (haplotypes) in only 11 copies. Our fears that electrophoresis underestimates the amount of genetic variation are confirmed, indeed enhanced. It appears that much of the variation present in DNA sequences is not detected by protein electrophoresis.

We mention one other interesting point about this study. Of the 14 polymorphic sites within the coding region, 13 were silent; that is, they did not change the amino acid sequence of the protein. The other polymorphism accounted for the difference between the *Fast* and *Slow* electrophoretic alleles. We return to this issue in Chapters 7 and 10.

Single nucleotide differences such as Kreitman found are the most common kind of variation at the DNA level. These are called **single nucleotide polymorphisms**, or **SNPs**. They can be detected by sequencing the DNA from different individuals, but newer techniques using DNA microarrays, or "gene chips" (Section 2.11) allow rapid automated detection of SNPs. Over 1 million SNPs have been detected in the human genome, with an average of about one SNP per thousand bases.

Kreitman's study revealed astonishingly high levels of variation at the DNA level. We must ask if his results are typical. The answer is yes; many subsequent studies have confirmed high levels of DNA variation in many genes. These studies have revolutionized the way population geneticists think about genetic variation. We next look at ways in which this variation can be quantified.

Quantifying Variation at the DNA Level

If, in fact, Kreitman's results are typical and nearly every allele is unique, then heterozygosity becomes an uninformative quantity. Heterozygosity is always essentially 100 percent, as Wallace (1958a) predicted. So we need some alternative way to describe variation at the DNA level.

One way is simply to calculate the proportion of nucleotide sites that are polymorphic in a sample, or **nucleotide polymorphism**. Let n_t be the total number of nucleotides in a sequence (i.e., the length of the sequence), and n_p be the number of nucleotide sites that are polymorphic in a sample. Then, the proportion of polymorphic nucleotide sites is estimated by

$$\hat{P}_n = \frac{n_p}{n_t} \qquad (2.16)$$

The subscript n indicates that we mean nucleotide polymorphism. $\hat{P}_n$ is analogous to genetic polymorphism, P, of equation (2.14). For information about the statistical properties of $\hat{P}_n$, see Nei (1987).

Table 2.6 gives $\hat{P}_n$ for several subsets of Kreitman's data. The second, third, and fourth columns are n_t, n_p, and $\hat{P}_n$, respectively. The proportion of polymorphic nucleotide sites over the entire region is about 0.016. In human genes studied so far, the average proportion of polymorphic sites is about 0.0033.

Another useful measure of variation at the DNA level is **nucleotide diversity**, usually symbolized by π. This is the average proportion of nucleotides that are different between any two copies of a gene: Pick two copies at random. Then look at the same nucleotide position in each. The probability that they are different is π.

Nucleotide diversity is the weighted average of the proportion of nucleotide differences among all haplotypes (different sequences) in the population. The weights are the frequencies of the different haplotypes. Let π_{ij} be the proportion of nucleotide differences between the ith and jth haplotypes, and let p_i and p_j be the frequencies of the two haplotypes. There are k different haplotypes in the population. The nucleotide diversity of the population is defined as

$$\pi = \sum_{i=1}^{k} \sum_{j=1}^{k} p_i p_j \pi_{ij} \qquad (2.17)$$

(Nei 1987), where the summation is over all possible pairs of haplotypes. This is the nucleotide diversity *of the population*. You might think that you could estimate it by taking the analogous quantities from a sample and substituting them into this equation. In fact, that would be a biased estimate (Box 2.2). An unbiased estimate of nucleotide diversity from a sample is

$$\hat{\pi} = \frac{n}{n-1} \sum_{i=1}^{k} \sum_{j=1}^{k} \hat{p}_i \hat{p}_j \hat{\pi}_{ij} \qquad (2.18)$$

where $\hat{p}_i$ and $\hat{p}_j$ are the observed haplotype frequencies in the sample, $\hat{\pi}_{ij}$ is the observed proportion of differences between the ith and jth haplotype, and n is the sample size, that is, the number of copies examined (not the number of different haplotypes). Nei (1987) describes some of the statistical properties of $\hat{\pi}$. Box 2.5 illustrates how to estimate nucleotide diversity using equation (2.18). There are several computer programs available that calculate $\hat{\pi}$ and its standard error (see Appendix B).

Box 2.5   Estimating nucleotide diversity from DNA sequence data

The following unique DNA sequences (alleles, or haplotypes) were found in a sample of five individuals. Allele *a* appeared twice. The total sequence length was 500 nucleotides. Only polymorphic sites are shown.

Allele *a*   ...... A ...... A ...... A ...... A ...... A ......
Allele *b*   ...... A ...... T ...... A ...... C ...... T ......
Allele *c*   ...... G ...... A ...... A ...... A ...... A ......
Allele *d*   ...... G ...... A ...... T ...... A ...... A ......

The proportion of polymorphic sites is, from equation (2.16),

$$P_n = \frac{n_p}{n_t} = \frac{5}{500} = 0.01$$

In the sample of five copies ($n = 5$), two had allele *a*, and one each had alleles *b*, *c*, and *d*. Therefore, $k = 4$, $p_1 = 0.4$, and $p_2 = p_3 = p_4 = 0.20$.

Alleles *a* and *b* differ by 3/500 nucleotides, or 0.006. Alleles *a* and *c* differ by 1/500 nucleotides, or 0.002. Similarly, the proportion of differences can be calculated for all pairs. The results are

	a	*b*	*c*	*d*
a	0			
b	0.006	0		
c	0.002	0.008	0	
d	0.004	0.010	0.002	0

Each entry represents $\hat{\pi}_{ij}$; the table is symmetrical about the diagonal.

We can now construct a table to calculate the summation:

i	j	$\hat{p}_i$	$\hat{p}_j$	$\hat{\pi}_{ij}$	$\hat{p}_i\hat{p}_j\hat{\pi}_{ij}$
1	1	0.40	0.40	0	0
1	2	0.40	0.20	0.006	0.00048
1	3	0.40	0.20	0.002	0.00016
1	4	0.40	0.20	0.004	0.00032
2	1	0.20	0.40	0.006	0.00048
2	2	0.20	0.20	0	0
2	3	0.20	0.20	0.008	0.00032
2	4	0.20	0.20	0.010	0.0004
3	1	0.20	0.40	0.002	0.00016
3	2	0.20	0.20	0.008	0.00032
3	3	0.20	0.20	0	0
3	4	0.20	0.20	0.002	0.00008
4	1	0.20	0.40	0.004	0.00032
4	2	0.20	0.20	0.010	0.0004
4	3	0.20	0.20	0.002	0.00008
4	4	0.20	0.20	0	0
					0.00352

The sum of the last column is 0.00352. Note the sum is over all possible combinations of *i* and *j*. We can now calculate $\hat{\pi}$ from equation (2.18):

$$\hat{\pi} = \frac{n}{n-1} \sum_{i=1}^{k} \sum_{j=1}^{k} \hat{p}_i\hat{p}_j\hat{\pi}_{ij} = \frac{5}{4}(0.00352) = 0.0044$$

This method works well on a spreadsheet for small problems. For large data sets, a specialized computer program is usually used.

Table 2.7 gives the pairwise differences among the 11 copies of *Adh* sequenced by Kreitman. From this table, one can calculate $\hat{\pi} = 0.0066$ (Problem 3).

Most estimates of nucleotide diversity have been done in *Drosophila* species or in humans. Moriyama and Powell (1996) and Powell (1997) reviewed *Drosophila* studies. In general, nucleotide diversity is usually in the range of 0.004 to 0.02. Przeworski et al. (2000) summarized nucleotide variation studies in humans. The average nucleotide diversity over all loci studied was about 0.00081, at least an order of magnitude lower than in *Drosophila*. In both species, the amount of variation depends on the locus studied and on the function of the sequence examined; for example, variation in noncoding regions is almost always higher than in coding regions.

TABLE 2.7 Nucleotide differences among 11 sequences of *Adh* in *Drosophila melanogaster.*
Entries in the table are $\hat{\pi}_{ij}$. The number of nucleotides compared is 2379. (This differs from the number in the text because insertions and deletions have been disregarded.) Sequences 8, 9, and 10 are identical. *S* and *F* refer to the electrophoretic alleles *Slow* and *Fast*. *Source*: Based on Kreitman (1983).

Allele	1(S)	2(S)	3(S)	4(S)	5(S)	6(S)	7(F)	8(F)	9(F)	10(F)	11(F)
1(S)	0										
2(S)	0.0013	0									
3(S)	0.0059	0.0055	0								
4(S)	0.0067	0.0063	0.0025	0							
5(S)	0.0080	0.0084	0.0055	0.0046	0						
6(S)	0.0080	0.0067	0.0038	0.0046	0.0059	0					
7(F)	0.0084	0.0071	0.0050	0.0059	0.0063	0.0021	0				
8(F)	0.0113	0.0110	0.0088	0.0097	0.0059	0.0059	0.0038	0			
9(F)	0.0113	0.0110	0.0088	0.0097	0.0059	0.0059	0.0038	0.00	0		
10(F)	0.0113	0.0110	0.0088	0.0097	0.0059	0.0059	0.0038	0.00	0.00	0	
11(F)	0.0122	0.0118	0.0097	0.0105	0.0084	0.0067	0.0046	0.0042	0.0042	0.0042	0

2.11 Indirect Methods of Inferring DNA Sequence Variation

DNA sequencing gives the most information about molecular variation in a population. However, sequencing one or more genes in many individuals can be a time-consuming and expensive process (although increasingly less so as techniques become faster and cheaper). An alternative is to use various molecular techniques that provide indirect information about DNA sequence variation. These methods are less sensitive than DNA sequencing, but are usually faster and cheaper. They are good for initial surveys of large numbers of individuals. Variants discovered by indirect methods can be sequenced later if necessary.

Restriction Fragment Length Polymorphisms

One of the earliest methods for studying DNA sequence variation was to use restriction enzymes and agarose gel electrophoresis to study variation in the location of restriction enzyme recognition sites. This is usually done for a single gene or other easily isolated piece of DNA, such as mitochondrial DNA. If two individuals have sequence differences that create or destroy recognition sites for a particular restriction enzyme, that enzyme will cut the DNA at different places, producing restriction fragments of different lengths, which can separated in an agarose or polyacrylamide gel and visualized by any of several methods. Because the presence or absence of restriction sites results in different sized restriction fragments, these polymorphisms are frequently called **restriction fragment length polymorphisms**, or **RFLPs**.

Figure 2.11a shows a simple example. A 10-kb piece of DNA has been isolated from five individuals, and cut with the restriction enzyme *Eco*RI. We define a site as a position in the DNA that is cut in at least one individual. There are

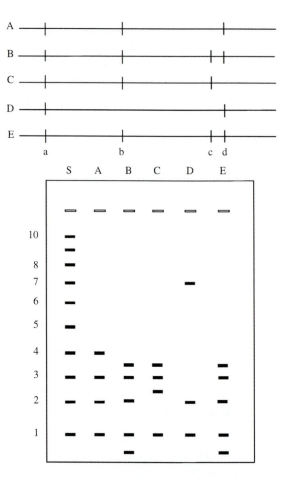

Figure 2.11 **Visualization of restriction fragment length polymorphisms.** Above, restriction maps of five individuals, A, . . . , E. Vertical lines indicate presence of the *Eco*RI recognition sequence at a site. Below, autoradiogram showing fragments produced after cutting with *Eco*RI and probing with a sequence that hybridizes to the region being studied. The lane labeled "S" contains size standards; in this example, each standard band is 1 kb larger than the one below it.

four sites represented by the sample, but not all individuals are cut at all sites. For example, individual B is cut at all four sites, but individual A is not cut at site c.

After cutting with the restriction enzyme, the fragments can be separated in a gel and visualized by, for example, autoradiography. Figure 2.11b shows a diagrammatic example. The first lane is a set of size standards, fragments of known sizes. Each subsequent lane contains the digested DNA of one individual. The sizes of the fragments can be estimated by comparison to the size standards. From such "fragment data" it is possible to draw a restriction map (as in Figure 2.11a) by comparing the fragment sizes when the DNA is cut by one or more restriction enzymes in various combinations. Each restriction site on the map can be considered a single locus, and each unique pattern of the presence or absence of sites is a haplotype. For example, in Figure 2.11 there are four haplotypes (individuals B and E have the same haplotype).

There are various ways to quantify the amount of variation revealed by restriction site polymorphisms, depending on whether we consider variation at the haplotype, restriction site, or nucleotide level. Nei (1987) discusses these ideas in detail, and several computer programs are available to do the calculations.

We consider one practical application of RFLP analysis. Sea turtles typically spend much of their long lives at sea. At reproductive maturity, females lay their eggs on sandy beaches. Hatchlings then enter the sea and stay there for several years until they are reproductively mature, at which time the females return to the beach of their hatching to lay their eggs in the sand. The extent to which females actually return to the same beach is not well known, and presumably varies among species. Many sea turtle populations are threatened by habitat destruction of their nesting beaches. Bowen et al. (1993) used RFLP analysis of mitochondrial DNA (mtDNA) to estimate variation in several populations of loggerhead turtles (*Caretta caretta*) in the Atlantic Ocean and Mediterranean Sea (Figure 2.12). Mitochondrial DNA is inherited maternally; thus it can serve as an indicator of female gene flow. Estimates of nucleotide diversity, π, are shown in Table 2.8. Turtles from the Florida beaches were the most variable, with turtles from Greece and South Carolina showing no variation. Using techniques described in Chapter 9, Bowen et al. were also able to estimate the amount of gene flow among populations. They estimated that about one or two individuals per generation move from one population to another. They concluded that this is sufficient to reestablish locally extinct populations over evolutionary time, but not over short periods of time. For conservation purposes they recommended that local nesting populations of loggerhead turtles be managed as independent populations.

Recently developed techniques for analyzing minisatellite and microsatellite DNA (see next section) have largely replaced RFLP analysis for studying genetic variation in natural populations. However, RFLPs are still useful in studying human genetic diseases and in gene mapping (Chapters 4 and 13).

TABLE 2.8 Nucleotide diversity ($\hat{\pi}$) in five populations of loggerhead turtle (*Caretta caretta*) as estimated by RFLP analysis of mitochondrial DNA.

Source: From Bowen et al. (1993).

Population	Nucleotide Diversity
Greece	0.0000
South Carolina	0.0000
Georgia	0.0002
Eastern Florida	0.0018
Western Florida	0.0018
Overall	0.0018

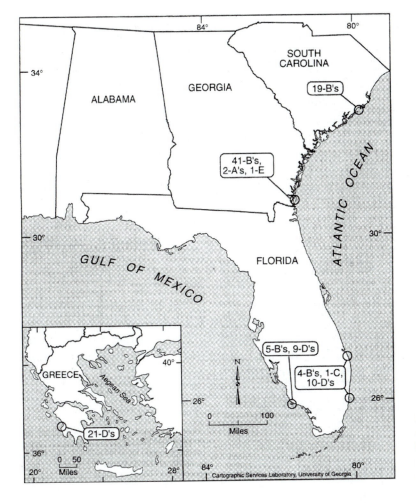

Figure 2.12 **Loggerhead turtle collection sites on Atlantic and Mediterranean beaches.** Numbers and letters in boxes represent numbers of each haplotype (A–E) collected at each site. *Source:* From Bowen et al. (1993).

Amplified Fragment Length Polymorphisms

There is another technique that takes advantage of the polymerase chain reaction (PCR) to amplify specific regions of DNA. The general procedure is to simultaneously cut genomic DNA with two restriction enzymes, creating some fragments that have, for example, an *Eco*RI site at one end and a *Bam*HI site at the other. Hundreds or thousands of such fragments will be created. Adapter molecules of known sequence are then added to these fragments. Finally, primers that recognize the adapter sequences are used to amplify a subset of these fragments by PCR. DNA sequence variation that creates or destroys restriction sites will create fragments of different sizes. Like RFLPs, these **amplified fragment length polymorphisms (AFLPs)** can be separated in an agarose or polyacrylamide gel and visualized.

AFLPs are scored as the presence or absence of a specific band on a gel. Bands usually cannot be associated with specific loci. For that reason, AFLPs are not very useful for estimating heterozygosity, or other studies that require locus specific information.

AFLPs are useful in identifying individuals and determining relationships, and in estimating gene flow and dispersal. See Mueller and Wolfenbarger (1999) for a summary of applications and a good overview of the procedures.

Random Amplified Polymorphic DNA

Primers of random sequences can be used to amplify some DNA sequences. Two primers, called forward and reverse, are simultaneously hybridized to genomic DNA. If the target sequences are close enough together, the DNA between them can be amplified by PCR and visualized on an agarose or polyacrylamide gel. Variation in the target sequences or in the length of the DNA between them results in different sized amplification products. Variation seen by this technique is called **random amplified polymorphic DNA (RAPD)**.

Like AFLPs, RAPDs are scored as the presence or absence of a specific band on a gel. Bands cannot be associated with specific loci, so RAPDs have some of the same disadvantages as AFLPs. In addition, repeatability is a greater problem with RAPDs than with either AFLPs or RFLPs.

Single Strand Conformational Polymorphisms

Like RNA, single stranded DNA tends to fold up on itself and form complex three-dimensional structures by intramolecular hydrogen bonding. Sequence variations can cause differences in this three-dimensional structure, which in turn can affect electrophoretic mobility. Thus, variation can be detected by denaturing small DNA fragments and separating them in a polyacrylamide gel. Differences in mobility indicate differences in three-dimensional structure (assuming fragments are the same size). Variations detected in this way are called **single strand conformational polymorphisms (SSCPs)**.

SSCP procedures are relatively simple, and are quite sensitive for DNA fragments up to about 200 base pairs in length. Such fragments are usually PCR products. SSCP techniques are useful for screening large numbers of individuals for variation in a particular region of DNA. They do not indicate the position or number of sequence differences that cause the conformational differences.

SSCP screening is especially useful in medical genetics and conservation biology. It has been used to diagnose diseases such as phenylketonuria (Dockhorn-Dworniczak et al. 1991), Tay-Sachs disease (Ainsworth et al. 1991), and cystic fibrosis (DesGeorges et al. 1993). In conservation and evolutionary biology, SSCPs are particularly useful for rapid screening of large numbers of individuals within or among populations. See Girman (1996) for a good overview of SSCP techniques and applications.

DNA Microarrays

The hottest new technology in molecular biology is the **DNA microarray** (sometimes called a DNA chip or gene chip). These are glass or silicon slides, about the size of a microscope slide, which contain a grid of up to a million or more cells. Single stranded DNA probes, about 20 nucleotides in length, are attached to the grid. Each cell in the grid receives a slightly different probe. For example, four probes might have A, T, G, or C at a particular position in the sequence. The test DNA is denatured and labeled with a fluorescent dye. It is then added to the DNA chip and hybridized under high stringency so that only a perfect match to the probe will hybridize. Unhybridized DNA is washed away and the pattern of fluorescence among cells in the grid is read by laser scanning and fed to a computer for analysis.

DNA chips with up to 1 million cells can be made, each containing a different probe. Using such chips, any possible variation in the DNA sequence of the test DNA can be detected. DNA chips have been used to screen for mutations in genes associated with cancer, such as the *p53* gene and the *BRCA*1 gene.

DNA chips have been used to identify single nucleotide polymorphisms in the human genome. More than 1 million SNPs have been identified and mapped so far. It is expected that these will be useful in mapping genes associated with complex diseases.

DNA chips are also used to study gene expression. Each cell in the grid contains a probe for a different gene. The test DNA, usually cDNA made from RNA isolated from a particular cell type or developmental stage, is hybridized to the grid, and the pattern of fluorescence indicates which genes are expressed in that particular cell or stage.

Gene chips have not yet been used much in population genetics, but as the cost comes down, they will inevitably be adopted and used extensively by population geneticists and evolutionary biologists, as were protein electrophoresis, RFLP analysis, DNA sequencing, and other molecular techniques.

2.12 Variation in Number of Tandem Repeats

Eukaryotic chromosomes contain regions in which short sequences of DNA are repeated over and over again (tandem repeats). Much of this "satellite DNA"[2] is

---

[2]It is called satellite DNA because it was initially observed as a separate (satellite) band when DNA was purified by density gradient centrifugation.

localized near the centromeres and telomeres of chromosomes, but certain kinds of satellite DNA are dispersed more or less randomly throughout the genome. Each cluster of repeats is considered a locus, and the number of repeats at a particular locus is variable. Such loci are called **VNTR** loci, for **variable number of tandem repeats**.

Much of satellite DNA is found in noncoding regions of eukaryotic chromosomes, thus variation in satellite DNA is more likely to be neutral with respect to natural selection than variation in coding regions. This makes some forms of satellite DNA useful as genetic markers, as we shall see. There are two kinds of VNTR loci, depending on the length of the repeat sequence.[3]

Minisatellite Loci and DNA Fingerprinting

Minisatellite loci are VNTR loci with a repeating unit of about 15 to 50 nucleotides. The number of repeats at a locus typically ranges from two or three to 20 or more (occasionally, many more). These clusters of repeats (loci) are scattered throughout the genome of many eukaryotic organisms. At each locus there may be dozens of different alleles corresponding to different numbers of repeats.

Minisatellite loci are studied by cutting the DNA with a restriction enzyme that cuts outside of the repeat cluster. The fragments are then separated by electrophoresis and Southern blotted to a nylon membrane. The membrane is then probed with a sequence that hybridizes with a conserved sequence (usually about 10–15 base pairs) within the repeat unit. The bands are then visualized by autoradiography or other methods. At a given minisatellite locus, individuals with different numbers of repeats will show different sized bands. Because most minisatellite sequences are present several times throughout the genome of an individual, many different bands typically show up on the autoradiogram (Figure 2.13).

Typically, this procedure reveals about 20 or so distinct bands per individual. Each band represents one cluster of repeats. Each locus produces one band, if homozygous, or two, if heterozygous. Note that we cannot tell which bands correspond to which loci. Minisatellite loci are so variable that no two individuals are likely to have the same number of repeats at every locus. Thus, if several loci are visualized with the same probe, each individual will show a unique pattern of bands. This is the basis of **DNA fingerprinting**.

Since the bands at each locus are inherited in Mendelian fashion, an individual will receive half its bands from its mother and half from its father. (There may be some ambiguity due to comigrating bands.) Every band must come from either one parent or the other. This makes minisatellite loci useful in determining parentage. Figure 2.14 shows the banding patterns of a mother and her child, and two possible fathers. Any band of the child's DNA that is not present in the mother had to have come from the father. Analysis of these bands indicates that individual F2 must be the father of the child.

DNA fingerprinting is now commonly used in criminal investigations. Figure 2.15 shows DNA fingerprints of a blood sample taken from the scene of a crime, and from several suspects. The blood sample matches the DNA fingerprint

---

[3]Some authors use the term VNTR loci to refer only to minisatellite loci. We use it in its more general meaning, referring to both minisatellite and microsatellite loci.

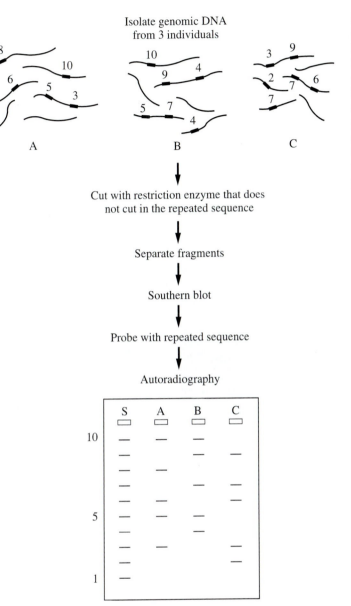

Isolate genomic DNA
from 3 individuals

Cut with restriction enzyme that does
not cut in the repeated sequence

Separate fragments

Southern blot

Probe with repeated sequence

Autoradiography

Figure 2.13 **Procedure for multilocus DNA fingerprinting.** In this case, genomic DNA is isolated from three individuals. Small boxes on the DNA represent minisatellite loci and the number above each box indicates repeat number. The DNA is then cut with a restriction enzyme that does not cut within the repeat unit, separated on an agarose gel, blotted to a nylon filter, probed with a sequence that hybridizes to the repeat sequence, and visualized by autoradiography. The autoradiograph shows one band for each allele (repeat number) present in the individual.

Figure 2.14 **Multilocus DNA fingerprints of a mother (M), her child (C), and two men (F1 and F2) claiming to be the child's father.** The child's bands that are not shared with the mother must have come from the father. On this basis, F1 can be excluded as the father of the child. *Source:* From Griffiths et al. (1996).

of suspect 3. Note, we did not say suspect 3 is guilty, or even that suspect 3 was present at the scene of the crime. Before we can say either of those things, we must establish beyond reasonable doubt that no one else is likely to have the same DNA fingerprint as suspect 3. We consider this issue in Section 4.3.

DNA fingerprinting can be useful in studying mating patterns in natural populations. For example, Gibbs et al. (1990) showed that in red-winged blackbirds, 45 percent of broods contained at least one individual derived from an extra-pair copulation. Westneat (1990) obtained similar results for indigo buntings. These results are surprising, because most Passerine birds have traditionally been thought to be monogamous. Burke (1989) and Petrie and Kempenaers (1998) review applications of DNA fingerprinting to studies of mating behavior in natural populations.

Figure 2.15 **DNA finger-prints from a blood sample collected at the crime scene and from several suspects.**
Source: From Griffiths et al. (2000).

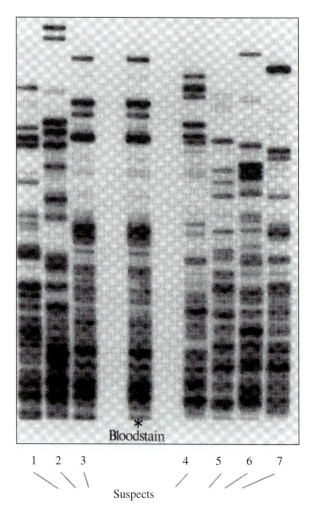

The multilocus nature of most minisatellite DNA sequences prevents assignment of specific bands to specific loci. Therefore, the number of loci, allele frequencies, heterozygosity, and other important population parameters usually cannot be estimated from multilocus DNA fingerprints. Advances in PCR technology and stringent hybridization conditions sometimes permit analysis of single locus minisatellite DNA. In this case, an individual will show one or two bands at a locus. Single locus DNA fingerprinting allows us to estimate allele frequencies and heterozygosity, (among other things), at individual loci, but another class of VNTR loci is more useful for these applications.

Microsatellite Loci

Some population genetic studies, such as gene mapping or estimating levels of gene flow, are most efficiently done with highly variable neutral genetic markers. Minisatellite loci are highly variable, but specific bands on a gel usually cannot be associated with specific loci. Allozyme and RFLP loci are single locus markers, but are not nearly as variable as minisatellite loci. These facts limit the usefulness of these loci in studies that require highly variable single locus marker loci. An ideal marker for such studies would be an easily identified single locus that is highly variable and neutral with respect to natural selection. It turns out that a second kind of satellite DNA fulfills these requirements.

Microsatellite loci are similar to minisatellites, except that the repeating unit is only two to five nucleotides in length (e.g., CACACA . . .). The average number of repeats in a cluster is usually about ten to 20, but some clusters have many more. Each cluster is considered a locus. There are thousands of microsatellite loci, densely scattered throughout the genome. For example, the human genome contains at least 50,000 microsatellite loci. Like their bigger siblings, microsatellite loci are highly variable, often with ten or more alleles (different numbers of repeats) per locus. Most alleles are relatively rare; therefore, most individuals are heterozygous at many, if not most, microsatellite loci. Most microsatellite loci do not code for proteins, and thus variation is often assumed to be neutral with respect to natural selection. However, high copy numbers in microsatellite loci flanking protein coding genes have been associated with several human diseases, for example, Fragile-X syndrome (Sutherland and Richards 1995). In addition, several diseases, including Huntington's disease, appear to be caused by high copy numbers of trinucleotide repeats *within* the coding sequence of a gene (Strachan and Read 1999).

Microsatellite loci are studied by using PCR to amplify an individual locus. Primers are designed from unique sequences flanking the locus. Because only that locus will be amplified, microsatellite loci can be analyzed one at a time; in other words, genotypes at each locus can be unambiguously determined. Typically, 20 or so loci are analyzed. This is similar to the number of loci usually studied by protein electrophoresis, but because microsatellites are much more variable than allozymes, the same number of loci will give much more information.

Allozymes frequently show little or no variation in small inbred populations, and thus are of little use for inferring population structure, history, and so on. However, microsatellite loci usually show variation even in populations with little or no allozyme variation. For example, Hughes and Queller (1993) surveyed six microsatellite loci in the social wasp *Polistes annularis*. All six loci were polymorphic, with an average observed heterozygosity of 0.62. By comparison, only three of 33 allozyme loci were polymorphic, with an average heterozygosity of 0.035.

Eldridge et al. (1998) describe an exception to the generalization that even small populations show substantial microsatellite variation. They surveyed ten microsatellite loci in two Australian mainland populations and one island population of the black-footed rock wallaby (*Petrogale lateralis*). Mainland populations had typical levels of heterozygosity of about 0.5 to 0.6. However, heterozygosity of the island population was only 0.05, the lowest value so far reported for microsatellite heterozygosity. They attribute this to long-term isolation and small population size. We will see in Chapter 7 how these factors are expected to reduce genetic variation.

Microsatellites are becoming increasingly important in mapping human disease genes and loci that affect quantitative traits. These applications are discussed in Chapters 4 and 13.

Microsatellite data can be analyzed using the traditional techniques. However, they contain much more information than can be extracted by these techniques alone. For example, there are genealogical relationships among alleles. Mutation usually increases or decreases the copy number by one, so alleles with similar copy number are more closely related than alleles with very different copy numbers. This allows far more detailed analysis than for other kinds of loci. There has been much recent activity in developing new statistical methods for analysis of microsatellite data. We will see some of these techniques in later chapters as we study population subdivision, genetic drift, and other topics. For reviews of some of these techniques, see Jarne and Lagoda (1996), Cavalli-Sforza (1998), and Luikart and England (1999).

2.13 Insertion/Deletion Variation

Minisatellite and microsatellite variation are examples of the more general phenomenon of length variation. Homologous DNA segments can vary in length as well as in nucleotide sequence. For example, Kreitman (1983) found six sites at which insertions or deletions of nucleotides occurred in the *Adh* region of *Drosophila melanogaster*. The length of the insertions or deletions ranged from one to 37 nucleotides. Such small insertions or deletions are often called **indels**, short for insertion/deletion.

Short indels are usually caused by errors in DNA replication. Their effects may range from neutral (e.g., if they occur within an intron) to lethal (e.g., if they cause a frameshift in an essential gene).

Longer length variation is frequently caused by transposable elements, at least in *Drosophila*. **Transposable elements** are DNA sequences that are capable of moving from one place to another in the genome. Barbara McClintock was the first to suggest that genetic elements might move from one place to another. She observed high mutation rates and high reversion rates in corn, and attempted to map the genes responsible. She found that they mapped to different places in different individuals within the same strain. She hypothesized that these genes were capable of moving from place to place, a radical idea at the time. Most scientists thought that either she was wrong or that this was an aberrant gene, and not of general interest. It was not until the development of molecular techniques in the 1970s and 1980s that transposable elements were discovered and understood. McClintock was proved right, and in 1983 she belatedly received a Nobel prize for her work.

Transposable elements are very common in both prokaryotes and eukaryotes. For example, in humans and *Drosophila*, about 25 percent or more of the genome consists of transposable elements. In corn, it is about 50 percent.

Transposable elements move (transpose) at relatively high rates. There are two kinds of transposition: Replicative transposition occurs when a transposable element makes a copy of itself, and the copy inserts itself somewhere else in the genome. The original stays in the original location. This leads to accumulation of transposable elements, which may explain why they are so frequent in some species. Conservative replication occurs when a transposable element "jumps" from one place to another, without making a copy of itself. This has led to the colorful but often inaccurate phrase "jumping genes" (transposable elements are not always genes, in the normal sense).

When a transposable element inserts itself into the middle of a protein coding gene, it causes a mutation in that gene. Similarly, if a transposable element moves by conservative transposition, the original DNA sequence is restored, and a reversion or back mutation occurs. Frequently though, excision is imperfect, and the transposable element takes some of the gene with it, leading to small deletions.

Transposition can also affect gene regulation. Some transposable elements contain promoter sequences so that insertion can affect regulation of nearby genes. Transposition can also disrupt existing promoter or protein binding sites.

As with random union of gametes, the allele frequency under random mating is unchanged in the offspring zygotes.

This may seem like a lot of algebraic manipulation to demonstrate something we already knew, but it is good practice, and the procedure will be useful in other contexts.

To summarize, the Hardy-Weinberg principle says that, if the frequencies of A_1 and A_2 are p and q, then, under the assumptions previously stated, the frequencies of the A_1A_1, A_1A_2, and A_2A_2 genotypes will be p^2, $2pq$, and q^2. These genotype frequencies will be reached after one generation of random mating, and will remain constant thereafter, as long as the stated assumptions hold in the population. Such a population is said to be in Hardy-Weinberg equilibrium. The assumptions define the conditions under which a population will not change, and thus implicitly define the processes which can cause a population to evolve.

Testing Population Samples for Hardy-Weinberg Conditions

The Hardy-Weinberg principle serves as a kind of null hypothesis. It tells us what to expect if all of the stated assumptions are true in the population. If we sample a population and find that the genotype frequencies are different from the Hardy-Weinberg predictions, then we can conclude that one or more of these assumptions is violated; that is, at least one evolutionary process is operating. This gives us motivation to study the population in more detail.

The usual way to compare a set of observed values to a set of expected values (based on some null hypothesis) is to use a goodness of fit test. The most commonly used goodness of fit test for Hardy-Weinberg conditions is the chi-square test. For those unfamiliar with the chi-square test, details of how to use it to test for Hardy-Weinberg conditions are explained in Box 3.1.

The chi-square test is commonly used to test population samples for deviation from Hardy-Weinberg expectations. However, three cautions must be emphasized:

Caution: If we find a population does not deviate from the Hardy-Weinberg expectations, we cannot conclude that no evolutionary processes are operating.

We illustrate with a numerical example. Start with a population of 100 adults, 25 A_1A_1, 50 A_1A_2, and 25 A_2A_2 (Table 3.4). In these adults, $p = 0.50$. Assume that A_1A_1 and A_1A_2 individuals produce two gametes each, but that A_2A_2 individuals are sterile. Then there will be 100 A_1 gametes and 50 A_2 gametes. In the gametes, $p = 0.75$. If these gametes unite at random, the frequencies in the zygotes will be (approximately) $p^2 = 0.56$, $2pq = 0.37$, and $q^2 = 0.06$. Since there were 150 gametes, there will be 75 zygotes. Therefore, the numbers of zygotes will be (rounded to the nearest whole number) 42, 28, and 5. Assume that survival is 100 percent, so that the adult numbers are the same. The allele frequency in these adults is $p(t + 1) = 0.75$, and the adult numbers match the Hardy-Weinberg expectations exactly (work it out to convince yourself). But natural selection is acting, since A_2A_2 individuals are sterile. Note that the allele frequency changed from one generation of adults to the next, but if you saw only one generation, you would be unable to detect natural selection. In general, a goodness of fit test is completely incapable of detecting deviations due to differential fertility among the adults.

(Text continues on middle of page 76)

Box 3.1   Using the chi-square test to test for Hardy-Weinberg conditions

The usual way to compare a set of observed values to a set of expected values (based on some null hypothesis) is to use a goodness of fit test. The most commonly used goodness of fit test for Hardy-Weinberg conditions is the chi-square test. The general formula for the chi-square statistic is

$$\chi^2_{calc} = \sum_{\substack{all \\ classes}} \frac{(O - E)^2}{E}$$

where O and E are the observed and expected *numbers* for each class (genotype) and the summation is over all classes (genotypes). The calculated value is then compared to a critical value obtained from a chi-square table (Table 3.2). If the calculated value is greater than the critical value, then we conclude that the observed values differ significantly from the expected values; that is, we reject the null hypothesis.

We illustrate with an example. Table 3.3 gives the observed genotypic frequencies for three genotypes at the MN blood group locus in a sample of U.S. Caucasians. At this locus there are two alleles, designated L^M and L^N. Let p be the frequency of the L^M allele. We first estimate the allele frequencies based on the observed genotype numbers. From equation (2.4) we get

$$p = \frac{2(1787) + 3039}{2(6129)} = 0.539$$

$$q = 1 - p = 0.461$$

We now use these allele frequencies to calculate the expected genotype frequencies, assuming all of the conditions of the Hardy-Weinberg principle hold. Using equations (3.1), (3.2) and (3.3), the expected genotype frequencies are

$$P_{MM} = (0.539)^2 = 0.291$$

$$P_{MN} = 2(0.539)(0.461) = 0.497$$

$$P_{NN} = (0.461)^2 = 0.212$$

We now calculate the expected numbers by multiplying the expected frequency by the sample size. The results are shown in the third column of the table. Finally, for each genotype, we calculate the difference between the observed number and the expected number, square it, and divide by the expected number for that genotype. These numbers are shown in the last column of the table. The sum of the last column is our calculated chi square value. We get $\chi^2_{calc} = 0.034$.

In general, the degrees of freedom for a chi-square test is the number of classes minus the number of constraints. The number of constraints is the number of things we must know to be able to calculate all of the expected values under the null hypothesis. In our two allele example, there are three classes (genotypes). There are two constraints; if we know N and p, we can calculate all of the expected genotypic values. Therefore the degrees of freedom is one.

There is always one constraint for the total (if we know the numbers in all classes but one, we can get the last by subtraction) and one for each parameter that must be estimated:

$$df = \left(\begin{array}{c} \text{number of} \\ \text{classes} \end{array}\right) - 1 - \left(\begin{array}{c} \text{number of} \\ \text{parameters estimated} \end{array}\right)$$

In a chi-square test for Hardy-Weinberg conditions, the number of parameters estimated is one less than the number of alleles (if we estimate all but one allele frequency, we can get the last by subtraction). If n is the number of genotypes and k is the number of alleles, this simplifies to

$$df = n - k$$

See Problem 3.1.

We look up the critical value of chi-square statistic in Table 3.2, and find it to be 3.84 for one degree of freedom and 0.05 level of significance. The calculated value is less than the critical value; therefore we fail to reject the null hypothesis. There is no evidence that the sample numbers differ significantly from the numbers expected if all of the Hardy-Weinberg conditions held. Note, we cannot say that no evolutionary processes are acting.

The chi-square test is not necessarily the best way to test for Hardy-Weinberg conditions, especially if some alleles are present in low frequencies. See the main text for references to alternative tests.

TABLE 3.2 Chi-square critical values.

Level of significance is the probability of rejecting the null hypothesis when it is actually true. Conventional level of significance is 0.05. Degrees of freedom is calculated as

$$df = \binom{number}{of\ classes} - 1 - \binom{number\ of}{parameters\ estimated}$$

To use the table, look up the critical value corresponding to level of significance desired and correct degrees of freedom. If calculated chi-square value is greater than critical value from table, reject the null hypothesis.

	Level of Significance		
Degrees of Freedom	**0.05**	**0.01**	**0.001**
1	3.84	6.64	10.83
2	5.99	9.21	13.82
3	7.82	11.34	16.27
4	9.49	13.28	18.47
5	11.07	15.09	20.52
6	12.59	16.81	22.46
7	14.07	18.48	24.32
8	15.51	20.09	26.13
9	16.92	21.67	27.88
10	18.31	23.21	29.59
11	19.68	24.72	31.26
12	21.03	26.22	32.91
13	22.36	27.69	34.53
14	23.68	29.14	36.12
15	25.00	30.58	37.70
16	26.30	32.00	39.25
17	27.59	33.41	40.79
18	28.87	34.81	42.31
19	30.14	36.19	43.82
20	31.41	37.57	45.31

TABLE 3.3 Chi-square test comparing observed genotype frequencies at the *MN* blood group locus, and expected frequencies under Hardy-Weinberg equilibrium. See Box 3.1 for details. *Source:* Data from Mourant et al. (1976).

Genotype	Observed Number (O)	Expected Number (E)	Difference ($O - E$)	$(O - E)^2/E$
$L^M L^M$	1787	1783.6	3.4	0.007
$L^M L^N$	3039	3046.1	−7.1	0.017
$L^N L^N$	1303	1299.3	3.7	0.010
Totals	6129	6129.0	0	$0.034 = \chi^2_{calc}$

TABLE 3.4 An example showing natural selection against sterile A_2A_2 homozygotes.
Initial allele frequencies in adults are 0.5 and 0.5. Adults of genotype A_1A_1 or A_1A_2 produce two gametes each but A_2A_2 individuals are sterile. Gametes unite at random to produce zygotes that are in Hardy-Weinberg frequencies. Assuming all zygotes survive, adult frequencies are also in Hardy-Weinberg frequencies. The same result holds for less than 100 percent survival, as long as all genotypes have the same survival rate.

	Genotypes			*Gametes*		*Allele Frequencies*	
	A_1A_1	A_1A_2	A_2A_2	A_1	A_2	*p*	*q*
Adults (number)	25	50	25			0.50	0.50
Adults (frequency)	0.25	0.50	0.25				
Adults (expected)	25	50	25				
Gametes produced per adult	2	2	0				
Gametes (number)				100	50	0.75	0.25
Zygotes (frequency)	0.56	0.37	0.06			0.75	0.25
Zygotes (number)	42	28	5			0.75	0.25
Adults (number)	42	28	5			0.75	0.25
Adults (expected)	42	28	5				

This is only one example; see Problem 3.5 for another. There are many other cases in which one or more evolutionary processes may be acting, but in ways that are impossible to detect with a goodness of fit test.

Caution: The power of the chi-square test to detect significant deviations from Hardy-Weinberg expectations may be weak.

Again, we illustrate with an example. Assume you have a population of 200 zygotes, 50 A_1A_1, 100 A_1A_2, and 50 A_2A_2. The frequency of A_1 is $p = 0.50$, and the zygote genotype frequencies match the Hardy-Weinberg expected frequencies. Assume that survival of the A_1A_1 and A_2A_2 genotypes is 100 percent, but that only 80 percent of the heterozygotes survive. Most evolutionary biologists would consider this fairly strong selection against the heterozygotes. The adult numbers will be (on the average) 50 A_1A_1, 80 A_1A_2, and 50 A_2A_2. The frequency of the A_1 allele in the adults is 0.50, unchanged because natural selection eliminated an equal number of A_1 alleles and A_2 alleles. The expected numbers for each genotype are 45, 90, and 45. The calculated chi-square value is 2.22, not significant according to Table 3.2.

Here, we have an example of fairly strong natural selection, yet we are unable to detect it with the goodness of fit test, even with a fairly large sample size. The problem is even worse for smaller sample sizes.

There are other statistical problems with using the chi-square test to detect deviations from the Hardy-Weinberg expectations, especially for small sample sizes. Weir (1996) discusses these problems and alternative tests.

Caution: Deviations from the Hardy-Weinberg expectations may give us no information about the kinds or directions of the evolutionary processes operating.

For example, it is sometimes claimed that an observed excess of heterozygotes over the Hardy-Weinberg expectation is evidence that natural selection is favoring heterozygotes. This is not necessarily true. There are many other situations that can give rise to an excess of heterozygotes, (for example, certain kinds of nonrandom mating). In general, if we are lucky enough to detect deviations from Hardy-Weinberg expectations, these deviations do not necessarily tell us anything about their causes. (See the example in Section 3.3.)

The conclusion to be drawn from these examples is that we must be careful in interpreting goodness of fit tests for Hardy-Weinberg equilibrium. Detecting significant deviations requires large sample sizes and strong disturbing forces. *Lack of significance cannot be interpreted to mean that no evolutionary processes are operating.* Different processes may be acting in ways that are not detectable with a goodness of fit test, or they may be too weak to be detectable with the given sample size. The theory behind these problems requires an understanding of probability theory, and of many topics we have not yet covered. However, the following papers are relatively easy to understand and give interesting examples: Wallace (1958b), Lewontin and Cockerham (1959), Workman (1969), and Smith (1970).

Using the Hardy-Weinberg Principle to Estimate Frequencies of Recessive Alleles

There is another common use for the Hardy-Weinberg principle: It sometimes allows us to estimate frequencies of recessive alleles. To do this, we must assume that the genotype frequencies are close to the Hardy-Weinberg expected frequencies. We illustrate with another example.

Albinism is a trait caused by homozygosity for a recessive allele (call it A_2). In humans, the frequency of albino individuals is about one in 20,000 or about 0.00005. This is an estimate of P_{22}, the frequency of A_2A_2 homozygotes. If we are willing to assume that the genotype frequencies are close to the Hardy-Weinberg expectations, we can estimate the frequency of the A_2 allele. The expected frequency of homozygous recessives is

$$P_{22} = q^2$$

Since we have an estimate of P_{22}, we can estimate q as

$$\hat{q} = \sqrt{\hat{P}_{22}} \cong 0.007$$

Therefore, the estimated frequencies of the three genotypes are

$$\hat{P}_{11} = (0.993)^2 \cong 0.986$$

$$\hat{P}_{12} = 2(0.993)(0.007) \cong 0.014$$

$$\hat{P}_{22} = (0.007)^2 \cong 0.00005$$

(The last calculation is, of course, circular.) Note, we cannot do a goodness of fit test, because in order to estimate q, we had to *assume* that the population was in Hardy-Weinberg equilibrium.

These kinds of calculations should make you nervous. Does it seem reasonable, for example, that natural selection is not operating at the albinism locus? Allele frequency estimates obtained in this way are appropriate only if the observed genotype frequencies are near the Hardy-Weinberg expectations. As we shall see in Section 3.3, significant deviations from Hardy-Weinberg expectations are relatively rare in large outbreeding natural populations. As emphasized above, this does not mean that no evolutionary processes are acting, but it does mean that we can estimate recessive allele frequencies by the method described here.

This example illustrates an important implication of the Hardy-Weinberg principle. If an allele is rare, it is almost always present in heterozygous genotypes. In the albinism example, the ratio of heterozygotes to recessive homozygotes is 0.014/0.00005, or about 280 to 1; there are nearly 300 heterozygotes for every homozygote. If natural selection works only against recessive homozygotes, it will be ineffective when the recessive allele is rare, because almost all copies of that allele will be in heterozygotes, which are not subject to natural selection. We return to this issue in Chapter 5.

Multiple Alleles

Many loci have more than two alleles. The Hardy-Weinberg principle easily extends to such loci. Using the notation of Chapter 2, we let p_i be the frequency of the ith allele, and P_{ii} or P_{ij} be the frequencies of homozygous or heterozygous genotypes A_iA_i or A_iA_j. Then, under the Hardy-Weinberg conditions, the expected frequencies of the homozygotes are

$$P_{ii} = p_i^2 \tag{3.7}$$

and the expected frequencies for the heterozygotes are

$$P_{ij} = 2p_ip_j \tag{3.8}$$

Figure 3.2 illustrates this graphically for three alleles. Another way to remember the expected frequencies is explained in Box 3.2.

We illustrate the calculations using the example in Box 2.1. There, we calculated the allele frequencies for each of the four alleles. For the A_1A_1 homozygote the Hardy-Weinberg expected frequency is

$$P_{11} = p_1^2 = (0.3925)^2 = 0.1541$$

For the A_1A_2 heterozygote, the expected frequency is

$$P_{12} = 2p_1p_2 = 2(0.3925)(0.3050) = 0.2394$$

You should calculate all of the other expected frequencies and do a chi-square test to see if the observed genotype frequencies are consistent with the Hardy-Weinberg expectations (Problem 3.3).

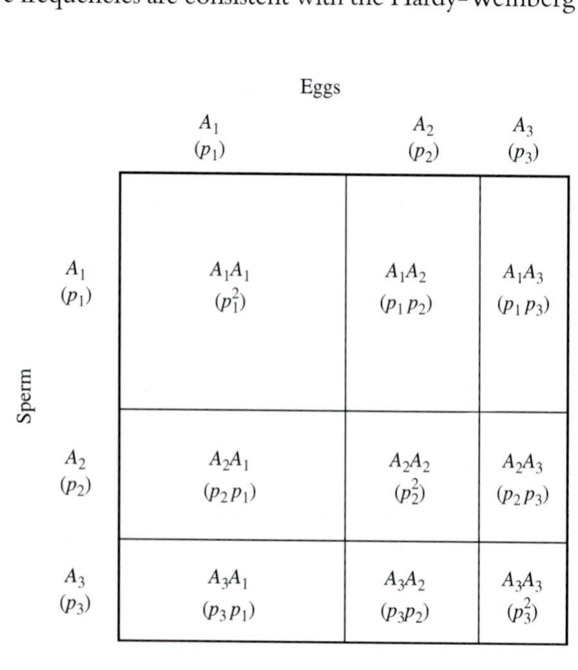

Eggs

		A_1 (p_1)	A_2 (p_2)	A_3 (p_3)
A_1 (p_1)		A_1A_1 (p_1^2)	A_1A_2 (p_1p_2)	A_1A_3 (p_1p_3)
A_2 (p_2)		A_2A_1 (p_2p_1)	A_2A_2 (p_2^2)	A_2A_3 (p_2p_3)
A_3 (p_3)		A_3A_1 (p_3p_1)	A_3A_2 (p_3p_2)	A_3A_3 (p_3^2)

Sperm

Figure 3.2 **Random union of gametes, A_1, A_2, and A_3, with allele frequencies p_1, p_2, and p_3.** Entries in the table represent zygote genotypes and frequencies.

Box 3.2 The binomial expansion and the Hardy-Weinberg principle

You may remember the binomial expansion from high school algebra:

$$(p + q)^2 = p^2 + 2pq + q^2$$

We can interpret this as

$$\left(\begin{array}{c} sum\ of \\ allele\ frequencies \end{array} \right)^2 = \left(\begin{array}{c} sum\ of \\ genotype\ frequencies \end{array} \right)$$

where the terms on the right-hand side represent the Hardy-Weinberg expected genotype frequencies. This easily extends to multiple alleles, giving an easy way to remember the expected genotype frequencies. For example, for three alleles,

$$(p_1 + p_2 + p_3)^2 =$$
$$p_1^2 + p_2^2 + p_3^2 + 2p_1p_2 + 2p_1p_3 + 2p_2p_3$$

where the squared terms on the right represent the frequencies of the homozygotes and the cross products represent the frequencies of the heterozygotes.

Caution: All of the previously mentioned problems about detecting deviations from Hardy-Weinberg expectations are magnified with multiple alleles.

Highly variable loci, such as microsatellite loci, frequently have numerous alleles, many of which are rare. This means that expected numbers of some genotypes will be very small. This inflates the value of the chi-square statistic (expected numbers are in the denominator), making it unreliable. One approach to the problem is to combine genotypes into classes with higher expected numbers, but this throws away information. Alternative tests have been proposed that try to avoid problems associated with goodness of fit tests. Rousset and Raymond (1995, 1997) and Weir (1996) discuss some of these tests. The computer program GenePop (see Appendix B) implements some of them.

3.2 Heterozygosity and Homozygosity

We will often use H to symbolize the frequency of heterozygotes at a locus. In Section 3.1 we saw that the **observed heterozygosity** at an autosomal locus with two alleles is given by equation (2.2):

$$H_{obs} = P_{12} = \frac{N_{12}}{N} \tag{3.9}$$

where N_{12} is the observed number of heterozygotes. The **expected heterozygosity**, under Hardy-Weinberg conditions is

$$H_{exp} = 2pq \tag{3.10}$$

For multiple alleles, observed heterozygosity at a locus is

$$H_{obs} = \sum_{j \neq i} P_{ij} \tag{3.11}$$

where the summation is over all heterozygotes, and P_{ij} is the observed frequency of the A_iA_j heterozygote, as calculated from equation (2.9). The expected heterozygosity is

$$H_{exp} = 1 - \sum p_i^2 \tag{3.12}$$

This is just one minus the expected frequencies of all of the homozygotes, where p_i^2 is the expected frequency of the A_iA_i homozygote.

Expected heterozygosity is sometimes called **gene diversity** (Nei 1987). It can be interpreted as the probability that two randomly chosen copies of a gene will be different alleles, and can be applied to haploid organisms such as bacteria. For example, Selander and Levin (1980) estimated gene diversity in *E. coli* to be about 0.47 based on 20 allozyme loci.

In Section 2.10, we defined **nucleotide diversity** as the probability that two randomly chosen homologous nucleotides are different. It should be clear that this is analogous to gene diversity (expected heterozygosity) as defined above; nucleotide diversity is expected heterozygosity at the nucleotide level.

We will sometimes be interested in the **homozygosity** of a locus (or population). Homozygosity is frequently symbolized by G, and is just one minus heterozygosity. For two alleles

$$G_{obs} = P_{11} + P_{22} \tag{3.13}$$

and

$$G_{exp} = p^2 + q^2 \tag{3.14}$$

For multiple alleles, we have

$$G_{obs} = \sum P_{ii} \tag{3.15}$$

and

$$G_{exp} = \sum p_i^2 \tag{3.16}$$

All of these are population parameters (no "hats"). Their values can be estimated from the corresponding sample values. For small sample sizes, Nei (1987) suggests corrections to adjust for sample bias. For random mating diploid populations, an unbiased estimate of heterozygosity is

$$\hat{H}_{exp} = \frac{2N}{2N - 1}\left(1 - \sum \hat{p}_i^2\right) \tag{3.17}$$

where N is the number of individuals sampled. Note that if N is reasonably large, the correction is negligible. Box 3.3 illustrates how to use this equation.

For mtDNA or Y-chromosome loci, the estimate is

$$\hat{H}_{exp} = \frac{n}{n - 1}\left(1 - \sum \hat{p}_i^2\right) \tag{3.18}$$

where n is the number of females for mtDNA or the number of males for Y-chromosome loci. If there are equal numbers of males and females, then $n = N/2 = 2N/4$. To understand why this equation has n instead of $2N$, recall that $2N$ is the number of copies of a diploid autosomal locus (two in each individual, male or female). However, a Y-chromosome locus is present only once in males and zero times in females. If the sex ratio is even, there are only 1/4 as many copies of a Y-chromosome locus as of an autosomal locus. The same is true for mtDNA, because mtDNA is inherited through the female only.

We will use both heterozygosity and homozygosity frequently in future chapters, and must be careful to indicate whether we mean the observed quantity or

Box 3.3   Calculating expected heterozygosity

The following table gives observed allele frequencies at a human microsatellite locus (Litt and Luty 1989). Both alleles were typed in 37 individuals ($2N = 74$ alleles examined). Allele size is the size in base pairs of the amplified fragment; it is not the number of repeats.

Allele Size	Observed Number	p_i	p_i^2
96	1	0.0135	0.0002
92	4	0.0541	0.0029
90	10	0.1351	0.0183
88	7	0.0946	0.0089
86	16	0.2162	0.0467
84	3	0.0405	0.0016
82	15	0.2027	0.0411
80	1	0.0135	0.0002
74	1	0.0135	0.0002
72	8	0.1081	0.0117
70	6	0.0811	0.0066
68	2	0.0270	0.0007
Sum	$2N = 74$	1.0000	0.1392

Using equation (3.17), we can estimate expected heterozygosity as

$$\hat{H}_{exp} = \frac{2N}{2N - 1}(1 - \sum \hat{p}_i^2)$$

$$= \frac{74}{73}(1 - 0.1392) = 0.8726$$

Litt and Luty do not give genotype frequencies, but they write that 32 of the 37 individuals were heterozygous. Thus, observed heterozygosity at this locus is 0.8649.

the expected quantity. To be sure you understand these concepts, you should calculate observed and expected heterozygosity and homozygosity for the blood group data in Table 2.3, and the example in Box 2.1 (Problems 3.2 and 3.3).

3.3   Deviations from Hardy-Weinberg Expectations

For an autosomal locus with two alleles, the expected heterozygosity in an idealized population (one satisfying all of the Hardy-Weinberg assumptions) is

$$H_{exp} = 2pq$$

However, real populations are not ideal. We can have a deficiency, or excess, of heterozygotes due to many factors: natural selection, genetic drift, inbreeding, nonrandom mating, and population subdivision, among others. Wright (1922) introduced the parameter F to describe the deviation from expected heterozygosity.[1] F can be defined as the proportionate reduction in heterozygosity compared to a population in Hardy-Weinberg equilibrium:

$$F = \frac{H_{exp} - H_{obs}}{H_{exp}} \tag{3.19}$$

---

[1]Wright used f instead of F, but F has become more common in modern usage.

Solving this equation for H_{obs} will show how F describes the deviation in heterozygosity. We get (using $H_{exp} = 2pq$)

$$H_{obs} = 2pq - 2pqF \tag{3.20}$$

So heterozygosity is reduced by the amount $2pqF$.

F is usually a fraction between zero and one. If $F = 0$, we get the Hardy-Weinberg expectations. If $F = 1$, there are no heterozygotes in the population. An intermediate value indicates a deficiency of heterozygotes. F can also be negative if there is an excess of heterozygotes.

We can see how this deviation affects the frequencies of homozygotes by using equation (2.6)

$$p = P_{11} + \frac{1}{2}P_{12}$$

where P_{11} and P_{12} are the observed frequencies of A_1A_1 and A_1A_2. Substituting equation (3.20) into this, we have

$$p = P_{11} + \frac{1}{2}(2pq - 2pqF)$$

Solving for P_{11}, we get (after a bit of algebra)

$$P_{11} = p^2 + pqF$$

Similarly,

$$P_{22} = q^2 + pqF$$

where P_{22} is the observed frequency of A_2A_2 homozygotes.

To summarize, the observed genotype frequencies can be described as deviations from the expected (Hardy-Weinberg) frequencies as

$$P_{11} = p^2 + pqF \tag{3.21}$$

$$P_{12} = 2pq - 2pqF \tag{3.22}$$

$$P_{22} = q^2 + pqF \tag{3.23}$$

Table 3.5 summarizes an interesting example of Hardy-Weinberg deviations. In an early study of allozyme variation in alcohol dehydrogenase (*Adh*), Vigue and Johnson (1973) surveyed 42 populations of *Drosophila melanogaster* in eastern

TABLE 3.5 Observed frequencies of *Adh* genotypes in four populations of *Drosophila melanogaster*.

F and *S* represent the *Fast* and *Slow* alleles, respectively; *N* is the sample size. Forty-two populations were studied; only those showing significant deviations from Hardy-Weinberg expectations are shown here. The last column indicates either a deficiency or excess of heterozygotes. *Source:* From Vigue and Johnson (1973).

Location	N	FF	FS	SS	χ^2	Deviation
Miami, Fla.	86	4	12	70	8.86	Deficiency
Miami, Fla.	128	6	25	97	5.65	Deficiency
Portland, Maine	307	45	174	88	7.49	Excess
Raleigh, N.C.	322	44	127	151	4.13	Deficiency

North America, ranging from Maine to Florida. *Adh* typically shows two alleles in *D. melanogaster*, designated *Fast* and *Slow*. The frequency of the *Fast* allele varied from about 0.04 to about 0.56, and was higher in northern populations than in the southern ones. Of the 42 populations, only four showed significant deviation from Hardy-Weinberg expectations; those four are shown in Table 3.5. Three populations showed a significant deficiency of heterozygotes, and one showed an excess. You should calculate *F* for each of these four populations (Problem 3.9). These data alone give us no information about why these populations (and not the others) deviate from the Hardy-Weinberg expectations. It is widely believed that the *Adh* locus is subject to natural selection, although the mechanism is not well understood. If this is true, then 38 of the 42 populations may be under natural selection at this locus, but do not deviate significantly from the Hardy-Weinberg expectations. These results are not unusual; significant deviation from Hardy-Weinberg expectations is the exception, rather than the rule, in most outbreeding natural populations.

Many plants and a few animals undergo extensive inbreeding (mating among relatives). As we shall see in Chapter 8, inbreeding reduces heterozygosity compared to Hardy-Weinberg expectations. Thus, we would expect inbred populations to have a positive *F*. For example, Marshall and Allard (1970) estimated allele frequencies and heterozygosity at several allozyme loci in *Avena barbata*, a grass with about 98 percent self-fertilization. Table 3.6 shows their results for three polymorphic loci. Observed heterozygosities were much less than Hardy-Weinberg expectations, and *F* ranged from 0.70 to 0.76. There is more to this story; we return to it in Section 8.1.

Extension of *F* to multiple alleles is straightforward as long as we consider H_{obs} and H_{exp} to be the overall heterozygosities at the locus, as defined by equations (3.11) and (3.12). Then equation (3.19) still defines *F*, and solving (3.19) for H_{obs} gives

$$H_{obs} = H_{exp}(1 - F)$$

Now, substitute equation (3.12) into this equation and we get

$$H_{obs} = (1 - \sum p_i^2)(1 - F) \qquad \textbf{(3.24)}$$

Wright originally defined *F* as a way to measure loss of heterozygosity due to inbreeding. However, *F* has acquired a variety of interpretations, which we shall meet in later chapters. For now, think of it in its most general interpretation: *F* is

---

TABLE 3.6 Allele frequencies and heterozygosities at two polymorphic esterase loci (*E4* and *E10*) and one phosphatase locus (*P5*) in the Calistoga, California, population of *Avena barbata*.

No sample size correction was made in estimating H_{exp}, but the correction would have been negligible. *Source:* From Marshall and Allard (1970).

Locus	p_1	H_{obs}	H_{exp}	F
E4	0.36	0.11	0.461	0.76
E10	0.53	0.13	0.498	0.74
P5	0.48	0.15	0.499	0.70

a measure of the deviation of observed heterozygosity from the Hardy-Weinberg expected heterozygosity, due to any of several factors.

To summarize, large outbreeding populations typically have F near zero for most loci. This does not necessarily mean that no evolutionary processes are operating, as the cautions in Section 3.1 and the *Adh* example suggest. Inbred populations, such as self-fertilizing plants, some snails, and fungi, typically have positive values for F. A positive F may also indicate genetic drift in a small population (Chapter 7).

3.4 Sex and the Hardy-Weinberg Principle

Until now, we have assumed that allele frequencies are the same in males and females, and that the locus being studied is autosomal. In this section, we examine what happens when these assumptions are relaxed.

Different Allele Frequencies in Males and Females

Allele frequencies may not always be the same in males and females. For example, allele frequencies may be different in two populations, and males from one population may migrate and mate with females in the other population. How does this affect the Hardy-Weinberg principle?

We will answer this question for a locus with two alleles. Let p_m and p_f be the initial frequencies of the A_1 allele in males and females, with q_m and q_f the frequencies of the A_2 allele. Assuming equal numbers of males and females, the overall allele frequency in the population will be the average of the two sexes:

$$\bar{p} = \frac{p_m + p_f}{2}$$

After mating, the genotype frequencies in the zygotes of the next generation will be (see Figure 3.3)

$$P_{11}(t + 1) = p_m p_f$$
$$P_{12}(t + 1) = p_m q_f + q_m p_f$$
$$P_{22}(t + 1) = q_m q_f$$

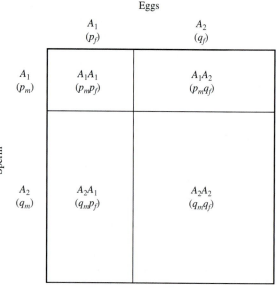

Eggs

	A_1 (p_f)	A_2 (q_f)
A_1 (p_m)	A_1A_1 ($p_m p_f$)	A_1A_2 ($p_m q_f$)
A_2 (q_m)	A_2A_1 ($q_m p_f$)	A_2A_2 ($q_m q_f$)

Sperm

Figure 3.3 **Random union of gametes at an autosomal locus with two alleles, A_1 and A_2.** The allele frequencies are initially different in males and females. Entries in the table represent zygote genotypes and frequencies.

These frequencies will be the same in males and females, since we are still considering an autosomal locus, but they are not necessarily the Hardy-Weinberg frequencies. Since the genotype frequencies are the same in males and females, the allele frequencies will be the same also. Using equation (2.6), the frequency of A_1 in the new zygotes will be

$$\overline{p}(t + 1) = P_{11}(t + 1) + \frac{1}{2}P_{12}(t + 1)$$

$$= p_m p_f + \frac{1}{2}(p_m q_f + q_m p_f)$$

$$= \cdots$$

$$= \frac{p_m + p_f}{2} = \overline{p}$$

(You should fill in the $\cdots$)

After one generation, the allele frequency is the same in males and females, and is equal to the average of the initial frequencies. At this point, if all the other conditions of the Hardy-Weinberg principle hold, the genotype frequencies in the next generation will be the Hardy-Weinberg expected frequencies of $\overline{p}^2$, $2\overline{p}\overline{q}$, and $\overline{q}^2$.

To summarize, if the allele frequencies are initially different in males and females, it will take two generations for the Hardy-Weinberg expected frequencies to occur. In the first generation, the allele frequencies in males and females are equalized, and in the second, the Hardy-Weinberg frequencies are reached, based on the average of the male and female frequencies.

Sex-Linked Loci

Many important genes are located on the X chromosome, for example, genes causing color blindness and hemophilia in humans. Does the Hardy-Weinberg principle hold for X-linked loci? We will consider a single X-linked locus with two alleles, designated A_1 and A_2, and will assume an even sex ratio (equal numbers of males and females). Let $p_m(t)$ and $p_f(t)$ be the frequencies of the A_1-bearing X chromosome in males and females in generation t. Because males have one X chromosome and females have two, the allele frequency in the population as a whole is the weighted average of the frequencies in males and females

$$\overline{p}(t) = \frac{1}{3}p_m(t) + \frac{2}{3}p_f(t) \tag{3.25}$$

Another way to think about this is that one-third of all X chromosomes in the population are found in males and two-thirds in females.

Now, in the next generation, sons get their X chromosome from their mothers, so their allele frequency will be that of their mothers:

$$p_m(t + 1) = p_f(t) \tag{3.26}$$

Similarly, the daughters get one X chromosome from their fathers and one from their mothers. Therefore, the allele frequency in the daughters will be the average of the parents:

$$p_f(t + 1) = \frac{p_m(t) + p_f(t)}{2} \qquad\qquad \text{(3.27)}$$

The average allele frequency in the next generation will be [analogous to equation (3.25)]

$$\bar{p}(t + 1) = \frac{1}{3}p_m(t + 1) + \frac{2}{3}p_f(t + 1) = \frac{1}{3}p_f(t) + \frac{2}{3}\left[\frac{p_m(t) + p_f(t)}{2}\right]$$

$$= \frac{1}{3}p_m(t) + \frac{2}{3}p_f(t)$$

$$= \bar{p}(t)$$

Therefore, the allele frequency in the population does not change.

What about the allele frequency in each sex? These are given by the pair of recursion equations (3.26) and (3.27). Rather than solve them, we can see what happens by considering a simple example. Assume that the allele frequencies in males and females are initially 0.80 and 0.20. Then $p_m(t) = 0.80$, $p_f(t) = 0.20$, and $\bar{p}(t) = 0.40$. Applying (3.26) and (3.27) we get

$$p_m(t + 1) = p_f(t) = 0.20$$

$$p_f(t + 1) = \frac{p_m(t) + p_f(t)}{2} = \frac{0.80 + 0.20}{2} = 0.50$$

In the next generation, we get

$$p_m(t + 2) = p_f(t + 1) = 0.50$$

$$p_f(t + 2) = \frac{p_m(t + 1) + p_f(t + 1)}{2} = \frac{0.20 + 0.50}{2} = 0.35$$

and so forth. Figure 3.4a shows the allele frequencies for the first ten generations. We see that p_m and p_f oscillate above and below $\bar{p}$, getting closer each generation. The approach to $\bar{p}$ is rapid, and after a few generations p_m and p_f are very nearly equal.

We can also follow the changes in genotype frequencies. Since males have only one allele, the genotype frequencies and allele frequencies are the same, and the solid line in Figure 3.4a also indicates the change in male genotype frequencies. The genotype frequencies in females are calculated from the male and female allele frequencies in the previous generation, since females get one X chromosome from their mother and one from their father. The genotype frequencies in females are thus

$$P_{11f}(t + 1) = p_m(t)p_f(t)$$

$$P_{12f}(t + 1) = p_m(t)q_f(t) + q_m(t)p_f(t)$$

$$P_{22f}(t + 1) = q_m(t)q_f(t)$$

These are illustrated in Figure 3.4b. We can note two things from these figures: (1) The female genotype frequencies approach the Hardy-Weinberg expected frequencies based on $\bar{p}$ and $\bar{q}$. (2) The genotype frequencies in males and females will remain different, even as equilibrium is approached.

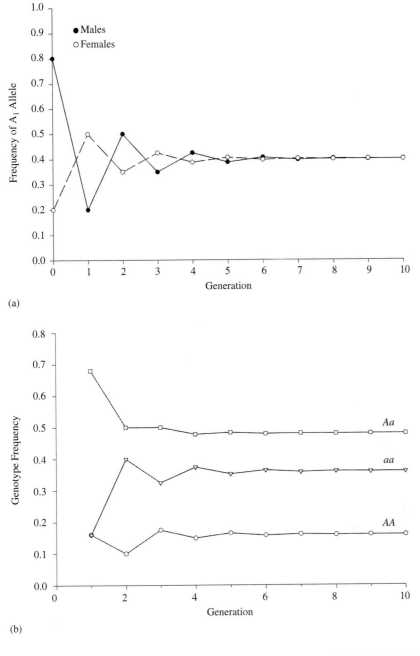

(a)

(b)

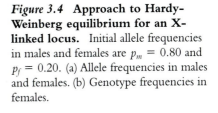

Figure 3.4 **Approach to Hardy-Weinberg equilibrium for an X-linked locus.** Initial allele frequencies in males and females are $p_m = 0.80$ and $p_f = 0.20$. (a) Allele frequencies in males and females. (b) Genotype frequencies in females.

Summary

1. The main point of this chapter is that Mendelian inheritance alone does not cause genetic variation to decay. Genotype frequencies and allele frequencies in a population will remain constant (at equilibrium) unless some evolutionary process is operating to change them.

2. The relationship between genotype frequencies and allele frequencies at Hardy-Weinberg equilibrium is given by equations (3.1) through (3.3) or (3.7) and (3.8). This equilibrium is reached after one generation of random mating. These equations are based on a number of simplifying assumptions.

3. For X-linked loci, the equilibrium is approached gradually, but the essential conclusion is the same: Genetic variation is maintained in the absence of disturbing forces.

4. The main evolutionary processes that can cause deviation from the Hardy-Weinberg equilibrium frequencies are mutation, genetic drift, natural selection, nonrandom mating, and gene flow.

5. The Hardy-Weinberg principle can be used to detect the influence of these evolutionary processes. The statistical tests are sometimes not very powerful, and cannot detect certain kinds of processes.

6. Deviations from Hardy-Weinberg conditions indicate the effect of at least one evolutionary process. Because in the absence of these processes the population will return to the Hardy-Weinberg equilibrium frequencies in one or a few generations, deviation indicates recent or ongoing evolutionary change.

7. Genotype frequencies are often close to the Hardy-Weinberg expectations in large outbreeding diploid populations. When this is true, the Hardy-Weinberg principle can be used to estimate the frequencies of recessive alleles.

8. Deviations from Hardy-Weinberg expectations are measured by the parameter F, which is the proportional reduction in heterozygosity compared to the Hardy-Weinberg expectation [equation (3.19)].

9. F is frequently near zero in large outbreeding populations, but significantly positive in inbred or very small populations.

The Hardy-Weinberg principle is fundamental to population genetics. Nearly all theories of evolutionary change start with the Hardy-Weinberg conditions and then consider one or more of the various evolutionary processes. In later chapters, we will study each of these processes, individually and in combination, almost always starting with the Hardy-Weinberg conditions. As we study these processes, we will try to answer a fundamental question: How do they affect the amount of genetic variation in a population?

Problems

3.1. a. For three alleles and four alleles:

How many homozygous genotypes are there?

How many heterozygous genotypes are there?

How many degrees of freedom in a chi-square test for Hardy-Weinberg frequencies are there?

  b. If k is the number of alleles and n is the number of genotypes, show that

$$n = \frac{k(k + 1)}{2}$$

  c. Show that the degrees of freedom in a chi-square test for Hardy-Weinberg conditions is

$$df = n - k$$

3.2. Calculate the expected heterozygosity for each population for the ABO, Lutheran, and Duffy loci in Table 2.3. Ignore the adjustment for sample size.

3.3. For the data in Box 2.1, calculate the following: allele frequencies, observed genotype frequencies, expected genotype frequencies, χ^2 test for genotype frequencies, observed heterozygosity, expected heterozygosity, F.

3.4. Make the same calculations for each locus in the two *Drosophila* populations of Problem 2.1. Within each population, does F differ between the two loci? Discuss.

3.5. Consider a population with three genotypes. You sample zygotes and get the following results:

$$A_1A_1 \qquad A_1A_2 \qquad A_2A_2$$
$$100 \qquad\quad 200 \qquad\quad 100$$

Assume the zygotes have different survival rates and imagine you can follow these zygotes to adults. You get the following:

$$A_1A_1 \qquad A_1A_2 \qquad A_2A_2$$
$$100 \qquad\quad 180 \qquad\quad 81$$

   a. Calculate the allele frequencies in both zygotes and adults. Is natural selection acting?

   b. Compare the observed and expected genotype frequencies in zygotes.

   c. Compare the observed and expected genotype frequencies in the adults, based only on adult data.

   d. Could you detect natural selection if you sampled only the adults? Discuss.

3.6. Cystic fibrosis is a severe, usually fatal, disease in humans, caused by homozygosity for a recessive allele. The frequency of the disease is about 0.0006. What is the estimated frequency of the *cf* allele? What is the expected frequency of individuals who are heterozygous for this allele? Does your answer surprise you? What assumptions did you have to make to get your answer? Why has this disease not been eliminated by natural selection?

3.7. Show that, for a single locus with two alleles, the ratio of A_1A_2 to A_2A_2 is $2(1 - q)/q$. Calculate this ratio for $q = 0.1$, 0.01, and 0.001. Are these results consistent with your answer to the previous problem?

3.8. Show that the expected frequency of heterozygotes at a locus with two alleles is maximized at $p = 0.5$. (*Hint:* Recall from calculus how to maximize a function.) Then plot the frequency of heterozygotes versus p for values of p ranging from 0 to 1. (Use a spreadsheet or graphing program.)

3.9. Calculate the observed and expected genotype frequencies, and F, for each of the four *Drosophila* populations in Table 3.5.

3.10. Show that for more than two alleles, the observed homozygosity is

$$G_{obs} = \sum p_i^2 + F(1 - \sum p_i^2)$$

3.11. Consider two alleles, A_1 and A_2, at an X-linked locus. Allele A_1 has an initial allele frequency of 1.0 in females and 0 in males. Use a spreadsheet and its graphing capabilities to calculate the allele frequencies in males and females for 25 generations. Also, plot the expected genotype frequencies in females for 25 generations (as in Figure 3.4b).

Recombination, Linkage, and Disequilibrium

The presumption that an adequate picture of evolutionary dynamics is contained in a space whose sole dimensions are the frequencies of alternative alleles at single loci is justified only if the various assumptions about independence of fitness relations and independence of genic assortment are correct or nearly so.

—*R. C. Lewontin (1974)*

Until now, we have considered a single locus in isolation. That is unrealistic, of course; any organism has thousands of loci, linked and unlinked, which may interact with one another to a greater or lesser degree. In this chapter, we introduce some of the fundamental concepts of multilocus population genetics. We derive a two-locus analog of the Hardy-Weinberg principle, and see how it can be applied to real-world problems such as DNA fingerprinting and mapping genes associated with human diseases.

Throughout most of this chapter, we assume that natural selection is not operating on the loci being considered. Multilocus models of natural selection and gene interaction will be considered in future chapters.

4.1 Recombination and Linkage

Whenever we consider two or more loci simultaneously, we must consider **recombination**. The most general definition of recombination is any process that creates new combinations of alleles in the offspring. A second essential concept is **linkage**. Two loci are linked if they are on the same chromosome, close enough together that the frequency of recombination between them is less than 50 percent.

First, consider recombination in diploid, sexually reproducing organisms. Recombination between unlinked loci occurs during meiosis I, when nonhomologous chromosome pairs align and separate randomly. This is the chromosomal basis of **independent assortment**. Figure 4.1 shows the situation for two pairs

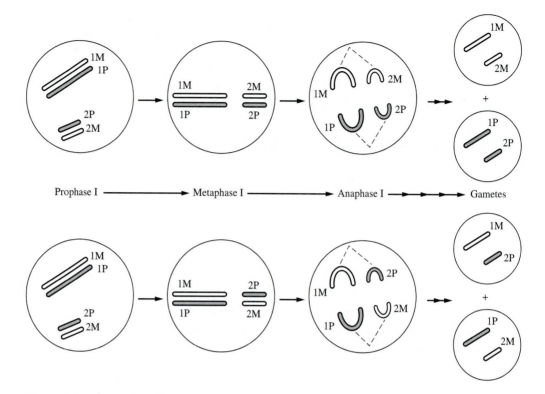

Figure 4.1 **Alternative alignments of two pairs of chromosomes during meiosis.**   Symbols 1M, 1P, 2M, and 2P represent maternally and paternally derived copies of chromosomes 1 and 2. In one possibility (*top*), maternal chromosomes align the same way in metaphase I and segregate together during anaphase I, ultimately producing gametes containing chromosomes 1M and 2M or 1P and 2P. Alternatively (*bottom*), the chromosomes can align so that 1M and 2P segregate together in anaphase I, ultimately producing gametes containing chromosomes 1M and 2P or 1P and 2M. Both alternatives are equally likely. In general, if *n* is the number of homologous pairs of chromosomes, the number of possible alignments is 2^{n-1}. Each alignment produces two combinations, so the number of possible maternal and paternal combinations is 2^n.

of chromosomes. In general, if there are *n* pairs of chromosomes, then there are 2^{n-1} ways in which maternally and paternally derived chromosomes can align during metaphase I. Each alignment produces two different combinations after anaphase I; therefore, there are 2^n possible combinations of maternal and paternal chromosomes after meiosis. For humans ($n = 23$), this number is more than 8 million!

Linked loci undergo recombination when **crossing over** occurs during meiosis. During prophase I, non-sister chromatids of homologous chromosomes exchange pieces, producing chromosomes with a combination of paternally and maternally derived genetic material. The probability that two linked loci will participate in crossing over depends in a general way on the distance between them; the closer together they are, the less likely crossing over will occur. However, the frequency of crossing over varies in different parts of the genome. For example, crossing over is very rare among loci on the small fourth chromosome of *Drosophila melanogaster*. In humans, crossing over in autosomes is more frequent in females than in males. Local hot spots exist in many species. Crossing over is entirely absent on the Y-chromosome of mammals and in male *Drosophila*.

In general, alleles at tightly linked loci are likely to stay together longer than alleles at unlinked loci. Some loci are so tightly linked that crossing over between them almost never occurs. Recall that a locus is just a position on a chromosome, so two nearby nucleotides in a single gene can be considered as two loci. Loci on the Y chromosome of mammals or on the mtDNA molecule of most animals are haploid and do not recombine except in very unusual circumstances. Therefore, each of these genetic elements can be considered a single locus in population genetics models.

When we think of recombination, we generally think of diploid sexual eukaryotes. But recombination can occur in other organisms as well. Bacteria undergo recombination by the processes of transformation, transduction, or conjugation. In fact, recombination can actually occur between different species of bacteria. This has allowed the acquisition of antibiotic resistance genes by many species of bacteria, a serious medical problem. Similar processes result in recombination in other organisms such as fungi and some protozoa.

The importance of recombination is this: If maternal and paternal chromosomes carry different alleles, as they often do, then both independent assortment and crossing over will produce new combinations of alleles in the offspring. If two loci interact in some way (for example, by coding for different parts of a multimeric enzyme) then natural selection may act on *combinations* of alleles at the two loci. Recombination will act both to create and to break up these combinations.

We saw in Chapter 2 that most populations contain large amounts of genetic variation. Therefore, recombination can be an important source of new allelic combinations. Natural selection may act on these combinations of alleles, as opposed to acting on each locus independently. This, and other factors such as mutation, genetic drift, and population subdivision, can create nonrandom associations between alleles at different loci. These nonrandom associations can be useful in gene mapping and in inferring information about natural selection, population subdivision, or other important processes. In the next section, we will see how to quantify the association between alleles at different loci.

4.2 Gametic and Linkage Disequilibrium

It is possible that the alleles at one locus may be randomly associated with one another as predicted by the Hardy-Weinberg principle, and the alleles at a second locus may also be randomly associated, but alleles at the first locus may be nonrandomly associated with alleles at the second locus. This is the concept of **gametic disequilibrium**. Table 4.1 illustrates an example. There are only three genotypes, *AABB*, *aabb*, and double heterozygotes of the form *AB/ab*. (Here, the notation means that the alleles before the slash came from one gamete and the alleles after the slash came from the other.) The frequency of the *A* allele and the frequency of the *B* allele are each 0.5, and the genotype frequencies at each locus match the Hardy-Weinberg expectations. But the *A* allele is always associated with the *B* allele.

Gametic disequilibrium can be generated in many ways, including hybridization, mutation, population subdivision, genetic drift, and some kinds of natural selection. But, whatever its cause, gametic disequilibrium will be temporary in the absence of forces acting to maintain it. Recombination will eventually break up nonrandom allelic associations. In the above example, recombination will occur in the double heterozygotes to produce gametes *Ab* and *aB*.

TABLE 4.1   An example of gametic disequilibrium.

All individuals are either *AABB, aabb*, or double heterozygotes of form *AB/ab*. Numbers refer to frequencies of the two-locus genotypes. Each locus is in Hardy-Weinberg equilibrium, but alleles at the *A* locus are nonrandomly associated with alleles at the *B* locus.

Genotype at A Locus	Genotype at B Locus		
	BB	**Bb**	**bb**
AA	0.25	0	0
Aa	0	0.50	0
aa	0	0	0.25

Gametic disequilibrium is frequently attributed to linkage. Tightly linked loci frequently show some gametic disequilibrium, but linkage is not necessary; epistatic interactions between alleles at unlinked loci can generate gametic disequilibrium, as can the other forces mentioned above. When disequilibrium appears to be an artifact of tight linkage, we often call it **linkage disequilibrium**. We will use the more general term *gametic disequilibrium*, unless linkage is obviously a factor.

The preceding discussion assumes a sexually reproducing diploid population. But disequilibrium can occur in haploid species as well. Different strains of *E. coli*, for example, frequently show nonrandom associations of alleles. This is usually interpreted to mean that recombination among strains of *E. coli* is very infrequent. However, DNA sequence data sometimes challenge this interpretation (Guttman 1997). Other species of bacteria sometimes show much less disequilibrium, for example *Bacillus subtilis* (Istock et al. 1992). For a good reviews of recombination in bacteria, see Maynard Smith et al. (1993) and Cohan (2000).

As mentioned previously, recombination will break up nonrandom allelic associations, causing disequilibrium to decay in the absence of forces acting to maintain it. The rate of decay should depend on the recombination rate between loci. Our goal here is to quantify the degree of nonrandom association and the rate of decay. We will assume a diploid, randomly mating sexual population in which neither mutation, genetic drift, nor natural selection is acting. In future chapters, we will examine what happens when these assumptions are relaxed. We will consider two loci, with alleles *A* and *a* at the first, and *B* and *b* at the second. It may help to think of these as linked loci, but keep in mind that linkage is not necessary for disequilibrium.

Notation can be a problem when working with two loci. We will try to make it as intuitive as possible. We define the allele frequencies as

$$p_A = \text{observed frequency of } A \text{ allele}$$

$$p_a = \text{observed frequency of } a \text{ allele}$$

$$p_B = \text{observed frequency of } B \text{ allele}$$

$$p_b = \text{observed frequency of } b \text{ allele}$$

We must also define the gamete frequencies:

g_1 = observed frequency of AB gamete

g_2 = observed frequency of Ab gamete

g_3 = observed frequency of aB gamete

g_4 = observed frequency of ab gamete

This assumes that we can actually tell how the alleles are associated in gametes. This is sometimes problematic, and will be discussed below.

From these definitions, the allele frequencies are related to the gamete frequencies as follows:

$$p_A = g_1 + g_2 \tag{4.1}$$

$$p_a = g_3 + g_4 = 1 - p_A \tag{4.2}$$

$$p_B = g_1 + g_3 \tag{4.3}$$

$$p_b = g_2 + g_4 = 1 - p_B \tag{4.4}$$

Now, if alleles at the first locus are randomly associated with alleles at the second locus, the expected frequency of a gamete is the product of the frequencies of the alleles in that gamete:

$$E_{AB} = p_A p_B \tag{4.5}$$

$$E_{Ab} = p_A p_b \tag{4.6}$$

$$E_{aB} = p_a p_B \tag{4.7}$$

$$E_{ab} = p_a p_b \tag{4.8}$$

where E denotes the expected frequency of the gamete indicated by subscripts.

We now define a deviation for each gamete as the difference between the observed and expected frequencies:

$$D_{AB} = g_1 - p_A p_B \tag{4.9}$$

$$D_{Ab} = g_2 - p_A p_b \tag{4.10}$$

$$D_{aB} = g_3 - p_a p_B \tag{4.11}$$

$$D_{ab} = g_4 - p_a p_b \tag{4.12}$$

It turns out that these D's are related in a very simple way. To see this, substitute equations (4.1) and (4.3) into (4.9):

$$D_{AB} = g_1 - (g_1 + g_2)(g_1 + g_3)$$

With some algebraic simplification this becomes

$$D_{AB} = g_1 g_4 - g_2 g_3$$

Now, substitute equations (4.1) and (4.4) into (4.10) and simplify. The result is

$$D_{Ab} = g_2 g_3 - g_1 g_4 = -D_{AB}$$

Similarly, it can be shown that $D_{aB} = -D_{AB}$ and $D_{ab} = D_{AB}$. So let us define

$$D = g_1 g_4 - g_2 g_3 \tag{4.13}$$

D is called the **disequilibrium coefficient**; it is a measure of the degree of nonrandom association among alleles at the two loci. The gamete frequencies now become [from equations (4.9) through (4.12)],

$$g_1 = p_A p_B + D \tag{4.14}$$

$$g_2 = p_A p_b - D \tag{4.15}$$

$$g_3 = p_a p_B - D \tag{4.16}$$

$$g_4 = p_a p_b + D \tag{4.17}$$

So far, this is mostly just definition. What we are really interested in is how the gamete frequencies and the disequilibrium coefficient change over time. The gamete frequencies will change each generation because of recombination during gamete formation. It is only by recombination in double heterozygotes that gamete frequencies will change. Note, there are two kinds of double heterozygotes, AB/ab (**coupling phase**) and Ab/aB (**repulsion phase**).[1] Recombination in the first will create Ab and aB gametes; recombination in the second will create AB and ab gametes (Figure 4.2). The gamete frequencies after recombination will be the gamete frequencies in the next generation, assuming no other processes are acting to change them. If we designate these new frequencies with subscript $t + 1$, the new frequency of AB will be

$$g_{1,t+1} = g_1 - \left(\begin{array}{c} \text{proportion lost} \\ \text{by recombination in} \\ \text{coupling heterozygotes} \end{array} \right) + \left(\begin{array}{c} \text{proportion gained} \\ \text{by recombination in} \\ \text{repulsion heterozygotes} \end{array} \right)$$

Let r be the frequency of recombination, defined as the proportion of recombinant gametes produced by a double heterozygote $(0 \leq r \leq 0.5)$. Then,

$$g_{1,t+1} = g_1 - r(g_1 g_4) + r(g_2 g_3)$$

The quantities in parentheses come from the assumption of random union of gametes, that is, the frequency of AB/ab is $g_1 g_4$ and the frequency of Ab/aB is $g_2 g_3$. Using the definition of D in equation (4.13), this equation simplifies to

$$g_{1,t+1} = g_1 - rD \tag{4.18}$$

Similarly,

$$g_{2,t+1} = g_2 + rD \tag{4.19}$$

$$g_{3,t+1} = g_3 + rD \tag{4.20}$$

$$g_{4,t+1} = g_4 - rD \tag{4.21}$$

We can now calculate the disequilibrium coefficient in generation $t + 1$:

$$D_{t+1} = g_{1,t+1} g_{4,t+1} - g_{2,t+1} g_{3,t+1}$$

Substituting equations (4.18) through (4.21) and simplifying, we get

$$D_{t+1} = (1 - r)D_t \tag{4.22}$$

---

[1]Note that, since allele symbols are arbitrary, the definitions of coupling and repulsion gametes are also arbitrary.

COUPLING PHASE

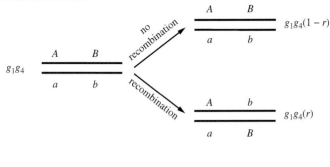

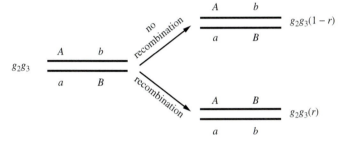

REPULSION PHASE

Figure 4.2 **Change in gamete frequencies as a result of recombination in double heterozygotes.** The frequency of coupling heterozygotes is g_1g_4 (assuming random union of gametes). Recombination creates repulsion gametes with frequency r. The frequency of repulsion heterozygotes is g_2g_3. Recombination creates coupling gametes with frequency r. The two loci are shown on the same chromosome for convenience only; linkage is not necessary.

where we have added the subscript t to D on the right side to emphasize that this is a one-generation recursion equation. Equation (4.22) has exactly the same form as equation (1.2), which we solved in Section 1.4. By analogy to equation (1.3), the solution to equation (4.22) is

$$D_t = (1 - r)^t D_o \qquad \textbf{(4.23)}$$

where D_o is the initial disequilibrium coefficient.

This is a very important result. It says that gametic disequilibrium will decay to zero if no other processes are acting except recombination. For independent assortment ($r = 0.5$), the decay to $D \cong 0$ will occur within a few generations, but if r is small (usually implying tight linkage), the decay will occur slowly. Figure 4.3 shows the decay of D for several values of r.

Most randomly mating natural populations typically show little disequilibrium. Table 4.2 shows an allozyme example from killifish (*Fundulus heteroclitus*). Pairwise disequilibrium coefficients were calculated for five polymorphic loci. Of the ten possible pairwise comparisons, only one had D significantly different from zero.

If two loci experience limited recombination, disequilibrium will decay slowly. Examples are tightly linked nucleotides, loci on mitochondrial or chloroplast DNA, or loci within inversions in certain *Drosophila* species. All of these situations frequently show significant gametic disequilibrium. Bacteria, which undergo limited recombination, frequently show high disequilibrium, although there are important exceptions (Maynard Smith et al. 1993).

Inbreeding (e.g., self-fertilization in many plant species) also slows the decay of gametic disequilibrium. The reason is that inbreeding decreases heterozygosity (Chapter 8), and therefore there are fewer double heterozygotes in which recombination can occur. Allard et al. (1972) found high disequilibrium among allozyme loci in a primarily self-fertilizing population of slender wild oats (*Avena*

Figure 4.3 **Decay of gametic disequilibrium (*D*) as a function of recombination frequency (*r*), as calculated from equation (4.23).** Shown is the magnitude (absolute value) of *D*. Positive and negative disequilibrium values decay at the same rate.

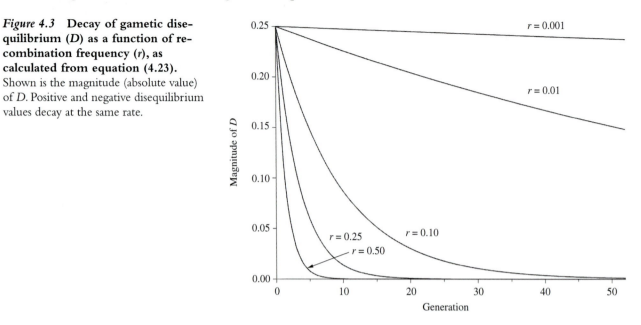

barbata). Similarly, Viard et al. (1997) found extensive disequilibrium among microsatellite loci in populations of a self-fertilizing snail (*Bulinus truncatus*).

Factors that can generate gametic disequilibrium are mutation, genetic drift, gene flow, nonrandom mating, population bottlenecks, or natural selection. In most cases, this disequilibrium will decay as described by equation (4.23). Processes that slow down the decay are limited recombination, inbreeding, and some kinds of natural selection. The interactions among these processes determine the amount of gametic disequilibrium observed between two loci.

TABLE 4.2   Pairwise gametic disequilibrium (*D*) among five polymorphic allozyme loci in the killifish, *Fundulus heteroclitus*.

The MDH-PGI pair is highly significant ($P < 0.001$); no other pairs are significant. D' is a standardized measure of gametic disequilibrium, discussed on pages 100–102. *Source:* From Mitton (1997).

Loci	D	D'
SERE-PGM	−0.010	−0.109
LDH-PGM	−0.009	−0.062
MDH-PGM	0.009	0.061
PGI-PGM	−0.004	−0.052
SERE-PGI	−0.016	−0.139
LDH-PGI	0.007	0.063
MDH-PGI	−0.024*	−0.228
SERE-MDH	−0.001	−0.002
LDH-MDH	0.006	0.096
SERE-LDH	0.003	0.050

*$P < 0.001$

Natural Selection and Gametic Disequilibrium

As we have seen, gametic disequilibrium will decay in the absence of processes acting to maintain it. One important process that is capable of slowing the decay of disequilibrium, or even of maintaining a stable nonzero disequilibrium, is natural selection.

Why do we care how much disequilibrium exists? We shall see in Chapter 5 that, if we know something about how natural selection acts at a locus, we can predict future changes in allele frequencies at that locus. This theory is valid only if different loci evolve independently, that is, if there is no gametic disequilibrium. If, in fact, gametic disequilibrium usually decays to zero and is a minor factor in most populations, then we can use one-locus selection theory to predict changes in allele frequencies. On the other hand, if gametic disequilibrium is common and stable, then predictions based on one-locus theory will be invalid, because allele frequencies at one locus will be affected by selection at other loci in disequilibrium with it. As suggested earlier, gametic disequilibrium seems to be relatively rare in large outbreeding populations, but it is important to understand the conditions under which exceptions can exist.

Alleles at different loci may interact in ways that favor certain gamete types. This gene interaction is potentially significant because it may allow natural selection to maintain a stable disequilibrium, that is, disequilibrium will not decay to zero. Thus, when gametic disequilibrium is found, and evidence suggests that it is long-standing and stable, natural selection is frequently invoked as an explanation.

Natural selection can create gametic disequilibrium by favoring certain gamete types. Recombination will reduce disequilibrium by breaking them up. The relative strength of these processes determines whether stable disequilibrium can exist. The higher the recombination rate between two loci, the stronger natural selection must be to overcome it. The mathematical details are complex, but the general principle is that linkage must be very tight, or selection very strong, in order for stable disequilibrium to exist. We will consider the details in Chapter 12.

The interactions between natural selection, linkage, and disequilibrium can be illustrated with an example. *Cepaea nemoralis* is a land snail found throughout much of western Europe. It is highly polymorphic for shell color and banding patterns on the shell, and the genetics and ecology of these characteristics have been extensively studied. Shell color ranges from yellow to pink to brown, with shades of each. These are due to several alleles at a single locus, with brown dominant over pink and yellow, and pink dominant over yellow. The number of bands on the shell can be from zero (dominant) to five, or rarely, six. Modifier loci affect the number of bands and how they are expressed. The loci for shell color and banding, along with other loci controlling expression of the bands, are tightly linked, with very little recombination among them in most populations. This complex of genes has been called a **supergene**, because the tightly linked genes effectively act as a single gene.

Natural selection acts on shell color and banding patterns in complex ways. Cryptic coloration is important in protecting snails from predators, primarily thrushes. Yellow snails are camouflaged on greenish backgrounds, and darker-colored snails are camouflaged on leaf litter and dark soil. Banded snails are camouflaged against a heterogeneous background, and unbanded against a uniform background. Thrushes take fewer snails that are better camouflaged, as judged by human eyes (Cain and Sheppard 1954a). Selective pressures change with the seasons. In early spring, woodland floors are covered with brownish leaf litter and soil. During this time, thrushes take more yellow than pink or brown snails. In late spring, the woodland floors are green and the thrushes take more pink and brown snails (Sheppard 1951; Jones et al. 1977).

In some habitats, there seems to be strong selection favoring combinations of shell color and banding pattern. For example, in beechwoods most snails are brown and unbanded. In hedgerows, most are yellow and banded. Yellow unbanded snails have higher survival rates in warm, dry conditions, and darker banded snails do better in cool, humid conditions.

These polymorphisms thus appear to be maintained by a variety of selection pressures that vary in space and time. Natural selection acts on both shell color and banding patterns, often on combinations of the two characters. If this is true, then we might predict significant gametic disequilibrium between loci controlling shell color and banding. In fact, disequilibrium is usually strong between these loci, and other loci within the supergene (Jones et al. 1977). For example, dark-brown shell color is almost always associated with the unbanded form.

If natural selection favors nonrandom association of alleles at different loci, one might expect that it would also favor reduced recombination between these loci, since recombination would break up favorable combinations. Any chromosomal rearrangement that brought the loci into tighter linkage might be favored. This is thought to be how the *Cepaea* supergene originated (Ford 1975). The genes within the supergene were probably brought together by selection favoring chromosomal changes that tightened linkage among them. For general reviews of polymorphism, selection, and disequilibrium in *Cepaea*, see Ford (1975) and Jones et al. (1977).

Other supergenes are thought to have evolved in the same way, for example, the genes for heterostyly in several plant species, or the genes controlling mimicry in some butterflies. See Ford (1975) and Hedrick et al. (1978) for reviews of supergenes.

To summarize, if natural selection favors certain gamete types, stable disequilibrium may be possible if recombination is low enough. Selection will favor reduced recombination to avoid breaking up favorable gamete types, sometimes resulting in tightly linked supergenes.

A Standardized Measure of Disequilibrium

One problem with D is that not only its value, but also its range depends on the allele frequencies. For example, if all gametes are AB or ab and $p_A = p_B = 0.5$, then D has its maximum value of 0.25. If all gametes are Ab or aB with $p_A = p_B = 0.5$, then D has its minimum value of -0.25. For allele frequencies other than 0.5 and for mixes of both coupling and repulsion gametes, the range of D is less. This variable range makes it difficult to compare values of D among populations or loci with different allele frequencies. It is desirable to have some kind of standardized disequilibrium coefficient whose range is not dependent on allele frequencies.

Since all gamete frequencies must be nonnegative, it follows from equations (4.15) and (4.16) that

$$D \le p_A p_b$$

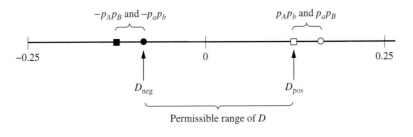

Figure 4.4   The possible range of D as a function of allele frequencies. All gamete frequencies must be positive by definition. Therefore, if D is positive, then from equations (4.15) and (4.16) D must be less than $p_A p_b$ and less than $p_a p_B$. The lesser of these is D_{pos}. Similarly, if D is negative, then from equations (4.14) and (4.17), D must be greater than $-p_A p_B$ and greater than $-p_a p_b$. The greater (least negative) of these is D_{neg}. Then D must be between D_{pos} and D_{neg}. D_{max} is defined as whichever of these is greatest in absolute value.

and

$$D \le p_a p_B$$

Let D_{pos} be the lesser of either $p_A p_b$ or $p_a p_B$. Then

$$D \le D_{pos}$$

See Figure 4.4.

Similarly, from equations (4.14) and (4.17), we see that

$$D \ge -p_A p_B$$

and

$$D \ge -p_a p_b$$

(Keep in mind that "greater than" means "to the right of" on the number line.) We define D_{neg} as the greater of (least negative of) $-p_A p_B$ or $-p_a p_b$. Then $D \ge D_{neg}$ (Figure 4.4).

To summarize, the range of D is

$$D_{neg} \le D \le D_{pos}$$

(Figure 4.4). We now define D_{max} as the absolute value of the greatest possible disequilibrium for a given set of allele frequencies. In other words,

$$D_{max} = D_{pos} = \min(p_A p_b, p_a p_B) \qquad \text{if } D > 0$$
$$D_{max} = |D_{neg}| = |\max(-p_A p_B, -p_a p_b)| \quad \text{if } D < 0$$

But $|\max(-p_A p_B, -p_a p_b)|$ is the same as $\min(p_A p_B, p_a p_b)$. So we can write

$$D_{max} = \min(p_A p_b, p_a p_B) \quad \text{if } D > 0$$
$$D_{max} = \min(p_A p_B, p_a p_b) \quad \text{if } D < 0$$

Finally, we define a new measure of disequilibrium, D' (D-prime) as

$$D' = \frac{D}{D_{max}} \qquad\qquad \textbf{(4.24)}$$

Defined in this way, D' is the proportion of the maximum disequilibrium possible for a given set of allele frequencies. It has the same sign as D, and has a range of -1 to $+1$, independent of allele frequencies. Note that the *value* of D' is still allele frequency dependent, but its *range* is not.

Table 4.3 illustrates the usefulness of D'. In the first two examples, D has the same positive value of 0.018. In the first it is only about 10 percent of its maximum value, but in the second it is about 90 percent of its maximum. Similarly, in the third

TABLE 4.3 Numerical examples showing identical values of D, but very different values of D'.
The latter has the same sign as D, and represents the proportion of the maximum disequilibrium possible for a given set of allele frequencies.

p_A	p_B	D	D'
0.40	0.30	0.018	0.100
0.10	0.80	0.018	0.900
0.60	0.40	−0.08	−0.333
0.90	0.10	−0.08	−0.889

and fourth examples, D has the same negative value but D' is very different. For the killifish example discussed earlier, Table 4.2 gives estimates of both D and D'.

D' makes it easy to understand how much disequilibrium is present compared to how much is possible, and is useful for comparing disequilibrium in different populations. However, it is more difficult to incorporate into theoretical studies, and D remains more useful in that context. See Hedrick (1987) and Lewontin (1988) for discussions of theoretical issues.

Estimating D

We have been discussing D as a population parameter. How can we estimate D from a sample? Sometimes this can be a problem. The reason is that the phase of the gamete or chromosome types must be determined. For haploid organisms such as bacteria, phase can be determined directly, but for diploids there are two kinds of double heterozygotes, coupling phase (AB/ab) and repulsion phase (Ab/aB), and they are indistinguishable most of the time. It is sometimes possible to determine the phase of a double heterozygote by pedigree analysis or by progeny testing. For some organisms, for example some species of *Drosophila*, it is possible to extract single chromosomes and determine the phase from extracted chromosomes, but this is laborious.

One approach to estimating D in diploids is to simply ignore the double heterozygotes. The phase of all other two-locus genotypes can be determined, and gamete frequencies estimated. Table 4.4 summarizes the notation for genotype numbers. If we ignore the double heterozygotes, then an estimate of g_1 is

$$\hat{g}_1 = \frac{2N_{11} + N_{12} + N_{21}}{2(N - N_{22})} \tag{4.25}$$

with analogous formulas for the other gamete frequencies (Problem 4.2). D can then be estimated using equation (4.13). These may be reasonable estimates if variation is low and there are few double heterozygotes. However, for highly variable loci in outbreeding populations, these estimates are not satisfactory.

Hill (1974) developed an iterative method for obtaining a better estimate of D. He showed that g_1 can be estimated by

TABLE 4.4 Notation for numbers of individuals for two-locus genotypes.
The first subscript refers to the row (genotype at the A locus) and the second refers to the column (genotype at the B locus). For example, the number of individuals of genotype $aaBb$ is N_{32}. The total number of individuals is N (without a subscript).

Genotype at A Locus	Genotype at B Locus		
	BB	**Bb**	**bb**
AA	N_{11}	N_{12}	N_{13}
Aa	N_{21}	N_{22}	N_{23}
aa	N_{31}	N_{32}	N_{33}

$$\hat{g}_1 = \frac{1}{2N}\left[(2N_{11} + N_{21} + N_{12}) + \frac{N_{22}\hat{g}_1(1 - \hat{p}_A - \hat{p}_B + \hat{g}_1)}{\hat{g}_1(1 - \hat{p}_A - \hat{p}_B + \hat{g}_1) + (\hat{p}_A - \hat{g}_1)(\hat{p}_B - \hat{g}_1)}\right] \quad \textbf{(4.26)}$$

The only unknown in this equation is $\hat{g}_1$; all other quantities can be estimated from genotypic data. This equation cannot be solved explicitly for $\hat{g}_1$, but can be iterated to any desired accuracy. The procedure is to use the estimate from equation (4.25) as an initial estimate, substitute this into the right side of equation (4.26) to obtain a new estimate, then substitute this new estimate into the right side of (4.26) again. Repeat until the estimate converges at the desired number of decimal places. Once a sufficiently precise estimate of g_1 is obtained, D can be estimated from equation (4.14). Box 4.1 illustrates the procedure.

See Hedrick et al. (1978) and Weir (1996) for more on estimating D.

Assuming we have an estimate of D, how do we know whether it is significantly different from zero? One approach is to use a chi-square test, with one degree of freedom. The chi-square table with four classes (gamete types) can be constructed in the usual way; alternatively, it can be shown (Problem 4.4) that an equivalent formula is

$$\chi^2_{calc} = \frac{2N\hat{D}^2}{p_A p_a p_B p_b} \quad \textbf{(4.27)}$$

where $\hat{D}$ is the estimate of D, and $2N$ is the number of gametes in the sample.

As when testing for deviations from Hardy-Weinberg equilibrium, the chi-square test is not always the best test for disequilibrium. Weir (1996) describes an exact test for the significance of disequilibrium.

To summarize, the main conclusion of this section is that if significant gametic disequilibrium is detected, it cannot be attributed solely to linkage. Some other evolutionary process must be operating to generate or maintain disequilibrium. This is analogous to our conclusion regarding significant departure from Hardy-Weinberg equilibrium; it indicates the action of one or more evolutionary processes. Keep in mind, however, that in the absence of these processes, the

Box 4.1   Iterative method for estimating D

The following table gives the genotype numbers from a hypothetical sample:

	BB	*Bb*	*bb*	Row Total
AA	430	120	75	625
Aa	145	170	180	495
aa	75	20	30	125
Column total	650	310	285	1245

Using the row and column totals, the allele frequencies at each locus can be estimated using equations (2.4) and (2.5). The results are

$$\hat{p}_A = 0.701$$

$$\hat{p}_a = 0.299$$

$$\hat{p}_B = 0.647$$

$$\hat{p}_b = 0.353$$

Using equation (4.25), an initial estimate of g_1 is

$$\hat{g}_1 = 0.523$$

We now substitute these values into the right hand side of equation (4.26). The result is 0.507, our

new estimate of g_1. We now substitute $\hat{g}_1 = 0.507$ into the right side of (4.26) again, and get a new $\hat{g}_1 = 0.503$, and repeat the process. D is estimated from equation (4.14).

The following table shows the spreadsheet calculations for the first ten iterations. It usually takes only a few iterations to converge at the third or fourth decimal place.

Iteration	$\hat{g}_1$	$\hat{D}$
0	0.5233	0.0538
1	0.5070	0.0498
2	0.5030	0.0488
3	0.5019	0.0485
4	0.5016	0.0484
5	0.5015	0.0484
6	0.5015	0.0484
7	0.5015	0.0484
8	0.5015	0.0484
9	0.5015	0.0484
10	0.5015	0.0484

Hardy-Weinberg equilibrium is reached in one or two generations (for autosomal genes), but the approach to gametic equilibrium may take much longer.

4.3 DNA Fingerprinting and Individual Identification

DNA fingerprinting has become routine in forensic investigations. Figure 4.5 shows an example using a single minisatellite locus. The lanes labeled V and P are DNA from a rape victim and the alleged perpetrator. Lanes marked E and J are evidence (DNA from a semen or blood sample at the scene of the crime) and a blood stain from the alleged perpetrator's jacket. The other lanes are control lanes and size standards. Some of the DNA obtained at the crime scene (E) does not belong to the victim, and matches that of the alleged perpetrator. Moreover, the blood stain on the perpetrator's jacket contains a small amount of DNA that matches the victim's (a very faint band in lane J, which may not show up in the figure). This evidence helped to convict the perpetrator.

This raises an important question. The DNA at the crime scene matched that of the alleged perpetrator. But could someone else have actually committed the crime, someone whose minisatellite DNA genotype was the same as the alleged perpetrator's? In other words, what is the probability that the alleged perpetrator's

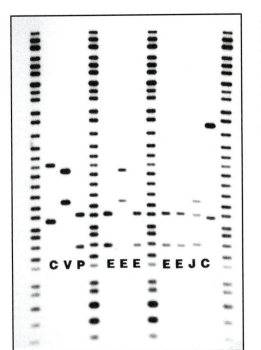

Figure 4.5 A DNA fingerprint used as evidence in a rape case. Lanes labeled V and P are samples from the victim and alleged perpetrator. Lanes labeled E represent evidence (DNA from a semen or blood sample at the scene of the crime). J represents a blood stain from the alleged perpetrator's jacket. Other lanes are control lanes and size standards. Some of the DNA obtained at the crime scene (E) does not belong to the victim, but matches that of the alleged perpetrator. Moreover, the blood stain on the perpetrator's jacket contains a small amount of DNA that matches the victim's. The bands in lane J have been enhanced for clarity. They were very faint on the original autoradiogram. *Source:* From Guilfoile (2000).

DNA was a random match to the real rapist's DNA? The answer depends on the amount of variability at the particular locus, and on the frequencies of the genotypes. For example, if the frequency of the real rapist's genotype was about 10 percent in the population, then one person out of ten would be a random match to the real rapist's DNA. Would you convict the alleged perpetrator based on a match?

The problem of random matches is essentially eliminated by analyzing several highly variable loci separately. Consider, for example, three loci, designated *A, B,* and *C.* Assume the rapist's genotype is $A_1A_5B_3B_4C_2C_7$. What is the probability that someone else has that same multilocus genotype? If we can estimate the genotype frequencies at each locus, we can estimate the probability of a match as

$$\mathrm{Pr}(match) = \mathrm{Pr}(A_1A_5) \times \mathrm{Pr}(B_3B_4) \times \mathrm{Pr}(C_2C_7)$$

where the probabilities on the right are the estimated frequencies of the single locus genotypes. By multiplying the single locus probabilities, we are assuming that the genotype at one locus is independent of the genotype at another locus, that is, we assume there is no disequilibrium among loci. In general, if *n* loci are examined, the probability of a random match is the product of the estimated frequencies of the *n* single locus genotypes.

$$\mathrm{Pr}(match) = \prod_{i=1}^{n} \mathrm{Pr}(G_i) \qquad \textbf{(4.28)}$$

where G_i is the genotype at the *i*th locus. This assumes that there is no disequilibrium among loci. If several highly variable loci are surveyed and the frequencies of individual genotypes are relatively low, the probability of a random match quickly becomes vanishingly small. For example, if the frequency of each of the single locus genotypes above is about 1 percent, then the probability of a match is about one in a million. That is probably not low enough to convict someone beyond reasonable doubt, but actual criminal cases use more loci, resulting in much lower probabilities of a match.

If we have estimates of allele frequencies in the relevant population, we can estimate the appropriate genotype frequencies by applying the Hardy-Weinberg principle. Recall that the expected frequency of a homozygous genotype is

$$\Pr(A_iA_i) = p_i^2$$

and the expected frequency of a heterozygous genotype is

$$\Pr(A_iA_j) = 2p_ip_j$$

where the left side is the expected probability (frequency) of the genotype, and p_i and p_j are the allele frequencies.

Table 4.5 gives estimated frequencies of different alleles at four minisatellite loci. These are based on FBI data for U.S. Caucasians. Suppose a blood sample

TABLE 4.5   Allele frequencies in U.S. Caucasians at four minisatellite loci.
Source: From Avise (1994); based on FBI data.

Allele Designation	Locus D1S7	Locus D2S44	Locus D1S79	Locus D4S139
1	0.004	0.005	0.010	0.004
2	0.006	0.003	0.003	0.010
3	0.009	0.016	0.007	0.006
4	0.012	0.024	0.004	0.014
5	0.011	0.046	0.015	0.033
6	0.014	0.034	0.223	0.024
7	0.010	0.123	0.199	0.040
8	0.029	0.106	0.263	0.047
9	0.021	0.084	0.200	0.054
10	0.014	0.049	0.029	0.071
11	0.028	0.083	0.032	0.108
12	0.031	0.039	0.010	0.190
13	0.046	0.041	0.006	0.129
14	0.067	0.039		0.095
15	0.057	0.087		0.036
16	0.061	0.089		0.036
17	0.069	0.075		0.103
18	0.055	0.022		
19	0.060	0.018		
20	0.063	0.008		
21	0.079	0.008		
22	0.077			
23	0.077			
24	0.032			
25	0.019			
26	0.050			

from the scene of a murder is genotype $A_3A_{11}B_5B_{17}C_1C_{10}D_3D_{13}$, where A, B, C, and D represent the four loci in Table 4.5. Assume this genotype does not match the victim's blood, but does match a suspect's. What is the probability that the suspect's multilocus genotype is a random match to the real murderer's? Applying equation (4.28), and using the allele frequencies in Table 4.5, we get

$$\Pr(match) = 2(0.009)(0.028) \times 2(0.046)(0.075) \times 2(0.010)(0.029)$$
$$\times 2(0.006)(0.129)$$
$$\cong 3.12 \times 10^{-12}$$

This is a very low number, lower than the probability of picking a specific individual from the entire human population. In fact, criminal convictions are usually based on probabilities many orders of magnitude lower. The essential point is that the probability of another person having the same multilocus genotype is essentially zero. The two major assumptions in these calculations are that each locus is in Hardy-Weinberg equilibrium, and that there is no disequilibrium among loci.

The use of DNA fingerprinting as a tool in criminal investigation has been controversial.[2] Critics have questioned both the reliability of the technique and the interpretation of the results. It is now generally (but not universally) accepted that with proper oversight and controls, the laboratory procedures are accurate and repeatable. Interpretation of results remains somewhat controversial. The main issues are potential differences in allele frequencies among different ethnic groups, and the assumptions of Hardy-Weinberg equilibrium within loci and of gametic equilibrium among loci. Some loci do show differences among ethnic groups, and this information can be considered in estimating the probability of a random match. The two assumptions of equilibrium seem to be approximately valid based on large samples of numerous loci, but exceptions do exist. Modifications have been proposed that favor the accused in estimating the probability of a random match. Most experts believe that these objections can be fairly addressed, and that DNA fingerprinting is a useful tool in forensic science. Evidence based on DNA fingerprinting has become widely accepted in criminal trials. In about a third of such trials, the suspect has been freed based on DNA fingerprinting.

DNA fingerprinting and individual identification are also useful in studies of mating behavior, ecological genetics, conservation biology, and other areas of ecology and evolutionary biology. In an early application of DNA fingerprinting, Burke et al. (1989) studied mating behavior and parental care in dunnocks (*Prunella modularis*). In these birds, two males share a single territory and a single female. One male (the alpha male) is dominant over the other (the beta male). Both males may mate with the female and may share in feeding her offspring. DNA fingerprinting showed that a male is more likely to help feed the offspring if he has sired some of them. Field studies showed that the probability of siring some of the offspring is related to the degree of exclusive access to the female, and that a beta male is more likely to share in feeding the offspring if he had some exclusive access to the female during her mating period. Apparently, the males use their degree of access to the female to judge whether the offspring are

---

[2]So have been (and are) "real" fingerprinting, lie detector tests, and other technological innovations.

theirs, and, therefore, whether to help feed them. For reviews of other studies of multiple paternity in birds, see Burke (1989) and Petrie and Kempenaers (1998).

In a similar study, Vigilant et al. (2001) used microsatellite variability to identify individuals in communities of chimpanzees. By comparing the multilocus genotypes of females, their offspring, and males in the group, they were able to assign paternity of most offspring to individual males in the group. They estimated that the proportion of offspring sired by males from outside the group was between 2.4 and 7.1 percent.

4.4 Mapping Genes Associated with Human Diseases

The Human Genome Project has revealed thousands of RFLPs, microsatellite and minisatellite polymorphisms, single nucleotide polymorphisms, and other kinds of molecular polymorphisms in the human genome. These are scattered, more or less randomly, on all the human chromosomes, and the positions of many have been mapped rather precisely. These **molecular markers** can be used to map genes causing simple Mendelian diseases such as Huntington's disease or sickle cell anemia, or to find genes associated with increased susceptibility to complex diseases that have a hereditary component, such as coronary artery disease or schizophrenia.

Why do we wish to map disease genes? Mapping can aid characterization of the disease-causing allele, which can lead to more accurate diagnosis and estimation of risk. It can also lead to isolation and cloning of the gene, which in turn can lead to characterization of the gene product and inferences about its biochemical function. For example, mapping the gene for cystic fibrosis led to an understanding of its role in transporting chloride ions across cell membranes. The disease causing allele creates defective transport channels, resulting in abnormal control of osmotic pressure and buildup of mucus outside the cells. This knowledge has led to treatments to regulate chloride transport, and the possibility of gene therapy to introduce normal alleles into the appropriate cells.

The general procedure for mapping disease genes is to examine families in which the disease occurs, type family members for many molecular markers, and to look for an allele or restriction site that is highly correlated with the presence of the disease. Assuming the correlation is due to tight linkage, you then know that a disease gene is near that marker. Before we look at the details, we need to review several statistical concepts.

The Binomial Distribution

The **binomial distribution** describes the possible outcomes of an experiment that consists of n independent and identical trials. Each trial has two possible outcomes, designated success or failure. The probability of success in any individual trial is p and the probability of failure is $1 - p$. Let X be a random variable indicating the number of successes out of n trials. Then X can take on values of $0, 1, 2, \ldots, n$. The binomial distribution describes the probability that X will take on each of these values:

$$\Pr(X = x) = \binom{n}{x} p^x (1 - p)^{n-x} = \frac{n!}{x!(n - x)!} p^x (1 - p)^{n-x} \quad \textbf{(4.29)}$$

where x is the realized value of the random variable X, and can be any integer between 0 and n. For example, the probability of getting one boy in a family of three children is

$$\Pr(X = 1) = \frac{3!}{1! \times 2!}(0.5)^1(0.5)^2 = 0.375$$

For more information about the binomial distribution, see Appendix A. Below, we will use the binomial distribution to determine the probability of a particular sibship in a family.

Conditional and Total Probability

Let A be a particular outcome of some experiment. Then the probability that A will occur is written as $\Pr(A)$. Let B be some event which affects whether or not A will occur. Then we define the **conditional probability** of A, assuming that B has already occurred as $\Pr(A|B)$, read as the probability of A given B. It can be shown from probability theory that

$$\Pr(A|B) = \frac{\Pr(A \text{ and } B)}{\Pr(B)} \qquad \textbf{(4.30)}$$

where $\Pr(A \text{ and } B)$ is the probability that both A and B occur.

If two events, B and C, both affect whether A will occur, and either B or C must occur, then the probability that A will occur is

$$\Pr(A) = \Pr(A \text{ and } B) + \Pr(A \text{ and } C)$$

Rearranging equation (4.30) and substituting, we can write the **law of total probability** as

$$\Pr(A) = \Pr(A|B) \times \Pr(B) + \Pr(A|C) \times \Pr(C) \qquad \textbf{(4.31)}$$

This is the weighted average of the two conditional probabilities.

For example, assume you have two different dice, designated D_1 and D_2 (assume they are different colors so you can tell them apart). Let T be the total shown when you roll both dice simultaneously. What is the probability that the total will be 4? There are 36 possible outcomes, of which only 3 sum to $4(1 + 3, 2 + 2, \text{ or } 3 + 1)$. So the probability of rolling a 4 is

$$\Pr(T = 4) = 3/36$$

Now, what is the probability of rolling a 4, given that the first die has already been rolled and shows a 1? In this case, the only possibility is that the second die shows a 3, and the probability of that is 1/6. We can write

$$\Pr(T = 4|D_1 = 1) = 1/6$$

which is not the same as $\Pr(T = 4)$.

From the law of total probability,

$$\begin{aligned}
\Pr(T = 4) &= \Pr(T = 4|D_1 = 1) \times \Pr(D_1 = 1) \\
&\quad + \Pr(T = 4|D_1 \neq 1) \times \Pr(D_1 \neq 1) \\
&= (1/6)(1/6) + (2/30)(5/6) \\
&= 3/36
\end{aligned}$$

as we showed initially. You should convince yourself that $\Pr(T = 4|D_1 \neq 1) =$ 2/30 in the above calculation (Refer to Figure A.2).

Appendix A explains more about conditional probabilities and how to calculate them. For what follows, the essential thing is to understand the meaning of conditional probability and how to apply the law of total probability.

Two-Point Mapping of Disease-Associated Genes

Let D be a disease causing allele (not necessarily dominant) and d be the normal allele. (Note that we are now using D to indicate an allele, and not the disequilibrium coefficient.) D can be an allele causing a simple Mendelian disease, or an allele causing increased susceptibility to a complex disease. We assume that the presence of D can be determined by phenotypic symptoms of the disease, or by a genetic or clinical test. Let M be a previously mapped molecular marker associated with the disease in a particular family, and m be an alternative allele (it can be any allele other than M). In order to determine whether the disease locus and the marker locus are linked, we need a family in which D is coupled with M in one parent. Figure 4.6a shows an example; here we can infer that the female in generation II carries D and M in coupling, and that only the third child in generation III is recombinant. We wish to know whether the information in the pedigree gives significant evidence for linkage between D and M. The general idea is to compare the probability of a given sibship assuming linkage with the probability of that sibship assuming no linkage. We can use the binomial distribution to estimate the probability of getting a particular set of recombinant and nonrecombinant offspring in the sibship under each hypothesis. Define success as a recombination event, and the probability of success as r.

First, consider the probability of getting the sibship in Figure 4.6a, assuming D and M are unlinked ($r = 0.5$). We can use the binomial distribution to get the

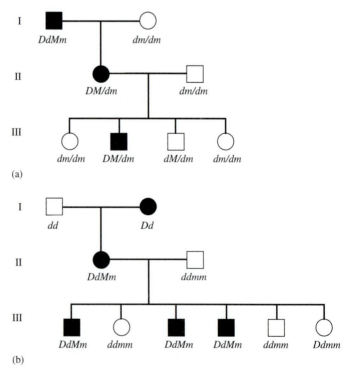

Figure 4.6 **Two families showing association between a marker allele (M) and the presence of a disease allele.**   Squares represent males and circles, females; shaded individuals have the disease allele, D. Genotypes are shown as (for example) DM/dm if linkage phase is known, or as $DdMm$ if linkage phase is unknown. (a) The woman in generation II must have inherited both D and M from her father. Therefore, phase can be determined unambiguously. Of her children in generation III, only the third is recombinant. (b) The marker genotypes of individuals in generation I are unknown so the woman in generation II has unknown linkage phase between D and M. Both coupling and repulsion phase have to be considered in determining whether the children in generation III are recombinant or nonrecombinant. If coupling, the first five children are nonrecombinant and the sixth is recombinant. If repulsion, the opposite is true.

probability of one recombinant in a sibship of four. Let $n = 4$ trials (siblings), $x = 1$ success (recombinant), and $p = 0.5$ (probability of recombination with no linkage). Expressing the probability as conditioned on $r = 0.5$, we get

$$\Pr(\text{sibship}|r = 0.5) = \frac{4!}{1! \times 3!}(0.5)^1(0.5)^3 = 0.25 \qquad \textbf{(4.32)}$$

Now consider the probability of getting the sibship assuming D and M are linked, with recombination rate r. Again, $n = 4$ and $x = 1$, but p is now the unknown recombination rate r, whose estimate is $\hat{r}$. Analogous to equation (4.32), we have

$$\Pr(\text{sibship}|r = \hat{r}) = \frac{n!}{x!(n - x)!}(\hat{r})^x(1 - \hat{r})^{n-x} \qquad \textbf{(4.33)}$$

One of four offspring is recombinant, so our best estimate of the recombination rate is $\hat{r} = 0.25$. This gives

$$\Pr(\text{sibship}|r = 0.25) = \frac{4!}{1! \times 3!}(0.25)^1(0.75)^3 = 0.4219$$

We now define the **likelihood ratio** as the probability of getting the sibship assuming linkage, divided by the probability of getting the sibship assuming no linkage

$$L = \frac{\Pr(\text{sibship}|r = \hat{r})}{\Pr(\text{sibship}|r = 0.5)} \qquad \textbf{(4.34)}$$

A likelihood ratio of greater than one indicates that the two loci are more likely linked than unlinked. The likelihood ratio is sometimes called the odds. In this example,

$$L = \frac{0.4219}{0.25} = 1.6875$$

Finally we define the **LOD score** as the logarithm of L

$$LOD = \log(L) \qquad \textbf{(4.35)}$$

LOD stands for logarithm of the odds. The statistical convention is that a LOD score of 3 or greater is significant evidence of linkage, and a LOD score of -2 or less is significant evidence of no linkage. (These numbers will be explained below.) In this example, $LOD = \log(1.6875) = 0.227$, so the pedigree in Figure 4.6a does not provide significant evidence of linkage between D and M. This is not surprising. It is analogous to flipping a coin four times and getting three heads; that result would probably not lead you to suspect that the coin was biased. As we shall see below, large sample sizes are required to detect linkage.

In the preceding example, we could infer that D and M were in coupling phase. Often the phase is unknown. In these cases, we must calculate the probability of the sibship given coupling or repulsion separately, and weight these probabilities accordingly.

Figure 4.6b illustrates a family pedigree in which the phase of D and M is unknown. The woman in generation II can be either DM/dm (coupling) or Dm/dM (repulsion). If the former, then the offspring are five parental types and one recombinant type. If the latter, the offspring are five recombinant types and one parental type. Intuitively, the former seems more likely, but we cannot ignore the latter possibility. If the former is true, then one estimate of r is $1/6 = 0.167$. But we must adjust this estimate based on the probability that D and M are actually in repulsion phase. Since repulsion seems unlikely given the sibship, we might expect our modified estimate to be near 0.167.

First, assume that D and M are in coupling phase. If so, the sibship consists of one recombinant and five parental types. Then the probability of the sibship is, from equation (4.33),

$$\Pr(\text{sibship}|r = \hat{r}) = \frac{6!}{1!(5)!}(\hat{r})^1(1 - \hat{r})^5$$

Now assume D and M are in repulsion phase. If so, the sibship consists of five recombinant types and one parental type. The probability of the sibship is

$$\Pr(\text{sibship}|r = \hat{r}) = \frac{6!}{5!(1)!}(\hat{r})^5(1 - \hat{r})^1$$

If we assume that either coupling or repulsion phase is equally likely, we can calculate the probability of the sibship assuming linkage, using the law of total probability

$$\Pr(\text{sibship}|\text{linkage}) = \Pr(\text{sibship}|\text{coupling}) \times \Pr(\text{coupling})$$
$$+ \Pr(\text{sibship}|\text{repulsion}) \times \Pr(\text{repulsion})$$
$$= \left[\frac{6!}{1!5!}\hat{r}^1(1 - \hat{r})^5\right](0.5) + \left[\frac{6!}{5!1!}\hat{r}^5(1 - \hat{r})^1\right](0.5)$$

which simplifies to

$$\Pr(\text{sibship}|\text{linkage}) = (0.5)(6)[(\hat{r})^1(1 - \hat{r})^5 + (\hat{r})^5(1 - \hat{r})^1]$$

Now consider the probability of the sibship assuming the disease locus and the marker locus are unlinked. We can go through the same procedure as for linkage; the final result is

$$\Pr(\text{sibship}|r = 0.5) = 6(0.5)^6$$

We can now calculate L; from equation (4.34)

$$L = \frac{(0.5)[(\hat{r})^1(1 - \hat{r})^5 + (\hat{r})^5(1 - \hat{r})^1]}{(0.5)^6}$$

The problem now is to find the value of $\hat{r}$ that maximizes L. In this example, L can be maximized using calculus, but typically the maximum is estimated graphically by plotting L against different values of r. Figure 4.7 shows graphs of both L and the LOD score against values of r, ranging from 0 to 0.5. There are two things to note: First, L is maximized near $r \cong 0.167$, as predicted above. Second, the maximum LOD score is between the critical values of 3 and -2. Thus, linkage between the disease locus and the marker locus can be neither concluded nor excluded from the pedigree.

If the maximum LOD score is greater than 3, we can conclude that the disease locus and the marker locus are linked, with the most likely recombination rate between them being the value of r at which the LOD score is maximized. If we know the position of the marker locus, we then know the approximate position of the disease locus. This can help in cloning the disease gene or in studying its

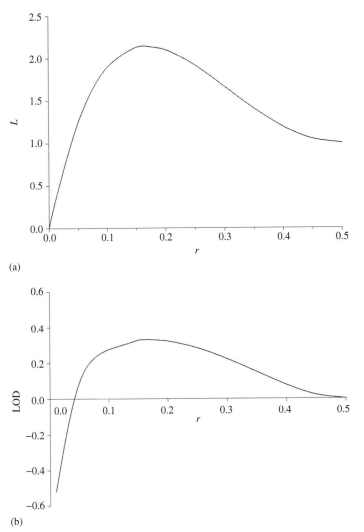

Figure 4.7 **Likelihood ratio (L) and LOD score as functions of recombination rate (r) for the family shown in Figure 4.6b.** (a) The likelihood ratio peaks near $r \cong 0.167$, indicating the most likely recombination rate. (b) The LOD score never goes above 3.0 or below -2.0, so the family in Figure 4.6b does not provide significant evidence either for or against linkage between the marker locus and the disease locus.

molecular effects. If linkage is very tight, it may also help in diagnosis of the disease (Section 4.5).

In neither of the above examples could we conclude linkage between the disease locus and the marker locus. In fact, single families can rarely provide significant evidence of linkage. An exception is shown in Figure 4.8; it shows a pedigree of a Venezuelan family in which Huntington's disease is common (Gusella et al. 1983). Below some individuals are their genotypes at a marker locus, G8, on chromosome 4. Gusella et al. were able to map the gene for Huntington's disease to a location very near this marker locus. The maximum LOD score for this family was 6.72 at a distance of $r = 0$. In other words, no recombination was observed between the marker locus and the disease locus.

Because information from individual families is usually limited, statistical power is weak. However, families can be combined, and the LOD scores for different families can be added. The reason for this is that the combined probabilities are multiplicative (law of multiplication) and therefore their logarithms are additive. (See Box 4.2 for a detailed explanation.) The increased sample sizes achieved by combining families make it easier to detect linkage between a disease locus and a marker locus.

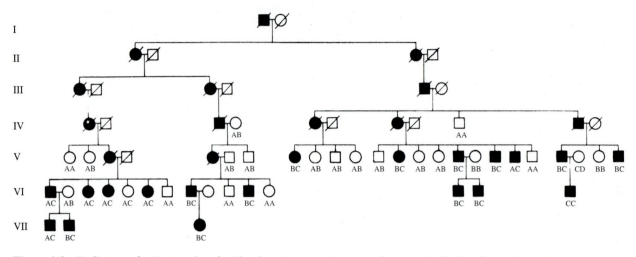

Figure 4.8 **Pedigree of a Venezuelan family showing Huntington's disease (shaded individuals).** Letters below some individuals indicate genotypes at the G8 locus (four alleles designated *A, B, C,* and *D*). The LOD score for this family was 6.72 at a distance of 0 centimorgans from the G8 locus. *Source:* From Gusella et al. (1983).

BOX 4.2   Combining LOD scores

Assume we have sibship data for two families. Let r_1 and r_2 be the estimates of recombination rate in family 1 and family 2. Similarly, use subscripts to indicate likelihood ratios and LOD scores for families 1 and 2. Then

$$L_1 = \frac{\Pr(\text{sibship}_1 | r_1)}{\Pr(\text{sibship}_1 | r_1 = 0.5)}$$

Similarly, for family 2,

$$L_2 = \frac{\Pr(\text{sibship}_2 | r_2)}{\Pr(\text{sibship}_2 | r_2 = 0.5)}$$

The denominators are numbers that depend only on family size, and not on r_1 or r_2. Call them k_1 and k_2. Then

$$L_1 = \frac{\Pr(\text{sibship}_1 | r_1)}{k_1}$$

and

$$L_2 = \frac{\Pr(\text{sibship}_2 | r_2)}{k_2}$$

Now, the combined probability of obtaining both sibships, given r_1 and r_2, is the product of the individual probabilities:

$$\Pr(\text{both sibships} | r_1, r_2) = \Pr(\text{sibship}_1 | r_1)$$
$$\times \Pr(\text{sibship}_2 | r_2) = (k_1 L_1) \times (k_2 L_2)$$
$$= k_1 k_2 L_1 L_2$$

Similarly, the combined probability of obtaining both sibships under no linkage is

$$\Pr(\text{both sibships} | r_1 = r_2 = 0.5)$$
$$= \Pr(\text{sibship}_1 | r_1 = 0.5) \times \Pr(\text{sibship}_2 | r_2 = 0.5)$$
$$= k_1 k_2$$

Next, the combined likelihood ratio is

$$L_{12} = \frac{\Pr(\text{both sibships} | r_1, r_2)}{\Pr(\text{both sibships} | r_1 = r_2 = 0.5)}$$
$$= \frac{k_1 k_2 L_1 L_2}{k_1 k_2} = L_1 L_2$$

Finally, the combined LOD score is

$$LOD_{12} = \log(L_{12}) = \log(L_1 L_2)$$
$$= \log(L_1) + \log(L_2)$$
$$= LOD_1 + LOD_2$$

Therefore, LOD scores from independent families can be added to obtain a combined LOD score.

Familial juvenile hyperuricemic nephropathy is a disease characterized by progressive kidney failure beginning at an early age. It is inherited as an autosomal dominant characteristic. Stiburkova et al. (2000) studied three Czech families and were able to map the gene for this disease to a small region on chromosome 16. The LOD score between the disease locus and microsatellite locus D16S3036 was 4.70 at a distance of $r = 0$. Stiburkova et al. also detected genetic heterogeneity for the disease, as only two of the families showed linkage between the disease locus and the marker locus.

Multipoint Mapping of Disease Genes

The preceding examples explain how to detect association between a particular marker locus and a disease causing allele. In mapping studies, dozens or even hundreds of marker loci are simultaneously tested for association with a disease allele. The general procedure is to assume the disease locus is at various points along the chromosome, and to calculate a composite LOD score, based on the marker loci, for each position. The chromosome position with the highest LOD score indicates the most likely position of the disease locus.

Infantile-onset spinocerebellar ataxia is a rare disease caused by an autosomal recessive allele, and is characterized by progressive degeneration of parts of the brain and spinal cord. Nikali et al. (1995) analyzed 213 DNA markers in 13 Finnish families, and were able to map the disease gene to a small region on chromosome 10. Figure 4.9 summarizes their results. The maximum LOD score of 5.83 occurred between marker loci F10S1267 and AFMb001wb9, but varied little in a span of about 4 centimorgans.

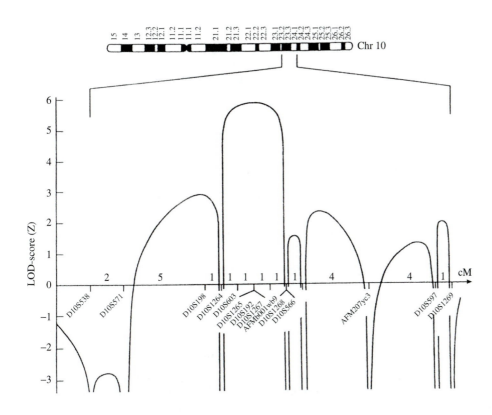

Figure 4.9 **Multilocus LOD score map for infantile-onset spinocerebellar ataxia.** The map of a small region of chromosome 10 is shown. Marker loci are indicated below the horizontal axis; numbers above the horizontal axis are distances in centimorgans between adjacent marker loci. Vertical axis is LOD score (often indicated by Z). The maximum LOD score of 5.83 occurs between marker loci F10S1267 and AFMb001wb9, but was nearly as high in an interval of about 4 centimorgans around this point. *Source:* From Nikali et al. (1995).

Multipoint mapping is done with specialized computer programs, such as GeneHunter or Fastlink (Appendix B). These programs perform the complex calculations, can step through multiple possibilities of ambiguous genotypes, and can allow for incomplete penetrance and other complications (see below). Many families have enough ambiguities that even single-marker mapping becomes unmanageably complex without a computer program.

Multipoint mapping procedures have been used to map genes for other single-gene diseases, such as Huntington's disease, cystic fibrosis, Duchenne muscular dystrophy, retinoblastoma, neurofibromatosis type 1, and many others. They have also been used to find genes causing increased susceptibility to complex diseases such as breast cancer and coronary artery disease, which are caused by multiple genetic and environmental factors.

Multipoint mapping is not limited to humans. Modern molecular techniques allow us to map hundreds of marker loci for virtually any organism we are interested in. We can then use these markers to map genes for simple Mendelian characteristics, or genes affecting quantitative traits. The procedure for mapping quantitative traits is essentially the same as for mapping genes affecting complex human diseases. Discovery and analysis of loci affecting quantitative traits (called **quantitative trait loci**) is a very active area of research. This subject is discussed in Chapter 13.

Significance Thresholds and False Positives

We stated above that the conventional threshold for significance of LOD scores is 3.0. This converts to a likelihood ratio of 1000:1, that is, we require that the probability of the sibship given linkage be 1000 times the probability of the sibship given independent assortment. This may seem like a very rigorous criterion for significance. The explanation is twofold. First, if you pick any two autosomal loci at random from the human genome, the a priori probability that they are linked is low; a common estimate is about 0.02. Thus, if the probability of linkage is low, we require strong evidence to convince us that two loci are in fact linked. Second, given this low prior probability (as it is sometimes called) of linkage, it can be shown using Bayesian statistics (Appendix A) that a LOD score of 3.0 corresponds to a posterior probability (i.e., the probability of linkage given the data) of about 0.95. In other words, a LOD score of 3.0 corresponds to the conventional significance level of about 0.05. (The significance level is the probability of concluding linkage when in fact the loci assort independently, i.e., the probability of a Type I error.)

When testing hundreds of markers against a disease gene, a few are likely to show strong associations by chance alone. This problem of false positives can lead to erroneous conclusions about linkage, and the severity of the problem increases with the number of marker loci examined. When a disease locus is simultaneously tested against many marker loci, we wish the *overall* probability of a false positive (erroneously concluding linkage) to be 0.05 or less. This requires higher LOD scores for significance. The threshold value has been debated, but Lander and Schork (1994) suggest a LOD score of about 3.3. This assumes the disease is caused by a single gene with no other genetic or environmental complications. For complex diseases such as schizophrenia, breast cancer, and others whose genetic basis is less clear, LOD scores of 4.0 or higher are required for significance.

Strachan and Read (1999) suggest that LOD scores of less than 5 be regarded as provisional.

The problem of false positives is illustrated by several widely publicized claims to have found genes associated with alcoholism, homosexuality, schizophrenia, and other complex characteristics. Initial studies found statistically significant LOD scores, but these have been mostly irreproducible, suggesting that the original results were false positives. Clearly, any claim to have mapped a disease gene must be independently verified before it can be accepted.

Problems and Cautions with LOD Score Analysis

The preceding examples were relatively simple, with few ambiguities and complications. Real-world analysis is complicated by many factors. One was hinted at in the example of Figure 4.6b, in which the linkage phase of the doubly heterozygous parent was unknown. More complex pedigrees may have more than one such ambiguity and may require many alternative probabilities to be calculated.

A second potential problem is incomplete penetrance of the disease. A disease is incompletely penetrant if not all individuals with the disease genotype show the disease phenotype. Many human diseases are incompletely penetrant, for example, breast cancer and retinoblastoma. Mapping programs can allow for incomplete penetrance.

A third problem is genetic heterogeneity for the disease. For example, tuberous sclerosis can be caused by disease alleles at genes on either chromosome 9 or 16. For many kinds of cancer, the genetic basis is complex or unknown. The problem of mapping genetically heterogeneous diseases was illustrated in the example of familial juvenile hyperuricemic nephropathy; only two of the three families studied by Stiburkova et al. (2000) showed significant association between the disease gene and the marker locus. The disease appeared to be due to a different cause in the third family.

Fourth, environmental factors may affect presence or absence of the disease. Lung cancer is an obvious example.

Complex diseases are those whose cause is not a simple Mendelian gene. They can be affected by many genes or environmental factors, and the relationships are obscure and varied. For example, at least two genes, *BRCA*1 and *BRCA*2, cause increased susceptibility to breast cancer. Other genes are probably involved, as are various environmental factors. Penetrance depends on the age of the woman. Obviously, it is harder to map genes associated with these kinds of diseases than to map simple Mendelian diseases. Strachan and Read (1999) and Maroni (2001) provide introductions to specialized techniques. See also Weiss and Clark (2002) for a review.

Useful Kinds of Molecular Markers and Informative Families

It should be obvious that some kinds of marker loci are more useful for gene mapping than others. The most useful markers have three characteristics: First, they must be codominant. It must be possible to distinguish homozygotes from heterozygotes; otherwise, linkage phase cannot be determined. Second, marker loci should be highly polymorphic. The more variation, the easier it is to determine linkage phase with the disease allele. In general, the marker allele associated with the disease allele should be relatively rare compared to other alleles in the population, so that most chromosomes are either *DM* or *dm*. Third, marker loci should be numerous and densely packed along the chromosomes. The techniques described in this section are effective only if the marker locus and the disease locus are fairly tightly linked; otherwise, LOD scores are inconclusive.

Microsatellite loci satisfy these criteria very well, and are currently the loci of choice for most gene mapping. Single nucleotide polymorphisms are even more numerous, but usually have only two alleles, making them potentially less useful than microsatellites. However, their greater numbers may overcome this disadvantage. Allozymes and RFLPs were initially used in gene mapping, but only a few allozyme loci have been mapped, and many are not very polymorphic. RFLPs share with SNPs the problem of having only two alleles.

Some families are more informative than others for gene mapping studies. The most informative families have one parent heterozygous at both the disease locus and the marker locus, and the other parent homozygous for the nondisease allele and heterozygous for different marker alleles from the parent carrying the disease allele. Figure 4.10a shows an example of a fully informative marker locus. Linkage phase can be established with certainty; the male in generation II must be DM_1/dM_2. Knowing this, the first five children in generation III are nonrecombinants, and only the sixth is recombinant.

Figure 4.10b shows the same family typed for a different marker locus. Here, the male in generation II is DA_1/dA_1, and *D* will be associated with A_1 in all of the offspring, whether or not recombination occurs. Such a family is sometimes called an uninformative mating. Note, however, that the pedigrees in Figure 4.10a

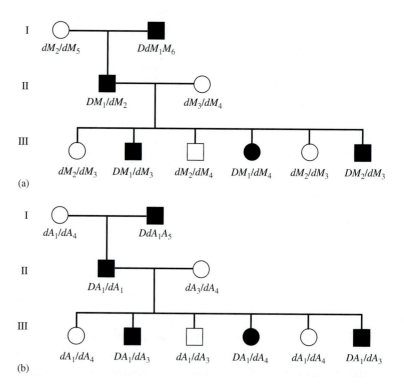

Figure 4.10   A family typed for two different marker loci. (a) A fully informative marker locus. The disease allele *D* and marker allele M_1 are in coupling phase in the male in generation II, and all of his children can be unambiguously typed as either recombinant or nonrecombinant. (b) An uninformative marker locus. The man in generation II is homozygous for marker allele A_1, so the disease allele is associated with A_1 in both recombinant and nonrecombinant children. Symbols and genotypes are as in Figure 4.6.

and b are the same family typed for different market loci. It is the marker locus that is uninformative in Figure 4.10b, and not the mating.

Mapping by Population Associations

In the preceding discussion, to map the disease gene we required an association within a family between a disease allele and a marker allele. This is essentially linkage disequilibrium on a small scale (within the family). Sometimes, the same principles can be used to map genes based on population level associations between the disease allele and a marker allele. If these associations are due to linkage disequilibrium, then the disease gene must be near the marker locus.

To illustrate, imagine a population that is polymorphic for marker alleles M and m, but in which the disease allele is absent. Then all gametes will be either Md or md. Now, assume that a single copy of the disease allele, D, is introduced into the population by, for example, mutation or gene flow. Assume that the single copy of D is associated with M. Then there will be three gamete types in the population, Md, md, and a single copy of MD. Disequilibrium between M and D is initially 100 percent. Over time, recombination will cause the disequilibrium to decay, but if linkage is very tight, the decay will be very slow, and disequilibrium between D and M may be detectable for a long time. Thus, any disease allele that arose as a single mutation might still be in strong linkage disequilibrium with an allele at a nearby marker locus. This may be detectable as a population level association between the disease allele and the tightly linked marker allele.

For example, idiopathic hemochromatosis is an autosomal recessive disease in which excessive iron is deposited in organs such as the heart and liver. In one association study, about 78 percent of individuals with idiopathic hemochromatosis had allele A_3 at the HLA-A locus on chromosome 6, but only 27 percent of normal individuals had this allele (Jorde et al. 1995). This statistically significant association led to mapping of the disease gene to chromosome 6.

Two requirements must be met for mapping by population association to be successful. First, the disease allele should have arisen only once in the population, making initial disequilibrium between the disease allele and a marker allele complete. This makes it much easier to detect association than if the disease allele has independently arisen multiple times, in association with different alleles at the marker locus. This situation is most likely found in small isolated populations, and human geneticists have invested much effort in studying relatively closed religious groups and populations whose history indicates long-term isolation, such as Iceland and Finland (e.g., Kere 2001).

The second requirement for successful association mapping is that the disease locus be tightly linked to the marker locus. How tight? A common estimate of the relation between physical distance and recombination frequency in humans is that about 1 megabase physical distance corresponds to about 1 percent recombination frequency (with much variation). Therefore, a distance of 1000 base pairs corresponds to a recombination rate of only $r \cong 10^{-5}$. Such a low recombination rate will preserve linkage disequilibrium for many generations. From equation (4.23), after 1000 generations, disequilibrium will still be about 99 percent of its original value. On the other hand, for a physical distance of 10^5 base pairs, disequilibrium will be about 37 percent of its original value after 1000 generations, and essentially zero for a physical distance of 10^6 base pairs. The conclusion is that the disease locus and the marker locus must be within a few thousand base pairs of one another. This is only a rough estimate, as the relationship between recombination rate and physical distance varies in different parts of the genome. Until recently, this requirement for such tight linkage was a major problem, but with completion of the human genome project, thousands of polymorphic markers (RFLPs, SNPs, microsatellites, etc.) have been mapped, creating a dense map of marker loci on all the human chromosomes. This will undoubtedly lead to mapping of many disease genes in the near future. For recent reviews of linkage disequilibrium and human gene mapping, see Cardon and Bell (2001), Reich et al. (2001), Stephens et al. (2001), and Weiss and Clark (2002).

Keep in mind that nonrandom associations between a disease allele and a marker allele can be due to things other than linkage disequilibrium. Population history, subdivision, natural selection, and other factors can also cause associations between alleles at different, perhaps unlinked, loci. Any of these would create a population association, but any attempt to map the disease locus to a position near the marker locus would fail. Thus, caution is needed when interpreting association studies. They can, however, lead to successful mapping of disease genes, as in the case of idiopathic hemochromatosis, mentioned above.

4.5 Diagnosing Human Genetic Diseases

Molecular markers can be used to diagnose some human genetic diseases, or to tell if a person is a carrier for a recessive disease. We must distinguish between two kinds of diagnosis. **Direct diagnosis** examines the disease-causing gene itself, and determines whether the disease allele is present. **Indirect diagnosis** predicts the presence of the disease allele based on nonrandom association with a marker allele. Direct diagnosis is definitive; it tells whether or not the tested individual carries the disease allele. Indirect diagnosis carries some degree of uncertainty, due to the possibility of recombination between the disease locus and the marker locus.

Direct Diagnosis

Figure 4.11 shows an example of direct diagnosis. Sickle cell anemia is due to a single nucleotide change in the gene coding for the β-chain of the hemoglobin molecule. The normal allele, designated *A*, has an *Mst II* site slightly upstream from the beginning of the coding sequence, one near the beginning, and one further downstream in the coding sequence. The sickle cell mutation destroys the middle *Mst II* site, so that the mutant allele, designated *S*, has only two sites. Thus, when cut with *Mst II*, an *A* allele will produce two fragments (0.2 and 1.1 kb) and an *S* allele will produce a single fragment (1.3 kb). Heterozygotes (carriers) will produce all three fragments. The fragments can be visualized by running the DNA on an agarose gel, Southern blotting, and probing with the appropriate sequences (Figure 4.11b).

The advantage of this technique is that it is frequently easier to test for the disease-causing allele than it is to test for the disease itself. For example, sickle cell anemia can be detected much more easily in a fetus by RFLP analysis than by examining red blood cells, because β-hemoglobin is present in very small amounts in the fetus. The test allows us to detect heterozygous carriers for recessive,

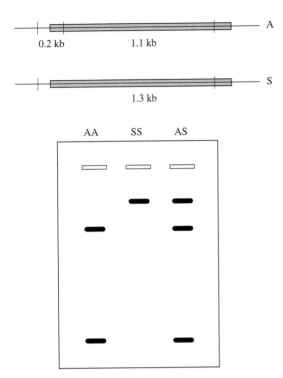

Figure 4.11 **Restriction site differences between the *A* and *S* alleles of the β-hemoglobin gene.**   The shaded bar represents the coding region of the gene; vertical lines represent *Mst II* sites. The middle site is destroyed by the mutation in the *S* allele, resulting in a single fragment instead of two. The bottom diagram illustrates an autoradiogram showing the bands seen in the three genotypes. Open rectangles at top are sample wells. Band sizes, from top to bottom, are 1.3, 1.1, and 0.2 kb.

disease-causing alleles. This is especially useful in some ethnic groups, where the frequency of a disease-causing allele may be high enough that both parents may be carriers. The parents can be tested, and if both are carriers, then the fetus can be tested and the parents can be counseled. Some dominant diseases, such as Huntington's disease, do not show their symptoms until relatively late in life. Individuals with a family history of the disease can be tested and can know whether they have the disease allele or not. This may affect their decisions on whether to have children, possibly passing the disease on to them.

Direct diagnosis requires that the disease-causing allele be identified and something known about its DNA sequence. The difference between the disease allele and the normal allele must be identifiable by a molecular marker, usually an RFLP or SNP. Unfortunately, relatively few diseases satisfy these criteria. Some examples are sickle cell anemia (above), Huntington's disease, and cystic fibrosis.

These techniques work well when the disease is caused by only one or a few known mutations in the disease gene. Tests can be devised to screen for these known mutations. Unfortunately, many diseases exhibit extensive allelic heterogeneity, that is, mutations at many places in the gene can all create a disease-causing allele. Diagnosis of these diseases requires a quick and easy way to screen for any mutation in the disease gene.

DNA sequencing is one obvious solution to this problem, and as the cost of sequencing continues to decrease, it will become more and more useful for disease diagnosis. However, comparing the DNA sequence of a disease allele with a normal allele does not indicate which differences actually cause the disease and which may be normal variation.

Another solution lies in the emerging technology of DNA microarrays (gene chips). Test DNA can be applied to a specially made chip and any possible mutation in the disease gene can be detected (Section 2.11). Gene chips have been used to screen for mutations in the *BRCA1* gene, which may cause increased susceptibility to breast cancer. As the technology improves and costs decrease, gene chips will become the main way to screen for mutations in known disease-causing genes.

Indirect Diagnosis

In the sickle cell anemia example, the disease allele causes an identifiable change in the RFLP pattern. The presence of the *S* allele is perfectly correlated with the presence of the 1.3 kb band. Sometimes we have a molecular marker that is not within the disease-causing DNA sequence, but is tightly linked to it. We saw in Section 4.4 how nonrandom associations between a disease allele and a marker allele can be used to map disease genes. Can this kind of nonrandom but imperfect association be useful in diagnosis also?

As before, let *D* be the disease-causing allele (not necessarily dominant) and *d* be the normal allele. Assume we have a tightly linked marker locus with alleles *M* and *m*, and that we have pedigree data with parents of the following genotypes:

$$DM/dm \times dm/dm$$

In other words, *D* and *M* are tightly linked in coupling phase in one parent. Then the presence of *M* in the children will indicate the presence of *D*, as long as recombination has not separated *M* and *D*. In order for these kinds of associations to be useful in diagnosis, the disease site and the marker site must be very tightly linked. Otherwise, recombination will break up the association between *D* and *M*. We are interested in the probability that an offspring with the marker *M* will also carry the disease allele *D*. It is just the probability that *D* and *M* are not separated by crossing over, or $1 - r$. We write this as

$$\Pr(D|M) = 1 - r$$

The left side is the conditional probability of *D* given *M*. Clearly, as recombination increases, the usefulness of the marker *M* decreases.

The preceding discussion assumes we have pedigree data, that is, we know the marker genotypes of both parents and offspring. Sometimes we know only the genotype of the individual in question. Then we must resort to inferences based on frequencies of *D* and *M* in the population as a whole.

We will describe a very general way of indirect diagnosis, using a technique called **Bayesian inference**.[3] As above, let *D* indicate the presence of the disease-causing allele and *d* the normal allele. Let *M* be a marker associated with the disease. It can be either a molecular marker or the results of a biochemical test that (imperfectly) indicates the presence of the disease allele. The alternative marker, associated with the normal allele is *m*. If *M* is always associated with *D* (as the 1.3 kb band was always associated with the sickle cell allele, above), then *M* is a perfect indicator of the disease. However the association between *D* and *M* is almost always imperfect, so that *M* only indicates presence of the disease allele with some probability. We wish to estimate this probability, designated $\Pr(D|M)$. This is the

---

[3]After Thomas Bayes (1744–1809), an English theologian and mathematician.

conditional probability that the individual has the disease allele, given that he or she has the marker M.

First, we need to define several probabilities:

$Pr(D)$ = probability that a person has the disease allele before we know whether he or she has M or m. This is sometimes called the prior probability, and can be obtained from population data or from pedigree analysis.

$Pr(M|D)$ = probability that a person with the disease allele has marker M. This is obtained from analysis of many family pedigrees.

$Pr(m|d)$ = probability that a person without the disease allele has marker m. This is also obtained from analysis of many family pedigrees.

$Pr(M)$ = probability (frequency) of M in the population as a whole. This can be obtained either from population data directly, of from the above two probabilities, as explained below.

$Pr(D|M)$ = probability that a person with M actually has the disease allele. This is what we want to estimate.

If we don't know $Pr(M)$ from population data, we can estimate it. Using the law of total probability,

$$Pr(M) = Pr(M|D) \times Pr(D) + Pr(M|d) \times Pr(d) \qquad \textbf{(4.36)}$$

We can now estimate the probability that a person with M actually has the disease allele, using **Bayes' theorem** (see Appendix A). In the present context, Bayes' theorem says

$$Pr(D|M) = \frac{Pr(D) \times Pr(M|D)}{Pr(M)} \qquad \textbf{(4.37)}$$

We will illustrate with two examples:

Example 1

Consider a disease caused by the allele D. The frequency of D in the population is 10^{-4}. Assume that the D allele is associated with marker allele M 99 percent of the time, and that the normal allele is associated with marker allele m 98 percent of the time. You might expect that such a strong association would provide a useful way of diagnosing the disease, but, as we shall see, that is not necessarily the case. We want to estimate $Pr(D|M)$.

From the information given, we know

$$Pr(D) = 10^{-4}$$

$$Pr(M|D) = 0.99$$

$$Pr(m|d) = 0.98$$

We first need to estimate $Pr(M)$. Using equation (4.36),

$$Pr(M) = 0.99 \times 10^{-4} + (1 - 0.98) \times (1 - 10^{-4}) \cong 0.02$$

Now, applying Bayes' theorem, the probability that a person with M has the disease allele is

$$Pr(D|M) = \frac{Pr(D) \times Pr(M|D)}{Pr(M)} = \frac{10^{-4} \times 0.99}{0.02} \cong 0.005$$

Only about one-half of 1 percent of individuals with M actually carry the disease allele! Even a near-perfect association has essentially no predictive value. Why? Because even though most people with the disease allele have M, most people with M do not have the disease allele. The frequency of the disease allele is only 0.0001, whereas the frequency of M is much higher, 0.02.

If you look at equation (4.37), you can see that if the test is to have any real predictive value, the association between the marker allele and the disease allele must be strong, that is, $Pr(M|D)$ must be high; *and* the frequency of the associated marker must be of the same order of magnitude as the frequency of the disease allele, that is, $Pr(M) \cong Pr(D)$. In other words, nearly all M alleles must be associated with D. The fact that nearly all D alleles are associated with M is not enough.

Example 2

Duchenne muscular dystrophy (DMD) is caused by an X-linked recessive allele. A woman, whose brother and uncle both have the disease, has had two normal sons. What is the probability that she carries the disease allele?

The relevant pedigree is shown in Figure 4.12. The woman in question is the female in generation III. Since the woman's brother has the disease, her mother must have been a carrier. The prior probability (before knowing anything about her children) that the woman is a carrier is therefore $Pr(D) = 0.5$. It is intuitively reasonable to believe that the real probability is somewhat less than this if we know that she has had two normal sons. It is this probability that we want to estimate.

We will consider the genetic test, or marker M, to be the presence of one or more sons with the disease. Then m indicates two normal sons. Now, if the

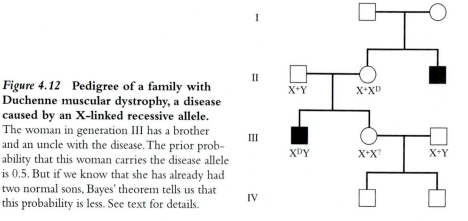

Figure 4.12   Pedigree of a family with Duchenne muscular dystrophy, a disease caused by an X-linked recessive allele. The woman in generation III has a brother and an uncle with the disease. The prior probability that this woman carries the disease allele is 0.5. But if we know that she has already had two normal sons, Bayes' theorem tells us that this probability is less. See text for details.

woman has one or more affected sons, then she obviously must carry the disease allele. Similarly, if she does not carry the disease allele, none of her sons will. In other words,

$$Pr(D|M) = 1$$
$$Pr(M|d) = 0$$

What we are really interested is $Pr(D|m)$, the probability that she carries the disease allele, given that she has already had two normal sons. Applying Bayes' theorem, we have

$$Pr(D|m) = \frac{Pr(D) \times Pr(m|D)}{Pr(m)}$$

We need to estimate $Pr(m|D)$ and $Pr(m)$. First, $Pr(m|D)$ is the probability that she will have two normal sons, given that she carries the disease allele. If she has the disease allele, then the probability that she will pass the normal allele to a son is 0.5. The probability that she will pass the normal allele to both sons is therefore 0.25 (using the law of multiplication).

Next, we must estimate $Pr(m)$. By symmetry with equation (4.36), we have

$$Pr(m) = Pr(m|D) \times Pr(D) + Pr(m|d) \times Pr(d)$$
$$= (0.25)(0.5) + (1)(0.5)$$
$$= 0.625$$

Finally, applying Bayes' theorem, we have

$$Pr(D|m) = \frac{Pr(D) \times Pr(m|D)}{Pr(m)} = \frac{(0.5)(0.25)}{0.625}$$
$$= 0.20$$

The probability that the woman is a carrier is substantially less than 0.5 if we know that she has already had two normal sons. This is intuitively reasonable since, if she had the disease allele, the probability that at least one of her sons would have it is 0.75.

Bayesian inference can also be used with clinical tests. Let M represent a positive result of the test, that is, the test indicates the individual has the disease allele. Similarly, m indicates a negative test for the disease allele. Most clinical tests are imperfect, with some normal individuals testing positive (false positive) and some disease individuals testing negative (false negative). The probabilities of these are $Pr(M|d)$ and $Pr(m|D)$ and can be estimated from clinical records. See Problems 4.9 and 4.10.

Summary

1. Recombination can be defined as any process that creates new combinations of alleles. Independent assortment leads to recombination among unlinked genes; crossing over leads to recombination among linked genes.

2. Alleles may be in Hardy-Weinberg equilibrium at each of two different loci, but alleles at one locus may be nonrandomly associated with alleles at the second locus. This is the concept of gametic disequilibrium.

3. Gametic, or linkage, disequilibrium can be produced by many processes, including hybridization, mutation, population subdivision, genetic drift, and natural selection.

4. Gametic disequilibrium will decay at a rate that depends on the recombination rate [equation (4.23)] unless some force acts to maintain it.

5. Factors that retard the decay of gametic disequilibrium are tight linkage, inbreeding, and some kinds of natural selection that involve gene interaction.

6. Large, randomly mating populations typically show little disequilibrium. Inbred populations or populations that have recently been through a bottleneck often show significant disequilibrium.

7. If natural selection favors certain gamete types, stable disequilibrium may be possible if selection is strong enough or recombination is low enough. Under these circumstances, natural selection will favor reduced recombination; sometimes creating supergenes.

8. DNA fingerprinting depends on the existence of extensive genetic variation at loci such as minisatellite or microsatellite loci. If one assumes gametic equilibrium among loci, the probability of a random match between two individuals can be estimated.

9. Molecular markers can be used to map genes associated with human diseases. The procedure depends on the existence of linkage disequilibrium between a disease-causing allele and a marker allele.

10. A statistical test of linkage between a disease locus and a marker locus is based on the LOD score, which is the logarithm of the likelihood ratio [equations (4.34) and (4.35)].

11. Gene mapping studies typically examine many marker loci simultaneously. This leads to the possibility of false positives (concluding linkage when it does not really exist). This possibility must be accounted for by modifying the significance thresholds for LOD scores.

12. In many families, gene mapping studies are complicated by the possibilities of unknown linkage phase, incomplete penetrance, genetic heterogeneity of the disease, environmental factors, and uninformative markers. Genes associated with complex diseases caused by combinations of many genetic and environmental factors are especially difficult to map.

13. In the absence of family pedigrees, disease genes can sometimes be mapped by examining population-level associations between disease alleles and marker alleles, but such nonrandom associations can be due to factors other than linkage disequilibrium.

14. Molecular markers can be used to diagnose human diseases. Direct diagnosis examines a molecular polymorphism caused by the disease allele itself. Indirect diagnosis depends on nonrandom association between a marker allele and the disease allele, and is an imperfect predictor of whether an individual actually has the disease allele.

15. Bayesian inference can be used to estimate the probability that an individual with the marker allele actually carries the disease allele. The incidence of false positives can be very high if the marker allele has a much higher frequency than the disease allele.

Problems

4.1. Table 4.6 shows restriction site variation at four polymorphic restriction sites in *Drosophila melanogaster*. Estimate D and calculate the chi-square statistics for all possible pairwise comparisons.

TABLE 4.6   Variation at four restriction sites in 17 lines of *Drosophila melanogaster*.

Presence or absence of the recognition sequence is indicated by + or −. Each line is monomorphic for the pattern shown, and can be considered a single individual for the purpose of disequilibrium analysis. *Source:* From Langley et al. (1982).

Line	BamHI	HindIII(1)	HindIII(2)	XhoI
R1	+	−	−	+
R2	+	−	−	+
M1	+	−	−	+
R3	+	−	−	+
N1	+	−	−	−
N2	+	−	−	−
R4	+	−	−	−
R5	+	−	−	−
R6	+	−	+	−
K1	+	−	+	−
K2	+	−	+	+
K3	−	−	−	+
K4	−	−	−	+
R8	−	+	−	+
M2	−	+	−	+
R9	−	−	−	+
R10	−	−	−	+

4.2. Derive formulas analogous to equation (4.25) for the other gamete frequencies.

4.3. Show that

$$p_A p_B + p_A p_b + p_a p_B + p_a p_b = 1$$

4.4. Derive equation (4.27) from the usual formula for the chi-square statistic. (*Hint:* use the result of Problem 4.3.)

4.5. Why is there only one degree of freedom for the chi-square test for disequilibrium?

4.6. The following table gives genotype frequencies at two blood group loci in humans (Mourant et al: 1976):

	MM	MN	NN
SS	91	32	5
Ss	147	78	17
ss	85	75	7

a. Estimate the allele frequencies at each locus.

b. Test for Hardy-Weinberg equilibrium at each locus.

c. Estimate D and D'.

d. Test for significant gametic disequilibrium.

4.7. For the family in Figure 4.10a calculate and graph the LOD score for various values of r from 0 to 0.5 in intervals of 0.02. Where does the LOD score seem to peak? Is that what you expect from the pedigree? What do you conclude about linkage between the marker locus and the disease locus?

4.8. Neurofibromatosis type 1 is an autosomal dominant disease characterized by dark patches on the skin, growths on the iris, and nonmalignant nerve tumors (Jorde et al. (1995). It is highly variable in its expression, but is essentially 100 percent penetrant. Figure 4.13 shows the pedigree of a family with this disease. Individuals have also been genotyped for marker locus 1F10 on chromosome 17. Alleles are designated 1 or 2 and genotypes are given below each individual.

a. Indicate the linkage phase between the disease allele and the marker allele for each individual in the pedigree. How many recombinant offspring in generation III?

b. Graph the LOD score for different values of r from 0 to 0.5 in intervals of 0.02. Where does the LOD score seem to peak? Is the maximum where you would expect it to be from the pedigree?

c. What can you conclude about linkage between the marker locus and the disease locus?

4.9. Porphyria is an autosomal dominant disease that can be imperfectly diagnosed by low levels of the enzyme porphobilinogen deaminase (Motulsky 1995). Laboratory records indicate that about 82 percent of individuals with the disease allele test positive, while about 3.7 percent of individuals without the disease allele test positive. The incidence of the disease allele in the population is 10^{-4}. What is the probability that a person who tests positive, but has no family history of the disease, actually has the disease?

4.10. A man whose sister has porphyria tests positive for the disease. What is the probability that he actually has the disease? Use the information in the previous problem.

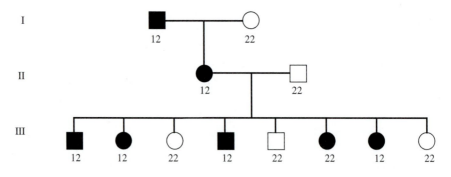

Figure 4.13 Pedigree of a family with neurofibromatosis type 1 an autosomal dominant disease. Shaded individuals have the disease allele. Genotypes at marker locus 1F10 (alleles designated 1 and 2) are shown below each individual. *Source:* From Jorde et al. (1995).

Natural Selection I: Basic Models

Can it, then, be thought improbable, seeing that variations useful to man have undoubtedly occurred, that other variations useful in some way to each being in the great and complex battle of life, should sometimes occur in the course of thousands of generations? If such do occur, can we doubt (remembering that many more individuals are born than can possibly survive) that individuals having any advantage, however slight, over others, would have the best chance of surviving and of procreating their kind? On the other hand, we may feel sure that any variation in the least degree injurious would be rigidly destroyed. This preservation of favourable variations, and the rejection of injurious variations, I call Natural Selection.

—*Charles Darwin (1859)*

If we wade through Darwin's Victorian prose, we see that Darwin viewed natural selection as a process that causes favorable variations to increase in frequency in a population and unfavorable variations to decrease. Much of *The Origin of Species* is Darwin's documentation of individual variations and his argument that natural selection of those individual variations is the main mechanism of evolutionary change.

A modern definition of natural selection is more restricted: Natural selection is the differential survival or reproduction, on the average, of different phenotypes in a population. It will lead to changes in frequencies of those phenotypes within a generation, that is, different age classes will have different phenotype frequencies. This differs from Darwin's definition; he apparently thought of natural selection as a process that occurs over multiple generations.

It is useful to keep in mind the distinction between natural selection and evolution. We define evolution as a change in the phenotypic (or genetic) composition of a population over time. Evolution occurs *between* generations. Natural selection can lead to evolution if the phenotypic changes it produces are maintained in the next generation. However, natural selection does not necessarily lead to evolution, as we shall see.

Evolution can be due to any of several causes, and natural selection is only one of them. Genetic drift, mutation, and gene flow can all cause changes over time. However, natural selection is the primary mechanism of *adaptive* evolution, defined as the process by which a population becomes better adapted to its environment over time.

Natural selection acts on the phenotype. Phenotypes can be controlled by single genes, groups of genes, or by genes interacting with environmental factors. Thus, natural selection can affect genotype frequencies and allele frequencies at single genes or groups of genes, and can vary in its effect, depending on genetic and environmental background.

To summarize, natural selection requires three things:

1. There must be phenotypic variation among individuals.

2. These variations must cause differences in survival or reproduction.

3. These variations must be heritable (at least to some degree).

If these three conditions are met, then natural selection will occur, and changes in phenotype frequencies will occur within a generation. If these changes are maintained in the next generation, evolution between generations will occur.

In this chapter, we consider the effect of natural selection on a phenotype determined by a single gene. It turns out that this is a reasonable approximation for many circumstances, and thus is a good place to begin. We will consider more complex forms of natural selection and interactions with other evolutionary processes in later chapters.

5.1 Fitness

Imagine a diploid sexual organism with a simple life cycle as follows:

$$zygote_{t} \xrightarrow{survival} adult_{t} \xrightarrow{fecundity} gametes_{t} \xrightarrow{mating} zygotes_{t+1}$$

where t and $t + 1$ indicate generations. An individual's contribution to the next generation depends on zygote-to-adult survival, on the ability to produce gametes (fecundity), and on the ability to obtain a mate. This contribution to the next generation is called **fitness**, and can be expressed in absolute or relative terms. If different phenotypes are due to different genotypes, and have different fitnesses, then natural selection will act and the phenotype and genotype frequencies will change.

The **absolute fitness** of a genotype is the average reproductive rate of individuals with that genotype. It is the average number of offspring (zygotes) contributed to generation $t + 1$ by a zygote of generation t. It can depend on all of the preceding factors. We will designate absolute fitnesses by W, with subscripts to indicate the genotypes. For example, genotypes A_1A_1, A_1A_2, and A_2A_2 will have absolute fitnesses of W_{11}, W_{12}, and W_{22}. Absolute fitnesses determine whether a population will increase or decrease in size. If the average absolute fitness of all individuals in the population is greater than one, the population will increase; if the average fitness is less than one, it will decrease. Many populations appear to be relatively constant in size, implying that average absolute fitness is near one. Obvious exceptions exist.

Population geneticists usually consider only **relative fitness**. The relative fitness of a genotype is its ability to survive and reproduce, compared to other genotypes in the population. It is obtained by specifying a relative fitness of one for one of the genotypes (call it the standard) and scaling the other fitnesses appropriately. Relative fitnesses are usually symbolized by w (lowercase), and are obtained from the absolute fitnesses as follows:

$$w_{ij} = \frac{W_{ij}}{W_{std}} \qquad \textbf{(5.1)}$$

where w_{ij} and W_{ij} are the relative and absolute fitnesses of genotype A_iA_j, and W_{std} is the absolute fitness of the genotype used to scale (standardize) the fitnesses. Relative fitnesses are commonly scaled by giving the genotype with the highest absolute fitness a relative fitness of one.

We illustrate with an example; refer to Table 5.1. Assume a population of zygotes, with 100 individuals each of genotypes A_1A_1, A_1A_2, and A_2A_2. Call this the zygote stage of generation t. The genotype frequencies in the zygotes are 0.33 for each genotype, and the frequency of the A_1 allele is $p = 0.5$. Assume that the zygote-to-adult survival rates for the three genotypes are 0.5, 0.4, and 0.3. Then there will be 50, 40, and 30 adults of genotypes A_1A_1, A_1A_2, and A_2A_2, respectively. The genotype frequencies in the adults are 0.42, 0.33, and 0.25, and the allele frequency is 0.58. Natural selection has occurred; the genotype and allele frequencies have changed because of different survival rates among the genotypes. Now assume that each adult leaves four, three, or two offspring, depending on genotypes. Then 50 A_1A_1 adults will produce a total of 200 zygotes in generation $t + 1$. Similarly, A_1A_2 and A_2A_2 adults will produce 240 and 120 zygotes, respectively. For now, ignore the genotypes of the offspring; they depend on the mating system and other factors irrelevant to the current discussion.

TABLE 5.1   An example of how to calculate absolute and relative fitnesses based on survival and fertility differences. See text for details.

	A_1A_1	A_1A_2	A_2A_2	N	p
Zygotes (t)	100	100	100	$N_t = 300$	$p_t = 0.50$
Survival probability	0.5	0.4	0.3		
Adults (t)	50	40	30	$N_{ad} = 120$	$p_{ad} = 0.58$
Offspring per adult	4	3	2		
Offspring per genotype[*]	200	120	60	$N_{t+1} = 380$	$p_{t+1} = 0.68$
Absolute fitness	2	1.2	0.6		
Relative fitness (scaled to A_1A_1)	1	0.6	0.3		
Relative fitness (scaled to A_1A_2)	1.67	1	0.5		
Relative fitness (scaled to A_2A_2)	3.33	2	1		

[*]These are the numbers of offspring produced by each genotype, not the numbers of each genotype in the offspring.

We can now calculate the absolute fitnesses: 100 A_1A_1 zygotes in generation t produced 200 zygotes in generation $t + 1$, for an average of two offspring per zygote. Similarly, 100 A_1A_2 zygotes in generation t produced 120 zygotes in generation $t + 1$, and 100 A_2A_2 zygotes in generation t produced 60 zygotes in generation $t + 1$, giving 1.2 and 0.6 offspring per zygote, respectively. Thus, the absolute fitnesses of the three genotypes are $W_{11} = 2$, $W_{12} = 1.2$, and $W_{22} = 0.6$.

To obtain relative fitnesses, we can scale to the best genotype, A_1A_1. Using equation (5.1) and the absolute fitnesses above, we get

$$w_{11} = \frac{W_{11}}{W_{std}} = \frac{2}{2} = 1$$

$$w_{12} = \frac{W_{12}}{W_{std}} = \frac{1.2}{2} = 0.6$$

$$w_{22} = \frac{W_{22}}{W_{std}} = \frac{0.6}{2} = 0.3$$

Table 5.1 also shows how the fitnesses can be scaled to the other genotypes.

Note that in this example, the number of zygotes changed from generation t to $t + 1$. If we are interested in population size, we must consider absolute fitnesses. However, population genetics models usually ignore population size. The reason is that we are usually interested only in the relative proportions of the genotypes, and not their absolute numbers. Density dependent regulating mechanisms frequently keep the population size relatively stable. If these mechanisms act independently of genotype, then we can ignore population size and consider only relative fitnesses.

Population Fitness

We can define the average or mean fitness of the population as the weighted average of the fitnesses (weighted by genotype frequencies). The **mean absolute fitness** of a population is

$$\overline{W} = P_{11}W_{11} + P_{12}W_{12} + P_{22}W_{22} \tag{5.2}$$

where Ps are the frequencies of the three genotypes, and the Ws are their absolute fitnesses. $\overline{W}$ is the average reproductive rate of all individuals in the population.

Mean absolute fitness is related to the parameter λ of population dynamics. In discrete generation models of unrestricted growth, λ describes how the population size changes from one generation to the next:

$$N_{t+1} = \lambda N_t \tag{5.3}$$

λ can be interpreted as the average number of offspring left by all individuals in the population: that is, the per capita reproductive rate of the population. Models of population growth usually assume that λ is the same for all individuals. If different genotypes have different reproductive rates (absolute fitnesses), then λ is simply a weighted average of the absolute fitnesses.

$$\lambda = P_{11}W_{11} + P_{12}W_{12} + P_{22}W_{22} \tag{5.4}$$

In other words, $\lambda = \overline{W}$.

We can similarly define the **mean relative fitness** of the population, symbolized by $\overline{w}$, as the weighted average of the relative fitnesses of the genotypes:

$$\overline{w} = P_{11}w_{11} + P_{12}w_{12} + P_{22}w_{22} \qquad \textbf{(5.5)}$$

If natural selection is acting, then not all of the relative fitnesses are equal, and the mean relative fitness is not one.

Unless indicated otherwise, when we say mean fitness of a population, we mean relative fitness. As we shall see in the next section, mean fitness plays an important role in mathematical models of natural selection.

Components of Fitness

We have illustrated how survival and fertility differences can change genotype and allele frequencies in a population. Survival and fertility are examples of **components of fitness**. If we consider everything that affects an individual's ability to leave offspring, each factor is a component of fitness. Figure 5.1 illustrates a generalized animal life cycle, identifying four major components of fitness:

Viability (survival). This is just the probability of surviving to reproductive maturity, as we have already discussed.

Fecundity. This is the number of eggs produced by females, or sperm produced by males. Females vary widely in their fecundity, so this can be an important component of fitness in females. Since most males produce an excess of sperm, this component is usually unimportant in males; sterility is an obvious exception. Keep in mind that what matters is not the total number of gametes produced, but the number that actually produce zygotes. And that depends also on mating ability.

Mating ability. Some genotypes may be better able to obtain mates than others, due either to discrimination by the opposite sex, or to behavioral differences. Human courtship and mating behavior, with all its complexities, is a familiar example.

Gamete competition. Some gametes may be more likely to produce a zygote than others. For example, a heterozygous (A_1A_2) male produces two

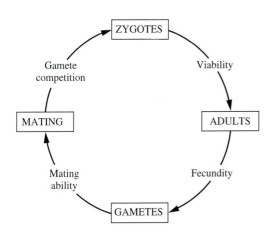

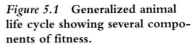

Figure 5.1 **Generalized animal life cycle showing several components of fitness.**

sperm types (A_1 and A_2). If one type is more likely to fertilize an egg, then it has a fitness advantage over the other. The classic examples of gamete competition occur in mice and in *Drosophila melanogaster*, but recent studies suggest that it may be more common than previously thought. We will discuss gamete competition when we discuss sexual selection in Chapter 12.

Fecundity and mating ability are sometimes combined into a single component of fitness called **fertility**. This refers to the number of offspring (zygotes) produced by an individual adult, and includes both gamete production and mating ability.

All of these components combine to determine fitness, which, as described above, is expressed as either absolute fitness or relative fitness. Overall fitness is what matters in an evolutionary sense. If a genotype has a high survival, but very low fertility, its fitness will be low and natural selection will work against that genotype.

Relatively few studies have attempted to estimate all components of fitness separately. In one important study, Christiansen et al. (1977) estimated components of fitness at an esterase locus in the live-bearing fish *Zoarces viviparus*. They found significant differences in zygote-to-adult viability, with the relative viabilities of 1.07, 1.00, and 1.04 for genotypes A_1A_1, A_1A_2, and A_2A_2, respectively. They found no evidence of gamete competition or differences in mating ability among the genotypes. Thus, viability may be a good estimate of net fitness in this case. An example showing fecundity and fertility differences is discussed in Section 5.10 (Table 5.6). Recent studies with *Drosophila* suggest that both gamete competition and mating ability may be important in some species.

The problems of estimating fitness and components of fitness will be discussed in Section 5.10.

5.2 A Diploid Model of Natural Selection at a Single Locus

Natural selection is the driving force of adaptive evolution. If we are to understand how populations adapt to their environment, we must develop a quantitative understanding of how natural selection works. In this section, we develop a simple model of natural selection that will aid in understanding and predicting the evolution of populations.

Natural selection can lead to rapid evolution. A classic example, discussed in detail in the next section, is the rapid increase in the frequency of the dark form of the peppered moth after the industrial revolution. Other examples are the rapid evolution of antibiotic resistance by bacteria, and the evolution of DDT resistance in mosquitoes and other insects.

On the other hand, natural selection can lead to stasis. The hermit crab looks essentially the same today as it did 200 million years ago. Some laboratory (and natural) populations will return to their original state if artificially disturbed.

Natural selection is a complex and variable process. It acts on phenotypes that may be determined by many genes interacting with one another in complex ways. It can act differently in different stages of the life cycle, in different environments, or in different genetic backgrounds. In order to understand some of this complexity, we must begin by simplifying. In this section, we consider the simplest form of natural selection, a model of selection acting at a single locus with two alleles. This is a reasonable approximation of many situations, and will lead to insights that will help in understanding more complex forms of natural selection.

Our goal is to gain some insight into the following questions:

1. Under what conditions will natural selection lead to evolutionary change, rather than evolutionary stability? Some populations change rapidly; other remain unchanged for long periods of time. Can we understand these differences in terms of natural selection?

2. Under what conditions will natural selection maintain genetic variation in a population? We have seen that most populations contain large amounts of genetic variation. What is the role of natural selection in preserving or eliminating this variation?

The Basic Model

Consider the following simplified life cycle of a hypothetical diploid organism:

$$zygotes_t \xrightarrow{natural\ selection} adults_t \xrightarrow{random\ mating} zygotes_{t+1}$$

We wish to understand the effect of natural selection acting on a single locus in this organism. We will keep track of changes in allele frequency only.

Assume two alleles, A_1 and A_2, with frequencies p and q, respectively. Let w_{11}, w_{12}, and w_{22} represent the relative fitnesses of the genotypes A_1A_1, A_1A_2, and A_2A_2. We will keep track of allele frequencies at the zygote stage of the life cycle. We wish to describe the allele frequencies in generation $t + 1$ (p_{t+1} or q_{t+1}) in terms of the allele frequencies in generation t (p_t or q_t). To minimize subscripts we will write p_t or q_t as simply p or q.

Initially, we make several assumptions:

1. Fitness differences are due only to differences in survival. All other components of fitness are equivalent in the three genotypes. Thus, this is a model of viability selection only.

2. All fitness differences can be attributed to genotypic differences at a single autosomal locus.

3. Mating is random with respect to the locus being studied.

4. No mutation, genetic drift, or gene flow occurs.

5. Fitnesses are constant over time and independent of allele frequencies or population size.

We will discuss the effects of these assumptions after we have developed and analyzed the model.

We can now construct the following table:

Genotype	A_1A_1	A_1A_2	A_2A_2
Frequency of zygotes	p^2	$2pq$	q^2
Relative fitness	w_{11}	w_{12}	w_{22}
Product	p^2w_{11}	$2pqw_{12}$	q^2w_{22}
Frequency of adults	$p^2w_{11}/\overline{w}$	$2pqw_{12}/\overline{w}$	$q^2w_{22}/\overline{w}$

The second row comes from the assumption of random mating, which yields the Hardy-Weinberg expected frequencies in the zygotes. The relative fitness of a genotype is the relative proportion of zygotes of that genotype that survive to adulthood. The Product row is the frequency of each genotype multiplied by the fitness of that genotype. It gives a fraction analogous to the frequency of that genotype in the adults, but these fractions are *not* frequencies, because they do not add up to one, as all sets of frequencies must. Convince yourself that

$$p^2w_{11} + 2pqw_{12} + q^2w_{22} < 1$$

To obtain the genotype frequencies in the adult population, we must divide each entry in the Product row by the sum of the entries in that row. This sum is the mean fitness of the population:

$$\overline{w} = p^2w_{11} + 2pqw_{12} + q^2w_{22} \tag{5.6}$$

Dividing each entry in the Product row by $\overline{w}$ gives the genotype frequencies in the adult population, listed in the fifth row of the preceding table. This process of dividing by $\overline{w}$ is sometimes called "normalizing" the frequencies (making them add up to one). One way of looking at this is that there are fewer adults than there were zygotes, because natural selection has eliminated some individuals as they grew up. We must compensate for this smaller number, and we do so by normalizing.

We can now calculate the allele frequencies in the adult population. Call them p_{ad} and q_{ad}. Applying equations (2.6) and (2.7), we get

$$p_{ad} = \frac{p^2w_{11}}{\overline{w}} + \frac{1}{2}\left(\frac{2pqw_{12}}{\overline{w}}\right) = \frac{p^2w_{11} + pqw_{12}}{\overline{w}}$$

$$q_{ad} = \frac{1}{2}\left(\frac{2pqw_{12}}{\overline{w}}\right) + \frac{q^2w_{22}}{\overline{w}} = \frac{pqw_{12} + q^2w_{22}}{\overline{w}}$$

Check this: Verify that $p_{ad} + q_{ad} = 1$.

In Chapter 3 we showed that if there are no fertility differences, and if mating is random, then the allele frequencies in the zygotes of generation $t + 1$ will be the same as the allele frequencies in the adults of generation t. Using subscripts to indicate generation $t + 1$, we have

$$p_{t+1} = \frac{p^2w_{11} + pqw_{12}}{\overline{w}} \tag{5.7}$$

$$q_{t+1} = \frac{pqw_{12} + q^2w_{22}}{\overline{w}} \tag{5.8}$$

Recall that in these equations, p and q really mean p_t and q_t, the allele frequencies in generation t. These recursion equations describe how the allele frequencies change from one generation to the next. Note that since $p + q = 1$ at any time, we need consider only one of these equations.

We have encountered recursion equations in Chapters 1 and 4. In general, they describe how some quantity (here, an allele frequency) changes from one time period to the next (usually one generation). Recall that to solve a recursion equation means to find some expression that gives the value of the quantity at any future time in terms of its initial value and the other parameters of the equation. We were able to do that in the recurrent mutation example of Section 1.4 [equation (1.2)]. That is difficult or impossible to do for most recursion equations, including (5.7) and (5.8). We can, of course, iterate these equations on a computer for any set of fitnesses and initial allele frequencies, and see what will happen to the population. But we wish to find a more general way to predict the future. One approach is to ask where the population will be after many generations. Will it approach some equilibrium point, and, if so, can we predict what it will be?

Recall that an equilibrium point is a point at which the allele frequency will remain constant; i.e., $p_{t+1} = p_t$. A **stable equilibrium** is a point of attraction; the allele frequency will approach the stable equilibrium from a short distance away. An **unstable equilibrium** is a point of repulsion; the allele frequency will move away from the unstable equilibrium point unless it is exactly on it. Even if the population is exactly on the unstable equilibrium, it will not stay there because random processes such as genetic drift or mutation will push it off. It is important to identify the stable equilibria of models because we expect natural populations to be near stable equilibria if our models are realistic. Box 5.1 gives an introduction to the ideas of equilibrium and stability.

We wish to determine whether equation (5.7) has one or more equilibrium points, and which one or ones will be approached over time. It is convenient to consider the difference in allele frequency from one generation to the next; i.e., the difference between p_{t+1} and p. We use the symbol Δp (delta p) for this difference.

$$\Delta p = p_{t+1} - p = \frac{p^2 w_{11} + pq w_{12}}{\overline{w}} - p \qquad \textbf{(5.9)}$$

A recursion equation written in this form is sometimes called a **difference equation**. The symbol Δ is usually used to indicate a change, or difference. Some messy algebra will show that equation (5.9) is equivalent to

$$\Delta p = \frac{pq[p(w_{11} - w_{12}) - q(w_{22} - w_{12})]}{\overline{w}} \qquad \textbf{(5.10)}$$

At equilibrium, the allele frequency does not change; i.e., $\Delta p = 0$. To find the equilibrium allele frequencies, we assume the population is at equilibrium and determine what allele frequencies result. In mathematical jargon, we set $\Delta p = 0$ and solve for p. If $\Delta p = 0$, then the numerator in equation (5.10) must be zero:

$$\Delta p = 0 = pq[p(w_{11} - w_{12}) - q(w_{22} - w_{12})]$$

The product of three numbers can be zero if and only if at least one of those numbers is zero. Therefore, there are three allele frequencies which can make $\Delta p = 0$; that is, there are three possible equilibrium frequencies:

either $p = 0$

or $q = 0$

or $[p(w_{11} - w_{12}) - q(w_{22} - w_{12})] = 0$

The first two are unsurprising and not very interesting. If mutation and gene flow do not occur (we assumed this at the beginning), and if the population consists of all

Box 5.1   Equilibrium and stability

One of the most important concepts of time-dependent systems is the idea of equilibrium. An equilibrium point, sometimes called a critical point, is a point where the variables of the system (e.g., allele frequencies of a population) do not change with time. Consider the general recursion equation

$$x_{t+1} = f(x_t) \qquad (1)$$

where f is any function of x. An equilibrium point of this equation is a value of x for which $x_{t+1} = x_t$. There may be more than one equilibrium point. In this book we indicate equilibrium points by putting a $\sim$ (tilde) over the variable; for example, $\tilde{x}$ is an equilibrium point of (1).

Finding the equilibrium points is straightforward in principle. Set $x_{t+1} = x_t = \tilde{x}$ and solve for $\tilde{x}$. For example, consider a linear recursion equation

$$x_{t+1} = ax_t + b \qquad (2)$$

where a and b are constants. Setting $x_{t+1} = x_t = \tilde{x}$ and solving for $\tilde{x}$, we get

$$\tilde{x} = \frac{b}{1 - a} \qquad (3)$$

In this example, there is only one equilibrium. For nonlinear equations, such as equation (5.7) or (5.8), there may be more than one. Once we have found the equilibrium points, we wish to know whether the system (population) actually approaches one of them.

An equilibrium point is **asymptotically stable** if it is approached whenever the system is in a neighborhood (region of parameter space) surrounding it. An equilibrium is just **stable** if, when the system enters the neighborhood of stability, it never leaves that neighborhood, but does not asymptotically approach the equilibrium point. When we say stable equilibrium, we usually mean asymptotically stable. A stable equilibrium is **globally stable** if it is approached regardless of initial conditions. An equilibrium is **locally stable** if it is approached only from a certain region of the parameter space.

An equilibrium point is **unstable** if there is a neighborhood of repulsion around it. If the system is in the neighborhood of repulsion, it leaves that neighborhood and never returns. If the system is *exactly* at an unstable equilibrium point, it will remain there until perturbed by some random factor. It will then move away.

To summarize, an asymptotically stable equilibrium is a point of attraction; the system will move toward it from anywhere in the neighborhood of stability. An unstable equilibrium is a point of repulsion; the system will move away from it from anywhere in the neighborhood of repulsion. These ideas are illustrated graphically in the figures here.

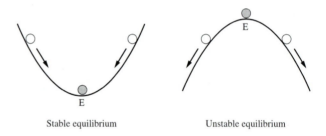

Stable equilibrium   Unstable equilibrium

The rule for determining stability of an equilibrium point is easy to state, but sometimes difficult to apply. For the general recursion equation (1), an equilibrium is locally stable if

$$|f'(\tilde{x})| < 1$$

where $f'(\tilde{x})$ is the derivative of f evaluated at the equilibrium point. For the linear equation (2), $f(x_t) = ax_t + b$, and the derivative of f is the constant a. Therefore, the linear recursion equation is stable if $|a| < 1$.

Differential equations are handled similarly. For the general differential equation

$$\frac{dx}{dt} = f(x, t)$$

the equilibria are found by setting $dx/dt = 0$ and solving for x. There may be more than one. An equilibrium, $\tilde{x}$, is asymptotically locally stable if

$$f'(\tilde{x}) < 0$$

where $f'(\tilde{x})$ is the derivative of f evaluated at the equilibrium point.

For complicated equations, equilibrium points and their stability are usually determined using a computer program such as Mathematica or MathCad.

A_1 alleles or all A_2 alleles; then the allele frequency cannot change. We have mathematically confirmed the obvious.

The third possibility is more interesting and informative:

$$[p(w_{11} - w_{12}) - q(w_{22} - w_{12})] = 0$$

Recall that $q = 1 - p$, and solve this equation for p. The result is

$$\tilde{p} = \frac{w_{22} - w_{12}}{w_{11} - 2w_{12} + w_{22}} \qquad \textbf{(5.11)}$$

The $\sim$ over the p indicates an equilibrium value.

This gives us the third equilibrium allele frequency. What does it tell us? First, we need to remember that $\tilde{p}$ represents an allele frequency. Mathematically, it can be any value, but its biological meaning requires it to be between zero and one. In other words, an equilibrium allele frequency of, say, $\tilde{p} = 5$ is meaningless in a biological model.

Skipping for now the mathematical details (see Problem 5.2), it is possible to show that $\tilde{p}$ will be between zero and one only if either of two conditions holds:

$$w_{12} > w_{11} \text{ and } w_{12} > w_{22}$$

$$\text{or} \qquad \textbf{(5.12)}$$

$$w_{12} < w_{11} \text{ and } w_{12} < w_{22}$$

This tells us that the third equilibrium will exist (in the biological sense) only if the heterozygote has the highest fitness or the lowest fitness. This third equilibrium is sometimes called the internal equilibrium or the polymorphic equilibrium, because it is between zero and one and both alleles are present.

We now know that there are three potential equilibrium points and how to find their values. Which ones are stable, and which are unstable? The easiest way to approach this question is to consider the relationships between the fitnesses.

Directional Selection

First consider the case in which A_1A_1 is the best genotype and A_2A_2 is the worst:

$$w_{11} > w_{12} > w_{22}$$

This is sometimes called directional selection because natural selection consistently acts in one direction, in this case favoring the A_1 allele.[1] It should be obvious that A_1A_1 will eventually take over the population and the allele frequency

---

[1] Strictly speaking, directional selection refers to selection in which fitness consistently increases or decreases with the phenotypic value of a trait, e.g., selection favoring larger body size.

will approach $p = 1$. In other words, $p = 1$ is an asymptotically stable equilibrium. Similar thinking will show that $p = 0$ is an unstable equilibrium. There is no internal equilibrium, because the necessary conditions of (5.12) are not satisfied.

Selection can work in the opposite direction, with A_2A_2 being the best genotype and A_1A_1 being the worst.

$$w_{11} < w_{12} < w_{22}$$

In this case, A_2A_2 will take over the population and the allele frequency will approach $q = 1$ $(p = 0)$. Therefore, $p = 0$ is asymptotically stable and $p = 1$ is unstable. Again, the internal equilibrium does not exist.

Figure 5.2 shows the results of a laboratory experiment on directional selection. Dawson (1970) studied a recessive lethal allele in the flour beetle, *Tribolium castaneum*. This allele is lethal when homozygous, but apparently has no detectable effect when heterozygous. Dawson examined several components of fitness and found no significant differences between homozygous normal and heterozygous genotypes. Therefore, the relative fitnesses are approximately 1, 1, and 0. This is strong selection, and the model predicts rapid decrease of the lethal allele. Dawson established two laboratory populations, initially consisting only of heterozygotes ($p_o = 0.5$), and followed the frequency of the lethal allele for twelve generations. The results are shown in Figure 5.2, along with the predicted change based on equation (5.8). There are minor differences, but both replicates follow the prediction of the one locus model reasonably well.

Heterozygote Superiority

Sometimes the heterozygote has the highest fitness of the three genotypes.

$$w_{12} > w_{11} \text{ and } w_{12} > w_{22}$$

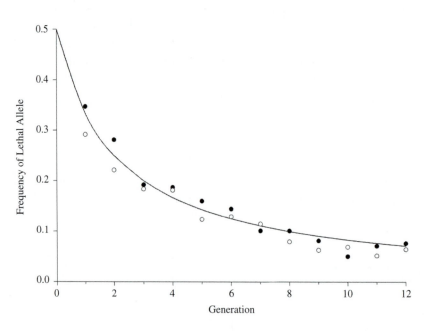

Figure 5.2 **Decrease in frequency of a recessive lethal allele in two laboratory populations of *Tribolium castaneum*.** The allele is lethal when homozygous, but heterozygotes have normal fitness. Solid line is the predicted change based on equation (5.8). *Source:* Data from Dawson (1970).

Heterozygote superiority is also called overdominance or single locus heterosis. The first thing to notice is that the conditions of (5.12) are satisfied, so the internal equilibrium exists. Therefore, there are three equilibrium points. In order to predict how the population will evolve, we must determine the stability of each of them. It turns out that the internal equilibrium is asymptotically stable and both $p = 0$ and $p = 1$ are unstable. To formally prove this, it is necessary to use the mathematical techniques of stability analysis outlined in Box 5.1. The general procedure is to take the derivative of the right side of the recursion equation (5.7) and evaluate it at $\widetilde{p}$. If you try this for the internal equilibrium, you will find yourself mired in algebraic quicksand.

Alternatively, we can approach the problem graphically by plotting Δp versus p (Figure 5.3). Note that the curve crosses the p-axis at the internal equilibrium, $\widetilde{p}$. We see that if $p < \widetilde{p}$, then Δp is positive and therefore p increases. But if $p > \widetilde{p}$, then Δp is negative and p decreases. In either case, p will approach $\widetilde{p}$. Therefore, $\widetilde{p}$ is an asymptotically stable equilibrium. Both $p = 0$ and $p = 1$ are unstable equilibria. (Why?)

This approach overlooks some mathematical subtleties; we have not rigorously proved that the internal equilibrium is stable for all relevant fitness values. However, the graphical analysis does lead to the correct conclusion: The internal equilibrium is asymptotically stable whenever the heterozygote has the highest fitness.

Figure 5.4 shows an example of heterozygote superiority. Prout (1971b) studied two loci on the fourth chromosome of *Drosophila melanogaster*. The *eyeless* allele (*ey*) causes a much reduced eye, and the *shaven* allele (*sv*) causes near absence of bristles on the thorax. Each of these alleles causes reduced fitness when homozygous.

The fourth chromosome of *Drosophila melanogaster* is very small, and essentially no recombination occurs among loci on this chromosome. Thus, *ey* and *sv* act as if they were alleles at a single locus. We can let A_1 represent the *ey sv*$^+$ (*eyeless*) chromosome type, and A_2 represent the *ey*$^+$ *sv* (*shaven*) chromosome type. Thus, A_1A_1 flies are eyeless, A_2A_2 flies are shaven, and A_1A_2 flies are phenotypically normal (double heterozygotes). Both *eyeless* and *shaven* homozygotes have reduced fitness compared to the double heterozygote, a condition that mimics single-locus overdominance (but note that neither locus exhibits overdominance by itself).

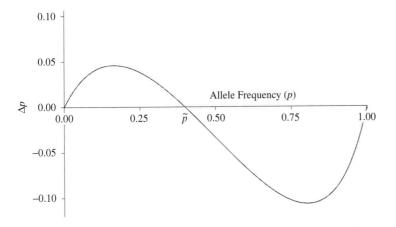

Figure 5.3 **Change in allele frequency (Δp) vs. p, for a locus showing heterozygote superiority.** The relative fitnesses are $w_{11} = 0.4$, $w_{12} = 1$, and $w_{22} = 0.6$.

Figure 5.4 **Convergence to a stable equilibrium of *eyeless* and *shaven* mutations in *Drosophila melanogaster*.**   The main solid line represents the mean of five populations; the main dashed line is the predicted equilibrium value. In generations 9 and 21, the population was perturbed away from the apparent equilibrium. Short solid and dashed lines represent observed and predicted responses to perturbation of individual populations. *Source:* From Prout (1971b).

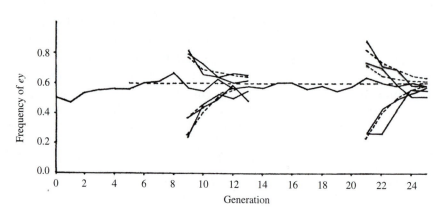

Prout started several experimental populations with different initial frequencies of *shaven* flies. All lines converged toward an equilibrium frequency of about 0.56 (Figure 5.4). In generations 9 and 21, he perturbed the allele frequency, raising it above or below the equilibrium. All perturbed lines again approached the equilibrium frequency within a few generations. The dashed lines represent predictions based on fitness estimates from independent experiments (Section 5.10).

This experimental system behaves exactly as our model predicts. Note, though, that we cannot use equation (5.11) to estimate the fitnesses of the homozygotes. (Why not?) Note also that the experimental situation consists of two loci that affect fitness. This illustrates the idea that a one-locus model can sometimes be more generally applicable than its literal meaning. For example, it also helps explain the dynamics of chromosomal inversions and translocations under some conditions.

Heterozygote Inferiority

The final possibility is that the heterozygote has the lowest fitness:

$$w_{12} < w_{11} \text{ and } w_{12} < w_{22}$$

This is sometimes known as underdominance or negative heterosis. The conditions of (5.12) are satisfied, so the internal equilibrium exists.

As with overdominance, we can determine stability of the equilibria by examining the graph of Δp versus p (Figure 5.5). The results are the opposite of overdominance. If $p < \tilde{p}$, then Δp is negative and p decreases. But if $p > \tilde{p}$, then Δp is positive and p increases. Therefore, $\tilde{p}$ is an unstable equilibrium. Both $p = 0$ and $p = 1$ are asymptotically stable equilibria. The long-term outcome depends on which side of $\tilde{p}$ the initial frequency lies (Figure 5.5). Again, formal proof of these conclusions requires local stability analysis.

If the initial allele frequency is *exactly* $\tilde{p}$, we might expect it to stay there. It would in the absence of other evolutionary processes. However, mutation, genetic drift, or gene flow will perturb the population from the unstable equilibrium point, and from then on the population will move away from it. Thus, an unstable equilibrium is not sustainable, even in the unlikely event that a population is exactly at that point. We do not expect to see populations near unstable equilibria.

We saw an example of underdominance for viability in the previous section. Christiansen et al. (1977) estimated the relative viabilities of three esterase genotypes in the fish *Zoarces viviparus* as 1.0, 0.94, and 0.97. If these are net fitnesses,

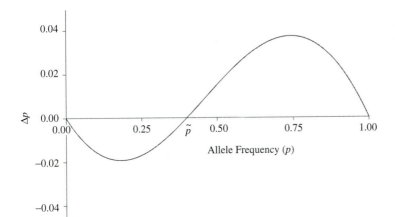

Figure 5.5 **Change in allele frequency (Δp) vs. p, for a locus showing heterozygote inferiority.** The relative fitnesses are $w_{11} = 1$, $w_{12} = 0.7$, and $w_{22} = 0.9$.

our model predicts an unstable equilibrium, and one allele or the other should go to fixation, depending on the initial allele frequencies. Since this has not happened, other factors are probably involved.

Summary of the Model

We have derived a simple model of natural selection at a single locus with only two alleles. Selection results from differences in zygote-to-adult viability only; all other components of fitness are assumed to be equal in the three genotypes. This model makes explicit predictions about the long-term outcome of natural selection. These predictions are summarized in Table 5.2 and Figure 5.6. The basic conclusion is that the allele frequency will move toward a stable equilibrium point, but will move away from an unstable equilibrium, unless it is *exactly* on that point (in which case, it will remain there only until it is moved off by random processes). This allows us to predict the long-term result of natural selection for any set of fitnesses, as seen in Figure 5.6.

We can now answer another important question: Under what conditions will genetic variation be maintained in the population? This is important because, without genetic variation, no future evolution can take place. If natural selection causes a population to become fixed for the A_1 allele, and the environment changes in a way that favors A_2, the population cannot adapt to the new environment because A_2 is no longer present. But if both alleles are present, the population can evolve in response to a changing environment. Under the assumptions of the model, genetic variability can be maintained indefinitely only if the heterozygote has the highest fitness.

TABLE 5.2   Stability of equilibria in the one-locus model of natural selection.
$\tilde{p}$ is the internal equilibrium given by equation (5.11). See also Figure 5.6.

Condition		$p = 0$	$p = \tilde{p}$	$p = 1$
Directional selection (A_1)	$w_{11} > w_{12} > w_{22}$	Unstable	Does not exist	Stable
Directional selection (A_2)	$w_{11} < w_{12} < w_{22}$	Stable	Does not exist	Unstable
Heterozygote superiority	$w_{12} > w_{11}$ and $w_{12} > w_{22}$	Unstable	Stable	Unstable
Heterozygote inferiority	$w_{12} < w_{11}$ and $w_{12} < w_{22}$	Stable	Unstable	Stable

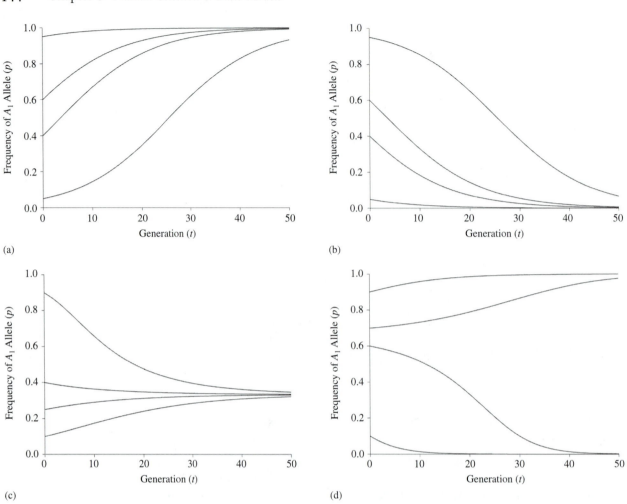

(a)

(b)

(c)

(d)

Figure 5.6 **Summary of allele frequency changes due to natural selection.** (a) Directional selection favoring A_1; (b) Directional selection favoring A_2; (c) Heterozygote superiority; (d) Heterozygote inferiority.

How Realistic Are the Assumptions?

The examples described above suggest that our one-locus selection model seems to be a reasonable guide to interpreting experimental results. But this model is based on many simplifying assumptions. How realistic are these assumptions? Can this simple model be useful in understanding real populations?

We should first clarify a common misconception. Many students believe that the main function of a model is to allow us to make precise predictions or to obtain precise estimates of relative fitnesses. In fact, this is usually not the case. We must not insist that the model give us precise estimates of fitness; that will never happen, because the assumptions of the model are never precisely met. But if the assumptions are approximately true, are the predictions approximately true? Like all models, this model allows us to evaluate possibilities. Does natural selection seem to be a reasonable explanation of what we see? Can we gain insight into what processes are acting, or not acting, in a population? If the model helps us obtain answers to these questions, it has been useful, even if we cannot obtain estimates of the parameters involved. This insight may lead to experiments that can help us to obtain those estimates.

We now consider some of the main assumptions of the basic model, and evaluate their validity and the possible effects of violating them.

We assumed random mating and absence of gene flow. These assumptions seem reasonable in many natural populations; moreover, they can be tested. We saw in Chapter 3 how to test for departure from the Hardy–Weinberg conditions. An estimate of F that is significantly different from zero is evidence that one or more of the Hardy–Weinberg assumptions is violated. If this test is performed on presumably neutral loci such as microsatellite loci, then a significant F value indicates the possibility of nonrandom mating or gene flow. The use of F statistics to estimate the amount of nonrandom mating and gene flow is discussed in Chapters 8 and 9.

We also assumed that mutation and genetic drift are negligible. This may be true in the short term, but cannot be true in the long term. Natural selection requires genetic variation. Mutation creates genetic variation and genetic drift eliminates it; clearly, we must consider all three processes in the long term. The interactions among mutation, genetic drift, and natural selection are complex and will be discussed in future chapters. For now, we can probably assume that the effects of mutation and genetic drift are negligible over the few generations involved in most field studies of natural populations. Important exceptions are small populations in which genetic drift may pose a short-term danger (Section 7.9).

One critical assumption is that all fitness differences can be attributed to a single locus. This is obviously unrealistic. Natural selection acts on the phenotype of the organism, which is affected by hundreds or thousands of genes. What we really want to know is whether natural selection at a single locus can be strong enough to have a detectible effect, separate from selection acting on other aspects of the individual's phenotype.

If alleles at a locus under selection are in gametic equilibrium with alleles at other loci, then the effects of natural selection at one locus should be relatively independent of selection at other loci. We saw in Chapter 4 that large outbreeding populations typically show little or no gametic disequilibrium. This suggests that if natural selection acts strongly on a locus, it should in principle be possible to design experiments to demonstrate it and estimate its strength. For example, Martinez and Levinton (1996) found evidence for a single gene of large effect conferring resistance to heavy metal poisoning in the oligochaete worm *Limnodrilus hoffmeisteri*.

Genes do not always act independently; natural selection can and does act on multilocus allelic combinations. Any time two genes interact, selection at one locus may affect allele frequencies at the other. The one locus model will be inappropriate. Two locus models of natural selection are very complex (Chapter 12). The general conclusion is that genes will evolve independently unless interaction is strong and/or recombination is very restricted (usually implying tight linkage).

Blocks of genes can sometimes be treated as a single locus, as in the *eyeless/shaven* example above. Chromosomal inversions can also sometimes be treated as alleles at a single locus because there is very little recombination between different gene arrangements. The many studies of Dobzhansky and his colleagues on chromosomal inversions in *Drosophila pseudoobscura* provide strong evidence that natural selection acts on different gene arrangements, and to a first approximation, these different arrangements can be treated as different alleles at a single locus.

In self-fertilizing or other highly inbred organisms, blocks of genes are often inherited together with very little recombination among genes within a block. We saw examples of extensive gametic disequilibrium in self-fertilizing species in Chapter 4. Again, these blocks of genes in strong disequilibrium can sometimes be treated as single loci in models of natural selection. For example, Marshall and Allard (1970) found evidence of natural selection on chromosomal blocks marked by allozyme loci in self-fertilizing wild oats, *Avena barbata*.

A final important assumption is that natural selection acts only on viability differences. There is increasing evidence that other components of fitness may be at least as important as viability. Models of natural section that consider other components of fitness, for example, sexual selection, will be considered in Chapter 12.

The preceding discussion suggests that we need to exercise caution when using the basic model to interpret observations of natural populations. Despite these reservations, the model is surprisingly useful. It is possible to detect and estimate the strength of natural selection at a single locus, if selection is strong at that locus and it evolves more or less independently of other loci. Several examples will be reviewed in the next few sections. However, we must recognize that consistency between predictions of the model and observations of natural populations does not prove that the model is correct and that we understand the whole story.

5.3 Two Classic Examples of Natural Selection

Natural selection is ubiquitous. Hundreds of studies have been published documenting natural selection in laboratory and natural populations (reviews in Endler 1986 and Kingsolver et al. 2001). Many of these studies are of quantitative traits whose genetic basis is complex or unknown; studies of selection at single loci are less common. In this section, we examine two classic examples of natural selection at a single locus, and ask how our model can help us to understand what we see in nature.

Industrial Melanism

Many species of moths have a light, speckled form and a dark, melanic form. The dark form is usually due to a dominant allele. These moths are inactive during the day. Their main predators appear to be birds that take them while they are resting on tree trunks or branches. The light form is well camouflaged on trees covered with lichens. The dark form is well camouflaged on trees covered with soot (Figure 5.7). Before about 1850, the dark forms were very rare in Great Britain, about 1 percent or less for the peppered moth (*Biston betularia*) the most thoroughly studied species. After that time, the frequency of the dark forms increased rapidly. By 1900, the frequency of the dark form of *Biston betularia* was 90 percent or greater in many areas of Great Britain. This is probably due, at least in part, to a selective advantage created by soot covering tree trunks as a result of the industrial

Figure 5.7 **Dark and light forms of the peppered moth (*Biston betularia*) on dark and light backgrounds.** *Source:* From Weaver and Hedrick (1992).

revolution. Beginning about 1950, air pollution controls were implemented, and emissions of soot were drastically reduced. Since that time, the frequency of the dark forms has decreased. This general pattern has been observed in several species, in both Great Britain and North America (Grant et al. 1996).

The standard explanation for these observations is this (Kettlewell 1973): Before about 1850, the light form was favored by natural selection, and the dark form remained at a low frequency, perhaps due to some form of frequency-dependent selection (Chapter 12) or mutation-selection equilibrium (Chapter 6). As the industrial revolution progressed, factory soot covered the lichens on tree trunks, creating a selective advantage for the dark form. Selection was strong and the dark form rapidly increased in frequency. After pollution controls were initiated, lichens again covered tree trunks and the light form was again favored. The dark form began to decrease in frequency.

There is much evidence to support this explanation, although recently some of that evidence has become controversial (see below). Kettlewell (1973) showed that birds do indeed take the conspicuous form at higher rates on both dark and light backgrounds. Mark-release-recapture experiments have confirmed that dark forms have a higher survival rate in polluted areas and light forms have higher survival in unpolluted areas. The fact that the increase and decrease in dark forms paralleled the increase and decrease of air pollution in many species simultaneously is persuasive evidence that the mechanism of selection is indeed related to air pollution.

Can our model of natural selection add to this story? We will examine some of the data for *Biston betularia*, the most well-studied species of the melanic moths. These moths produce one generation per year, and the dark form is apparently controlled by a dominant allele C (but see below) and the light form by a recessive allele c. Table 5.3 shows data from a typical experiment (Clarke and Sheppard

TABLE 5.3   Estimation of survival rates and relative viabilities for *Biston betularia* on dark and light backgrounds. *Source:* Data from Clarke and Sheppard (1966).

	Light Background		Dark Background	
	Light Form	**Dark Form**	**Light Form**	**Dark Form**
Number exposed	40	40	70	70
Number recovered	32	24	39	58
Survival rate	$\frac{32}{40} = 0.80$	$\frac{24}{40} = 0.60$	$\frac{39}{70} = 0.56$	$\frac{58}{70} = 0.83$
Relative viability (fitness)	$\frac{0.80}{0.80} = 1.00$	$\frac{0.60}{0.80} = 0.75$	$\frac{0.56}{0.83} = 0.67$	$\frac{0.83}{0.83} = 1.00$

1966). Dead light and dark moths were pinned to light or dark backgrounds and the numbers taken by birds were recorded. From these numbers, we can estimate survival on both backgrounds. Table 5.3 explains how. We can then estimate the relative fitnesses of the three genotypes:

Genotype	CC	Cc	cc
Phenotype	dark	dark	light
Fitness on light background	0.75	0.75	1.0
Fitness on dark background	1.0	1.0	0.67

Figure 5.8 shows the predicted changes on the dark background. The dark form should increase very rapidly and approach fixation within 100 years. This is consistent with observations, although the predicted increase in frequency may be somewhat faster than actually observed. Given the uncertainty of the fitness

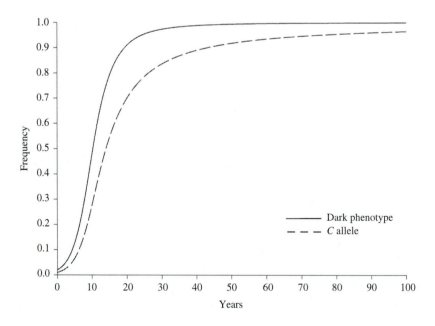

Figure 5.8 **Predicted frequency of the dark phenotype and the dark allele on a dark background, based on fitnesses as estimated in Table 5.3.**

estimates, it appears that the observed frequency changes are satisfactorily explained by a model of differential survival.

This is the classic story of the peppered moth. The real story is more complicated. First, anomalies of distribution suggest that more than air pollution is involved. For example, the dark form has been found in high frequencies (up to 80 percent) in East Anglia, U.K., where there has been little air pollution (Creed et al. 1973). Second, several lines of evidence suggest that the dark form may have a selective advantage independent of cryptic coloration. Creed et al. (1980) summarized laboratory experiments on viability and concluded that the dark form has a significant viability advantage over the light form (relative viabilities of 1 and 0.7, respectively). Third, the genetic basis of melanism is not as simple as once believed. Apparently, there are at least three dark alleles, and dominance is not always complete. Fourth, the importance of cryptic coloration and bird predation has been questioned (Sargent et al. 1998). It appears that the moths frequently rest on the underside of branches high in the trees, and not on the trunks, as has usually been assumed. Unfortunately, we know little of the behavior of the moths or of the birds that prey on them. This makes the relevance of the predation experiments suspect.

There has been a recent resurgence of interest in industrial melanism. Brakefield (1987), Sargent et al. (1998), Majerus (1998), and Grant (1999) all provide good reviews of our current understanding. Sargent et al. and Majerus are especially critical, and Coyne (1998) has responded by writing "for the time being we must discard *Biston* as a well-understood example of natural selection in action...." Grant, on the other hand, argues that the evidence is still strong that natural selection is causing the observed changes, even though the details are more complex than we initially thought. There is still much to learn about industrial melanism.

To summarize, in the most thoroughly studied species, *Biston betularia*, the dark form has increased dramatically in frequency, probably as a result of natural selection at the *C* locus. However, the mechanism of selection is not well understood. Cryptic coloration protecting the moths from bird predation probably plays a role, but the *C* allele may have other advantages unrelated to predation.

This example illustrates an important point. We must not take our models too literally, even if they *appear* to explain observations in nature. We may be overlooking significant complicating factors.

Sickle Cell Anemia

Every biology student has heard the story of sickle cell anemia. Individuals with the disease have abnormally shaped ("sickle shaped") red blood cells (Figure 5.9a) and are unable to transport oxygen efficiently. As a result, they suffer a variety of physiological problems (Figure 5.9b) and nearly always die before reaching reproductive age.

Sickle cell anemia is due to homozygosity for a mutation that causes a single amino acid substitution in the β-chain of the hemoglobin molecule. The normal allele is designated *A* and the sickle cell allele *S*. Thus, *AA* individuals are normal and *SS* individuals have sickle cell anemia. Heterozygotes do not have sickle cell anemia and are phenotypically normal under most circumstances, but their red blood cells sickle under low oxygen concentration. Heterozygotes are said to have sickle cell *trait*.

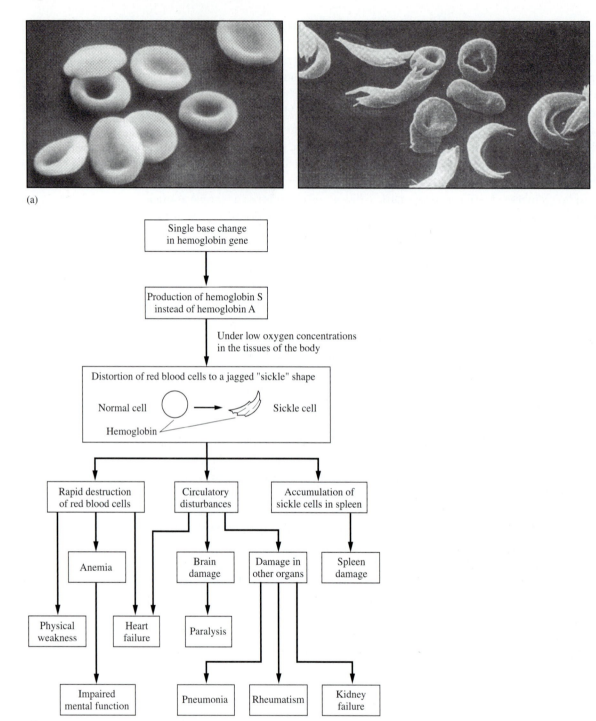

Figure 5.9 (a) Normal red blood cells (left) and sickle cells (right). (b) Pleiotropic effects of sickle cell anemia. *Source:* (a) From Freeman and Herron (2001); (b) From Griffiths et al. (1996).

From this we expect strong natural selection against the *S* allele, and can predict that it will either be eliminated or be present in very low frequency. However, the *S* allele is present in surprisingly high frequencies in some populations, especially in Africa. Allison (1954a,b) noted the correlation between frequency of the *S* allele and frequency of malaria (Figure 5.10), and suggested that the *S* allele might be relatively common in such areas because heterozygotes have some resistance to malaria. He showed that children with the sickle cell trait (heterozygotes) had significantly fewer infections of the malaria parasite (*Plasmodium falciparum*) than did normal (*AA*) children. Much subsequent work has confirmed this conclusion. The mechanism seems to be that when the parasite enters a red blood cell, it causes the cell to sickle and be destroyed, killing the parasite along with the cell.

Here we have the classic example of heterozygote superiority. The heterozygote has a fitness advantage over either homozygote *in environments where malaria is common*. The last phrase must be emphasized because this advantage does not exist in malaria-free environments. Our model thus predicts a stable equilibrium allele frequency that depends on the relative fitnesses of the *AA* and *SS* homozygotes.

There have been many attempts to estimate the relative viabilities of the two homozygotes. The *SS* individuals usually die before reaching reproductive age,

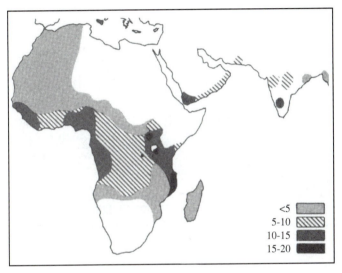

<5	
5-10	
10-15	
15-20	

(a)

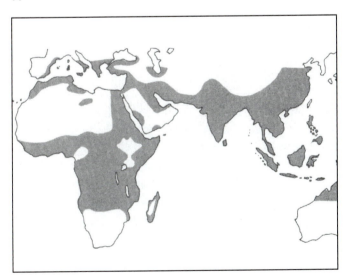

(b)

Figure 5.10   (a) Distribution of the sickle cell allele in Africa. Legend indicates frequency of the *S* allele. (b) Distribution of malaria in Africa. *Source:* Both from Cavalli-Sforza and Bodmer (1971).

and relative viability estimates range from 10 to 20 percent of normal. This is independent of the environment, except that it can be increased with good medical care. Fitness of the AA homozygotes has been estimated at about 85 percent that of the heterozygotes in malarial regions. We can use these fitness estimates to estimate the equilibrium allele frequency from equation (5.11). If we let p be the frequency of the A allele and q the frequency of the S allele, and estimate $w_{AA} = 0.85$, $w_{AS} = 1$, and $w_{SS} = 0.15$, and substitute these values into equation (5.11), we get a predicted equilibrium frequency of $\tilde{p} \cong 0.85$, or $\tilde{q} \cong 0.15$. This corresponds reasonably well to the observed frequencies of the S allele in central and west Africa (Figure 5.10a). Considering the difficulty of estimating fitnesses, the agreement is good. However, we must not forget two main assumptions of this calculation: We have assumed that all fitness differences are due to viability differences, that is, no fertility differences exist. The second assumption is that the African populations are actually in equilibrium.

5.4 Evolution of Pesticide and Antibiotic Resistance

Insects and other pests cause tremendous economic loss and widespread illness. It is estimated that about 20 percent of all agricultural products are lost to insects and other pests. Malaria currently kills more than 3 million persons each year, as well as being responsible for hundreds of millions of nonfatal cases. Tuberculosis, influenza, AIDS, and numerous other bacterial and viral diseases are major health problems.

Beginning in the 1940s, humans began developing sophisticated chemical weapons against these pests and pathogens. At first, wonder drugs such as DDT and penicillin were remarkably effective, but it took only a few years for resistance to develop. Today, resistance to pesticides and antibiotics is a serious and growing economic and medical problem.

Drug resistance is frequently due to a mutation in a single gene. Selection for resistance can be very strong, causing the rapid spread of resistance within a population or among populations. In this section, we examine several examples.

Insecticide Resistance

Our first line of defense against insect pests and disease vectors has been the use of insecticides. The first organic insecticides were introduced in the early 1940s and were initially quite effective. However, many species of insects rapidly developed resistance to one or more insecticides. Today, more than 400 species of arthropods are resistant to one or more insecticides, and some species, for example, the Colorado potato beetle, are resistant to all major classes of insecticides. The pests appear to be winning the arms race: In the 1940s about 7 percent of agricultural crops were lost to insects, compared to about 20 percent today. Malaria cases declined dramatically after the initial use of DDT, but have risen again as mosquitoes have developed resistance to DDT and other insecticides. In the early 1960s, there were about 75 million cases of malaria worldwide; today, the number is about 300 million. Evolution of resistance to insecticides is one of the most dramatic and important examples of natural selection.

Resistance to insecticides is frequently due to mutation at one or two genes, with resistance being at least partially dominant. For example, in the Australian sheep blowfly (*Lucilia cuprina*) resistance to organophosphate insecticides is primarily due to two genes, *Rop*-1 and *Rop*-2, with different physiological mechanisms. Resistance to dieldrin and malathion is due primarily to single genes *Rdl* and *Rmal*, respectively.

Examples of partial dominance of resistance are shown in Figure 5.11. Resistance to DDT in the mosquito *Culex quinquefasciatus* is almost completely dominant (Figure 5.11a), whereas in the mosquito *Aedes aegypti*, the heterozygote is about midway between the resistant and susceptible homozygotes for pyrethroid resistance (Figure 5.11b).

Selection for insecticide resistance is typically very strong. Doses are typically high enough so that only a quantum jump in resistance is likely to be beneficial to the population. Curtis et al. (1978) estimated relative fitnesses in two species of mosquito, *Anopheles culicifacies* and *An. stephensi*; both species are important malaria vectors in Asia. Curtis et al. did not know the genetic basis of resistance, but assumed it to be due to a single gene. They estimated fitnesses of the three

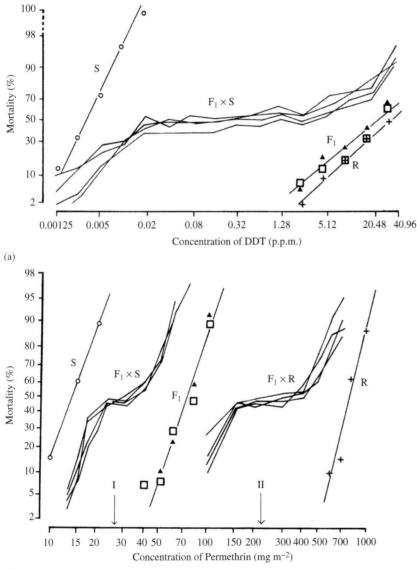

(a)

(b)

Figure 5.11 **Insecticide resistance due to a single gene in mosquitoes.** S = sensitive homozygote; R = resistant homozygote; F_1 = heterozygote; $F_1 \times S$ = backcross between F_1 and the sensitive homozygote; $F_1 \times R$ = backcross between F_1 and the resistant homozygote. (a) DDT resistance in *Culex quinquefasciatus*. The resistance allele is almost completely dominant, with heterozygotes showing almost the same degree of resistance as the resistant homozygotes. (b) Pyrethroid resistance in *Aedes aegypti*. The resistance allele shows partial dominance, with heterozygotes about half way between sensitive and resistant homozygotes. *Source:* From Wood (1981).

genotypes *SS, SR,* and *RR* (where *S* is susceptible and *R* is resistant) under exposure to several insecticides based on assumptions that the resistance is dominant, intermediate, or recessive. The fitnesses were scaled so that *SS* had a fitness of one and resistant genotypes had fitness greater than one. The estimates of Curtis et al. are summarized in Table 5.4. Resistant genotypes are frequently two to three times, or more, as fit as *SS*. This is strong natural selection.

The frequency of resistance often declines when insecticide treatment is stopped. This suggests that the resistant genotypes are at some disadvantage in the absence of the insecticide. Curtis et al. (1978) also estimated fitnesses of resistant genotypes in the absence of DDT selection. In *Anopheles stephensi*, fitness of the *RR* genotypes, compared to *SS*, was 0.91 to 0.93 after DDT treatment was stopped. In *Anopheles culicifacies*, fitness of *RR* was about 0.57, assuming resistance is dominant.

These fitness estimates are only rough estimates based on assumptions about the genetics of resistance, but they do give some idea of the magnitude of selection for and against insecticide resistance. Selection can be strong in both directions.

The role of gene flow in spreading insecticide resistance is illustrated by organophosphate resistance in the mosquito *Culex pipiens*. Organophosphates inhibit the enzyme acetylcholinesterase, resulting in overstimulation of the nervous system and eventual death. In *Culex pipiens*, the main mechanism of resistance is to degrade the insecticide before it reaches its target. This degradation is accomplished by chemical compounds called esterases, produced by the *estB* genes. Organophosphate resistance in *Culex pipiens* is due to overproduction of these esterases due to gene amplification of the $estB_1$ and $estB_2$ genes.

TABLE 5.4   Relative fitnesses of sensitive (*S*) and resistant (*R*) phenotypes in two species of mosquitoes.
Curtis et al. assumed resistance is due to a single gene and considered three hypotheses about dominance. They used two methods of estimating fitnesses, and estimated that the true values were in-between the two estimates. Values given here are the arithmetic means of their two estimates. Fitnesses are scaled so that *SS* has fitness of one. *Anopheles culicifacies*/DDT estimates are means of six villages in India. *Source:* Based on Curtis et al. (1978).

Species	Insecticide	Hypothesis	Genotypes		
			SS	*SR*	*RR*
Anopheles culicifacies	DDT	*R* dominant	1	1.28	1.28
		R partially dominant	1	1.25	1.51
		R recessive	1	1	1.43
	Dieldrin	*R* dominant	1	4.51	4.51
		R partially dominant	1	2.44	3.88
		R recessive	1	1	3.70
Anopheles stephensi	Dieldrin	*R* dominant	1	1.77	1.77
		R partially dominant	1	1.87	2.73
		R recessive	1	1	1.74
	Malathion	*R* dominant	1	1.32	1.32
		R partially dominant	1	1.30	1.60
		R recessive	1	1	1.33

Gene amplification is a process whereby specific genes are duplicated and expressed at very high levels. The amplified genes $estB_1$ (about 250 copies) and $estB_2$ (about 60 copies) are very closely related (about 96 percent similarity in DNA sequence) and are derived from the unamplified $estB$ gene. Raymond et al. (1991) compared the DNA sequences (indirectly by using restriction enzymes) of the coding region and the upstream and downstream flanking regions of the $estB$ and $estB_2$ genes. The coding sequences were very similar in all strains examined, indicating selective constraints. The flanking sequences of the $estB$ gene showed variation among sensitive strains. However, the flanking sequences of the $estB_2$ genes were identical in resistant strains from Africa, Asia, and North America. Raymond et al. interpret these results to mean that the amplification of the $estB_2$ gene has occurred only once, recently, and has spread worldwide by migration and gene flow. *Culex pipiens* is frequently associated with human international movements. The combination of extensive gene flow and strong selection has made populations of *Culex pipiens* throughout the world resistant to organophosphates.

The evolution of insecticide resistance is an important medical and economic problem. Resistance has evolved independently in many species, and by many different physiological mechanisms. For good reviews of the subject, see Wood (1981), Wood and Bishop (1981), May (1985), Roush and McKenzie (1987), and McKenzie and Batterham (1994).

Warfarin Resistance in Rats

Warfarin is a rat poison, first used in the United Kingdom in 1953 and used extensively thereafter. It prevents blood clotting, causing the rats to die of internal hemorrhaging or of minor cuts and wounds. Resistance to warfarin first appeared in 1958, and has since evolved in many regions of Great Britain, continental Europe, and the United States.

In order to understand the evolution of resistance to warfarin, it is necessary to understand some of the biochemistry of blood clotting. Clotting is the result of a complex cascade of chemical reactions. One essential component of this cascade is prothrombin (Factor II). Mature prothrombin is produced in the liver and requires vitamin K. During the process, vitamin K is converted to its oxidized form, dihydro vitamin K (Figure 5.12). The enzyme vitamin K reductase converts dihydro vitamin K back to vitamin K. Warfarin inhibits this latter reaction, preventing restoration of vitamin K (Bell and Matschiner 1972). The result is that mature prothrombin is not produced, and blood clotting fails.

Resistance to warfarin is due to a mutation in the gene producing the vitamin K reductase enzyme. The modified enzyme is less sensitive to inhibition by warfarin, but is also less efficient in its normal activity (converting dihydro vitamin K back to vitamin K). As a result, resistant rats are susceptible to vitamin K deficiency and require much more vitamin K in their diets than do susceptible rats.

Several resistance alleles have appeared. The best studied is designated Rw^2; the sensitive allele is Rw^1. We will simplify the notation and call these alleles R, for resistant, and S, for sensitive. The R allele is essentially fully dominant for resistance, but is nearly completely recessive in vitamin K requirement. The differences among the genotypes are summarized in Table 5.5.

Figure 5.12 **Biochemical mechanism of warfarin resistance.** (a) Immature prothrombin is converted to mature prothrombin in a coupled reaction that converts vitamin K to its oxidixed form, dihydro vitamin K. The latter is converted back to vitamin K by the enzyme vitamin K reductase. Warfarin inhibits this enzyme, which prevents the production of mature prothrombin, and ultimately prevents blood clotting. Warfarin resistance is due to a modified vitamin K reductase, which is less inhibited by warfarin, but also is less efficient in converting vitamin K to dihydro vitamin K, resulting in increased vitamin K requirements in resistant rats. (b) Chemical structures of vitamin K and its oxidized form, dihydro vitamin K. *Source:* From Bell and Matschimer (1972).

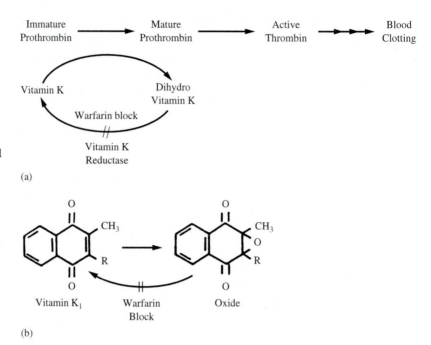

TABLE 5.5 Warfarin resistance and vitamin K requirements in warfarin-resistant and warfarin-sensitive rats.
Vitamin K requirements are in µg/100 g body weight in males. Males have higher requirements than females. *Source:* From Bishop (1981).

	Genotype		
Characteristic	***SS***	***SR***	***RR***
Warfarin sensitivity	Sensitive	Resistant	Resistant
Vitamin K requirement (Wistar lab strain)	6	4–6	80
Vitamin K requirement (Welsh wild strain)	0.5	1	5–7

Given that *SS* individuals are sensitive to warfarin, and *RR* individuals are subject to vitamin K deficiency, we might predict that, in the presence of warfarin, the heterozygote would have the highest fitness and a stable polymorphism would be established for the *R* and *S* alleles. That is exactly what has occurred. Figure 5.13 shows that the frequency of the resistant phenotype has remained stable at about 40 percent in an area where warfarin has been used extensively. Greaves et al. (1977) estimated the relative fitnesses of *RR, SR,* and *SS* genotypes to be about 0.68, 1.0, and 0.37, respectively. Substituting these values into equation (5.11) gives a predicted equilibrium frequency for the *R* allele of about 0.66.

When warfarin treatment is stopped, the frequency of the *R* allele declines rapidly, as expected due to the vitamin K deficiency of *RR* genotypes. Estimated

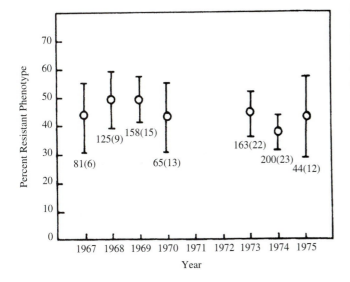

Figure 5.13 **A stable polymorphism for warfarin resistance in English-Welsh populations of rats.** Error bars represent ±2 standard errors. Numbers represent sample sizes and (in parentheses) number of infestations from which the samples are drawn. *Source:* From Bishop (1981); original from Greaves et al. (1977).

fitnesses for *SS, SR,* and *RR* in the absence or warfarin are 1, 0.77, and 0.46, respectively (Partridge 1979). This suggests that the *R* allele would be eliminated in the absence of warfarin.

What can we learn from this example? First, a detailed understanding of resistance requires a knowledge of both the biochemistry and genetics of the system. Second, resistance can evolve rapidly. Third, there appears to be a cost of resistance; resistant individuals are at a disadvantage when the selective agent is removed. These principles apply to the evolution of pesticide and antibiotic resistance in many organisms.

Antibiotic Resistance in Bacteria

When antibiotics were first used in the 1940s, they were effective against many of our most deadly diseases, for example, pneumonia and tuberculosis. These drugs saved millions of lives. But with time, more and more species of bacteria became resistant to one or more antibiotics. In the last few years, there has been an alarming increase in the number of multiple antibiotic-resistant strains. For example, *Staphylococcus aureus* is a major cause of hospital acquired infections. It can cause, among other things, blood poisoning and pneumonia, and is sometimes fatal, especially in already weakened patients. Some strains of *S. aureus* are resistant to all known antibiotics (more than 100 drugs) except vancomycin. In 1997, a vancomycin-resistant strain appeared, and it is just a matter of time before strains appear that are resistant to all known antibiotics.

At least three other species of bacteria that can cause potentially life-threatening diseases are resistant to more than 100 different drugs (Levy 1998): (1) *Enterococcus faecalis*, which causes blood poisoning and urinary tract and wound infections; (2) *Pseudomonas aeruginosa*, which causes blood poisoning and pneumonia; and (3) *Mycobacterium tuberculosis*, which causes tuberculosis. Tuberculosis death rates have recently begun to rise because of antibiotic-resistant strains of *M. tuberculosis*. Bacterial resistance to antibiotics is common and becoming more so.

Mechanisms of resistance vary widely, but resistance is frequently due to a mutation in a single gene. We can examine a simple haploid selection model to see if we can gain some insight into the potential rates of evolution of antibiotic resistance in bacteria.

Models of bacterial growth usually assume continuous growth. We will consider a discrete model analogous to our diploid model; the qualitative results are the same as in a continuous model. Assume two bacterial strains, designated A_1 and A_2. We usually think of these as single-locus haploid genotypes (e.g., resistant vs. sensitive), but the two strains may differ at many loci. The only important thing is that recombination is negligible, so that the two strains remain distinct. Let p and q ($= 1 - p$) be the frequencies of the two forms. We are interested only in the relative proportions of the two strains, so we can work with relative finesses. Scaling fitnesses to the A_2 strain, we have

Type	A_1	A_2
Frequency	p	q
Relative fitness	w	1

As before, we define the mean fitness as the weighted average of the fitnesses:

$$\overline{w} = pw + q \tag{5.13}$$

Using the same logic as in the diploid model (but without heterozygotes), the frequency of A_1 after one generation will be

$$p_{t+1} = \frac{pw}{\overline{w}} \tag{5.14}$$

So A_1 will increase if $w > \overline{w}$. Using equation (5.13), it is easy to show that this is equivalent to $w > 1$. In other words, A_1 will increase if it has a higher relative fitness than A_2. This is not a surprising result; the form with the higher reproductive rate simply outruns the other. No polymorphic equilibrium is possible if the relative fitnesses remain constant.

We can use this equation and some reasonable approximations to estimate how long it might take for an antibiotic-resistant strain of bacteria to replace a sensitive strain. Let A_2 be the sensitive type and A_1 be the resistant type. Assume that the initial frequency of the resistant type is about 1×10^{-8}. This is approximately the mutation rate to antibiotic resistance in some strains of bacteria. Also assume that the resistant type has a growth rate of about ten times the growth rate of the sensitive type in the presence of the antibiotic. This is very conservative; it assumes that the antibiotic kills about 90 percent of sensitive cells. So we have $p_o = 1 \times 10^{-8}$ and $w = 10$. How long will it take for the resistant type to take over?

Figure 5.14 plots p versus t for 20 generations. By about 12 generations, the resistant form has effectively taken over the population. Recall that for some species of bacteria, generation time may be measured in minutes, and you get some idea of the magnitude of the antibiotic-resistance problem. This simple model is unrealistic in many ways; real bacteria will not behave exactly as our model tells them to. But it illustrates the point that strong selection favoring resistance will lead to very rapid evolution.

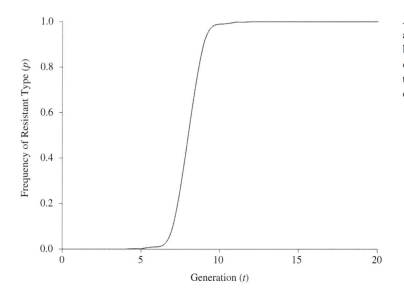

Figure 5.14 **Evolution of antibiotic resistance in a bacterial population, as predicted by equation (5.14).** In presence of the antibiotic, the fitness of the resistant strain is ten times the fitness of the sensitive strain. Initial frequency of resistant strain is 10^{-8}.

The problem may be even worse than our analysis suggests. Resistance genes are frequently found on plasmids (small, circular, independently replicating DNA molecules in bacteria), and some specialized plasmids, called resistance transfer factors, contain several resistance genes, conferring resistance to several different antibiotics. These plasmids are not only heritable, they can be infectious; that is, they can be passed from cell to cell, *even among different species*. It does not take much theoretical population genetics to see that the evolution of antibiotic resistance is a serious problem. Perhaps the question should be, "Why are not all bacteria resistant to all antibiotics?" For recent reviews of the subject, see Spratt (1996), Baquero and Blazquez (1997), Levy (1998), and Levin (2000).

5.5 Selection Against Human Genetic Diseases and Rare Recessive Alleles

Many human genetic abnormalities are caused by homozygosity for a rare recessive allele, for example, Tay Sachs disease and albinism. Since these abnormalities are usually detrimental in some way, we would expect natural selection to eliminate them. However, these alleles frequently persist in a population at low frequencies. Our model of natural selection can help us to understand why. If a recessive allele is rare, nearly all copies are present in heterozygotes. Under random mating, the frequencies of heterozygotes and homozygous recessives before selection are $2pq$ and q^2, respectively. The ratio of heterozygotes to homozygotes is

$$\frac{2pq}{q^2} = \frac{2p}{q}$$

For example, the frequency of albinism is about 1 in 20,000, or about 0.00005. If we assume the genotype frequencies are close to the Hardy-Weinberg expectations, we can estimate the frequency of the albinism allele as $q \cong 0.007$. Then the frequency of heterozygotes is about 0.014. The ratio of heterozygotes to homozygotes is about 280 to 1. There are nearly 300 heterozygotes for every homozygote! If heterozygotes are phenotypically normal, natural selection will not act on the albinism allele most of the time.

The ratio gets higher as q gets smaller. If $q = 0.001$, the ratio is 1998 to 1. Figure 5.15 shows how slowly the allele frequency changes due to natural selection against a rare recessive allele. The fitnesses are 1, 1, and 0; this represents a recessive lethal—the most extreme case. For a starting frequency of $q = 0.01$, the frequency of the recessive allele is still about 25 percent of its starting value after 250 generations. For a starting frequency of 0.001, the frequency of the recessive allele is virtually unchanged after 250 generations.

We conclude that even very strong natural selection will be ineffective in eliminating rare deleterious recessive alleles. This is consistent with observations in several species of *Drosophila*. Many studies (summarized in Powell 1997 and references therein) have shown that about 10 to 20 percent of *Drosophila* chromosomes carry recessive lethal alleles at one or more loci. Powell (1997) estimated that the average frequency of a recessive lethal allele at any locus is about 0.001. This estimate is consistent with Figure 5.15, which suggests that natural selection will be ineffective at eliminating recessive lethal alleles near this frequency.

As a recessive allele gets rarer and rarer, it may reach a frequency at which the force of natural selection to eliminate it is balanced by the force of recurrent mutation to recreate that allele. We will quantify this idea of mutation-selection equilibrium in Chapter 6.

5.6 Selection on Allozyme Loci

In the 1970s, the mechanism for maintenance of allozyme polymorphisms was intensely debated. Neutralists claimed that allozyme variants are selectively neutral and that polymorphisms are maintained by the interactions between mutation and genetic drift, while selectionists claimed that allozyme polymorphisms are maintained by some form of natural selection.

If the null hypothesis is that natural selection is not acting, then two things must be done to demonstrate that allozyme polymorphisms are maintained by

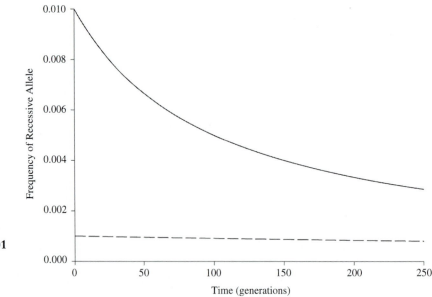

Figure 5.15 **Predicted change in frequency of a recessive lethal allele from initial frequencies of 0.01 (solid line) and 0.001 (dashed line).** Fitnesses are $w_{11} = w_{12} = 1$ and $w_{22} = 0$.

natural selection. First, it must be demonstrated that natural selection acts directly on allozyme loci; second, it must be shown that natural selection acts in such a way to maintain genetic variation, rather than to eliminate it.

While it is relatively easy to demonstrate deterministic changes in allozyme frequencies, it is much more difficult to demonstrate that these changes are due to natural selection at a particular locus. Two things are necessary: First, biochemical differences among different alleles must be demonstrated; second, it must be shown that these biochemical differences actually result in fitness differences among the genotypes. This requires a link between the function of the protein and the ecology or physiology of the organism.

Differences in catalytic efficiency and heat stability among alleles at allozyme loci have been documented many times. Gillespie (1991) and Mitton (1997) provide many examples. One well-studied example is the alcohol dehydrogenase (*Adh*) locus in *Drosophila melanogaster*. Alcohol dehydrogenase is an enzyme that detoxifies alcohol; based on electrophoretic mobility, there are two alleles, designated Adh^F (*Fast*) and Adh^S (*Slow*). The *F* protein is more efficient than *S* (Chambers 1988), but the *S* protein is more heat stable (Vigue and Johnson 1973).

Do these biochemical differences result in fitness differences among the genotypes? One approach to this question is to manipulate the environment in a way that is relevant to the differences in enzyme function. *Drosophila* larvae typically live in fermenting fruit, and the ability to detoxify ethanol is obviously advantageous. We can predict that the Adh^F allele would be favored in environments high in ethanol, because the *F* protein is more efficient in breaking down ethanol. Cavener and Clegg (1981) grew flies in medium supplemented with alcohol, and observed a consistent increase in the frequency of the Adh^F allele over many generations (Figure 5.16). In another observation, McKechnie and Geer (1993) found that files inside wineries sometimes had a higher frequency of Adh^F than nearby flies outside the wineries. Both of these results are consistent with the prediction that Adh^F should be favored in environments high in ethanol.

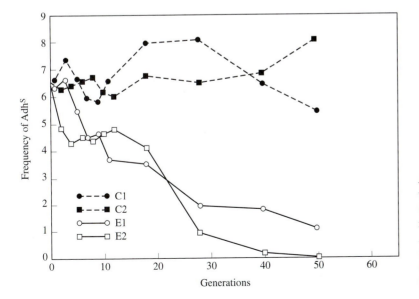

Figure 5.16 **Changes in the frequency of the** Adh^S **allele in normal medium (dashed lines) and in medium supplemented with ethanol (solid lines).** The Adh^S allele has lower enzymatic activity than the Adh^F allele, and was predicted to decline in frequency in the ethanol medium. *Source:* From Cavener and Clegg (1981).

Other evidence, including patterns of geographic variation and detailed analysis of DNA sequences, suggests that the alcohol dehydrogenase locus is subject to natural selection (reviews in Chambers 1988; Kreitman and Akashi 1995; Eanes 1999). However, the precise mechanism of selection remains unclear. For example, *Adh* is also involved in lipid synthesis, and this may be the trait subject to selection.

If selection does act on allozyme variants, does it act to maintain polymorphism? The one-locus, two-allele model predicts that genetic variation will be maintained only if the heterozygote has the highest fitness. Other possibilities for maintaining polymorphism include selection that varies in space or time, gene interaction, or others. Is there evidence that allozyme polymorphisms are maintained by any of these kinds of natural selection?

Ward Watt and his colleagues have conducted a remarkable series of studies on several species of *Colias* butterflies in western North America. These butterflies can fly only when their body temperature is high. In general, they cannot fly on cool or overcast days due to insufficient metabolic energy.

The enzyme phosphoglucose isomerase (*PGI*) is important in glucose metabolism and energy production. In *Colias* butterflies, each species has several alleles (designated by 1, 2, 3, etc.). In a series of in vitro experiments, Watt and his colleagues showed that there are differences in catalytic efficiency and heat stability among the different genotypes. For example genotypes 2/2 and 2/3 have higher activity at low temperatures, but lower activity at high temperatures. Genotypes containing alleles 4 and 5 have highest activity at high temperatures. Some heterozygotes, especially 3/4, showed overdominance for enzyme activity.

From these in vitro studies, Watt et al. hypothesized that *PGI* has a direct influence on flying ability in *Colias* and made several predictions about natural populations: (1) Alleles 2 and 3 should be more common in high-altitude (cooler) populations; (2) in insects flying during the coolest part of the day, the most common genotype should be 3/4; (3) genotypes containing alleles 4 and 5 should be favored during hot weather. These predictions were all confirmed by collecting butterflies in the field and genotyping them. In addition, Ward et al. showed that recently emerged butterflies did not differ from Hardy-Weinberg expectations, but older butterflies showed heterozygote excess, suggesting higher survival for heterozygotes. They also showed that the genotype 3/4 had the highest mating success and genotypes with lower enzymatic activity had lower mating success.

Taken together, these observations provide strong evidence that the *PGI* polymorphisms in *Colias* are maintained by a combination of heterozygote superiority and environmental heterogeneity. Here, we have only very briefly summarized these important studies. The original papers (Watt 1977, 1983; Watt et al. 1983, 1985, 1996; Carter and Watt 1988) will repay careful study.

These two examples provide convincing evidence that natural selection *can* act on allozyme variation and that allozyme polymorphisms *can* be maintained by natural selection. Whether allozyme polymorphisms are *generally* maintained by natural selection is still an unanswered, and perhaps not very meaningful, question. Evidence from DNA sequences (Chapter 10) suggests that natural selection, mutation, and genetic drift are all involved, and their relative effects vary for different loci. Mitton (1997) and Eanes (1999) provide good modern reviews of selection on allozyme polymorphisms.

5.7 Balancing Selection

We saw in Chapter 2 that most populations have extensive genetic variation for a variety of morphological and molecular characteristics. One of the major goals of population genetics is to understand how this variation is maintained. Two hypotheses (not mutually exclusive) are that genetic variation is maintained by natural selection, or is the result of interactions between mutation and genetic drift. These hypotheses are discussed briefly in Chapter 2 and in detail in Chapter 10.

We define **balancing selection** as any kind of natural selection that tends to preserve genetic variation. Heterozygote superiority (single-locus overdominance) is one example. The one-locus model of Section 5.2 predicts that genetic variation will be maintained only if the heterozygous genotype has the highest fitness. Relatively few examples of single-locus overdominance have been well documented, and it seems unlikely that this is the cause of much of the genetic variation seen in nature. Can more complex kinds of natural selection help preserve genetic variation?

One of the most important assumptions of our model is that fitnesses are constant. Several times we have suggested that natural selection can vary in space and time. In Section 4.2, we reviewed evidence that shell color and banding patterns of the snail *Cepaea nemoralis* are subject to different selection pressures in different environments and in different seasons. Establishment of air pollution controls changed the direction of selection in the peppered moth. Different *PGI* genotypes are favored at different times of the day and in different weather. Numerous other examples exist, and it is widely believed that selection often varies in both space and time.

Selection can act differently in different stages of the life cycle. It can act differently in males and females. Fitness of a genotype can depend on the frequency of that genotype, on genotypes at other loci, or on the size of the population. These are only a few examples of how the assumption of constant fitnesses can be violated.

The conclusion is that fitnesses are typically not constant, and that natural selection varies in many ways. Do these varying selection pressures lead to balancing selection? Sometimes. Varying selection can produce balancing selection, but it does not necessarily do so.

It is widely believed that much (but not all) of the genetic variation observed in natural populations is the result of balancing selection due to varying selection pressures. This does not invalidate the basic model discussed in this chapter, as the examples demonstrate. It only means that the model is applicable only some of the time.

This has been only a very brief introduction to the idea of balancing selection. Different kinds of natural selection and their relationship to balancing selection are discussed in Chapter 12.

5.8 Alternative Formulations of the Viability Selection Model

The viability selection model was formulated in terms of relative fitnesses, and we have usually scaled the fitnesses so that the best genotype has a relative fitness of one and the other genotypes have fitnesses less than one. Sometimes it is convenient to think of the fitnesses in other ways. In this section, we introduce alternative ways to parameterize relative fitnesses. The purpose is simply to establish some equations that will be useful later. These equations will sometimes make the algebra of future models more manageable.

Selection Coefficients

We originally formulated the viability model in terms of relative fitnesses, that is, the relative proportion of individuals that survive to reproductive age. Sometimes it is more convenient to work with selection coefficients. The **selection coefficient** of a genotype is related to its relative fitness as

$$w + s = 1$$

When using selection coefficients, we often scale the fitnesses to the heterozygote. The relative fitnesses then become

Genotype	A_1A_1	A_1A_2	A_2A_2
Relative fitness	$1 - s_1$	1	$1 - s_2$

If both s_1 and s_2 are positive, then the heterozygote has the highest fitness, and s_1 and s_2 can be interpreted as the relative proportions of the homozygotes that die. If both s_1 and s_2 are negative, the heterozygote has the lowest fitness, and s_1 and s_2 are measures of the selective advantages of the homozygotes.

Scaling the fitnesses this way, the mean fitness of the population is

$$\overline{w} = 1 - s_1 p^2 - s_2 q^2 \tag{5.15}$$

and the basic recursion equation is

$$p_{t+1} = \frac{p(1 - s_1 p)}{\overline{w}} \tag{5.16}$$

The corresponding difference equation is

$$\Delta p = \frac{pq(-s_1 p + s_2 q)}{\overline{w}} \tag{5.17}$$

Nothing in the model has changed, except our way of describing it. Therefore, there are still either two or three equilibria, depending on the fitnesses. The internal equilibrium is

$$\tilde{p} = \frac{s_2}{s_1 + s_2} \tag{5.18}$$

This equilibrium will exist only if s_1 and s_2 have the same sign. You should derive all of these equations from the basic model (Problem 5.6).

Dominance Formulation

There is yet another useful way of setting up the model. Sometimes we are interested in the degree of dominance between two alleles. We can set up the model with a dominance parameter, as follows:

Genotype	A_1A_1	A_1A_2	A_2A_2
Fitness	1	$1 - hs$	$1 - s$

We define A_1 as the allele with the highest homozygous fitness. Then s is the selection coefficient against the A_2A_2 genotype, and must be between zero and one. If $s = 0$, there is no selection; if $s = 1$, the A_2A_2 genotype is lethal. The other parameter, h, measures the degree of dominance. If $h = 0$, then A_1 is fully dominant (with respect to fitness); if $h = 1$, then A_2 is fully dominant. For h between zero and one, the heterozygote has intermediate fitness, and if $h = 0.5$, the heterozygote is exactly intermediate between the two homozygotes. If $h < 0$ or $h > 1$, we have overdominance or underdominance, respectively.

Using this setup, the basic equations of the model are

$$\overline{w} = 1 - 2pqhs - sq^2 \tag{5.19}$$

$$p_{t+1} = \frac{p^2 + pq(1 - hs)}{\overline{w}} \tag{5.20}$$

$$\Delta p = \frac{pqs[ph + q(1 - h)]}{\overline{w}} \tag{5.21}$$

$$\widetilde{p} = \frac{h - 1}{2h - 1} \tag{5.22}$$

Again, you should derive these for yourself (Problem 5.8). We will use this version of the selection model in Chapter 6 when we consider the effects of mutations on viability.

Marginal Fitnesses

Natural selection acts on phenotypes, which are determined by genotypes. (We are ignoring environmental effects.) Therefore, we have assigned relative fitnesses to the genotypes. But genotypes are broken up during Mendelian segregation, and only alleles are passed on to offspring. It is possible to consider a quantity analogous to fitness for an allele. First consider the A_1 allele. Its effect on the fitness of a genotype depends on what allele it is paired with. Under random mating, an A_1 allele will be paired with another A_1 allele a fraction p of the time (genotype A_1A_1 with fitness w_{11}), and will be paired with an A_2 allele a fraction q of the time (genotype A_1A_2 with fitness w_{12}). We can define the average effect (average fitness) of the A_1 allele as

$$w_1 = pw_{11} + qw_{12} \tag{5.23}$$

Similarly, the average effect of the A_2 allele is

$$w_2 = pw_{12} + qw_{22} \tag{5.24}$$

The average effect of an allele is sometimes called **marginal fitness** of that allele.

Using this notation, the mean fitness of the population becomes

$$\overline{w} = pw_1 + qw_2 \tag{5.25}$$

and the recursion equation and change in allele frequency become

$$p_{t+1} = \frac{pw_1}{\overline{w}} \tag{5.26}$$

and

$$\Delta p = p\left(\frac{w_1 - \overline{w}}{\overline{w}}\right) \tag{5.27}$$

These are much simpler-looking equations than the ones given earlier, and they will be useful in the future. You should derive all three from the corresponding equations (5.6), (5.7), and (5.10).

Equation (5.26) implies that if $w_1 > \bar{w}$, then $p_{t+1} > p_t$, and the frequency of the A_1 allele will increase. This can be especially useful in determining whether a new allele will survive when introduced into a population at a low frequency (for example, by mutation or gene flow). For example, assume a population initially consists of all A_2A_2 individuals. Introduce a very small number of A_1 alleles, so that the frequency of A_1 is very low, but not zero. Then $p \cong 0$ and $q \cong 1$. If the fitnesses of the genotypes are

A_1A_1	A_1A_2	A_2A_2
0.9	1	0.8

then $w_1 \cong qw_{12} \cong 1$ and $w_2 \cong qw_{22} \cong 0.8$. The mean fitness is $\bar{w} \cong qw_2 \cong 0.8$. Therefore, $w_1 > \bar{w}$, so A_1 will increase in the population. Note, this analysis shows only that A_1 will increase when rare; it will not increase to fixation. (Why not?)

You may not be impressed; you should have been able to predict that A_1 will increase as soon as you saw the fitnesses. But the real usefulness of this approach is that it works with multiple alleles and with more complex models of natural selection, where the fate of a rare allele is not obvious. We will see examples in Chapter 12.

The other advantage of using marginal fitnesses is that it sometimes makes algebraic manipulations much easier, as we shall see in the next section.

5.9 Evolution of Population Fitness

In our analysis of the basic viability selection model, we examined the different relationships among the fitnesses and predicted the long-term outcome of natural selection in each case. Is there some fundamental principle that can unify all of these cases? The answer is yes, and it should be obvious if you think about it. One of the fundamental principles of evolution is that populations evolve in response to their environment; they adapt to it. So we might expect the mean fitness of a population to increase with time.

It is fairly easy to show that mean fitness will always increase under the basic model. The direct approach would be to find an expression for the change in population fitness, $\Delta \bar{w}$, and show that $\Delta \bar{w}$ is always positive. This is possible, but difficult. We will take an indirect approach and show that for any set of fitnesses, p always changes in a direction that causes population fitness to increase.

We already know how p will change for any set of fitnesses. We want to see how $\bar{w}$ changes as p changes. First, we need the derivative of $\bar{w}$ with respect to p. The details are in Box 5.2; the result is, using marginal fitnesses,

$$\frac{d\bar{w}}{dp} = 2(w_1 - w_2)$$

Note, this is the derivative of $\bar{w}$ with respect to p, not with respect to time. It is much more difficult to obtain an expression for the change in $\bar{w}$ with respect to time.

Now, recall from equation (5.27) that the change in p is given by

$$\Delta p = \frac{p(w_1 - \bar{w})}{\bar{w}} \tag{5.28}$$

Box 5.2   The derivative of $\overline{w}$

From, equation (5.6), $\overline{w}$ is

$$\overline{w} = p^2 w_{11} + 2pq w_{12} + q^2 w_{22}$$

Differentiate with respect to p, using the product rule on the second term and the chain rule on the second and third terms:

$$\frac{d\overline{w}}{dp} = 2p w_{11} + 2w_{12}\left[p\frac{dq}{dp} + q\right] + 2q w_{22}\frac{dq}{dp}$$

But

$$\frac{dq}{dp} = \frac{d(1-p)}{dp} = -1$$

So we have

$$\frac{d\overline{w}}{dp} = 2p w_{11} + 2w_{12}[-p + q] - 2q w_{22}$$

$$= 2[(p w_{11} + q w_{12}) - (p w_{12} + q w_{22})]$$

Using equations (5.23) and (5.24), the last equality can be written as

$$\frac{d\overline{w}}{dp} = 2(w_1 - w_2)$$

which is the form used in the main text.

Substitute equation (5.25) for $\overline{w}$ and simplify. The result is

$$\Delta p = \frac{pq}{\overline{w}}(w_1 - w_2) \qquad \textbf{(5.29)}$$

The part in parentheses is one-half of $d\overline{w}/dp$. So we can rewrite equation (5.29) as

$$\Delta p = \left(\frac{pq}{2\overline{w}}\right)\frac{d\overline{w}}{dp} \qquad \textbf{(5.30)}$$

Finally, solving for the derivative,

$$\frac{d\overline{w}}{dp} = \left(\frac{2\overline{w}}{pq}\right)\Delta p \qquad \textbf{(5.31)}$$

All of the quantities in parentheses are positive. Therefore, the sign of the derivative (slope of $\overline{w}$ versus p) is the same as the sign of Δp. If Δp is positive, the slope of $\overline{w}$ versus p is positive and $\overline{w}$ will increase as p increases (Figure 5.17a). If Δp is negative, the slope of $\overline{w}$ versus p will be negative, and $\overline{w}$ will increase as p decreases (Figure 5.17b).

Under directional selection favoring A_1, p will always increase; Δp is always positive, therefore $\overline{w}$ always increases. Under directional selection favoring A_2, Δp is always negative. But $\overline{w}$ increases as p decreases, so directional selection in either direction always causes $\overline{w}$ to increase. Similarly, it can be shown that with overdominance or underdominance, p always changes in the direction that increases $\overline{w}$ (Problem 5.4).

We have shown that, under the many assumptions of the model, a population will evolve in such a way that its mean relative fitness will continually increase. Under more complex models of natural selection, this conclusion does not always hold; conflicting selection pressures, gene interaction, and changing environments sometimes prevent mean fitness from increasing.

It is useful to keep in mind the difference between absolute fitness and relative fitness. If population size is to remain relatively stable, mean absolute fitness must remain near one; we do not necessarily expect absolute fitness to increase in a

Figure 5.17   Change in $\overline{w}$ related to change in p.   From equation (5.31), the slope of $\overline{w}$ vs. p is the same sign as Δp. (a) If Δp is positive, the slope of $\overline{w}$ vs. p is positive, and $\overline{w}$ increases as p increases (arrow). (b) If Δp is negative, the slope of $\overline{w}$ vs. p is negative, and $\overline{w}$ increases as p decreases (arrow).

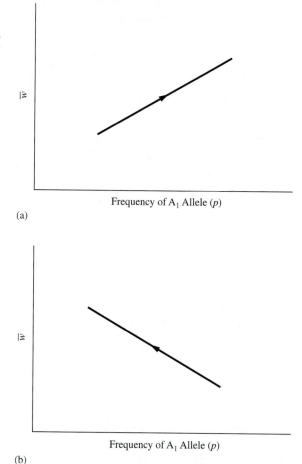

(a)

(b)

stable environment. Relative fitness may or may not increase, depending on the way natural selection is acting. However, it is possible that absolute fitness of a population will decrease because of a deteriorating environment, *even as relative fitness is increasing.* The best genotypes may be increasing in frequency, even as the absolute reproductive rates of all genotypes are decreasing. Consider that nearly all species that have ever existed are extinct. They were probably unable to evolve fast enough to survive in a changing environment; in other words, their absolute fitness decreased as the environment changed.

The Fundamental Theorem of Natural Selection

The idea that population fitness should always increase has been attributed to R. A. Fisher. In 1930, Fisher published what he called The Fundamental Theorem of Natural Selection, which he stated as, "The rate of increase in fitness of any organism at any time is equal to its genetic variance in fitness at that time" (Fisher 1958, p. 37). Fisher attached great significance to his theorem. He considered it to be exact and of great generality. He wrote about its "supreme position among the biological sciences" (Fisher 1958, p. 39).

Like most things Fisherian, the explanation and derivation are very hard to follow. Population geneticists have been puzzling over its meaning since it was first published. The usual interpretation of Fisher's theorem has been that the rate of increase of population fitness is *approximately* equal to the additive genetic variance for fitness at that time. Since a variance must be positive, this means that population fitness will always increase with time. Thus, Fisher has been credited with the first statement that population fitness will always increase.

By this interpretation, Fisher's theorem is only approximate, and the maximization principle holds only for selection at a single locus. It is known that in some circumstances population fitness may decrease (Chapter 12). This has led many to downgrade the importance of Fisher's theorem. Gillespie (1998), for example, has claimed that the fundamental theorem is neither fundamental nor a theorem. Either Fisher was wrong about the theorem's exactness and generality, or he has been misinterpreted. Both possibilities have been argued.

Fisher has indeed been widely misinterpreted. According to some authors, the fundamental theorem does *not* claim that population fitness will always increase. Fisher's definition of fitness was unorthodox and cryptic, and in fact he never claimed that population fitness, as we have defined it, will always increase. He claimed only that a particular component, the additive genetic component (to be discussed in Chapter 13) will increase. The maximization principle that $\bar{w}$ will always increase has been erroneously attributed to Fisher. It is not generally true, and he apparently never claimed that it was. See Price (1972), Ewens (1989), Edwards (1994), and Crow (2002) for important papers on the fundamental theorem.

Adaptive Landscapes

The intuitive (though not always accurate) idea that $\bar{w}$ should increase suggests a metaphor for thinking about evolution. This idea is due to Sewall Wright, who first suggested it in 1932. Imagine a landscape full of peaks and valleys, and a mountain climber who cannot go downhill. The peaks of this "adaptive landscape" represent points of maximum fitness. Each peak is a local maximum, and the highest peak is the global maximum, the Mount Everest of the adaptive landscape. The climber represents the population whose fitness (almost) always increases. The climber wants to climb Mount Everest. But if he starts up the wrong peak, he will be stuck on a lesser peak, because he cannot cross the valleys between peaks.

We illustrate with an example from our one-locus model. Imagine a population with the following fitnesses:

A_1A_1	A_1A_2	A_2A_2
1.0	0.7	0.8

Imagine that the population initially consists of all A_2A_2 individuals, but a few A_1 alleles are introduced by mutation or gene flow. Clearly, the ideal situation for this population would be for the A_1 allele to become fixed, resulting in the maximum possible mean fitness of the population. Can this happen? We can use the principle (for the one-locus model) that mean fitness always increases to find out.

First, notice that the heterozygote is the least fit genotype. Using equation (5.11) the unstable equilibrium is $\tilde{p} = 0.25$. So if the initial frequency of the A_1 allele is less than 0.25, it will be lost. The population cannot reach its ideal condition.

Figure 5.18 illustrates this graphically. It plots Δp versus p and $\overline{w}$ versus p. Note that $\overline{w}$ is minimum at $\widetilde{p}$ and that there are local maxima[2] at $p = 0$ and $p = 1$. If the population starts out below $\widetilde{p}$, then p will go to 0 (because $\Delta p < 0$) and $\overline{w}$ will go to 0.8 and not to 1. We conclude that a potentially beneficial mutation will be lost if it creates a heterozygous disadvantage.

We can think of the graph of $\overline{w}$ versus p as a simple two-dimensional adaptive landscape. The fitness maxima at $p = 0$ and $p = 1$ are adaptive peaks separated by an adaptive valley. In order for the population to move from the ridge leading to the peak at $p = 0$ to the ridge leading to the higher peak at $p = 1$, it must cross the adaptive valley between them. Since fitness cannot decrease, this cannot happen under the assumptions of the model. The population cannot reach its global maximum for fitness.

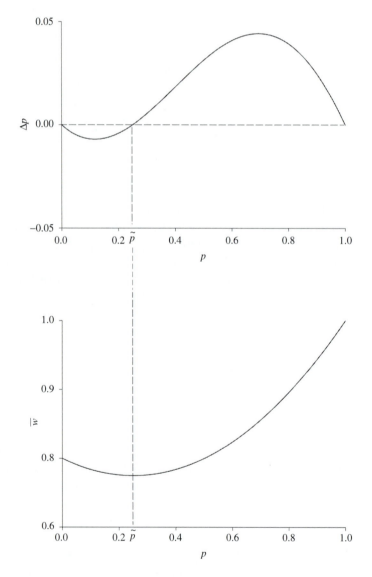

Figure 5.18 Δp **vs. p (above) and $\overline{w}$ vs. p (below) when fitnesses of A_1A_1, A_1A_2, and A_2A_2 are 1, 0.7, and 0.8.** $\overline{w}$ is a minimum at $\widetilde{p}$. The population will always evolve to increase fitness. If the initial frequency is greater than $\widetilde{p}$, the population will move toward $p = 1$ and $\overline{w} = 1$. If the initial allele frequency is less than $\widetilde{p}$, the population will move toward $p = 0$ and $\overline{w} = 0.8$; it will be unable to reach the global maximum for fitness at $p = 1$.

---

[2]The mathematically astute reader will notice that $p = 0$ and $p = 1$ are not really maxima of $\overline{w}$, but they are, considering the restriction that p must be between zero and one.

This is a very simple example of Wright's concept of the adaptive landscape. Wright imagined a complex adaptive landscape where fitness is determined by many genes and their interactions. The landscape has many adaptive peaks and valleys. The peaks are of different heights, with only one representing the global maximum for fitness. If fitness cannot decrease, the population will climb to a peak depending on its initial position in the landscape (i.e., the initial allele frequencies). It will then be stuck on this local maximum, even though it may be less fit than the global maximum. Figure 5.19 illustrates this idea for two interacting genes. For more than two genes, visualization is difficult, but the same principle holds.

Now imagine that the peaks and valleys are constantly changing positions and heights because the environment is changing. The mountain climber may start up toward a local peak, but end up going downhill a few generations later. The conclusion is that a population may be unable to evolve to its maximum fitness, even if mean fitness always increases.

Wright was very interested in how a population could cross adaptive valleys in order to reach the global maximum fitness. This led to his shifting balance theory of evolution, which proposed that genetic drift, population subdivision, and gene interaction allow populations to cross valleys and reach higher fitness peaks (Wright 1931, 1932, and many other papers; see Wright 1977 for a summary). For a recent controversy over the shifting balance theory see Coyne et al. (1997, 2000), Peck et al. (1998), and Wade and Goodnight (1998).

5.10 Estimating Fitnesses

Natural selection is the most important process of adaptive evolution. Therefore, it is important to be able to detect it and to estimate its strength. Detecting natural selection is not difficult; any systematic change over time is likely due to natural selection, at least in part. Endler (1986) discusses methods of detecting natural selection and lists many examples from natural populations. Estimating the strength of natural selection is a more difficult problem, and estimating the strength of selection on different genotypes *at a single locus* is one of the most difficult problems

Allele Frequency at First Locus

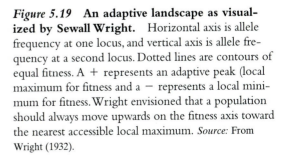

Figure 5.19 **An adaptive landscape as visualized by Sewall Wright.** Horizontal axis is allele frequency at one locus, and vertical axis is allele frequency at a second locus. Dotted lines are contours of equal fitness. A + represents an adaptive peak (local maximum for fitness and a − represents a local minimum for fitness. Wright envisioned that a population should always move upwards on the fitness axis toward the nearest accessible local maximum. *Source:* From Wright (1932).

in population genetics. Here we discuss a few examples, and point out some of the problems.

It is sometimes suggested that deviations from Hardy-Weinberg expectations can be used to estimate the direction and strength of natural selection. That is usually *not* possible for at least two reasons: First, natural selection may act in a way that does not cause deviations from Hardy-Weinberg expectations. For example, fertility differences will not create deviations from Hardy-Weinberg expectations in adults, because selection has not yet acted (see also Problems 5.17 and 5.18). Second, other evolutionary processes, for example, nonrandom mating, may cause deviations from Hardy-Weinberg expectations that mimic the effects of natural selection. Review all of the cautions in Chapter 3 about interpreting Hardy-Weinberg deviations or lack thereof.

There are two basic approaches to estimating fitnesses: estimating components of fitness, and estimating net fitnesses based on allele frequency changes. Each method has many experimental and statistical difficulties.

Estimating Components of Fitness

Estimating components of fitness is relatively straightforward in the laboratory. For example, put 100 *Drosophila* eggs in a vial and count the number of adults that emerge. That gives an estimate of egg-to-adult viability. Similarly, it is possible to estimate other components of fitness with properly designed experiments. Prout (1971a) discussed the design of experiments to estimate various components of fitness.

Dawson (1970) studied the *Sa* mutation (short antennae) in the flour beetle, *Tribolium castaneum*. *Sa* is phenotypically dominant, but is lethal when homozygous. Dawson started four populations with only *Sa*/+ heterozygotes. The decrease in the *Sa* allele is shown in Figure 5.20. The decrease is best explained by assuming that heterozygotes have reduced fitness, about 0.8 compared to the ++ homozygotes, although this is not a strong conclusion. Dawson also estimated various components of fitness in independent experiments. The results are summarized in Table 5.6. The results are consistent with the preceding hypothesis that heterozygotes suffer some

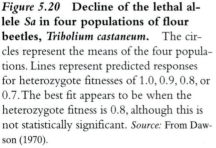

Figure 5.20 **Decline of the lethal allele** *Sa* **in four populations of flour beetles,** *Tribolium castaneum.*   The circles represent the means of the four populations. Lines represent predicted responses for heterozygote fitnesses of 1.0, 0.9, 0.8, or 0.7. The best fit appears to be when the heterozygote fitness is 0.8, although this is not statistically significant. *Source:* From Dawson (1970).

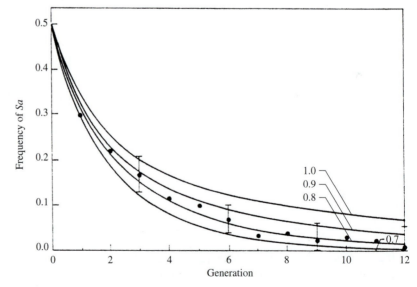

TABLE 5.6   Components of fitness for *Sa/+* heterozygotes in *Tribolium castaneum*, relative to +/+ homozygotes.

Sa/Sa homozygotes are lethal. *Source:* From Dawson (1970).

Component	Relative Fitness of *Sa/+* Compared to +/+
Male fertility (percent of males that produced offspring)	0.82
Female fertility (percent of females that produced offspring)	0.95
Female fecundity (eggs/female/day)	0.71
Larval survival	0.96
Developmental time (days from egg to pupa)	0.99

reduction in fitness. Note, however, that the most important fitness components seem to be male fertility and female fecundity. This should inspire caution about using a selection model that assumes viability selection only.

Estimating fitness components from field data is more complicated. We give one example. Assume we have estimates of genotype frequencies before and after mortality (for example, in immatures and in adults). From these we can estimate relative viabilities for the time period examined. Consider the one-locus model with two alleles. Let P_{11}, P_{12}, and P_{22} be the frequencies of A_1A_1, A_1A_2, and A_2A_2, respectively, before selection, and P'_{11}, P'_{12}, and P'_{22} be the frequencies after viability selection. We use P' instead of P_{t+1} to emphasize that selection may be incomplete at this stage of the life cycle. Note also that we are not assuming the Hardy-Weinberg proportions. This is to allow for possible selection in some earlier stage of the life cycle. We scale the viabilities to the heterozygote, so the viabilities are v_{11}, 1, and v_{22}. Proceeding exactly as we did when we constructed our selection model (it *is* a viability model, remember), we can construct the following table:

Genotype	A_1A_2	A_1A_2	A_2A_2
Frequency before selection	P_{11}	P_{12}	P_{22}
Viability	v_{11}	1	v_{22}
Product	$P_{11}v_{11}$	P_{12}	$P_{22}v_{22}$
Frequency after selection	$P_{11}v_{11}/\bar{v}$	$P_{12}/\bar{v}$	$P_{22}v_{22}/\bar{v}$

where $\bar{v}$ is the weighted average of the viabilities,

$$\bar{v} = P_{11}v_{11} + P_{12} + P_{22}v_{22}$$

This should look familiar. It is exactly like the setup of the original selection model. The last row gives P'_{11}, P'_{12}, and P'_{22}. We have three equations:

$$P'_{11} = P_{11}v_{11}/\bar{v}$$

$$P'_{12} = P_{12}/\bar{v}$$

$$P'_{22} = P_{22}v_{22}/\bar{v}$$

From these, we want to estimate v_{11} and v_{22}. We can eliminate $\bar{v}$ by dividing P'_{11} and P'_{22} by P'_{12}:

$$\frac{P'_{11}}{P'_{12}} = \frac{P_{11}v_{11}/\bar{v}}{P_{12}/\bar{v}} = \frac{P_{11}v_{11}}{P_{12}}$$

$$\frac{P'_{22}}{P'_{12}} = \frac{P_{22}v_{22}/\bar{v}}{P_{12}/\bar{v}} = \frac{P_{22}v_{22}}{P_{12}}$$

Solve the first equation for v_{11} and the second for v_{22}:

$$\hat{v}_{11} = \frac{P'_{11}P_{12}}{P_{11}P'_{12}} \tag{5.32}$$

$$\hat{v}_{22} = \frac{P'_{22}P_{12}}{P_{22}P'_{12}} \tag{5.33}$$

The $^\wedge$ over the v's indicates that they are estimates. If we have estimates of the genotype frequencies before and after selection, we can estimate the relative viabilities. For example,

Genotype	A_1A_1	A_1A_2	A_2A_2
Frequency before selection	0.50	0.30	0.20
Frequency after selection	0.60	0.25	0.15

We can estimate v_{11} and v_{22} as

$$\hat{v}_{11} = \frac{P'_{11}P_{12}}{P_{11}P'_{12}} = \frac{(0.60)(0.30)}{(0.50)(0.25)} = 1.44$$

$$\hat{v}_{22} = \frac{P'_{22}P_{12}}{P_{22}P'_{12}} = \frac{(0.15)(0.30)}{(0.20)(0.25)} = .90$$

If desired, we can rescale so that the maximum viability is one, by dividing all the viabilities by the highest:

A_1A_1	A_1A_2	A_2A_2
1	0.69	0.62

This technique assumes that the genetic basis of the trait is known and that heterozygotes can be distinguished from homozygotes. Usually we are not so lucky. For example, even though resistance to insecticides is often controlled by a single gene, we do not always know this. Recall that Curtis et al. (1978) estimated relative fitnesses of resistant and sensitive mosquitoes (Table 5.4) under the *assumption* that resistance is controlled by a single gene. As another example, consider industrial melanism in *Biston betularia*, where the heterozygotes cannot be distinguished from the dark homozygotes in the field. In Table 5.3 we showed how to estimate the relative viabilities of the dark and light forms from experimental data on bird predation. Here, we generalize this procedure to cases where the genetics of the trait may be unknown. Consider two phenotypes; call them A and B (e.g., resistant versus sensitive, or light versus dark). Let P_A and P_B be the frequencies of A and B before selection, and

P'_A and P'_B be the frequencies after viability selection, but before other kinds of selection. Assume we have estimates of these four frequencies. The relative viabilities (to be estimated) are v_A and v_B, scaled so that $v_A = 1$. To summarize, we have

Phenotype	A	B
Frequency before selection	P_A	P_B
Relative viability	$v_A = 1$	v_B
Product	P_A	$P_B v_B$
Frequency after selection	$P'_A = P_A / \bar{v}$	$P'_B = P_B v_B / \bar{v}$

where $\bar{v}$ is the weighted average of the viabilities:

$$\bar{v} = P_A v_A + P_B v_B \tag{5.34}$$

We want to estimate v_B from the phenotype frequencies. That's easy. Substitute $\bar{v}$ from equation (5.34) into the equation for P'_B:

$$P'_B = \frac{P_B v_B}{\bar{v}} = \frac{P_B v_B}{P_A v_A + P_B v_B}$$

The only unknown in this equation is v_B. Solving for v_B we get

$$\hat{v}_B = \frac{P_A P'_B}{P_B (1 - P'_B)} \tag{5.35}$$

Again, we use $^\wedge$ to indicate an estimate. You should use this equation to verify the viability estimates in Table 5.3 (Problem 5.14).

These techniques give estimates of relative viability during the particular period of the life cycle examined. They *cannot* be taken as estimates of relative fitnesses, unless you are willing to assume that all other components of fitness are equal in the various forms. There is no reason to assume this unless you have data from independent experiments, so estimates of net fitness based only on viability estimates must be interpreted cautiously.

Estimating Fitnesses from Allele Frequency Changes

It is difficult to get reliable estimates of fitness from allele frequency changes. We illustrate with a simple example. Consider a population with fitnesses of A_1A_1, A_1A_2, and A_2A_2 being 1, 0.50, and 0.40 (directional selection favoring A_1). If the initial frequency of A_1 is 0.10, then using equation (5.8), the frequency in the next generation is 0.13. Now assume the fitnesses are 0.70, 1, and 0.82 (overdominance). The allele frequency in the next generation is again 0.13. It does not get much better in the second generation. You should verify that the frequencies in the second generation are 0.169 and 0.165 for the two different fitness sets. Only after several generations do the allele frequency curves diverge enough so that there is some reasonable chance of distinguishing between them (Figure 5.21). If you had only the first two or three generations of data, it would be virtually impossible to distinguish between these very different selection regimes.

More generations of data make the problem somewhat easier. Dobzhansky and Pavlovsky (1953) followed the frequencies of two chromosomal inversions in *Drosophila pseudoobscura* in laboratory population cages, starting with an initial frequency of $p = 0.80$ for CH (Chiricahua) and $q = 0.20$ for ST (Standard).

Figure 5.21 **Predicted allele frequency change over ten generations for two different kinds of natural selection.**   The solid line is the change when fitnesses of A_1A_1, A_1A_2, and A_2A_2 are 1, 0.5, and 0.4 (directional selection favoring A_1. The dashed line is the change when fitnesses are 0.82, 1, and 0.70 (overdominance). The allele frequency dynamics are essentially indistinguishable for the first few generations.

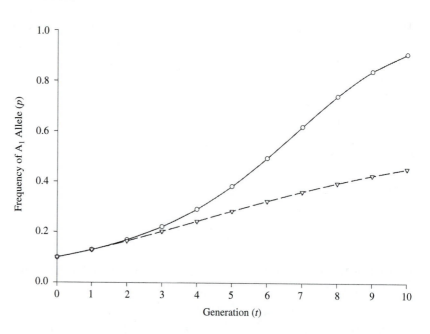

The observed frequencies of CH are indicated by the circles in Figure 5.22. Dobzhansky and Pavlovsky estimated fitnesses using a method devised by Dobzhansky and Levene (1951) that examined Δp in adjacent generations. Standardizing to the heterozygote, the individual estimates ranged from 0.27 to 0.76 (mean 0.413) for CH homozygotes and 0.73 to 1.42 (mean 0.895) for ST homozygotes, suggesting overdominance. The solid line in Figure 5.22 shows the predicted frequency of CH based on the mean fitness estimates of the homozygotes. The fit is reasonably good, but keep in mind that it is based on 15 generations of data that took a year to collect. Also note the variation in individual estimates. Finally, we must ask how relevant these estimates are to the flies in nature.

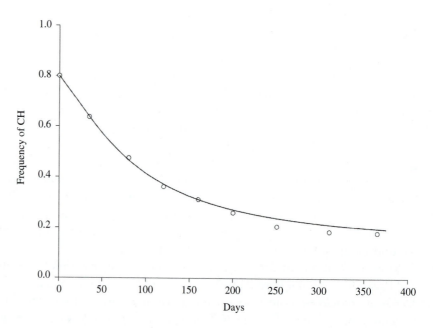

Figure 5.22 **Decline in frequency of the CH inversion in laboratory populations of *Drosophila melanogaster*.** Circles represent means of four replicate populations. Solid line represents predicted response when fitnesses were estimated by the method of Dobzhansky and Levene (1951). *Source:* Data from Dobzhansky and Pavlosky (1953).

If we are willing to assume complete dominance or complete recessiveness for fitness, the problem becomes easier; there is only one fitness to estimate. Consider the dominance formulation of fitnesses:

A_1A_1	A_1A_2	A_2A_2
1	$1 - hs$	$1 - s$

If A_1 is completely dominant or completely recessive with respect to fitness, then $h = 0$ or $h = 1$, and we only have to estimate s. Similarly, for a lethal, $s = 1$, and we only have to estimate h.

Wallace (1963) followed the frequency of an eye color mutation in *Drosophila melanogaster* for 10 generations (Figure 5.23). The mutation is lethal when homozygous. Therefore, $s = 1$ and we can use equation (5.20) to predict the frequency of the lethal mutation if it is completely recessive ($h = 0$). The solid line shows this predicted change. The allele frequency seems to decline faster than predicted, suggesting that there is some selection against heterozygotes. The dashed line shows the predicted change if $h = 0.1$; the fit appears to be better.

If we do not know anything about the nature of selection, there are serious problems in estimating fitnesses from allele frequency changes. Prout was the first to address these issues, in a series of classic papers (Prout 1965, 1969, 1971a, 1971b).

Prout (1969) showed that if fitnesses are estimated at a stage in the life cycle when selection is only partially completed (for example in adults after viability selection and before fertility selection), then four generations of allele frequency data are required. If there are four independent components (two viabilities and two fertilities) then four generations of data are needed to generate the four required equations. Prout also claimed that estimates of fitness obtained in this way are not very reliable. The estimates have large variances, and are sometimes negative.

Prout's solution to the problem was to estimate the various components of fitness in independent experiments. Prout (1971a, 1971b) described experimental procedures to estimate different components of fitness. He then derived a recursion

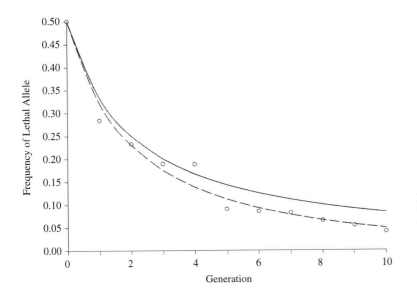

Figure 5.23 **Decline in frequency of a lethal allele in laboratory populations of *Drosophila melanogaster*.** Circles represent observed decline. Solid line represents predicted response if $h = 0$; dashed line represents predicted response if $h = 0.1$. *Source:* Data from Wallace (1963).

equation that incorporated all those components, and used that equation to predict allele frequency change in a population. We have already seen some of his experimental results in the *eyeless/shaven* experiment discussed in Section 5.2. In one set of experiments, Prout estimated fitness components of the three genotypes. He then used these estimates to predict the course of selection in a separate set of experiments. In Figure 5.4, the dashed lines are the predictions and the solid lines are the observed results. The agreement is reasonably good, suggesting that the method is capable of giving reliable estimates of net fitnesses.

The problems with this approach (other than the time and labor involved) are that it is not possible in all experimental organisms, and it is completely inapplicable to natural populations that cannot be studied in the laboratory. Nevertheless, these important papers have pointed out the problems in estimating fitnesses, and possible solutions in some cases.

To summarize, there are several serious problems with any attempt to estimate fitnesses: First, we do not always know the genetic basis of the phenotype under selection (for example, in the studies of insecticide resistance in *Anopheles* mosquitoes). Second, variances of allele frequency estimates can be large, and variances of fitness estimates even larger. Third, selection is frequently (although not always) weak. It takes large samples and careful techniques to detect fitness differences of less than 10 percent or so. Fourth, it takes several generations of allele frequency data to estimate net fitnesses. This is often not possible in natural populations, and if it is, the effect of fitnesses varying in time becomes a realistic consideration. Fifth, genes don't act in a vacuum. Fitness is determined by the interactions among all genes in the organism. Fitness of a genotype at a particular locus may well depend on genotypes at other loci. Finally, consider the basic assumptions of the model. Much evidence suggests that fertility differences among genotypes can be important, and that males and females of the same genotype sometimes have different fitnesses. Perhaps the most questionable assumption of all is that fitnesses are constant.

It is no wonder, then, that even though natural selection is pervasive and easily demonstrated, we have relatively few good estimates of the strength of natural selection at individual loci in natural populations. Most studies of natural selection estimate the strength of selection on phenotypic variation whose genetic basis is complex or unknown. This creates a whole new set of problems, in addition to those mentioned earlier. We will consider natural selection on these kinds of traits in Chapter 13.

Summary

1. Natural selection is the differential survival or reproduction of different phenotypes in a population. It can lead to adaptive evolution.
2. Absolute fitness is the net reproductive rate of an individual (or genotype). Relative fitness is the relative reproductive rate, compared to other individuals (or genotypes) in the population.
3. The main components of fitness are viability, fecundity, mating ability, and gamete competition (Figure 5.1).
4. A simple model of natural selection at a single locus with two alleles was derived and analyzed. Selection was assumed to be due to differential viability only. The

predicted change in allele frequency is given by equation (5.7). This model has many simplifying assumptions that we will begin to relax in future chapters.

5. The viability selection model predicts that genetic variation will be preserved only if the heterozygote has the highest relative fitness.

6. Examples from laboratory experiments and natural populations suggest that in many cases, the model is a reasonable approximation of reality.

7. Several examples of natural selection were examined from the perspective of the viability selection model. The model generally increases our understanding of selection in nature, but does not explain the entire story.

8. The evolution of pesticide resistance and antibiotic resistance are serious problems. Both are frequently due to strong viability selection at a single locus.

9. Natural selection is ineffective against rare deleterious recessive alleles, including those causing some human diseases, because these alleles are almost always present in heterozygous genotypes.

10. Natural selection can act on allozyme variants, but the relative importance of natural selection in maintaining allozyme polymorphisms is variable and not well understood except for a few specific loci.

11. Balancing selection is any kind of natural selection that preserves genetic variation. Variable and contradictory selection pressures often lead to balancing selection, which is thought to be an important factor in maintaining genetic variation in natural populations.

12. The one-locus selection model can be formulated in terms of selective coefficients and dominance effects. The equations are sometimes easier to work with.

13. In the one-locus viability selection model, mean fitness will always increase with time. This is not always true for more complex models.

14. Estimating components of fitness and net fitnesses in natural populations is a difficult and complex problem.

Problems

5.1. Show that $\bar{w} < 1$, unless $w_{11} = w_{12} = w_{22} = 1$.

5.2. Show that $\tilde{p}$ will be between zero and one only if the heterozygote has the highest fitness or the lowest fitness.

5.3. For each of the following fitness sets, plot p versus time and $\bar{w}$ versus time for 50 generations, starting with the initial p.

Think before you start, and set up a spreadsheet so that you only have to change the fitnesses and the initial p.

	w_{11}	w_{12}	w_{22}	Initial p
a.	1	0.9	0.8	0.01
b.	0.8	0.9	1.0	0.99
c.	0.8	1	0.6	0.01
d.	0.8	1	0.6	0.99
e.	1	0.6	0.8	0.30
f.	1	0.6	0.8	0.35

5.4. For fitness sets (a), (b), (c), and (e) of Problem 5.3, plot Δp versus p and $\overline{w}$ versus p for values of p from zero to one in increments of 0.02. On each graph, draw arrows showing the direction the population will move.

5.5. For each of the fitness sets in Problem 5.3, calculate the marginal fitnesses of each allele at generation 0 and at generation 50.

5.6. Derive equations (5.15) through (5.18).

5.7. What is the relationship between w_1 and w_2 at equilibrium?

5.8. Derive equations (5.19) through (5.22).

5.9. Derive equations (5.25) through (5.27).

5.10. For each of the fitnesses sets in Problem 5.3, rewrite the fitnesses in terms of s and h.

5.11. Write an equation expressing h as a function of w_{12} and s.

5.12. For each of the following fitness sets, plot p versus t for 100 generations, starting from $p = 0.50$. Calculate h for each fitness set. What is the effect of partial dominance on the rate of evolution? Explain in biological terms.

w_{11}	w_{12}	w_{22}
1	1	0.95
1	0.995	0.95
1	0.98	0.95
1	0.95	0.95

5.13. Prout (1971a) crossed *ey/sv* double heterozygotes with one another and recorded the numbers and genotypes of adult flies that emerged. Following are his results:

Phenotype	*eyeless*	wild type	*shaven*
Number of males	1064	2535	981
Number of females	1022	2352	1057

Estimate egg-to-adult viability for males and females separately. Can you think of fitness components where males and females might differ dramatically?

5.14. Using the method described in Section 5.10, confirm the relative viability estimates for the peppered moth data in Table 5.3.

5.15. Starting with equation (5.14), derive an expression for the time required for a given change in frequency of the resistant strain of bacteria. (*Hint:* Let $x = p/q$ and work with x instead of p.)

5.16. Consider the example of antibiotic resistance discussed in the text. Assume that with use of antibiotics, the resistant strain has almost taken over the population, so that the frequency of the sensitive strain is only 0.0001. Now, discontinue use of the antibiotic and assume the sensitive strain has a 1 percent advantage in the absence of the antibiotic. Use your answer to Problem 5.15 to determine how many generations it will take for the sensitive strain to effectively take over the population (i.e., how long it will be until the frequency of the sensitive strain is at least 0.99).

5.17. Assume you have the following series of adult genotype frequencies:

Generation	A_1A_1	A_1A_2	A_2A_2
1	0.1811	0.4889	0.3300
2	0.2038	0.4953	0.3009
3	0.2282	0.4990	0.2728
4	0.2540	0.5000	0.2460
5	0.2812	0.4982	0.2206
6	0.3096	0.4936	0.1968

Plot allele frequency versus generation. Is there evidence of natural selection from allele frequency changes? For each generation, calculate the observed allele frequencies, and expected genotype frequencies under the Hardy-Weinberg expectations. For each generation, compare the observed and expected genotype frequencies. Is there evidence of selection based on deviations from expected frequencies? Discuss.

5.18. Assume the fitnesses of genotypes A_1A_1, A_1A_2, and A_2A_2 are 1, w, and w^2, respectively. Show that the adult genotype frequencies will not deviate from Hardy-Weinberg expectations by the following steps:

a. Show that $\bar{w} = (p + qw)^2$.

b. Using the same procedure as in the initial derivation of the model, show that the adult genotype frequencies will be

$$P_{11} = \frac{p^2}{\bar{w}}, \quad P_{12} = \frac{2pqw}{\bar{w}}, \quad \text{and} \quad P_{22} = \frac{q^2w^2}{\bar{w}}$$

c. Show that the adult allele frequencies will be

$$p_{ad} = \frac{p(p + qw)}{\bar{w}}, \quad \text{and} \quad q_{ad} = \frac{qw(p + qw)}{\bar{w}}$$

d. Finally, show that in the adults

$$P_{11} = (p_{ad}^2), \quad P_{12} = 2(p_{ad})(q_{ad}), \quad \text{and} \quad P_{22} = (q_{ad})^2$$

In other words, the adult genotype frequencies are the Hardy-Weinberg functions of the adult allele frequencies. See Lewontin and Cockerham (1959) for more on this set of fitnesses.

5.19. Assume the following fitnesses:

A_1A_1	A_1A_2	A_2A_2
1.0	0.9	1.0

Graph p versus t for 50 generations from initial frequencies of $p = 0.49$, $p = 0.50$, and $p = 0.51$. Does it seem reasonable that such small differences in the initial conditions can cause such large differences in the outcome of natural selection? Discuss.

5.20. Calculate the equilibrium allele frequency for each of the following fitness sets:

A_1A_1	A_1A_2	A_2A_2
0.70	1	0.80
0.40	1	0.60
0.90	1	0.933
0.10	1	0.40
1.30	1	1.20

What can you say about determining fitnesses from the equilibrium allele frequencies? Discuss.

5.21. Assume the following fitness set:

A_1A_1	A_1A_2	A_2A_2
1.0	0.9	0.80

Find the minimum sample size necessary to detect a significant deviation from Hardy-Weinberg expectations due to natural selection. Use a chi-square test with level of significance of 0.05, and assume $p = 0.5$. (See Table 3.2 for critical values of the chi-square test.)

Mutation

*Some authors believe it to be as much the function of the
reproductive system to produce individual differences, or very slight
deviations of structure, as to make the child like its parents.*

—*Charles Darwin (1859)*

Mutation is the ultimate source of all genetic variation. Without mutation, there would be no genetic variation, hence no evolution. However, as we saw in the example in Section 1.4, mutation alone is an insignificant force for changing allele frequencies in the short term. The importance of mutation lies in its long-term effects and interactions with other processes, such as natural selection and genetic drift. However, before we can understand these interactions, we must understand how mutation acts in isolation, that is, if no other processes are acting.

Mutations are relevant to many problems in population genetics and evolution, for example, levels and patterns of genetic variation, rates of evolution, evolution of recombination, risk of extinction in small populations, evolution of breeding systems, and the maintenance of sexual reproduction. In order to understand the evolutionary effects of mutation, we need to answer two major questions: (1) How frequently do they occur? (2) What are their effects on the fitness of an organism? The answers to these questions will help us to understand the evolutionary importance of mutation and build a foundation for understanding the interactions between mutation and other evolutionary processes.

In this chapter, we review the kinds of mutations that are important in evolution, discuss their effects on the fitness of individuals and populations, and summarize attempts to estimate mutation rates and frequencies in natural populations. We will discuss interactions with other evolutionary processes in future chapters.

6.1 Kinds of Mutations

Mutation can be defined as any heritable change in the genetic material. Mutations can be due to a variety of processes, for example, errors in DNA replication, unequal crossing over, chromosomal breakage, or meiotic nondisjunction. Traditionally, they have been divided into two major classes. Gene mutations are mutations that map to a single gene locus. Chromosomal mutations are changes in the number or structure of chromosomes. In modern usage, these terms are inadequate. For example, if we define a gene as a DNA sequence that codes for a protein, then most mutations that map to a single locus (position on a chromosome) are not gene mutations. This is because much of the genome in many eukaryotes does not code for proteins.

Alternatively, we can classify mutations into four general kinds. First, we retain the term **chromosomal mutations** to mean changes in the number or structure of chromosomes. These changes are frequently visible under the microscope, and usually affect blocks of genes. Changes in chromosome structure are usually classified as duplications, deletions, inversions, or translocations.

Point mutations are changes in a single nucleotide in a DNA sequence. Usage of this term is inconsistent. Two leading genetics textbooks (Griffiths et al. 2000; Weaver and Hedrick 1997) consider point mutations as changes in one or a few base pairs of DNA. In molecular biology, the term is often restricted to changes in a single nucleotide (e.g., Li 1997; Lewin 2000). We will follow this latter usage.

Indels, short for insertions/deletions, are (usually small) insertions or deletions in a DNA sequence. The inserted or deleted sequence can be as small as a single base pair or as large as several hundred base pairs. See Section 2.13.

Gene duplications are duplications of one or more protein-coding genes. Gene duplications have been important in evolution because they create new genetic material on which natural selection can act. They sometimes lead to new cellular functions.

There is some overlap in these categories. For example, the *Bar* mutation (a dominant mutation affecting the structure of the compound eye) in *Drosophila melanogaster* is a gene duplication that is visible as a small insertion of a few bands in the polytene chromosomes. It is both a chromosomal duplication and a gene duplication. Insertion of a single nucleotide can be considered either a point mutation or a small indel.

We can also consider mutations based on where in the organism they occur. **Germline mutations** occur in gametes or cells that produce gametes. They can be passed on to the next generation. **Somatic mutations** occur in somatic cells and normally are not passed on to offspring. (Exceptions can occur by vegetative propagation of some plants.) Somatic mutations are important causes of cancer, ageing, and other problems in individuals, but generally are not important in evolution.

Finally, we must distinguish between a mutation and a substitution. A mutation is a change in the genetic material. It usually occurs in a single individual, and produces a mutant allele. A **substitution** is the fixation of a mutant allele in a population or species, replacing previous alleles. Fixed differences between two groups are the result of substitutions in one or both lineages.

The terminology for mutations is inconsistent and contradictory. The above definitions are useful in population genetics and evolutionary biology, but different researchers define these terms somewhat differently or use different terms.

Chromosomal Mutations

In meiosis I, homologous chromosomes usually separate into opposite cells. On rare occasions nondisjunction occurs, and a pair of homologous chromosomes fails to separate properly. The result is a gamete with an extra chromosome or a missing chromosome. If this gamete combines with a normal gamete, the resulting zygote will have an extra or missing chromosome. This is known as **aneuploidy**. Aneuploid individuals usually have low fitness and frequently do not survive to reproduce. A well-known example in humans is Down syndrome, which is due to three copies of chromosome 21. Because natural selection usually eliminates aneuploidy, it has probably not been very important in evolution.

Polyploidy occurs when entire sets of chromosomes are duplicated, so that an individual has three or more sets of chromosomes instead of the usual two sets. The usual cause is meiosis without cytokinesis in a reproductive cell, resulting in an unreduced gamete with two sets of chromosomes.

About half of all angiosperm plants are polyploid or derived from polyploid ancestors. Many agriculturally important crops are polyploid, for example, cotton (tetraploid), wheat (hexaploid), strawberries (octoploid), and bananas (triploid).

Polyploidy is rare in most animal groups, and seems to be mostly limited to groups with unusual reproductive systems, for example, parthenogenetic salamanders. The reason for the rarity appears to be related to disruption of dosage compensation of genes on the sex chromosomes (Orr 1990).

Sometimes a chromosome breaks and a piece is flipped end-for-end. The result is a chromosomal **inversion**. Inversions are quite common in many *Drosophila* species, and have been studied extensively by Dobzhansky and his colleagues (reviews in Dobzhansky 1970 and Powell 1997).

In an inversion heterokaryotype, recombination within the inversion is suppressed. This is because the recombinant gametes are either unbalanced (carrying duplications and/or deletions) and nonfunctional, or put into the polar bodies (e.g., *Drosophila* females). Therefore, the genes within an inversion will tend to be inherited as a unit. Dobzhansky (e.g., 1970) called these inversions "coadapted gene complexes" and believed that different inversions carry different alleles at homologous loci. Allozymes and DNA sequences provide some support for this view.

There is strong evidence that these inversion polymorphisms are maintained by some form of balancing selection and that inversion heterokaryotypes have higher fitness than homokaryotypes (review in Powell 1997). There are two possible explanations for this phenomenon. The first is single-locus overdominance. Dobzhansky believed that heterozygotes frequently have higher fitness than homozygotes. Therefore, if different inversions carry different alleles, then inversion heterokaryotypes will be heterozygous at many loci, and thus will have higher fitness than homokaryotypes. An alternative explanation is suggested by the viability studies described in Section 2.7. Flies that are chromosomally homozygous generally have lower fitness than flies that are chromosomally heterozygous. This is due to recessive viability reducing alleles at one or more loci on most chromosomes. If inversions carry different alleles, then the probability that a locus is homozygous for a viability-reducing allele is less in a heterokaryotype than in a homokaryotype. This is apparently common, at least in *Drosophila*. Single-locus overdominance is probably not as common.

Translocations occur when a chromosome breaks and a piece attaches to a nonhomologous chromosome. Translocations can be reciprocal, in which nonhomologous chromosomes exchange pieces, or nonreciprocal. Another kind of translocation is a fusion, in which parts of two nonhomologous chromosomes combine to create a single chromosome. The opposite, chromosomal fission, can also occur. Translocations are almost always deleterious because translocation heterozygotes produce a high proportion of abnormal gametes, and because moving genes around can disrupt their regulation. Several kinds of cancer in humans are caused by translocations that disrupt normal gene regulation.

Closely related species often have karyotypes that differ by one or a few translocations. Several *Drosophila* species differ by chromosomal fusions or fissions (Dobzhansky 1970). The great apes provide another example. Chimpanzees, gorillas, and orangutans have 24 pairs of chromosomes, whereas humans have 23 pairs. Two acrocentric chromosomes in the other primates have fused to produce chromosome 2 in humans (Figure 6.1). Humans and the other apes also differ by several translocations and inversions.

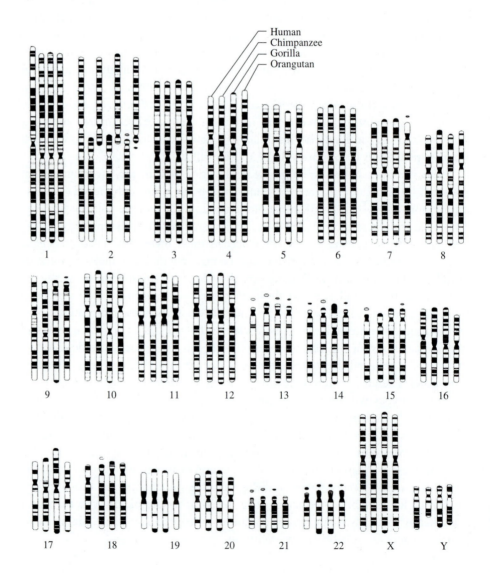

Figure 6.1 **Chromosomal homologies in humans, chimpanzees, gorillas, and orangutans.** Chromosome 2 in humans is a fusion of two chromosomes in the other primates. *Source:* From Strickberger (2000).

Point Mutations

We define point mutations as changes in a single nucleotide of a DNA sequence. The effect can vary from lethal or near lethal to no physiological or phenotypic effect at all, depending on the location of the mutation. For example, sickle cell anemia results from a mutation that causes a single amino acid change in the β-hemoglobin of humans. Other mutations in the same gene seem to have no physiological effect.

Point mutations can be classified biochemically as either transitions or transversions. Recall that adenine and guanine are purines, and cytosine and thymine are pyrimidines. A **transition** is a mutation replacing one purine with the other or one pyrimidine with the other. A **transversion** is a replacement of a purine by a pyrimidine or vice versa. Transitions are observed much more frequently than transversions, for reasons not well understood.

In protein-coding sequences, point mutations can also be classified based on their effect on the amino acid sequence of the protein. **Replacement mutations** replace one amino acid with another. They can be further divided into missense mutations, which change a single amino acid, with no effect on the rest of the amino acid sequence, and nonsense mutations, which insert a premature stop codon. **Silent mutations** do not change the amino acid sequence, due to redundancy of the genetic code. From the standard genetic code, we can predict that about 25 percent of all single base changes will be silent (assuming all changes occur with equal probability). However, at the third codon position, about 70 percent of changes will be silent. In fact, in coding sequences, the majority of observed differences between closely related species are silent, as are most single-nucleotide polymorphisms within a species. This has important implications for population genetics and molecular evolution, as we shall see.

Insertions and Deletions

Insertions and deletions can be caused by several factors. Small insertions and deletions (indels) are usually caused by errors in DNA replication. The extensive length variation found at minisatellite and microsatellite loci (Section 2.12) is due to the high mutation rate caused by polymerase slippage during DNA replication.

Frameshift mutations are insertions or deletions of one or a few nucleotides that cause the reading frame of an mRNA molecule to be shifted. As a result, all amino acids downstream from the mutation are changed. This almost always results in a nonfunctional protein. Therefore, most frameshift mutations are deleterious, and frameshift polymorphisms are rare in protein-coding genes in natural populations. However, short indels are relatively common in noncoding DNA sequences.

Longer insertions and deletions can be caused by insertion or removal of transposable elements (Section 2.13). Transposable elements are common in both prokaryotes and eukaryotes, and transpose (move) at relatively high rates. Whenever a transposable element moves into or out of a protein-coding gene, a mutation occurs. Many of the classical mutations studied by *Drosophila* geneticists are due to insertion or removal of transposable elements, and transposable elements have been associated with phenotypic variation in quantitative characters, for example, bristle number in *Drosophila*.

Finally, insertions or deletions can be caused by chromosomal breakage. These chromosomal duplications and deletions are often large and can be seen under the microscope. Large deletions are almost always severely deleterious when heterozygous, and lethal when homozygous. One example is cri du chat syndrome in humans, caused by heterozygosity for a small deletion on chromosome 5. Affected individuals have severe physiological and mental abnormalities and never live to reproductive age.

Gene Duplications

Sometimes, individual genes, pieces of chromosomes, entire chromosomes, or entire genomes can be duplicated. These duplications are important because they create new copies of a gene or a group of genes. This is thought to be an important source of evolutionary novelty, an idea first suggested by Ohno (1970). Gene duplications can result from many causes; here, we only consider their evolutionary effects on populations. See Lewin (2000) for a good discussion of gene duplication from a molecular viewpoint.

Once a gene is duplicated, there are three possible pathways: First, both copies may remain functional and continue to produce identical gene products. Natural selection favors multiple copies of a gene when large amounts of the gene product are required. Examples are the multiple genes coding for histone proteins and genes coding for rRNA molecules.

Second, one copy of the gene can retain its original function and the other can accumulate mutations, degenerate, and ultimately become nonfunctional. Such nonfunctional, duplicated genes are called **pseudogenes**, and they are very common. We discuss pseudogenes further in Chapters 7 and 10.

The third possibility is the most interesting from an evolutionary point of view: One copy of a newly duplicated gene may maintain its original function, and the other may evolve to acquire a new function. This appears to have happened many times during the course of evolution.

The globin gene family in vertebrates is probably the most well-known group of duplicated genes. The globin family consists of myoglobin (a muscle protein), and several forms of hemoglobin, the oxygen transport molecule in the blood of vertebrates. Hemoglobin is a tetramer, usually consisting of two α-subunits and two β-subunits in adults. In humans, fetal hemoglobin consists of two α-subunits and two γ (β-like) subunits. There are also several other α-like and β-like subunits. All of these subunits have slightly different physiological properties and their production is precisely regulated in different cell types. The pattern is similar in other vertebrates. The genes coding for all of these proteins almost certainly originated by gene duplication. The evidence is strong: (1) The amino acid sequences of the subunits are similar; (2) all of the hemoglobin genes share a similar intron/exon structure; and (3) the genes for the α-like and β-like subunits occur in two linked clusters. Figure 6.2 shows the chromosomal locations and inferred evolutionary history of these genes. The original gene duplication event produced two genes, one of which evolved a myoglobin-like function, and the other a hemoglobin-like function. Then the hemoglobin gene duplicated again to produce genes that evolved α-like and β-like functions. Either the gene duplication event itself or a subsequent translocation separated the α-like and β-like genes onto different chromosomes. Both the α-like and β-like genes have further duplicated to produce the

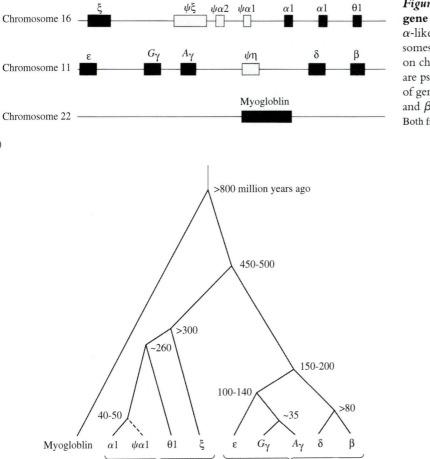

Figure 6.2 Evolution of the globin gene family. (a) Arrangement of human α-like and β-like globin genes on chromosomes 16 and 11, and the myoglobin gene on chromosome 22. Genes prefixed with ψ are pseudogenes. (b) Hypothesized sequence of gene duplications that led to the α-like and β-like globin genes in humans. *Source:* Both from Graur and Li (2000).

clusters of genes in Figure 6.2a. The figure also shows genes labeled ψα, ψβ, and so on. These are duplicated pseudogenes that have lost their function.

Langley et al. (1982) discovered a duplicate of the *Adh* gene that occurs only in *Drosophila teissieri* and *D. yakuba*. All other species of *Drosophila* examined so far lack this duplicate gene. It was originally thought to be a pseudogene. Long and Langley (1993) sequenced several copies of this gene in both species and compared them to the functional *Adh* gene. Based on detailed molecular analysis, they concluded that this gene is an active gene and has evolved a new, still unknown, function. Long and Langley named this new gene *jingwei*, after a Chinese legend of reincarnation.

Another recent gene duplication has occurred in the plant genus *Clarkia*. The *PgiC* gene[1] apparently duplicated early in the evolution of the genus. Today, some species express one copy of the gene and some species express two (Gottlieb and Weeden 1979). In species with two functional genes, the nucleotide sequences are about 96 percent identical and the enzymes apparently function identically. In other species, one gene (*PgiC2*) has been inactivated several times independently

---

[1]Phosphoglucose isomerase, an enzyme that catalyzes the isomerization of glucose-6-phosphate and fructose-6-phosphate.

(Gottlieb and Ford 1996). In *C. mildrediae*, the inactivation has apparently been very recent (Gottlieb and Ford 1997). Thus, two of the three possible evolutionary outcomes of gene duplication have been observed (so far) for this gene.

These examples suggest that gene duplication may have played an important role in evolution by providing new genetic material for natural selection to act on. Lynch and Conery (2000) surveyed gene duplication in a variety of species (four vertebrates, two invertebrates, two plants, and one yeast). They estimated the average rate of gene duplication to be about 0.01 duplications per gene per million years, with a range of about 0.002 to 0.02. This is a surprisingly high rate, the same order of magnitude as the average mutation rate per nucleotide. Lynch and Conery extrapolate to estimate that about half of all genes in a genome are expected to duplicate and increase to high frequency at least once on a time scale of 35 to 350 million years.

The vast majority of duplicated genes are fated to become pseudogenes. Most duplicated genes are inactivated soon after duplication. Lynch and Conery estimated the average half-life of a duplicated gene to be about 4 million years. However, because the rate of gene duplication is high, the few duplicate genes that survive can still provide an important source of new functions or patterns of gene expression, as illustrated by the examples above.

6.2 Molecular Models of Mutation

Before we discuss rates and effects of mutations, we should consider different ways of describing mutations at the molecular level. These concepts have coevolved with increasingly sophisticated molecular techniques.

Recurrent Mutation

The simplest concept of mutation is the classical concept of a "wild type" and a "mutant type." The wild type is the common form, found in almost all individuals in a population, and the mutant type is a rare variant. Examples are red eyes versus white eyes in *Drosophila*, or normal versus sickle cell hemoglobin in humans. Single-gene human genetic diseases are typically thought of in these terms. Note that wild type and mutant refer to the phenotype, and nothing is implied about the underlying molecular details.

The term recurrent mutation comes from the idea that the wild type is constantly mutating to the mutant type at a very low but measurable rate. This has given rise to the classical genetic notation of, say, *A* for wild type and *a* for mutant. The implication is that both the wild type and the mutant are homogeneous, at least as far as their phenotypic effects are concerned.

In this two-allele concept of mutation, we can visualize two different kinds of mutation: a forward mutation, from wild type to mutant, and a back mutation, from mutant to wild type. In Section 1.4 we developed a simple model that ignored natural selection and considered forward mutation only. The result was that the mutant allele should eventually take over the population, but that this would take a very long time, on the order of tens of thousands of generations (Figure 1.2). Still ignoring natural selection, we then considered back mutation and derived a recursion equation for the change in allele frequency [equation (1.4)]. This equation is easily solved for an equilibrium allele frequency, which is

$$\tilde{q} = \frac{u}{u + v} \qquad \textbf{(6.1)}$$

where $\tilde{q}$ is the equilibrium frequency of the mutant allele, u is the forward mutation rate (A to a), and v is the backward mutation rate (a to A). This equation predicts that the mutant allele should reach an equilibrium frequency that depends only on the forward and back mutation rates. For reasons to be explained in Section 6.4, forward mutations are expected to occur more frequently than back mutations. If this is true, the mutant allele should be much more common than the wild type allele at equilibrium.

These predictions are contrary to observation and common sense. Red-eyed flies are much more common than mutant white-eyed flies; the normal hemoglobin allele is more common than the sickle cell allele in all populations. The problem, of course, is that we have not considered natural selection. We consider the interactions between mutation and selection in Section 6.5.

Another important omission is that we have not considered the diversity of possible mutations *within* a gene.

The Infinite Alleles Model

At the molecular level, the two-allele recurrent mutation model is unrealistic. A gene is a sequence of nucleotides in a DNA molecule. Depending on how one defines a gene, it can consist of several hundred to several thousand nucleotides. Each of these nucleotides can mutate to three alternatives. The result is that any given gene can mutate in thousands of different ways. This considers only single base changes; insertions and deletions within a gene increase the possibilities essentially infinitely.

These facts have led to the development of the infinite alleles model, first proposed by Kimura and Crow (1964). It assumes that each mutation creates a unique allele, an approximation of the finite but very large number of possible DNA sequences of a gene.

Every copy of a gene will eventually mutate. Therefore, if mutation were the only evolutionary process acting, then all alleles in a population would eventually be unique. In other words, heterozygosity would approach 100 percent—every individual would contain two different alleles at every locus. However, mutation is counteracted by genetic drift, which eliminates alleles from the population. We shall see in Chapter 7 how these two opposing processes are expected to reach an equilibrium in the absence of natural selection.

The infinite alleles model is fundamental to molecular population genetics. Many of the models we discuss in future chapters are based on this model of mutation.

The Stepwise Mutation Model

At the allozyme level, the infinite alleles model is sometimes inappropriate. It was recognized early that protein electrophoresis does not reveal all variation at the protein level, let alone all variation at the DNA level. Electrophoresis detects variation based primarily on the net charge of proteins. The stepwise mutation model, sometimes called the charge-state model, was developed to account for this (Ohta and Kimura 1973). It assumes that an electrophoretically detectable mutation will change the net charge of a molecule by plus or minus one unit (Figure 6.3) and that mutations that do not change the net charge are undetectable. The main

Figure 6.3   **The stepwise mutation model as it was originally proposed for allozymes.** A_{-1}, A_0, A_1, etc. represent alleles that differ in net charge by one unit. Only mutations that change the net charge of a protein are detectable, and they change the net charge by ± 1. The mutation rate, v, is the mutation rate to electrophoretically detectable alleles, not the total mutation rate at a locus. Each allele mutates to an adjacent allele with a frequency of $v/2$ in each direction. The model can also apply to stepwise changes in repeat number at microsatellite and minisatellite loci.
Source: Based on Kimura (1983a).

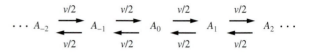

difference between the infinite alleles model and the stepwise mutation model is that the latter allows back mutation and recurrent mutation.

It turns out that the stepwise mutation model is not a particularly good model for allozymes, but is a reasonably good description of the mutation process at minisatellite and microsatellite loci. At these loci, mutation is assumed to increase or decrease the number of repeats by one.[2] Thus, stepwise changes in repeat number show up as incremental differences of mobility on an agarose gel. Again, the stepwise mutation model allows for recurrent mutation, as repeat numbers wander up and down.

As in the infinite alleles model, mutation and genetic drift affect heterozygosity in opposing directions, and eventually reach an equilibrium in the absence of natural selection.

The Infinite Sites Model

As DNA sequences became available, it became obvious that under some circumstances neither of the two models described above is appropriate. When considering a DNA sequence, the infinite alleles model is appropriate, but sometimes not very useful. For example, we saw in Section 2.10 that heterozygosity for many genes may be near 100 percent if we consider each unique sequence as a different allele. The infinite sites model was developed to circumvent this and similar problems. Under this model, each nucleotide position (site) within a DNA sequence is considered independently. It assumes that the mutation rate per nucleotide is low enough that most nucleotides do not mutate, and those that do, mutate only once. Thus, any given nucleotide site will show at most only two alleles (different nucleotides) in the population. Single-nucleotide polymorphisms support this assumption; it is very rare to find polymorphisms consisting of more than two nucleotides at a given site.

There are two main versions of the infinite sites model. The first was proposed by Kimura (1969) and assumes linkage equilibrium among nucleotide sites. A more realistic model that considers linkage disequilibrium among tightly linked sites was proposed by Watterson (1975). We will not discuss the details of either model here.

Once again, mutation and genetic drift act in opposing directions, and eventually reach an equilibrium in the absence of natural selection. The infinite sites

---

[2]Some stepwise mutation models allow steps of greater than one, but most steps are assumed to be up or down by one.

model makes predictions about the number of segregating (polymorphic) sites under certain assumptions.

Most models of DNA sequence evolution assume the infinite sites model; it is critical to interpretation of DNA sequence data from natural populations, as we shall see in Chapters 10 and 11.

6.3 Mutation Rates

We must not confuse the *frequency* of a mutant allele with the mutation *rate*. Since most mutations that affect fitness are deleterious (Section 6.4), natural selection tends to decrease the frequency of mutant alleles, while mutation tends to increase their frequency. These opposing processes will reach a balance, at which point the mutant allele will be maintained in the population at a low equilibrium frequency. This frequency depends on the mutation rate and on the strength of natural selection, primarily against heterozygotes (Section 6.5). The point is, the frequency of a mutant allele tells us nothing about the mutation rate to that allele unless we know how natural selection acts at that particular locus.

Mutation rates can be described in a variety of ways. We can describe the number of mutations per locus, or per nucleotide, or per chromosome, or per genome. We must also be clear whether we are talking about the number of mutations per cell division or per generation or per year. For example, in *Drosophila* there are about 36 germline cell divisions between a zygote and a gamete produced by the adult that developed from that zygote. So, a mutation rate of, say, 2×10^{-6} mutations per locus per cell division is equivalent to 72×10^{-6} mutations per locus per generation (or per zygote).

Two aspects of mutation rates are important: (1) the per locus (or per nucleotide) mutation rate; (2) the overall genomic rate of deleterious mutations. We discuss the first here, and defer the second until Section 6.6.

Mutation rates at individual loci vary over several orders of magnitude. Table 6.1 shows a few examples. Mutation rates at loci causing human genetic diseases are typically in the range of 10^{-5} to 10^{-6} mutations per locus per generation, and this seems to be typical of many genes in eukaryotes. Microsatellite loci have much higher mutation rates; many estimates give rates of about 10^{-3} (Jarne and Lagoda 1996; Goldstein and Pollock 1997). This high rate is probably due to polymerase slippage during DNA replication, and is what makes microsatellite loci so variable (Section 2.12).

Mutation rates per nucleotide are typically estimated to be about 10^{-8} to 10^{-10} per generation. For example, Nachman and Crowell (2000) estimated the rate in humans to be about 2.5×10^{-8} mutations per nucleotide site per generation, a higher estimate than in previous studies. This is consistent with per locus estimates if we assume that a protein-coding gene typically has about 1000 coding nucleotides.

These are very low rates. The reason is that there are proofreading mechanisms that act during DNA replication, and multiple DNA repair pathways in most cells. Interestingly, mitochondria, which lack most DNA repair pathways, have higher mutation rates per nucleotide in most organisms. Denver et al. (2000) estimated the mutation rate in the mitochondrial genome of the nematode worm *Caenorhabditis elegans* to be about 1.6×10^{-7} mutations per nucleotide site per

TABLE 6.1 Mutation rates in various organisms.
Estimates are based on a variety of techniques. Some references are secondary sources.

Organism	Character	Mutation Rate	Units	Reference
Bacteriophage T2	rapid lysis ($r^+ \rightarrow r$)	1×10^{-8}	mutations/locus/replication	Griffiths et al. (2000)
	host range ($h^+ \rightarrow h$)	3×10^{-9}	mutations/locus/replication	Griffiths et al. (2000)
HIV	reverse transcription	10^{-4} to 10^{-3}	mutations/nucleotide/rev. trans.	Nowak (1990)
Escherichia coli	streptomycin resistance	4×10^{-10}	mutations/locus/cell division	Strickberger (2000)
	T1 resistance	2×10^{-8}	mutations/locus/cell division	Sager and Ryan (1961)
	arabinose dependence	2×10^{-6}	mutations/locus/cell division	Strickberger (2000)
	$his^+ \rightarrow his^-$	2×10^{-6}	mutations/locus/cell division	Griffiths et al. (2000)
	$his^- \rightarrow his^+$	4×10^{-8}	mutations/locus/cell division	Griffiths et al. (2000)
	$lac^- \rightarrow lac^+$	2×10^{-7}	mutations/locus/cell division	Griffiths et al. (2000)
	total mutation rate	3×10^{-3}	mutations/genome/cell division	Drake (1991)
Salmonella typhimurium	*his* dependence	2×10^{-6}	mutations/locus/cell division	Strickberger (2000)
Chlamydomonas reinhardi	streptomycin resistance	1×10^{-6}	mutations/locus/cell division	Griffiths et al. (2000)
Caenorhabditis elegans	total mtDNA mutations	1.6×10^{-7}	mutations/nucleotide/generation	Denver et al. (2000)
	point mutations in mtDNA	9.7×10^{-8}	mutations/nucleotide/generation	Denver et al. (2000)
Drosophila melanogaster	yellow ($y^+ \rightarrow y$)	1.2×10^{-4}	mutations/locus/generation	Strickberger (2000)
	brown ($bw^+ \rightarrow bw$)	3×10^{-5}	mutations/locus/generation	Strickberger (2000)
	ebony ($e^+ \rightarrow e$)	2×10^{-5}	mutations/locus/generation	Strickberger (2000)
	visible mutations	6.5×10^{-6}	mutations/locus/generation	Simmons and Crow (1977)
	electrophoretic mobility	1.28×10^{-6}	mutations/locus/generation	Voelker et al. (1980)
	electrophoretic null alleles	3.86×10^{-6}	mutations/locus/generation	Voelker et al. (1980)
	transposition	0.8	transpositions/zygote	Nuzhdin and Mackay (1995)
	recessive lethals	0.006	mutations/chromosome/generation	Mukai et al. (1972)
	recessive lethals	0.02 to 0.05	mutations/zygote	Lynch et al. (1999a)
	point mutations	1.6×10^{-8}	mutations/nucleotide/year	Sharp and Li (1989)
	point mutations	1.3	mutations/zygote	Lynch et al. (1999a)
	total mutations	>2.1	mutations/zygote	Lynch et al. (1999a)
	microsatellite repeat number	9.3×10^{-6}	mutations/locus/generation	Schug et al. (1998)

Organism	Trait/Mutation	Rate	Units	Reference
Honeybee	microsatellite repeat number	4.75×10^{-4}	mutations/locus/generation	Estoup et al. (1995)
Corn	$Sh \to sh$	1.2×10^{-6}	mutations/locus/generation	Stadler (1942)
	$Pr \to pr$	1.1×10^{-5}	mutations/locus/generation	Stadler (1942)
	$Su \to su$	2.4×10^{-6}	mutations/locus/generation	Stadler (1942)
	$I \to i$	1.06×10^{-4}	mutations/locus/generation	Stadler (1942)
Durum wheat	microsatellite repeat number	2.4×10^{-4}	mutations/locus/generation	Thuillet et al. (2002)
Mouse	recessive coat color mutations	1.1×10^{-5}	mutations/locus/generation	Schlager and Dickie (1971)
	dominant coat color mutations	2.5×10^{-6}	mutations/locus/generation	Schlager and Dickie (1971)
Humans	achondroplasia	1×10^{-5}	mutations/locus/generation	Strickberger (2000)
	aniridia	4×10^{-6}	mutations/locus/generation	Strickberger (2000)
	Huntington's chorea	5×10^{-6}	mutations/locus/generation	Strickberger (2000)
	neurofibromatosis	7.5×10^{-5}	mutations/locus/generation	Strickberger (2000)
	retinoblastoma	8.5×10^{-6}	mutations/locus/generation	Strickberger (2000)
	hemophilia	3×10^{-5}	mutations/locus/generation	Sager and Ryan (1961)
	albinism	3×10^{-5}	mutations/locus/generation	Sager and Ryan (1961)
	electrophoretic mobility	3×10^{-6}	mutations/locus/generation	Neel et al. (1986)
	point mutations	1.0×10^{-8}	mutations/nucleotide/generation	Neel et al. (1986)
	point mutations	2.5×10^{-8}	mutations/nucleotide/generation	Nachman and Crowell (2000)
	microsatellite repeat number	1×10^{-3}	mutations/locus/generation	Jarne and Lagoda (1996)
	microsatellite repeat number	1.2×10^{-3}	mutations/locus/generation	Weber and Wong (1993)
	microsatellite repeat number	5×10^{-4}	mutations/locus/generation	Chakraborty et al. (1997)

generation, at least an order of magnitude higher than most estimates for nuclear DNA.

Retroviruses have RNA as their genetic material. The first step in replication of a retrovirus is to make a DNA copy of its RNA molecule, a process known as reverse transcription. The enzyme that controls this process is called reverse transcriptase and is coded for by the retroviral RNA molecule. In most retroviruses, reverse transcriptase is error prone and there are no error correction mechanisms. Consequently, the mutation rate in retroviruses is much higher than in other organisms. For example, consider HIV, the retrovirus that causes AIDS. Every time the HIV reverse transcriptase makes a DNA molecule, it makes from one to ten mistakes (Nowak 1990). The RNA molecule is about 10,000 nucleotides, so this is a mutation rate of about 10^{-3} to 10^{-4} mutations/nucleotide site/reverse transcription. This is several orders of magnitude higher than per nucleotide mutation rates in humans. This incredibly high mutation rate has led to rapid evolution of drug resistance, and has (so far) allowed HIV to evade all attempts at long-term control. For an excellent discussion of the interaction between mutation rate and natural selection in HIV, see Freeman and Herron (2001).

Keep in mind that all of these numbers are only rough estimates. There are severe experimental and statistical problems with estimating mutation rates.

6.4 Fitness Effects of Mutations

The fitness effect of a mutation can range continuously from lethal to deleterious to neutral to advantageous. Most mutations with large effects are deleterious. The proportion of mutations that are neutral or nearly neutral is unknown and somewhat controversial. What is agreed on is that advantageous mutations are much rarer than deleterious mutations.

Deleterious Mutations

If a gene produces an efficiently functioning protein, then a random change in the coding sequence is unlikely to improve the protein. The change will most likely be deleterious or have no effect. This is the basis for the often stated, but somewhat inaccurate, generalization that most mutations are deleterious. It may be more accurate to say that most mutations that affect fitness are deleterious, since an unknown proportion of mutations may have no significant effect on fitness (see below).

Similarly, loss of function mutations are expected to be more common than gain of function mutations. There are many changes in a DNA sequence that can destroy the function of a protein. On the other hand, once a gene has mutated to produce a nonfunctional protein, the options to restore function are much more limited. Most changes will have either no effect or a further negative effect. Only a few changes, perhaps only a precise reversion to the original sequence, can restore function. Thus (usually deleterious) mutations that destroy the function of the gene product are expected to be more common than (usually advantageous) mutations that restore or improve function. For example, the mutation rate of *his*$^+$ (histidine producing) to *his*$^-$ (histidine deficient) is about 2×10^{-6} in *E. coli* (Table 6.1). The back mutation from *his*$^-$ to *his*$^+$ is only about 4×10^{-8}, about 2 percent of the forward rate.

That mutations are more often deleterious than advantageous has been confirmed by mutation accumulation experiments, to be described in Section 6.6. The general idea of these experiments is to let experimental populations evolve and accumulate mutations under conditions that minimize the effects of natural selection. After many generations, individuals are made homozygous for entire chromosomes, and viability, growth rate, or some other fitness component is compared to the original population. The results of these experiments vary in detail, but as mutations accumulate, the fitness of homozygotes almost always decreases. This is attributed to the accumulation of slightly deleterious mutations that are exposed in the homozygotes.

One might expect natural selection to effectively eliminate deleterious mutations from a population. However, for two reasons this is not always the case. First, recurrent mutation may continually regenerate mutant alleles. Natural selection and recurrent mutation may reach an equilibrium at which the mutant allele is kept in the population at a low frequency that depends on the mutation rate and the strength of selection. We consider the details of mutation–selection equilibrium in Section 6.5.

The second reason is genetic drift, which causes allele frequencies to vary randomly from one generation to the next. The impact of genetic drift is inversely proportional to population size, that is, drift is greater in small populations. In a small population, genetic drift may overwhelm natural selection, and a deleterious allele may actually become fixed. We consider the details of genetic drift in Chapter 7; for now, it is sufficient to understand that genetic drift may occasionally allow a deleterious mutation to reach a high frequency or even become fixed in small populations.

The elimination of deleterious mutations by natural selection is called **purifying selection**. Its effectiveness depends on the mutation rate, the strength of selection against the mutant allele, and on the population size. We can expect purifying selection to eliminate most severely deleterious mutations or to keep them at very low frequencies, at least in moderately large populations. On the other hand, mildly deleterious mutations may reach frequencies that are high enough to affect population fitness (Section 6.6)

Neutral and Nearly Neutral Mutations

Given that a protein-coding gene can mutate in thousands of different ways, it is possible that some of these mutations will have no effect on the protein or on the fitness of the organism. Others may have such small effects that selection for or against them may be overpowered by genetic drift. Such mutations are called neutral or nearly neutral mutations. A nearly neutral mutation is one whose effect on fitness is so small that its fate may be primarily determined by either genetic drift or natural selection, depending on population size.

In protein-coding genes, the proportion of mutations that are neutral is unknown, and probably varies greatly among different genes and different organisms. Figure 6.4 illustrates two hypotheses about the proportion of neutral mutations in coding sequences. The horizontal axis is the marginal fitness (Section 5.8) of a new mutation. We can think of it as the relative fitness of a heterozygote carrying the mutation, compared to the genotype before mutation. If the marginal fitness of a mutation is greater than one, the mutation improves the

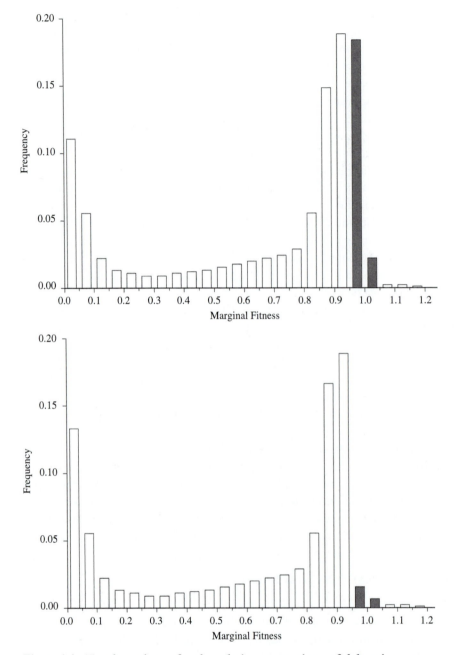

Figure 6.4 **Two hypotheses for the relative proportions of deleterious, neutral, and advantageous mutations.** The horizontal axis is the marginal fitness of a mutation and the vertical axis is the relative frequency of that kind of mutation. Fitness effects are grouped into increments of 0.05 for convenience only. Both hypotheses assume that mutations with small effect are more common than mutations with large effect, that there is a relatively large group of lethal or nearly lethal mutations, and that advantageous mutations are rare. The shaded bars indicate neutral or nearly neutral mutations, arbitrarily defined here as those with marginal fitness between 0.95 and 1.05. The top graph assumes there is a relatively high proportion of neutral or nearly neutral mutations (about 21 percent in the figure). The bottom graph assumes there are relatively few neutral or nearly neutral mutations (about 2 percent in the figure). The graphs are intended to show general relationships only, and are not representative of real data. The hypotheses are not mutually exclusive; some loci may have many neutral mutations, while other loci have few.

fitness of its bearer. Figure 6.4a assumes a relatively large proportion of mutations are neutral or nearly neutral (fitness very close to one), whereas Figure 6.4b assumes that nearly all mutations are deleterious, with few being neutral or nearly neutral. In both figures, there is a mode near neutrality and another mode near lethality.

Before DNA sequence data became available, Kimura (1983b) developed a method for estimating the proportion of neutral alleles among segregating allozyme variants. In a survey of five populations, he estimated that about 14 percent of electrophoretically detectable mutations are neutral.

As mentioned previously, not all mutations in the coding sequence of a gene result in electrophoretically detectable variation. What proportion of point mutations in the coding sequence are neutral? To answer this question, we need to consider silent and replacement mutations separately.

From the standard genetic code, we can calculate that about 25 percent of all point mutations in the coding sequence of a gene will be silent, assuming all changes are equally likely. Since these mutations do not change the amino acid sequence of the protein, it is tempting to assume they must be neutral. That is not necessarily true, as these mutations can affect translational efficiency, RNA stability, and other characteristics that may affect fitness in minor ways. Silent mutations may be subject to weak natural selection. Its effectiveness will depend on population size; in small populations, selection may be overpowered by genetic drift, but in large populations, alternative codons may be efficiently selected for or against. Silent mutations appear to be nearly neutral mutations. We address this issue in more detail in Chapter 10.

What proportion of replacement mutations in the coding sequence are neutral? One approach to this question is to compare sequences in different species. A substitution is a replacement of one nucleotide by another in a population or species. It results in a fixed difference between two groups. A **silent**, or **synonymous**, substitution is one that does not change the amino acid sequence of the protein; a **replacement**, or **nonsynonymous**, substitution is one that does change the amino acid sequence. By comparing silent and replacement substitution rates, we can get some idea of the proportion of replacement mutations that are neutral. From the genetic code, we can estimate that the ratio of replacement to silent mutations should be about 3 to 1 (assuming all changes are equally likely). If all point mutations in a coding sequence are neutral and have an equal probability of becoming fixed, then we expect to see a similar ratio of replacement to silent substitutions. In practice, we compare substitution rates per site by adjusting for the numbers of silent (synonymous) and replacement (nonsynonymous) sites in the sequence. (We will see how to do this in Chapter 11.) Thus, if all mutations are neutral, we expect a per site ratio of replacement to silent substitutions to be about one. A ratio of less than one indicates that some replacement mutations have been eliminated by natural selection.

For example, Ohta (1995) compared the rates of replacement to silent substitutions among 49 genes in several groups of mammals. For primates, the per site ratio of replacement to silent substitutions was about 0.27. This is typical of many studies (Chapters 7 and 10, Table 7.1). If we assume that all substitutions are fixations of neutral mutations, this means that about 27 percent of replacement mutations are neutral, and the rest have been eliminated by natural selection. This ignores the possibility that some rare advantageous mutations have been fixed by natural selection.

Other estimates of the proportion of neutral replacement mutations are generally in the range of 0.20 to 0.30, but sometimes higher (Eyre-Walker et al. 2002; Fay et al. 2002; Eyre-Walker and Keightley 1999). These estimates are based on a variety of techniques and assumptions, and must be interpreted cautiously. They do suggest, however, that most replacement mutations in coding sequences are deleterious, but a significant minority are neutral.

The above estimates consider only the coding sequences of a gene. The proportion of neutral mutations in noncoding DNA sequences is generally believed to be higher than in coding sequences. For example, mutations in pseudogenes or introns are probably neutral most of the time. Confirmation of this hypothesis comes from sequence comparisons: Substitution rates are usually higher in pseudogenes and introns than in most protein-coding genes. However, some noncoding sequences are clearly functional. For example, many noncoding sequences, such as promoters, are involved in gene regulation and expression. Mutations in these sequences would probably not be neutral. Fay et al. (2001) estimate that about 50 percent of noncoding sites are under some selective constraint.

We must consider another factor: The definition of neutral depends on population size. In small populations, genetic drift may overpower weak selection. A mildly deleterious mutation may become fixed by genetic drift in a small population, but be efficiently eliminated by natural selection in a large population. More on this in Chapter 10.

Advantageous Mutations

It was argued above that most mutations that affect fitness are deleterious. That does not mean that advantageous mutations do not occur, or that they are not evolutionarily important. Rare advantageous mutations are the raw material of adaptive evolution.

Consider the fate of an advantageous mutation. It will originate as a single copy (ignoring recurrent mutation), and will most likely be lost by genetic drift within a few generations (Section 6.7). If it avoids this fate, it will increase in frequency due to natural selection. The initial rate of increase will be very slow because the mutant allele is rare and natural selection seldom gets a chance to act on it. However, once the mutant allele reaches a frequency of more than a few percent, it will increase in frequency, with dynamics determined by equation (5.7) in randomly mating diploids, or equation (5.14) in haploids. Eventually the mutant allele will replace other alleles in the population and become fixed. This is known as **positive selection**, as opposed to purifying selection.

Bacterial populations, because of their enormous population sizes and short generation times, provide an unparalleled opportunity to observe and study rare events such as the fixation of advantageous mutations. Richard Lenski and his colleagues have maintained populations of *E. coli* for thousands of generations, and have learned much about the long-term effects of mutation, natural selection, and genetic drift in bacterial populations (e.g., Lenski and Travisano 1994; Elena and Lenski 1997a; Papadopoulos et al. 1999).

Lenski's *E. coli* populations were started from a single cell, and were serially transferred every day for several years. Each day, the population goes through about 6.6 cell divisions, and increases from about 5×10^6 to about 5×10^8 cells. These populations are isolated from one another and cannot undergo recombination, so mutation is the only source of new genetic variation within a population. Also, such large population sizes mean that genetic drift must have been a minor factor in their evolution. Complications of diploidy and dominance are also avoided with haploid bacteria populations. As of 1994, these populations had been maintained for more than 10,000 generations (cell divisions). The experiment is ongoing; as of 2002, the populations had been maintained for more than 20,000 generations (e.g., Lenski et al. 2003).

Lenski and his colleagues monitored fitness and cell size through these years. They found that both fitness and cell size increased rapidly at the beginning of the experiment, and increased in a stepwise fashion. Figure 6.5 shows the increase in cell size over the first 3000 generations (Elena et al. 1996). A statistical model of stepwise increases fits the data much better than a model of continuous increase. Mean fitness was highly correlated with cell size ($r \cong 0.95$) and showed similar dynamics. Elena et al. conclude that the short periods in which both fitness and cell size increased rapidly were due to fixation of rare advantageous mutations that increased fitness. Whether increased cell size was directly selected or was a pleiotropic effect of mutations affecting both fitness and cell size is unknown.

How rare were these advantageous mutations? Elena et al. estimate that about 10^6 mutations occurred every day in each population. The 3000 generations correspond to 3000/6.6, or about 450 days. Thus, about 4.5×10^8 mutations occurred, of which only a half dozen or so caused the rapid changes that Elena et al. observed. These must have been mutations with large advantages. Mutations that produce a small advantage would be more common but would take much longer to go to fixation, and their effects would have been overwhelmed by the effects of the few mutations with large effects. We do not know how often these slightly advantageous mutations might occur, but these experiments suggest that mutations with fairly large fitness benefits are very rare.

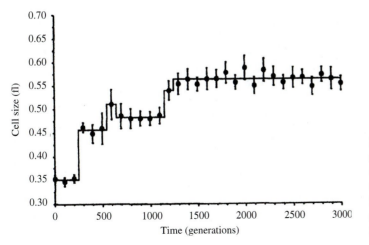

Figure 6.5 **Change in cell size in an experimental population of E. coli.** Circles are the means of 10 replicates, and error bars are 95 percent confidence intervals. The solid line is the best fit of a step function that assumes cell size increases in discrete steps. (fl = femtoliter = 10^{-15} liter). *Source:* From Elena et al. (1996).

Even though advantageous mutations are rare, they may account for a large proportion of the sequence differences between species. For example, Smith and Eyre–Walker (2002) estimated that about 45 percent of amino acid substitutions between *Drosophila simulans* and *D. yakuba* have been fixed by natural selection.

The Magnitude of Mutational Effects

Mutations that produce obvious morphological effects are rare. Not all genes are capable of producing visible changes, and those changes that do occur must be fairly obvious to be noticed. There must be many more mutations that produce small changes that go unnoticed by experimenters.

Similarly, recessive lethal mutations seem to be rarer than mutations that have minor effects on viability. In one large chromosome extraction experiment (Section 2.7), Mukai et al. (1972) estimated that the mutation rate to recessive lethals was about 0.006 per second chromosome per generation in *Drosophila melanogaster*. The mutation rate to viability-reducing alleles was about 0.06 to 0.12 per chromosome, 10 to 20 times higher. The average reduction in viability due to a single mutation (excluding lethals) was about 4 to 8 percent when homozygous. Simmons and Crow (1977) summarized similar experiments in *Drosophila* and estimated that the mutation rate to lethal alleles was about 2.5×10^{-6} per locus, and to viability-reducing alleles was a minimum of about 6.2×10^{-5} per locus, about 25 times higher. They estimated the average reduction in viability to be about 3 percent per mutation when homozygous, excluding lethals and semilethals (those with relative viability less than about 0.5 when homozygous).

Other studies give similar results (review in Lynch et al. 1999a). For example, Langley et al. (1981) found that mutations to null electrophoretic alleles (no detectable activity) decreased viability by less than 0.2 percent in heterozygotes. Figure 6.6 shows the proportionate reduction in fitness (selection coefficient), as estimated by growth rate, plotted against cumulative frequency in three experiments with microorganisms. Nearly all mutations had a selection coefficient of less than about 0.05.

To summarize, it is generally believed that mutations with small effect on fitness are more common than those with large effects. The average effect of a

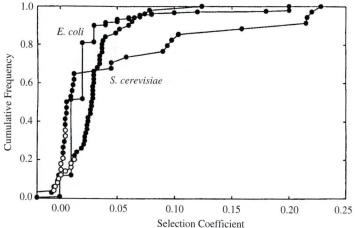

Figure 6.6 **Cumulative frequency distribution of mutational effects in three experiments with microorganisms (*E. coli* and *Saccharomyces cerevisiae*).** The selection coefficient is the proportional reduction in fitness, as estimated by growth rate. Cumulative frequency is the frequency of mutations with a given selection coefficient or smaller. In *E. coli* (top two lines), about 80 percent of mutations had a selection coefficient of 0.05 or smaller. *Source:* From Lynch et al. (1999a).

slightly deleterious mutation in homozygous state is a few percent reduction in viability. Lynch et al. (1999a) review studies leading to these conclusions.

Heterozygous Effects of Mutations

Most mutant alleles are found in heterozygous genotypes. This is a direct result of the Hardy-Weinberg principle; if a mutant allele is rare, the ratio of heterozygotes to homozygotes is very high (Section 5.5). Therefore, it is important to know the effects of mutations on the fitness of heterozygotes.

In order to understand the relationship between the effect of a mutant allele when homozygous and its effect when heterozygous, we need to consider the relative viabilities of the three genotypes. If A_1 is the normal allele and A_2 is a mutant allele, then, from Section 5.8, the genotypes and their relative viabilities can be written as

$$A_1A_1 \quad A_1A_2 \quad A_2A_2$$
$$1 \quad 1-hs \quad 1-s$$

where s is the selection coefficient and h indicates how much the viability of the heterozygote is reduced.

We are interested in the relationship between h and s. Greenberg and Crow (1960) were the first to suggest an inverse relationship, based on a complex analysis of chromosome extraction experiments with *Drosophila melanogaster*. Simmons and Crow (1977) reviewed the results of several experiments that attempted to estimate heterozygous effects of mutations. They estimated that a recessive lethal lowered the viability of heterozygotes by about 3 percent, or $hs \cong 0.03$. But $s = 1$ for a recessive lethal; therefore, $h = 0.03$ for recessive lethal mutations. At the other extreme, heterozygotes for mildly deleterious mutations suffer a reduction that is about 30 to 50 percent of the reduction in homozygotes. In other words, $h \cong 0.3$ to 0.5 for mildly deleterious mutations. The evidence strongly supports an inverse relationship between s and h. In the words of Simmons and Crow, "the milder the effect of a mutant, the greater its dominance." One possible shape of this relationship is illustrated in Figure 6.7. In general, mutations with severe effects when homozygous ($s \cong 1$) are nearly completely recessive ($h \cong 0$), while mutations with small homozygous effects ($s \cong 0$) have intermediate effects when heterozygous ($h \cong 0.4$). Lynch and Walsh (1998) reviewed other *Drosophila* studies and estimated that h for all deleterious alleles is about 0.1, and for mildly deleterious alleles, h is 0.15 to 0.30.

Figure 6.8 shows the dominance relationships for *P*-element insertions affecting viability in *Drosophila melanogaster* (Mackay et al. 1992). The graph shows only moderately to severely deleterious mutations. For s between 0.3 and 0.6, the average value of h is about 0.53; for $s > 0.6$, the average h is about 0.17 (Caballero and Keightley 1994). These estimates of both s and h are higher than those of Simmons and Crow, but the inverse relationship holds.

Garcia-Dorado and Caballero (2000) have suggested that these estimates may be too high. They reanalyzed the data from an earlier *Drosophila* experiment (Ohnishi 1977; see Section 6.6) and estimated h to be about 0.065 for deleterious mutations with $s \leq 0.37$.

How about organisms other than *Drosophila*? Houle et al. (1996) reviewed a number of recent experiments with other organisms and estimated an overall

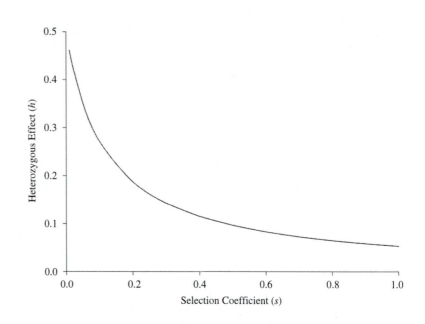

Figure 6.7 **One possible relationship between heterozygous effect (*h*) and selection coefficient (*s*).** Relative viabilities of *AA, Aa,* and *aa* are 1, 1 − *hs*, and 1 − *s*, respectively. The curve is intended to show only the general shape of the relationship.

average heterozygous effect (*hs*) of about 0.02 for mutations affecting life history characteristics. Similarly, Lynch et al. (1999a) concluded that data from many organisms are "broadly compatible with the existence of a common pool of deleterious mutations with heterozygous effects on the order of 0.1% to 1.0%." Both estimates suggest that mildly deleterious mutations will reduce viability in heterozygotes by about 1 to 2 percent, consistent with most of the *Drosophila* data.

Clearly, there is much variation in the heterozygous effects of slightly deleterious mutations, but it seems reasonable to conclude that most are only partially recessive and that some deleterious effects are expressed in heterozygotes (a few percent reduction in viability). This has important implications for evolution, because these rare, slightly deleterious alleles may contribute to lowered fitness of populations. It is important to know how frequently these slightly deleterious mutations occur. We consider this question in Section 6.6

6.5 Mutation-Selection Equilibrium

We saw in Section 5.5 that, as a deleterious allele becomes rarer, natural selection is less effective in eliminating it. Eventually, selection is so ineffective that it is balanced by recurrent mutation. New copies of the allele are generated by mutation as fast as old copies are eliminated by natural selection. An equilibrium is reached,

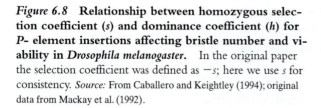

Figure 6.8 **Relationship between homozygous selection coefficient (*s*) and dominance coefficient (*h*) for *P*- element insertions affecting bristle number and viability in *Drosophila melanogaster*.** In the original paper the selection coefficient was defined as −*s*; here we use *s* for consistency. *Source:* From Caballero and Keightley (1994); original data from Mackay et al. (1992).

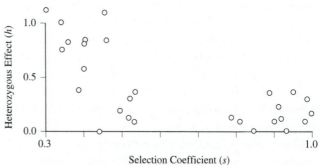

at which the mutation is maintained in the population at a constant low frequency. The equilibrium frequency of a mutant allele will depend primarily on the mutation rate and the degree of selection against heterozygotes. Selection against homozygotes will usually have a minor effect because if the mutant allele is rare, homozygotes will be very rare.

We can combine our models of mutation and selection to gain some insight into this equilibrium. We consider recurrent mutation of A_1 to A_2 only and ignore back mutation. The mutation rate from A_1 to A_2 is u, and the fitnesses of the three genotypes are

$$\begin{array}{ccc} A_1A_1 & A_1A_2 & A_2A_2 \\ 1 & 1 - hs & 1 - s \end{array}$$

In Chapter 1, we derived the recursion equation for the change in allele frequency due to mutation. From equation (1.2),

$$p_{t+1} = p_t(1 - u)$$

It is more convenient to work with Δp, the change in p, which is

$$\Delta p_u = -up$$

where the subscript u indicates the change due to mutation. As usual, we have suppressed the subscript t because it is unnecessary.

Similarly, in Section 5.8 [equation (5.21)] we showed that the change in allele frequency due to selection is

$$\Delta p_s = \frac{pqs[ph + q(1 - h)]}{\overline{w}} \tag{6.2}$$

where the subscript s indicates the change due to selection.

The total change in allele frequency is the change due to mutation plus the change due to selection:

$$\Delta p = \Delta p_u + \Delta p_s$$

At equilibrium, the total change is zero, and we have

$$\Delta p = -up + \frac{pqs[ph + q(1 - h)]}{\overline{w}} = 0$$

We can solve this equation for q to get the equilibrium frequency of the mutant allele. At this point we have two choices: We can plow through the algebra and find that it gets ugly very quickly (remember that $\overline{w}$ is quadratic in q). Or we can make some approximations and simplify. We will take the second approach.

If a mutation is deleterious, we expect that natural selection will keep the mutant allele at a low frequency; it will be kept in the population only by recurrent mutation, and we know that mutation rates per locus are low. Therefore, at equilibrium we expect the frequency of the mutant allele to be low. In other words, we can assume that $q \cong 0$ and $p \cong 1$. If we make this approximation, the change due to mutation becomes

$$\Delta p_u \cong -u$$

Similarly, if $q \cong 0$, $\bar{w}$ is approximately one (convince yourself) and the change due to selection becomes

$$\Delta p_s \cong qhs$$

Finally, the total change becomes

$$\Delta p \cong -u + qhs$$

We can easily solve this equation for the equilibrium value. Setting $\Delta p = 0$ and solving for q, we get

$$\tilde{q} \cong \frac{u}{hs} \qquad \qquad (6.3)$$

as long as $h \neq 0$. The $\sim$ indicates equilibrium, as usual. Note, the equilibrium frequency depends on the mutation rate and on the strength of selection against heterozygotes (hs), as suggested above.

If $h = 0$, equation (6.3) does not apply. But equation (6.2) reduces to

$$\Delta p_s \cong q^2 s$$

(Again, this assumes that $\bar{w} \cong 1$.) At equilibrium, $\Delta p = 0$, and we have

$$\Delta p = -up + q^2 s = 0$$

Solving for q, we get

$$\tilde{q} \cong \sqrt{\frac{u}{s}} \qquad \qquad (6.4)$$

There are two important applications of equations (6.3) and (6.4). They can be used to estimate mutation rates, and to estimate the heterozygous effects of lethal mutations. We consider an example of each.

Estimating Mutation Rates

If we know the frequency of a mutation in a population, and are willing to assume the population is in mutation-selection equilibrium, we can use equation (6.3) or (6.4) to estimate the mutation rate.

For example, achondroplasia is a form of dwarfism in humans. It is due to a dominant mutation. In one study mentioned by Crow (1986), 108 individuals with achondroplasia had 27 offspring, while 457 siblings without achondroplasia left 582 offspring. The achondroplastics left 27/108 or 0.25 offspring per individual. The nonachondroplastics left $582/457 = 1.27$ offspring per individual. Scaling to the nonachondroplastics, the relative fitness of the achondroplastics is $0.27/1.27 \cong 0.20$. Since the mutation is dominant, the relative fitnesses of the genotypes are (letting A_1 be the normal allele and A_2 be the mutant)

A_1A_1	A_1A_2	A_2A_2
1	$1 - hs = 0.20$	$1 - s = 0.20$

So, $s = 0.80$ and $h = 1$.

In an independent study, ten individuals with achondroplasia were observed in 94,075 births. They were all heterozygotes, so there were ten mutant alleles in $2 \times 94,075$ gametes. If we assume this represents the equilibrium frequency, then we can estimate $\tilde{q} = 10/(2 \times 94075) = 5.3 \times 10^{-5}$.

We can now use equation (6.3) to estimate the mutation rate. Solving equation (6.3) for u and substituting these numbers, we get

$$u = \tilde{q}sh = (5.3 \times 10^{-5})(0.80)(1) = 4.25 \times 10^{-5} \text{ mutations per locus per gamete}$$

Table 6.1 gives the estimated mutation rate, obtained by a different method, as about 1×10^{-5}. Considering the small sample size (only ten achondroplastics to estimate the equilibrium allele frequency) and the difficulty in estimating fitnesses, the agreement is satisfactory.

Heterozygous Effects of Lethal Mutations

One estimate of the mutation rate to recessive lethals in *Drosophila melanogaster* is about 0.006 lethal mutations per chromosome per generation (Mukai et al. 1972; Table 6.1). There are about 5000 genes on the second chromosome, but not all are capable of mutating to lethals. Assuming about 25 percent can do so, the lethal mutation rate per locus that is capable of mutating to a lethal allele is about

$$u = \frac{0.006}{(0.25)(5000)} = 4.8 \times 10^{-6}$$

If these mutations are completely recessive, we can use equation (6.4) to estimate their expected equilibrium frequency:

$$\tilde{q} \cong \sqrt{\frac{u}{s}} = \sqrt{u} = \sqrt{4.8 \times 10^{-6}} = 2.2 \times 10^{-3}$$

This is the predicted equilibrium frequency of recessive lethal alleles, assuming they are completely recessive ($h = 0$).

We can compare this prediction to results of chromosome extraction experiments described in Chapter 2. Dobzhansky and Spassky (1954) and Lewontin (1974) summarized experiments with several species of *Drosophila*. Lewontin estimated the number of lethal alleles per second chromosome to be about 0.386 in *Drosophila melanogaster*. If, as above we assume that about 1250 genes on the second chromosome are capable of mutating to lethals, then the observed frequency of lethal alleles is about $0.386/1250 = 3.1 \times 10^{-4}$ per locus. The observed frequency is about an order of magnitude less than expected if the alleles are completely recessive. This suggests that lethal alleles have some deleterious effect when heterozygous.

We can use equation (6.3) to estimate the heterozygous effect of lethal mutations. Solving (6.3) for h, letting $s = 1$, and substituting the estimated values for u and $\tilde{q}$, we have

$$h = \frac{u}{\tilde{q}} = \frac{4.8 \times 10^{-6}}{3.1 \times 10^{-4}} \cong 0.016$$

This is typical of many studies that have estimated the heterozygous effect of lethal mutations to be around 2 percent. Like all estimates of mutation rates, these estimates have large variances and must be taken with a grain of salt.

6.6 Genomic Mutation Rates and Mutational Load

Mutation-selection balance creates a **genetic load** on a population. In a general sense, genetic load is the reduction in mean fitness of the population due to the presence of genotypes that are less fit than the best possible genotype. Genetic load is a very general concept. It is due to the presence of genetic variation for fitness, and can be caused by any of several evolutionary processes (mutation, segregation, genetic drift, etc.). Here, we are concerned only with the genetic load due to mutation, called the **mutational load**. It is the reduction in population fitness due to segregating deleterious mutations, compared to a mutation-free genotype. Our goal is to determine whether mutational load constitutes a significant danger for most populations. We will see that it may be surprisingly high in some organisms (humans!).

Mutational Load

The problem of mutational load was first addressed by Haldane (1937) and then expanded on by H. J. Muller (1950) in his important paper; "Our Load of Mutations." Muller was a pioneer in the study of mutations and in using X-rays to induce mutations; he won a Nobel prize for his work. Muller's 1950 paper was motivated by the increasing exposure of humans to various kinds of radiation, for example, X-rays and above-ground nuclear testing. He feared that increased exposure to radiation would significantly increase the mutational load of the human species and would be manifest as a general deterioration in health of the population. The problem of mutation rates and mutational load is at least as important today as it was in 1950, because of increased use of radiation in medicine, nuclear energy, the deterioration of the ozone layer, and so on.

Under recurrent mutation, as in many human genetic diseases, deleterious mutations are kept in a population at low equilibrium frequency by a balance between mutation and selection. This causes the mean fitness of the population to be lower than it would be if mutation did not occur. We can easily quantify this mutational load.

Consider a partially recessive deleterious mutation, A_2, with fitnesses as follows:

$$
\begin{array}{ccc}
A_1A_1 & A_1A_2 & A_2A_2 \\
1 & 1 - hs & 1 - s
\end{array}
$$

From equation (5.19), the mean fitness of the population is

$$\overline{w} = 1 - 2pqhs - sq^2$$

If the mutation rate from A_1 to A_2 is u, then the equilibrium frequency of A_2 is, from equation (6.3),

$$\tilde{q} \cong \frac{u}{hs} \tag{6.5}$$

At equilibrium, then, the mean fitness of the population is

$$\overline{w} = 1 - 2\tilde{p}\tilde{q}hs - s\tilde{q}^2 \tag{6.6}$$

We can make some reasonable approximations. We have seen that the equilibrium frequency of the A_2 allele is very low. Therefore, we can assume that $\tilde{q}^2 \cong 0$, and ignore the third term on the right. Similarly,

$$\tilde{p}\tilde{q} = (1 - \tilde{q})\tilde{q} = \tilde{q} - \tilde{q}^2 \cong \tilde{q}$$

Making these approximations to equation (6.6) gives us a more manageable equation, without much loss in accuracy:

$$\overline{w} \cong 1 - 2\widetilde{q}hs$$

So, the mutational load is $2\widetilde{q}hs$. But from equation (6.5), $\widetilde{q}hs \cong u$. Therefore,

$$L \cong 2u \qquad \textbf{(6.7)}$$

For a completely recessive mutation, the mutational load turns out to be $L \cong u$ (Problem 6.12). But we will use equation (6.7), because most mutations are not completely recessive.

At mutation-selection equilibrium, the mutational load at a locus depends only on the mutation rate. It is independent of the severity of the mutation! This may be surprising. The reason is that more severe mutations come to a lower equilibrium frequency [see equation (6.3)] and this balances the stronger selection against them.

This is the mutational load at a single locus. What is the mutational load in the genome as a whole? In order to calculate the overall mean fitness due to many loci, we have to assume that the individual loci interact in some way to determine overall fitness. One assumption is that they combine multiplicatively. This means, for example, if the fitness due to the genotype at one locus is 0.8, and the fitness due to a second locus is 0.7, then the overall fitness is $0.8 \times 0.7 = 0.56$. This may be a reasonable assumption for loci affecting viability. If we assume that all loci in the genome are at mutation-selection equilibrium, and that overall fitness is determined multiplicatively, the overall mean fitness is approximately

$$\overline{w}_n \cong (1 - 2\overline{u})^n \qquad \textbf{(6.8)}$$

where $\overline{w}_n$ is the mean fitness as determined by all loci, $\overline{u}$ is the average mutation rate over all loci, and n is the number of loci in the haploid genome. This assumes that each locus has an independent effect on fitness, and that the average reduction in fitness at each locus is $2\overline{u}$.

Next, use the approximation $(1 + x)^n \cong 1 + nx$, if $|x| \ll 1$ (see any calculus text) to rewrite $\overline{w}_n$ as

$$\overline{w}_n \cong 1 - n(2\overline{u})$$

The total load due to n loci is then

$$L_n \cong 1 - \overline{w}_n = 1 - (1 - 2n\overline{u})$$

or,

$$L_n \cong 2n\overline{u}$$

Note that $n\overline{u}$ is the genomic mutation rate (the number of new mutations per haploid genome), so $2n\overline{u}$ is the total deleterious mutation rate per diploid zygote, usually symbolized by U. So, the total mutational load is

$$L_n = U = 2n\overline{u} \qquad \textbf{(6.9)}$$

Remember, our goal is to determine if mutation creates a significant genetic load for a population. So, if we can estimate the total mutation rate per zgyote, U, we can predict the mutational load of a population. We will summarize attempts to estimate U below; first we consider another reason why total deleterious mutation rates are important.

Muller's Ratchet

There is a second, more insidious aspect of mutational load. As we have seen, most mutations that affect fitness are deleterious, and mildly deleterious mutations are more common than those that are severely deleterious. Severely deleterious mutations will be efficiently eliminated by natural selection or kept at very low frequency by mutation-selection balance. But slightly deleterious mutations may be effectively neutral in small populations, and thus have a nonzero probability of becoming fixed by genetic drift. As these slightly deleterious mutations are fixed, the *absolute* fitness of every individual in the population is reduced slightly. It follows that long-term accumulation of slightly deleterious mutations may lower reproductive rates and increase the danger of extinction, especially in small populations.

Unless a population has some way of effectively eliminating deleterious mutations, they will inexorably accumulate over time. This idea is known as **Muller's ratchet**, after H. J. Muller, who first suggested it was a potential problem (Muller 1950, 1964). As slightly deleterious mutations accumulate, they continually decrease viability until the mean absolute fitness of the population becomes less than one and the population cannot replace itself. At this point, the population quickly declines to extinction. Lynch and Gabriel (1990) called this process **mutational meltdown**.

Muller's ratchet is particularly damaging in asexual populations. Because there is no recombination, no individual can produce an offspring with fewer mutations than itself, so the number of deleterious mutations in the population can only increase with time; it can never decrease. The mutational ratchet turns in only one direction. As the number of mutations increases, the average fitness of the population will decrease, ultimately to the point where the population cannot replace itself. Eventual extinction is inevitable.

Chao (1990) tested this prediction. He cultured bacteriophage $\phi6$, an RNA virus, for 40 cycles. Each cycle began with a single virus and increased to about 10^8. The average growth rate after 40 cycles was only about 78 percent of the original. Chao attributed the decrease to the accumulation of deleterious mutations as predicted by Muller's ratchet.

Muller (1964) suggested that recombination in sexual reproduction would slow down or stop the turning of the ratchet. Recombination during sexual reproduction can create some zygotes with fewer mutations than either parent, and some with more. Those with few mutations are favored by natural selection, and those with many are selected against. Natural selection will tend to eliminate mutations in groups, through the selective death of a single individual carrying many mutations.

In another experiment, Chao et al. (1997) tested whether sexual reproduction and recombination could free the $\phi6$ virus from Muller's ratchet. After culturing the virus as described above without recombination, they allowed recombination to occur through mixed infection of the host cell. They found that lost fitness was recovered faster in populations that were allowed recombination than in populations that were not. They concluded that recombination via sexual reproduction can decrease or eliminate the deleterious effects of Muller's ratchet, at least in these viruses.

It is still unclear whether Muller's ratchet is an important factor in the evolution of most sexual organisms. Its importance depends on the total deleterious mutation rate and on population size. The reason population size matters is that slightly deleterious mutations may become fixed by genetic drift in a small population. Hence, the ratchet should turn faster in small populations.

Estimates of Genomic Mutation Rates

The early studies of total deleterious mutation rates were done with *Drosophila*, using variations of the chromosome extraction procedure described in Chapter 2 (Mukai 1964; Mukai et al. 1972; Ohnishi 1977). The idea is to let mutations accumulate over many generations by keeping them in a heterozygous state and minimizing the effects of natural selection. At intervals, the experimental lines are made homozygous and viability estimated. If deleterious mutations accumulate, viability should decrease compared to control lines. From the rate of decrease in viability and the rate of increase in variation among replicate lines, it is possible to estimate the total deleterious mutation rate, U, and the mean decrease in homozygous viability per mutation. The statistical techniques were originally derived by Bateman (1959), and are summarized by Mukai et al. (1972), Simmons and Crow (1977), and Crow and Simmons (1983). Newer statistical methods are described by Garcia-Dorado (1997) and Garcia-Dorado et al. (1999).

Several of these mutation accumulation experiments have been done with *Drosophila*, and a few with other organisms. Viability does indeed decline. Figure 6.9 shows one example, from Mukai et al. (1972). The average rate of decrease in these experiments was about 0.4 percent per generation. Similar experiments have estimated an average rate of decrease of about 0.2 percent (Ohnishi 1977; Fry et al. 1999). These declines result from mutation accumulation on the second chromosome only; when extrapolated to the entire genome, they suggest rates of decrease of about 1 to 2 percent per zygote.

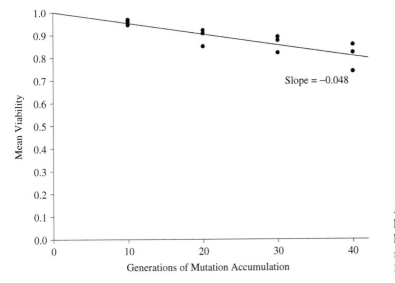

Figure 6.9 **Decrease in mean viability in lines of *Drosophila melanogaster* allowed to collect deleterious recessive mutations.** Lethal mutations are excluded. *Source:* Based on data in Mukai et al. (1972).

Simmons and Crow (1977) summarized the early *Drosophila* mutation accumulation experiments. They estimated that the probability that a second chromosome would acquire a recessive viability-reducing mutation (excluding lethal and severely deleterious mutations) ranged from 0.058 to 0.172 per chromosome per generation, with a mean of about 0.124. The second chromosome is about 40 percent of the *Drosophila* genome, so if the probability of a second chromosome acquiring a deleterious mutation is about 0.124, then the probability that the genome as a whole will acquire one is about 0.31. Doubling this number to account for diploidy, the estimated total deleterious mutation rate in *Drosophila* is $U \cong 0.62$ new viability reducing mutations per zygote.

Recent evidence suggests that these estimates may be too high. Keightley (1996) has reevaluated the classical *Drosophila* experiments, and claimed that they may overestimate the viability-reducing mutation rate by at least an order of magnitude. Similarly, Garcia-Dorado (1997) and Garcia-Dorado et al. (1999) used new statistical methods on Mukai's and Ohnishi's data and obtained much lower estimates, $U \cong 0.01$ to 0.05. Finally, Fry et al. (1999) have performed similar experiments and estimated $U \cong 0.10$ viability-reducing mutations per zygote. On the other hand, Lynch et al. (1999a) argue that the classical estimates of $U \cong 0.6$ are probably reasonably accurate.

What about other organisms? Are these *Drosophila* estimates typical of other organisms? Recently, we have begun to find answers to this question. Table 6.2 summarizes the results of several recent studies.

Vassilieva et al. (2000) conducted a 214 generation, mutation accumulation experiment with the nematode worm *Caenorhabditis elegans*. They studied several characteristics affecting fitness (e.g., survival, longevity, intrinsic rate of increase, and others). The overall decline in fitness was only about 0.1 percent per generation, about an order of magnitude lower than the original *Drosophila* estimates. The overall deleterious mutation rate was $U \cong 0.03$ deleterious mutations per zygote, again, much lower than the original *Drosophila* estimates.

Kibota and Lynch (1996) estimated the deleterious mutation rate in *E. coli* to be about 1.7×10^{-4} deleterious mutations per genome per cell division. This seems many times smaller than the *Drosophila* estimates, but we must compensate for two things: First, the *E. coli* genome is only about 2 percent of the size of the *Drosophila* genome. Thus, we have to multiply the *E. coli* rate by 50 to get comparable rates per *Drosophila* genome. Second, as mentioned in Section 6.3, there are about 36 germline cell divisions in the cell line leading from a newly formed zygote to a gamete in the adult. So we have to multiply the *E. coli* estimates by 36 to get a comparable estimate. Making these adjustments gives an estimate of about 0.31 in units of "*Drosophila* gametes."[3] This is almost identical to the early *Drosophila* estimates, but keep in mind that a much higher proportion of *E. coli* DNA codes for proteins than does *Drosophila* DNA.

Eyre-Walker and Keightley (1999) estimated deleterious mutation rates in humans, chimpanzees, and gorillas. They compared the observed rate (per site) of silent substitutions, which they assumed were neutral, to the observed rate of replacement

---

[3]Kibota and Lynch used an earlier estimate of 25 germline cell divisions in *Drosophila*, giving an *E. coli* estimate of 0.21 in terms of *Drosophila* gametes.

TABLE 6.2   Estimates of total deleterious mutation rates (*U*) in various organisms.

Most are based on mutation accumulation experiments and use viability as an estimate of fitness; some estimates are based on comparisons of rates of silent substitutions versus replacement substitutions. Most *Drosophila* estimates are based on second chromosome extraction experiments and have been converted to estimates per zygote. Some references are secondary sources.

Organism	Character	Mutation Rate (*U*) (new mutations/zygote)	Reference
E. coli	total fitness (growth rate) (mutations/genome/cell division)	0.00017	Kibota and Lynch (1996)
Yeast	total fitness	9.5×10^{-5}	Zeyl and DeVisser (2001)
Drosophila melanogaster	viability-reducing mutations	0.70	Mukai (1964)
	viability-reducing mutations	0.86	Mukai et al. (1972)
	viability-reducing mutations	0.29	Ohnishi (1977)
	fitness compared to control line	0.85 to 1.0	Houle et al. (1992)
	viability-reducing mutations	0.10	Fry et al. (1999)
	viability-reducing mutations	0.04	Garcia-Dorado et al. (1999)
Daphnia pulex	several fitness components	0.76	Lynch et al. (1998)
Caenorhabditis elegans	deleterious mutations	0.005	Keightley and Caballero (1997)
	deleterious mutations	0.03	Vassilieva et al. (2000)
Arabidopsis thaliana	total fitness	0.148	Lynch et al. (1999a)
Corn	several reproductive characters	0.06 to 0.09	Lynch et al. (1999a)
Humans	deleterious mutations	1.6	Eyre-Walker and Keightley (1999)
	deleterious mutations	3.0	Nachman and Crowell (2000)
Chimpanzees	deleterious mutations	1.7	Eyre-Walker and Keightley (1999)
Gorillas	deleterious mutations	1.2	Eyre-Walker and Keightley (1999)

substitutions in 46 protein–coding genes. If all replacement substitutions are neutral, the per site rates should be equal. Fewer replacement substitutions indicates that deleterious mutations have been eliminated. They estimated 143 replacement substitutions, compared to 231 expected if all were neutral. Using published estimates of divergence times among species, generation times, numbers of genes in each species, and other parameters, they converted these numbers to an estimate of $U \cong 1.6$ deleterious mutations per zygote in humans. Estimates for chimpanzees and gorillas were similar ($U \cong 1.7$ and 1.2, respectively). They judged that these estimates were very conservative and that the actual numbers were about three deleterious mutations per zygote. Nachman and Crowell (2000), using similar techniques, also obtained an estimate of about three deleterious mutations per zygote in humans.

These estimates are surprisingly high compared to the *Drosophila* estimates of about 0.6 or lower, and raise the question of how humans and other primates can sustain such a high rate of deleterious mutations. If a population is to avoid accumulation of deleterious mutations, every new mutation must be eliminated by one genetic death. A genetic death is a selective death or failure to reproduce of one individual, or proportionate reduction of fitness in several individuals. If $U \cong 3$, as estimated above for primates, then each individual must suffer three genetic deaths if the population is to avoid Muller's ratchet!

Something is clearly wrong. Under the assumptions above,

$$\overline{w}_n \cong 1 - L_n$$

But if $L_n = U$ and $U = 3$, then $\overline{w}_n$ is negative! The problem is that in equating the mutational load, L_n, with the total deleterious mutation rate, U, we assumed that $\overline{w}_n$ is a multiplicative function of single-locus fitnesses, as expressed in equation (6.8). If overall fitness is determined by epistatic interactions among loci, the mutational load may be much lower than the total deleterious mutation rate. Deleterious mutations may interact in such a way that each additional mutation that an organism carries decreases its fitness by a greater amount than the previous one. This was first suggested by Kimura and Maruyama (1966) and has been called **synergistic epistasis**. For example, fitness may decrease by an amount proportional to the square of the number of mutations. Under synergistic epistasis, an individual with many mutations is much more likely to be eliminated by natural selection than an individual with only a few. Natural selection will eliminate mutations in groups; each genetic death will eliminate many mutations at once. Whether synergistic epistasis is common in natural populations is still unknown.

As suggested above, recombination during sexual reproduction may aid synergistic epistasis. Recombination can create zygotes with more mutations than either parent. If mutations interact by synergistic epistasis, these zygotes will have low fitness and will have a high probability of being eliminated by natural selection. Natural selection will eliminate many mutations through a single genetic death. This hypothesis has received much recent attention (e.g., Kondrashov 1988, 1995; Crow 1997, 1999; Lynch et al. 1995, 1999a; Keightley and Eyre-Walker 2000). In one interesting study, Keightley and Eyre-Walker (2000) surveyed estimates of total deleterious mutation rates in many organisms, and suggested that in many sexual organisms these rates are much less than one. From this, they concluded that purging of deleterious mutations is probably not the main reason that sexual reproduction is so widespread. However, Kondrashov (2001) challenged this conclusion because it was based on the assumption that silent substitutions are completely neutral.

The main point of this analysis is that the mutational load depends on the mutation rate, and may be quite high. Increasing the mutation rate will lead to increased genetic load. This prediction has been tragically confirmed in studies of several species in the area of the Chernobyl nuclear disaster. Humans, for example, show significantly increased incidence of leukemia, thyroid cancer, and developmental abnormalities, compared to humans in unexposed regions. These are probably due to somatic mutations, but suggest that germline mutation rates may have increased also. Dubrova et al. (1996, 1997) found that the germline mutation rate at minisatellite loci was about twice as high in exposed individuals as in unexposed controls. Other loci probably show increased mutation rates also, with unknown, long-term effects. In a related study, Dubrova et al. (2002) found a similar increase in germline mutation rates in individuals exposed to radioactive fallout from a nuclear test site in the Soviet Union during the 1950s. Ellegren et al. (1997) and Møller and Mosseau (2001) found increases in germline mutation rates at microsatellite loci in populations of barn swallows (*Hirundo rustica*) breeding close to Chernobyl. These populations showed corresponding increases in partial albinism and decreases in fitness. There is no reason to think that barn swallows are unusual in this respect. Muller was right.

To summarize, the total deleterious mutation rate is surprisingly high in some organisms. For good reviews, see Lynch et al. (1999a) and Keightley and Eyre-Walker (1999). This raises the possibility of long-term genetic deterioration and eventual extinction, especially in small populations. Whether this is an important factor in the evolution of sexual populations is still not clear. Muller (1950) first suggested that it might be a problem for human populations; this has been reiterated by Crow (1997) and Lynch et al. (1999a).

6.7 Fate of a New Mutation

In theory, an infinite number of mutations is possible, but in real life, most are lost within a few generations. What is the probability that a new mutant allele will be lost in the first generation? Or the second? Or at any time in the future?

When a new (neutral) mutation (call it A_2) arises, the individual carrying it must be heterozygous (A_1A_2) and must mate with a normal (A_1A_1) individual. If population size is stable, the *average* number of offspring per mating (family size) will be two. If this $A_1A_2 \times A_1A_1$ mating produces two offspring, what is the probability that neither will get the A_2 allele? The probability that the A_1A_2 individual will pass the A_1 allele to the first offspring is one-half; independently, the probability that it will pass the A_1 allele to the second offspring is also one-half. Using the law of multiplication, the probability that both will get the A_1 allele, that is, the probability that A_2 will be lost, is

$$\Pr(loss) = \left(\frac{1}{2}\right)^2 = \frac{1}{4}$$

Not all matings produce two offspring, of course. If the mating produces zero offspring the probability of loss is 1; if it produces one offspring, the probability of loss is $1/2$. A little thought should convince you that if the family size is k, the probability of loss is $(1/2)^k$. We write this as

$$\Pr(loss|k) = \left(\frac{1}{2}\right)^k \tag{6.10}$$

where the left side is the conditional probability of loss, given a family size of k.

It is frequently assumed that family size is a random variable that follows a Poisson distribution (Box 6.1; Appendix A). If K is a random variable indicating family size, and the average family size is two, then, from the definition of the Poisson distribution, the probability that a family size will be k is

$$\Pr(K = k) = \frac{e^{-2}2^k}{k!} \tag{6.11}$$

where k is a nonnegative integer.

Now we can calculate the probability that the A_2 allele will be lost, using the law of total probability. It is the probability of loss for a given family size multiplied by the probability of that family size, summed over all possible family sizes:

$$\Pr(loss) = \sum_{k=0}^{\infty} [\Pr(loss|k) \times \Pr(K = k)]$$

where the summation is over all possible family sizes. Substituting equations (6.10) and (6.11) into this, we get

BOX 6.1   The Poisson distribution

A Poisson-distributed random variable has the following characteristics:

1. Any given experiment (trial) has two possible outcomes; call them success and failure.
2. The probability of success in any single trial (p) is low.
3. The number of trials (n) is large.
4. Each trial is independent of the others.

Then the average number of successes is np, and the Poisson distribution describes the distribution of the number of successes.

Imagine firing a gun randomly at a target. The probability of hitting the target is p and is assumed to be low. The number of bullets fired is n, and is assumed to be large. You can also think of n as the number of targets. This describes the **target theory** of mutation due to radiation. Ionizing particles are the bullets and genes are the targets. There are other causes of mutation, of course, but it is widely assumed that the number of new mutations is a random variable that follows a Poisson distribution.

If X is a Poisson distributed random variable, then the probability that X will have a realized value of x is

$$Pr(X = x) = \frac{e^{-\lambda}\lambda^x}{x!}$$

where $\lambda = np$ is the expected (average) number of successes in n trials.

What is the probability that a chromosome (or genome, or zygote) will have one, or two, or some other number of new mutations? We use the data of Mukai (1964) to illustrate how to estimate the distribution of new mutations. Mukai estimated the deleterious mutation rate per locus as 2.9×10^{-5}. The number of loci is about $13,600$. These are p and n, respectively. Therefore, mean number of deleterious mutations per genome is $\lambda = np = 0.39$. We can now use the formula for the Poisson distribution to calculate the probability of one, two, three, and so on mutations occurring:

$$Pr(X = 0) = \frac{e^{-0.39}(0.39)^0}{0!} \cong 0.68$$

$$Pr(X = 1) = \frac{e^{-0.39}(0.39)^1}{1!} \cong 0.26$$

$$Pr(X = 2) = \frac{e^{-0.39}(0.39)^2}{2!} \cong 0.04$$

etc.

Note that the most likely number of new mutations is zero and the probability of having more than two is about 2 percent.

Similarly, if K is a random variable indicating family size, and the average family size is two, then the probability of having k offspring is

$$Pr(K = k) = \frac{e^{-2}2^k}{k!}$$

This is equation (6.11) in the main text.

$$Pr(loss) = \sum_{k=0}^{\infty} \left(\frac{1}{2}\right)^k \left(\frac{e^{-2}2^k}{k!}\right)$$

This is easier than it looks. The 2^k and $(1/2)^k$ cancel, and we can move e^{-2} outside the summation. We are left with

$$Pr(loss) = e^{-2} \sum_{k=0}^{\infty} \frac{1}{k!}$$

You may recognize the summation. It is just e. (See any calculus textbook if you're skeptical.) So we have

$$Pr(loss) = e^{-1} \cong 0.37$$

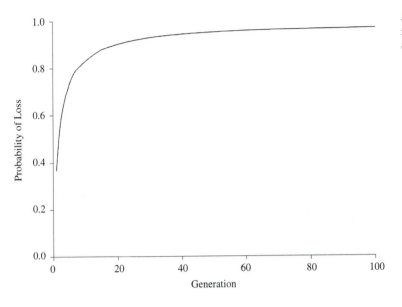

Figure 6.10 **Probability of loss for a new neutral mutation.** *Source:* Based on calculations of Fisher (1930).

What have we shown? The probability that a new mutation will be lost in the first generation is about 37 percent. This assumes a neutral mutation and a stable population with average family size of two.

What about future generations? R. A. Fisher (1930) calculated the probability of loss for successive generations. His results are summarized in Figure 6.10. The probability that the new mutation will be lost within ten generations is more than 80 percent. Clearly, a new mutation is unlikely to survive for more than a few generations.

Here we have our first introduction to the power of genetic drift. A new neutral mutation is very likely to be lost by chance alone in the first few generations. In Chapter 7, we examine the probability that a mutation will be lost or fixed in a finite population. We will see, surprisingly, that the probability that a favorable mutation will be lost is also high. Perhaps even more surprisingly, the probability that a deleterious mutation will be fixed is low, but not zero.

Summary

1. Mutation is the ultimate source of all genetic variation; without mutation, there would be no evolution.

2. Chromosomal mutations are changes in the number or structure of chromosomes. Point mutations are changes in a single nucleotide of a gene or other region of DNA.

3. Polyploidy has been important in plant evolution, but is rare in most animal groups.

4. Inversion polymorphisms are common in some *Drosophila* species. They are probably maintained by some form of balancing selection.

5. Point mutations are changes in a single nucleotide. Replacement mutations change the amino acid sequence of the protein. Silent mutations do not

change the amino acid sequence. Silent mutations are frequently assumed to be neutral or nearly neutral with respect to natural selection.

6. Small insertions and deletions are common in noncoding DNA. Insertions or deletions in coding DNA are usually deleterious because they disrupt the mRNA coding sequence.

7. Gene duplication has been an important source of new genetic material. Sometimes, duplicated genes evolve new functions; more often, they lose their function and become pseudogenes.

8. The infinite alleles model postulates that each new mutation creates a unique allele (DNA sequence). The stepwise mutation model assumes stepwise increases in the net charge of a protein, or the number of repeats in minisatellite or microsatellite DNA. The infinite sites model considers each nucleotide site independently.

9. Mutation rates per locus vary over several orders of magnitude (Table 6.1). Typical mutation rates for protein-coding genes in eukaryotic organisms are about 10^{-5} to 10^{-6} per locus per generation. Microsatellite mutation rates are much higher, about 10^{-3}.

10. Deleterious mutations are more common than advantageous mutations, and mutations with small effects are more common than those with large or lethal effects.

11. An unknown proportion of mutations is selectively neutral or nearly neutral. The proportion of neutral mutations is probably higher in noncoding DNA sequences than in coding sequences.

12. Advantageous mutations are rare, but can lead to adaptive evolution of a population.

13. On the average, there is an inverse relationship between the deleterious effect of a mutation when homozygous and its effect when heterozygous (Figure 6.7).

14. When recurrent mutation and natural selection act in opposing directions, the mutant allele comes to a low equilibrium frequency as given by equation (6.3) or (6.4).

15. Severely deleterious mutations are efficiently eliminated by natural selection, but slightly deleterious mutations may be effectively neutral in small populations, and persist for many generations or become fixed by chance alone.

16. The mutational load of a population is the reduction in population fitness caused by recurrent deleterious mutation. At a single locus, the mutational load is equal to about twice the mutation rate.

17. Muller's ratchet predicts that deleterious mutations will accumulate in a population, leading to long-term reduction in population fitness and possible extinction. It has been documented in asexual populations, but its importance in sexual populations is less well understood.

18. Total genomic deleterious mutation rates may be quite high in some species, including humans, and raise the question of how some populations can survive in the face of constant accumulation of slightly deleterious mutations.

19. Most new mutations will be lost within the first few generations after they arise. This is true for advantageous mutations, as well as for neutral and deleterious mutations.

Problems

6.1. Postgate (1994) estimates that humans produce about 200 grams of feces per day and each gram contains about 10^7 new *E. coli* cells. Assuming there are 5 billion individuals on the planet, how many new *E. coli* cells are produced each day worldwide?

Now, use the estimate of total mutation rate in *E. coli* from Table 6.1, and the fact that the *E. coli* genome contains about 4.6×10^6 nucleotides, to estimate the number of times that each nucleotide in the *E. coli* genome mutates each day, worldwide.

6.2. Postgate (1994) made similar calculations, using an estimate of 10^{-7} mutations per cell per cell division. He estimated that each gene in the *E. coli* genome mutates about 2.5×10^9 times each day, worldwide. How does this compare to your estimate in Problem 6.1? Assume that each gene contains about 1000 nucleotides. Why are the estimates so different? Which do you think is more reasonable? The point of these calculations is to illustrate that the total number of new mutations produced in microorganisms is unimaginably huge. Even if the calculations are off by several orders of magnitude, every possible base substitution is occurring thousands of time every day!

6.3. Assume the following deleterious mutation rates (based on Table 6.2):

E. coli	1.7×10^{-4} mutations/genome/cell division
Drosophila	0.6 mutations/zygote (average of several studies)
Humans	3.0 mutations/zygote

Convert these to mutations/gene/cell division and to mutations/nucleotide/cell division. Assume genome sizes as follows:

E. coli	4300 genes, 4.6×10^6 nucleotides
Drosophila	13,600 genes, 1.8×10^8 nucleotides
Humans	40,000 genes, 3×10^9 nucleotides

Assume that there are about 36 cell divisions from zygote to gamete in *Drosophila* and 24 in humans (females).

Are these estimates consistent with the typical per locus estimates in Table 6.1? Speculate on why the deleterious mutation rate per gene per cell division is lower in *E. coli*.

6.4. Assume that an adult human has 10^{12} cells and that these cells result from repeated doubling by mitosis, with no cells dying. Use the information in Problem 6.3 to estimate the total number of new deleterious mutations in your body. Why is this number probably a large underestimate? What are the possible phenotypic effects of these somatic mutations?

6.5. Assume that a typical human chromosome has about 2000 genes and that the average mutation rate per gene per cell division is about 2×10^{-6}. What is the probability that there will be zero mutations after one cell division? One mutation? Two or more mutations?

6.6. Plot the frequency of a mutant allele versus time from an initial frequency of 0.01 and a typical mutation rate of 10^{-5} per locus per generation, ignoring back mutation and natural selection. On the same graph, plot frequency versus time for a typical microsatellite mutation rate of 10^{-3}. Use a log scale for generations, and plot up to 100,000 generations. (*Hint*: See Section 1.4.)

6.7. Would you expect the change in allele frequency to be faster with one-way mutation or with two-way mutation? Explain.

6.8. For the two-way neutral mutation model, show that $\tilde{q} \cong 0.91$ for any values of u and v as long as $u = 10v$. What is the equilibrium value if $u = 100v$?

6.9. Assume $h = 0.03$ for a lethal mutation, and $h = 0.4$ for a mildly deleterious mutation with $s = 0.05$. Compare the reduction in heterozygous viability (hs) for the two mutations. Speculate on the long-term evolutionary consequences of heterozygote intermediacy for slightly deleterious mutations.

6.10. In deriving the equation for mutation-selection equilibrium, we ignored back mutation. Why is this a reasonable assumption?

6.11. Calculate the equilibrium allele frequency ($\tilde{q}$) for each of the following fitness sets. Assume $u = 1 \times 10^{-5}$ in each case. What is the effect of partial dominance on the mutation-selection equilibrium? Explain in biological terms.

	w_{11}	w_{12}	w_{22}
a.	1	1	0.90
b.	1	0.95	0.90
c.	1	0.90	0.90
d.	1	1	0.98
e.	1	0.99	0.98

6.12. Show that $L \cong u$ for a completely recessive mutation.

6.13. Assume that the average mutation rate per locus is about 1×10^{-5} in both *Drosophila* (assume 13,600 genes) and humans (assume 40,000 genes). Assume all mutations are partially recessive. Calculate $\bar{w}_n$ and L_n for both species. How do these numbers compare with the estimates in Table 6.2? Discuss.

6.14. Huntington's disease is a fatal neurological disease that frequently does not strike until after reproductive age. It is caused by a dominant mutation. In one population, the frequency of the disease is about 3×10^{-6}. Using the estimate of the mutation rate from Table 6.1, and assuming mutation-selection equilibrium, what is the relative fitness of individuals with Huntington's disease?

6.15. Cystic fibrosis (CF) is a fatal disease caused by a recessive mutation. Individuals homozygous for the CF allele die before reproducing. The frequency of the CF allele is about 0.02 in some populations. Assume these populations are at equilibrium.

   a. Assume that the CF allele is maintained in the population by mutation-selection equilibrium and that heterozygotes have no reduction in viability. Estimate the mutation rate to the CF allele. Does this seem reasonable given what you know about mutation rates?

   b. Alternatively, assume that the CF allele is maintained in the population by overdominance, and that mutation is negligible compared to natural selection (i.e., u is much smaller than s_1 or s_2 and can be ignored as a factor determining equilibrium allele frequency). Estimate the selection coefficient against normal homozygotes.

   c. Which of these hypotheses seems more reasonable? Can you design an experiment to choose between them?

Genetic Drift

Variations neither useful nor injurious would not be affected by natural selection, and would be left a fluctuating element, as perhaps we see in certain polymorphic species, or would ultimately become fixed, owing to the nature of the organism and the nature of the conditions.

—Charles Darwin (1872)

Evolution can be due to different causes. Natural selection leads to adaptive evolution, as we have seen. However, evolution sometimes appears to be random and unpredictable, uncorrelated with any environmental or other selective conditions. In this chapter, we examine how the random process of genetic drift can lead to unpredictable evolution, sometimes with important consequences.

Genetic drift is random variation in allele frequencies from one generation to the next. It has two causes: (1) Mendelian segregation—a diploid individual contains two copies of every gene, but produces gametes containing only one copy; (2) finite population size—any population is a finite (random) sample of the gametes produced by the parent generation.

The importance of genetic drift in natural populations has always been somewhat controversial. This is because its short-term effect is expected to be minor except in very small populations, and it is difficult to observe allele frequency changes over many generations. For these reasons, the study of genetic drift is heavily dependent on mathematical models and computer simulations. However, as we shall see, molecular markers sometimes provide evidence that we cannot ignore the long-term effects of genetic drift.

The theory of genetic drift was worked out primarily by Sewall Wright in a series of papers in the 1930s and 1940s. The mathematical theory is quite sophisticated, and requires knowledge of advanced probability theory for a full understanding. In this chapter, we only summarize the main conclusions. See Wright's papers or Wright (1969) for the mathematical details.

We begin with an analogy. Consider a bucket containing 500 white marbles and 500 red marbles. The frequency of white marbles in the bucket is 0.50. Now draw a sample of 100 marbles at random. You may get 54 white and 46 red. The frequency of white marbles in the sample is 0.54, slightly different from that of the population (bucket). Now, put all of the marbles back in the bucket and draw another sample. This time you may get 47 white marbles; the frequency of white in this sample is 0.47.

Now repeat the experiment again, but this time taking two random samples of only ten marbles each. You may get three white marbles in one sample, and seven in the other. Note that the large sample frequencies are closer to the original frequency than the small sample frequencies.

This simple analogy illustrates three important points about random sampling:

1. The sample frequency usually differs from the population frequency, and the direction of the change is unpredictable. In each of the two experiments, the frequency of white marbles went up in one sample, and down in the other.

2. The magnitude of the change is inversely related to the sample size; the change was larger for the smaller samples.

3. Different samples from the same population will differ not only from the original population, but will differ from each other. The two samples of 100 marbles differed slightly from each other; the two samples of ten differed substantially.

Now consider the effect of repeated sampling. Consider the first sample of ten marbles. Imagine reconstituting the population (bucket) with 1000 marbles with the same frequencies as in the sample, that is, 300 white and 700 red. Now draw another sample, reconstitute the bucket based on that sample, and repeat the process over and over again. It can be shown from probability theory that eventually the sample will contain all white or all red marbles. The long-term effect of repeated sampling is to reduce or eliminate variation.

Genetic drift is due to repeated random sampling of gametes from a population. Think of the bucket as all of the gametes produced by a population of adults; the two colors of marbles represent alternative alleles. The sample represents those alleles that are actually included in the next generation. Based on the above principles of random sampling, we can then identify the four main aspects of genetic drift:

1. The direction of genetic drift is unpredictable. In any generation, the allele frequency may increase or decrease with equal probability.

2. The magnitude of genetic drift depends on population size. The smaller the population, the larger the change from one generation to the next.

3. The long-term effect of genetic drift is to reduce genetic variation within a population.

4. Genetic drift causes populations to diverge from one another.

Note the difference between the last two points: Genetic drift will *reduce* variation *within* a population, but will *increase* variation *among* populations.

Figure 7.1 illustrates these points with computer simulations. In each graph, the initial allele frequency was 0.5 and six replicate populations were followed for 50 generations. In Figure 7.1a, b, or c, the population size was 10, 30, or 100, respectively. Study this figure until you understand all of the points above.

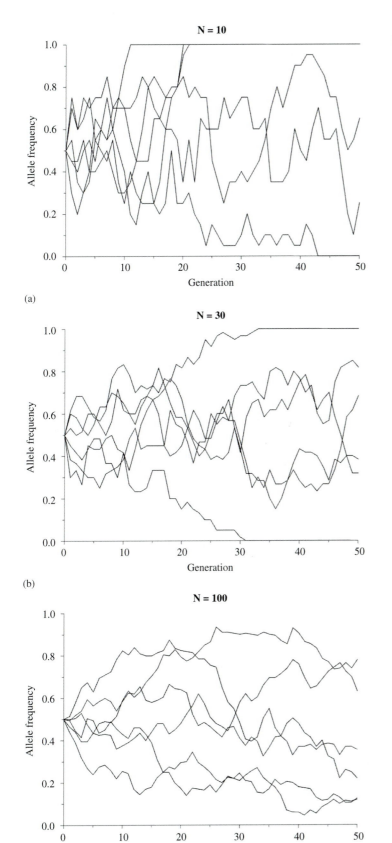

Figure 7.1 **Computer simulations of change in allele frequencies due to genetic drift.**
(a) $N = 10$. (b) $N = 30$. (c) $N = 100$.

In the rest of this chapter we consider each of these points in detail. We will formulate a model of genetic drift based on what we call an ideal population. Initially, we will make a number of assumptions about this ideal population:

1. We assume a diploid organism with sexual reproduction and Mendelian inheritance of an autosomal gene.
2. Generations are discrete and nonoverlapping.
3. Individuals mate at random, and self-fertilization is allowed.
4. Population size (N) is constant
5. The population is closed (no gene flow into or out of the population).
6. Mutation does not occur.
7. Natural selection does not occur.

We first examine how genetic drift affects a population under these simplifying assumptions. We will then see how genetic drift interacts with mutation and natural selection.

7.1 The Direction and Magnitude of Genetic Drift

Allele frequencies will change from one generation to the next. Let p_t be the frequency of an allele in generation t; then $1 - p_t$ is the frequency of all other alleles in the population. Look again at Figure 7.1. In any generation, the allele frequency increases about as often as it decreases. Now consider 100 identical populations, each with allele frequency of $p_t = 0.5$. Look at the allele frequency in each of these 100 populations one generation later; call this frequency p_{t+1}. Then p_{t+1} will not be the same in all 100 populations (Figure 7.2). In some it will be greater than 0.5 and in some will be less, but the average will be about the same as in the original population; in other words,

$$\overline{p}_{t+1} \cong p_t$$

where $\overline{p}_{t+1}$ is the average allele frequency in all populations in generation $t + 1$.

Another way to look at this is to consider p_{t+1} to be a random variable. Like any random variable, p_{t+1} has a mean (expected value) and a variance. From probability theory, it is easy to show that the expected value of p_{t+1} is p_t:

$$E(p_{t+1}) = p_t \tag{7.1}$$

and the variance of p_{t+1} is

$$V(p_{t+1}) = \frac{p_t(1 - p_t)}{2N} \tag{7.2}$$

where N is the population size. See Box 7.1 for the details.

Equation (7.2) allows us to get some idea of the magnitude of allele frequency changes from one generation to the next. To do so, we make use of the fact that about 95 percent of the time, p_{t+1} will be within two standard deviations of its mean. In other words, p_{t+1} will be between

$$p_t - 2\sqrt{\frac{p_t(1 - p_t)}{2N}} \quad \text{and} \quad p_t + 2\sqrt{\frac{p_t(1 - p_t)}{2N}}$$

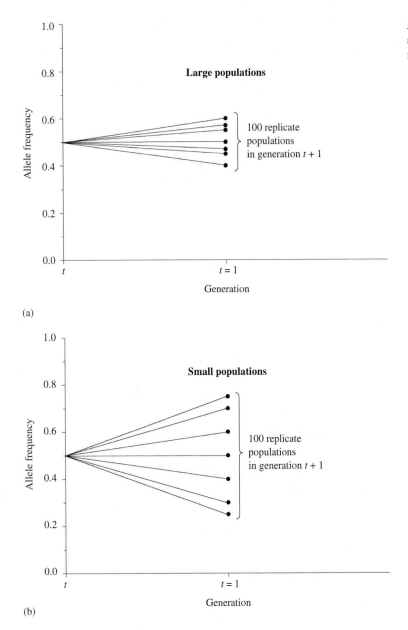

(a)

(b)

Figure 7.2 **Variation in allele frequency after one generation of genetic drift in 100 replicate populations (not all shown).** (a) Large populations. (b) Small populations.

about 95 percent of the time. For example, if the population size is 25 and $p_t = 0.5$, then the limits are approximately 0.36 to 0.64. Allele frequency change due to genetic drift alone will be outside this range only about 5 percent of the time.

Note that the magnitude of the change depends not only on the population size, but on the allele frequency as well.

7.2 Genetic Drift Reduces Genetic Variation Within a Population

Genetic drift reduces genetic variation in two ways: Heterozygosity is reduced, and alleles are lost. Before we can understand how heterozygosity is reduced, we must understand the concept of identity by descent.

Box 7.1   The mean and variance of p_{t+1}

First, review the binomial distribution (Section 4.4; Appendix A): If X is any binomially distributed random variable, then

$$\Pr(X = x) = \frac{n!}{x!(n-x)!} p^x (1-p)^{n-x}$$

where n is the number of trials, x is the the number of successes, and p is the probability of success in any individual trial. The mean (expected value) and variance of X are

$$E(X) = np$$

$$V(X) = np(1-p)$$

Now, consider a population of size N, with the frequency of an allele A_1 equal to p_t in generation t. To get alleles for generation $t + 1$, think of picking $2N$ alleles, one at a time. If Y is the number of A_1 alleles picked, then Y is a binomially distributed random variable with $2N$ trials and probability of success p_t. The mean and variance of Y are

$$E(Y) = 2Np_t$$

$$V(Y) = 2Np_t(1 - p_t)$$

The frequency of A_1 in generation $t + 1$ is

$$p_{t+1} = \frac{Y}{2N}$$

The expected value of p_{t+1} is

$$E(p_{t+1}) = E\left(\frac{Y}{2N}\right) = \frac{1}{2N}E(Y) = \frac{1}{2N}(2Np_t) = p_t$$

The variance of p_{t+1} is

$$V(p_{t+1}) = V\left(\frac{Y}{2N}\right) = \left(\frac{1}{2N}\right)^2 V(Y)$$

$$= \left(\frac{1}{2N}\right)^2 [2Np_t(1 - p_t)] = \frac{p_t(1 - p_t)}{2N}$$

(Recall that the variance of a constant times a random variable is the constant squared times the variance of the random variable; Appendix A.)

Identity by State and Identity by Descent

The classical definition of an allele is an alternative form of a gene. Examples are *round* versus *wrinkled* alleles in peas, *Fast* versus *Slow* alleles in *Drosophila Adh*, *A* versus *a* in generic genetic models, and so on. The implication is that these alleles are functionally different in some way. On the other hand, two copies of the same allele are assumed to be functionally equivalent. Such functionally equivalent copies are said to be **identical by state (ibs)**.

In a finite population, two individuals may have a common ancestor. If so, they may have received two copies of the same allele from that ancestor. Such copies are said to be **identical by descent (ibd)**. This means they are derived from the same ancestral DNA molecule by one or more cycles of DNA replication. Two alleles that are identical by descent will be identical in state unless mutation has altered the function of one.

Figure 7.3 illustrates these ideas. It follows the lines of descent for eight copies of a gene for three generations. All alleles labeled A (shaded) are identical by state; similarly, alleles marked a (unshaded) are identical by state, but different in state from A. In generation t, alleles A_1 and A_2 are identical by descent from generation $t - 1$. Similarly, a_2 and a_3 are identical by descent from generation $t - 1$, as are A_3 and A_4. Note that alleles A_4 and A_5 are *not* identical by descent from generation $t - 1$, but they *are* identical by descent from generation $t - 2$. This illustrates the point that whether two alleles are considered identical by descent depends on how far back in time we follow the population. If we go far enough back in time, all alleles in a finite population must be identical by descent. We discuss this idea in more detail later.

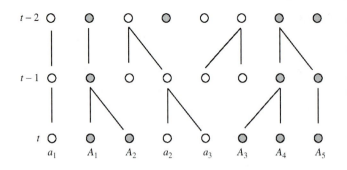

Figure 7.3 **Transmission of eight copies of a gene for three generations.** All alleles labeled *A* (shaded) are identical by state. Alleles marked *a* (unshaded) are identical by state to each other, but different in state from *A*. Pairs of alleles that trace to the same copy in the previous generation are identical by descent; for example, A_1 and A_2, or a_2 and a_3 in generation *t* are ibd from generation $t - 1$. Alleles A_4 and A_5 are not ibd from generation $t - 1$, but are ibd from generation $t - 2$.

Identity by state and identity by descent do not necessarily imply anything about identity of DNA sequences. Alleles that are identical by descent are identical by state and identical in DNA sequence only if no mutation has occurred since the common ancestor. However, mutation can change either the sequence or the function, so alleles that are identical by descent may not be identical in state or identical in sequence. Similarly, if two alleles are identical in state, this does not necessarily mean that their DNA sequences are identical, because mutations may have changed the DNA sequence without affecting the function.

The probability that two copies of a gene are identical by descent is called the **inbreeding coefficient**, symbolized by *f*. Consider a randomly mating diploid population of constant size *N*. The probability that two alleles chosen at random are copies of the same allele in the previous generation is $1/2N$. (Pick a marble from a bucket. Put it back. Pick again. If there are $2N$ marbles in the bucket, the probability of picking the same marble the second time is $1/2N$.) The probability that two alleles are identical by descent from the previous generation is therefore

$$f = \Pr(ibd) = \frac{1}{2N}$$

This describes the probability that two alleles will be identical by descent after one generation of random sampling, assuming no alleles in the previous generation were ibd. We should expect the probability of identity by descent to increase each generation, because each generation of sampling creates a new possibility of ibd to be added to the already existing probability. The recursion equation that describes how the inbreeding coefficient increases each generation is

$$f_{t+1} = \frac{1}{2N} + \left(1 - \frac{1}{2N}\right)f_t \qquad \textbf{(7.3)}$$

We will derive this equation and consider the inbreeding coefficient in detail in Chapter 8.

Reduction of Heterozygosity

We have defined homozygosity, *G*, as the frequency of homozygous individuals in the population. Alternatively, it can be interpreted as the probability that two alleles chosen at random are identical by state. Under random mating, these interpretations are equivalent. Genetic drift increases homozygosity because, in addition to being identical by state, two alleles may be identical by descent. As homozygosity increases, heterozygosity ($= 1 - G$) will decrease. We wish to quantify this decrease.

Under the assumptions of random mating, no mutation, no natural selection, and no gene flow into or out of the population, the only thing causing allele frequencies and genotype frequencies to change is genetic drift. Under these assumptions, genotype frequencies will be (approximately) the Hardy-Weinberg expected frequencies each generation. As the allele frequencies change, the genotype frequencies and heterozygosity will change, but on the average, the genotype frequencies will match the Hardy-Weinberg expected frequencies in each generation.

Under random mating, formation of genotypes is equivalent to sampling two alleles, with replacement, from the population of gametes. "With replacement" is because we have assumed that self-fertilization is possible; that is, an individual can receive both of its alleles from the same parent. The probability that the same allele is picked twice is the probability of identity by descent, or $1/2N$.

An individual can be homozygous in two ways (Figure 7.4). First, it can receive two alleles that are identical by descent from the previous generation. The probability of this is $1/2N$. Alternatively, it can receive two alleles that are not identical by descent from the previous generation (probability $= 1 - 1/2N$), but were identical by state in the previous generation. The probability that they were identical by state is just G_t, the homozygosity. Putting these together, we have

$$\text{Pr(homozygous in gen } t + 1) = \text{Pr(ibd from gen } t)$$
$$+ \text{Pr(not ibd from gen } t) \times \text{Pr(ibs in gen } t)$$

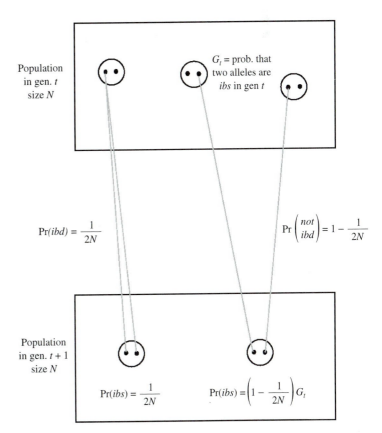

Figure 7.4 **Increase in homozygosity due to genetic drift.** Circles represent individuals in the population; dots within circles represent the two alleles of an individual. Light lines indicate Mendelian transmission. *ibd* = identical by descent; *ibs* = identical by state.

In symbols, this is

$$G_{t+1} = \frac{1}{2N} + \left(1 - \frac{1}{2N}\right)G_t$$

After some algebraic rearrangement, this becomes

$$G_{t+1} = \frac{1}{2N}(1 - G_t) + G_t$$

Now, convert to heterozygosity $(H = 1 - G)$ and simplify. The result is

$$H_{t+1} = H_t\left(1 - \frac{1}{2N}\right) \tag{7.4}$$

Equation (7.4) shows that heterozygosity decreases each generation by an amount that depends on N. This is a linear recursion equation, of the same form as several we have previously encountered. The solution is

$$H_t = H_o\left(1 - \frac{1}{2N}\right)^t \tag{7.5}$$

where H_o is the initial heterozygosity in the population. This equation describes how genetic drift reduces heterozygosity over time. Figure 7.5 illustrates the expected decay of heterozygosity for several values of N.

Equation (7.5) and Figure 7.5 may give the impression that heterozygosity should decrease monotonically in a predictable way. That would be misleading. In any given population allele frequencies will vary randomly; therefore heterozygosity will vary randomly. You must think of equation (7.5) as describing the *expected*, or *average*, decrease of heterozygosity in a large number of replicate populations.

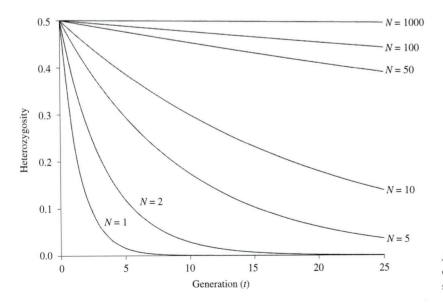

Figure 7.5 **Decay of expected heterozygosity for different population sizes, as predicted by equation (7.5).**

Loss of Alleles

Another way to view the reduction of genetic variation is to see that genetic drift causes alleles to be lost from a population. Look once again at Figure 7.1. In some populations, the allele became fixed; in others, it was lost. It can be shown that in all populations the allele frequency will eventually go to zero or one. If there are more than two alleles, a population will eventually lose all alleles but one.

We can ask two questions about this process: (1) What is the probability that a particular allele will be fixed (or lost) in a population? (2) How long is it likely to take?

Intuition should suggest that the more common an allele (state) is initially, the more likely it is to be randomly fixed in a population. Conversely, if an allele is initially rare, the probability that it will be fixed is low. In fact, the relationship is extremely simple. If p_o is the initial frequency of a particular allele, and $\Pr(fix)$ and $\Pr(loss)$ indicate the probabilities of fixation or loss of that allele, then

$$\Pr(fix) = p_o \tag{7.6}$$

and

$$\Pr(loss) = 1 - p_o \tag{7.7}$$

One way to understand the probability of fixation is to recognize that in each generation some alleles will not be replicated and passed on to the next generation. Eventually, all alleles will be identical by descent from a single allele that existed some time in the past. The probability that any particular allele (copy) in that past population will be the one that survives is $1/2N$. If there are n copies of that allele that are identical by state, the probability that one of them will be the survivor is $n/2N$. But this is just p_o, the frequency of that allele (state) in the initial population. Thus, the probability that a particular state will be fixed is the initial frequency of that state, p_o.

Consider the fate of a newly arisen neutral mutation. Assuming the mutation is not previously present in the population, its initial frequency is $1/2N$. In a population of 100 individuals, the probability that this new mutation will be fixed by genetic drift is $1/200 = 0.005$. Similarly, the probability that a new mutation will be lost is 0.995. For larger populations the probability of loss is even greater. Thus, new mutations are very unlikely to be fixed in any but the very smallest populations, unless favored by natural selection. In fact, even a favorable mutation is likely to be lost, as we shall see in Section 7.8.

How long will it take until an allele is fixed or lost? Let t_{fix} be the time to fixation for an allele that is destined for fixation and t_{loss} be the time to loss of an allele that is destined to be lost. Then t_{fix} and t_{loss} are random variables. Their means are approximately (Kimura and Ohta 1971)

$$\bar{t}_{fix} \cong -\frac{1}{p_o}[4N(1 - p_o)\ln(1 - p_o)] \tag{7.8}$$

and

$$\bar{t}_{loss} \cong -\frac{4Np_o}{1 - p_o}\ln p_o \tag{7.9}$$

For the example of a new mutation above, with $N = 100$ and $p_o = 1/200$, these work out to be approximately 399 and 10.6 generations for $\bar{t}_{fix}$ and $\bar{t}_{loss}$ respectively. (See also Problems 7.4 and 7.5.)

Note that $\bar{t}_{fix}$ and $\bar{t}_{loss}$ are *mean* times to fixation or loss, that is, they are the means of the random variables t_{fix} and t_{loss}. These random variables have large variances, so the actual times to fixation or loss may differ substantially from their means. Note also that $\bar{t}_{fix}$ and $\bar{t}_{loss}$ are the mean times for alleles that are actually fixed or lost, respectively.

Equations (7.6) through (7.9) assume a neutral allele and a constant population size. We will consider natural selection and changing population size in Section 7.8.

7.3 Genetic Drift Causes Populations to Diverge

Figure 7.1 illustrates the idea that genetic drift causes populations to diverge from one another. For each population size, all six of the replicate populations began with an allele frequency of 0.5, but had different allele frequencies after a few generations. Comparing Figure 7.1a, b, and c, we can guess that this divergence is faster for a smaller population, and that it increases with time.

One way to quantify this divergence is to look at the variance of allele frequency among replicate populations. In equations (7.1) and (7.2) we saw that, after one generation, the mean and variance of allele frequency among populations were

$$E(p_{t+1}) = p_t$$

$$V(p_{t+1}) = \frac{p_t(1 - p_t)}{2N}$$

It is easy enough to show that the average allele frequency over all populations will not change in the next generation. This makes intuitive sense. The allele frequency will increase in some populations and decrease in others, but there is no reason the change should favor any particular allele overall.

We can expect the variance to increase each generation. This is because some populations with high allele frequencies in generation 1 will produce even higher frequencies in generation 2, and some populations with low allele frequencies in generation 1 will produce even lower frequencies in generation 2. This divergence will continue each generation; see Figure 7.6. The increase in variance over time turns out to be

$$V_t(p) = p_o(1 - p_o)\left[1 - \left(1 - \frac{1}{2N}\right)^t\right] \qquad \textbf{(7.10)}$$

where $V_t(p)$ represents the variance of allele frequency among populations in generation t. See Crow and Kimura (1970) for a derivation of this equation.

What does equation (7.10) tell us? First, it confirms that variance increases each generation. (The part in square brackets approaches one as t increases.) The variance approaches a maximum of $p_o(1 - p_o)$. This equation also confirms our guess above that divergence will be faster among small populations. (The part in square brackets increases faster for smaller N.) Figure 7.7 illustrates the increase in variance for several population sizes.

Figure 7.6   **Increase in variation of allele frequency over three generations due to genetic drift.**

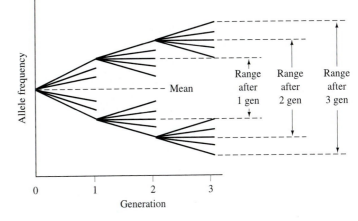

Equation (7.10) predicts that the allele frequencies in different populations will spread out from the initial frequency as time increases. But allele frequencies can spread only so far. When they reach 0 or 1, they are trapped (assuming no mutation). So we expect the number of populations that are fixed for one allele or the other to increase with time. Figure 7.8 shows an example of this. It shows the distribution of allele frequencies among 400 simulated populations after 1, 2, 4, 8, 16, and 32 generations, with $N = 8$ and $p_o = 0.5$. We indeed see "pileups" at $p = 0$ and $p = 1$. The distribution of allele frequencies under genetic drift was investigated by Sewall Wright in many papers. The problem was first solved completely by Kimura (1957, 1962). Figure 7.8 is a computer simulation based on their results, and illustrates the main conclusions. In this example, all populations have $N = 8$ and an initial allele frequency of 0.5. The distribution of allele frequencies spreads out each generation. By about N generations (8 in the figure) the distribution is essentially flat. By about $4N$ generations, the majority of populations have become fixed for one allele or the other. If the initial allele frequency

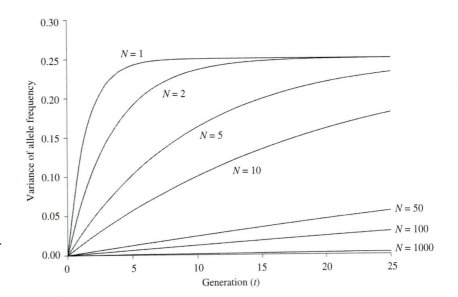

Figure 7.7   **Increase in variance of allele frequency for different population sizes, as predicted by equation (7.10).**   In each case, the initial allele frequency was $p_o = 0.5$.

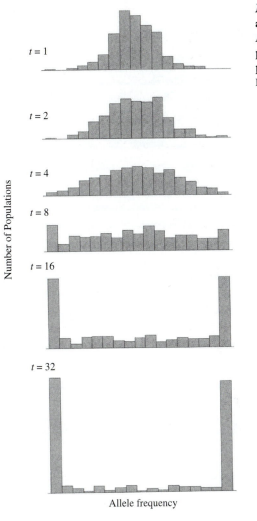

Number of Populations

$t = 1$

$t = 2$

$t = 4$

$t = 8$

$t = 16$

$t = 32$

Allele frequency

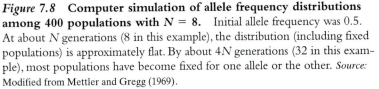

Figure 7.8 **Computer simulation of allele frequency distributions among 400 populations with $N = 8$.**   Initial allele frequency was 0.5. At about N generations (8 in this example), the distribution (including fixed populations) is approximately flat. By about $4N$ generations (32 in this example), most populations have become fixed for one allele or the other. *Source:* Modified from Mettler and Gregg (1969).

is near 0 or 1, the pileup is asymmetrical, as expected from equation (7.6), and the distribution never really becomes flat.

A famous experiment testing these predictions was conducted by Buri (1956). He studied 107 replicate laboratory populations of *Drosophila melanogaster* over 19 generations. Each population initially consisted of eight males and eight females, all heterozygous for two alleles (bw^{75}/bw) at a locus affecting eye color. These alleles appear to be neutral with respect to one another, so changes in frequency can be attributed to genetic drift. In each generation, eight males and eight females were chosen at random to produce the next generation. Thus, population size was maintained at 16 in each of the 107 lines. In each generation, the frequency of the bw^{75} allele was estimated in each line. The results are shown in Figure 7.9. Each of the 19 histograms plots the number of bw^{75} alleles (equivalent to frequency of that allele) versus the number of lines that had that many bw^{75} alleles. Recall that in generation 0, all 107 lines had 16 bw^{75} alleles ($p_o = 0.5$). After one generation, most lines were still close to this value. But as time increased, the lines diverged more and more from this initial value. By generation nineteen, about half the lines had lost one allele or the other.

Figure 7.9   **Dispersion of allele frequencies in laboratory populations of *Drosophila melanogaster* subject to genetic drift.**   Histograms represent generations 1 to 19, from top to bottom. Height of each bar is the number of populations. Population size was 16 in each generation. *Source:* From Ayala (1982), based on Buri (1956).

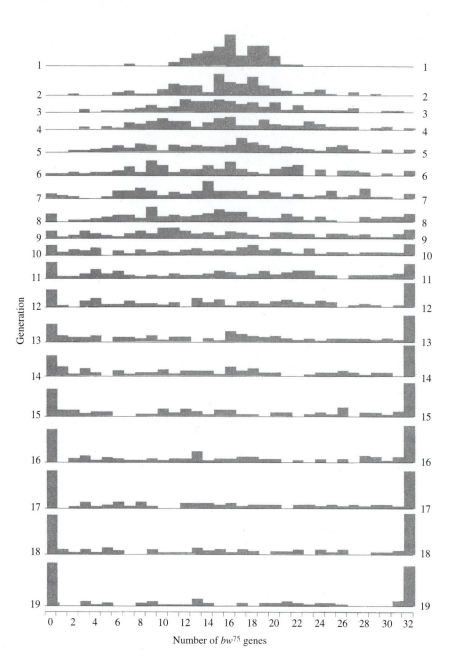

How do these experimental results compare with the theoretical predictions of genetic drift? First, equations (7.6) and (7.7) predict that the probability of fixation or loss of the bw^{75} allele is 0.5. After 19 generations, 28 lines had fixed the bw^{75} allele and 30 lines had lost it, so the results agree satisfactorily with the predictions. Second, the allele frequency averaged over all populations should remain constant at about 0.5, and the variance should increase as predicted by equation (7.10). Figure 7.10 compares the observed values to the predicted values. The mean stays reasonably close to the predicted value of 0.5 (Figure 7.10a), but the variance increases faster than predicted for the actual population size of 16 (Figure 7.10b, dashed line). The observed results seem to better fit the predicted increase in variance based on a population size of 9 (Figure 7.10b, dotted

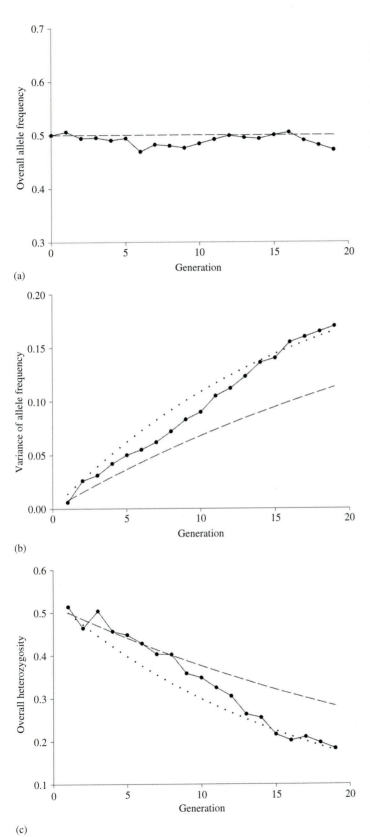

Figure 7.10 **Mean and variance of allele frequency, and heterozygosity over all 107 lines in Buri's (1956) experiment.** (a) Observed allele frequency (solid line) and expected allele frequency (dashed line) averaged over all lines. (b) Observed increase in variance (solid line) and expected increase, assuming $N = 16$ (dashed line) or $N = 9$ (dotted line). (c) Observed decrease in heterozygosity (solid line) and expected decrease, assuming $N = 16$ (dashed line) or $N = 9$ (dotted line). *Source:* Data from Buri (1956).

line), especially in the later generations. This is a fairly common observation: From the perspective of genetic drift, the population seems to be smaller than it actually is. The third prediction is that heterozygosity, averaged over all lines, should decrease as predicted by equation (7.5). Figure 7.10c compares the average heterozygosity with the prediction. Average heterozygosity is close to the predicted value for $N = 16$ in the earlier generations, but close to the predicted value for $N = 9$ in the later generations.

Both the observed increase in variance and observed decrease in heterozygosity suggest that, from the point of view of genetic drift, the population is behaving as if it were smaller than it actually is. This suggests that one or more of our initial assumptions is violated. In some situations, it is possible to compensate for this in such a way that we can still use our equations for genetic drift. This idea is discussed in the next section.

7.4 Effective Population Size

When comparing experimental outcomes with the predictions of the drift model, we frequently see that the population behaves as if it were smaller than it actually is. One example is the experiment of Buri (1956), described in the previous section.

We can sometimes allow for this by considering the **effective population size**, instead of the actual population size. The idea of effective population size, usually designated N_e, was introduced by Wright (1931). It is a number that, when substituted for N in our equations for an ideal population, describes the drift experienced by the actual (more complex) population.

For example, we have seen that in an ideal population, the decay of heterozygosity is described by equation (7.5). If one or more of the conditions of the ideal population are violated, the decay of heterozygosity will not be accurately described by this equation. We wish to find some N_e to substitute for N, so that equation (7.5) will describe the decay of heterozygosity in the actual population, at least to a good approximation.

The effective population size depends on the breeding structure of the population, the sex ratio, the mean and variance of family size, the history of population size, and other factors. The effective population size is usually smaller than the actual population size, a fact that is important in animal and plant breeding, and in conservation biology.

Why should we care if the effective population size is smaller than the actual size? Recall that heterozygosity decays faster, therefore homozygosity increases faster, for small populations. Most populations contain deleterious recessive alleles at low frequencies (Chapter 6). Homozygosity for these alleles can increase dramatically in small populations. The result is a higher proportion of unfit individuals and a lower mean fitness. This might have disastrous consequences for small endangered populations. In plant and animal breeding and in captive breeding programs, it is important to minimize these effects. We can sometimes do so by careful design of breeding schemes.

The effects of genetic drift can be described in three ways: change in allele frequency variance, equation (7.2); change in the inbreeding coefficient, equation (7.3); or change in heterozygosity, equation (7.4). There are various ways of determining the effective size of a population that deviates from an ideal population, depending on which of these aspects we are most interested in. The effective sizes that describe changes in variance, inbreeding, or heterozygosity are called variance effective number, inbreeding effective number, or eigenvalue effective number, respectively. Usually, the different methods give similar results if population size is relatively constant, but the differences can be important in small populations and in populations that are rapidly changing size. For example, in rapidly growing populations, the variance N_e is greater than the inbreeding N_e. We will not discuss these differences, but see Crow and Kimura (1970), Crow and Denniston (1988), Neigel (1996), and Crandall et al. (1999) for more information.

Here, we summarize a few examples of determining effective population size based on simple deviations from the ideal population. Formulas are given without derivation and their application described briefly. Most of the examples below are inbreeding effective numbers. For more thorough treatments of effective population size, see Wright (1969), Crow and Kimura (1970), or Falconer and Mackay (1996), in addition to the references above. Neigel (1996) summarizes methods of estimating N_e.

No Self-Fertilization

One of our initial assumptions was that an individual is capable of self-fertilization. We assumed this for arithmetic convenience only; it makes the algebra much simpler. However, most species of higher animals have separate sexes, and many plant species do not self-fertilize. The effect of self-fertilization on effective population size is trivial unless the population is very small. If self-fertilization is excluded, the effective population size is

$$N_e = N + \frac{1}{2} \qquad (7.11)$$

Note that the effective population size is larger than the actual population size, although trivially so. This makes intuitive sense because excluding self-fertilization reduces the probability that two alleles will be identical by descent.

Uneven Sex Ratio

Our model of genetic drift implicitly assumes equal numbers of males and females. If the sex ratio is uneven, we must adjust for this. If N_m and N_f are the numbers of breeding males and females, respectively, the effective population size is

$$N_e = \frac{4N_m N_f}{N_m + N_f} \qquad (7.12)$$

The effective population size depends heavily on the rarer sex. This is because the rarer sex contributes half the gametes to the next generation, even though it may comprise much less than half the population. This is important in many animal breeding programs where artificial insemination is practiced, and one male can fertilize a hundred or more females. For example, if $N_m = 1$ and $N_f = 100$, then $N_e \cong 4$.

Some species, for example, elephant seals, have a mating structure in which a few males have harems and mate with many females, while many males do not mate at all. Again, the effective population size is substantially less than the actual population size.

Variation in Population Size

Many populations fluctuate dramatically in size, due to either random or deterministic factors. This fluctuation can have important effects on the evolution of the population. If N_i represents the population size in the ith generation, then the effective population size after t generations is the harmonic mean of the population sizes:

$$N_e = \frac{t}{\sum\limits_{i=1}^{t} \frac{1}{N_i}} \tag{7.13}$$

The harmonic mean is always less than the arithmetic mean (unless all the N_is are equal). In fact, N_e is disproportionately affected by small population sizes (population bottlenecks). For example, if population sizes over four generations are 1000, 1000, 50, and 1000, the arithmetic mean is 762.5, whereas the effective population size is only 174.

Variation in Family Size

Even if the population size remains constant, every mating pair will not leave exactly two offspring. We suggested in Section 6.7 that family size can be modeled as a Poisson random variable. If K is the number of offspring left by a mating pair (family size), then in a stable population, the mean and variance of K are both 2. This assumes that all individuals have an equal probability of contributing to the next generation. Many factors, for example, fertility differences, can cause this assumption to be violated, making the variance of K greater than 2. The effective population size when the variance of family size is V_K is approximately

$$N_e \cong \frac{4N}{V_K + 2} \tag{7.14}$$

This reduces to $N_e = N$ if $V_K = 2$, as with a Poisson distribution of family size.

On the other hand, it is possible to increase the effective population size by reducing the variance of family size. This can be done in plant or animal breeding programs, or in captive breeding of endangered species. If, for example, all individuals in the breeding program contribute the same number of offspring, then $V_K = 0$, and $N_e = 2N$.

Variation in family size is one of the most important factors in determining effective population size. In natural populations, this variation usually reduces the effective population size to some number less than the actual size.

X-Linked, Y-Linked, and Mitochondrial DNA

Genes on the X and Y chromosomes of mammals are inherited differently from autosomal genes. Similarly, mitochondrial DNA is (usually) inherited only from the female parent. Genetic drift will affect these genes differently, and effective population sizes must account for these differences.

Many important genes in mammals are located on the X chromosome. The effective population size for X-linked genes is

$$N_e = \frac{9N_m N_f}{4N_m + 2N_f} \qquad \textbf{(7.15)}$$

where, as above, N_m and N_f are the numbers of males and females.

There are few genes on the Y chromosome of most mammals, but they can be useful for tracing gene flow (Chapter 9). For Y-linked genes, the effective population size is

$$N_e = \frac{N_m}{2} \qquad \textbf{(7.16)}$$

The difference between the forms of equations (7.16) and (7.15) is due to the difference in transmission of X-linked and Y-linked genes. Only males transmit Y-linked genes, but both males and females transmit X-linked genes, but at different rates. Equation (7.15) takes this into account.

Mitochondrial DNA (mtDNA) is the female analog of Y-chromosomal DNA. Mitochondrial DNA is inherited maternally, and there is no recombination among genes on the mtDNA molecule. This makes it useful for many kinds of studies. The effective population size for genes on the mtDNA molecule is

$$N_e = \frac{N_f}{2} \qquad \textbf{(7.17)}$$

There are $2N$ copies of an autosomal gene in a diploid population. If there are equal numbers of males and females, there are only $N/2$ copies of a Y-linked gene, because only half the population has a Y chromosome, and those that have it have only one copy. The same is true for mtDNA genes. Thus, there are only one-quarter as many copies of these genes in the population as there are copies of autosomal genes. Effective population size is only one-quarter as large as it is for autosomal genes (again, assuming equal numbers of males and females). Therefore, gene diversity (analogous to expected heterozygosity; see Section 3.2) should decay faster and allele frequency variance should increase faster than for autosomal genes.

Another way to look at it is that for an autosomal gene, the probability that two genes chosen at random are identical by descent is $1/2N$. But for Y-linked or mtDNA genes, the probability is $1/N_m$ or $1/N_f$. If there are equal numbers of males and females, these probabilities are four times higher than for autosomal genes.

We will continue to use N in our models, but whenever you see N, you should think effective population size.

7.5 Genetic Drift in Natural Populations

We have seen that the results of laboratory experiments are consistent with the theory of genetic drift. Can we extrapolate these results to natural populations?

In some ways, this is still a controversial question. As we shall see, it is frequently difficult to distinguish the effects of genetic drift from those of natural selection.

One of the first studies of genetic variation in natural populations was conducted by Dobzhansky and Queal (1938). They surveyed the frequencies of third chromosome inversions in 11 geographically isolated populations of *Drosophila pseudoobscura*. Frequencies ranged from 0.51 to 0.88 for the Arrowhead inversion, 0.02 to 0.20 for the Chiricahua inversion, and 0.08 to 0.39 for the Standard inversion. This geographic heterogeneity is statistically significant. The data show "no geographical trend or regularity in these variations, so that populations from adjacent localities may be more different from each other than populations from remote ones." (Dobzhansky and Queal 1938, p. 248). The authors reject mutation, gene flow, and natural selection as possible explanations of this pattern, and conclude that "By far the most probable explanation of the observed differences between populations of the separate mountain ranges is that the frequency of a gene or a chromosome structure is subject to random fluctuations." (p. 249).

This interpretation has not held up to subsequent studies. Dobzhansky and Queal's statistical techniques to detect natural selection were either ineffective or wrong, and many subsequent studies have demonstrated that natural selection can and does act on these inversions. Much evidence suggests that inversion polymorphisms in natural populations of *Drosophila* are maintained, at least in part, by some form of balancing selection (review in Powell 1997). Nevertheless, the apparent randomness of Dobzhansky and Queal's results remains suggestive. This study illustrates a common difficulty in studying natural populations: Observed patterns of variation can often be explained by either genetic drift or natural selection, and it can be difficult or impossible to choose between them without further study.

Other examples of the long-standing controversy over the importance of genetic drift in natural populations include shell color and banding patterns in the snails *Cepaea nemoralis* and *C. hortensis*, color morphs of the moth *Panaxia dominula*, wing patterns in the butterfly *Maniola jurtina*, and flower color in the desert plant *Linanthus parryae*. These early studies, and others, are summarized in the classic books by Mayr (1963) and Ford (1975).

As long ago as 1963, Ernst Mayr wrote: "Virtually every case quoted in the past as caused by genetic drift due to errors of sampling has more recently been reinterpreted in terms of selection pressures." (Mayr 1963, pp. 207–208). Coyne et al. (1997) reviewed the evidence for genetic drift in natural populations and came to essentially the same conclusion: "Our inability to cite many cases in which drift has affected characters beyond protein and DNA sequence does not mean that sampling events are unimportant, only that they are difficult to document" (Coyne et al. 1997, p. 656).

As Coyne et al. indicate, the best examples of genetic drift in natural populations come from studies of protein and DNA variation. Many patterns of molecular variation within and among species are best explained by the interactions between mutation and genetic drift. This is especially true for some DNA sequences.

Recall that silent (synonymous) mutations are changes in the coding sequence of a gene that do not change the amino acid sequence of the protein, and replacement (nonsynonymous) mutations do change the amino acid sequence. These can result in polymorphisms—variation among individuals within a species, or substitutions—fixed differences between species. From a table of the standard genetic code, one can calculate that 134 of 549, or about 24 percent of all possible single base changes (excluding stop codons) will be silent. Therefore, if all mutations in the coding sequence were neutral (not subject to natural selection), we would expect to find about 24 percent of all single nucleotide polymorphisms within a species to be silent, and about 24 percent of substitutions between species to be silent.

In virtually every protein-coding gene studied, silent polymorphisms are observed much more frequently than replacement polymorphisms. In a classic study, Kreitman (1983) sequenced 768 nucleotides in the coding sequence for alcohol dehydrogenase in *Drosophila melanogaster*. He compared 11 copies of this sequence from different individuals and found that 14 of the 768 sites were polymorphic (not the same in all 11 copies). Of those 14 variable sites, 13 were silent polymorphisms. These results suggest that most replacement mutations have been eliminated by natural selection, and the silent polymorphisms are neutral, with their frequencies determined primarily by genetic drift.

Li (1997) and Graur and Li (2000) summarized substitution rates between humans and rodents for 47 protein-coding genes (Table 7.1). The average rate of silent substitutions was 3.51×10^{-9} substitutions per nucleotide site per year, compared to 0.74×10^{-9} for replacement substitutions. The most widely held explanation for this pattern is that silent mutations are neutral or nearly neutral with respect to natural selection, and therefore silent substitutions represent random fixation of neutral mutations by genetic drift. On the other hand, replacement mutations are more likely to adversely affect the function of a protein and be eliminated by natural selection, resulting in lower substitution rates.

Pseudogenes are nonfunctional genes that have probably arisen as copies of functional genes. Pseudogenes do not code for functional proteins, and are almost certainly not subject to natural selection. When we compare levels of variation in pseudogenes with variation in their homologous functional genes, we find that pseudogenes are almost always more variable. Graur and Li (2000) estimated substitution rates in pseudogenes to be about 3.8×10^{-9} substitutions per nucleotide site per year. This is close to their estimate for silent substitutions in coding regions of functional genes (Table 7.1). Again, the accepted explanation for this pattern is that pseudogenes are not subject to natural selection, and their substitution rates are determined primarily by genetic drift.

We conclude that much of the observed variation in DNA sequences is best explained by genetic drift of neutral mutations. Note that the patterns described above provide evidence for both natural selection and genetic drift. If mutations *occur* randomly, but we *observe* silent polymorphisms and substitutions more frequently than replacement ones, then we must conclude that natural selection has eliminated most replacement mutations (purifying selection). Whether the replacement polymorphisms and substitutions *that we see* are due to natural selection is another question, to be addressed in Chapter 10.

TABLE 7.1   Rates of replacement and silent substitutions in protein-coding genes of mammals.

Rates are (substitutions $\times 10^{-9}$) per nucleotide site per year. Averages and standard deviations are based on 45 proteins. *Source:* Summarized from Li (1997).

Gene	Number of Codons	Replacement Substitution Rate	Silent Substitution Rate
Histone 3	135	0.00	4.52
Histone 4	102	0.00	3.94
Ribosomal protein S14	150	0.02	2.16
Robosomal protein S17	134	0.06	2.69
Actin α	376	0.01	2.92
Myosin β, heavy chain	1933	0.10	2.15
Erythropoietin	191	0.77	3.56
Growth hormone	189	1.34	3.79
Insulin	51	0.20	3.03
Insulin C-peptide	31	1.07	4.78
Interleukin I	265	1.50	3.27
Luteinizing hormone	140	1.05	2.90
Parathyroid hormone	90	1.00	3.47
Relaxin	53	2.59	6.39
Somatostatin-28	28	0.00	3.10
α-globin	141	0.56	4.38
β-globin	146	0.78	2.58
Myoglobin	153	0.57	4.10
Apolipoprotein E	291	1.10	3.72
Apolipoprotein A-I	235	1.64	3.97
Ig V_H	100	1.10	4.76
Ig k	106	2.03	5.56
Interferon α1	166	1.47	3.24
Interferon β1	159	2.38	5.33
Albumin	590	0.92	5.16
Aldolase A	363	0.09	2.78
Amylase	506	0.63	3.42
Creatine kinase M	380	0.15	2.72
Fibrinogen γ	411	0.58	4.13
Glucagon	29	0.00	2.36
Glyceraldehyde-3-phosphate dehydrogenase	332	0.20	2.30
Lactate dehydrogenase A	331	0.19	4.06
Prion protein	224	0.29	3.89
Thymidine kinase	232	0.43	3.93
Average ± standard deviation		0.74 ± 0.67	3.51 ± 1.01

7.6 Founder Effects and Population Bottlenecks

Mayr (1963) defined the founder effect as "the establishment of a new population by a few original founders (in an extreme case, by a single fertilized female) which carry only a small fraction of the total genetic variation of the parental population." For example, most island populations are established by only a few individuals from the mainland. Several religious groups in the United States were originally founded by only a few individuals from Europe.

The theory of genetic drift predicts two consequences of the founder effect: First, populations founded by a small number of individuals will be less variable than the original population. Second, allele frequencies in the founder population will be different from the original population.

Founder Effect Reduces Genetic Variation

One consequence of the founder effect is that rare alleles are likely to be lost. Let p be the frequency of some allele in the ancestral population, and N_f be the size of the founder population. The probability that this allele will not be represented in the founder population is the probability that it is not present in N_f individuals, or $2N_f$ gametes:

$$\Pr(loss) = (1 - p)^{2N_f}$$

Figure 7.11 illustrates this probability for several allele frequencies. Any allele with a frequency of 0.01 or less will probably be lost if the founder population is small.

This is the probability that the allele will be lost during the founder event. Even if an allele makes it into the founder population, the probability that it will be lost within a few generations is high if the allele is rare and the founder population remains small (Section 7.2).

Heterozygosity is also decreased by the founder effect. If we think of genetic drift as random sampling over time, the founder effect can be thought of as random

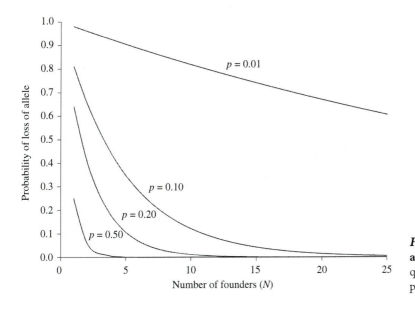

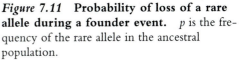

Figure 7.11 **Probability of loss of a rare allele during a founder event.** p is the frequency of the rare allele in the ancestral population.

sampling over space. In other words, the founder population is simply a new generation in a different place. We can therefore reinterpret equation (7.4) as

$$H_f = H_a\left(1 - \frac{1}{2N_f}\right)$$

where H_a and H_f represent heterozygosity in the ancestral and founder populations, respectively, and N_f is the size of the founder population. Heterozygosity in the founder population is reduced by an amount that depends on its size. However, the effect is minimal except for very small founder populations. If the founder population is ten or larger, the founder heterozygosity is at least 95 percent of the ancestral heterozygosity (Figure 7.12).

To summarize, heterozygosity is reduced by a founder effect, but the effect is minimal unless the founder population is very small. The most important consequence of the founder effect is that rare alleles are likely to be lost, thus decreasing allelic diversity in the founder population.

It is commonly assumed that island populations are frequently founded by a very small number of individuals. This creates an immediate decrease in genetic variation, as we have just discussed. Furthermore, once established, island populations are likely to remain smaller than mainland populations, so genetic variation in small island populations will also decay faster than in larger mainland populations. Therefore, for both reasons, we can predict that island populations will have lower genetic variation than similar mainland populations. Frankham (1997) tested this prediction by comparing levels of allozyme and DNA variation in island versus mainland populations of animals and plants. The results confirm the prediction. In 165 of 202 comparisons based on allozymes, island populations had lower heterozygosity than mainland populations of the same species. The average reduction was about 29 percent. In comparisons based on DNA sequences, 29 of 29 were in the predicted direction. Clearly, island populations tend to have less genetic variation than mainland populations. Both the founder effect and continued small population size probably contribute.

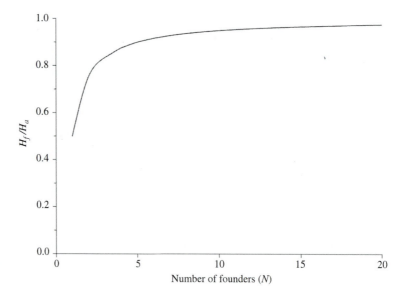

Figure 7.12 **Loss of heterozygosity during a founder event.** H_f/H_a **is the ratio of heterozygosity in the founder population to heterozygosity in the ancestral population.** For a founder population larger than ten, the heterozygosity is at least 95 percent of the ancestral population.

The founder effect is not necessarily limited to island populations. The European starling (*Sturnus vulgaris*) was introduced to North America in 1890–1891. The founding population is thought to be about 100 individuals (Cabe 1998). Given this relatively large founder size, one would expect to see little or no loss of heterozygosity in North American populations compared to European populations. That is exactly what Cabe (1998) found. Allozyme heterozygosity was 0.031 in both North American and British populations. However, North American populations had lost as many as 42 percent of alleles at variable loci, compared to British populations. This is consistent with the prediction above that rare alleles will be lost, and with computer simulations of Nei et al. (1975), who concluded that a high proportion of rare alleles will be lost, even with large founder sizes.

Drosophila buzzatii is native to parts of Brazil and Argentina, and has been introduced to other parts of the world. It was introduced to Australia, probably in the early 1930s, and the founder population was probably very small (Dodd 1940; Barker 1982). Several studies show evidence of the founder effect in Australian populations: Barker et al. (1985) found that Australian and South American populations did not differ significantly in allozyme heterozygosity, but that several rare alleles present in South America are absent in Australia. Knibb et al. (1987) found similar results for chromosomal inversions. Halliburton and Barker (1993) found no mtDNA variation in Australian populations, whereas Rossi et al. (1996) found normal amounts of mtDNA variation in South American populations. The mtDNA haplotype found in Australia was the most common one in the South American populations (Halliburton and Fontdevila, unpublished).

Founder Effect Changes Allele Frequencies

Even if an allele is represented in the founder population, its frequency is likely to be different from the ancestral population. We can use equation (7.2) to get some idea of how different. If we think of a founder population as simply a new generation in a different place, then according to equation (7.2), the variance of allele frequency in the founder population will be

$$V(p_f) = \frac{p_a(1 - p_a)}{2N}$$

where p_f represents the allele frequency in the founder population and p_a is the frequency in the ancestral population. If N is small, then the variance can be large. In other words, the allele frequencies in a small founder population can be very different from the allele frequencies in the ancestral population.

Human populations often show founder effects. Almost any review of the literature will turn up dozens of studies claiming the founder effect as a cause of unusual allele frequencies in isolated human populations. For example, the Amish religious group of Lancaster County, Pennsylvania, has a high frequency of Ellis–Creveld syndrome, sometimes called six-fingered dwarfism. In one study, 82 individuals with the disease all traced their ancestry to a single man and his wife (McKusick et al. 1964). Spiess (1977) describes several of the classic cases of founder effect in human populations. For more recent examples, see Laake et al. (1998), Vanschothorst et al. (1998), or almost any issue of any human genetics journal.

Finally, we summarize one last example that illustrates several important aspects of the founder effect and genetic drift. The common myna (*Acridotheres tristis*) is a bird related to the starling. It is native to India, and has been introduced to other parts of the world in the last 100 years or so. Baker and Moeed (1987) compared native Indian and introduced populations at 39 allozyme loci. Their results are not surprising: The average number of alleles per locus, number of polymorphic loci, and heterozygosity were all significantly lower in the introduced populations. The reduction of heterozygosity was slight (0.05 in introduced populations *vs.* 0.06 in natives), but statistically significant. Several alleles that were rare in India were absent in the introduced populations. All these results are predicted by the theory of genetic drift.

Perhaps the most interesting aspect of this study concerns the differentiation among populations. One way of describing differentiation among populations is with a quantity known as F_{ST}. We will discuss F_{ST} in detail in Chapter 9, but for the present context, you only need to understand that high values of F_{ST} indicate significant genetic differentiation among populations. Baker and Moeed found little differentiation among the Indian populations; F_{ST} averaged over all loci was 0.032. However, the introduced populations were genetically different from one another, as well as from the Indian populations. The average F_{ST} among introduced populations was 0.123. Baker and Moeed conclude: "The introduced populations, however, are characterized by significantly reduced levels of within-population variation and significantly enhanced among-population variation" (Baker and Moeed 1987, pp. 534–535). That is precisely what the theory of genetic drift predicts.

Population Bottlenecks

A bottleneck occurs when a population is suddenly reduced to a very small size. Think of a bottleneck as a founder event in which the founder population and the ancestral population are the same. The genetic consequences of a bottleneck are identical to those of a founder effect: Heterozygosity is reduced, rare alleles are lost, and allele frequencies are changed.

Several experimental studies have confirmed these predictions. In one experiment, McCommas and Bryant (1990) subjected houseflies to bottlenecks of 2, 8, or 32 individuals. They compared allozyme variation at four polymorphic loci before and after the bottlenecks. With a single exception, they detected significant decreases in heterozygosity only after the $N = 2$ bottlenecks. However, the number of alleles decreased significantly for all bottleneck sizes. In most cases, their results closely matched the predictions of genetic drift theory.

Packer et al. (1991) describe a bottleneck in a population of lions in the Ngorongoro Crater, Tanzania. This population is isolated from other lion populations, and in 1962 suffered an epidemic disease that nearly exterminated it. Packer et al. document that the population today is descended from seven females and eight males that survived the epidemic. They found less genetic variation in the Crater population than in the large "control" population in the Serengeti. Four polymorphic loci were detected in the Crater lions, compared to seven in the Serengeti population. Heterozygosity was 0.022 compared to 0.033 in the Serengeti lions. Furthermore, Yuhki and O'Brien (1990) found that, for the genes of the major histocompatibility complex (MHC), the Crater lions had only about one-third the variation of the Serengeti lions. Packer et al. (1991) also conducted simulation studies and concluded that the Crater population must have been through more than one bottleneck; a single bottleneck was insufficient to explain the observed reduction of genetic variation.

A low level of genetic variation has frequently been interpreted to indicate a bottleneck in the recent history of a population. For example, Bonnell and Selander (1974) found no allozyme variation in a sample of 159 northern elephant seals (*Mirounga angustirostris*). They attributed this to a bottleneck caused by hunting, which reduced the population to about 20 individuals by 1890. Similarly, O'Brien et al. (1983, 1985) found essentially no genetic variation in a large study of South African cheetahs (*Acinonyx jubatus jubatus*). They attributed this to one or more bottlenecks in the recent past, although this interpretation is somewhat controversial (Pimm 1991; Merola 1994; O'Brien 1994).

Just because a population has little or no genetic variation, we cannot necessarily conclude that it has been through a recent bottleneck or founder effect. A small isolated population of the eastern barred bandicoot (*Perameles gunnii*) in Australia showed no allozyme variation at 27 loci. However, two large populations also showed no variation at these loci (Sherwin et al. 1991).

Recovery from a Bottleneck or Founder Event

Equation (7.4) describes the immediate loss of heterozygosity when a population suddenly crashes from a large size to small. If the population remains small, heterozygosity will continue to decline, as described by equation (7.5) and Figure 7.5. However, if the population immediately begins to grow rapidly, the decay of heterozygosity may be much slower. We can rearrange equation (7.4), and solve for the change in heterozygosity, $\Delta H_t = H_{t+1} - H_t$:

$$\Delta H_t = -\frac{1}{2N_t}$$

(The negative sign indicates that heterozygosity will decrease each generation.) Now imagine a population growing rapidly, so that N_t increases each generation. The change in heterozygosity, ΔH_t, will rapidly approach zero and heterozygosity will stabilize at a new value, lower than the original. The faster the population grows, the sooner heterozygosity will stabilize, and the less the total loss will be. Finally, if there is no gene flow into the population, the only way heterozygosity can increase is by the appearance of new mutations. This will take a very long time.

Nei et al. (1975) have studied this process theoretically and with computer simulations. They considered a population that grows logistically after passing through a bottleneck, and assumed that all mutations are unique (infinite alleles model) and neutral with respect to natural selection. Their general results are summarized in Figure 7.13. The loss of heterozygosity depends on the size of the bottleneck and on r, the intrinsic rate of increase (a measure of population growth rate) after the bottleneck. The figure shows the three stages described above: (1) The loss of heterozygosity while the population size is small. The rate of loss (slope of the initial decline) is less for larger bottleneck sizes and higher r. Note that nearly all the loss of heterozygosity occurs in the first 10 or so generations. (2) Stabilization of heterozygosity. The new heterozygosity value (region

Figure 7.13 **Loss and recovery of heterozygosity after a population bottleneck.** Solid lines are $N = 2$ and dashed lines are $N = 10$. r is the intrinsic rate of increase. *Source:* From Nei et al. (1975).

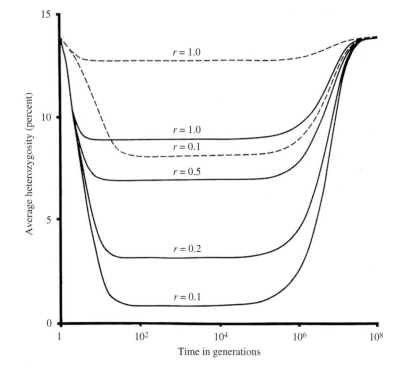

where the curves are approximately horizontal) also depends on N and r. Populations with higher r and higher N stabilize at higher heterozygosities (although still less than the original). (3) Finally, heterozygosity starts to increase again due to new mutations. This takes hundreds of thousands of generations, and the theory predicts that it will be millions of generations before the original level of heterozygosity is restored.

To summarize, there are two main conclusions from this model: (1) If a population grows rapidly after going through a bottleneck, the reduction in heterozygosity may be small, even if the bottleneck size is extremely small. (2) It will take a *very* long time to restore the initial level of heterozygosity. Nei et al. give a rule of thumb that the number of generations required is about the reciprocal of the neutral mutation rate, or about 10^8 generations (plus or minus an order of magnitude). They claim that this is frequently longer than the evolutionary history of a species. In other words, a population will essentially never return to its original heterozygosity.

7.7 Genetic Drift and Mutation

Mutation creates new alleles, increasing heterozygosity. Genetic drift eliminates alleles, decreasing heterozygosity. Under what conditions will these two processes balance one another? Is there an equilibrium value of heterozygosity determined by the interaction between mutation and genetic drift?

We assume the infinite alleles model, in which every mutation creates a unique allele (Section 6.2), and consider only mutations that are neutral with respect to natural selection. The logic is similar to that used to derive heterozygosity under drift alone, except we must allow for mutation. Refer to Figure 7.14.

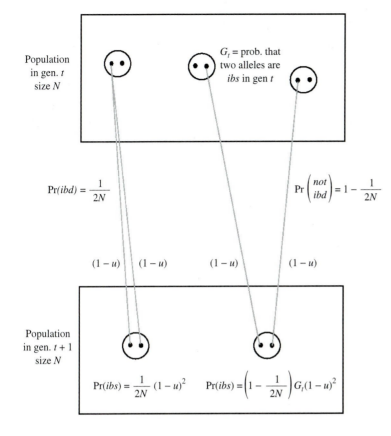

Figure 7.14 **Change in homozygosity due to interaction between mutation and genetic drift.**   Circles represent individuals in the population; dots within circles represent the two alleles of an individual. Light lines indicate Mendelian transmission. *ibd* = identical by descent; *ibs* = identical by state.

Two alleles can be identical by state in two ways: First, they may be identical by descent from the previous generation, and neither of them has mutated. Second, they may not be identical by descent, but were identical in state in the previous generation, and neither has mutated. In other words,

$$\Pr\!\begin{pmatrix} \text{ibs} \\ \text{in } t+1 \end{pmatrix} = \Pr\!\begin{pmatrix} \text{ibd} \\ \text{from } t \end{pmatrix} \times \Pr\!\begin{pmatrix} \text{neither} \\ \text{mutates} \end{pmatrix}$$

$$ + \Pr\!\begin{pmatrix} \text{not ibd} \\ \text{from } t \end{pmatrix} \times \Pr\!\begin{pmatrix} \text{ibs} \\ \text{in } t \end{pmatrix} \times \Pr\!\begin{pmatrix} \text{neither} \\ \text{mutates} \end{pmatrix}$$

If u is the mutation rate to neutral alleles, then the probability that one allele does not mutate is $(1 - u)$, and the probability that neither mutates is $(1 - u)^2$. So the recursion equation becomes

$$G_{t+1} = \left(\frac{1}{2N}\right)(1 - u)^2 + \left(1 - \frac{1}{2N}\right)(G_t)(1 - u)^2$$

Collecting terms, we get

$$G_{t+1} = \left[\frac{1}{2N} + \left(1 - \frac{1}{2N}\right)G_t\right](1 - u)^2 \qquad \textbf{(7.18)}$$

To find the equilibrium, set $G_{t+1} = G_t = \widetilde{G}$ in the above equation and solve for $\widetilde{G}$. After a bit of algebraic manipulation, we get

$$\widetilde{G} = \frac{(1-u)^2}{2N - (2N-1)(1-u)^2}$$

Now consider the factor $(1-u)^2 = 1 - 2u + u^2$. Recall that u is a small number (usually 10^{-6} or less). Therefore, u^2 is even smaller and can be neglected, and $(1-u)^2 \cong 1 - 2u$. So the above equation can be approximated as

$$\widetilde{G} \cong \frac{1-2u}{2N - (2N-1)(1-2u)}$$

which simplifies to

$$\widetilde{G} \cong \frac{1-2u}{4Nu + 1 - 2u}$$

We can make a further approximation, since $1 - 2u \cong 1$:

$$\widetilde{G} \cong \frac{1}{4Nu + 1}$$

(Note we could not make this second approximation earlier, since we had terms involving the product of N and u, and this product is not necessarily small.)

Finally, since $H = 1 - G$, the equilibrium heterozygosity is

$$\widetilde{H} \cong \frac{4N_e u}{4N_e u + 1} \tag{7.19}$$

Here, we have used N_e to make the dependence on effective population size explicit.

Figure 7.15 shows how equilibrium heterozygosity varies with $4N_e u$. If $4N_e u$ is small (less than about 0.01), then heterozygosity is near zero (less than about 0.01). Genetic drift predominates, and nearly all genetic variation is lost from the population. On the other hand, if $4N_e u$ is large (greater than about 4), then heterozygosity is near one and nearly all alleles in the population are different in state; mutation predominates.

Values of $4N_e u$ between about 0.04 and 0.2 correspond to heterozygosity values frequently found in allozyme studies of natural populations (Table 2.5). Figure 7.15b magnifies this area of the graph. Equation (7.19) suggests that we might use these heterozygosity estimates to estimate $N_e u$. This would give us some idea of effective population size or the neutral mutation rate if we knew the other (Problem 7.14). However, this would be assuming that all allozyme variation is neutral and that the population is at mutation–drift equilibrium.

In the above derivation, we assumed the infinite alleles model, in which every mutation creates a unique allele. For minisatellite and microsatellite loci, this assumption is unwarranted, and the stepwise mutation model (Section 6.2) is often assumed. Under the stepwise mutation model, the equilibrium heterozygosity is

$$\widetilde{H} = 1 - \frac{1}{(1 + 8N_e u)^{1/2}} \tag{7.20}$$

See Ohta and Kimura (1973) or Kimura (1983).

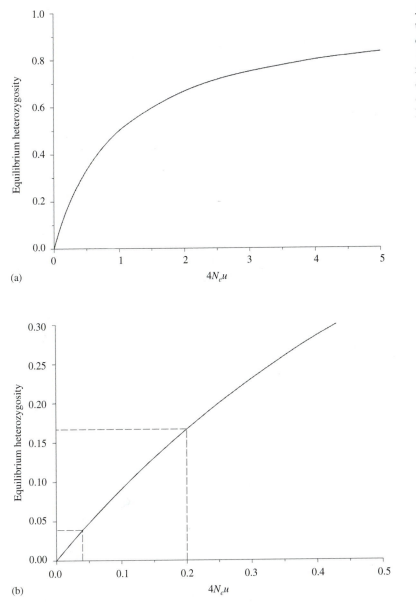

Figure 7.15 **Heterozygosity as a function of neutral mutation and genetic drift, as predicted by equation (7.19).** (a) Equilibrium heterozygosity. (b) A magnified view of the lower-left corner of (a). The dashed lines in (b) indicate approximate range of values of $4N_e u$ that correspond to observed heterozygosities from allozyme studies.

7.8 Genetic Drift and Natural Selection

Natural selection causes a systematic and predictable change in allele frequencies. Genetic drift causes allele frequencies to vary randomly from generation to generation. How do these opposing processes interact? Figure 7.16a illustrates natural selection in an infinite population with no genetic drift. The frequency of the favored allele increases each generation in a predictable way, as described by equation (5.7). Now consider what might happen in a finite population. If the population is relatively large, natural selection will still cause the favored allele to increase in frequency, but genetic drift will cause some variation around the predicted curve (Figure 7.16b). On the other hand, if the population size is small, the fluctuations around the predicted curve will be larger, and the allele frequency may even change in the wrong direction (Figure 7.16c). The idea is that we can

Figure 7.16 **Interaction between natural selection and genetic drift.**   (a) With no genetic drift, the favored allele increases in a predictable way. (b) In a large population, the favored allele increases predictably (solid line), but with some variation around the theoretical curve (dashed line). (c) In a small population, genetic drift overpowers natural selection and the favored allele changes unpredictably (solid line). It may even be lost.

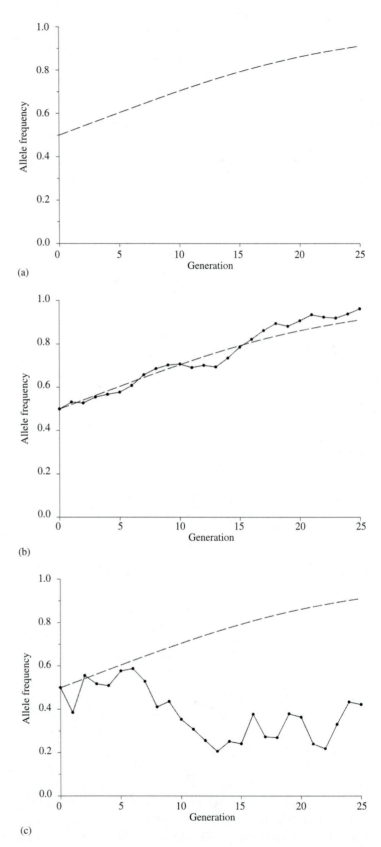

superimpose the effect of genetic drift on the effect of natural selection. If the population size is large, the effect of drift is minor and changes in allele frequency are determined primarily by natural selection. But if the population size is small, drift can be a major factor affecting allele frequency changes.

Rich et al. (1979) tested this prediction with laboratory populations of the flour beetle *Tribolium castaneum*. They established populations of sizes 10, 20, 50, and 100, with 12 replicates of each population size. Each population started with an initial frequency of 0.50 for the *b* allele. This allele produces a black body color; the corresponding wild type allele is b^+. Allele frequencies were followed for 20 generations. In nearly all lines, the b^+ allele increased in frequency. Statistical analysis indicated that this increase was due to natural selection against the *b* allele. Variation within lines from generation to generation and variation among lines were both greater for the smaller population sizes (Figure 7.17). The deleterious allele was actually fixed in one of the lines with $N = 10$, but not in the larger lines. These results agree well with the predictions of the previous paragraph.

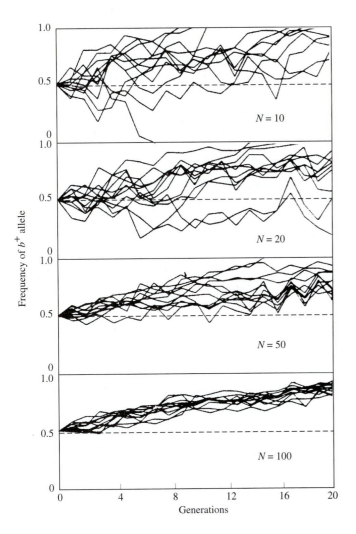

Figure 7.17 **Natural selection and genetic drift in different-sized laboratory populations of *Tribolium castaneum*.** *Source:* From Rich et al. (1979).

If populations are small enough, drift can overpower natural selection and cause major evolutionary changes. We saw this in the *Tribolium* experiment, in which one of the small lines became fixed for the "wrong" allele. Whether drift overpowers selection very often in natural populations is controversial. We describe one well-documented example.

Eichhornia paniculata is an aquatic annual plant that exhibits polymorphism for style length. The three phenotypes are designated *L* (long), *M* (mid), and *S* (short). Style length is controlled by two interacting loci. A stable equilibrium is maintained by nonrandom mating and frequency-dependent natural selection (Chapter 12), and at equilibrium, the three phenotypes are expected to be present in equal frequencies. Because of the nature of the gene interactions, the allele controlling the short phenotype is present in low frequency at equilibrium. The theory of genetic drift predicts that small populations will sometimes lose this allele, resulting in populations with only the *M* and *L* phenotypes. This has actually been observed. Husband and Barrett (1992) studied 167 populations in northeastern Brazil, ranging in size from one to about 8000. Only about 42 percent of populations with 25 or fewer individuals were trimorphic (all three phenotypes), whereas about 90 percent of populations with 100 or more individuals were trimorphic. Of populations that had lost one or more phenotypes, nearly all had lost the *S* phenotype, as predicted. Moreover, in populations that were sampled in two or more years, the magnitude of the change in frequency of the *S* phenotype was strongly inversely related to population size, again as predicted by the theory. Eckert and Barrett (1992) and Eckert et al. (1996) found similar patterns in another tristylous species, *Lythrum salicaria*.

We conclude that genetic drift *can* overpower natural selection in small populations. How often it actually does so remains an open question.

Probability of Fixation for a New Mutation

In Section 5.2, we saw that, ignoring genetic drift, a favorable mutation will increase in frequency and will eventually go to fixation, as long as it provides an advantage in both heterozygous and homozygous conditions. Conversely, a deleterious mutation will be eliminated if it reduces fitness in both heterozygotes and homozygotes. However, the above discussion suggests that a favorable mutation has some chance of being lost, in spite of natural selection. Similarly, we hinted in Section 6.6 that deleterious mutations may rarely become fixed in small populations, in spite of natural selection against them.

The probability of fixation for a new neutral mutation is $1/2N$. It is reasonable to expect that the probability of fixation should be higher for a favorable mutation and lower for a deleterious mutation. Moreover, the probability of fixation for a new mutation depends primarily on its fitness in heterozygotes, since it will initially be rare for many generations, and will be found almost exclusively in heterozygotes. Here, we examine how genetic drift affects the probability that a mutation subject to natural selection will become fixed.

We will scale the fitnesses of genotypes slightly differently than we have before. Let A_1 be the normal allele, and A_2 be a new mutant allele. Then we will scale the fitnesses to the normal homozygote

$$
\begin{array}{ccc}
A_1A_1 & A_1A_2 & A_2A_2 \\
1 & 1+s & 1+2s
\end{array}
$$

Here, s is the effect of the mutation when heterozygous, and $2s$ is its effect when homozygous. For a favorable mutation, $s > 0$ and for a deleterious mutation, $s < 0$. We are assuming that the fitness of heterozygotes is midway between the fitnesses of the homozygotes. This is approximately true for mutations with small effects (Section 6.4). Moreover, this assumption will have little effect on the probability of fixation, because the most critical time for a new mutation is in the early generations when it is very rare and present almost entirely in heterozygotes.

Kimura (1957, 1962) first derived a general expression for the probability of fixation of a mutation, although some special cases had been worked out by Haldane (1927b), Fisher (1930b), and Wright (1931). We will skip the derivation, which requires an understanding of diffusion equations, and consider only some special cases. For the above fitnesses, Kimura showed that the probability of fixation for a mutation with initial frequency p is approximately

$$\Pr(fix) \cong \frac{1 - e^{-4N_e s p}}{1 - e^{-4N_e s}} \qquad \textbf{(7.21)}$$

The initial frequency of a new mutation is $1/2N$. Substituting this for p gives

$$\Pr(fix) \cong \frac{1 - e^{-2\frac{N_e}{N} s}}{1 - e^{-4N_e s}}$$

Note that the probability of fixation depends on both the effective and actual population sizes. The reason is that drift depends on N_e, but the initial allele frequency depends on N. If we assume $N_e = N$,

$$\Pr(fix) \cong \frac{1 - e^{-2s}}{1 - e^{-4Ns}} \qquad \textbf{(7.22)}$$

First consider an advantageous mutation $(s > 0)$. If s is small, then $e^{-2s} \cong 1 - 2s$, and the numerator becomes $2s$. (See Box 7.2 for an explanation of this approximation.) Furthermore, if $4Ns$ is large, then $e^{-4Ns} \cong 0$, and the denominator is approximately one. The probability of fixation becomes

$$\Pr(fix) \cong 2s \qquad \textbf{(7.23)}$$

As long as $4Ns$ is large, the probability of fixation is approximately independent of population size, and depends primarily on the strength of natural selection. Figure 7.18a shows the probability of fixation for a favorable mutation, calculated from equation (7.22) for $s = 0.01$ and $s = 0.05$ (solid lines). The dashed line is the probability of fixation for a neutral mutation $(= 1/2N)$. For relatively large populations, the probability of fixation is approximately $2s$, as suggested by equation (7.23).

Look again at equation (7.22). If both $2s$ and $4Ns$ are small, the numerator becomes approximately $2s$, and the denominator becomes approximately $4Ns$ (using the approximation in Box 7.2). The probability of fixation is then

$$\Pr(fix) \cong \frac{2s}{4Ns} \cong \frac{1}{2N}$$

In other words, if $4Ns$ is small, a new advantageous mutation has about the same probability of fixation as a neutral mutation. If the population size is small

---

BOX 7.2   A useful approximation

The irrational number e can be written as an infinite sum

$$e = \sum_{n=0}^{\infty} \frac{x^n}{n!}$$

$$= 1 + x + \frac{x^2}{2} + \frac{x^3}{3!} + \cdots$$

This sum converges to the value $e = 2.71828\ldots$ for all real values of x (see any calculus text).

If the absolute value of x is near zero, the terms containing x^2 and higher are very near zero, and we can write

$$e^x \cong 1 + x \qquad \textbf{(1)}$$

For example, if $x = 0.1$, $e^x = 1.105$ and $1 + x = 1.100$; if $x = -0.1$, $e^x = 0.905$ and $1 + x = 0.900$. For values of x closer to zero, the approximation is better.

If we let $y = -x$, then

$$e^y \cong 1 + y$$

or,

$$e^{-x} \cong 1 - x \qquad \textbf{(2)}$$

which is the approximation used to obtain equation (7.23) from (7.22).

---

enough, or selection is weak enough, the mutation will be effectively neutral. Figure 7.18b plots the probability of fixation against $4Ns$ for $s = 0.01$ and $s = 0.05$. If $4Ns$ is much less than one, the probability of fixation depends primarily on population size. If $4Ns$ is much greater than one, the probability of fixation is approximately equal to $2s$. So, a useful guideline is this: If $4Ns \ll 1$, genetic drift is the most important process acting on a new mutation, and the probability of fixation is approximately $1/2N$. If $4Ns \gg 1$, natural selection is predominant, and the probability of fixation is approximately $2s$.

If a new advantageous mutation is completely recessive, then neither equation (7.22) nor (7.23) is applicable, because heterozygotes have the same fitness as normal homozygotes. We might expect that such a mutation will be more likely to be lost by drift than one that shows some advantage in heterozygotes. Kimura (1962) showed that the probability of fixation for a completely recessive advantageous mutation is approximately

$$\Pr(fix) \cong 1.13\sqrt{\frac{s}{2N}} \qquad \textbf{(7.24)}$$

where s is the fitness advantage of the mutant homozygote. In general, this is a much lower probability than for a mutation with some heterozygote advantage. For example, if $N = 10^4$ and fitnesses are 1.00, 1.01, and 1.02, the probability of fixation is about 0.02, but if fitnesses are 1.00, 1.00, and 1.02, the probability of fixation is only about 8×10^{-4}.

It may surprise you that favorable mutations will usually be lost. If a mutation gives a 1 percent advantage to heterozygotes, the probability that it will become fixed in the population is only about 2 percent. Even if the mutation gives a 10 percent advantage to heterozygotes, considered strong selection, the probability of fixation is only about 20 percent. The conclusion is that even very favorable mutations are likely to be lost much of the time. As Gillespie (1998) says, "Think of all the great mutations that failed to get by the quagmire of rareness!"

These results have important implications for evolutionary biology. First, evolution is inefficient. Most mutations are lost, even if they would be valuable to the

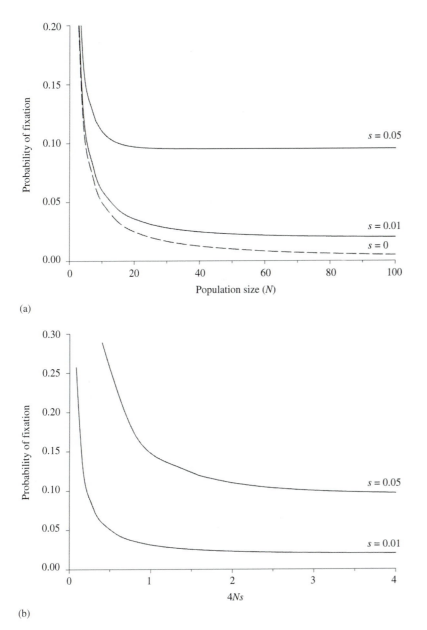

Figure 7.18 **Probability of fixation for a favorable mutation.**   (a) Probability of fixation as a function of population size for $s = 0.01$ or $s = 0.05$, as predicted by equation (7.22). Dashed line is the probability of fixation for a neutral mutation ($= 1/2N$). (b) Probability of fixation as a function of $4Ns$. If $4Ns$ is much greater than 1, the probability of fixation is approximately $2s$.

population. A population may be unable to maximize its adaptation because most beneficial mutations are lost by drift. In the language of adaptive landscapes, a population may be unable to reach a high fitness peak because the mutations allowing it to get there have been lost. Adaptation is generally *not* maximized. Second, evolution is unpredictable. Mutations occur randomly, and are fixed or lost randomly. A favorable mutation may be fixed largely because it was lucky, and not because it was better than some other favorable mutation. If we were to "rewind the tape," as Gould (1989) envisions, and replay the history of life, the outcome might be very different from what we see today because different adaptive mutations would be fixed over the course of evolution. Adaptive evolution is not always a deterministic trend, as is frequently thought; it can sometimes be a random, unpredictable process. Genetic drift can restrict adaptive evolution even in large populations. Its importance is not limited to small populations, as we often assume.

The above discussion has implicitly assumed the infinite alleles model; that is, an advantageous mutation will occur only once. However, in very large populations the same mutation may appear many times, substantially increasing the probability that one copy will escape the quagmire of rareness and increase to fixation. Under these conditions, adaptive evolution may in fact be predictable and repeatable. For example, HIV inevitably acquires resistance to the drug AZT. The evolution of resistance in different patients is often due to one of a few replacement substitutions (e.g., lys to arg at codon 70) in the viral gene coding for reverse transcriptase (Larder and Kemp 1989; Mohri et al. 1993). Similarly, Bull et al. (1997) found that, as the virus ϕX174 adapted to high temperature, more than half of the observed substitutions occurred in more than one replicate line. The common characteristics of these studies are very large population sizes (on the order of 10^8 or greater) and strong natural selection. We do not know how often this kind of convergent molecular evolution occurs in eukaryotes with much smaller population sizes, but recurrent mutation must be much rarer, and therefore genetic drift is probably more important in restricting adaptive evolution.

Now consider a new deleterious mutation. We saw in Chapter 6 that slightly deleterious mutations are surprisingly common in some species, and suggested that they can become fixed in small populations due to genetic drift. Here, we investigate this idea more quantitatively. Intuitively, we expect the probability of fixation to be low and inversely related to the severity of the mutation in heterozygotes. For mutations with intermediate fitness in heterozygotes, equation (7.22) is still approximately valid, with $s < 0$. For example, if $s = -0.005$, a weakly deleterious effect, the probability of fixation is about 0.0016 in a population of 100. For mutations with more severe effects or for larger populations, the probability of fixation rapidly becomes vanishingly small, because the denominator of (7.22) quickly becomes very large. For example, if $|4Ns| = 8$, the probability of fixation is only about $(7 \times 10^{-4})|s|$. Figure 7.19 plots the probability of fixation against population size for $s = -0.01$ and $s = -0.05$.

We conclude that deleterious mutations with large effect have near zero probability of fixation, while mutations with small effect (which are probably much more common) can become fixed by genetic drift, especially in small populations. The same general guidelines apply as for a favorable mutation: If $|4Ns| \ll 1$, the probability of fixation is primarily determined by population size, and weakly deleterious mutations may become fixed in small populations. If $|4Ns| \gg 1$, the deleterious mutations will almost certainly be eliminated by natural selection.

The probability that any given deleterious mutation will be fixed is very low, but we saw in Section 6.6 that genomic deleterious mutation rates may be high enough that a few mutations will inevitably become fixed. It is like buying a lottery ticket: The probability that any given individual will win the lottery is very low, but *someone* eventually wins the big prize.

The above conclusions assume a constant population size. Many populations fluctuate in size, either randomly or by steadily increasing or decreasing. This can have a dramatic effect on the probability of fixation of a new mutation (Otto and Whitlock 1997). Favorable mutations are more likely to be fixed in a growing population. Recall that a favorable mutation is most likely to be lost soon after it occurs, when it is very rare. In a stable population, each individual leaves, on

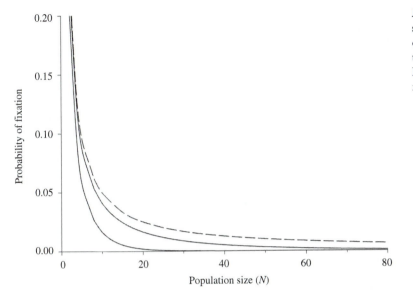

Figure 7.19 **Probability of fixation for a slightly deleterious mutation ($s = -0.01$ or $s = -0.05$) as a function of population size, as predicted by equation (7.22).** Dashed line is the probability of fixation for a neutral mutation.

the average, one offspring, but in a growing population, each individual will leave more than one on the average. This means that a rare allele is less likely to be lost because its carriers did not reproduce or did not transmit the allele to their offspring. Favorable alleles are more likely to get through the initial danger period of rarity and increase to a frequency at which loss by genetic drift is unlikely. The opposite is true in a declining population. Each individual will leave an average of less than one offspring, and the rare allele is more likely to be lost through lack of transmission.

Deleterious alleles, on the other hand, are more likely to be fixed in a declining population. As the population size gets smaller, drift becomes more important, and the probability of fixation increases.

The general principle is the same for both favorable and deleterious alleles. In a large population, natural selection is more effective and genetic drift less so than in a small population. So, as the population size increases, selection will be more effective in fixing favorable alleles and eliminating deleterious alleles. The reverse is true as the population size decreases. Drift becomes more important, increasing the probability that favorable alleles will be lost and deleterious alleles fixed. Otto and Whitlock (1997) have studied this process for various models of population growth and decline. They develop formulas for the probability of fixation analogous to those presented here, and propose formulas for effective population size which take into account the past and future demographic patterns.

The fact that deleterious mutations are more likely to be fixed as a population declines has important implications for small endangered populations. As these mutations become fixed, they lower the absolute fitness of the population, which may increase the danger of extinction.

7.9 Genetic Drift, Population Dynamics, and Extinction

An important problem in conservation biology is to estimate the probability of extinction for threatened or endangered species. Small populations are subject to increased risks of extinction because of various demographic, environmental, and genetic factors. Stochastic (random) effects are much greater in small populations, and can greatly increase the risk of extinction. In this section, we briefly consider some of the factors that increase the danger of extinction, and ways in which their effects can be estimated.

Stochastic Population Dynamics

Imagine a sexually reproducing population that consists of two males and two females ($N_t = 4$). Each female mates and has three offspring, then all four parents die ($N_{t+1} = 6$). All six offspring are males. What will N_{t+2} be?

The answer, of course, is zero. The males will have no one to mate with, and the population will go extinct. This is a simple example of how random events can affect population size. It is not as unrealistic as you might expect. The dusky seaside sparrow went extinct in 1987 (Ehrlich et al. 1988). However, in 1980, the last six surviving birds were all males, a cruel twist of probability that assured extinction years before the last bird actually died. (See Problem 7.18.)

Many populations, especially on islands, begin with only a very few individuals, sometimes only a single fertilized female. The probability of extinction for such a population can be very high, especially in the first few generations. Imagine a predator discovering the first cohort of eggs or young and eating them all.

These are examples of stochastic (random) effects on population size. Any real population is subject to these effects, and we must always ask ourselves if they are likely to be important in the population we are studying. The smaller the population, the more important these effects are likely to be.

Any population increases or decreases by individual births and deaths. We can assign average rates and probabilities to births and deaths, but each individual birth or death is a random event. One sequence of births and deaths might be

$$BBDBDDBDBBBDD\ldots$$

where B and D represent individual births and deaths in the population. Another sequence might be

$$DBDDBBBDDDDD\ldots$$

This randomness in the order of births and deaths is known as **demographic stochasticity**. Clearly, a long sequence of deaths can cause a small population to go extinct, even though the *average* reproductive rate is positive. Demographic stochasticity is a significant cause of extinction only in very small populations, but all extinct populations were once very small.

Another source of danger to small populations is **environmental stochasticity**. In natural populations, there are always random fluctuations in weather, food supply, predators and parasites, sex ratio, and so forth. These cause random fluctuations in reproductive rate and population size. A few consecutive bad years can endanger even a fairly large population; they are likely to be fatal to a small one.

A third danger might be called **genetic stochasticity**, to be consistent with the above terminology. Here we refer to the deleterious effects of genetic drift in small populations. We saw in the previous section that slightly deleterious mutations are more likely to become fixed in a small population. Every time this happens, the mean absolute fitness (reproductive rate) of the population decreases slightly. This may decrease the population size, which further increases the probability that deleterious mutations will be fixed, which further decreases absolute fitness, which further decreases population size, which further increases the probability that deleterious mutations will become fixed, ... If this continues, the mean absolute fitness may eventually fall below one, and the population will rapidly decline to extinction. Lynch and Gabriel (1990) called this process "mutational meltdown." We saw in Section 6.6 that sexual reproduction and recombination may slow or stop this process in most sexual populations, but it may be a significant danger in very small populations.

Lynch et al. (1995) performed computer simulations to estimate the effect of mutation accumulation on the time to extinction in small sexual populations. They concluded that for genomic deleterious mutation rates of about one (i.e., one deleterious mutation per genome per generation; Section 6.6) and weak selection, mutation accumulation alone could lead to extinction within a few thousand generations in small populations (64 or fewer individuals).

Small populations are also subject to other genetic effects that increase the likelihood of extinction. Genetic drift reduces genetic variation, especially in small populations. This decreases evolutionary flexibility, limiting a population's ability to respond to environmental change, and may increase susceptibility to parasitism and disease. Moreover, small populations are more subject to inbreeding depression in which individuals suffer a variety of deleterious genetic abnormalities (Chapter 8). Thus, reduced genetic variation and inbreeding depression both lower reproductive rates; in other words, mean absolute fitness is likely to be lower in small populations than in large ones.

Stochastic models of population dynamics and extinction can be very complicated; for example, Lacy (1987) simulated the interacting effects of mutation, genetic drift, natural selection, population subdivision, and gene flow. Renshaw (1991), Foley (1994), Gillman and Hails (1997), and Roughgarden (1998) discuss some simple models. We will not examine stochastic extinction models in detail, but the general conclusions are straightforward:

- *Any* finite population is doomed to eventual extinction. The important question is when. What is the mean persistence, or expected time to extinction?
- Predicted time to extinction depends on population size. The larger the population, the greater its persistence.
- Predicted time to extinction depends on the variance of reproductive rate. Large variance increases the probability of extinction (decreases persistence).
- Extinction is more likely in a variable environment than in a constant one.
- Environmental stochasticity is more important than demographic stochasticity except in very small populations.
- Density dependent population regulation decreases the expected time to extinction.

We conclude that demographic, environmental, and genetic effects all conspire against persistence in small populations. Furthermore, these factors interact in positive feedback loops, creating what Gilpin and Soule (1986) have called extinction vortices. As the population becomes smaller, each of these factors becomes more deleterious, further decreasing the population size, and the population spirals down to extinction.

Berger (1990) described a dramatic example of extinction in small populations. More than 100 populations of bighorn sheep (*Ovis canadensis*) have been followed for 50 years or more. The probability of persistence was directly related to population size (Figure 7.20). After 50 years, all populations with fewer than 50 individuals were extinct, whereas all populations with more than 100 individuals still existed. Berger concluded that both environmental and genetic effects contributed to shorter persistence of small populations. Some of the conclusions of this study have been challenged; see Wehausen (1999) and Berger (1999).

Jones and Diamond (1976) studied persistence of bird populations in the Channel Islands. Their results are summarized in Figure 7.21. Of populations with fewer than ten breeding pairs, 39 percent went extinct over an 80-year period. At the other extreme, no populations with more than 1000 breeding pairs went extinct. Pimm et al. (1988) obtained similar results in a study of birds on small islands of Great Britain.

Minimum Viable Population Size

What is the minimum population size necessary to prevent extinction? The answer depends on many biological factors, of course, and on the time frame being considered, since all populations go extinct sooner or later. In the above examples, a population of 50 bighorn sheep was clearly too small to persist, whereas 100 or more breeding pairs allowed most bird populations in the Channel Islands to persist, at least in the short term. Shaffer (1981) introduced the concept of a minimum viable population size, and defined it as "the smallest isolated population having a 99 percent chance of remaining extant for 1000 years despite the foreseeable effects of demographic, environmental, and genetic stochasticity, and natural catastrophes." Other definitions are possible, for example, the population size necessary to assure a 95 percent probability of survival for 100 years.

Estimating the minimum viable population size is a difficult problem. It requires detailed knowledge of the ecology, genetics, breeding structure, and reproductive biology of the population in question. It helps to have a crystal ball to predict future environmental variation. Computer simulations are useful for estimating the effects of environmental and demographic stochasticity, but must be

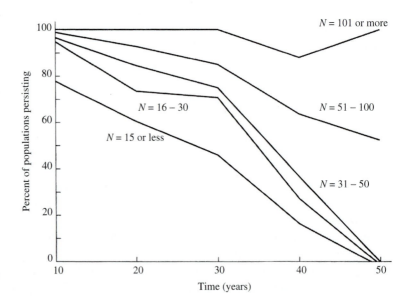

Figure 7.20 **Relationship between population size and persistence in bighorn sheep populations.** *Source:* From Primack (1993); data from Berger (1990).

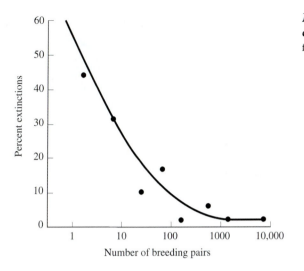

Figure 7.21 **Probability of extinction for bird populations of different sizes in the Channel Islands, California.** *Source:* Modified from Jones and Diamond (1976).

based on reliable estimates of survival, reproductive rates, gene flow, and so on. We will consider one example.

The grizzly bear (*Ursus arctos*) once ranged widely over western North America, but in the continental United States it is now confined to six isolated populations, some of which have fewer than ten individuals (Allendorf and Servheen 1986). The largest populations are in Yellowstone and Glacier National Parks. Shaffer (1983) and Shaffer and Samson (1985) estimated the minimum viable population (95 percent chance of survival for 100 years) of grizzly bears to be about 50 to 90 individuals. This estimate considered only demographic and environmental stochasticity; it did not consider genetic effects. Harris and Allendorf (1989) estimated that the effective population size in grizzly bears is about 25 percent of the actual size. If their estimate is correct, the actual population size needed to avoid genetic problems is about 200 to 360 individuals. The population of grizzly bears in the Greater Yellowstone Ecosystem (Yellowstone National Park and surrounding national forests and other lands; Figure 7.22) is about 200 (Allendorf 1997; National Park Service 1998). In another study, Dennis et al. (1991) considered a model of environmental stochasticity and estimated the average growth rate ($\bar{r}$) to be about −0.003 per year with large variance. The estimated mean time to extinction was about 200 years, with large variance, but the most likely time to extinction was only 79 years. Foley (1994) considered a model that included density-dependent population regulation and obtained a conservative estimate of the mean time to extinction of about 1200 years. However, the variance of extinction time was large, so that the population had only about 50 percent probability of persisting for 800 years or more. These estimates are less pessimistic than those of Dennis et al., but should still be worrisome to those who care about grizzly bears in Yellowstone.

These estimates must be taken with a grain of salt. They depend on rather crude estimates of important parameters, and assume that the future will be more or less the same as the present. However, they give some way of quantifying the risk of extinction. They suggest that the long-term future of grizzly bears in the Greater Yellowstone Ecosystem is uncertain.

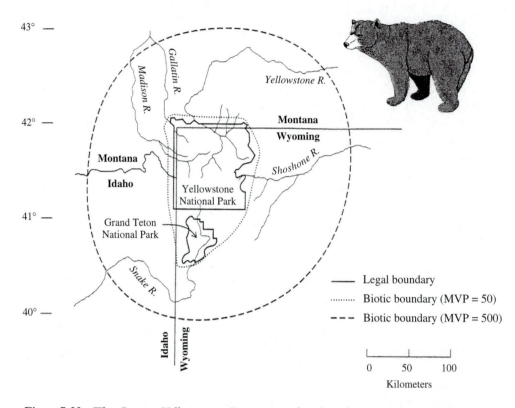

Figure 7.22 **The Greater Yellowstone Ecosystem showing the range for a minimum viable population of 50 (dotted line) and 500 (dashed line) grizzly bears.** *Source:* From Krebs (1994); original from Newmark (1985).

Some general guidelines have come from theoretical and empirical research. The minimum population size to avoid the short-term deleterious effects of genetic drift and inbreeding is about 50 to 100 individuals. To avoid extinction due to long-term environmental stochasticity and catastrophes, the population should be about 500. This 50/500 rule should be considered a general guideline only, and it has been criticized as being much too simplistic. Lande (1988) suggested a rule of thumb of about 1000 individuals for long-term persistence of vertebrate populations. More recently, Lande (1995) has considered the effects of deleterious mutations and revised this estimate upward to 5,000–10,000 individuals.

To summarize, demographic, environmental, and genetic effects interact in positive feedback loops to increase the danger of extinction in small populations. The emerging science of population viability analysis can help us to understand the relative importance of these factors, and can aid in formulating management policies for endangered populations.

7.10 How Important Is Genetic Drift in Nature?

That gene frequencies will be affected by random factors is evident and not questioned by any author. What is disputed is the evolutionary significance of such random fluctuations of genes in local populations.

—*Ernst Mayr (1963)*

There is no doubt that the theory of genetic drift is correct. Small, isolated, and controlled laboratory populations behave approximately as the theory predicts. There is also little doubt that genetic drift explains many patterns of DNA variation in natural populations. However, the short-term importance of genetic drift in many natural populations remains unclear. The problems are many: First, it is difficult to obtain accurate estimates of population sizes in nature, and even more difficult to obtain histories of population sizes. Second, genetic drift is usually such a slow process that it would take many generations to confirm its effects, and we have very little data on how allele frequencies change in natural populations. Third, population structure (subdivision) complicates the effects of genetic drift, and subdivision is a subtle and difficult thing to study in natural populations. Fourth, the effects of drift are inevitably confounded with the effects of mutation, natural selection, gene flow, and so on. We see this when terms like $Nu, Ns,$ and Nm appear in our models (m is the rate of gene flow, analogous to the mutation rate; see Chapter 9). Wright (1940, 1969) suggested some useful guidelines: A population is "small" when $4Nu$, $4Ns,$ or $4Nm$ is less than one; genetic drift is a significant evolutionary process. A population is "large" when all of these quantities are much greater than one; drift is minor and the deterministic processes predominate. The difficulty, of course, is in estimating $N, u, s,$ and m.

The long-term of importance of genetic drift is more clear. All populations will eventually lose genetic variation and accumulate slightly deleterious mutations. Many advantageous mutations will be lost, restricting long-term adaptive evolution. How rapidly these effects occur depends on the effective population size, and on other variables.

Summary

1. Genetic drift is random variation in allele frequencies due to Mendelian segregation and random sampling in a finite population.

2. The direction of genetic drift is unpredictable in the short term. In any given generation, the allele frequency is equally likely to increase or decrease.

3. The magnitude of genetic drift is inversely proportional to population size. The variance of allele frequency change in one generation is given by equation (7.2).

4. The long-term effect of genetic drift is to reduce genetic variation within a population. This is manifested as decreased heterozygosity, described by equation (7.5), and loss of alleles, described by equation (7.7).

5. Genetic drift causes populations to diverge from one another. The increase in variance of allele frequency among populations is given by equation (7.10).

6. Experimental results are generally consistent with the predictions of the theory of genetic drift, but frequently heterozygosity decreases faster and allele frequency variance increases faster than predicted by actual population size.

7. The effective population size describes how heterozygosity decreases or allele frequency variance increases in an actual population that violates the assumptions of the theory.

8. The effective population size depends on breeding structure of the population, sex ratio, the history of population size, the variance of family size, and other factors. The effective size is usually smaller than the actual size.

9. Genetic drift is often difficult to document in natural populations because natural selection can frequently explain observations equally well.

10. Genetic drift explains observed patterns of silent and replacement polymorphisms and substitutions in protein-coding genes, and in pseudogenes. Purifying selection to preserve protein function is also seen in coding sequences.

11. A founder effect occurs when a new population is established by a few individuals from the original population. The new population has less genetic variation than the original, and different allele frequencies; many examples have been documented.

12. A population bottleneck occurs when a population is suddenly reduced to a very small size. The consequences are the same as for a founder effect.

13. Genetic drift and mutation have opposing effects on the level of heterozygosity in a population. Under the infinite alleles, neutral mutation model, an equilibrium is reached, as described by equation (7.19).

14. Genetic drift can sometimes oppose natural selection in small populations. This can lead to loss of favorable alleles or fixation of slightly deleterious mutations.

15. Accumulation of slightly deleterious mutations can reduce absolute fitness of a population and increase the risk of extinction in small populations.

16. Demographic, environmental, and genetic effects all interact to increase the risk of extinction in small populations. Population viability analysis can estimate the minimum population size necessary to avoid extinction in the short or near term, and can help evaluate the relative effects of various factors contributing to extinction.

Problems

7.1. In a population of size 50, the frequencies of A_1 and A_2 are 0.7 and 0.3. What is the expected heterozygosity after ten generations?

7.2. In an isolated island population of 500 individuals, the frequency of a recessive allele changed from 0.75 to 0.70 in a single generation. What is the most likely cause of this change, mutation, natural selection, gene flow, or genetic drift? Estimate u or s, as appropriate. How confident are you with this estimate?

7.3. Calculate the expected time to fixation for initial allele frequencies of 0.1, 0.5, and 0.9 for population sizes of 10 and 100. Does the pattern you see make biological and mathematical sense? Explain.

7.4. Show that for large N, equation (7.8) is approximately

$$\bar{t}_{fix} \cong 4N$$

for a new mutation. (*Hint*: Use the approximation in Box 7.2.)

7.5. Find the probability of fixation and the expected time to fixation for a new neutral mutation in populations of size 5, 50, and 500. Use both equation (7.8) and the approximation of the previous problem. How large does N have to be for the approximation to be acceptable?

7.6. Plot expected heterozygosity for 100 generations for population sizes of 5, 50, and 500 (all on the same graph). Assume initial heterozygosity is 0.5.

7.7. Plot variance of allele frequency for 100 generations for population sizes of 5, 50, and 500 (all on the same graph). Assume initial allele frequency of 0.5.

7.8. Assume that an allele has a frequency of p in generation t, and that the population size is N. What is the probability that that allele will be represented in the next generation?

7.9. Assume that the allele in the previous problem is a new mutation, so $p = 1/2N$. Show that, for large N, the probability that the allele will be represented in the next generation is approximately independent of N, and is about 0.73. (*Hint*: Use the approximation in Box 7.2.)

7.10. What is the effective population size if a population grows exponentially for five generations, with initial population size of 100 and $\lambda = 1.3$. (*Hint*: First find the actual population size each generation by using the equation $N_{t+1} = \lambda N_t$.)

7.11. Buri's observed variance among lines and overall heterozygosity fit the predicted values for an effective population size of 9, whereas the actual population size was 16. Why might the effective population size have been reduced in these experiments?

7.12. How many generations would it take for heterozygosity to decay to half its initial value in a population size of N? Express your answer in generations as a function of N. Obtain an exact answer and an approximation using the approximation in Box 7.2. What else is going on during this time? Does loss of heterozygosity due to genetic drift seem to be much of a problem for large populations? Define large.

7.13. Laboratory populations are sometimes inbred by making single-pair matings between siblings each generation. What is the expected heterozygosity in such a laboratory population after ten generations? Express your answer as a multiple of the initial heterozygosity.

7.14. The neutral mutation rate is frequently estimated (guessed) to be about 1×10^{-7} per locus per generation. Use this value and equation (7.19) to estimate the average effective population sizes for invertebrates, vertebrates, and plants based on the heterozygosity estimates in Table 2.5. Do these numbers seem reasonable? Discuss why or why not. What assumptions went into these calculations? Do they seem reasonable?

7.15. Consider the following possible fitnesses for a new mutation *A2*.

	A_1A_1	A_1A_2	A_2A_2
a. neutral	1	1	1
b. favorable resessive	1	1	1.02
c. favorable semidominant	1	1.01	1.02
d. favorable dominant	1	1.02	1.02
e. deleterious semidominant	1	0.99	0.98

Calculate the probability of fixation for each set of fitnesses. Assume $N = 500$ in each case. Repeat for $N = 50$. What changes?

7.16. What is the minimum population size necessary for an allele with a heterozygous advantage of 1 percent to have a greater probability of fixation than a neutral allele.

7.17. Dobzhansky and Pavlovsky (1957) established 20 experimental populations of *Drosophila pseudoobscura*. Each population initially contained the inversions PP (Pikes Peak) and AR (Arrowhead) in equal frequencies. Ten populations were established with 4000 founders ("large" populations) and ten were established with 20 founders ("small" populations). The frequency of PP was estimated in each population after 5 months and 18 months. Following are their results:

Large Populations		Small Populations	
5 Months	**18 Months**	**5 Months**	**18 Months**
.393	.317	.377	.180
.423	.290	.307	.320
.293	.347	.310	.460
.380	.340	.323	.467
.333	.227	.343	.327
.360	.203	.417	.473
.403	.320	.373	.163
.410	.223	.253	.343
.370	.257	.377	.320
.420	.220	.253	.220

a. Plot the frequency of PP versus time for each line. Plot large populations and small populations on separate graphs with the same scale.

b. Calculate the mean and variance for large populations and for small populations at 5 months and at 18 months.

c. Interpret the results in terms of natural selection and genetic drift.

7.18. Assume that N fertilized females reach an island uninhabited by that species, and that each female produces n offspring.

a. Derive an expression for the probability that the population will go extinct in the first generation due to lack of one sex. In other words, what is the probability that the offspring of these N females will be either all males or all females?

b. What is this probability if a single fertilized female reaches the island and leaves four offspring? Assuming the offspring are not all the same sex, what other problems will the population have to overcome if it is to become established on the island?

Inbreeding and Nonrandom Mating

It often occurred to me that it would be advisable to try whether seedlings from cross-fertilised flowers were in any way superior to those from self-fertilised flowers.

—Charles Darwin (1876)

In order for the Hardy-Weinberg principle to hold, individuals in a population must mate at random. Until now, we have nearly always assumed that this condition is met as we examined the effects of various evolutionary processes. In this chapter, we examine what happens when this assumption is violated; that is, we will study the consequences of nonrandom mating.

Nonrandom mating is actually rather common: Many plants regularly undergo extensive self-fertilization. Other plants have mechanisms that ensure they will mate with a plant with a different phenotype. Mating beetles are frequently more similar in size than two beetles chosen at random from the population. Marriages between cousins are relatively common in some human cultures.

In the broadest sense, there are three kinds of nonrandom mating: **Inbreeding** is mating between related individuals; that is, between individuals having a common ancestor. **Assortative mating** occurs when two mating individuals are phenotypically more alike than two individuals chosen at random. **Disassortative mating** occurs when mates are phenotypically less alike than two individuals chosen at random. We shall examine each of these in turn.

8.1 Inbreeding

Inbreeding is a form of nonrandom mating based on genetic relationship. It occurs when two mating individuals have a common ancestor. As we shall see, one of the consequences of inbreeding is an increase in the frequency of homozygous genotypes. Deleterious recessive alleles are more frequently expressed in homozygotes, resulting in a variety of developmental and morphological defects, or reduced viability or fertility. Thus, inbreeding can lower the mean fitness of the population. Natural populations must either avoid extensive inbreeding or find ways to avoid its deleterious consequences. Plant or animal breeding programs and captive breeding programs must also take into account the consequences of inbreeding and attempt to minimize these potential problems. Therefore, it is important to be able to quantify the degree of inbreeding and its effects on population fitness. That is the main goal of this section. Throughout, we will assume a diploid, sexually reproducing organism, and autosomal loci, unless explicitly stated otherwise.

Consanguinity and Inbreeding

Two individuals are said to be **consanguineous** if they have a common ancestor. Full siblings or first cousins are examples. As we saw in Section 7.2, individuals with a common ancestor may have received copies of the same allele from that ancestor; such copies are called identical by descent (ibd). Alleles that are identical by descent are copies of the same ancestral DNA molecule. One definition of inbreeding, then, is the mating of consanguineous individuals, and one consequence of inbreeding is that their offspring have a nonzero probability that two alleles at a locus are identical by descent.

In any finite population, two individuals chosen at random must have a common ancestor if we go back far enough into the past. Any individual has two parents, four grandparents, eight great-grandparents, and so on. In general, the number of ancestors n generations in the past is 2^n. This number increases rapidly with n. Ten generations back, an individual has 1024 ancestors; 30 generations back, the number of ancestors is more than 10^9. Clearly, it does not take many generations before the number of ancestors is more than the population size (Problem 8.1). Therefore, any two individuals *must* have a common ancestor somewhere in the distant or not-so-distant past.

The genetic consequences of consanguinity and inbreeding decrease as the number of generations to the common ancestor increases. At some point, these consequences become negligible, and common ancestry can be ignored. We can arbitrarily define an ancestral population in which it is assumed that all individuals are unrelated and no alleles are identical by descent. Such an ancestral population is called the **reference population**, or base population. We shall see below that both inbreeding and the reference population can sometimes be defined differently, but inbreeding is always a relative concept, compared to a stated or implied reference population.

A simple way of measuring the degree of relatedness between two individuals is the **coefficient of relationship**. This is the expected proportion of alleles that are identical by descent in two individuals. For full sibs of unrelated parents, the coefficient of relationship is 1/2; if you have allele A_1, the probability that your sibling also has it is 1/2. For first cousins, the coefficient of relationship is 1/8.

The coefficient of relationship is intuitively attractive and conceptually clear, but it is not the most useful measure of relationship. A more useful measure is called the **coefficient of consanguinity**. It is the probability that two alleles, one chosen at random from each of two individuals, are identical by descent. Pick an allele at random from individual X. Pick an allele at random from the same locus in individual Y. The probability that they are identical by descent is the coefficient of consanguinity. Make sure you understand why the coefficient of relationship and the coefficient of consanguinity are not the same. Pick an allele from X. The probability that Y *has* that allele is the coefficient of relationship; the probability that you will *pick* that allele is the coefficient of consanguinity. Since you can either pick that allele or the other with equal probability, the coefficient of consanguinity is half the coefficient of relationship.

The coefficient of consanguinity has also been called the coefficient of kinship or the coefficient of coancestry. These three terms are identical, but the coefficient of relationship is not. We will use coefficient of consanguinity, and symbolize it by g, usually with subscripts to indicate the individuals involved.

A useful formula for calculating the coefficient of consanguinity of two individuals, X and Y, from a pedigree is

$$g_{XY} = \sum_{\substack{\text{all} \\ \text{alleles}}} \frac{1}{2} \Pr(X = A_i) \times \frac{1}{2} \Pr(Y = A_i) \qquad (8.1)$$

where g_{XY} is the coefficient of consanguinity between individuals X and Y, and A_i is the ith allele that is possibly identical by descent from the common ancestor. $\Pr(X = A_i)$ and $\Pr(Y = A_i)$ are the probability that individuals X or Y have inherited A_i from the common ancestor. These can be calculated from the pedigree. The factors of $1/2$ are because, given that X or Y has allele A_i, the probability that you will pick it is $1/2$. The summation is over all alleles that are possibly identical by descent. Equation (8.1) also assumes that none of the alleles in the common ancestor are identical by descent.

When studying consanguinity and inbreeding, we usually simplify the pedigrees and include only individuals that can contribute to identity by descent. It is usually easier to trace lines of descent in these simplified pedigrees, sometimes called path diagrams. Figure 8.1 illustrates both ways of diagramming relationships. Individuals X and Y are second cousins. Figure 8.1a shows all individuals in the pedigree; Figure 8.1b shows only the paths through which X and Y can receive alleles that are identical by descent. For complex relationships, path diagrams are usually easier to interpret.

Figure 8.2 illustrates the pedigrees of several kinds of consanguineous individuals. We will illustrate how to calculate the coefficient of consanguinity for full siblings and for first cousins.

Look at Figure 8.2c. Individuals X and Y are full siblings with common ancestors (parents) A and B. We label the alleles in the parents with different subscripts so we can keep track of them individually. This does not necessarily mean they are different in state. The probability that X receives A_1 from A is $1/2$, and the probability that Y receives A_1 is (independently) $1/2$. The same is true for each of the other alleles A_2, A_3, and A_4. Therefore, the probability that you will pick two alleles that are identical by descent is

$$g_{XY} = \sum_{\substack{\text{all} \\ \text{alleles}}} \frac{1}{2} \Pr(X = A_i) \times \frac{1}{2} \Pr(Y = A_i)$$

$$= \sum_{\substack{\text{all} \\ \text{alleles}}} \frac{1}{2} \left(\frac{1}{2}\right) \times \frac{1}{2} \left(\frac{1}{2}\right)$$

***Figure 8.1* Alternative ways of showing genealogical relationships.**   Individuals X and Y are second cousins. (a) Full pedigree, showing all individuals. (b) Path diagram, showing only lines of descent that may lead to identity by descent in consanguineous individuals.

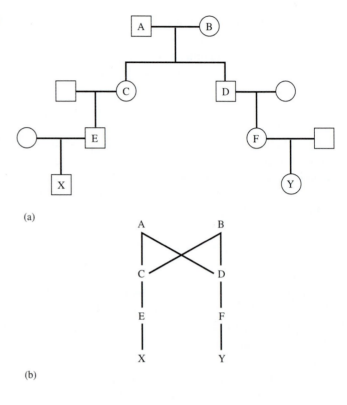

(a)

(b)

$$= 4\left[\frac{1}{2}\left(\frac{1}{2}\right) \times \frac{1}{2}\left(\frac{1}{2}\right)\right]$$

$$= \frac{1}{4}$$

Now consider first cousins (Figure 8.2e). The probability that X will receive A_1 is 1/4, and the probability Y will receive A_1 is 1/4. The same is true for alleles A_2, A_3, and A_4. Therefore, using equation (8.1),

$$g_{XY} = \sum_{\substack{\text{all} \\ \text{alleles}}} \frac{1}{2} \Pr(X = A_i) \times \frac{1}{2} \Pr(Y = A_i)$$

$$= \sum_{\substack{\text{all} \\ \text{alleles}}} \frac{1}{2}\left(\frac{1}{4}\right) \times \frac{1}{2}\left(\frac{1}{4}\right)$$

$$= 4\left[\frac{1}{2}\left(\frac{1}{4}\right) \times \frac{1}{2}\left(\frac{1}{4}\right)\right]$$

$$= \frac{1}{16}$$

You should work out the coefficients of consanguinity for each of the other relationships in Figure 8.2 (Problem 8.3).

Now consider the offspring of two consanguineous individuals. Such an individual is said to be **inbred**. Note that consanguinity refers to relatedness of two individuals (who may or may not mate), whereas inbreeding refers to the mating of consanguineous individuals.

(a) Self-Fertilization

(b) Parent-Offspring

Figure 8.2 **Examples of consanguineous individuals.** In
each pedigree, individuals marked
X and Y are consanguineous; that
is, they have one or more common ancestors.

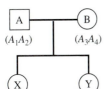

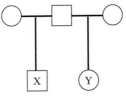

(c) Full Siblings

(d) Half Siblings

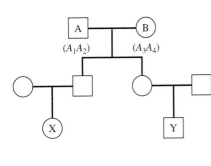

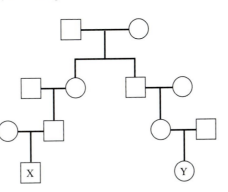

(e) First Cousins

(f) Second Cousins

The Inbreeding Coefficient

Because of its consanguineous parents, an inbred individual may have two alleles
at a locus that are identical by descent. We define the **inbreeding coefficient** as
the probability that an individual has two alleles at a locus that are identical by
descent. We introduced the inbreeding coefficient in Section 7.2 in the context
of genetic drift; we will consider the relationship between inbreeding and genetic drift below. Since the alleles of an inbred individual are chosen at random, one
from each of its parents, the inbreeding coefficient of an individual is the same as
the coefficient of consanguinity of its parents.

It is possible to generalize equation (8.1) to calculate the inbreeding coefficient of any individual from its pedigree. We will use f (lowercase) to symbolize
the inbreeding coefficient, usually with a subscript to indicate which individual.
The procedure is as follows:

1. List all common ancestors.
2. List all pathways through each common ancestor. Start with one parent, work
   back to the common ancestor, then forward to the other parent. Any individual can be present in a pathway only once, but can be present in more than one
   pathway. The number of pathways is p, and is not necessarily the same as the
   number of common ancestors (see example below). The number of individuals
   in the ith pathway (counting both parents of the inbred individual) is n_i.

3a. If the common ancestor is not inbred, the partial inbreeding coefficient due to the ith pathway is

$$f_i = \left(\frac{1}{2}\right)^{n_i}$$

The logic behind this equation is as follows: The probability that a given allele will be transmitted to the next generation by Mendelian segregation is $1/2$, and there are $n_i - 1$ generations of segregation from the common ancestor to both parents (count them in a few pedigrees to convince yourself). Therefore, the probability that a given allele is ibd in the inbred individual is $(1/2)^{n_i-1}$. There are two alleles that are possibly ibd from the common ancestor, so the probability that the inbred individual receives alleles that are ibd through the ith pathway is $(1/2)^{n_i-1} \times 2 = (1/2)^{n_i}$. Note that here, the subscript i represents a pathway, and not an individual.

3b. If the common ancestor is itself inbred, multiply the partial inbreeding coefficient by one plus the inbreeding coefficient of the common ancestor:

$$f_i = \left(\frac{1}{2}\right)^{n_i} (1 + f_{ai})$$

where f_{ai} is the inbreeding coefficient of the common ancestor in the ith pathway. The inbreeding coefficient of the common ancestor is usually easily determined by applying step 3a to its parents (see example below).

4. Repeat step 3a or 3b for all possible pathways through all common ancestors.

5. Sum the partial inbreeding coefficients over all possible pathways through all common ancestors. The final equation is

$$f = \sum_{i=1}^{p} \left(\frac{1}{2}\right)^{n_i} (1 + f_{ai}) \qquad \textbf{(8.2)}$$

where p is the number of possible pathways and f_{ai} is the inbreeding coefficient of the common ancestor in the ith pathway.

To illustrate the use of equation (8.2), consider the examples in Figure 8.3. First, consider Figure 8.3a. Individuals G and H are second cousins, and I is their offspring. There are two common ancestors, A and B. Neither is inbred, and there is only one pathway through each. Table 8.1 gives the steps in calculating the inbreeding coefficient of I.

Figure 8.3b shows a more complicated example. Common ancestors are A, C, F, and G. G is itself inbred, and there are four pathways through A. The lower part of Table 8.1 shows the calculations for the inbreeding coefficient of individual I.

Effect of Inbreeding on Heterozygosity

We define an inbred population as one in which at least some individuals are inbred. Inbreeding increases the frequency of homozygotes and decreases the frequency of heterozygotes, compared to a reference population. In the reference population, some alleles may be identical by state, but by definition are not identical by descent; homozygosity is due to identity by state only. In an inbred population, alleles are not

TABLE 8.1 Calculation of the inbreeding coefficient for the individuals indicated in Figure 8.3.
In each path the common ancestor is indicated in boldface. See text for details.

Common Ancestor	Path	n	f_{ai}	$\left(\dfrac{1}{2}\right)^{n}(1 + f_{ai})$
From Figure 8.3a:				
A	GECADFH	7	0	0.0078
B	GECBDKH	7	0	0.0078
Total				0.0156
From Figure 8.3b:				
F	HFJ	3	0	0.125
G	HGJ	3	$(1/2)^3$	0.125 × 1.125
C	HFCGJ	5	0	0.0313
A	HEBACFJ	7	0	0.0078
A	HEBACGJ	7	0	0.0078
A	HEBADGJ	7	0	0.0078
A	HFCADGJ	7	0	0.0078
Total				0.3281

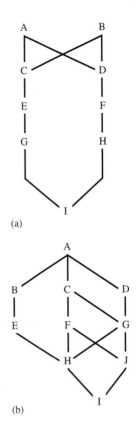

(a)

(b)

Figure 8.3 **Sample pedigrees for calculating the inbreeding coefficient, *f*.** In each case, the inbreeding coefficient of the individual marked I is calculated. See Table 8.1 and text for details of calculations.

only identical by state, but can also be identical by descent due to inbreeding. Assuming the reference population is in the relatively recent past, mutation can be ignored and alleles that are identical by descent must be identical by state. Therefore, in an inbred population homozygosity is due to both identity by state inherited from the reference population and to identity by descent due to inbreeding since the reference population.

Let G_f be the observed frequency of homozygotes in an inbred population, and G_r be the frequency of homozygotes (identity by state) in the reference population. Then the homozygosity (probability of identity by state) in the inbred population is (using the law of total probability—see Appendix A),

$$G_f = \Pr(\text{ibs}) = \Pr(\text{ibs}|\text{ibd}) \times \Pr(\text{ibd}) + \Pr(\text{ibs}|\text{not ibd}) \times \Pr(\text{not ibd})$$

Barring mutation, $\Pr(\text{ibs}|\text{ibd}) = 1$. Also, $\Pr(\text{ibd}) = f$, and $\Pr(\text{ibs}|\text{not ibd}) = G_r$. Substituting, we get

$$G_f = 1 \times f + G_r(1 - f)$$

Rearranging slightly,

$$G_f = G_r + f(1 - G_r) \qquad (8.3)$$

Inbreeding increases homozygosity by an amount that depends on *f*. We can see the effect on heterozygosity using the relationship $H = 1 - G$. Substituting and simplifying, we get

$$H_f = H_r(1 - f) \qquad (8.4)$$

where H_f and H_r are the frequencies of heterozygotes in the inbred and reference population, respectively. Rearranging equation (8.4) gives

$$f = \frac{H_r - H_f}{H_r} \tag{8.5}$$

So f can be interpreted as a measure of the proportionate reduction of the frequency of heterozygotes, compared to the reference population.

In Section 3.3, we defined F as the proportionate deviation of observed heterozygosity in a population compared to the heterozygosity expected under Hardy-Weinberg equilibrium:

$$F = \frac{H_{exp} - H_{obs}}{H_{exp}} \tag{8.6}$$

F was defined as a measure of deviation from the Hardy-Weinberg expectation due to unspecified causes. If inbreeding is the only process affecting the locus being considered, then genotype frequencies in the reference population will be the Hardy-Weinberg expected frequencies, and any deviation from Hardy-Weinberg expected heterozygosity in the inbred population will be due solely to inbreeding. Therefore, $H_r = H_{exp}$ and $H_f = H_{obs}$, and we can write

$$f = \frac{H_{exp} - H_{obs}}{H_{exp}} \tag{8.7}$$

Equation (8.7) is valid only if no other processes are affecting genotype frequencies. Under these conditions, f is a measure of deviation from Hardy-Weinberg expected heterozygosity due to inbreeding in the population. We will continue to use F to indicate a deviation due to unspecified causes, and f to indicate the deviation due to inbreeding.

Note the two different interpretations of f. At the individual level, it is the probability that two alleles in an individual are identical by descent. At the population level, it is the proportionate reduction of heterozygosity in an inbred population compared to a noninbred reference population. These interpretations are equivalent if no other evolutionary processes are operating, but keep in mind that a population can deviate from Hardy-Weinberg expected heterozygosity for many other reasons besides inbreeding.

If we are willing to assume that no other evolutionary processes are acting on a locus, we can use observed and expected heterozygosities at that locus to estimate the average inbreeding coefficient in a population. For example, Table 8.2 gives the observed and expected heterozygosities at seven microsatellite loci in two populations of snails (*Physa acuta*). The estimates of f were calculated from equation (8.7). Monsutti and Perrin (1999) attribute these high values of f to self-fertilization, an extreme form of inbreeding (see below).

Effect of Inbreeding on Genotype and Allele Frequencies

We have seen that inbreeding increases homozygosity and decreases heterozygosity in a population. We can determine the relationship between the genotype

TABLE 8.2 Observed and expected heterozygosities at seven microsatellite loci, and estimates of f in two populations of snails, *Physa acuta*. *Source:* From Monsutti and Perrin (1999).

Locus	Lake Population			Pond Population		
	H_{obs}	H_{exp}	$\hat{f}$	H_{obs}	H_{exp}	$\hat{f}$
32-B	0.438	0.604	0.276	0.286	0.712	0.598
61	0.133	0.357	0.627	0.313	0.507	0.383
19	0.500	0.813	0.385	0.000	0.635	1.000
83	0.000	0.362	1.000	0.200	0.856	0.766
27	0.250	0.760	0.671	0.286	0.708	0.596
59-B	0.250	0.492	0.492	0.313	0.544	0.425
9	0.000	0.250	1.000	0.000	0.179	1.000

frequencies and f, as follows: Assume that a genotype is formed by picking two alleles at random (with replacement) from the gene pool. Then the frequency of A_iA_i homozygotes is

$$P_{ii} = \Pr(A_iA_i|\text{not ibd}) \times \Pr(\text{not ibd}) + \Pr(A_iA_i|\text{ibd}) \times \Pr(\text{ibd})$$

The only term in this equation that may not be obvious is $\Pr(A_iA_i|\text{ibd})$. Given that two alleles are ibd, what is the probability that they are A_i? It is just the frequency of that allele in the population, or p_i. So we get

$$P_{ii} = p_i^2(1 - f) + p_if$$

Rewriting this as

$$P_{ii} = p_i^2 + p_i(1 - p_i)f \tag{8.8}$$

we see how inbreeding increases the frequency of A_iA_i homozygotes compared to the frequency under random mating.

Similarly, the frequency of the A_iA_j heterozygote under inbreeding is

$$P_{ij} = \Pr(A_iA_j|\text{not ibd}) \times \Pr(\text{not ibd}) + \Pr(A_iA_j|\text{ibd}) \times \Pr(\text{ibd})$$

Two alleles must be the same if they are ibd; therefore, $\Pr(A_iA_j|\text{ibd}) = 0$, and we get

$$\Pr(A_iA_j) = 2p_ip_j(1 - f) \tag{8.9}$$

For two alleles, equations (8.8) and (8.9) reduce to

$$P_{11} = p^2 + pqf \tag{8.10}$$
$$P_{12} = 2pq(1 - f) \tag{8.11}$$
$$P_{22} = q^2 + pqf \tag{8.12}$$

How does inbreeding affect allele frequencies? First, for two alleles let p be the frequency of A_1 under random mating and p_f the frequency under inbreeding.

Likewise for the genotype frequencies. Then, using equation (2.6)

$$p_f = P_{11f} + \frac{1}{2}P_{12f}$$

$$= (p^2 + pqf) + \frac{1}{2}[2pq(1 - f)]$$

$$= \cdots$$

$$= p$$

You should convince yourself that the same conclusion holds for multiple alleles (Problem 8.7).

To summarize, inbreeding alone does not affect allele frequencies. Only the genotype frequencies change, with an increase in homozygotes and decrease in heterozygotes. This assumes that natural selection is not acting on the locus; we will consider the effects of selection later.

Inbreeding in Finite Populations

In any finite population, there is a nonzero probability that an individual receives two alleles that are identical by descent, even if individuals in the population mate at random. If we start with the reference population, in which no alleles are identical by descent, the probability that two randomly chosen alleles in the first generation are ibd is $1/2N$ (Section 7.2). This probability must increase with time, because each generation creates a new possibility that alleles are identical by descent, to be added to the already existing probability. To see this, refer to Figure 8.4. Two alleles in generation $t + 1$ can be identical by descent in two different ways. First, they can come from the same allele in generation t, with probability $1/2N$ (solid lines). This is new identity by descent. Second, two alleles in generation $t + 1$ can come from different alleles in generation t, with probability $(1 - 1/2N)$; but these two alleles can be identical by descent from a previous generation, with probability f_t (dashed lines). So the probability that two alleles are identical by descent in generation $t + 1$ is

$$f_{t+1} = \frac{1}{2N} + \left(1 - \frac{1}{2N}\right)f_t \qquad (8.13)$$

If the initial inbreeding coefficient in the reference population is $f_o = 0$, the solution to this equation is

$$f_t = 1 - \left(1 - \frac{1}{2N}\right)^t \qquad (8.14)$$

(See Box 8.1 for the algebraic details.) Figure 8.5 illustrates how fast the inbreeding coefficient increases for various population sizes.

From equation (8.4), it is obvious that as the inbreeding coefficient increases, the frequency of heterozygotes must decrease. Therefore, heterozygosity must decrease each generation in a finite population. That should come as no surprise; one of the fundamental conclusions from Chapter 7 was that expected heterozygosity will decrease according to equation (7.5)

$$H_t = H_o\left(1 - \frac{1}{2N}\right)^t \qquad (8.15)$$

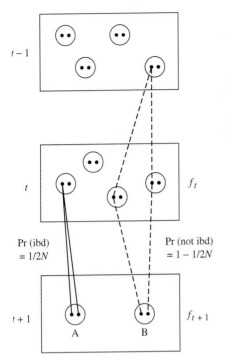

Figure 8.4 Identity by descent in a finite population. In generation $t + 1$, an individual can receive two alleles that are identical by descent (ibd) in two ways. Individual A received two copies of the same allele from generation t. (This assumes self-fertilization is possible.) Individual B received copies of two different alleles from generation t, but those alleles are ibd from generation $t - 1$. See text for calculation of f_{t+1}.

BOX 8.1 The solution to the recursion equation for f

We start with the recursion equation (8.13)

$$f_{t+1} = \frac{1}{2N} + \left(1 - \frac{1}{2N}\right)f_t \qquad \textbf{(1)}$$

The algebra is easier if we work with $1 - f$ instead of f. Multiply both sides by (-1) and then add 1. The result is

$$1 - f_{t+1} = 1 - \frac{1}{2N} - \left(1 - \frac{1}{2N}\right)f_t$$

Factor the right side and rewrite as

$$1 - f_{t+1} = \left(1 - \frac{1}{2N}\right)(1 - f_t) \qquad \textbf{(2)}$$

Now, let $x = 1 - f$. Equation (2) then becomes

$$x_{t+1} = \left(1 - \frac{1}{2N}\right)x_t$$

We already know how to solve this equation. The solution is

$$x_t = \left(1 - \frac{1}{2N}\right)^t x_o$$

Now convert back to f:

$$1 - f_t = \left(1 - \frac{1}{2N}\right)^t (1 - f_o)$$

Solving for f_t gives

$$f_t = 1 - \left(1 - \frac{1}{2N}\right)^t (1 - f_o) \qquad \textbf{(3)}$$

This is the general solution to equation (1) for any initial inbreeding coefficient f_o. If $f_o = 0$, then

$$f_t = 1 - \left(1 - \frac{1}{2N}\right)^t$$

This is equation (8.14) in the main text.

Figure 8.5 **Increase in the average inbreeding coefficient for different population sizes.**

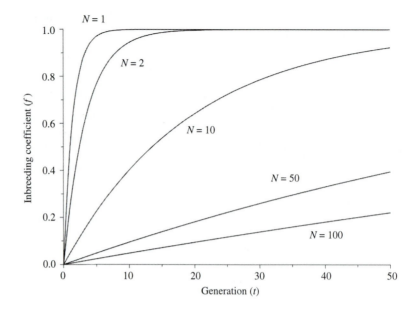

Combining equations (8.14) and (8.15), we get

$$H_t = H_o(1 - f_t)$$

which shows how the frequency of heterozygotes in generation t is related to the initial frequency in the reference population, H_o.

Clearly, inbreeding is similar to genetic drift, in that both decrease the frequency of heterozygotes. There are differences. Genetic drift changes allele frequencies; ultimately, one allele becomes fixed, and as allele frequencies drift toward zero or one, heterozygosity must approach zero even if mating is random. However, in any given generation, observed heterozygosity may not deviate from Hardy-Weinberg expected heterozygosity, based on allele frequencies in that generation. Inbreeding, on the other hand, does not change allele frequencies, but creates a deviation between observed and expected heterozygosity each generation.

There is another important difference: Genetic drift ultimately causes all genetic variation at a locus to be lost (ignoring mutation). Inbreeding resulting from consanguineous matings in a large population does not change allele frequencies; therefore, a large inbred population retains alleles, and a single generation of random mating will restore the Hardy-Weinberg expected frequencies.

We have seen that the probability that two alleles are identical by descent (f) increases each generation. There are two reasons for this: First, f will increase due to random sampling in a finite population, as predicted by equation (8.13). This occurs in all populations. Second, f will increase if consanguineous matings occur. This will be in addition to the increase due to finite population size, and will cause f to increase faster than predicted by equation (8.13).

We have defined inbreeding as mating among consanguineous individuals. The reference population is a population in the past in which all individuals are assumed to be unrelated. However, in any finite population, random mating includes the possibility of mating with relatives (or with oneself, for that matter).

Thus, some inbreeding, that is, mating among consanguineous individuals, will occur in any randomly mating finite population. An alternative definition of inbreeding is mating between individuals that are more closely related than two individuals chosen at random. In this sense, inbreeding is a kind of nonrandom mating based on relationship. It refers to matings among consanguineous individuals over and above what would be expected due to chance alone. The reference population is a randomly mating population of the same size.

The song sparrow population on Mandarte Island, British Columbia, has been studied extensively for many years. Since 1974 all individuals have been banded and all mating pairs observed. The population is small; the median number of breeding birds has been about 89, with much variation. Because of extensive pedigree data, Keller and Arcese (1998) were able to calculate the inbreeding coefficient of the offspring of each breeding pair from 1981 through 1995. Because of the small population size, mating between consanguineous individuals was common; 59 percent of all matings were between known relatives. For each year, Keller and Arcese compared the average inbreeding coefficient based on pedigrees with the expected inbreeding coefficient if all available adults were mated at random. They found no significant differences in either the means or the distributions of observed versus expected inbreeding coefficients. Figure 8.6 illustrates some of their results. Keller and Arcese conclude that mating among related individuals was no higher than it would be by chance, and that the observed inbreeding coefficients are due to finite population size alone. The song sparrow population was inbred in the sense that consanguineous matings occurred, but not in the sense that consanguineous matings occurred no more often than expected due to chance in a small population. The difference is in definition and choice of a reference population.

Self-Fertilization

An extreme form of inbreeding is self-fertilization. Some flowering plants are almost entirely self-fertilizing, while others have various mixtures of self-fertilization

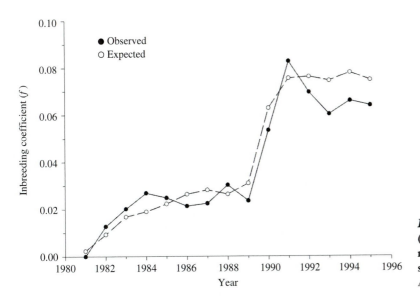

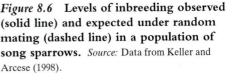

Figure 8.6 **Levels of inbreeding observed (solid line) and expected under random mating (dashed line) in a population of song sparrows.** *Source:* Data from Keller and Arcese (1998).

and outcrossing. Many of the world's most important crops are primarily self-fertilizing, for example, wheat, barley, rice, peas, and beans. Some of these have self-fertilization rates of 90 percent or higher. Self-fertilization is relatively rare in animals, but occurs in some snails, worms, many parasites, and a few other groups.

The frequency of heterozygotes will decrease rapidly under self-fertilization. Homozygotes will produce only homozygotes, and heterozygotes will produce half homozygotes and half heterozygotes. Therefore, under self-fertilization heterozygosity will decrease by half each generation

$$H_{t+1} = \frac{1}{2}H_t \qquad \text{(8.16)}$$

and heterozygosity will approach zero very quickly.

If we start with a large noninbred reference population ($f_o = 0$), the inbreeding coefficient after one generation of self-fertilization is $1/2$. It is easy to show (Problem 8.8) that the recursion equation for the inbreeding coefficient under self-fertilization is

$$f_{t+1} = \frac{1}{2}(1 + f_t) \qquad \text{(8.17)}$$

Figure 8.7 shows how heterozygosity and the inbreeding coefficient change under self-fertilization. Both approach their limiting values very quickly. After 10 generations, heterozygosity is essentially zero at any given locus.

The prediction is that the frequency of heterozygotes should be zero in self-fertilizing plants. In fact, the observed heterozygosity is frequently very low, for example, 0.00024 in one study of *Arabidopsis thaliana* (Abbott and Gomes 1989). However, most self-fertilizing plants retain some heterozygosity. One reason is that self-fertilizing organisms are usually not exclusively self-fertilizing; most undergo some degree of outcrossing.

We can predict the amount of inbreeding and the frequency of heterozygotes in an infinite population with a mating system of mixed self-fertilization

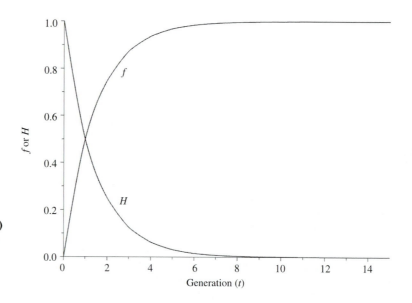

Figure 8.7 **Decrease in heterozygosity (*H*) and increase in the inbreeding coefficient (*f*) under continued self-fertilization.**
The initial population is assumed to be completely heterozygous.

and outcrossing. Let S be the proportion of the population that undergoes self-fertilization. Then, $1 - S$ is the proportion that undergoes outcrossing. We assume that outcrossing is equivalent to random mating. If an individual self-fertilizes, the inbreeding coefficient of its offspring is, from equation (8.17), $(1/2)(1 + f_t)$. If an individual outcrosses, the inbreeding coefficient of its off-spring is zero. So the average inbreeding coefficient in the offspring generation is the weighted average

$$f_{t+1} = S\left(\frac{1}{2}\right)(1 + f_t) + (1 - S) \times 0$$

$$= \frac{1}{2}S(1 + f_t)$$

At equilibrium, $f_t = f_{t+1} = \tilde{f}$, which gives the inbreeding coefficient at equilibrium

$$\tilde{f} = \frac{S}{2 - S} \qquad \textbf{(8.18)}$$

To see how heterozygosity changes, first recall that allele frequencies do not change with inbreeding. Therefore, the proportion $1 - S$ that mates randomly each generation will produce offspring whose heterozygosity is the same as the initial population, H_o. The proportion S that self fertilizes, will produce offspring whose heterozygosity is reduced according to equation (8.16). So the offspring heterozygosity will be the weighted average:

$$H_{t+1} = (1 - S)H_o + S\left(\frac{1}{2}H_t\right) \qquad \textbf{(8.19)}$$

At equilibrium, $H_{t+1} = H_t = \tilde{H}$, which gives

$$\tilde{H} = 2\left(\frac{1 - S}{2 - S}\right)H_o \qquad \textbf{(8.20)}$$

The initial heterozygosity is decreased by a fraction that depends on S. Note that if $S = 0$, the heterozygosity does not change, and if $S = 1$, the equilibrium heterozygosity is zero, as we previously showed. Figure 8.8 shows how heterozy-gosity approaches equilibrium for several values of S.

Most animals do not undergo self-fertilization, but the freshwater snail *Bulinus truncatus* is an exception. Viard et al. (1997) studied variation at four microsatellite loci in 38 populations of this species in Africa and on several Mediterranean islands. One advantage of studying microsatellite loci is that they are frequently polymor-phic, even in populations that show no variation for other kinds of genetic markers (Section 2.12). Over all 38 populations, Viard et al. found from 6 to 55 alleles per locus, with the average number of alleles per locus within a population ranging from 1.6 to 6.8. Table 8.3 gives observed and expected heterozygosities for several populations. The observed heterozygosities are remarkably low for microsatellite loci. Based on previous studies of the mating system, Viard et al. assumed this was due to a high degree of self-fertilization. The deviations from expected heterozy-gosity were assumed to be due to self-fertilization, and f was estimated from these

Figure 8.8 **Decrease in heterozygosity under different mixtures of self-fertilization and random mating.** S is the proportion of the population that undergoes self-fertilization each generation. The initial population is assumed to be completely heterozygous.

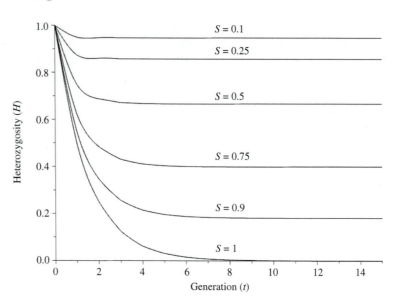

deviations using equation (8.7). S was then estimated from equation (8.18). Considering only populations with expected heterozygosities greater than 0.01, estimates of S ranged from 0.51 to 1.00, with an average of 0.92.

If we have a reasonable estimate of S from an independent source, we can estimate the predicted equilibrium value of f from equation (8.18), and compare it to an estimate obtained from equation (8.7). Any difference suggests

---

TABLE 8.3 Observed and expected heterozygosities, and estimates of inbreeding coefficients and self-fertilization rates in several populations of snails (*Bulinis truncatus*).

Heterozygosities are means over four microsatellite loci. In all, 38 populations were studied, of which four were monomorphic at all loci. Only the first ten polymorphic populations are listed here. The last row is the arithmetic average of the 34 polymorphic populations. *Source:* From Viard et al. (1997).

Population	$H_{obs} \pm$ s.e.	$H_{exp} \pm$ s.e.	$\hat{f}$	$\hat{S}$
Travignano	0.00 ± 0.00	0.30 ± 0.29	1.00	1.00
Orbo	0.00 ± 0.01	0.12 ± 0.22	0.96	0.98
Budoni	0.02 ± 0.02	0.35 ± 0.41	0.95	0.98
Lotzorai	0.02 ± 0.02	0.42 ± 0.31	0.96	0.98
Pula	0.00 ± 0.00	0.14 ± 0.24	1.00	1.00
Djanet	0.01 ± 0.01	0.05 ± 0.06	0.75	0.85
RD4RD	0.01 ± 0.03	0.05 ± 0.07	0.74	0.85
Fint	0.05 ± 0.09	0.07 ± 0.14	0.35	0.51
Sidi Chikh	0.01 ± 0.01	0.01 ± 0.01	0.00	0.00
Mada	0.05 ± 0.04	0.56 ± 0.25	0.91	0.95
Mean (34 populations)			0.84	0.89

the presence of evolutionary processes other than inbreeding. Marshall and Allard (1970) did this for two populations of wild oats, *Avena barbata*, in California. They estimated the outcrossing rate $(1 - S)$ by looking for heterozygous progeny of known homozygotes. For the population near Calistoga, California, they estimated S to be about 0.986. This gives a predicted equilibrium value of $\tilde{f} \cong 0.97$ from equation (8.18). From allele frequencies at four polymorphic allozyme loci, estimates of f based on equation (8.7) ranged from 0.70 to 0.78, with an average of about 0.75. This is significantly lower than the predicted value of 0.97, corresponding to higher heterozygosity than predicted from the estimated level of self-fertilization. Marshall and Allard attributed this excess heterozygosity to overdominance for chromosomal regions marked by the allozyme loci.

Repeated Inbreeding and Systems of Mating

Repeated self-fertilization is an extreme form of inbreeding. However, many plant and animal breeding programs use less extreme forms, for example, repeated matings between full or half siblings (or more distant relatives). Many laboratory cultures repeatedly involve full-sib matings, and many experiments require repeated half-sib or parent-offspring matings. Such systems of mating lead to repeated inbreeding and a continual increase in the inbreeding coefficient and decrease in heterozygosity, although at a slower rate than with repeated self-fertilization.

We will derive the recursion equations for f and H under repeated full sib mating. Look at Figure 8.9a. Individual I in generation $t + 1$ has parents X and Y (full siblings) in generation t and grandparents A and B (also full siblings) in generation $t - 1$. We wish to find the inbreeding coefficient of I, designated f_{t+1}.

Individual I could have received both of its alleles from grandparent A (probability $= 1/4$), or both from grandparent B (probability $= 1/4$), or one allele from A and one from B (probability $= 1/2$); see Figure 8.9b. If both alleles of I came from A, the probability that they are both A_1 or both A_2 is 1/2. But A is inbred, which means that A_1 and A_2 may be ibd, so to get the probability that both alleles of I are ibd, we must apply step 3b from page 274 and multiply by $(1 + f_A)$. This gives the conditional probability that the two alleles of I are ibd, given that both came from grandparent A as $(1/2)(1 + f_A) = (1/2)(1 + f_{t-1})$. Similarly, the conditional probability that the two alleles of I are ibd, given that both came from grandparent B is $(1/2)(1 + f_B) = (1/2)(1 + f_{t-1})$. Finally, if one allele came from A and one from B, the probability that they are ibd is g_{t-1}, which is equal to f_t. Using the law of total probability to put all of this together, we get

$$f_{t+1} = \frac{1}{4}\left[\left(\frac{1}{2}\right)(1 + f_{t-1})\right] + \frac{1}{4}\left[\left(\frac{1}{2}\right)(1 + f_{t-1})\right] + \frac{1}{2}[f_t]$$

which simplifies to

$$f_{t+1} = \frac{1}{4}(1 + 2f_t + f_{t-1}) \qquad (8.21)$$

Figure 8.9   **Continued full-sib mating.**   (a) Path diagram showing repeated full-sib mating. Alleles in individuals A and B are labeled differently only to distinguish them; they may or may not be identical by state. (b) Identity by descent in full-sib mating. Alleles along the top represent alleles I possibly received from parent X; alleles along the left side represent alleles I possibly received from parent Y. The areas outlined in bold represent the probabilities that individual I received two alleles from grandparent A or both from grandparent B. The conditional probability that two alleles are ibd, given that they both came from the same grandparent, is $(1/2)(1 + f_{t-1})$. If one allele came from A and one came from B, the conditional probability of ibd is g_{t-1}. See text for details.

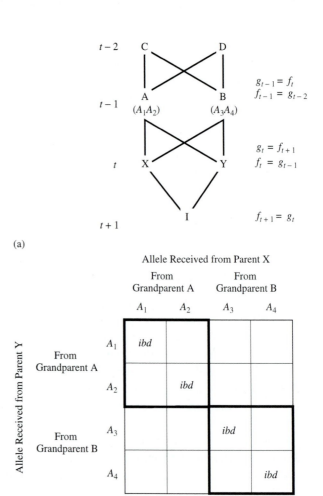

(a)

(b)

Once we have the recursion for f, it is easy to find the recursion for H (Problem 8.11). It is

$$H_{t+1} = \frac{1}{2}H_t + \frac{1}{4}H_{t-1} \qquad (8.22)$$

Figure 8.10 shows how f and H change under repeated full sib mating.

For more complicated systems of inbreeding, the derivations are more complex. Table 8.4 summarizes the recursion equations for f under several systems of mating. For details, see Crow and Kimura (1970) or Falconer and Mackay (1996).

Inbreeding Depression

Since long before Mendelian genetics, it has been known that matings between close relatives tend to produce offspring with morphological, developmental, or mental defects, or with lowered viability or fertility. In humans, for example, it is well established that the children of consanguineous parents have a higher than normal frequency of deleterious abnormalities such as birth defects, mental retardation, and rare genetic diseases. Most human societies forbid parent-offspring or

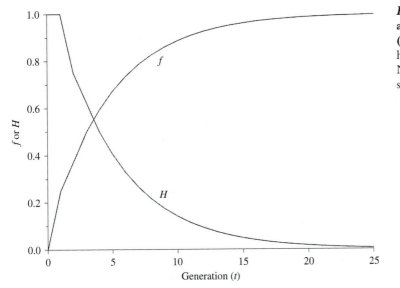

Figure 8.10 **Decrease in heterozygosity (*H*) and increase in the inbreeding coefficient (*f*) under repeated full-sib mating.** Initial heterozygosity is assumed to be 100 percent. Note that *H* does not begin to decrease until the second generation.

brother-sister matings, and this taboo is probably based on prescientific observations of abnormal children from such matings. Darwin (1875) found that cattle, pig, and bird breeders generally believed that close inbreeding leads to a decrease in vigor and fertility. He studied many plant species and found that most showed deleterious effects of close inbreeding. For example, he conducted experiments demonstrating that self-fertilized morning glories (*Ipomoea purpurea*) produced fewer seed capsules and fewer seeds per capsule than outcrossed plants (Darwin 1876). Several nineteenth-century experiments documented a decline

TABLE 8.4 Recursion equations for the inbreeding coefficient under several systems of repeated inbreeding.

Mating System	Equation for *f*
Self-fertilization	$f_{t+1} = \dfrac{1}{2}(1 + f_t)$
Mixed self-fertilization and random mating	$f_{t+1} = \dfrac{1}{2}S(1 + f_t)$
Parent-offspring	$f_{t+1} = \dfrac{1}{4}(1 + 2f_t + f_{t-1})$
Full siblings	$f_{t+1} = \dfrac{1}{4}(1 + 2f_t + f_{t-1})$
Half siblings	$f_{t+1} = \dfrac{1}{8}(1 + 6f_t + f_{t-1})$
Repeated backcrossing to individual with inbreeding coefficient of f_I	$f_{t+1} = \dfrac{1}{4}(1 + f_I + 2f_t)$
Repeated backcrossing to inbred line	$f_{t+1} = \dfrac{1}{2}(1 + f_t)$

in vitality, weight, fertility, and viability after close inbreeding in rats and mice (Wright 1977).

This phenomenon is known as **inbreeding depression**, and is a direct consequence of increased homozygosity due to inbreeding. Rare recessive (or nearly recessive) alleles, whose deleterious effects are usually masked in heterozygotes, are expressed more frequently as homozygosity is increased by inbreeding. Overdominance, in which heterozygotes have higher fitness than either homozygote, may also contribute to inbreeding depression since the frequency of heterozygotes is reduced under inbreeding.

First, consider rare recessive disease alleles. Let A_2 be a rare deleterious allele that, when homozygous, causes a specific disease. Under random mating, the frequency of the disease is q^2. But with inbreeding, the frequency becomes $q^2 + fpq$. Consider first cousins, for example. The inbreeding coefficient is $1/16 = 0.0625$. If the recessive allele has a frequency of 0.01, the frequency of the disease genotype among children of unrelated parents is 10^{-4}. Among children of first cousins, the frequency of the disease is about 7.2×10^{-4}, more than seven times higher. The rarer the disease allele, the greater the ratio; for example, if $q = 0.001$, the disease is about 63 times more frequent among children of first cousins. Table 8.5 shows the frequencies of several kinds of genetic abnormalities in the general population, and among children of first cousins.

We saw in Chapter 2 (Section 2.7) that there is much hidden variation for fitness in most *Drosophila* populations. When chromosomes are made homozygous, the viability of chromosomal homozygotes is nearly always less than the viability of chromosomal heterozygotes. This is almost certainly due to rare deleterious recessive or nearly recessive alleles whose effects are expressed when a chromosome is made homozygous.

We can predict the conditions under which inbreeding depression will occur by considering the mean fitness of the population under random mating and under inbreeding. Designate these fitnesses by $\overline{w}_o$ and $\overline{w}_f$, respectively. Using the notation of Section 5.8, the fitnesses of the genotypes A_1A_1, A_1A_2, and A_2A_2 are $1, 1 - hs$, and $1 - s$. We assume that A_2 is a deleterious, partially recessive allele,

---

TABLE 8.5 Percent of genetic abnormalities in children of unrelated parents and in children of parents who are first cousins.

Numbers in parentheses indicate sample sizes. *Source:* From Ayala (1982); original data from Stern (1973).

Country	Children of Unrelated Parents (%)	Children of First Cousins (%)
United States	9.8 (163)	16.2 (192)
France	3.5 (833)	12.8 (144)
Sweden	4.0 (165)	16.0 (218)
Japan	8.5 (3570)	11.7 (1817)
Average	6.5	14.2

and that the heterozygote has intermediate fitness $(0 < h < 1)$. Then, the mean fitness under random mating is

$$\overline{w}_o = p^2 + 2pq(1 - hs) + q^2(1 - s) \qquad \textbf{(8.23)}$$

To get the mean fitness under inbreeding, we must multiply the genotype fitnesses by the genotype frequencies under inbreeding, instead of under random mating. This gives

$$\overline{w}_f = (p^2 + pqf) + 2pq(1 - f)(1 - hs) + (q^2 + pqf)(1 - s)$$

Rewrite this as

$$\overline{w}_f = [p^2 + 2pq(1 - hs) + q^2(1 - s)] + pqf[1 - 2(1 - hs) + (1 - s)]$$

or,

$$\overline{w}_f = \overline{w}_o + pqf[1 - 2(1 - hs) + (1 - s)] \qquad \textbf{(8.24)}$$

Inbreeding depression means $\overline{w}_f < \overline{w}_o$. This will be true if the quantity in brackets on the right side of equation (8.24) is negative. In other words, inbreeding depression will occur if

$$1 - 2(1 - hs) + (1 - s) < 0$$

which simplifies to

$$h < 0.5 \qquad \textbf{(8.25)}$$

This means that inbreeding depression will occur if the heterozygote is closer in fitness to A_1A_1 than it is to A_2A_2.

We saw in Section 6.4 that this condition is usually met for mutant alleles, at least in *Drosophila*, where we have the most data. Lynch and Walsh (1998) estimated that for an average deleterious allele, $h \cong 0.1$. For mildly deleterious mutations, $h \cong 0.15$ to 0.30, and for recessive lethals (much rarer than mildly deleterious mutations), $h \cong 0.02$ to 0.04.

We conclude from this analysis that inbreeding depression should be common, to the extent that partially recessive deleterious alleles are common. The available data suggest that such alleles are surprisingly common (Section 6.6).

Inbreeding depression is commonly observed in laboratory, domesticated, and captive populations. However, it is difficult to document in natural populations. It requires estimates of f for individuals in the population, along with estimates of their fitness, or major components thereof. Both require intensive and long-term field work. In small, isolated populations of birds, individuals can be banded and followed throughout their lives, making it possible to estimate survival and breeding success, and to estimate inbreeding coefficients.

Keller (1998) studied inbreeding depression in the Mandarte Island population of song sparrows mentioned earlier. Pedigree information was available for most individuals going back to 1975. The average coefficient of consanguinity (g) between the male and female of 671 breeding pairs was 0.031. About 7.6 percent of all matings were between individuals with $g \geq 0.125$. Offspring of these consanguineous matings suffered dramatic fitness losses. For individuals with $f = 0.25$, survival from egg to breeding age was reduced by 49 percent. For females with $f = 0.25$, or $f = 0.0625$, lifetime reproductive success (number of offspring that survived to parental independence) was reduced by 48 percent or 13 percent, respectively, compared to noninbred females. Surprisingly, inbreeding seemed to have no significant effect on lifetime reproductive success of males.

If pedigree information is unavailable or incomplete, the degree of inbreeding can sometimes be estimated from heterozygosity of neutral molecular markers. Slate et al. (2000) used this approach to document inbreeding depression in a natural population of red deer (*Cervus elaphus*). They estimated heterozygosity at several microsatellite loci, and assumed that individuals with fewer heterozygous loci were more inbred than individuals who were more heterozygous. For both males and females, they found significant positive correlations between a standardized measure of multilocus heterozygosity and lifetime breeding success (number of offspring produced, usually estimated from field observations). Less heterozygous (presumably more inbred) individuals had significantly lower lifetime breeding success. Slate et al. conclude that, if their results are typical of other vertebrates, inbreeding depression can significantly lower population fitness, especially in small, isolated populations.

We discussed the increased risk of extinction for small populations in Section 7.9. One conclusion was that the increased probability of fixation of slightly deleterious mutations can reduce absolute fitness in an already stressed population, possibly increasing the probability of extinction. Is inbreeding depression an additional factor in increasing the danger of extinction in small populations? This has been a controversial question. One view is that inbreeding depression will decrease the reproductive rate of an already small and endangered population, and may tip the scales toward extinction. Alternatively, if a population is so small that inbreeding depression is a serious problem, then other factors such as demographic and environmental stochasticity (Section 7.9) may be more important in determining survival or extinction.

Frankham (1998) addressed this question by surveying the literature on island populations. His logic was (1) island populations historically have had a higher risk of extinction than mainland populations; (2) island populations are smaller than mainland populations, and therefore should have higher inbreeding coefficients; and (3) laboratory and domestic populations show increased risk of extinction when inbreeding increases beyond $f \cong 0.5$. Frankham asked whether inbreeding coefficients in island populations are in this range. He calculated an effective inbreeding coefficient for island populations by comparing observed heterozygosities of island populations with observed heterozygosities in mainland populations, based on allozymes and microsatellites. Endemic island populations had a mean effective inbreeding coefficient of 0.57, with 64 percent having $f > 0.50$. Frankham concluded that increased risk of extinction occurs when inbreeding coefficients reach about $f \cong 0.5$, and that island populations frequently have inbreeding coefficients in this range. Inbreeding depression can increase the risk of extinction in island populations, in addition to demographic and environmental stochasticity and environmental catastrophes discussed in Section 7.9.

Saccheri et al. (1998) obtained direct evidence to support this conclusion. They took advantage of the metapopulation structure of the Glanville fritillary butterfly (*Melitaea cinxia*) in Finland. A metapopulation consists of numerous local populations of varying size, with some migration and gene flow between them. Local populations are subject to relatively high rates of extinction and re-colonization. Saccheri et al. studied 42 local populations, estimating various eco-logical parameters such as population size, distance to other local populations, patch size, and abundance of host plants. They estimated levels of inbreeding from heterozygosity at seven allozyme loci and one microsatellite locus. Between 1995 and 1996, seven of the 42 local populations had gone extinct. Saccheri et al. de-veloped a complex multivariate statistical model with which they were able to separate the effect of heterozygosity from the other variables. They found a strong inverse relationship between observed heterozygosity and extinction; pop-ulations with low frequencies of heterozygotes (thus, inferred high levels of in-breeding) were more likely to go extinct ($P < 0.005$). Inbreeding depression was manifest as decreased larval survival, adult longevity, and egg-hatching rate. Saccheri et al. concluded that inbreeding depression is a significant factor in-creasing the risk of extinction in small populations.

It has been suggested that inbreeding depression may be more pronounced in stressful environments (e.g., Hoffmann and Parsons 1991). If, as seems reasonable, small populations are frequently stressed in some way, then inbreeding depression may contribute even more to the risk of extinction. The population of song spar-rows described previously crashed to only 11 individuals during the severe win-ter of 1988–1989. Inbreeding coefficients were significantly higher for birds that died ($\hat{f} = 0.0312$) than for survivors ($\hat{f} = 0.0065$) (Keller et al. 1994). Even though most mortality was probably due to density-independent factors, birds that were only slightly inbred suffered more than those that were less so.

Keller and Waller (2002) reviewed studies of inbreeding and inbreeding de-pression in many natural plant and animal populations. They concluded that both inbreeding and inbreeding depression are widespread. The effects of inbreeding vary among different groups, but inbreeding can have strong effects on both in-dividual and population fitness. The effects of inbreeding on individual fitness have been well documented, but the effects on population dynamics are less well understood. These effects depend on the kind of selection involved, and on the mechanisms of density-dependent population regulation. However, the examples summarized here and in Keller and Waller (2002) suggest that inbreeding can sig-nificantly increase the risk of extinction in small populations.

Young et al. (1997) provided an example of how information about inbreed-ing can be useful in conservation biology. The plain pigeon (*Columba inornata wetmorei*) is an endangered subspecies endemic to Puerto Rico. Its population size has declined dramatically, primarily due to habitat destruction and hunting. A captive breeding program was begun in 1983 in an attempt to save the sub-species from extinction. Young et al. (1997) compared several measures of repro-ductive success to the proportion of minisatellite bands shared in several captive breeding pairs of plain pigeons. The proportion of bands shared is an indication of degree of relationship between them, analogous to the coefficient of consan-guinity. Parents and offspring or full siblings will share at least half their bands; more distant relatives will share fewer bands. Young et al. found a statistically sig-nificant inverse relationship between the proportion of bands shared by the male and female, and the number of eggs they produced (Figure 8.11). Based on these and similar results in a study of Puerto Rican parrots, they concluded that

Figure 8.11   **Relationship between proportion of minisatellite bands shared and four measures of reproductive success in six breeding pairs of the Puerto Rican plain pigeon.**   A nonparametric statistical test indicated that only the relationship between total eggs and proportion of bands shared was significantly negative. *Source:* Data from Young et al. (1997).

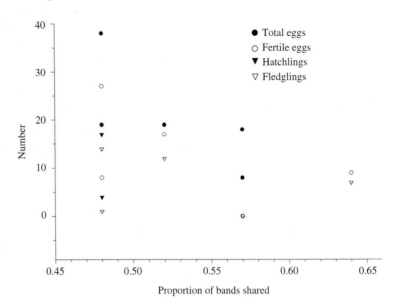

inbreeding depression is a potential problem in captive breeding programs and that DNA fingerprinting can help identify closely related individuals. They suggested that birds with the fewest shared bands should be paired for maximum breeding success.

It is useful to have some standard way of quantifying inbreeding depression, for comparison among different populations. The simplest model of inbreeding depression is to assume that population fitness (or some major component) declines linearly with f:

$$\overline{w}_f = \overline{w}_o - \beta f \tag{8.26}$$

where $\overline{w}_f$ and $\overline{w}_o$ are the fitnesses of inbred and noninbred populations, f is the average inbreeding coefficient of the population, and β is the regression coefficient. An alternative model, which sometimes provides a better fit to data, is to assume that fitness declines exponentially with f:

$$\overline{w}_f = \overline{w}_o e^{-\beta f} \tag{8.27}$$

In both models, a positive β indicates inbreeding depression ($\overline{w}_f < \overline{w}_o$). More complicated models can also be fit to existing data.

The **cost of inbreeding** can be defined as the proportionate decrease in fitness of an inbred population (with specified f) compared to a noninbred population. It is often symbolized by δ (delta)

$$\delta = \frac{\overline{w}_o - \overline{w}_f}{\overline{w}_o} \tag{8.28}$$

For the model of equation (8.27), this turns out to be

$$\delta = 1 - e^{-\beta f} \tag{8.29}$$

Equation (8.29) is useful for comparing the cost of inbreeding in different organisms. For example, Ralls et al. (1988) estimated the cost of inbreeding in 40 captive mammal populations with various levels of inbreeding. Estimates of β

were quite variable, but were positive for 36 of the 40 populations, indicating that inbreeding depression is widespread. For $f = 0.25$ (offspring of full-sib or parent-offspring matings) estimates of δ ranged from -0.19 (an inbreeding advantage) to 0.98, with a mean of 0.33. Ralls et al. considered these to be underestimates because they were based only on reduced juvenile survival; other components of fitness probably also contribute to the cost of inbreeding.

Inbreeding and Purging of Deleterious Recessive Alleles

In Section 6.6, we saw that the total deleterious mutation rate is surprisingly high in many organisms, including humans. This mutational load potentially contributes to reduced population fitness, and may increase the probability of extinction in small populations.

On the other hand, inbreeding allows deleterious recessive mutations to be expressed in homozygous genotypes, and thus be subject to elimination by natural selection. This elimination of deleterious recessive alleles in inbred populations is called **purging** (Byers and Waller 1999).

Whether purging will be effective in eliminating the mutational load in natural populations depends on many things: mutation rates and effects, degree of dominance, homozygous and heterozygous selection coefficients, effective population size, breeding system, and so on. For example, recall that most mutations are not completely recessive; i.e., there is some heterozygous effect. If so, natural selection can act on the heterozygote, and the purging advantage of inbreeding is much reduced. Similarly, recall that most mutations have small effects, and that selection is less effective in small populations because genetic drift can overpower weak natural selection. Since inbreeding reduces effective population size, there is some question as to whether natural selection can effectively eliminate slightly deleterious mutations in small populations. We summarize two studies that ask how often purging occurs in plants and animals.

Byers and Waller (1999) surveyed the plant literature, asking the question, "Do plant populations purge their genetic load?" They examined 52 studies that attempted to answer the question using a variety of techniques. They predicted that populations with a history of inbreeding would show less inbreeding depression (because deleterious mutations had been purged) than populations with little or no prior inbreeding. The history of inbreeding was inferred by several methods, including average inbreeding coefficients, flower morphology, population size, and others. Inbreeding depression was usually estimated by comparing fitness components in experimentally inbred populations to noninbred populations and estimating δ from equation (8.28). Overall, 20 studies showed evidence of purging, 29 did not, and three were ambiguous. Byers and Waller concluded that "although populations are capable of purging under some circumstances, purging appears neither consistent nor effective enough to reliably reduce ID [inbreeding depression] in small and inbred populations" (Byers and Waller 1999, p. 505).

Ballou (1997) conducted a similar survey of 25 captive mammal populations. The hypothesis was similar to that of Byers and Waller (1999): If purging occurs, populations with a history of inbreeding should show less inbreeding depression than populations with no history of inbreeding, because some deleterious alleles have been purged. Most of the observed inbreeding depression was in the form of reduced neonatal survival; of the 19 populations with sufficient data, 17 showed inbreeding depression for neonatal survival. Populations with greater history of inbreeding showed less inbreeding depression for neonatal survival, but not for other components of fitness. Because of small sample sizes, the reduction was significant in only one species, but the overall trend was consistent and highly significant. Ancestral inbreeding tended to reduce, but not eliminate, inbreeding depression. In other words, purging appears to be weak but consistent with respect to neonatal survival, but not with respect to other components of fitness.

The conclusion from both surveys is that purging does occur, and it can reduce inbreeding depression, but its effects are weak and inconsistent unless inbreeding is substantial.

It has been suggested that intentional inbreeding might sometimes be used as a strategy to purge genetic load in captive breeding programs (e.g., Templeton and Read 1984). The results summarized here suggest that this is probably not a good idea for most populations. It appears that the weak beneficial effects of purging are usually outweighed by the deleterious effects of the intensive inbreeding necessary for it to be effective. Moreover, the natural selection involved in the purging process may select for the captive environment, which will be very different from the natural environment into which captive-bred individuals will eventually be released.

Inbreeding and Gametic Disequilibrium

Recall from Section 4.2 that the coefficient of gametic disequilibrium is defined as

$$D = g_1 g_4 - g_2 g_3$$

where the g's are the frequencies of the two-locus gamete types. Recall also that in the absence of selection, gametic disequilibrium decays at a rate that depends on the recombination rate:

$$D_{t+1} = (1 - r)D_t$$

As discussed in Section 4.2, most outbreeding populations show little or no gametic disequilibrium for allozyme or microsatellite loci.

Recombination occurs only in double heterozygotes. Inbreeding reduces heterozygosity at all loci; therefore, the frequency of double heterozygotes should be low under intensive inbreeding (e.g., self-fertilization). This means that initial gametic combinations will rarely be broken up, and decay to gametic equilibrium will be much slower than in a randomly mating population. We would therefore expect to find high levels of gametic disequilibrium in predominantly self-fertilizing organisms. This prediction was confirmed in the self-fertilizing snail populations described previously. Viard et al. (1997) found extensive two-locus gametic disequilibrium in many populations, a result that is rare in outbreeding populations.

Similarly, if natural selection favors particular allelic combinations, then gametic disequilibrium can be very strong in inbred populations, because the favored types increase due to natural selection, and are rarely broken up by recombination. Allard et al. (1972) found just this situation in a population of slender wild oats, *Avena barbata*. This species has a self-fertilization rate of $S \cong 0.98$. Allard et al. surveyed five polymorphic allozyme loci and found a great excess of 12221 and 21112 gamete types (where the five positions represent the five loci, and 1 or 2 represent the slow or fast allele, respectively). Moreover, they found great excess of 21112/21112 and 12221/12221 homozygotes, with the former occupying mostly mesic sites and the latter mostly xeric sites. They interpreted these results as indicating natural selection favoring 21112 in mesic habitats and 12221 in xeric habitats. This does not necessarily mean that natural selection acts directly on the loci surveyed; they may themselves be in strong gametic disequilibrium with the loci under direct selection. Favored allelic combinations such as these were called "coadapted gene complexes" by Dobzhansky (1970). Because of the high rate of self-fertilization and consequent low recombination, these complexes have reached high frequencies in their favored habitats, with corresponding deficiencies in the frequencies of other gamete types.

Outbreeding, Hybrid Vigor, and Outbreeding Depression

The opposite of inbreeding is sometimes called **outbreeding**: mating of individuals who are less related than would be expected by chance. It occurs in laboratory populations when two separate lines (usually previously inbred) are crossed to one another. In natural populations, it can occur when two previously isolated populations come into contact and breed with one another.

The main consequence of outbreeding is the opposite of inbreeding: The mean fitness of the outbred population is higher than the fitness of either of the initial lines. To see this, imagine a large, randomly mating population ($f = 0$). Split this population into two lines and inbreed each. In each inbred line, some deleterious alleles will become fixed due to genetic drift, but different alleles are likely to be fixed in each line. Homozygosity will increase due to both genetic drift and inbreeding, and inbreeding depression will occur due to expression of deleterious recessive alleles. Now cross these two lines. Heterozygosity will be restored, and deleterious recessive alleles will again be masked. The hybrid (outbred) population will have a higher fitness than either inbred line.

This phenomenon of increased fitness in outbred populations is called **hybrid vigor**, or heterosis.[1] It is the opposite of inbreeding depression. Hybrid vigor is due primarily to masking of deleterious recessive alleles; overdominance (heterozygote advantage) is not necessary, although it may occur. If heterozygotes are intermediate in fitness, then the hybrid population will have higher fitness than the inbred population because fewer homozygotes for deleterious recessive alleles are present.

Hybrid vigor is commonly observed when inbred lines of laboratory organisms are crossed to one another. Outbreeding is also commonly used in plant and animal breeding. For example, hybrid corn is the result of a cross between two or more inbred lines. Yield of hybrid corn is 20 percent or more greater than the yield of inbred lines, and nearly all corn planted in the United States is hybrid corn.

---

[1]The word heterosis has been used in several, sometimes contradictory, ways. We will use the more descriptive and less ambiguous phrase hybrid vigor.

Hybrid vigor may not occur when distantly related populations are crossed. If two populations have been isolated for a long time, they may have diverged substantially, both as a result of random factors (mutation and genetic drift) and natural selection. When such populations are crossed, genetic or developmental incompatibilities may occur, and the hybrids may show reduced fertility or viability. This is sometimes called **outbreeding depression**, and is due, at least in part, to disruption of favorable gene interactions. As isolated populations adapt to their environments, selection may favor different multilocus genotypes (coadapted gene complexes) that interact to increase fitness in the local environment. Hybridization disrupts these favorable interactions, leading to decreased fitness in the hybrids.

Outbreeding depression is a potential concern in captive breeding and reintroduction programs. Conservation biologists sometimes wish to supplement an endangered population by introducing individuals from another population. The danger is that when these introduced individuals mate with the natives, outbreeding depression may occur. For example, ibex (*Capra ibex*) were hunted to extinction in the mountains of Czechoslovakia. Ibex from the ecologically similar mountains of Austria were introduced and established themselves. Later, biologists attempted to supplement the population by introducing ibex from Turkey and the Sinai. These areas are much warmer and drier than the mountains of Czechoslovakia, and the reproductive biology of ibex from these regions is very different. These introduced ibex mated with the Austrian-derived ibex, but the hybrids reproduced at the wrong time of the year for the mountainous environment. They mated in the fall, and young were born in February, in the middle of the cold winter. Reproduction failed, and the population went extinct (Greig 1979; Templeton 1997).

The ibex example is dramatic, but unusual. Conservation biologists sometimes introduce individuals from related populations into very small endangered populations that have low fitness. The result is more often hybrid vigor rather than outbreeding depression. For example, a small isolated population of greater prairie chicken (*Tympanuchus cupido pinnatus*) with low fertility and hatching success showed dramatically increased fertility when birds from a large distant population were introduced (Westemeier et al. 1998). Hedrick and Kalinowski (2000) review several examples, and the cautions associated with these kinds of breeding programs.

8.2 Assortative Mating

Assortative mating (sometimes called positive assortative mating) is nonrandom mating based on phenotype. It occurs when mating individuals are more alike phenotypically than two individuals chosen at random from the population. **Disassortative mating** (sometimes called negative assortative mating) is the opposite; individuals tend to mate with individuals that are phenotypically different from themselves. We consider assortative mating in this section, and disassortative mating in the next.

Assortative mating is fairly common in human populations. It occurs with respect to height, weight, skin pigmentation, IQ score, and other characteristics. Assortative mating also occurs among individuals with certain disabilities or diseases, for example, deafness, blindness, or albinism. The degree of assortative mating varies for different traits and among different populations.

In animals, assortative mating sometimes occurs based on body size or pigmentation. Light and dark phases of the Arctic skua (*Stercorarius parasiticus*) and the snow goose (*Chen caerulescens*) mate assortatively (O'Donald 1959; Cooch and Beardmore 1959). Fruit flies from different geographic regions sometimes mate assortatively under laboratory conditions.

Plants can also mate assortatively. Plants that flower early in the season tend to pollinate and be pollinated by other early-flowering plants, and the same is true for late-flowering plants. Insect pollinators attracted to flowers of a specific color or structure or height will transfer pollen to other flowers with similar phenotypic characteristics.

Most traits for which assortative mating occurs are complex traits affected by many genes. Realistic models can be complex. In this section, we consider only simple models in which the trait is controlled by a single gene. The insights we gain will generally be applicable to more complex traits.

Effect of Assortative Mating on Genotype and Allele Frequencies

The main effect of assortative mating is to decrease the frequency of heterozygotes at loci controlling the trait on which assortative mating is based. The easiest way to see this is to consider complete assortative mating at a single locus with two alleles and incomplete dominance. Under complete assortative mating, the only mating types are $(A_1A_1 \times A_1A_1)$, $(A_1A_2 \times A_1A_2)$, and $(A_2A_2 \times A_2A_2)$. This is genetically equivalent to self-fertilization and the same conclusions hold: Heterozygosity is reduced by half each generation, and the allele frequencies do not change.

Under full dominance, the situation is more complicated. Under complete assortative mating by phenotype, all matings are (dominant × dominant) or (recessive × recessive), giving four possible mating types. Table 8.6 summarizes the mating types, their frequencies, and the offspring genotype frequencies. From these, the recursion equations for the genotype frequencies can be derived (see Box 8.2). For heterozygosity, we get

$$H_{t+1} = \frac{2pH_t}{2p + H_t} \tag{8.30}$$

Note that $2p/(2p + H_t)$ is always less than one, so heterozygosity at this locus decays to zero, but at a rate slower than for incomplete dominance. Figure 8.12 compares the rates of decay for full and incomplete dominance. As with incomplete dominance, allele frequencies do not change (Problem 8.12).

Assortative mating is rarely complete, so we must consider the effects of partial assortative mating. Partial assortative mating with incomplete dominance is equivalent to partial self-fertilization, and the same conclusion holds: Heterozygosity decreases to an equilibrium value that depends on the proportion of assortative mating.

For partial assortative mating with full dominance, let A be the proportion of the population that mates assortatively (analogous to S for self-fertilization), and $1 - A$ the proportion that mates randomly. Then heterozygosity will be a

TABLE 8.6 Mating frequencies and offspring genotypes under complete assortative mating and full dominance.

Genotype frequencies in generation t are P, H, and Q (with subscript t omitted). Mating frequencies are the product of the genotype frequencies divided by the frequency of dominant $(1 - Q)$ or recessive (Q) phenotypes. See Box 8.2 for details.

Mating Type	Product	Mating Frequency	Offspring		
			A_1A_1	A_1A_2	A_2A_2
$A_1A_1 \times A_1A_1$	P^2	$P^2/(1 - Q)$	$P^2/(1 - Q)$		
$A_1A_1 \times A_1A_2$	$2PH$	$2PH/(1 - Q)$	$(1/2)[2PH/(1 - Q)]$	$(1/2)[2PH/(1 - Q)]$	
$A_1A_2 \times A_1A_2$	H^2	$H^2/(1 - Q)$	$(1/4)[H^2/(1 - Q)]$	$(1/2)[H^2/(1 - Q)]$	$(1/4)[H^2/(1 - Q)]$
$A_2A_2 \times A_2A_2$	Q^2	Q^2/Q			Q
Sum	<1	1	P_{t+1}	H_{t+1}	Q_{t+1}

BOX 8.2 Complete assortative mating with full dominance

We will derive the recursion equations for genotype frequencies under complete assortative mating with full dominance. To simplify notation, let P, H, and Q be the genotype frequencies of A_1A_1, A_1A_2, and A_2A_2 in generation t. The frequency of dominant phenotypes is then $P + H = 1 - Q$ and the frequency of recessive phenotypes is Q. Under complete assortative mating by phenotype, all matings are (dominant $\times$ dominant) or (recessive $\times$ recessive), with frequencies $1 - Q$ and Q, respectively. The four possible mating types and their frequencies are shown in Table 8.6. The entries in the product column are the product of the frequencies of the male and female genotypes. These products do not sum to one; to make them do so, we divide each product by either $1 - Q$ or Q, depending on whether it is a (dominant $\times$ dominant) or (recessive $\times$ recessive) mating. Convince yourself that the entries in the product column sum to less than one and the entries in the mating frequency column sum to one. Once the mating frequencies are determined, the offspring genotypes follow the rules of Mendelian segregation. These are the entries in the last three columns.

To get the genotype frequencies in the next generation, we sum entries in each of the last three columns. For example,

$$P_{t+1} = \frac{P^2 + (\frac{1}{2})2PH + (\frac{1}{4})H^2}{1 - Q}$$

$$= \frac{(P + H/2)^2}{1 - Q}$$

$$= \frac{p^2}{1 - Q}$$

For heterozygosity we get

$$H_{t+1} = \frac{(\frac{1}{2})2PH + (\frac{1}{2})H^2}{1 - Q}$$

which simplifies to

$$H_{t+1} = \frac{2pH}{2p + H}$$

Once we have the recursion equations for genotype frequencies, it is easy to get the equations for allele frequencies (Problem 8.12).

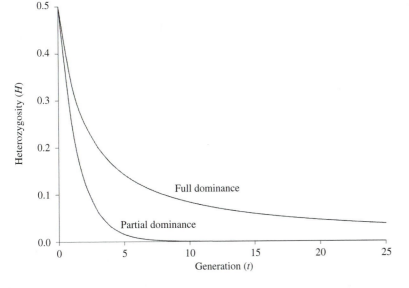

weighted average of the two parts of the population. In the random mating part, heterozygosity will be $2pq$; in the assortative mating part, heterozygosity will be as in equation (8.30). The overall heterozygosity will therefore be

$$H_{t+1} = (1 - A)2pq + A\left(\frac{2pH_t}{2p + H_t}\right) \tag{8.31}$$

Note that the first term on the right is constant; some heterozygotes will be produced every generation. Thus, heterozygosity will not decay to zero, as in complete assortative mating, but should reach some equilibrium value. To find the equilibrium, set $H_{t+1} = H_t = \tilde{H}$. After much algebraic manipulation, we get

$$\tilde{H}^2 + \tilde{H}(2p^2)(1 - A) - 4p^2q(1 - A) = 0$$

This is the standard form of a quadratic equation. Using the quadratic formula, and choosing the positive root, the solution is (after much algebraic manipulation)

$$\tilde{H} = p^2(A - 1) + p\sqrt{p^2(1 - A)^2 + 4q(1 - A)} \tag{8.32}$$

Figure 8.13 shows how heterozygosity approaches equilibrium for various levels of assortative mating. The essential conclusions are that equilibrium is approached rapidly, and fairly high levels of heterozygosity are maintained, even with high levels of assortative mating.

To summarize, under complete assortative mating, heterozygosity decays to zero, but under partial assortative mating, heterozygosity reaches an equilibrium value that depends on the amount of assortative mating in the population. Allele frequencies do not change under either complete or partial assortative mating.

The consequences of assortative mating are similar to those of inbreeding, but with one important difference: Under inbreeding, heterozygosity is decreased at

Figure 8.13 **Change in heterozygosity under partial assortative mating with full dominance.** A is the proportion of the population that undergoes assortative mating. Initial conditions are $p = q = 0.5$, and $H = 0.5$.

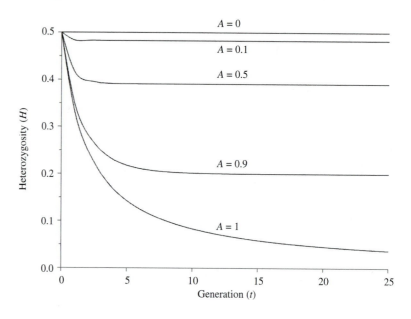

all loci. Under assortative mating, heterozygosity is decreased only at the locus for which assortative mating occurs (although loci in gametic disequilibrium with it may be affected).

Assortative Mating and Population Fitness

Assortative mating changes genotype frequencies. If the genotypes have different fitnesses, then assortative mating will affect the mean fitness of the population. Clearly, if heterozygotes have the highest fitness, then assortative mating will reduce the population fitness, compared to random mating.

Conversely, if heterozygotes have the lowest fitness, then assortative mating will reduce the frequency of heterozygotes and thus increase population fitness. Consider again the example of the ibex introduced to the mountains of Czechoslovakia. When the introduced animals bred with the natives, the hybrids mated at the wrong time of the year and failed to reproduce. Population fitness was reduced so much that the population went extinct. Clearly, if assortative mating had occurred so that natives mated with each other and not with the introduced animals, population fitness may not have declined so disastrously. Obviously, a character such as reproductive behavior is controlled by more than one gene, but the same principle holds for complex traits.

We can gain some insight about the effect of assortative mating when the heterozygote has intermediate fitness by using an approach similar to the one we used for inbreeding. As before, let 1, $1 - hs$, and $1 - s$ be the relative fitnesses of genotypes A_1A_1, A_1A_2, and A_2A_2, respectively. Then $\overline{w}_o$ is the population mean fitness under random mating, as given by equation (8.23). Let $\overline{w}_{AM}$ be the fitness under assortative mating. Recall that the genotype frequencies (P_{11}, P_{12}, and P_{22}) can be expressed as deviations from Hardy-Weinberg expectations using F, as in equations (3.21), (3.22), and (3.23). Then

$$\overline{w}_{AM} = P_{11}(1) + P_{12}(1 - hs) + P_{22}(1 - s)$$
$$= (p^2 + pqF) + 2pq(1 - F)(1 - hs) + (q^2 + pqF)(1 - s)$$

which can be rewritten as

$$\overline{w}_{AM} = \overline{w}_o + pqF[1 - 2(1 - hs) + (1 - s)]$$

Under assortative mating, F is positive. Therefore, assortative mating will increase the mean fitness of the population, compared to random mating, if the quantity in brackets is positive, which is true if $h > 0.5$. In other words, the population is better off under assortative mating if the heterozygote fitness is closer to the deleterious homozygote (A_2A_2) than to the better homozygote (A_1A_1). Heterozygote inferiority $(h > 1)$ is a special case.

Conversely, mean population fitness is decreased if the quantity in brackets is negative. This corresponds to $h < 0.5$; that is, the heterozygote fitness is closer to the better homozygote. Heterozygote superiority $(h < 0)$ is a special case. This is the same condition as for inbreeding depression. Once again, we see the similarity between assortative mating and inbreeding.

Note that we have not considered the dynamics of natural selection with assortative mating, but only considered population fitness within a single generation. Models of natural selection with assortative mating get very complicated. For example, there are five variables to keep track of: three genotype frequencies plus p and F, of which any three are independent (three degrees of freedom). The recursion equations for these variables are nonlinear and coupled, and results depend on initial conditions, as well as on the selection and dominance coefficients, s and h.

Assortative Mating and Speciation

Speciation is the process in which a single population splits into two different populations that are reproductively isolated from one another. Assortative mating frequently plays an important role in this process.

Before we discuss the process of speciation and the role of assortative mating, we need to review several terms and concepts. First is the **biological species concept**. Mayr (1940, 1963) defined species as "groups of actually or potentially interbreeding natural populations which are reproductively isolated from other such groups" (Mayr 1963, p. 19). There are other definitions of a species, but the biological species is the most useful for our purposes. The essential idea is reproductive isolation. Two groups are reproductively isolated if they are unable to produce viable and fertile offspring when they mate with one another.

Reproductive isolation can be due to a number of factors, called **reproductive isolating mechanisms**. These are usually grouped into two major kinds. **Premating isolating mechanisms** prevent mating between different groups. One kind is ecological isolation, in which related groups occupy different ecological niches, so they rarely or never come into contact with each other. Temporal isolation occurs when two groups mate at different times, for example, pine trees that shed their pollen at different times of the year, or *Drosophila* species that mate at different times of the day. Behavioral isolation is an important kind of premating isolation in many groups of animals; assortative mating based on mating calls or courtship rituals limits or prevents mating between different groups.

Postmating isolating mechanisms[2] act after mating to prevent the production of viable offspring (hybrid inviability) or to cause the offspring to be sterile (hybrid sterility), or even to cause the offspring of the hybrids to be inviable (F_2 breakdown). An example of hybrid inviability occurs in the frog genus *Rana*. Hybrids between different species usually die at some stage of development, depending on which species hybridize. A well-known example of hybrid sterility is the mule, which is a viable but sterile hybrid between a horse and a donkey.

Two populations can be either allopatric or sympatric. **Allopatric** populations are geographically isolated; their distributions do not overlap. Immigration between populations is prevented by some kind of geographical barrier. The main consequence of allopatry is that there is no gene flow between allopatric populations, and they follow independent evolutionary paths. **Sympatric** populations have overlapping geographical distributions. Gene flow can occur between sympatric populations unless prevented by reproductive isolating mechanisms.

The essential problem of speciation is how to develop reproductive isolation so that two groups have no gene flow between them, even if they are sympatric. The most common theoretical model of speciation is the model of **allopatric speciation**. It hypothesizes three major steps in the process of speciation:

Step 1: Allopatric Divergence

Imagine a large, randomly mating population with much genetic variation. Imagine a mountain range, river, or other geographic barrier arising to separate the population into two groups having essentially no gene flow between them. What will happen? First, the two groups may initially be different because of the founder effect (Section 7.6). Each subpopulation will have a random subset of the genetic variation present in the original population. Second, the two subpopulations will diverge over time because evolutionary processes will act differently in the two groups. Mutation and genetic drift will occur randomly and independently in the two groups. Natural selection will cause the two groups to diverge if their environments are different. Eventually, the two groups may become morphologically, developmentally, or behaviorally different from one another. Complete or partial reproductive isolation may develop as a by-product of this divergence.

Step 2: Secondary Sympatry

Now assume that the geographic barrier disappears, and the two groups expand their ranges so that they overlap once again (secondary sympatry). What will happen? There are three possibilities: First, the two groups may not have diverged enough to have developed any reproductive isolation between them. They will successfully breed with one another and the two groups will merge into a single population again. No speciation has occurred. Second, they may have diverged so much that they do not breed with one another at all. Speciation has occurred during allopatry. The third possibility is most interesting. The two groups may still mate with one another, but genetic differences may have developed so that the hybrids are somehow less fit than their parents; for example, they may have lower viability or fertility.

---

[2]Premating and postmating isolating mechanisms are often called prezygotic and postzygotic, respectively. But premating and postmating are more precise, because an isolating mechanism can be postmating but prezygotic, for example sperm-egg incompatibilities.

Step 3: Reinforcement

Assuming the third result has occurred, there will be natural selection favoring premating isolating mechanisms. Individuals that mate with their own kind will produce more viable and fertile offspring than individuals that mate with the other kind. Eventually, the two groups will not mate with one another and will become separate species. Speciation has occurred partly in allopatry and partly because of reinforcement of isolating mechanisms during secondary sympatry.

This is actually two models of speciation. In the first, purely allopatric model, all divergence and evolution of reproductive isolation occurs during allopatry of stage 2. In the second, reinforcement model, much of the evolution of reproductive isolation occurs during secondary sympatry. These models describe how speciation is thought to occur most of the time. We consider an alternative model below.

During the second stage, postmating reproductive isolation is thought to develop as a by-product of divergence for other characteristics. This may be due to pleiotropic effects of genes affecting development, morphology, and so on. It is important to realize that natural selection cannot favor postmating reproductive isolation directly, because it would be favoring reduced viability or fertility. Postmating isolation can arise only as an accidental by-product of divergence in other characters. Some premating isolation may arise during the stage of allopatric divergence, but this too is a by-product of divergence in other characters, and is not selected for.

It is during the third stage of reinforcement that assortative mating becomes important. If some postmating isolation has developed during allopatry, there will be strong selection favoring assortative mating. Individuals that mate with their own kind will have higher fitness than individuals that mate with the other kind, and natural selection will favor discrimination. Any preexisting or newly arising tendency toward assortative mating will be strengthened, provided it has some genetic basis.

Assortative mating can be manifest as divergence of flowering time in plants, different courtship behaviors in animals, differences in time of mating, and so forth. The important thing is that these characteristics will be favored if they reduce the frequency of matings with the other group. As these premating isolating mechanisms accumulate, there will eventually be no gene flow between the two groups, and speciation will be complete.

The idea of reinforcement was first suggested by Dobzhansky (1937). Although it has sometimes been controversial, current consensus seems to be that reinforcement is not only possible, but plausible (e.g., Coyne and Orr 1998; Noor 1999; Turelli et al. 2001). Much evidence supports the idea that premating isolating mechanisms can be reinforced by natural selection. For example, two species of frogs, *Hyla ewingi* and *H. verrauxi*, occur sympatrically and produce inviable hybrids. Mating calls are more different in regions of sympatry than in regions of allopatry. Coyne and Orr (1989, 1997) summarized studies of 171 sympatric and allopatric pairs of *Drosophila*. Divergence times were estimated from allozyme data. The found that recently diverged sympatric pairs had far higher levels of assortative mating than recently diverged allopatric pairs. In other words, assortative mating evolves faster in sympatric populations than in allopatric ones. We will see how to quantify the degree of assortative mating below.

Another model of speciation depends heavily on assortative mating. The theory of **sympatric speciation** hypothesizes that divergence and reproductive isolation can occur without geographic isolation. One kind of observation that suggests sympatric speciation is the fish fauna in several large lakes. For example, Lake Victoria in Africa has more than 200 species of cichlid fish (family cichlidae). The lake apparently dried up completely about 12,000 years ago and all of these species have probably arisen since then (Johnson et al. 1996). It is hard to imagine that all of them could have arisen by allopatric speciation within a single lake in so short a time.

There are several versions of the sympatric speciation model. One postulates ecological isolation instead of geographical isolation as the barrier preventing gene flow between two groups. Imagine a large randomly mating population. Assume that mutations in a small number of genes cause a shift in, for example, the ecological habitat, or host plant preference of phytophagus insects. Selection may favor each group in its own habitat with selection against hybrids, a process known as **disruptive selection**. If habitat preference or assortative mating is strong, natural selection may cause divergence between the two groups and ultimately speciation without geographical isolation.

Sympatric speciation has been very controversial. Disruptive selection alone will not necessarily cause divergence if mating is random. Theoretical models suggest that, in addition to strong disruptive selection, habitat preference or assortative mating is also required. Reasonable models of competition or sexual selection have been proposed, but they also depend heavily on some kind of assortative mating. For a good review of the controversy and recent work on sympatric speciation see Via (2001). Current consensus seems to be that sympatric speciation is plausible under some circumstances.

This has been the barest introduction to the problem of speciation, with emphasis only on the role of assortative mating. In the last decade or so, speciation has been one of the most active areas of research in evolutionary biology. There has been renewed interest in ecological and genetic aspects of speciation. Population genetic models have become much more sophisticated and plausible, and statistical methods for analysis of DNA sequences have given us the ability to test some of their predictions. For general introductions to speciation, see textbooks on evolutionary biology (e.g., Futuyma 1998; Freeman and Herron 2001). For review papers and introductions to the current literature on speciation, see Coyne and Orr (1998) and the many excellent papers in the special issue of *Trends in Ecology and Evolution* (Vol. 16, No. 7; July 2001).

Quantification of Assortative Mating

It is frequently useful to be able to quantify the degree of assortative mating between two types. For example, as described above, assortative mating can be an important aspect of the process of speciation. If two populations have diverged, will they prefer to mate with their own kind? Answering this question requires some way to quantify the degree of assortative mating.

One measure of the degree of assortative mating is the **isolation index** (Malogolowkin-Cohen et al. 1965). Designate the two groups by A and B. Let n_{AA} be the number of matings between A females and A males, n_{AB} be the number of matings between A females and B males, and so on. Let n be the total number of matings. We call matings between members of the same group **homogamic matings**, and matings between members of different strains **heterogamic matings**. Then the isolation index, I, is defined as

$$I = \frac{n_{AA} + n_{BB} - n_{AB} - n_{BA}}{n} \tag{8.33}$$

I ranges from -1 (all heterogamic matings), through 0 (random mating) to $+1$ (all homogamic matings). An estimate of I, designated $\hat{I}$, is obtained by putting experimental data into equation (8.33). Then $\hat{I}$ is a random variable, and we must consider its variance in order to interpret its significance.

Let p be the frequency of homogamic matings and q the frequency of heterogamic matings. Then $p = (n_{AA} + n_{BB})/n$ and $q = 1 - p$. The variance of I is

$$\sigma_I^2 = \frac{4pq}{n}$$

An approximate 95 percent confidence interval for the true value of I can be constructed as

$$\hat{I} - 2\sigma_I \leq I \leq \hat{I} + 2\sigma_I$$

where I is the (unknown) true value, and $\hat{I}$ is its estimate. If the confidence interval does not include zero, the conclusion is that mating is significantly nonrandom (either assortative or disassortative, depending on whether $\hat{I}$ is positive or negative).

Drosophila paulistorum is widespread throughout much of tropical South America. It is composed of six taxa that are morphologically very similar but that differ in chromosomal inversions. Different groups show partial reproductive isolation with each other. For this reason, they have been called semispecies, or incipient species, and they represent a classic example of speciation in process. The geographic distributions of several semispecies overlap, and some semispecies occur both sympatrically and allopatrically. In regions of sympatry, there appears to be little hybridization among different semispecies. Laboratory hybrids usually produce fertile females and sterile males, relatively strong postmating reproductive isolation. Laboratory experiments also show strong assortative mating based on differences in courtship behavior or other mating preferences (sexual isolation).

Clearly, there is reduced gene flow among different semispecies, but some gene flow may occur by way of fertile hybrid females. Whether or not these semispecies are considered separate species is arbitrary. Since some gene flow probably exists, they are usually considered incipient species, not yet fully reproductively isolated.

Ehrman (1965) tested the reinforcement hypothesis by comparing levels of assortative mating among sympatric and allopatric populations of *D. paulistorum* semispecies. She predicted that sympatric populations of two different semispecies would show more assortative mating than allopatric populations of the same two semispecies. Mating choice was observed directly by introducing males and females of two semispecies into a mating chamber and recording the kinds of matings that occurred. Ehrman's results are summarized in Table 8.7. The average

TABLE 8.7 The estimated isolation index and its standard error from mating choice experiments between sympatric and allopatric semispecies of *Drosophila paulistorum*.
Hybrid males from crosses between different semispecies are usually sterile. *Source:* From Ehrman (1965).

Semispecies	Source	Number of Matings	$\hat{I}$	σ_I
Amazonian × Andean	Sympatric	108	0.86	0.049
	Allopatric	100	0.66	0.074
Amazonian × Guianan	Sympatric	104	0.94	0.033
	Allopatric	109	0.76	0.061
Amazonian × Orinocan	Sympatric	106	0.75	0.065
	Allopatric	124	0.61	0.070
Andean × Guianan	Sympatric	109	0.96	0.026
	Allopatric	102	0.74	0.066
Orinocan × Andean	Sympatric	100	0.94	0.033
	Allopatric	111	0.46	0.084
Orinocan × Guianan	Sympatric	104	0.85	0.053
	Allopatric	100	0.72	0.069
Centro-American × Amazonian	Sympatric	102	0.68	0.072
	Allopatric	103	0.71	0.070
Centro-American × Orinocan	Sympatric	110	0.85	0.052
	Allopatric	103	0.73	0.069
Average	Sympatric	843	0.85	
	Allopatric	852	0.67	

isolation index between sympatric strains was 0.85, compared to 0.67 among allopatric strains. The difference is highly significant. The results are consistent with the reinforcement prediction; premating isolation is higher in sympatric populations than in allopatric ones.

8.3 Disassortative Mating

Disassortative mating is the opposite of assortative mating: Mating individuals are less alike phenotypically than two individuals chosen at random from the population. As you might expect, heterozygosity is increased, relative to random mating, at loci controlling the phenotype on which disassortative mating is based.

One obvious and widespread example of disassortative mating is sexual reproduction in most animals; males mate with females. With this exception, disassortative mating appears to be relatively rare in most animal populations; however, certain mating systems in plants and fungi are disassortative. We will consider examples as we discuss the theory.

Disassortative mating is inextricably bound to selection. Therefore, it can cause allele frequencies to change. In Section 5.1, we discussed several components of fitness, including viability, fecundity, mating ability, and gamete competition (Figure 5.1). Disassortative mating frequently leads to differences in mating ability or to gamete competition. This is one form of sexual selection. Similarly, under some forms of disassortative mating, rare genotypes or rare alleles have an advantage during reproduction. This is a kind of frequency-dependent selection. Sexual selection and frequency dependent selection will be discussed in more detail in Chapter 12. Our main goal here is to consider a few simple models of disassortative mating, along with some important examples from natural populations.

Disassortative Mating with Dominance

The simplest system of disassortative mating is complete disassortative mating between two phenotypes. As usual, let P_{11}, P_{12}, and P_{22} represent the frequencies of genotypes A_1A_1, A_1A_2, and A_2A_2, respectively, and p the frequency of A_1. Assume that A_1 is completely dominant, so that A_1A_1 and A_1A_2 have the same phenotype.

If mating is completely disassortative based on phenotype, all matings are dominant × recessive, and there are only two possible mating types, $(A_1A_1 \times A_2A_2)$ and $(A_1A_2 \times A_2A_2)$. The frequencies of these matings, and their offspring are listed in Table 8.8. Adding the columns under offspring genotypes, we get the genotype frequencies in the next generation:

$$P_{11}(t + 1) = 0 \qquad (8.34)$$

$$P_{12}(t + 1) = \frac{2P_{11}P_{22} + P_{12}P_{22}}{T} \qquad (8.35)$$

$$P_{22}(t + 1) = \frac{P_{12}P_{22}}{T} \qquad (8.36)$$

where $T = 2P_{11}P_{22} + 2P_{12}P_{22}$, and the products of the genotype frequencies are divided by T to make the mating frequencies add to one. The latter two equations can be rewritten as (Problem 8.14)

$$P_{12}(t + 1) = \frac{p}{P_{11} + P_{12}} \qquad (8.37)$$

TABLE 8.8 **Mating frequencies and offspring genotypes under complete disassortative mating and full dominance.**
See text for details.

Mating Type	Product	Mating Frequency	A_1A_1	A_1A_2	A_2A_2
				Offspring	
$A_1A_1 \times A_2A_2$	$2P_{11}P_{22}$	$2P_{11}P_{22}/T$		$2P_{11}P_{22}/T$	
$A_1A_2 \times A_2A_2$	$2P_{12}P_{22}$	$2P_{12}P_{22}/T$		$P_{12}P_{22}/T$	$P_{12}P_{22}/T$
Sum	T	1	$P_{11}(t + 1)$	$P_{12}(t + 1)$	$P_{22}(t + 1)$

$$P_{22}(t + 1) = \frac{(1/2)P_{12}}{(P_{11} + P_{12})} \qquad \textbf{(8.38)}$$

The first thing to notice is that there are no A_1A_1 homozygotes in generation $t + 1$! If all the matings are $A_1A_1 \times A_2A_2$ or $A_1A_2 \times A_2A_2$, then all the offspring (the $t + 1$ generation) must be either A_1A_2 or A_2A_2. These offspring will mate disassortatively ($A_1A_2 \times A_2A_2$ matings only), and the $t + 2$ generation will be half A_1A_2 and half A_2A_2. These genotype frequencies will continue indefinitely; a stable equilibrium has been reached in two generations. The equilibrium genotype frequencies are

$$\widetilde{P}_{11} = 0$$

$$\widetilde{P}_{12} = 0.5$$

$$\widetilde{P}_{22} = 0.5$$

This gives equilibrium allele frequencies of $\widetilde{p} = 0.25$ and $\widetilde{q} = 0.75$, regardless of the starting allele frequencies. To see that allele frequencies must change initially, consider the allele frequency in generation $t + 1$:

$$p_{t+1} = P_{11}(t + 1) + \frac{1}{2}P_{12}(t + 1) = \frac{p}{2(P_{11} + P_{12})} \qquad \textbf{(8.39)}$$

which, in general, is not equal to p.

There are two important examples of this kind of complete disassortative mating. The first is heterostyly in plants. Some species of plants produce two flower types, called pin and thrum (Figure 8.14). In pin flowers, the style is long so that the stigma (pollen receptacle) is high in the flower, and the filaments of the stamens are short, so that the anthers (pollen producers) are well below the stigma. In thrum flowers, the filaments are long and the style short, so that the anthers are well above the stigma. An individual plant produces only one kind of flower, and a population contains approximately half of each type of plant. The flowers are insect pollinated. Insects that tend to work high on the flower transfer pollen from thrum flowers to pin flowers; insects that work lower in the flower transfer pollen from pin to thrum. Thus, disassortative mating is caused by pollinator preferences.

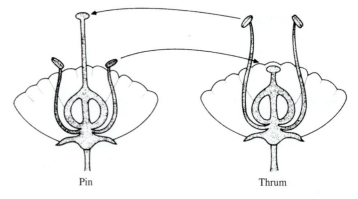

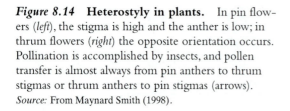

Figure 8.14 **Heterostyly in plants.** In pin flowers (*left*), the stigma is high and the anther is low; in thrum flowers (*right*) the opposite orientation occurs. Pollination is accomplished by insects, and pollen transfer is almost always from pin anthers to thrum stigmas or thrum anthers to pin stigmas (arrows). *Source:* From Maynard Smith (1998).

Pin   Thrum

The genetics of pin and thrum flowers are best understood in the primrose (*Primula vulgaris*). One gene controls style length, with G producing short style and g producing long style. A tightly linked gene controls stamen length, with A producing long stamens and a producing short stamens. Pin flowers (long style, short stamen) are ga/ga. Thrum flowers (short style, long stamens) are GA/ga. Because the two genes are very tightly linked, recombinant gametes are very rare. Disassortative mating produces matings of type $GA/ga \times ga/ga$. Thus, we have the situation described in the model above, with GA/ga as the A_1A_2 genotype and ga/ga as A_2A_2. The predicted result is a stable equilibrium, with half of each genotype in the population, which is approximately what is seen in *Primula vulgaris* and several other species.

A significant degree of assortative mating could upset this equilibrium. However, tightly linked to the pin/thrum loci are one or more incompatibility loci (see below). These assure that the rare assortative matings produce few seeds, effectively eliminating the effects of assortative mating.

The other important example of complete disassortative mating is the usual mode of sexual reproduction in animals, with disassortative mating between males (XY) and females (XX).[3] Again a stable equilibrium exists, with approximately half of each phenotype (sex).

Self-Incompatibility Systems

Many plants and fungi have genetic mechanisms that prevent self-fertilization, or fertilization by another plant with the same genotype at a particular locus. Such loci are called self-incompatibility loci.

In one system, called **sporophytic incompatibility**, the pollen tube does not grow (pollen elimination) or the zygote does not develop (zygote elimination) when the pollen donating plant has any allele in common with the pollen receiving plant. Under this system, a minimum of four alleles is necessary to ensure compatibility between mating plants. If some dominance occurs, so that incompatibility is based on similar phenotypes, fewer alleles are necessary. In reality, most plants with sporophytic incompatibility have many alleles at the incompatibility locus.

A more common system is **gametophytic incompatibility**. Here, the pollen tube does not grow if the (haploid) pollen lands on a (diploid) stigma whose genotype contains the same allele as the pollen. Obviously, this requires at least three alleles, and heterozygosity will be 100 percent at this locus. Typically, many alleles are present.

To model gametophytic incompatibility, let P_{ij} be the frequency of genotype A_iA_j, and p_i be the frequency of the ith allele in generation t. We will consider a three-allele system. Assume that excess pollen is produced, so that every stigma receives pollen. Under this model, there are only three successful mating types, with mating frequencies equal to the genotype frequencies of the stigma-containing plants. These mating types, their frequencies, and their offspring are shown in Table 8.9. From the table, the recursion equations for the genotypes are

$$P_{ij}(t + 1) = \frac{1}{2}(1 - P_{ij}) \qquad \textbf{(8.40)}$$

---

[3]In some animals, the female is the heterogametic sex, but the principle is exactly the same.

TABLE 8.9 Mating frequencies and offspring genotypes under a gametophytic self-incompatibility system.

The pollen tube does not grow if the pollen lands on a stigma that contains the same allele as the pollen grain. It is assumed that pollination is random and all stigmas are pollinated. Therefore, the successful mating frequencies are the same as the genotype frequencies of the stigmas, which sum to one. Only successful matings are shown.

Pollen	Stigma	Mating Frequency	Offspring A_1A_2	Offspring A_1A_3	Offspring A_2A_3
A_1	A_2A_3	P_{23}	$(1/2)P_{23}$	$(1/2)P_{23}$	
A_2	A_1A_3	P_{13}	$(1/2)P_{13}$		$(1/2)P_{13}$
A_3	A_1A_2	P_{12}		$(1/2)P_{12}$	$(1/2)P_{12}$
Sum		1	$P_{12}(t+1)$	$P_{13}(t+1)$	$P_{23}(t+1)$

This is really three equations, where P_{ij} represents either P_{12}, P_{13}, or P_{23}. To find the equilibrium genotype frequencies, set $P_{ij}(t+1) = P_{ij} = \widetilde{P}_{ij}$ and solve for $\widetilde{P}_{ij}$. The result is

$$\widetilde{P}_{ij} = \frac{1}{3}$$

In other words, at equilibrium each of the three heterozygous genotypes is present in equal frequency.

Now consider allele frequencies at equilibrium. Using equation (2.11) (with no homozygotes), we get

$$\widetilde{p}_1 = \frac{1}{2}(\widetilde{P}_{12} + \widetilde{P}_{13}) = \frac{1}{3}$$

The same is true for the other alleles, so at equilibrium all three alleles are present in equal frequency. The result can be generalized; if there are k alleles, the equilibrium frequency of each is $1/k$. It can be shown that the equilibrium is locally stable. This is an example of frequency-dependent selection. If an allele is rare (for example, a new mutation) it will favored by natural selection because most matings that involve that allele will be compatible. But if the allele is common, it will be incompatible in a high frequency of matings, and will be selected against.

Self-incompatibility systems like this usually have many alleles present. If there were only a few alleles, a high proportion of pollinations would produce no offspring, and the population fitness would be low. In one classic example, Emerson (1939) found 34 self-incompatibility alleles in 134 individuals of *Oenothera organensis*, a rare plant found only in the Organ Mountains of New Mexico. The expected equilibrium frequency of each of these alleles is $1/34 \cong .029$. Actual frequencies ranged from 0.007 to 0.052 (Figure 8.15). This is reasonable agreement, considering the small sample size (Problem 8.15).

If a self-incompatibility system is stable and alleles are not lost by genetic drift, polymorphism at the self-incompatibility locus can be maintained for a very long time. As time passes, alleles will accumulate mutations in their DNA sequences; the longer the system has been in existence, the more divergent the allelic sequences will become. Ioerger et al. (1990) compared DNA sequences

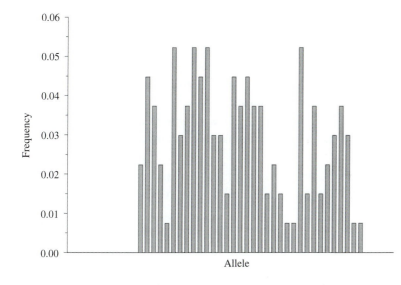

Figure 8.15 **Frequency distribution of self-incompatibility alleles in *Oenothera organensis*.** Each bar represents one allele. The expected frequency of each allele is 0.029. *Source:* Data from Emerson (1939).

of 11 alleles at the gametophytic self-incompatibility locus in three species of the family Solanaceae. They found extraordinarily high sequence diversity. Amino acid sequence similarities were as low as 40 percent between alleles in the same species. Moreover, some alleles in different species were more closely related than alleles in the same species, indicating that the polymorphism at this locus predates the divergence of these species, about 27 to 36 million years ago.

On the other hand, if a population goes through repeated bottlenecks, genetic variation at all loci will be reduced and alleles will be lost. If self-incompatibility alleles are lost, the population may suffer reduced fertility and increased danger of extinction. *Aster furcatus* is a rare self-incompatible plant found in fewer than 50 populations in the midwestern United States. Variation at allozyme loci is near zero in most populations (Les et al. 1991), suggesting severe bottlenecks or founder effects. Les et al. suggested that dispersal mechanisms and fruit structure contribute to severe founder effects. Reinartz and Les (1994) found little variation at the sporophytic incompatibility locus, only four or five alleles in the most well-studied population (Sheboygan Falls, Wisconsin). Note that four alleles are the minimum required if compatibility requires no shared alleles between mates. Populations showed significant differences in seed set, ranging from 0 to 53 percent. Six populations had seed set less than 10 percent. Reinartz and Les suggested that low seed set was due to the small number of incompatibility alleles in those populations, a result of severe founder effects. In such a situation, one might expect natural selection to favor self-compatibility to increase fertility. Indeed, Reinartz and Les found varying degrees of self-compatibility and concluded that the self-incompatibility system was breaking down due to loss of alleles, and consequent low fertility.

To summarize, all of the disassortative mating systems we have considered have several things in common: (1) Allele frequencies change; (2) a stable equilibrium is reached at which allele frequencies and genotype frequencies remain constant; (3) heterozygosity at equilibrium is greater than it would be under random mating; and (4) there are differences in fertility or mating ability, and selection often favors rare alleles or genotypes.

Summary

1. Two individuals are consanguineous if they have a common ancestor.

2. Two alleles are identical by descent if they are derived from the same copy of an allele one or more generations back. Consanguineous individuals may have alleles that are identical by descent.

3. Inbreeding is the mating of consanguineous individuals. It is always defined relative to some reference population in which all individuals are assumed to be unrelated.

4. The inbreeding coefficient (f) is the probability that the two alleles in an individual are identical by descent. The inbreeding coefficient can be calculated from pedigrees.

5. Inbreeding increases the proportion of homozygotes and decreases the proportion of heterozygotes in a population. Inbreeding alone does not change allele frequencies.

6. An alternative interpretation of the inbreeding coefficient is the proportionate reduction in heterozygosity compared to a reference population; equation (8.5).

7. In a finite population, the average inbreeding coefficient increases each generation. The rate of increase depends on the population size; equation (8.13).

8. Self-fertilization decreases heterozygosity by half each generation. With a mixture of self-fertilizing and randomly mating individuals, heterozygosity reaches an equilibrium that depends on the initial heterozygosity and on the proportion of self-fertilizing individuals; equation (8.20).

9. Repeated self-fertilization, repeated full-sib matings, and other regular systems of inbreeding are common in experimental design and in plant and animal breeding. Table 8.4 summarizes how f changes under several regular systems of inbreeding.

10. Inbreeding depression occurs when inbred individuals have lower fitness than noninbred individuals. It is common in natural populations and can contribute to increased risk of extinction in small populations.

11. Purging occurs when deleterious alleles are eliminated from a population by natural selection after they have been made homozygous by inbreeding. Purging is relatively weak and inconsistent in most populations.

12. Inbreeding retards the decay of gametic disequilibrium. Significant gametic disequilibrium is frequently seen in self-fertilizing species.

13. Outbreeding occurs when mating individuals are less closely related than expected by chance. A frequent result of outbreeding is hybrid vigor, but outbreeding depression can occur if matings occur between populations that have diverged substantially.

14. Assortative mating occurs when mating pairs are more similar in phenotype than would be expected under random mating.

15. Assortative mating decreases heterozygosity compared to random mating, but does not change allele frequencies. In a population with a mixture of assortative and random mating, heterozygosity reaches a stable equilibrium.

16. Assortative mating sometimes plays an important role in the process of speciation by preventing mating between groups that have evolved some postmating reproductive isolation.

17. The degree of assortative mating in a population can be quantified by using an isolation index; equation (8.33).

18. Disassortative mating occurs when mating pairs are less similar in phenotype than would be expected under random mating. Several disassortative mating systems occur in plants (self-incompatibility systems).

19. Disassortative mating increases heterozygosity compared to random mating. Under various models of disassortative mating, heterozygosity and allele frequencies generally approach a stable equilibrium.

20. Disassortative mating frequently leads to differences in mating ability or to gamete competition (sexual selection). Frequency-dependent selection often favors rare alleles or genotypes.

Problems

8.1. How many years back must you go before any two humans *must* have a common ancestor, assuming no consanguinity? Assume the world population is about 6×10^9 and an average generation time of about 20 years.

8.2. Draw path diagrams for each of the relationships in Figure 8.2.

8.3. Find the coefficient of consanguinity for individuals X and Y in each of the relationships shown in Figure 8.2.

8.4. Calculate the inbreeding coefficient for individual I in Figure 8.16a and 8.16b.

8.5. For each of the snail populations in Table 8.2 estimate the frequency of self-fertilization based on each locus separately. Discuss possible reasons why the estimates are different for different loci.

8.6. Cystic fibrosis is a usually fatal disease caused by homozygosity for a recessive allele. The frequency of the allele in some human populations is about 0.0006. What is the expected frequency of the disease in the population at large, assuming random mating? What is the frequency of the disease among children of first cousin marriages? How much more likely are first cousins to have a child with cystic fibrosis than two unrelated individuals?

8.7. Show that for multiple alleles, the allele frequencies do not change under inbreeding.

8.8. Derive the recursion equation for f under self-fertilization.

8.9. Assume that a locus has the following fitnesses and genotype frequencies before selection:

Genotype	A_1A_1	A_1A_2	A_2A_2
Frequency	0.6	0.3	0.1
Fitness	1	0.9	0.75

Assume that the deviation from Hardy-Weinberg expected heterozygosity is due to inbreeding only. Estimate f and the cost of inbreeding for this locus in this population.

8.10. What is the expected proportion of heterozygotes after ten generations of full-sib mating? Assume an initial heterozygosity of 0.5.

8.11. Use equation (8.21) to derive equation (8.22).

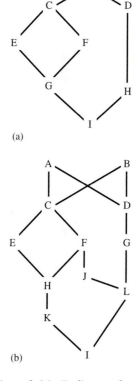

Figure 8.16 **Pedigrees for calculating inbreeding coefficients.** See Problem 8.4.

8.12. Use the results in Box 8.2 to show that allele frequencies do not change under complete assortative mating with full dominance.

8.13. Dodd (1989) tested for assortative mating between strains of *Drosophila pseudoobscura* adapted to starch (S) or maltose (M). The results, combining tests of several M strains and several S strains, are given below. Calculate the isolation index and its standard error. Interpret.

Kind of mating (♀ × ♂)	S × S	S × M	M × S	M × M
Number	290	149	153	312

8.14. Show that equations (8.37) and (8.38) are equivalent to equations (8.35) and (8.36), respectively.

8.15. Table 8.10 gives the observed allele frequencies at the self-incompatibility locus of *Oenothera organensis*. (This is the data on which Figure 8.15 is based.) Perform a chi-square test for the null hypothesis that the observed numbers are consistent with the predicted numbers based on the gametophytic incompatibility model.

TABLE 8.10 Observed numbers of self-incompatibility alleles in a population of *Oenothera organensis*.
See Figure 8.15 and Problem 8.15. *Source:* From Emerson (1939).

Allele Designation	Observed Number	Allele Designation	Observed Number
2	3	21	5
3	6	22	5
4	5	23	2
5	3	24	3
6	1	26	2
7	7	27	1
9	4	28	1
11	5	29	7
12	7	30	2
13	6	31	5
14	7	33	2
15	4	34	3
16	4	35	4
17	2	36	5
18	6	37	4
19	5	38	1
20	6	39	1

Population Subdivision and Gene Flow

A certain degree of partial isolation of local populations within a large species seems to provide the most favorable conditions for evolutionary advance.

—*Sewall Wright (1939)*

In the previous chapter, we examined the consequences of nonrandom mating based on relationship or phenotype. Nonrandom mating can also occur based on geography. If two individuals are far apart, they may be less likely to mate with each other than two individuals that are close together. Populations may be subdivided into neighborhoods (subpopulations) with random mating within subpopulations but limited mating between them.

If two populations are geographically isolated from each other, they will tend to diverge. Mutation, natural selection, and genetic drift will act independently in each population, and after some time they will be genetically different from each other. Gene flow slows down this divergence by moving alleles from one population to the other.

In this chapter, we consider the opposing processes of subdivision (partial geographic isolation), which leads to divergence among populations, and gene flow, which retards divergence. We have three major goals: First, we examine ways to describe the extent of subdivision within a population. Second, we examine several models of gene flow to see how it retards divergence. Third, we examine ways to estimate the amount of gene flow among subpopulations.

9.1 The Wahlund Effect

Sometimes subdivision and isolation are not obvious. For example, the distributions of *Drosophila pseudoobscura* and *D. persimilis* overlap. These two species look alike to human eyes, but they rarely mate with each other in nature. Similarly, there are several species of *Anopheles* mosquitoes that look alike but rarely breed with one another; initially, they were thought to be a single species. It is important to recognize this isolation because some species carry the malaria parasites and some do not.

What is the effect of sampling what appears to be a single population, but is actually two or more subpopulations with limited gene flow between them? As we shall see, failure to discriminate between partially isolated subpopulations leads to a perceived deficiency of heterozygotes. This may suggest that one or more evolutionary processes are acting, when in fact the deficiency is due solely to population subdivision. We first illustrate with an example, and then consider the general situation.

Consider a population that is divided into two subpopulations of equal size. Random mating occurs within each subpopulation, but individuals from one subpopulation do not mate with individuals from the other. You sample 100 individuals from each subpopulation and get the following results:

Genotype	A_1A_1	A_1A_2	A_2A_2
Subpopulation 1	64	32	4
Subpopulation 2	4	32	64
Total	68	64	68

The frequency of the A_1 allele is 0.80 in the first subpopulation, and 0.20 in the second. The genotype frequencies in each subpopulation match the Hardy-Weinberg expectations (convince yourself).

Now imagine that you do not know about the subdivision, and lump all individuals in your sample together. The frequency of A_1 in the combined sample is $\bar{p} = 0.50$. The expected frequency of heterozygotes is $2\bar{p}\bar{q} = 0.50$, and the observed frequency is $64/200 = 0.32$. There is a deficiency of heterozygotes compared to the Hardy-Weinberg expectations. This *perceived* deficiency of heterozygotes is due to our treating two different populations as one. This effect is called the **Wahlund effect**, after its discoverer, S. Wahlund (1928).

Estoup et al. (1995) described a probable example of the Wahlund effect. In a study of microsatellite variation in honeybees, they collected several populations throughout Europe and Africa. Only one population, from Cape Town, South Africa, showed significant differences between observed and expected heterozygosities (0.784 and 0.839, respectively). It turns out that the Cape Town sample was actually combined from two populations about 150 km apart.

We can visualize the Wahlund effect for two subpopulations of equal size, each satisfying the Hardy-Weinberg conditions. Figure 9.1 plots heterozygosity versus allele frequency. p_1 and p_2 are the allele frequencies in subpopulations 1 and 2, and H_1 and H_2 are the observed heterozygosities. They fall on the curve because each subpopulation is in Hardy-Weinberg equilibrium. $\bar{p}$ is the observed allele frequency in the total population (the average of p_1 and p_2), and H_{obs} is the observed heterozygosity in the total population (the average of H_1 and H_2). H_{exp} is the expected heterozygosity in the total population ($= 2\bar{p}\bar{q}$). Because of the concavity of the curve, H_{exp} is always greater than H_{obs}. If the two subpopulation sizes are unequal, the position of $\bar{p}$ is changed, but the relationship between H_{exp} and H_{obs} is not.

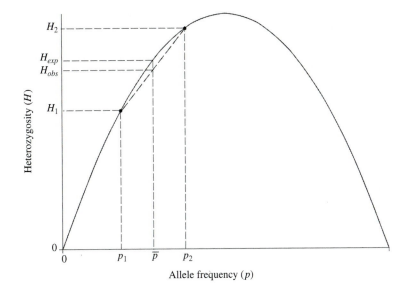

Figure 9.1 Observed and expected heterozygosity at a single locus with two alleles in a subdivided population. The curve is expected heterozygosity $(H_{exp} = 2pq)$. p_1 and p_2 are allele frequencies in two subpopulations, with expected heterozygosities H_1 and H_2, respectively. The overall allele frequency is $\bar{p}$ (the average of p_1 and p_2). H_{obs} is the observed heterozygosity in the entire population (the average of H_1 and H_2). H_{exp} is the expected heterozygosity for the entire population $(= 2\bar{p}\bar{q})$. Because the curve is concave downward, H_{obs} is always less than H_{exp}.

The Wahlund effect is true for more than two populations, but the demonstration is more complicated. Box 9.1 explains the algebraic details.

We have shown that if subpopulations are completely isolated and differ in allele frequencies, but are sampled without distinguishing between them, there will be a perceived deficiency of heterozygotes in the sample, even though there is no such deficiency in any subpopulation. Does the same principle hold if subpopulations are only partially isolated? The answer is yes, as we shall see in the next section.

9.2 Population Subdivision and F Coefficients

In a subdivided population, the overall deviation from Hardy-Weinberg expected heterozygosity has two components: the deviation due to factors acting within subpopulations, and the deviation due to subdivision (the Wahlund effect). Wright (1951) defined three F coefficients that describe these deviations. His original definition was in terms of correlations among gametes; we will use Nei's (1977) equivalent definitions in terms of deviations from expected heterozygosities. These F coefficients then take the general form that we saw in Chapter 3 (Section 3.3):

$$F = \frac{H_{exp} - H_{obs}}{H_{exp}}$$

where H_{obs} is the observed frequency of heterozygotes and H_{exp} is the Hardy-Weinberg expected frequency. A positive F means a deficiency of heterozygotes compared to Hardy-Weinberg expectations.

Before we define these F coefficients, we must carefully describe the model and its parameters. Consider a single large population subdivided into numerous subpopulations, and single locus with two alleles, A_1 and A_2 (Figure 9.2). We are concerned with the allele frequencies and the observed and expected heterozygosities in each subpopulation, and in the total population. The parameters of the model are as follows (refer to Figure 9.2):

Box 9.1   The Wahlund effect for more than two subpopulations

---

For more than two subpopulations, visualization of the Wahlund effect fails and we must resort to algebraic demonstration. The following may look complex, but it is really fairly straightforward if you pay close attention to what the summations mean. Consider a single locus with two alleles, A_1 and A_2. Assume a population is divided into n subpopulations. The relative proportion of the jth subpopulation is c_j, and the frequency of A_1 in that subpopulation is p_j. Assume there is random mating within each subpopulation, but no mating between subpopulations. Then we have the following:

	A_1A_1	A_1A_2	A_2A_2
jth subpopulation	p_j^2	$2p_jq_j$	q_j^2
Total population	$\sum c_j p_j^2$	$\sum c_j(2p_jq_j)$	$\sum c_j q_j^2$

The first row is just the Hardy-Weinberg expected genotype frequencies within each subpopulation. Since we are assuming random mating within subpopulations and ignoring other forces, these are also the observed genotype frequencies for each subpopulation. In the second row, the population genotype frequencies are the weighted averages (weighted by proportion) of the frequencies in each subpopulation. Similarly, the allele frequency in the total population is the weighted average of the allele frequencies in the subpopulations:

$$\bar{p} = \sum c_j p_j$$

The observed heterozygosity in the total population is, again, the weighted averages of the heterozygosities in the subpopulations

$$H_{obs} = \sum c_j(2p_jq_j) = 2\sum c_j p_j(1 - p_j)$$

$$= 2\left[\sum c_j p_j - \sum c_j p_j^2\right]$$

The expected heterozygosity in the total population, ignoring subdivision, is

$$H_{exp} = 2\bar{p}\bar{q} = 2\left(\sum c_j p_j\right)\left(1 - \sum c_j p_j\right)$$

$$= 2\left[\sum c_j p_j - \left(\sum c_j p_j\right)^2\right]$$

The difference is

$$H_{exp} - H_{obs} = 2\left[\sum c_j p_j - \left(\sum c_j p_j\right)^2\right]$$

$$- 2\left[\sum c_j p_j - \sum c_j p_j^2\right]$$

$$= 2\left[\sum c_j p_j^2 - \left(\sum c_j p_j\right)^2\right]$$

The part in brackets is just the variance of the ps (see Appendix A). So the difference is

$$H_{exp} - H_{obs} = 2Var(p_j)$$

Since a variance is always positive, H_{exp} is always greater than H_{obs}.

---

n = number of subpopulations

c_j = relative size (proportion) of jth subpopulation

p_j = frequency of A_1 in jth subpopulation

q_j = frequency of A_2 in jth subpopulation = $1 - p_j$

H_{oj} = observed heterozygosity in jth subpopulation

H_{ej} = expected heterozygosity in jth subpopulation = $2p_jq_j$

H_I = mean of observed heterozygosities over all subpopulations

$\quad = \sum c_j H_{oj}$

H_S = mean of expected heterozygosities over all subpopulations

$\quad = \sum c_j H_{ej} = \sum c_j(2p_jq_j)$

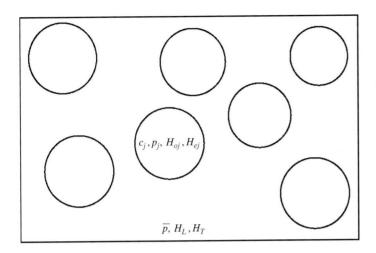

Figure 9.2 A large population subdivided into many partially isolated subpopulations.
c_j, p_j, H_{oj}, and H_{ej} represent the relative size, allele frequency, observed, and expected heterozygosity of the jth subpopulation. $\bar{p}$ is the allele frequency in the total population (average of all the p_j), and H_I is the observed heterozygosity in the total population (average of all the H_{oj}). H_T is the expected heterozygosity of the total population based on $\bar{p}$. See text for details.

$$\bar{p} = \text{mean of } p_j \text{ over all subpopulations} = \sum c_j p_j$$

$$\bar{q} = \text{mean of } q_j \text{ over all subpopulations} = 1 - \bar{p}$$

$$H_T = \text{expected heterozygosity in total population, ignoring subdivision}$$

$$= 2\bar{p}\,\bar{q}$$

Note that for the jth subpopulation, c_j is the relative proportion of the total population, and does not refer to a sample proportion. We usually assume that the subpopulations are equal size unless we have good reason to believe otherwise. Therefore, we usually specify $c_j = 1/n$ for all subpopulations.

Make sure you understand the meaning of each of these parameters before you read any further.

The average deviation in heterozygosity within subpopulations is

$$F_{IS} = \frac{H_S - H_I}{H_S} \qquad (9.1)$$

where H_I is the average of the observed heterozygosities, and H_S is the average of the expected heterozygosities. The subscript IS indicates deviation among individuals relative to their subpopulation. The factors acting within subpopulations (natural selection, genetic drift, inbreeding, assortative mating, etc.) can either increase or decrease heterozygosity over Hardy-Weinberg expectations. Therefore, F_{IS} can be either positive or negative, ranging from -1 to $+1$.

The deviation in heterozygosity due to subdivision alone is

$$F_{ST} = \frac{H_T - H_S}{H_T} \qquad (9.2)$$

where H_T is the expected heterozygosity in the total population, and H_S is the average of expected heterozygosities. The subscript ST indicates deviation among subpopulations relative to the total population. Subdivision always causes a perceived deficiency of heterozygotes, as we saw in the previous section. Therefore, F_{ST} is always zero or positive, ranging from 0 to 1.

Finally, the overall deviation in heterozygosity in the total population is

$$F_{IT} = \frac{H_T - H_I}{H_T} \qquad (9.3)$$

where H_T is the expected heterozygosity in the total population, assuming no subdivision, and H_I is observed heterozygosity in the total population (i.e., the average of the observed heterozygosities). The subscript *IT* indicates deviation among individuals relative to the total population. F_{IT} can range from -1 to $+1$, depending on the values of F_{IS} and F_{ST}.

These three quantities are *not* additive, as will be explained below.

Usually, we are most interested in F_{ST}, which is a measure of how different the subpopulations are. It can be interpreted as the proportion of the total heterozygosity that is due to differences in allele frequencies among subpopulations. Note that F_{ST} can be calculated from allele frequencies alone, whereas F_{IS} and F_{IT} require observed genotype frequencies.

Calculating F coefficients can be tedious. Box 9.2 and Table 9.1 illustrate the procedure as you would do it on a spreadsheet. This example makes no correction for possible bias introduced by small sample sizes. Real data are usually analyzed with specialized computer programs. These vary in their approach to sampling bias; we will discuss this issue later in this section.

Your intuition might suggest that the three F coefficients are additive; that is, F_{IT} might be the sum of the other two. Because of the way they are defined, that is usually *not* the case. They are related by the expression

$$(1 - F_{IT}) = (1 - F_{IS})(1 - F_{ST}) \tag{9.4}$$

See Box 9.3 for the algebraic details. This equation can be rewritten as

$$F_{IT} = F_{IS} + F_{ST} - F_{IS}F_{ST} \tag{9.5}$$

Box 9.2 Calculating F coefficients for loci with two alleles

We use the sample data in Table 9.1 to illustrate how to calculate F_{IS}, F_{IT}, and F_{ST} from sample genotype and allele frequencies. The allele frequencies in each subpopulation are calculated in the normal way from equations (2.6) and (2.7). We usually assume each subpopulation is equal in size, unless we have strong evidence to the contrary. From the table, the calculation of heterozygosities is straightforward. H_{oj} is the observed frequency of heterozygotes in the *j*th subpopulation, and H_{ej} is the expected frequency ($= 2p_jq_j$). The allele frequencies in the total population are

$$\bar{p} = \sum c_j p_j = 0.510$$

and

$$\bar{q} = \sum c_j q_j = 0.490$$

Then, from the table we have

$$H_I = \sum c_j H_{oj} = 0.447$$

$$H_S = \sum c_j H_{ej} = 0.449$$

$$H_T = 2\bar{p}\bar{q} = 2(0.510)(0.490) = 0.500$$

Finally, using equations (9.1), (9.2), and (9.3), the F coefficients are

$$F_{IS} = \frac{H_S - H_I}{H_S} = \frac{0.449 - 0.447}{0.449} = 0.0052$$

$$F_{ST} = \frac{H_T - H_S}{H_T} = \frac{0.500 - 0.449}{0.500} = 0.1016$$

$$F_{IT} = \frac{H_T - H_I}{H_T} = \frac{0.500 - 0.447}{0.500} = 0.1063$$

We conclude that each subpopulation is consistent with the Hardy-Weinberg expectations, but that there appears to be moderate differentiation among subpopulations.

TABLE 9.1 Sample data for calculating F coefficients from genotype and allele frequency samples.
P_{11}, P_{12}, and P_{22} are the observed genotype frequencies. See Box 9.2.

j	c_j	P_{11}	P_{12}	P_{22}	p_j	q_j	$c_j p_j$	$c_j q_j$	H_{oj}	$c_j H_{oj}$	H_{ej}	$c_j H_{ej}$
1	0.333	0.10	0.42	0.48	0.31	0.69	0.103	0.230	0.420	0.140	0.428	0.143
2	0.333	0.27	0.50	0.23	0.52	0.48	0.173	0.160	0.500	0.167	0.499	0.166
3	0.333	0.49	0.42	0.09	0.70	0.30	0.233	0.100	0.420	0.140	0.420	0.140
							0.510	0.490		0.447		0.449
							$\bar{p}$	$\bar{q}$		H_I		H_S

The last term on the right can be considered a kind of interaction term modifying the additive relationship.

F Coefficients with Multiple Alleles

Extension to multiple alleles is straightforward in principle, although we must be very clear about notation and subscripts. Assume there are k alleles and n subpopulations. We use the subscript i to indicate alleles $(i = 1, \ldots, k)$ and j to indicate subpopulations $(j = 1, \ldots, n)$. The frequency of the ith allele in the jth subpopulation is p_{ij}.

Using the notation of Chapter 2, we let P_{ii} be the observed frequency of genotype $A_i A_i$. Then P_{iij} is the frequency of $A_i A_i$ in the jth subpopulation. The observed heterozygosity in the jth subpopulation is then

$$H_{oj} = 1 - \sum_{i=1}^{k} P_{iij}$$

BOX 9.3   The relationship among F coefficients

Contrary to what you might expect, F_{IT} is not the sum of F_{IS} and F_{ST}. To see the actual relationship, first consider equation (9.1)

$$F_{IS} = \frac{H_S - H_I}{H_S}$$

With some algebraic rearrangement, this can be rewritten as

$$1 - F_{IS} = \frac{H_I}{H_S} \qquad (1)$$

Similarly, equation (9.2) can be rewritten as

$$1 - F_{ST} = \frac{H_S}{H_T} \qquad (2)$$

Now, multiply the left side of equation (1) by the left side of equation (2) and the right side of (1) by the

right side of (2) to get

$$(1 - F_{IS})(1 - F_{ST}) = \left(\frac{H_I}{H_S}\right)\left(\frac{H_S}{H_T}\right) = \frac{H_I}{H_T}$$

Now add and subtract 1 from the right side:

$$(1 - F_{IS})(1 - F_{ST}) =$$

$$\left(1 - \frac{H_T}{H_T}\right) + \frac{H_I}{H_T} = 1 - \left(\frac{H_T - H_I}{H_T}\right)$$

But the quantity in parentheses on the far right is F_{IT}. Therefore, we have

$$(1 - F_{IS})(1 - F_{ST}) = 1 - F_{IT}$$

which is equation (9.4) in the main text.

This is just one minus the frequencies of all the homozygotes in that subpopulation. As for the two-allele case, H_I is the average of H_{oj} over all subpopulations:

$$H_I = \sum_{j=1}^{n} c_j H_{oj}$$

The expected heterozygosity in the jth subpopulation is, using equation (3.12),

$$H_{ej} = 1 - \sum_{i=1}^{k} p_{ij}^2$$

where p_{ij} is the frequency of the ith allele in the jth subpopulation. Next, H_S is the average of the H_{ej} over all subpopulations:

$$H_S = \sum_{j=1}^{n} c_j H_{ej}$$

To calculate H_T, we need the average allele frequencies over all subpopulations. For the ith allele, its average frequency is

$$\bar{p}_i = \sum_{j=1}^{n} c_j p_{ij}$$

Once we have the average allele frequencies, we can calculate H_T using equation (3.12):

$$H_T = 1 - \sum_{i=1}^{k} \bar{p}_i^2$$

We can now use equations (9.1), (9.2), and (9.3) to calculate the F coefficients. Box 9.4 and Table 9.2 illustrate the calculations. Again, no corrections for sampling bias have been made.

F Coefficients for Multiple Loci

Most population genetic surveys include data for more than one locus. We can estimate F_{ST} for each locus separately, or we can try to find some analogous statistic that summarizes all loci. Nei (1977) suggested calculating H_I, H_S, and H_T separately for each locus, and then averaging them. Call these averages $\bar{H}_I$, $\bar{H}_S$, and $\bar{H}_T$, respectively. Nei then defined the multilocus version of F_{ST} as

$$\bar{F}_{ST} = \frac{\bar{H}_T - \bar{H}_S}{\bar{H}_T} \tag{9.6}$$

See Nei (1977, 1987) for more information. Weir and Cockerham (1984) suggested an alternative way of combining data from several loci. When using a computer program to make these calculations, read the documentation to determine which method is used.

Alternative Interpretations of F_{ST}

We have defined F_{ST} in terms of expected heterozygosities within and among subpopulations [equation (9.2)]. It can be interpreted as the proportion of the total heterozygosity in the population that is due to differences in allele frequencies among subpopulations. There are various alternative interpretations of F_{ST}, and several closely related parameters. Here, we introduce some of them.

BOX 9.4   Calculating F coefficients for loci with multiple alleles

We illustrate how to calculate F coefficients for a locus with multiple alleles using the data in Table 9.2. The observed frequencies of homozygotes (P_{ij}) and the allele frequencies (p_{ij}) are given in the table for each subpopulation. The observed heterozygosity in each subpopulation is calculated from

$$H_{oj} = 1 - \sum_{i=1}^{k} P_{iij}$$

where the P_{iij} are the observed homozygosities in the jth subpopulation. For example, in the first subpopulation,

$$H_{o1} = 1 - (0.139 + 0.165 + 0 + 0.038) = 0.658$$

H_I is then calculated as

$$H_I = \sum_{j=1}^{n} c_j H_{oj} = 0.644$$

Expected heterozygosity in the jth subpopulation is now calculated by

$$H_{ej} = 1 - \sum_{i=1}^{k} p_{ij}^2$$

For example,

$$H_{e1} = 1 - [(0.33)^2 + (0.38)^2 + (0.09)^2 + (0.20)^2]$$
$$= 0.699$$

From this, H_S is calculated as

$$H_S = \sum_{j=1}^{n} c_j H_{ej} = 0.666$$

Finally, the overall frequency of the ith allele is

$$\bar{p}_i = \sum_{j=1}^{n} c_j p_{ij}$$

This is just the average of the ith allele over all subpopulations. From Table 9.2, $\bar{p}_1 = 0.363$, $\bar{p}_2 = 0.406$, and so on.

From this, we calculate H_T as

$$H_T = 1 - \sum_{i=1}^{k} \bar{p}_i^2 = 1 - [(0.363)^2$$
$$+ (0.406)^2 + (0.083)^2 + (0.147)^2] = 0.675$$

Finally, calculate F_{IS}, F_{ST}, and F_{IT} from equations (9.1), (9.2), and (9.3):

$$F_{IS} = \frac{H_S - H_I}{H_S} = \frac{0.666 - 0.644}{0.666} = 0.032$$

$$F_{ST} = \frac{H_T - H_S}{H_T} = \frac{0.675 - 0.666}{0.675} = 0.014$$

$$F_{IT} = \frac{H_T - H_I}{H_T} = \frac{0.675 - 0.644}{0.675} = 0.046$$

There appears to be little genetic differentiation among subpopulations.

TABLE 9.2   Blood group frequencies in three populations of South American Indians.

The frequencies of the four homozygous genotypes (P_{iij}) and the four allele frequencies (p_{ij}) are given. See Box 9.4 for explanation of the calculations. *Source*: From Nei (1987); data from Gershowitz et al. (1967).

i	c_j	P_{11j}	P_{22j}	P_{33j}	P_{44j}	p_{1j}	p_{2j}	p_{3j}	p_{4j}	$c_j p_{1j}$	$c_j p_{2j}$	$c_j p_{3j}$	$c_j p_{4j}$	H_{oj}	$c_j H_{oj}$	H_{ej}	$c_i H_{ej}$
1	0.333	0.139	0.165	0.000	0.038	0.33	0.38	0.09	0.20	0.110	0.127	0.030	0.067	0.658	0.219	0.699	0.233
2	0.333	0.181	0.119	0.017	0.004	0.37	0.36	0.15	0.12	0.123	0.120	0.050	0.040	0.679	0.226	0.697	0.232
3	0.333	0.164	0.222	0.000	0.018	0.39	0.48	0.01	0.12	0.130	0.160	0.003	0.040	0.597	0.199	0.603	0.201
										0.363	0.406	0.083	0.147		0.644		0.666
										$\bar{p}_1$	$\bar{p}_2$	$\bar{p}_3$	$\bar{p}_4$		H_I		H_S

Wright (1951 and later papers) initially defined the F coefficients in terms of correlations among gametes. He also showed that for two alleles, F_{ST} can be expressed in terms of the variance of allele frequency among subpopulations. If p is the frequency of the A_1 allele, then

$$F_{ST} = \frac{V(p)}{\overline{p}(1 - \overline{p})}$$

where $V(p)$ is the variance of p among subpopulations, and $\overline{p}$ is the average over all subpopulations. This definition is equivalent to the definition based on heterozygosities. See Box 9.5 for the algebraic details.

Weir and Cockerham (1984) defined a parameter θ, which is, loosely speaking, a ratio of the variance of allele frequencies among subpopulations to the overall variance in allele frequencies. θ is essentially the same as F_{ST}, although they differ in their assumptions about the sampling process (Cockerham 1969, 1973; Weir and Cockerham, 1984). We shall see in Section 9.5 that the variance interpretation leads to a very flexible approach to estimating F_{ST} and related parameters.

Box 9.5 F_{ST} in terms of allele frequency variances

Consider the simplest situation, with two alleles. Let p_j and $q_j = 1 - p_j$ be the frequency of A_1 and A_2 in the jth population. Then we have defined H_S as

$$H_S = \sum_{j=1}^{n} c_j(2p_j q_j)$$

If we assume equal weighting for all subpopulations, then $c_j = 1/n$, and H_S simplifies to

$$H_S = 2\left(\frac{1}{n}\right)\sum_j p_j(1 - p_j) = 2\left(\frac{1}{n}\right)\sum_j (p_j - p_j^2)$$

Separating the summations,

$$H_S = 2\left[\frac{1}{n}\sum_j p_j - \frac{1}{n}\sum_j p_j^2\right]$$

But $(1/n)\sum p_j$, the average allele frequency over all subpopulations, $\overline{p}$. Similarly, $(1/n)\sum p_j^2$ is the average of the squared allele frequencies, which we will designate by $\overline{(p^2)}$. Therefore,

$$H_S = 2[\overline{p} - \overline{(p^2)}]$$

Next, recall that H_T is defined as

$$H_T = 2\overline{p}\,\overline{q}$$

which is equivalent to

$$H_T = 2\overline{p}(1 - \overline{p}) = 2(\overline{p} - \overline{p}^2)$$

We can now write F_{ST} as

$$F_{ST} = \frac{H_T - H_S}{H_T} = \frac{2\overline{p}(1 - \overline{p}) - 2[\overline{p} - \overline{(p^2)}]}{2\overline{p}(1 - \overline{p})}$$

$$= \frac{\overline{p} - \overline{p}^2 - \overline{p} + \overline{(p^2)}}{\overline{p}(1 - \overline{p})}$$

$$= \frac{\overline{(p^2)} - \overline{p}^2}{\overline{p}(1 - \overline{p})}$$

Note that $\overline{p}^2$ and $\overline{p^2}$ are not the same; $\overline{p}^2$ is the square of the average allele frequency, whereas $\overline{p^2}$ is the average of the squared allele frequencies. Since we are considering the entire population, we can write these as expectations; $\overline{p} = E(p)$ and $\overline{(p^2)} = E(p^2)$. F_{ST} then becomes

$$F_{ST} = \frac{E(p^2) - [E(p)]^2}{\overline{p}(1 - \overline{p})}$$

But the numerator is the variance of p; see Appendix A. Therefore,

$$F_{ST} = \frac{V(p)}{\overline{p}(1 - \overline{p})}$$

For more than two alleles, the situation is more complicated, because the expected heterozygosity is not equal to $2p_i(1 - p_i)$.

Wright's original definition of F_{ST} assumed a diploid population with only two alleles at a locus, and no natural selection at the locus being considered. He also made various other assumptions about the population. Our definition based on expected heterozygosities also assumes diploidy. It is inapplicable for haploid populations or for haploid loci (e.g., mitochondrial DNA). Nei (1973) introduced a parameter closely related to F_{ST}, which is applicable to haploid loci, and which makes no assumptions about natural selection, gene flow, or the number of alleles at a locus. He called this parameter the **coefficient of gene differentiation**, and designated it by G_{ST}. G_{ST} is related to F_{ST} in the same way that gene diversity is related to expected heterozygosity. In equation (3.12) we defined expected heterozygosity as

$$H_{exp} = 1 - \sum p_i^2$$

The right side of this equation can be calculated for any population. In the context of randomly mating diploids, it is the expected heterozygosity, but it can apply to any population, and in other contexts is called gene diversity (Section 3.2). Similarly, F_{ST} is based on expected heterozygosities, while G_{ST} is based on gene diversities.

Nei defined gene identity at a locus as

$$J = \sum_{i=1}^{k} p_i^2$$

where the summation is over the k alleles at a locus. Similarly, he defined gene diversity at a locus as

$$H = 1 - J = 1 - \sum_{i=1}^{k} p_i^2$$

These are analogous to expected homozygosity and expected heterozygosity, respectively, in a randomly mating diploid population; see equations (3.12) and (3.16). In a subdivided population, J_S is the average of the within subpopulation gene identities and J_T is the gene identity in the total population. Similar meanings apply to H_T and H_S. The latter are analogous to H_T and H_S as we have defined them, but are based on gene diversity instead of heterozygosity. Nei then defined the coefficient of gene differentiation as

$$G_{ST} = \frac{H_T - H_S}{H_T}$$

This *looks* just like equation (9.2), but there are subtle differences in the definitions of H_S and H_T. F_{ST} is based on expected heterozygosities within and among subpopulations, while G_{ST} is based on gene diversities. This has led to confusion and disagreement over the relationship between G_{ST} and F_{ST} (Weir and Cockerham 1984, Nei 1986, Slatkin and Barton 1989). F_{ST} and G_{ST} are often used interchangeably in the literature, adding to the confusion. Nei (1977) showed that for a single locus, F_{ST} and G_{ST} are identical when F_{ST} is defined as we have done. For multiple loci, they differ in their assumptions about the sampling process and the method of averaging over loci.

Whittam et al. (1983) surveyed allozyme variation among 178 strains of *E. coli* isolated from different individuals in three human populations (United

States, Sweden, and Tonga). Variation was high. Of the 12 loci they studied, all were polymorphic, and overall gene diversity was 0.518. However, differentiation among populations was minimal; their estimate of G_{ST} was only 0.020. (They called it F_{ST}, but it was equivalent to Nei's G_{ST}.) This suggests that geographically distant populations are very similar in allele frequencies, but Whittam et al. found significant differentiation when they considered multilocus associations. They concluded that human-mediated gene flow is high enough to prevent differentiation of allele frequencies among *E. coli* populations, but that both random and selective factors generate linkage disequilibrium that is different in different populations, and persists due to the near absence of recombination in *E. coli*.

To summarize, we have three ways to describe differentiation among subpopulations. They are F_{ST}, G_{ST}, and θ, and can be interpreted as the proportion of total heterozygosity, total gene diversity, or total allele frequency variance that is due to differences in allele frequencies among subpopulations. These parameters differ in minor ways, but all three describe the degree of differentiation among subpopulations. For most practical purposes, they are equivalent. From now on, when we mention F_{ST} we shall implicitly include G_{ST}, and θ, unless stated otherwise. We next consider how to estimate these parameters.

Estimating F_{ST} from Sample Data

Until now, we have defined F coefficients in terms of population parameters, and in the examples we have estimated their values by applying the appropriate population equations to sample data. As discussed in Box 2.2, this procedure sometimes leads to biased estimates of parameters. In theory, estimates of F coefficients as calculated in the above examples are biased and must be adjusted to account for the numbers of individuals sampled in each subpopulation and the number of subpopulations. There are various approaches to correcting for sampling bias; for discussions of this issue, see Weir and Cockerham (1984), Nei (1986), and Slatkin and Barton (1989).

As we have seen, there are various other parameters related to F_{ST}, and there are corresponding ways to estimate them. Nei and Chesser (1983) proposed an unbiased estimator based on heterozygosities, and one based on gene diversities. We will call these estimators $\hat{F}_{ST}$ and $\hat{G}_{ST}$, respectively. Weir and Cockerham (1984) proposed an unbiased estimator based on allele frequency variances. We will call this estimator $\hat{\theta}$. Still other estimators have been proposed.

Thus, we have three different estimates of subdivision: $\hat{F}_{ST}$, $\hat{G}_{ST}$, and $\hat{\theta}$. They differ in their assumptions about sampling, their approaches to correcting for sampling bias, their methods of averaging over multiple loci, and other subtleties. Since their corresponding parameters are all essentially identical, we shall consider them alternative ways to estimate F_{ST}. Weir and Cockerham did computer simulations and concluded that $\hat{\theta}$ (designated $\hat{\theta}_W$ in their paper) was overall the best of several estimators under a wide range of conditions. Most papers today present two or more of $\hat{F}_{ST}$, $\hat{G}_{ST}$, or $\hat{\theta}$ in their results.

The calculations for all of these estimators are computationally complex and best handled with a specialized computer program. There are several programs available (see Appendix B); most give the option of calculating one or more of these estimates. One very easy-to-use program is called FSTAT (Goudet 1995, 2000); see Appendix B.

Knowledge about the degree of differentiation among populations can be useful in conservation biology. The button wrinklewort (*Rutidosis leptorrhynchoides*) is a grassland daisy found in southeastern Australia. Its numbers have declined dramatically since the 1870s, primarily due to habitat destruction; its current habitat is only about 5 percent of what it was in the 1850s. The species is endangered, and is known in only 24 populations, about half of which consist of fewer than 200 individuals. The plants are insect pollinated, and the seeds are wind dispersed, but dispersal distances are short, usually less than 0.5 m. As part of an attempt to save this species, Young et al. (1999) studied allozyme diversity in 16 populations, 13 from the northern part of the range, and 3 from the southern part. The northern and southern populations are separated by several hundred kilometers. Genetic diversity was surprisingly high; all nine loci they surveyed were polymorphic and overall observed heterozygosity was about 0.20, ranging from 0.12 to 0.25 in different populations. Smaller populations were less heterozygous than larger populations. Divergence among populations was moderate. The overall F_{ST} was about 0.17; among 13 northern populations it was 0.14, and among 3 southern populations it was 0.17.

As we have seen (Sections 7.9 and 8.1), habitat fragmentation frequently leads to reduced genetic diversity within populations, which in turn leads to increased risk of extinction. Young et al. conclude that this is not an imminent threat to this species, since genetic variation in most populations is high. They suggest that, since genetic differences among populations are relatively small, the major conservation efforts should go toward preserving the five largest populations in the north and the largest one in the south.

Table 9.3 gives estimates of F_{ST} based on several kinds of molecular markers in a variety of organisms. In general, mobile populations such as humans have relatively low values of F_{ST}. On the other hand, the kangaroo rat, which is known from field studies to disperse very short distances, has a high F_{ST}. Most of these studies are based on allozyme surveys; we will consider alternative estimates of subdivision based on other kinds of genetic markers in Section 9.5.

Statistical Significance of Subdivision

Once you have an estimate F_{ST}, how do you know whether it represents significant subdivision? Most computer programs will calculate a confidence interval for the estimate, but sometimes they will not. Here, we describe a simple test that applies to any number of subpopulations and any number of alleles. The test is called a heterogeneity chi-square (or contingency chi-square) test. It seeks to answer the question, Do allele frequencies differ significantly among subpopulations? The null hypothesis is that they do not.

The numbers of interest are observed numbers (not frequencies) of each allele in each subpopulation. These can easily be calculated from the observed genotype numbers. The data are arranged as in Table 9.4. As before, the subscript i indicates allele and j indicates subpopulation. There are k different alleles (rows) and n subpopulations (columns). Let O_{ij} be the observed number of the ith allele in the jth subpopulation. The row totals are designated by R_i and the column totals by C_j. T is the total number of alleles (not individuals) in the entire sample. Let

TABLE 9.3   Estimates of F_{ST} ($\hat{G}_{ST}$ or $\hat{\theta}$) for various organisms.

Where a range is given, it indicates different kinds of comparisons.

Organism	Kind of Marker	Number of Loci	Number of Populations	F_{ST} Estimate	Reference
E. coli	allozymes	12	3	0.036	Whittam et al. (1983)
Horseshoe crab (*Limulus polyphemus*)	allozymes	25	4	0.072	Selander et al. (1970)
Drosophila aldrichi	allozymes	3	10	0.021	Krebs and Barker (1993)
Drosophila simulans	microsatellites	4	16	0.138	Irvin et al. (1998)
Snails (*Bembicium vittatum*)	allozymes	47	13	0.122 to 0.342	Johnson and Black (1998)
Squid (*Loligo opalescens*)	microsatellites	11	6	0.0028	Reichow and Smith (2001)
Atlantic cod (*Gadus morhua*)	RFLP	8	10	0.014	Pogson et al. (2001)
Brown trout (*Salmo trutta*)	allozymes	35	38	0.292	Ryman (1983)
Water snakes (*Nerodia sipedon*)	allozymes	7	7	0.005 to 0.093	King and Lawson (1995)
Shrew (*Sorex araneus*)	microsatellites	24	6	0.032	Wyttenbach et al. (1999)
House mouse (*Mus musculus*)	allozymes	40	4	0.119	Selander et al. (1969)
Kangaroo rat (*Dipodomys ordii*)	allozymes	18	9	0.674	Johnson and Selander (1971)
Gray wolf (*Canis lupis*)	allozymes	8	5	0.074	Kennedy et al. (1991)
Canada lynx (*Lynx canadensis*)	microsatellites	17	9	0.033	Schwartz et al. (2002)
Humans (major races)	allozymes	62	3	0.088	Nei and Roychoudhury (1982)
Humans (Yanomama Indian villages)	allozymes	15	37	0.069	Weitcamp et al. (1972)
Club moss (*Lycopodium lucidulum*)	allozymes	13	4	0.284	Levin and Crepet (1973)
Coulter pine (*Pinus coulteri*)	allozymes	8	25	0.165	Ledig (2000)
Aspen (*Populus tremuloides*)	RAPDs	132	23	0.18 to 0.34	Stevens et al. (1999)
Daisy (*Rutidosis leptorrhynchoides*)	allozymes	16	9	0.17	Young et al. (1999)

E_{ij} be the expected number of the ith allele in the jth subpopulation, assuming no subdivision. The E_{ij} are calculated as

$$E_{ij} = \frac{R_i C_j}{T} \tag{9.7}$$

This is just the overall frequency of the ith allele (R_i/T) multiplied by the total number of alleles (copies) in the jth subpopulation (C_j).

TABLE 9.4   Arrangement of data for the chi-square test for significant subdivision.
There are k different alleles ($i = 1, \ldots, k$) and n subpopulations ($j = 1, \ldots, n$). Each entry in the table is the observed number of a particular allele in a particular subpopulation; O_{ij} is the observed number of the ith allele in the jth subpopulation. R and C are the row totals and column totals, respectively. T is the total number of alleles (copies) in the entire population.

Alleles	Subpop 1	Subpop 2	. . .	Subpop j	. . .	Subpop n	Row Totals
				Subpopulations			
A_1	O_{11}	O_{12}				O_{1k}	R_1
A_2							R_2
. . .							
A_i	O_{i1}	O_{i2}		O_{ij}		O_{in}	R_i
. . .							
A_k	O_{k1}	O_{k2}				O_{kn}	R_k
Column totals	C_1	C_2		C_j		C_n	T

The chi-square test statistic is then

$$\chi^2 = \sum_{i=1}^{k} \sum_{j=1}^{n} \frac{(O_{ij} - E_{ij})^2}{E_{ij}}$$

The degrees of freedom for the test is $(k - 1)(n - 1)$. The null hypothesis of homogeneity (no differences among subpopulations) is rejected if the calculated chi-square value is greater than the appropriate critical value in Table 3.2. A significant chi-square value indicates significant allele frequency differences among subpopulations, that is, significant subdivision. It gives no information about the causes of the differences. Box 9.6 illustrates the calculations. See also Problem 9.13.

This may not be the best way to determine whether subdivision is significant, especially if there are rare alleles present in one or more subpopulations, a common occurrence with microsatellite loci. If expected numbers are few, the chi-square statistic is inflated and unreliable (see the cautions in Section 3.1). Rousset and Raymond (1997) review the use of so-called "exact tests," which may be preferable in some situations. Computer programs that estimate F_{ST} will usually provide some estimate of significance. You should read the program's documentation to see how it does this.

Keep in mind that F_{ST} simply *describes* the differentiation among subpopulations. We have made no assumptions about the causes of differentiation or the actual structure of the population. In Section 9.5, we shall see that, under certain assumptions, we can use estimates of F_{ST} to estimate levels of gene flow among subpopulations.

F_{ST} and G_{ST} were initially developed to describe levels of subdivision based on allele frequencies at classical morphological and allozyme loci. Most population genetic studies today use microsatellite or DNA markers. In Section 9.5, we consider ways to describe population subdivision based on these kinds of data, and methods that allow us to estimate levels of gene flow under assumptions that are more appropriate to them. But first we need to discuss models of gene flow.

Box 9.6   Chi-square test for significant subdivision

Sample data are usually summarized as genotype frequencies. The number of copies of each allele in each subpopulation can easily be calculated from these genotype frequencies. For example, if a subpopulation has 50 A_1A_1 homozygotes, 20 A_1A_2 heterozygotes, and 30 A_1A_3 heterozygotes, there are 150 A_1 alleles in that subpopulation.

Consider an example with three subpopulations and three alleles. The observed numbers of alleles are

	Subpopulation 1	Subpopulation 2	Subpopulation 3	Row Totals
Allele 1	20	25	160	$R_1 = 205$
Allele 2	20	50	20	$R_2 = 90$
Allele 3	160	25	20	$R_3 = 205$
Column totals	$C_1 = 200$	$C_2 = 100$	$C_3 = 200$	$T = 500$

The expected numbers are calculated from equation (9.7). For example, the expected number of A_1 alleles in subpopulation 3 is $(205)(200)/500 = 82$. The expected numbers are

	Subpopulation 1	Subpopulation 2	Subpopulation 3	Row Totals
Allele 1	82	41	82	$R_1 = 205$
Allele 2	36	18	36	$R_2 = 90$
Allele 3	82	41	82	$R_3 = 205$
Column totals	$C_1 = 200$	$C_2 = 100$	$C_3 = 200$	$T = 500$

Finally, we set up a table for the calculations as follows:

i	j	O_{ij}	E_{ij}	$\dfrac{(O_{ij} - E_{ij})^2}{E_{ij}}$
1	1	20	82	46.88
1	2	25	41	6.24
...	...	...	...	...
3	2	25	41	6.24
3	3	20	82	46.88
Sum				325.75

The sum of the last column is the chi-square statistic. The degrees of freedom is 4. Clearly, the chi-square value is highly significant.

You should confirm for yourself that the uncorrected estimate of F_{ST} is 0.33. This represents highly significant subdivision according to the chi-square test.

9.3 Gene Flow

We now examine how gene flow homogenizes populations. It should be intuitively obvious that if two populations are different, then gene flow between them will tend to reduce those differences. Our goal is to quantify that reduction.

Gene flow is defined as movement of individuals (or gametes) from one population to another *and subsequent breeding*. The latter part is necessary, because if individuals move, but do not leave any genes in the new population, then they have had no genetic effect on that population. The level of gene flow is usually symbolized by m, and is defined as the proportion of alleles in a population that has come from another population in that generation. The symbol m comes from the older term migration, which can mean different things to geneticists or ecologists. We shall follow Endler's (1977) recommendation and use the term gene flow, but will retain the conventional symbol m. (Note that m is still frequently referred to as the migration rate in the literature.)

We can imagine various ways in which populations are structured, and several models of gene flow between subpopulations. Figure 9.3 illustrates several examples. In the continent-island model (Figure 9.3a), there is one-way movement from a single large population (the continent) to a single smaller population (the island). In the island model (Figure 9.3b), there are numerous subpopulations, each exchanging individuals with all the others. We will examine these models to quantify the rate at which gene flow homogenizes populations. Initially, we assume that population sizes are effectively infinite and that no other evolutionary processes are acting on the loci being studied. We will relax these assumptions in Section 9.4.

The Continent-Island Model

The simplest model of gene flow is a continent-island model in which one-way gene flow occurs from a large continental (mainland) population to a smaller island population (Figure 9.3a). Let p_m be the frequency of an allele on the mainland, and p_t be the frequency of the same allele on the island in generation t. If no other evolutionary processes are operating on the mainland, we can assume that p_m is constant. The rate of gene flow, m, is the proportion of alleles on the island that have come from the mainland each generation. Then in the next generation, a proportion m on the island will have just come from the mainland, and a proportion $(1 - m)$ will be from the island. Therefore, the allele frequency on the island will be

$$p_{t+1} = (m)(p_m) + (1 - m)(p_t) \tag{9.8}$$

This is just the weighted average of the allele frequencies among the immigrants and the residents. We can rearrange this equation to get

$$p_{t+1} = p_t + m(p_m - p_t) \tag{9.9}$$

The change in allele frequency is then

$$\Delta p = p_{t+1} - p_t = m(p_m - p_t) \tag{9.10}$$

The allele frequency on the island changes at a rate that depends on the rate of gene flow and on the difference in allele frequency between the island and the mainland.

The equilibrium frequency on the island can be found by setting $p_{t+1} = p_t = \widetilde{p}$ and solving for $\widetilde{p}$. The result is

$$\widetilde{p} = p_m$$

Figure 9.3   **Several models of population structure with gene flow.**   (a) Continent-island model,
with one large mainland population and one island population; gene flow occurs from the mainland to the island only. (b) Island model; gene flow occurs from any
island to any other island. (c) One-dimensional stepping
stone model; gene flow occurs from an island to an adjacent island only. (d) Two-dimensional stepping stone
model; same as (c), except the islands are arranged in
two dimensions.

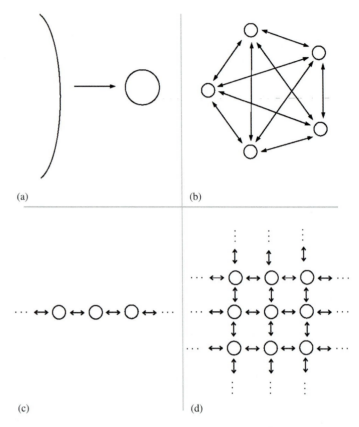

(a)

(b)

(c)

(d)

By plotting Δp versus p, it is easy to show that this equilibrium is asymptotically
stable (Problem 9.14; see also Box 9.7 for a more formal approach). That should
be obvious in retrospect. If the island receives immigrants from the mainland
every generation, then eventually all alleles will have come from the mainland
one or more generations back. To see the rate of approach to this equilibrium, we
can solve equation (9.9). The details are in Box 9.7; the solution is

$$p_t = p_m + (1 - m)^t(p_o - p_m) \tag{9.11}$$

where p_o is the initial allele frequency on the island. As t increases, $(1 - m)^t$ approaches zero, and p_t approaches p_m. Figure 9.4 plots equation (9.11) for several
levels of gene flow.

This model may be realistic for literal continent-island systems where gene
flow is essentially one-way, such as from the west coast of South America to the
Galapagos Islands, or from the European mainland to Great Britain. It is also useful in describing the effects of gene flow in wildlife management programs such
as captive breeding programs or stocking of river systems with hatchery fish. For
example, Hansen et al. (1995) studied restriction site variation in mtDNA of natural and hatchery populations of brown trout (*Salmo trutta*) in a river system in
Denmark. The rivers were extensively stocked with hatchery trout during the
1980s. We can view the natural populations as islands and the hatchery population as a mainland (source) population. Hansen et al. used a more complex model
than the one described here, incorporating the characteristics of mtDNA and age

Box 9.7   The general linear recursion equation

Consider equation (9.8):

$$p_{t+1} = mp_m + (1 - m)p_t \qquad (1)$$

This is an example of the general linear recursion equation

$$x_{t+1} = a + bx_t \qquad (2)$$

where a and b are constants. Here, we find the general solution to equation (2) and analyze the stability of its equilibrium. We first find the equilibrium value of x by setting $x_{t+1} = x_t = \tilde{x}$. Solving for $\tilde{x}$ we get

$$\tilde{x} = \frac{a}{1 - b} \qquad (3)$$

We now need to determine whether the equilibrium is stable or unstable. To do that, we need the general solution to equation (2). We can find the pattern by brute force. In the first generation.

$$x_1 = a + bx_o$$

In the second generation,

$$x_2 = a + bx_1 = a + b(a + bx_o)$$
$$= a + ab + b^2 x_o$$

In the third generation

$$x_3 = a + bx_2 = a + b(a + ab + b^2 x_o)$$
$$= a(1 + b + b^2) + b^3 x_o$$

And, in general,

$$x_t = a(1 + b + \cdots + b^{t-1}) + b^t x_o$$
$$= a\left(\sum_{i=0}^{t-1} b^i\right) + b^t x_o$$

This is the general solution to the recursion equation; a formal proof would require the use of mathematical induction. We now use the fact that if $b < 1$, then the summation is equal to

$$\sum_{i=0}^{t-1} b^i = \frac{1 - b^t}{1 - b}$$

Substituting, we get

$$x_t = a\left(\frac{1 - b^t}{1 - b}\right) + b^t x_o$$
$$= \left(\frac{a}{1 - b}\right)(1 - b^t) + b^t x_o$$

But $a/(1 - b) = \tilde{x}$, so

$$x_t = \tilde{x}(1 - b^t) + b^t x_o$$

Finally, a slight rearrangement gives

$$x_t = \tilde{x} + b^t(x_o - \tilde{x}) \qquad (4)$$

Now, if $|b| < 1$, then b^t approaches zero, and x_t approaches $\tilde{x}$. So the condition for asymptotic stability of the equilibrium is that $|b| < 1$.

Comparing equations (1) and (2) above, we see that $a = mp_m$ and $b = 1 - m$. Since $(1 - m) < 1$, the equilibrium $\tilde{p} = p_m$ is stable. From equation (4), we can write the solution to equation (1) as

$$p_t = p_m + (1 - m)^t(p_o - p_m)$$

which is equation (9.11).

---

structure (different reproductive rates in different age classes). Based on estimated population sizes and stocking records, the contribution of the hatchery population to present-day natural populations was predicted to be about 70 percent. The actual contribution was estimated to be about 2 percent. They attributed the difference to natural selection against hatchery trout. Note, this does not mean that natural selection is acting on the RFLP polymorphisms. These are almost certainly neutral markers whose frequency is being affected by natural selection at other loci. This example also emphasizes the fact that gene flow depends not just on movement of individuals, but also on their survival and reproduction in the new population.

Figure 9.4 **Change of allele frequency on the island in a continent–island model of gene flow.**   Initial allele frequency on the island is 0.3 and allele frequency on the mainland is 0.8. The curves show change in frequency on the island for four different levels of gene flow (*m*).

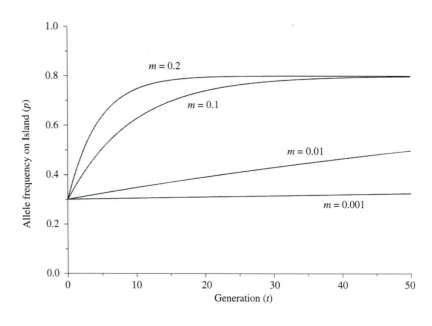

The Island Model

Many populations *exchange* immigrants with one another. Imagine a large number of islands, each randomly exchanging individuals with all other islands (Figure 9.3b). This might approximately describe gene flow among a literal group of islands, such as the Hawaiian islands. It might also be useful for describing "habitat islands," such as fragmented rain forests, or mountaintops separated by desert, as in the Great Basin region of the United States. Any time we have many populations, each exchanging individuals with other populations, we have an island model of gene flow.

The gene flow rate, *m*, is the proportion of alleles on any island that have come from elsewhere in that generation. The islands differ in allele frequency, and the average allele frequency among all the islands is $\bar{p}$. Now consider a single island. From the point of view of that island, all the other islands are equivalent to a "continent" with allele frequency $\bar{p}$. So, if p_t is the allele frequency on a given island in generation *t*, the recursion equation for the allele frequency on that island is

$$p_{t+1} = (m)(\bar{p}) + (1 - m)(p_t) \tag{9.12}$$

This is the same as equation (9.8), except that $\bar{p}$ replaces p_m. So, all of the conclusions of the continent–island model hold for any single island. The allele frequency on each island will approach the mean frequency, $\bar{p}$. Figure 9.5 illustrates this for four islands with initial allele frequencies of 0, 0.3, 0.7, and 1.0, and $m = 0.1$. In this case, $\bar{p} = 0.50$, and all of the islands converge to this frequency.

Stepping Stone Models and Isolation by Distance

Figure 9.3 shows other models of population structure and gene flow. Figure 9.3c and d illustrate one- and two-dimensional **stepping stone models**. These models allow subpopulations to exchange individuals only with adjacent subpopulations, and can be formulated in one or two (or more!) dimensions. The

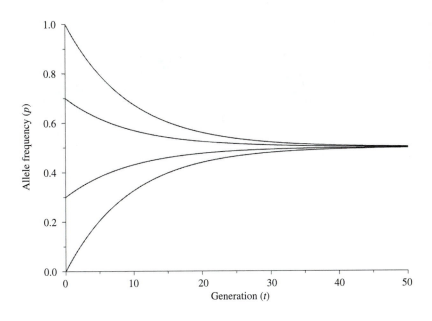

Figure 9.5 **Change of allele frequency on four islands in an island model of gene flow, with** $m = 0.1$**.** Initial allele frequencies on the islands are 0, 0.3, 0.7, and 1.0. The four curves show all islands converging to an allele frequency of 0.5.

mathematics of stepping stone models can be difficult, but the main conclusions are straightforward. Two subpopulations that are far apart will experience little of the homogenizing effects of gene flow, and thus will be more different than two subpopulations that are closer together.

A variant of the two-dimensional stepping stone model assumes that individuals are distributed continuously over a large geographical area, but the probability of two individuals mating decreases with their distance apart. Thus, individuals chosen from locations close together are likely to be more similar genetically than individuals chosen from locations far apart. This idea is known as **isolation by distance**, and was first discussed by Sewall Wright (1940, 1943), and extended by Slatkin and Maddison (1990), Slatkin (1993), Rousset (1997), and others.

The Arctic brown bear (*Ursus arctos*) provides an interesting example of isolation by distance. These bears are distributed more or less continuously throughout much of extreme northern North America, and there are no major barriers to gene flow. Paetkau et al. (1997) surveyed eight microsatellite loci in bears from six regions throughout their distribution (Figure 9.6a). From their results, they constructed several measures of genetic distance. Genetic distance is a measure of how different two populations are genetically. There are several ways to estimate it, which need not concern us now. One way to think of it is that genetic distance is analogous to a pairwise estimate of F_{ST} between two subpopulations. Most measures of genetic distance range from zero (two populations are exactly alike) to infinity (two populations share no alleles at any of the loci studied). Paetkau et al. plotted several different measures of genetic distance versus geographic distance. Figure 9.6b shows their results for one commonly used measure of genetic distance (Nei 1972). Clearly, there is a strong relationship between geographic distance and genetic distance. The regression line explains about 87 percent of the variation.

Whether local differentiation occurs in a large, continuously distributed population depends primarily on the population density and the dispersal distances

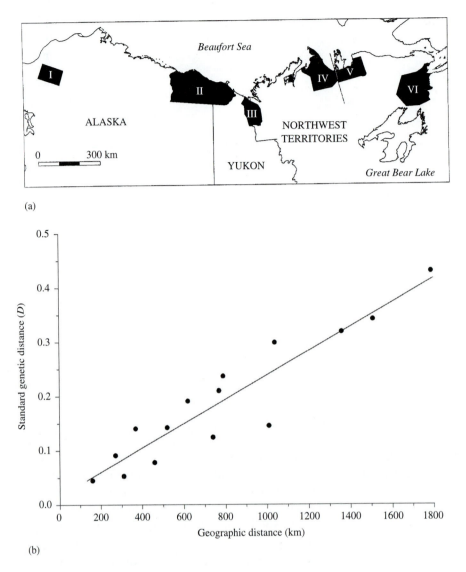

Figure 9.6 **Isolation by distance in the Arctic brown bear** *(Ursus arctos).* (a) Regions sampled from a large, more or less continuous population. (b) Isolation by distance, as measured by Nei's (1972) measure of genetic distance. Each point represents a pairwise comparison of two subpopulations. The regression line explains about 87 percent of the variation. *Source:* (a) From Paetkau et al. (1997).

of individuals. If we define dispersal distance as the distance from an individual's place of birth to its place of reproduction, then dispersal distance is a random variable with standard deviation σ. The number of individuals within a circle of radius σ is designated by N_σ. If $N_\sigma > 12$, there will be essentially no regional differentiation within the population (Kimura and Ohta 1971c). In other words, if the "neighborhood size" is 12 or more individuals, the entire population will be homogeneous with respect to allele frequencies. This condition must surely be met by organisms such as continental *Drosophila* species.

Drosophila willistoni is a widespread neotropical species, found from southern Florida, through Mexico, the West Indies, Central America, and much of South America. The species is more or less continuously distributed throughout much of tropical South America. Francisco Ayala and his colleagues (Ayala et al. 1971, 1972; Ayala 1972) conducted extensive studies of allozyme variation throughout the range of this species. They studied 27 loci and as many as 80 populations for some loci. Sampled populations were as far apart as 6000 km. The species is highly variable; the average observed heterozygosity per locus was about 0.18. The remarkable thing about this study is the consistency from population to population. Ayala et al. found that most populations are very similar in both the amount and pattern of genetic variation at most loci. There are two possible explanations for this observation: First, the observed patterns may be maintained by some kind of balancing selection that maintains similar allele frequencies in different populations. Second, the loci may be selectively neutral, and the observed uniformity is the result of extensive gene flow, as hypothesized in the previous paragraph. Ayala et al. argued for the role of balancing selection, but the issue has never been satisfactorily resolved. Jeffrey Powell, one of the participants in the original study, wrote, many years later, "… surely migration [gene flow] must be playing a large role" (Powell 1997, p. 40).

The main qualitative conclusion of all these models of gene flow is the same: Gene flow tends to homogenize populations. The rate of homogenization depends on the amount of gene flow and on the population structure, but the eventual outcome is homogeneity, regardless of the details of the model or the starting point. This assumes that population sizes are effectively infinite and that no other evolutionary processes are affecting the loci being studied. In the next section, we examine the quantitative relationship between the homogenizing effects of gene flow and the diversifying effects of other processes.

9.4 Gene Flow and Differentiation

Genetic drift causes populations to diverge; gene flow tends to prevent divergence. How much gene flow is required to counteract the divergent effects of genetic drift? The specific answer to this question depends on the particular model of gene flow, genetic drift, and mutation, but the general answer is surprising.

First, consider a finite islands model, in which each subpopulation has a finite population size, and there is a finite number of subpopulations (islands). Ignore mutation for now. If there is no gene flow, drift acts independently in each subpopulation and each will become fixed for one allele. Different alleles will likely become fixed in different subpopulations. If there is gene flow among subpopulations, the tendency toward fixation will be retarded, but since the total population size is finite, all subpopulations will eventually become fixed for the same allele. Gene flow only slows down the process.

Now consider a slightly different model. Assume that each subpopulation is finite, but there are so many subpopulations that the total population size is effectively infinite (infinite islands model). Again, without gene flow, each subpopulation will eventually become fixed for a single allele. But with gene flow, the outcome is very different. If gene flow is low, each subpopulation will go its own way, and allele frequencies will be different in different subpopulations; some will be fixed for one allele, some fixed for another, and some will remain polymorphic. Drift will be the dominant process in each subpopulation, with gene flow only slowing the rate of differentiation. The reason that fixation does not always occur is that any given subpopulation is receiving alleles from the population as a whole, where there is no overall tendency toward fixation (because the *total* population size is infinite). On the other hand, if gene flow is high enough, it will be the dominant process and all subpopulations will be similar. How high is "high enough"?

Wright (1940) first studied this model quantitatively. He assumed an infinite islands model with no mutation or natural selection. The average allele frequency over all islands is $\bar{p}$. As shown in the previous section, the *expected* frequency on any single island will approach $\bar{p}$. However, the *distribution* of allele frequencies over all the islands depends on m and N, more specifically, on their product, Nm. Here, and in all that follows, N is the effective population size on each island. Note that Nm is the actual number of immigrant individuals in an island population each generation.

Wright showed that if Nm is large enough, the distribution will be humped, and nearly all islands will have an allele frequency near $\bar{p}$. Gene flow dominates genetic drift, and differentiation among islands will be minimal. On the other hand, if Nm is small, the distribution of allele frequencies will be U–shaped and there will be significant differentiation among islands. Genetic drift dominates gene flow, and allele frequencies will approach fixation on some islands.

Figure 9.7 illustrates these ideas for $\bar{p} = 0.5$. If $Nm \geq 5$, there is a large peak at $\bar{p}$. Recall that Nm is the actual number of individuals that contribute genes from elsewhere each generation. Therefore, an average gene flow rate of five individuals per generation, *irrespective of population size*, is enough to prevent substantial differentiation and keep nearly all subpopulations near $\bar{p}$. Furthermore, for any $Nm > 0.5$, the distribution is at least somewhat humped, implying that nearly all subpopulations are polymorphic at this locus. In other words, one migrant individual every other generation is enough to prevent fixation of alternative alleles in different subpopulations. These are surprisingly low numbers. We conclude that very little gene flow is necessary to counteract the effect of genetic drift. It may also be surprising that this effect is independent of population size. The reason is that in large populations drift is weak, so it takes a less gene flow (expressed as proportion of immigrants each generation) to counteract it.

Wright formulated his model in terms of two alleles, but it can apply to any number of alleles by simply letting p be the frequency of one allele and $1 - p$ be the frequency of all other alleles. Mutation has so far been ignored. However, any time we consider long-term processes we must consider mutation. Under the infinite alleles model with no genetic drift (infinite population size), all alleles will eventually be unique and at equilibrium, heterozygosity will be 100 percent (Section 6.2). In a finite population (as in a single subpopulation) equilibrium heterozygosity will be as given by equation (7.19);

$$\tilde{H} \cong \frac{4Nu}{4Nu + 1} \tag{9.13}$$

Consider the infinite islands model again, this time with mutation according to the infinite alleles model. In any given subpopulation, equilibrium heterozygosity is expected to be as in equation (9.13). From the point of view of any given subpopulation, the rest of the population is infinite, with equilibrium heterozygosity

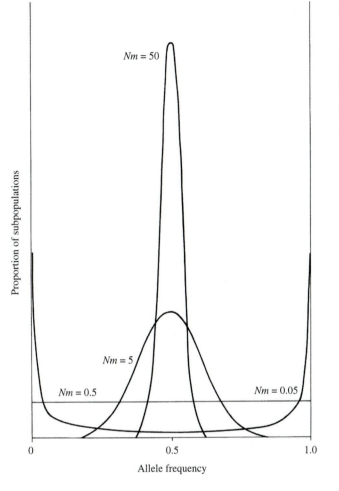

Figure 9.7 **Distribution of allele frequencies among subpopulations with gene flow and genetic drift under Wright's island model of migration.** Horizontal axis is allele frequency, p. Vertical axis is probability density function; it can be interpreted as the relative proportion of populations with allele frequencies near p. Overall allele frequency in the entire population is $p_m = 0.5$. *Source:* Modified from Crow (1986).

of 100 percent. Every allele coming into the subpopulation is a new allele. Therefore, gene flow, which brings a new allele into a subpopulation from elsewhere, is genetically equivalent to mutation, which creates a new allele within the subpopulation. We can simply substitute m for u in equation (9.13) to get the equilibrium heterozygosity in the infinite islands model of gene flow:

$$\tilde{H} \cong \frac{4Nm}{4Nm + 1} \qquad \textbf{(9.14)}$$

This assumes that mutation is insignificant compared to gene flow; otherwise, we must replace m in equation (9.14) by $(m + u)$.

Equation (9.14) tells us how equilibrium heterozygosity depends on Nm, the number of immigrants that contribute to a subpopulation each generation. Heterozygosity changes rapidly over a short range of Nm (Figure 9.8). For example, as Nm goes from 0 to 1, heterozygosity goes from 0 to 0.8. A very small amount of gene flow among subpopulations, on the order of one individual per generation, is enough to assure high levels of heterozygosity.

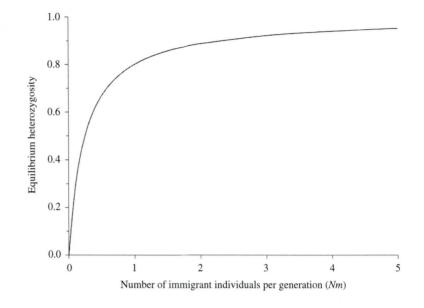

Figure 9.8 **Equilibrium heterozygosity ($\tilde{H}$) as a function of *Nm* for the island model, as predicted by equation (9.14).**

Now consider the issue of differentiation among subpopulations. Recall that differentiation can be measured by F_{ST}:

$$F_{ST} = \frac{H_T - H_S}{H_T}$$

where H_T is the expected heterozygosity in the total population, and H_S is the average of the expected heterozygosities in the subpopulations. Under the infinite alleles model, $H_T = 1$, and H_S is given by equation (9.14). Substituting, we get, at equilibrium,

$$\tilde{F}_{ST} \cong 1 - \left(\frac{4Nm}{4Nm + 1}\right)$$

or,

$$\tilde{F}_{ST} \cong \frac{1}{4Nm + 1} \tag{9.15}$$

This equation tells us how gene flow, as measured by Nm, affects population differentiation, as measured by F_{ST}. Keep in mind that N is really the effective population size, N_e.

Figure 9.9 plots equilibrium F_{ST} versus Nm. Differentiation among subpopulations decreases rapidly with only a few immigrant individuals per generation. For $Nm > 1$ (more than one immigrant per generation), F_{ST} is less than about 0.20; for $Nm > 5$, F_{ST} is less than about 0.05. Thus, one immigrant per generation is sufficient to prevent substantial differentiation, and five per generation are enough to assure that all subpopulations are essentially identical.

The qualitative conclusion from this model is the same as for the two-allele model discussed earlier: It takes very little gene flow to prevent differentiation among subpopulations. Other models of drift and gene flow differ in details, but the general conclusion is the same: Gene flow is very effective in counteracting the divergent effect of genetic drift. It takes only a very few immigrants each

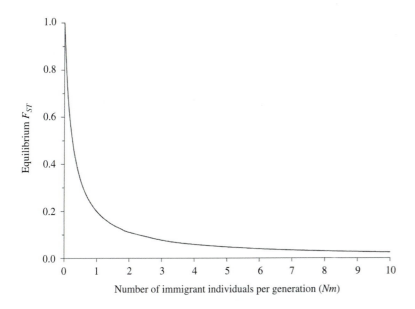

Figure 9.9 **Equilibrium F_{ST} as a function of Nm for the island model, as predicted by equation (9.15).**

generation to prevent populations from diverging. This, of course, assumes that natural selection is not acting on the locus being studied.

Gene Flow and Natural Selection

We have seen that it takes very little gene flow to prevent significant differentiation *in the absence of natural selection.* If natural selection is acting differently in different subpopulations, how much gene flow is necessary to counteract the effects of local selection and prevent differentiation among subpopulations? Alternatively, how strong must natural selection be to overcome the homogenizing effects of gene flow?

Consider an island model of gene flow. Assume that each subpopulation is large enough that genetic drift can be ignored in the short term, and that natural selection acts in a single subpopulation as

Genotype	A_1A_1	A_1A_2	A_2A_2
Fitness	1	$1 - hs$	$1 - s$

We showed in Section 5.8 that the change in allele frequency due to natural selection is

$$\Delta p_{sel} = \frac{pqs[ph + q(1 - h)]}{\overline{w}}$$

where $\overline{w} = 1 - 2pqhs - sq^2$. These are equations (5.21) and (5.19).

Under the island model of gene flow, the change in allele frequency due to gene flow alone is

$$\Delta p_{gf} = m(\overline{p} - p)$$

where p is the allele frequency in the subpopulation and $\overline{p}$ is the allele frequency in the entire population. This is equation (9.10) with $\overline{p}$ instead of p_m.

The total change in allele frequency in the subpopulation is the sum of the changes due to the two processes. At equilibrium, the total change is zero, and we have

$$\Delta p = \frac{pqs[ph + q(1 - h)]}{\overline{w}} + m(\overline{p} - p) = 0 \qquad \textbf{(9.16)}$$

If you try to solve this equation for the equilibrium value of p, you will obtain, after much algebra, a cubic equation that yields little biological insight. However, equation (9.16) allows us to plot Δp versus p, or to obtain the equilibrium value of p, for any specified values of s, h, m, and $\overline{p}$. Note that if $0 < \overline{p} < 1$, neither $p = 0$ nor $p = 1$ is an equilibrium, because every generation both alleles will be introduced into the subpopulation by gene flow.

Figure 9.10a illustrates an example. The fitnesses are $1, 0.95$, and 0.9 ($s = 0.1$ and $h = 0.5$). With no gene flow, A_1 would go to fixation ($p = 1$). With small

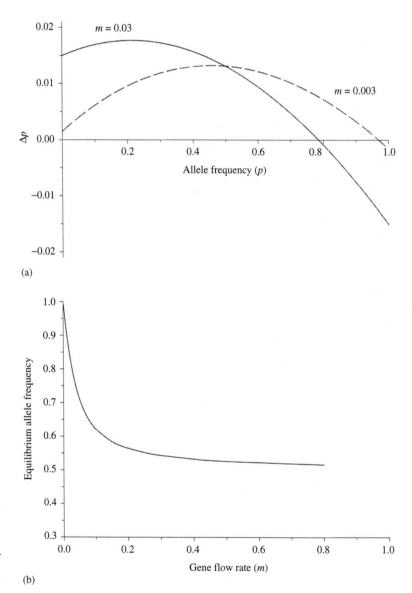

Figure 9.10 Allele frequencies under an island model of gene flow with natural selection.   (a) Change in allele frequency, Δp, as predicted by equation (9.16). Fitnesses are $1, 0.95$, and 0.90 ($s = 0.1$ and $h = 0.5$). A stable equilibrium occurs where Δp crosses the horizontal axis, going from positive to negative. With no gene flow, the population would go to fixation at $p = 1$. With low levels of gene flow (dashed line), the stable equilibrium is only slightly less than 1; with higher levels of gene flow (solid line), the equilibrium is shifted substantially. (b) Equilibrium allele frequency, $\tilde{p}$, as a function of gene flow, m. For high levels of gene flow, the allele frequency on the island approaches the average allele frequency (0.5 in this example).

amounts of gene flow, there is a stable equilibrium near $p = 1$. (Recall that a stable equilibrium occurs when Δp crosses the p-axis, going from positive to negative.) Natural selection is the dominant process and gene flow has little effect on the subpopulations. With higher levels of gene flow, the stable equilibrium is displaced further from $p = 1$. The general effect is illustrated in Figure 9.10b, which plots $\tilde{p}$ versus m, again for $s = 0.1$ and $h = 0.5$. For large values of m, the equilibrium allele frequency approaches the overall allele frequency, $\bar{p}$ (0.5 in this example). We can take as a general guideline that if m is much less than s, then natural selection will dominate and each subpopulation will approach an equilibrium determined mostly by selection in that subpopulation. If m is much greater than s, then gene flow will prevent local adaptation to a great extent; there will be little differentiation among subpopulations.

9.5 Estimating Levels of Gene Flow

Gene flow has important implications for many practical problems. For example, gene flow has been implicated in the worldwide spread of resistance to organophosphate insecticides in the mosquito *Culex pipiens* (Raymond et al. 1991; see Section 5.4).

One of the controversies over the release of genetically modified organisms into the environment is whether they are likely to pass their genes to natural populations. Ellstrand et al. (1999) reviewed data on the world's 13 most important food crops and concluded that 12 of them interbreed with their wild relatives to some extent. Quist and Chapela (2001) claimed to demonstrate the presence of DNA from transgenic corn in native races in Mexico. Based on DNA sequence diversity, they inferred that transfer had occurred multiple times, and that such gene flow may be common between genetically modified organisms and native crops. This work is controversial; see Mann (2002), Metz and Futterer (2002), Kaplinsky (2002), and Quist and Chapela (2002). However, if this kind of gene flow does occur to natural populations, it may have deleterious effects, such as the evolution of weeds resistant to pesticides or increased danger of extinction of wild relatives (Ellstrand et al. 1999). In order to evaluate these possible dangers, it is important to be able to estimate how much gene flow actually occurs.

Endangered populations are frequently small and isolated from other populations of the same species (if there are any). Small isolated populations suffer increased risk of extinction due to reduced genetic variation and a number of other genetic and demographic factors (Section 7.9). In order to evaluate these risks and make informed management decisions, we need to know how much genetic variation actually exists in an endangered population, and the degree to which it is connected to other populations through gene flow.

An intelligent approach to these issues requires quantitative estimates of the amount of gene flow among populations. In this section, we summarize several approaches to estimating gene flow, and discuss the assumptions behind them and potential problems with each method.

Direct Estimates of Gene Flow

Amounts of gene flow can sometimes be inferred from movements of individuals. We must be careful to distinguish between dispersal and gene flow. Dispersal is the movement of individuals from one region to another, and will result in gene flow (movement of *genes* from one region to another) only if dispersal is followed by reproduction (Endler 1977).

Many of the early estimates of dispersal involved mark and recapture experiments with various species of *Drosophila*. Slatkin (1985b) reviews some of these studies. Some of the general conclusions are: (1) Dispersal distances are typically short, on the order of 500 m or less; (2) occasional long-distance dispersal, on the order of several kilometers, occurs; and (3) individuals in a favorable habitat disperse less than individuals in an unfavorable habitat.

Experiments with other organisms generally confirm these results. Dispersal is usually limited. Organisms typically move much less than they are physically capable of moving, but a few individuals will disperse long-distances. This occasional long-distance dispersal may have significant long-term evolutionary effects if it is accompanied by gene flow. Whether such long-distance gene flow actually occurs is difficult to determine by direct observation.

Movement from one population to another does not necessarily result in gene flow. Physical movement may give a misleading picture of actual gene flow. Recall the study by Hansen et al. (1995) of brown trout in Denmark. During the 1980s the Karup River was extensively stocked with hatchery trout. One RFLP haplotype was present in very high frequency in hatchery trout, but rare in rivers stocked with the hatchery trout. Hansen et al. concluded that hatchery trout had low reproductive success and the genetic contribution of hatchery trout to the river population was much less than predicted from stocking numbers alone.

Highly variable loci can sometimes be used to obtain direct estimates of gene flow. For example, Vigilant et al. (2001) used noninvasive sampling (feces, hairs, etc.) to determine the genotypes at nine microsatellite loci in 108 chimpanzees in three communities. By comparing the genotypes of females and their offspring to the genotypes of males in the group, they were able to assign paternity to 34 of 41 offspring by excluding all males but one as possible fathers. In all 34 cases, the male belonged to the same community as the female. Of the seven offspring whose paternity could not be assigned, Vigilant et al. concluded that only one represented a probable case of extra-group paternity. The others were ambiguous because some males within the community had died before being sampled. Because of these ambiguities, they estimated the frequency of extra-group paternity to be between 2.4 and 7.1 percent. This represents significant gene flow in an evolutionary sense. As microsatellites are increasingly used in population studies, these kinds of gene-flow estimates will become more common.

Direct estimates of gene flow are applicable only to the specific time and circumstances at the time of the estimate. They cannot provide any information about historic processes, nor are they likely to reveal the occasional long-distance gene flow that may be evolutionarily important. In this sense, indirect estimates complement direct estimates of gene flow. For discussions of the relationship between direct and indirect estimates of gene flow, dispersal, and differentiation, see Ehrlich and Raven (1969), Endler (1977), Slatkin (1985b), and Neigel (1997).

Indirect Estimates of Gene Flow

We have seen that gene flow retards divergence between populations. It would seem reasonable, therefore, that the degree of divergence between populations might give some indication of the amount of gene flow between them. Our goal is to quantify this relationship.

Indirect methods of estimating gene flow use the patterns of genetic variation within and among populations to infer the amount of gene flow between them. Different methods make different assumptions about population structure, mutation, modes of inheritance, and so forth. Neigel (1997) discusses four things that must be considered when using indirect estimates of gene flow. We will modify his classification slightly, to explicitly consider the mutation model. Thus, there are five factors to be considered when using indirect techniques to estimate gene flow:

Demographic Model: This is the model of population structure. Examples described previously are the island model, stepping stone models, and so on.

Genetic Markers: What kinds of genetic markers are being studied? The expectations are different for allozymes, microsatellites, nuclear DNA sequences, or mtDNA sequences.

Genetic Model: Is the population diploid or haploid? Are the marker loci inherited biparentally or uniparentally (as, for example, mtDNA)?

Mutation Model: What model of mutation is being assumed (infinite alleles model, stepwise mutation model, infinite sites model, etc.)?

Parameter Estimator: The most common parameter to be estimated is F_{ST} or G_{ST}. Others have been proposed. What is the best way to estimate the parameter from sample data? What is the bias and variance of the estimate?

The general approach is to develop a theoretical model based on the above considerations. From this model, a relationship between gene flow and genetic differentiation can be predicted. The appropriate measure of genetic differentiation is estimated from sample data, and gene flow is then estimated accordingly. We will consider this process for several kinds of genetic data. In all that follows, we will assume that natural selection is not acting on the loci being studied.

Estimating Gene Flow from Allozyme Data

Until recently, the standard way to estimate gene flow was to use allozyme surveys and estimate F_{ST} from the sample data. From equation (9.15), the relationship between F_{ST} and gene flow at equilibrium is

$$F_{ST} \cong \frac{1}{4Nm + 1}$$

Solving for Nm, we get

$$Nm \cong \frac{1 - F_{ST}}{4F_{ST}} \tag{9.17}$$

So, if we can estimate F_{ST} from allozyme data, we can get an estimate of gene flow, expressed as the number of immigrants per generation, Nm.

$$\widehat{Nm} \cong \frac{1 - \hat{F}_{ST}}{4\hat{F}_{ST}} \tag{9.18}$$

where the hats indicate estimates from a sample. Analogous equations can be written by substituting $\hat{G}_{ST}$ or $\hat{\theta}$ for $\hat{F}_{ST}$. Equation (9.18) has frequently been used to estimate gene flow in natural populations. From the studies summarized in Table 9.4, estimates of Nm range from 0.1 in kangaroo rats to 89 in squid.

There are many assumptions behind equation (9.18). It assumes a diploid population with biparental inheritance of the loci being examined. It assumes an island model of population structure, that mutation follows the infinite alleles model, and that the population is at equilibrium between genetic drift and mutation. It also assumes that natural selection is not affecting the loci being studied, and that population size is constant. It seems unlikely that all of these assumptions will ever hold. Nevertheless, equation (9.18) probably has been the most common way to estimate gene flow. Given the long list of restrictive assumptions, such estimates must be interpreted very cautiously. Slatkin and Barton (1989) and Neigel (1997) discuss the effects of these assumptions and alternative methods of estimating Nm. See also Bossart and Prowell (1998) and Whitlock and McCauley (1999).

These assumptions can sometimes be considered a set of null hypotheses that can be tested with experimental data. In a large study of allozyme variation in *Drosophila melanogaster*, Singh and Rhomberg (1987a,b) sampled 15 populations worldwide. They surveyed 117 loci, of which 61 were polymorphic. They estimated F_{ST} separately at each polymorphic locus; Figure 9.11 summarizes their results. Clearly, there is much variation from locus to locus; F_{ST} estimates ranged from about 0.03 to 0.58. Under the assumptions on which equation (9.18) is based, each locus ought to give an independent estimate of F_{ST}, and hence Nm. The great heterogeneity of estimates from different loci led Singh and Rhomberg to argue that natural selection must be affecting some of these loci. In Figure 9.11, about two-thirds of the polymorphic loci are clustered around $F_{ST} \cong 0.09$. They concluded that these loci were consistent with the model of divergence based on genetic drift and gene flow. From equation (9.17), the estimate of Nm based on these loci is about 2.5. The remaining loci were highly divergent in their F_{ST} values. Singh and Rhomberg argued that the high F_{ST}

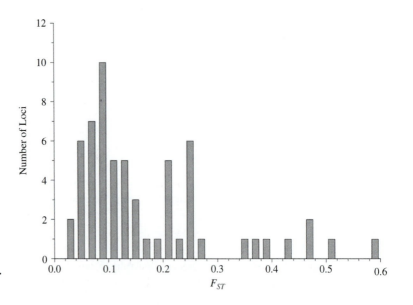

Figure 9.11   **Distribution of F_{ST} values among 61 polymorphic loci in 15 worldwide populations of *Drosophila melanogaster*.** *Source:* Data from Singh and Rhomberg (1987b).

values for these loci indicate selection in which different alleles are favored in different populations.

Estimating Gene Flow from Microsatellite Loci

Two potential problems with allozyme studies are the low resolution (usually few alleles detectable), and the possibility that some allozyme loci may be affected by natural selection. Microsatellite loci minimize both of these problems: They show much more variation than allozyme loci, and since they are in noncoding regions of the genome, are less likely to be subject to natural selection (but see below). Microsatellites have almost entirely replaced allozymes as markers for surveys of genetic variation.

It is possible to estimate gene flow from microsatellite loci, using equation (9.18) or one of its analogs. We summarize one example of this approach, then describe a different, potentially more powerful, approach.

Forbes and Boyd (1997) studied microsatellite variation in native and reintroduced populations of gray wolves (*Canis lupus*) in the Rocky Mountains of Canada and the United States. They studied three native Canadian populations, one population that had naturally recolonized northern Montana from Canada during 1985–1995, and two populations established by introduction of wolves from Canadian populations. Expected heterozygosity was high in all populations, including the reintroduced populations; over all six populations and ten microsatellite loci, the average was 0.641. For the natural populations (Canadian plus northern Montana) $\hat{\theta}$ was about 0.074, corresponding to $Nm \cong 3.1$. Estimates of Nm between pairs of Canadian populations ranged from 1.6 to 2.7, depending on which two populations were compared. These numbers indicate high gene flow and little subdivision, and are consistent with a previous study of allozyme variation in different Canadian wolf populations (Kennedy et al. 1991). Wolves are very mobile, and packs are known to disperse over hundreds of kilometers. The naturally recolonized Montana population has already expanded its range about halfway to the reintroduced population in Yellowstone National Park. Thus, the microsatellite-based, indirect estimates of gene flow are consistent with known dispersal abilities of the gray wolf. This study is good news for the reintroduced wolf population in Yellowstone National Park for two reasons: First, it indicates that the reintroduced population is not genetically depauperate, which would diminish its long-term evolutionary potential. Second, it indicates that this population is not isolated, and will probably experience gene flow with other wolf populations in the future.

Because microsatellite loci are much more variable than allozyme loci, they are potentially more powerful tools for detecting gene flow and divergence among populations. As illustrated above, we can use estimates of F_{ST} (or G_{ST} or θ) to estimate gene flow, but there is a potential problem with this approach: Microsatellite loci are thought to follow the stepwise mutation model, whereas the F_{ST} approach assumes the infinite alleles model.

Slatkin (1995) proposed a quantity, which he called R_{ST}, as an alternative method of estimating subdivision and gene flow from microsatellite data. R_{ST} differs from F_{ST} in that it considers allele sizes, which, according to the stepwise mutation model, contain information about the relationships among alleles (Sections 2.12 and 6.2). R_{ST} is, roughly speaking, a ratio of the variance of allele sizes (measured as number of repeat units) among subpopulations to the variance of allele sizes in the total population. This is analogous to Weir and Cockerham's (1984) definition of θ, described in Section 9.2. Ignoring sampling considerations, R_{ST} is related to Nm the same way as is F_{ST} in equation (9.17). Slatkin performed computer simulations that suggested R_{ST} is generally better at estimating gene flow than F_{ST} (but see below). He did not consider sampling bias in his derivation. Michalakis and Excoffier (1996) proposed an unbiased estimator of R_{ST}, which we will designate by $\hat{R}_{ST}$ to be consistent with our usual notation for estimates of parameters.

Gaggiotti et al. (1999) performed computer simulations to compare the performance of $\hat{\theta}$ and $\hat{R}_{ST}$ as estimators of gene flow based, on microsatellite data. Their simulations differed from Slatkin's (1995) in two ways: First, they assumed constraints on the number of repeats (allele size), whereas Slatkin did not. Second, they considered sampling properties of $\hat{R}_{ST}$, which Slatkin did not. They concluded that, for large sample sizes and many loci, $\hat{R}_{ST}$ is a better estimator of gene flow than $\hat{\theta}$, a result consistent with Slatkin's. However, for realistic experimental conditions of small sample sizes and few loci, the opposite was true. They concluded that the most conservative approach under most experimental conditions is to use $\hat{\theta}$. They also found that even moderate constraints on allele size led to large overestimates of gene flow, and that this was true for both $\hat{R}_{ST}$ and $\hat{\theta}$. For example, if the maximum allele size is 100 repeats and the true value of Nm is 10, the estimates of Nm were about 14 and 15 based on $\hat{R}_{ST}$ or $\hat{\theta}$, respectively. This is of concern because there is increasing evidence that microsatellite loci are subject to constraints on allele sizes (Garza et al. 1995; Jarne and Lagoda 1996; Pollock 1997; Gaggiotti et al. 1999).

Whether R_{ST} is a better estimate of gene flow than F_{ST} depends on whether microsatellite loci more closely follow the stepwise mutation model or the infinite alleles model. For each model, it is possible to predict the expected number of alleles at a locus, given the observed heterozygosity. The expected number of alleles can then be compared to the observed number at that locus, and tested for significance. Estoup et al. (1995) used this approach to compare the stepwise mutation model to the infinite alleles model in populations of honeybees (*Apis mellifera*). The calculations are complex, and are explained in Estoup et al. Out of 60 population × locus comparisons, the infinite alleles model could never be rejected, and the stepwise mutation model could be rejected only once. Clearly, the data were consistent with either model.

Wyttenbach et al. (1999) obtained similar results in a study of six microsatellite loci in the common shrew (*Sorex araneus*); neither model was consistently better than the other. Wyttenbach et al. compared gene-flow estimates based on both models. Over all six loci in 24 populations, the overall F_{ST} estimate ($\hat{\theta}$) was 0.032. The overall R_{ST} estimate was 0.016. These correspond to Nm estimates of about 8 and 15, respectively, indicating quite high gene flow among these populations. They also found a weak but statistically significant trend of isolation by distance. These results are consistent with evidence discussed by Wyttenbach et al. that these shrews are capable of long-distance dispersal through mountain passes that connect the populations. In a related study, Lugon-Moulin et al. (1999) estimated Nm to be about 0.5 or 1.1 (based on $\hat{R}_{ST}$ or $\hat{F}_{ST}$, respectively) between two chromosomal races of this species, indicating substantial reduction of gene flow between chromosomal races.

In conclusion, it seems appropriate to calculate both $\hat{R}_{ST}$ and $\hat{\theta}$, and to consider carefully the assumptions behind each estimator. We still have much to learn about mutation rates, size constraints, and so forth at microsatellite loci.

Goodman (1997) described a computer program, RSTCALC, that calculates $\hat{R}_{ST}$. It adjusts for unequal sample sizes and unequal variances in different populations. See Appendix B for information on how to obtain the program.

Assignment Tests for Microsatellite Loci

Because microsatellite loci are so variable, they can sometimes be used to infer which of several populations an individual has come from. For example, assume population 1 is nearly fixed for genotype $A_1A_1B_1B_1C_1C_1$, and population 2 is fixed for $A_2A_2B_2B_2C_2C_2$. Then if you found an individual with the latter genotype in population 1, you could reasonably infer that it was an immigrant from population 2. Note, this does not necessarily imply reproduction and gene flow. Similarly, if you found an individual of genotype $A_1A_2B_1B_2C_1C_2$ in either population, you could infer that it was the offspring of an immigrant that mated with a native. Such inferences are made by use of an assignment test, which determines the population with highest probability of being the source population for an individual.

Why is it useful to know which population an individual came from? For example, it is possible to test a hunter's claim that an animal was taken from a population where hunting is legal, and not from a protected population. Similarly, it is possible to determine the source populations of invasive species. The Mediterranean fruit fly ("medfly," *Ceratitis capitata*) is a serious economic pest that has recently invaded California and Florida. Knowing which of several possible populations were the source of the invasions could lead to improved quarantine measures directed at the presumed source (He and Haymer 1999). The assignment test also allows inferences about the history of gene flow. Nielsen et al. (1997) showed that Danish Atlantic salmon (*Salmo salar*) are indeed descendents of historic Danish populations, and not immigrants from other populations, as had been suggested. Finally, Reed et al. (1997) have written a paper with the intriguing title "Molecular Scatology: The Use of Molecular Genetic Analysis to Assign Species, Sex, and Individual Identity to Seal Faeces." Waser and Strobeck (1998) discuss several other interesting examples.

The assignment test first calculates the probability of an individual's multilocus genotype in its own population, based on estimated allele frequencies in that population, and assuming Hardy–Weinberg equilibrium at each locus and gametic equilibrium among loci. In other words, the probability of a given genotype is calculated for each locus, and the resulting probabilities multiplied to give the multilocus probability. The process is repeated for all other populations, using allele frequencies in each population to calculate the multilocus probability for that population. The population with the highest probability is assumed to be the population of origin for the individual. The more variable the loci are, the fewer loci are needed to make an assignment with high probability of being correct. Individuals sampled in one population but assigned to another may well be immigrants.

We illustrate with an example. Consider three loci; to simplify notation as much as possible, we will assume only two alleles at each of three loci. Let p_{A1}, p_{B1}, and p_{C1} represent the frequencies of A, B, and C and p_{a1}, p_{b1}, and p_{c1} represent the frequencies of a, b, and c at the three loci in population one. Similarly, let p_{A2}, p_{B2}, and p_{C2} represent the frequencies of A, B, and C, and p_{a2}, p_{b2}, and p_{c2} the frequencies of a, b, and c in population 2. Note the letter in the subscript indicates the allele and the number indicates the population. Assume the two populations have allele frequencies as shown in Table 9.5. Now consider an individual of genotype $AAbbCc$. The probability of that genotype in population one is

$$\Pr\left(\begin{array}{c} AAbbCc \\ \text{in pop 1} \end{array}\right) = p_{A1}^2 \times p_{b1}^2 \times (2p_{C1}p_{c1})$$

$$= (0.3)^2 \times (0.45)^2 \times (2 \times 0.75 \times 0.25)$$

$$= 0.0074$$

The probability of that genotype in population two is

$$\Pr\left(\begin{array}{c} AAbbCc \\ \text{in pop 2} \end{array}\right) = p_{A2}^2 \times p_{b2}^2 \times (2p_{C2}p_{c2})$$

$$= (0.6)^2 \times (0.9)^2 \times (2 \times 0.4 \times 0.6)$$

$$= 0.140$$

This individual is almost 20 times more likely to have come from population 2 than from population 1. If it was sampled in population 1, it is very likely an immigrant.

In practice, adjustments for sampling bias are made. For more on assignment tests, see Paetkau et al. (1995), Rannala and Mountain (1997), Waser and Strobeck (1998), and Davies et al. (1999). Brzustowski (2002) has written an easy-to-use, Web-based assignment calculator; see Appendix B.

TABLE 9.5 Allele frequencies at three loci in two hypothetical populations.

These data are used in the example of the assignment test.

	Allele Frequencies	
Allele	**Population 1**	**Population 2**
A	0.3	0.6
a	0.7	0.4
B	0.55	0.1
b	0.45	0.9
C	0.75	0.4
c	0.25	0.6

The assignment test was originally developed by Paetkau et al. (1995) in a study of microsatellite variation in polar bears (*Ursus maritimus*). They found that they could correctly assign only about 60 percent of polar bears to the population from which they were sampled. In contrast, Paetkau et al. (1998) could correctly assign 92 percent of brown bears (*Ursus arctos*) using the same loci. Bears that were misassigned were usually assigned to a nearby population, and were suspected of being immigrants.

Estimating Gene Flow from RFLP Data

Increasingly, population genetic data are coming in DNA sequences, either as complete sequences or as surveys of restriction site variation. We need to consider how to estimate gene flow from these kinds of data. The basic approach is similar for both.

If the sequence data are in the form of restriction site polymorphisms, one approach is to consider the region studied to be a single locus, and treat each unique haplotype as a different allele. F_{ST} can then be estimated by either $\hat{F}_{ST}$, $\hat{G}_{ST}$, or $\hat{\theta}$, as described previously. Equation (9.18) can then be used to estimate the amount of gene flow.

This method ignores information about the relationships among haplotypes. For example (letting 1 indicate presence of a restriction site and 0 indicate absence), the haplotype 11110 is more closely related to 11111 than it is to 10101. Considering these relationships allows for more powerful tests of subdivision. See Box 9.8 for an introduction to the relationships among sequences. We will consider these relationships in detail in Chapters 10 and 11.

Several F_{ST} analogs have been proposed that consider the relationships among haplotypes. For example, Lynch and Crease (1990) defined N_{ST} based on the average numbers of nucleotide differences between haplotypes. They defined

$\hat{v}_b$ = an estimate of the average proportion of nucleotide substitutions between subpopulations

$\hat{v}_w$ = an estimate of the average proportion of nucleotide substitutions within subpopulations.

From these, they defined

$$\hat{N}_{ST} = \frac{\hat{v}_b}{\hat{v}_w + \hat{v}_b} \qquad \textbf{(9.19)}$$

Lynch and Crease studied the sampling properties of $\hat{N}_{ST}$, and found that its standard error is usually quite large, due primarily to the limited number of nucleotides sampled in RFLP studies.

The details of calculating $\hat{N}_{ST}$ are complex. Computer programs that analyze restriction site variation usually calculate $\hat{N}_{ST}$ or one of the other F_{ST} analogs, and estimate gene flow from them.

Lynch and Crease (1990) reviewed 16 studies of restriction site variation in a variety of organisms. $\hat{N}_{ST}$ revealed significant subdivision in seven, whereas $\hat{G}_{ST}$ or $\hat{\theta}$ detected significant subdivision in only four, suggesting that N_{ST} is, in fact, a more powerful tool for detecting subdivision than G_{ST} or θ.

Box 9.8   Relationships among haplotypes

For multiple alleles of allozyme loci, it is not obvious which alleles are related to each other by mutation. However, for DNA polymorphisms, either restriction site polymorphisms or direct DNA sequences, it is possible to infer the evolutionary relationships among haplotypes. The main assumptions are that each nucleotide position is an independent site, there is no recombination between sites, and that mutation is rare enough that each site will mutate only once (infinite sites model; Section 6.2). For restriction site polymorphisms, it is assumed that gain or loss of a restriction site is due to a mutation in a single nucleotide within the recognition sequence.

Consider the following five restriction site haplotypes:

$$h_1: \quad 1 \quad 1 \quad 1 \quad 1 \quad 1$$
$$h_2: \quad 1 \quad 0 \quad 1 \quad 1 \quad 1$$
$$h_3: \quad 1 \quad 1 \quad 1 \quad 0 \quad 1$$
$$h_4: \quad 1 \quad 0 \quad 0 \quad 1 \quad 1$$
$$h_5: \quad 1 \quad 0 \quad 1 \quad 1 \quad 0$$

where 1 indicates presence of a restriction site and 0 indicates its absence.

These five haplotypes can be related to one another by single mutational steps in the following way:

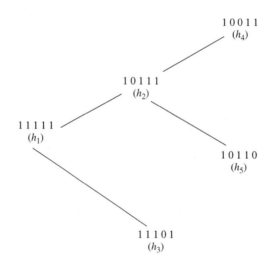

Note that without further information, we cannot indicate the direction of the changes; for example, we cannot say whether h_2 arose as a mutation from h_1 or the other way around. We can construct a table of the proportion of differences between haplotypes. For this example we get

	h_1	h_2	h_3	h_4	h_5
h_1	—	1	1	2	2
h_2	0.2	—	2	1	1
h_3	0.2	0.4	—	3	3
h_4	0.4	0.2	0.6	—	2
h_5	0.4	0.2	0.6	0.4	—

Above the diagonal are the number of site differences between two haplotypes; below the diagonal are the proportion of differences. We can use these differences in various ways, as we shall see later.

By considering evolutionary relationships among haplotypes, it is possible to construct powerful tests of subdivision or other genetic hypotheses. For example, N_{ST}, as discussed in the main text, is based on this principle. The same principle holds for microsatellite loci if we assume the stepwise mutation model and that mutation changes the number of repeats by plus or minus one unit.

This is a simple and unambiguous example. Real data usually contain ambiguities. For example, if a haplotype 1 0 0 1 0 existed, it could be related to either h_4 or h_5 by a single mutational step. Ambiguities such as this frequently mean that there is no single best relationship among haplotypes. Many computer programs are available to infer the most probable relationships. For an introduction to the principles of drawing "gene trees," as they are often called, see Li (1997) or Hall (2001). Two widely used computer programs are PHYLIP (Felsenstein 1993) and PAUP (Swofford 2000). See Appendix B.

Lynch et al. (1999b) compared estimates of subdivision and gene flow based on different kinds of molecular markers in *Daphnia pulex*, a small crustacean that inhabits temporary ponds throughout western Oregon. G_{ST} estimates based on allozyme or microsatellite loci were 0.307 or 0.293, respectively. Their N_{ST} estimate, based on mtDNA restriction site analysis, was 0.516, almost twice as high as the G_{ST} estimates. The higher value for mtDNA was attributed to the smaller effective population size for mitochondrial loci than for nuclear loci (Section 7.4).

Estimating Gene Flow from DNA Sequence Data

One of the problems of restriction site surveys is the small number of nucleotides surveyed. One way around this problem is to use direct DNA sequencing, which typically examines hundreds or thousands of nucleotides. The tradeoff, of course, is that fewer individuals are usually examined, and fewer loci studied (sometimes only one). As sequencing techniques have become faster and cheaper, DNA sequencing has become the standard technique in many areas of molecular population genetics.

The simplest approach to analyzing DNA sequence data is to treat each nucleotide site as a different locus, and estimate F_{ST} by one of the methods described above. However, as with restriction site data, this approach ignores information about relationships among different haplotypes (sequences). Also, polymorphic nucleotide sites within a single sequence usually violate the assumption of independence of loci; sites are tightly linked and often show linkage disequilibrium.

Several alternative approaches have been proposed for estimating gene flow from DNA sequence data (e.g., Lynch and Crease 1990; Nei 1982; Slatkin and Maddison 1989; Hudson et al. 1992). All of these methods use nucleotide diversity, π_{ij} (Section 2.10), or some analog as a measure of the difference between two sequences. They differ in the way they estimate π_{ij}, in the mutation model assumed, and in other subtleties. We will summarize a method described by Nei and Kumar (2000), which is very similar to that of Lynch and Crease (1990) described above.

Recall that π_{ij} is the proportion of nucleotide differences between haplotypes i and j (Section 2.10). If p_i and p_j are the frequencies of haplotypes i and j, then the nucleotide diversity in a population is defined as

$$\pi = \sum_{i=1}^{k} \sum_{j=1}^{k} p_i p_j \pi_{ij} \qquad (9.20)$$

where k is the number of different haplotypes. Nucleotide diversity is the average proportion of nucleotides that are different between two randomly chosen haplotypes in a population. It can be estimated by equation (2.18); see Box 2.5 for an example.

Now consider a population divided into n subpopulations. Let $\hat{\pi}_m$ be the estimate of π in subpopulation m ($m = 1, 2, \ldots, n$), as calculated from equation (2.18). Define $\hat{\pi}_S$ as the average of the $\hat{\pi}_m$

$$\hat{\pi}_S = \frac{1}{n} \sum_{m=1}^{n} \hat{\pi}_m)$$

(9.21)

Now consider the entire population, ignoring subdivision. Let $\bar{p}_i$ be the average frequency of the ith haplotype in the entire population

$$\bar{p}_i = \frac{1}{n} \sum_{m=1}^{n} p_{im}$$

Define $\bar{p}_j$ similarly for the average frequency of the jth haplotype. Then, from equation (2.18), an estimate of haplotype diversity in the entire population, ignoring subdivision, is

$$\hat{\pi}_T = \frac{S}{S-1} \sum_{i=1}^{k} \sum_{j=1}^{k} \bar{p}_i \bar{p}_j \pi_{ij}$$

(9.22)

where S is the total sample size.

Finally, define $\hat{N}_{ST}$ as

$$\hat{N}_{ST} = \frac{\hat{\pi}_T - \hat{\pi}_S}{\hat{\pi}_T}$$

(9.23)

Box 9.9 illustrates an example of the calculations. Nei and Kumar (2000) call N_{ST} the **coefficient of nucleotide differentiation**, and discuss the variance and sampling properties of $\hat{N}_{ST}$.

DNA sequences contain a wealth of information about the history and geography of different haplotypes (sequences) that we have only recently begun to learn how to extract. Recent advances in coalescent theory, nested clade analysis, statistical methods, and other techniques allow us to make inferences about populations that were impossible a few years ago. We will discuss some of these ideas further in Chapter 10 and 11.

Analysis of Variance Approach

As described in Section 9.2, subdivision can be described based on heterozygosities or on allele frequency variances. Cockerham (1969, 1973) and Weir and Cockerham (1984) defined θ as the ratio of the variance of allele frequencies among subpopulations to the overall variance in allele frequencies. They used analysis of variance (ANOVA) to partition the variance of allele frequencies in the entire population into components representing variance among individuals within subpopulations, and variance among subpopulations. This is similar to dividing the overall deviation from Hardy-Weinberg expectations (F_{IT}) into deviation within subpopulations (F_{IS}) and deviation due to subdivision (F_{ST}).

Excoffier et al. (1992) and Michalakis and Excoffier (1996) extended the ANOVA approach so that it can be applied to other kinds of population genetic data, including microsatellite data or DNA sequences. They call their method analysis of molecular variance (AMOVA). Understanding the details requires a

---

BOX 9.9 Calculating $\hat{N}_{ST}$ from nucleotide sequence data

In a sample of three subpopulations (X, Y, and Z), you find four haplotypes at a locus. Three nucleotides are variable in a sequence of 100. The haplotypes are as follows (only variable sites are shown):

Haplotype	Sequence
h_1	...A...A...A...
h_2	...A...A...T...
h_3	...C...A...T...
h_4	...C...T...T...

The numbers (frequencies) of the haplotypes in the three subpopulations are

Haplotype	Frequency in X	Frequency in Y	Frequency in Z	Total
h_1	16 (0.80)	14 (0.70)	10 (0.50)	40 (0.67)
h_2	2 (0.10)	3 (0.15)	4 (0.20)	9 (0.15)
h_3	1 (0.05)	2 (0.10)	2 (0.10)	5 (0.08)
h_4	1 (0.05)	1 (0.05)	4 (0.20)	6 (0.10)

The table of haplotype differences is

	h_1	h_2	h_3	h_4
h_1	—	1	2	3
h_2	0.01	—	1	2
h_3	0.02	0.01	—	1
h_4	0.03	0.02	0.01	—

Numbers above the diagonal are the numbers of nucleotide differences. Below the diagonal are the proportionate differences; these are the π_{ij}.

Next, we calculate $\hat{\pi}$ for each subpopulation separately, using equation (2.18) and following the procedure explained in Box 2.5. The results are

$$\hat{\pi}_X = 0.0063$$
$$\hat{\pi}_Y = 0.0081$$
$$\hat{\pi}_Z = 0.0131$$

From these, we calculate $\hat{\pi}_S$ using equation (9.21)

$$\hat{\pi}_S = \frac{1}{3}(0.0063 + 0.0081 + 0.0131) = 0.00914$$

To calculate $\hat{\pi}_T$, we treat the entire sample as a single population, using equation (9.22) and the average haplotype frequencies (the numbers in parentheses in the last column of the table above). Here S is 60, the total sample size. The result is

$$\hat{\pi}_T = 0.0094$$

Finally, we calculate $\hat{N}_{ST}$ from equation (9.23)

$$\hat{N}_{ST} = \frac{\hat{\pi}_T - \hat{\pi}_S}{\hat{\pi}_T} = \frac{0.0094 - 0.0091}{0.0094} = 0.0272$$

This example is to illustrate the calculations, and is not intended to be representative of real data. Calculations are usually done with specialized computer programs, which may consider sampling bias differently than done here.

---

knowledge of analysis of variance and matrix algebra, but the general idea is to choose some measure of the distance between two haplotypes. Haplotypes can be based on RFLP data, DNA sequences, microsatellite data, or potentially other kinds of genetic data. For example, if there are four haplotypes, the distance matrix would look like

$$\begin{bmatrix} \delta_{11}^2 & \delta_{12}^2 & \delta_{13}^2 & \delta_{14}^2 \\ \delta_{21}^2 & \delta_{22}^2 & \delta_{23}^2 & \delta_{24}^2 \\ \delta_{31}^2 & \delta_{32}^2 & \delta_{33}^2 & \delta_{34}^2 \\ \delta_{41}^2 & \delta_{42}^2 & \delta_{43}^2 & \delta_{44}^2 \end{bmatrix}$$

where δ_{ij}^2 is the measure of distance between the ith and jth haplotypes. Excoffier et al. (1992) describe several ways to define the δ_{ij}^2. For example, under certain assumptions, δ_{ij}^2 can be the number of restriction site differences between the ith and jth haplotypes, or the proportion of nucleotide differences between two sequences (π_{ij}).

Once the distance matrix has been determined, analysis of variance is used to partition the total variation of the δ_{ij}^2 into variation among haplotypes within subpopulations and variation among subpopulations. Further hierarchical levels of subdivision can also be included. From the analysis of variance, they define Φ_{ST} (phi), analogous to Weir and Cockerham's (1984) definition of θ.

The most useful thing about the AMOVA approach is its versatility. The distance matrix can be defined in many different ways, incorporating different assumptions about the mutation model, independence of loci, levels of subdivision, and so forth. Once this matrix has been defined, the standard analysis of variance techniques are available, and Φ_{ST} is estimated under the specific assumptions of the model.

Excoffier et al. (1992) applied the AMOVA approach to human mtDNA restriction site data. They studied 672 individuals in ten populations grouped into five regions (Figure 9.12a). They examined 34 polymorphic restriction sites and found a total of 56 haplotypes. The most parsimonious relationship among the haplotypes is shown in Figure 9.12b. Excoffier et al. (1992) performed the AMOVA procedure several times, based on alternative distance matrices. Some of their results are summarized in Table 9.6. In the first half of the table, the δ_{ij}^2 were the number of restriction site differences between two haplotypes. In the second half, all haplotypes were assumed to be equally distant from one another. This ignores the relationships among the haplotypes, and simply treats them as multiple alleles, as in the F_{ST} approach.

The table shows the amount of variation within populations, among populations but within regions, and among regions, along with estimates of the appropriate Φ statistics and their P-values. The structure of the table should be familiar to those who have studied analysis of variance. Note that the total variation can be partitioned into hierarchical levels.

In Table 9.6a, the variation within populations accounts for 75.4 percent of the total variation, but variation among populations within regions accounts for only 3.5 percent of the total. This suggests that there is relatively high gene flow among populations within regions, an inference reinforced by the relatively low estimate of Φ_{SR} ($\hat{\Phi}_{SR} = 0.044$). On the other hand, there is more variation among regions (21.1 percent of the total variation; $\hat{\Phi}_{RT} = 0.211$), suggesting less gene flow among different regions. The numbers are similar in Table 9.6b, but note that considering the relationships among haplotypes reveals more differentiation than ignoring those relationships. For variation above the population level, the percentages are 24.6 and 19.3, respectively. Note also that no explicit estimates of Nm are given (but they could be calculated from the appropriate $\hat{\Phi}$).

We have previously discussed the song sparrow (*Melospiza melodia*) in different contexts (e.g., Section 8.1). These birds are present throughout much of

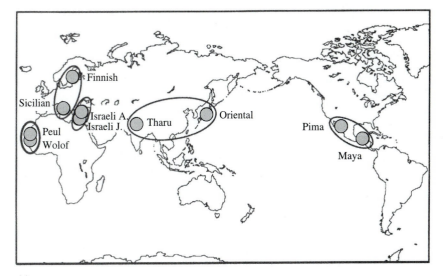

(a)

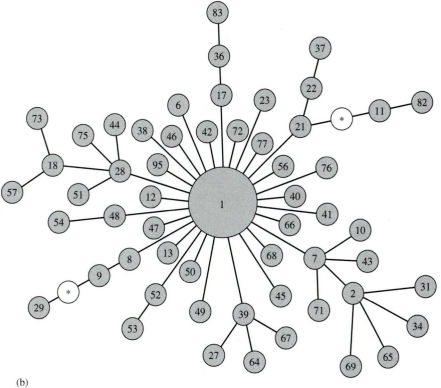

(b)

Figure 9.12 **Restriction site variation in human mitochondrial DNA.**   (a) Locations of populations surveyed, showing ten populations in five regions. (b) Relationships among mtDNA haplotypes found in these populations. The haplotypes marked with an asterisk have not been found, and are hypothetical intermediates. *Source:* Both from Excoffier et al. (1992).

TABLE 9.6   AMOVA of human mtDNA restriction site variation.

The total variation is partitioned into variation within populations, variation among populations within regions, and variation among regions. Φ statistics are F statistic analogs; subscripts S, R, and T, refer to (sub)populations, regions, and total, respectively. P is the P-value, the probability that the observed or higher value of Φ could occur by chance under the null hypothesis of no differentiation. *Source:* From Excoffier et al. (1992).

a.   Considering relationships among haplotypes

Source of Variation	Variance	Percent
Among regions	0.134	21.1
Among populations, within regions	0.022	3.5
Within populations	0.479	75.4
Total	0.635	100.0

$\hat{\Phi}_{SR} = 0.044$ $\hat{\Phi}_{RT} = 0.211$ $\hat{\Phi}_{ST} = 0.246$

$P < 0.0001$ $P \cong 0.002$ $P < 0.0001$

b.   Not considering relationships among haplotypes

Source of Variation	Variance	Percent
Among regions	0.055	15.7
Among populations, within regions	0.013	3.6
Within populations	0.281	80.7
Total		

$\hat{\Phi}_{SR} = 0.043$ $\hat{\Phi}_{RT} = 0.157$ $\hat{\Phi}_{ST} = 0.193$

$P < 0.0001$ $P \cong 0.008$ $P < 0.0001$

North America, and there is much variation among populations in morphology, plumage coloration, and other characteristics. Zink and Dittman (1993) asked whether patterns of mtDNA variation are congruent with patterns of morphological variation. They sampled 29 populations representing at least 19 subspecies, and analyzed mtDNA with 17 restriction enzymes. They treated the mtDNA molecule as a single locus, and each unique haplotype as a different allele. They found much genetic variation, with gene diversity estimated at 0.97, but essentially no subdivision. Overall G_{ST} was estimated at 0.09, not significantly different from zero, suggesting high levels of gene flow. Geographic patterns of mtDNA variation did not match patterns of morphological variation or subspecies designation. In a follow-up study, Fry and Zink (1998) used AMOVA to analyze both RFLP and sequence data from mtDNA. They used the numbers of restriction site differences or the proportions of nucleotide differences as their distance matrices. Their estimates of Φ_{ST} were 0.294 and 0.271 for RFLP and sequence data, respectively. Both are higher than Zink and Dittman's G_{ST} estimate of 0.09, suggesting again that AMOVA, by considering the relationships among haplotypes, is more powerful in detecting subdivision. There was no simple relationship between haplotype distances (π_{ij}) and geography or morphology.

Fry and Zink interpret their results as indicating a complex history of multiple refugia during Pleistocene glaciation, followed by rapid postglacial colonization and range expansion. The weakly structured mtDNA variation reflects past high levels of gene flow. Strongly structured morphological variation reflects recent adaptation to local environments, with less gene flow recently than in the past.

An excellent tutorial on AMOVA is found at http://www.bioss.sari.ac.uk/smart/unix/mamova/slides/intro.htm. Arlequin is a free collection of programs that perform a variety of analyses, including AMOVA, on many kinds of genetic data. See Appendix B for information on how to get it.

9.6 A Case Study: Lake Erie Water Snakes

The Lake Erie water snake, *Nerodia* (formerly *Natrix*) *sipedon*, lives in marshlands along the shores of Lake Erie. The snakes are also found on several islands in the lake (Figure 9.13a). The island habitat is rocky, with little vegetation.

These snakes exhibit variation in color pattern, ranging from gray unbanded forms to highly banded individuals (Figure 9.13b). Snakes along the shoreline (the mainland population) are nearly always banded. Snakes on the islands show a range of patterns. Detailed genetics of the color pattern variation are unknown, but there appears to be a single gene with large effect, with the banded pattern dominant over unbanded (King 1993a). Modifier loci are also involved, producing a range of banding patterns.

Camin and Ehrlich (1958) demonstrated significant differences between young and adult snakes in the frequency of banded individuals on the islands. They attributed the difference to natural selection favoring unbanded snakes on the islands.

Natural selection appears to favor banded individuals along the mainland shorelines. The banding pattern apparently provides cryptic coloration in the marshlands, protecting snakes from predators, primarily gulls. On the other hand, selection favors unbanded forms on the islands. Unbanded forms apparently exhibit cryptic coloration on the unvegetated rocky shores of the islands. Mark-recapture studies show that on the islands, unbanded young snakes have higher survival rates than banded snakes. Relative fitness of the banded forms on the islands is estimated to be about 0.8 to 0.9 compared to the unbanded forms (King 1993b; King and Lawson 1995).

Here, we have different populations, with natural selection acting in opposite directions on the banding patterns. There appears to be relatively strong selection favoring unbanded forms on islands, but banded forms persist. Why have they not been eliminated?

Conant and Clay (1937) first suggested that banding pattern polymorphism on the islands is maintained by dispersal of banded snakes from the mainland to the island. This hypothesis has been widely accepted. There is much anecdotal evidence to support it; for example, snakes have been seen swimming in the lake, miles from the mainland. As banded snakes reach the islands and establish themselves and breed, they continually introduce alleles for banding into the island populations. This gene flow opposes natural selection, which tends to eliminate the alleles for banding. A stable equilibrium is maintained due to the balance of these opposing processes. This is consistent with the analysis in Section 9.4: When

Figure 9.13 **Lake Erie water snakes.** (a) Map showing islands on which the snakes occur, along with the Ohio and Ontario mainlands. (b) Banding patterns of snakes, ranging from unbanded (*top*) to heavily banded (*bottom*). *Source:* (a) modified from King and Lawson (1995); (b) from Camin and Ehrlich (1958)

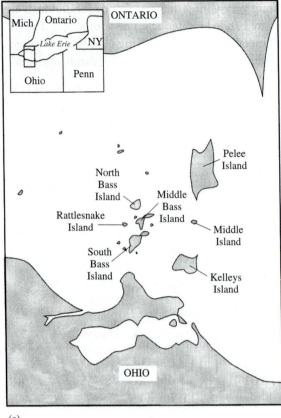

(a)

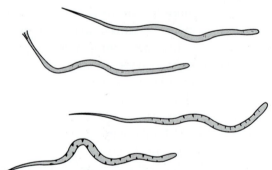

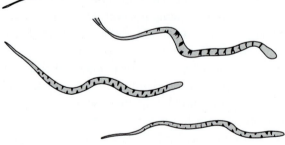

(b)

natural selection and gene flow oppose one another, the equilibrium allele frequency will be displaced from its value without gene flow. The magnitude of the displacement will depend on the amount of gene flow.

This hypothesis raises several questions: How strong is natural selection on the islands? How much gene flow is there from the mainland to the islands? Is it enough to prevent natural selection from eliminating the banded forms? Are other evolutionary processes, for example, genetic drift, likely to be important? In order to answer these questions, we need estimates of the strength of selection, rates of gene flow, and population sizes.

The selection story is more complicated than it appears. Camin and Ehrlich (1958) found significant differences in the frequency of unbanded forms between litters and adults on the islands, and concluded that the difference was due to selective elimination of banded forms. However, their litter and adult samples were taken several years apart, and the difference may have been due to other causes. King (1987) found no evidence of selection in a capture-recapture study on the islands. In both of these studies, snakes were classified simply as banded or unbanded. The pattern is much more complicated than this. King (1993b) performed a multivariate analysis of several aspects of color pattern, and compared color pattern scores with measures of relative crypsis, indicating how well a pattern was camouflaged against different backgrounds. From this, he was able to predict which color patterns should have greater survival. He tested these predictions by capture-recapture experiments. Results showed that, on the islands, color patterns producing reduced banding were favored during the first year of life, but not afterwards. The essential conclusion is that natural selection does act on banding patterns, but in a more subtle way than was assumed in earlier studies. These patterns are controlled by more than one gene, but a gene with major effect is involved. This means we can approximate the action of natural selection with a one-locus model, with banded dominant over unbanded. King and Lawson (1995) estimated the relative fitness of the banded pattern on the islands as 0.78 to 0.90, depending on how snakes were classified as banded or unbanded. They estimated the frequency of the banded forms to be about 0.46 on the islands.

King and Lawson (1995) studied subdivision and gene flow among the island and mainland populations. They estimated F_{ST} based on seven polymorphic allozyme loci. Pairwise estimates of F_{ST} among island populations were generally not significantly different from zero; therefore, they considered the islands as a single population. Estimates of Nm based on equation (9.18) were 3.6 between the Ohio mainland and the islands, and 9.2 between the Ontario mainland and the islands. They combined these to get an estimate of Nm of 12.8 from mainland to island populations. (Remember, this assumes that the allozyme loci are not subject to natural selection; that is, they are in gametic equilibrium with the banding locus.) An independent estimate of population size gave $N \cong 1262$ on the islands, corresponding to $m \cong 0.01$. The confidence interval for N was 523 to 4064, corresponding to m of 0.024 to 0.003.

We can use these estimates to predict an equilibrium frequency of the un-banded allele from equation (9.16). Let A_1 be the unbanded allele and p its frequency. Then the fitnesses are

Genotype	A_1A_1	A_1A_2	A_2A_2
Phenotype	unbanded	banded	banded
Fitness	1	$1 - s$	$1 - s$

Here, $h = 1$ and s ranges from 0.10 to 0.22 (fitnesses of 0.9 to 0.78). Recall that the frequency of the banded form on the mainland is 1, so $\bar{p} = 0$ in equation (9.16).

Figure 9.14 plots Δp versus p for the range of estimates of m and s. We are interested in the stable equilibrium, the point at which Δp crosses the p-axis, going from positive to negative. The predicted stable equilibria range from $\tilde{p} \cong 0.64$ (for $s = 0.10$ and $m = 0.024$) to $\tilde{p} \cong 0.99$ (for $s = 0.22$ and $m = 0.003$). Note that the predicted equilibrium has moved from $\tilde{p} = 1$, where it would be with no gene flow, to $\tilde{p} \cong 0.64$, with only 2.4 percent gene flow.

The actual frequency of the unbanded allele is about 0.73, within the predicted range and toward the lower end of it. This suggests that s may be toward the low end and m may be near the high end of their respective confidence intervals. The agreement seems satisfactory, but we have made many assumptions.

One assumption is certainly false. Using equation (9.18) to estimate Nm assumes that populations are at equilibrium and that population size on the islands has remained constant. Populations of snakes on the islands have decreased dramatically in the last hundred years or less. King and Lawson (1995) estimated that historical population size of island snakes was about 16,016, compared to their 1995 estimate of 1262. If this is typical of most of their history, gene flow must have had much less effect on island populations in the past than today. On the other hand, mainland population sizes appear not to have declined in historic times. Assuming that mainland populations have always sent a constant number of

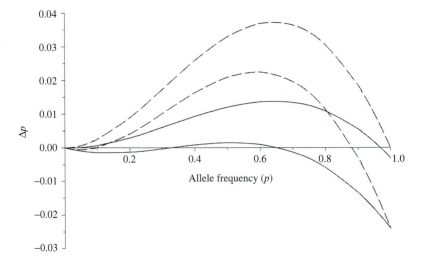

Figure 9.14 Predicted change of allele frequency in island populations of water snakes based on an island model of gene flow, with different selection coefficients and rates of gene flow. Stable equilibria occur where Δp crosses the p-axis, going from positive to negative. Solid lines are for $s = 0.1$; dashed lines for $s = 0.22$. In each pair the upper curve is for $m = 0.003$ and the lower curve is for $m = 0.024$. Stronger selection and less gene flow push the stable equilibrium toward $p = 1$. The actual observed allele frequency is about 0.73.

immigrants to the islands, we can use King and Lawson's estimate of $Nm = 12.8$ to estimate $m \cong 0.0008$ for historic population sizes. The predicted equilibrium allele frequency for $m = 0.0008$ and $s = 0.10$ is greater than 0.99; gene flow would have had essentially no effect on allele frequency until recent population declines.

This is consistent with data from museum specimens and early collection data. King (1987) and King and Lawson (1995) estimated the frequency of the un-banded allele to be greater than 0.90 before the 1950s. It seems reasonable to conclude that as island population sizes have decreased, a constant *number* of im-migrants has had a proportionately greater effect on the frequency of the un-banded forms, displacing the allele frequency equilibrium from a historic value of 0.90 or higher to its current value of about 0.70.

Keep in mind that this analysis is based on the assumption that banding patterns are controlled by a single gene. We know that this hypothesis is only approximate-ly true; therefore, these conclusions should be interpreted cautiously. However, the analysis does suggest that both current and historical patterns are due to interac-tion between natural selection, gene flow, and changing population sizes.

9.7 Cautions and Prospects

We have only begun to study the problems of estimating population subdivision and gene flow. Several methods have been introduced, which depend on a vari-ety of assumptions about the population structure, mutation model, genetic drift, and so forth. When using any of these methods, the assumptions must be careful-ly evaluated for the particular project at hand.

Keep in mind that population subdivision, as indicated by high values of $\hat{F}_{ST}$, $\hat{R}_{ST}$, $\hat{N}_{ST}$, etc. can be due to many factors other than limited gene flow. For example, natural selection may be strong enough to cause local differentiation, even with high levels of gene flow.

Demographic processes can also affect levels of subdivision as estimated by F_{ST}. Rapid increase or decrease in population sizes can affect heterozygosity and thus affect F_{ST}. Similarly, local extinction and recolonization of subpopulations will affect F_{ST}. If both occur relatively frequently, and if most recolonization is from one or two stable source populations, F_{ST} will be low, even if there is little gene flow among most subpopulations.

Lack of differentiation among subpopulations may be due to recent fragmen-tation of a large population into several (relatively large) subpopulations among which there may be no gene flow. If the fragmentation is recent, the subpopula-tions will not have had time to diverge significantly unless natural selection is very strong. This may be the case with the button wrinklewort example discussed earlier. The extent of subdivision was low, and estimates of gene flow were rela-tively high. But these estimates are based on the assumptions that subpopulations are at mutation–drift equilibrium and that population sizes have remained con-stant. Historical evidence suggests that this is not the case due to recent habitat destruction and fragmentation.

Similarly, recent range expansion followed by isolation of subpopulations may lead to little differentiation among subpopulations, even if gene flow is low. This appears to be the case with the song sparrow populations discussed in Section 9.5.

In all of these examples, using F_{ST} or one of its analogs to estimate gene flow would give misleading results. The problem is that using equation (9.18) to estimate gene flow is based on many assumptions. We review some of them here for emphasis:

1. Population structure is approximated by an island model with an infinite number of islands.
2. The population size is constant and the same on each island.
3. The actual population size, N, is the same as the effective population size, N_e.
4. Mutation is negligible compared to gene flow.
5. Gene flow occurs randomly between all islands, regardless of distance between them.
6. Gene flow occurs randomly with respect to genotype.
7. Natural selection is not affecting the loci being used to estimate gene flow.
8. Each subpopulation is at equilibrium with respect to mutation, gene flow, and genetic drift.

Bossart and Prowell (1998) and Whitlock and McCauley (1999) make compelling arguments that no real population is likely to meet all of these assumptions, and that deviations from them can lead to serious errors in estimating Nm from equation (9.18). Therefore, estimates of Nm based on F_{ST} or its analogs must be interpreted very cautiously.

Estimates of Nm should not be taken literally, but should be used as a basis of comparison, like EPA estimates of gas mileage.[1] You know the estimate is not accurate, but it allows you to compare gas mileages of different cars. Estimates of Nm do tend to be correlated with field-based estimates of dispersal abilities (Bohonak et al. 1998). Birds and insects, for example, generally have higher Nm estimates than sedentary organisms. Waples (1998) summarized F_{ST} estimates in marine and freshwater fishes. The median F_{ST} estimate in 57 species of marine fishes was 0.020, compared to 0.144 in 49 species of freshwater fishes, which have significantly greater barriers to gene flow.

As we have seen, several kinds of molecular markers are useful for estimating population subdivision and gene flow. Sunnocks (2000) reviews the advantages and disadvantages of different kinds of genetic markers. Ideally, two or more kinds should be used in a study, and the results compared. Differences may suggest violations of the assumptions. For example, differences in gene-flow estimates based on mtDNA and on microsatellites may indicate differences between male and female dispersal. Paetkau et al. (1998) estimated subdivision between several island and mainland populations of Arctic brown bears (*Ursus arctos*). Estimates based on mtDNA sequences were higher than estimates based on microsatellites. Paetkau et al. attribute the difference to lower gene flow in females than in males. This is consistent with behavioral observations, which suggest that females are much more sedentary than males. Mitton (1994) and Bossart and Prowell (1998) review other studies in which different kinds of genetic markers give different interpretations of gene flow.

Remember that F_{ST} is a *description* of the extent of subdivision. It makes no assumptions about gene flow or population structure. It is only when we use F_{ST} to estimate gene flow that the above assumptions become important. F_{ST}

---

[1]The U.S. Environmental Protection Agency provides official estimates of gas mileage for every car sold in the United States. These estimates consistently overestimate actual mileage.

remains a good description of subdivision, independent of its use (or misuse) in estimating gene flow.

Recent advances in statistical analysis of patterns of DNA variation promise improved estimates of gene flow. In particular, approaches using the relationships among haplotypes to construct gene genealogies or gene trees, as in Box 9.8, are very promising. For example, the nested clade analysis of Templeton (1998) is sometimes able to separate the effects of ongoing gene flow from historical processes such as habitat fragmentation or range expansion. Similar techniques exist for microsatellite loci (Luikart and England 1999). Gene genealogies and their applications are discussed in Chapter 11. For different approaches to using gene genealogies to estimate gene flow, see Slatkin and Maddison (1989), Lynch and Crease (1990), Kaplan et al. (1991), Templeton (1998), and Beerli and Felsenstein (1999, 2001). For recent reviews on the general topic of estimating gene flow, see Mitton (1994), Neigel (1997), Bossart and Prowell (1998), Waples (1998), Luikart and England (1999), Whitlock and McCauley (1999), Davies et al. (1999), and Sunnocks (2000).

Summary

1. Population subdivision occurs when a large population is divided into a number of smaller subpopulations, with individuals more likely to mate with an individual from the same subpopulation than from a different one.

2. If a subdivided population is treated as a single, randomly mating population, there will be a perceived deficiency of heterozygotes compared to Hardy-Weinberg expectations, even if every subpopulation shows the Hardy-Weinberg expected genotype frequencies (Wahlund effect).

3. The degree of subdivision is quantified by F_{ST}, which is the proportional reduction of expected heterozygosity in subpopulations compared to expected heterozygosity in the entire population ignoring subdivision; equation (9.2).

4. Quantities closely related to F_{ST} describe the degree of subdivision based on allele frequency variances (θ) or gene diversities (G_{ST}).

5. Estimating F_{ST} from sample data is a complex problem. Computer programs usually account for multiple loci and calculate one or more estimates of F_{ST}. Most adjust for sampling error and give some estimate of statistical significance.

6. Gene flow is the movement of individuals or gametes from one population to another, with subsequent breeding in the new population. The main effect of gene flow is to make populations more similar than they would be without it.

7. There are several models of population structure and gene flow (Figure 9.3). The island model is frequently used to approximate gene flow in subdivided natural populations.

8. Under the island model, all subpopulations will approach an equilibrium allele frequency equal to the average allele frequency among all the islands; equation (9.12) and Figure 9.5.

9. All gene flow models, if they exclude other evolutionary processes, lead to the same general conclusion: Subpopulations will become more alike over time, and allele frequencies in each subpopulation will converge to the average allele frequency in the population.

10. Isolated populations will tend to diverge due to mutation, genetic drift, natural selection, and so on. Gene flow retards this divergence. At equilibrium, the degree of subdivision is given by equation (9.15). This equation is based on many assumptions.

11. In general, it takes only a very few immigrant individuals per generation to prevent differentiation due to genetic drift alone.

12. Whether natural selection can cause differentiation in the face of gene flow depends on the relationship between the strength of selection in different populations and the rate of gene flow between them. In general, if m is much less than s, then natural selection will be the predominant factor determining allele frequencies in each subpopulation. If m is much greater than s, then gene flow will prevent local adaptation to a great extent.

13. Levels of gene flow can sometimes be estimated from movements of individuals or from genotyping of highly variable loci such as microsatellites.

14. Levels of gene flow are frequently estimated from equation (9.18), but this is based on many unrealistic assumptions, and estimates must be interpreted very cautiously.

15. Other estimates of subdivision and gene flow take into account the relationships among alleles (microsatellites) or haplotypes (RFLP or DNA sequence data). Analysis of molecular variance (AMOVA) allows one to consider different assumptions about population structure, mutation models, and so on and obtain estimates based on those specific assumptions.

16. The water snakes of Lake Erie provide a good example of the interactions between natural selection, population subdivision, gene flow, and changing population sizes. The observed allele frequencies match the predicted frequencies reasonably well, but the predictions are based on many assumptions of unknown validity.

17. Differentiation among subpopulations can be due to many factors other than restricted gene flow, for example, varying selection coefficients. Historical processes such as population fragmentation and range expansion must also be considered.

18. F_{ST} is a good measure of subdivision, but using it to estimate gene flow, as in equation (9.18) or analogous equations, is fraught with difficulties. Estimates of Nm should not be interpreted literally, but should be used as a basis of comparison.

Problems

For Problems 9.1 through 9.5, use a spreadsheet and ignore corrections for sampling bias.

9.1. Calculate F_{IS}, F_{IT}, and F_{ST} for the following subdivided population:

	Number of Individuals		
	A_1A_1	A_1A_2	A_2A_2
Subpopulation 1	10	180	810
Subpopulation 2	250	500	250

9.2. Now assume that the two subpopulations are indistinguishable and calculate the allele frequencies, H_{obs}, H_{exp}, and F for the combined population.

9.3. Calculate F_{ST} for the Lewis and Duffy loci in Table 2.3.

9.4. Calculate F_{ST} for the ABO locus in Table 2.3

9.5. Calculate F_{ST} for each locus in the *Drosophila* populations of Problem 2.1.

9.6. Show that $F_{ST} \leq 1 - H_S$. [*Hint:* Use the facts that $H_S < H_T$ (Why?) and $H_T < 1$.] If $H_S = 0.60$, what is the maximum value of F_{ST}? See Hedrick (1999a) for a discussion of this issue.

9.7. Assume that one population has five alleles at a particular locus, each with a frequency of 0.2. A second population has five *different* alleles at the same locus, each with a frequency of 0.2. Calculate F_{ST} for this locus, ignoring bias corrections. Does the result surprise you? Discuss.

9.8. The equations for change of allele frequency have the same form for the recurrent mutation model and for the continent-island model of gene flow. Show this by putting equations (1.4) and (9.8) into the general form of the linear recursion equation, $p_{t+1} = a + bp_t$. But we have concluded that mutation is an insignificant force in changing allele frequencies, whereas gene flow is a powerful force for homogenizing populations. Explain why mutation and gene flow can have such different effects on a population.

9.9. The table below gives genotype frequencies at the acid phosphatase locus in three Israeli populations of different origins (Mourant et al. 1976). Calculate F_{ST} for these populations. Use a spreadsheet and ignore corrections for sampling bias.

		Genotype Number				
Origin	**Sample Size**	A_1A_1	A_1A_2	A_1A_3	A_2A_2	A_2A_3
Iran	49	10	16	0	21	2
Iraq	82	11	33	1	35	2
Yemen	37	1	9	0	26	1

9.10. Download the program FSTAT (see Appendix B) and use it to calculate $\hat{F}_{ST}$ and $\hat{\theta}$ for the populations in Table 9.7. Compare to your answer for Problem 9.9.

9.11. Use FSTAT to calculate $\hat{F}_{ST}$ and $\hat{\theta}$ for the populations of Problem 2.1. Compare to your answer for Problem 9.5.

9.12. Based on your answers to Problems 9.9 through 9.11, do the bias corrections seem important? Discuss.

9.13. Do a chi-square test to test for significant subdivision in the data of Problem 9.9.

9.14. Plot Δp versus p for the continent-island model of gene flow, with $p_m = 0.6$ and $m = 0.01$ and 0.10 (plot both on the same graph). Before you start, try to predict the shape of the curves and how they will be different. Where is the stable equilibrium? Are $p = 0$ and $p = 1$ equilibria? Explain.

9.15. The following table gives genotypes at three loci from individuals in two populations. For each individual, calculate the probability that it came from population X and the probability that it came from population Y. Which individuals do you think might be immigrants?

Population X	Population Y
$A_2A_2B_1B_1C_1C_1$	$A_1A_1B_1B_2C_2C_2$
$A_1A_2B_2B_2C_1C_2$	$A_1A_2B_2B_2C_1C_2$
$A_1A_1B_1B_2C_1C_2$	$A_2A_2B_2B_2C_1C_1$
$A_1A_2B_1B_2C_1C_1$	$A_1A_2B_2B_2C_2C_2$
$A_2A_2B_1B_1C_1C_1$	$A_1A_1B_2B_2C_1C_2$
$A_1A_2B_1B_1C_2C_2$	$A_1A_2B_2B_2C_2C_2$
$A_2A_2B_1B_2C_1C_1$	$A_1A_1B_2B_2C_1C_1$
$A_2A_2B_1B_2C_1C_1$	$A_1A_2B_1B_2C_2C_2$
$A_1A_2B_1B_2C_1C_2$	$A_2A_2B_2B_2C_1C_2$
$A_2A_2B_2B_2C_1C_1$	$A_1A_1B_2B_2C_1C_2$

9.16. Use the assignment calculator found at www.biology.ualberta.ca/jbrzusto/Doh.php to repeat the analysis of Problem 9.15. Compare your results to the program's. Explain the difference.

9.17. Plot equation (9.11) for initial allele frequencies of 0, 0.3, 0.7, and 1.0 for $m = 0.01$. Compare to Figure 9.5.

9.18. Assume that natural selection acts at a single locus in an island population as follows:

Genotype	A_1A_1	A_1A_2	A_2A_2
Fitness	1	0.98	0.95

Assume that on a nearby mainland, the frequency of the A_2 allele is 0.7. Plot equation (9.16) for $m = 0.01$ and $m = 0.10$. Discuss.

9.19. Now assume the following fitnesses on the island:

Genotype	A_1A_1	A_1A_2	A_2A_2
Fitness	0.98	1	0.95

a. Rescale the fitnesses so that the fitness of A_1A_1 is 1; determine s and h.

b. What is the equilibrium frequency in the absence of gene flow?

c. Assume that on the mainland, the frequency of the A_2 allele is 0.8. Plot equation (9.16) for $m = 0.01$ and $m = 0.10$. How is the equilibrium allele frequency changed? Discuss.

9.20. Make a spreadsheet to work through all of the calculations in Box 9.9. Make it as general as possible, so you need to make minimal changes to use it for Problem 9.21. Test it with the data in Box 9.9.

9.21. Modify your spreadsheet from Problem 9.20 to calculate $\hat{N}_{ST}$ for the following data. The sequence was 100 nucleotides long, of which four sites were variable.

Haplotype	Sequence	Number in Subpopulation X	Number in Subpopulation Y	Number in Subpopulation Z
h_1	...G...C...A...A	25	22	18
h_2	...C...C...T...A	0	1	1
h_3	...G...C...T...A	0	2	1
h_4	...G...C...A...T	0	0	5

Molecular Population Genetics

The neutral theory asserts that the great majority of evolutionary changes at the molecular level, as revealed by comparative studies of protein and DNA sequences, are caused not by Darwinian selection but by random drift of selectively neutral or nearly neutral mutants.

—*Motoo Kimura (1983)*

In the 1960s, technological advances made it possible to compare the amino acid sequences of homologous proteins in different species. Similarly, in the 1980s, it became possible to compare homologous DNA sequences among different species. In parallel with these developments, electrophoretic techniques, PCR, and rapid DNA sequencing have allowed us to examine protein and DNA variation among individuals of the same species. We have already seen that protein electrophoresis revealed an unexpectedly high level of allozyme variation within species, and that DNA sequencing reveals even more. As a result of these developments, we now have much information about the degree of molecular differentiation among different species and the amount of molecular variation within species.

One of the most active areas of modern population genetics is the attempt to understand the patterns of molecular variation revealed by numerous studies over the last 30 years. There is no doubt that much genetic variation exists within and among most species. One of the most important problems of population genetics is to understand the causes of this variation. As we have seen in previous chapters, opposing processes such as mutation, genetic drift, natural selection, gene flow, and inbreeding interact to determine the amount and pattern of genetic variation within a population. Our goal in this chapter is to understand how these processes interact to determine patterns of variation at the protein and DNA levels.

The unifying theory of molecular population genetics is the neutral theory of molecular evolution, which we have discussed briefly in previous chapters. In this chapter, we examine the neutral theory in more detail, and look at the predictions it makes about the patterns of molecular variation within and among populations. We then use these predictions to look for evidence of natural selection.

10.1 Introduction to the Neutral Theory

Darwin himself recognized that not all variations would be subject to natural selection. He suggested that such neutral variations would either fluctuate in frequency in a population, or occasionally become fixed or lost (see the quotation at the beginning of Chapter 7). This seems a remarkable insight, given that he knew nothing of Mendelian genetics or the theory of genetic drift. However, after the rediscovery of Mendel's work and the early development of population genetics theory, natural selection was usually considered the main, if not the only, driving force in evolution. Kimura (1983) discussed what he called the "overdevelopment of the synthetic theory." He wrote: "By the early 1960s, consensus seems to have been reached that every biological character can be interpreted in the light of adaptive evolution by natural selection, and that almost no mutant genes are selectively neutral" (Kimura 1983, p. 21). This was the prevailing paradigm of evolution when the molecular data began to appear.

It was initially assumed that this molecular variation must be subject to natural selection. Differences among species were assumed to be the result of fixation of advantageous mutations, and variations within species are maintained by some form of balancing selection, probably overdominance. However, Kimura (1968) and King and Jukes (1969) independently proposed an alternative hypothesis: Most of the amino acid differences among different species, and most of the molecular polymorphisms within species, are the result of interaction between genetic drift and neutral mutations. The molecular variation that we see is selectively neutral and frequencies are controlled by genetic drift and not by natural selection. These are the core ideas of the neutral theory of molecular evolution, usually called just the neutral theory. It has been extensively developed by Motoo Kimura, and modified by Tomoko Ohta and others.

Two experimental observations in the 1960s motivated the development of the neutral theory. The first was the apparent constancy of amino acid substitution rates in a wide variety of species. The second was the unexpectedly high levels of protein variation revealed by allozyme surveys. We will consider each of these in turn, but first we review several basic ideas.

Basic Concepts

Before we examine the neutral theory in detail, we must clarify certain terms that are sometimes confused. A mutation is any change in the genetic material; in this chapter, we consider only single base changes. Mutations occur in individuals. They create differences in nucleotide sequences among individuals, and, sometimes, differences in the amino acid sequences of proteins. These differences can lead to either substitutions or polymorphisms. A **substitution** is a replacement in a population of one amino acid or nucleotide by another. Substitutions lead to fixed differences between groups. For example, humans have *met* at position 12 of cytochrome c, whereas horses have *gln* at the same position. Not all DNA substitutions produce amino acid substitutions, due to the redundancy of the genetic code. A **polymorphism** is the segregation of two or more variants within

a single population. Examples are the *Fast/Slow* polymorphism in alcohol dehydrogenase of *Drosophila melanogaster*, or the hemoglobin polymorphisms in
humans.

At the DNA level, we can distinguish between **silent**, or synonymous, differences (substitutions or polymorphisms), which do not change the amino acid sequence of the protein, and **replacement**, or nonsynonymous, differences, which
do change the amino acid sequence. About 25 percent of all single base changes
in the coding sequence of a gene will be silent, assuming all changes are equally
probable.

Substitutions can occur due to either natural selection or genetic drift. Each
generation, new mutations arise in a population. The number of new mutations
each generation is the mutation rate per locus, u, multiplied by the total number
of copies, or $2Nu$ new mutations each generation (assuming diploidy). Most of
these will be lost within a few generations, but a few will ultimately become
fixed in the population (Figure 10.1). For a new neutral mutation, the probability of ultimate fixation by genetic drift is $1/2N$, and for an advantageous mutation, it is approximately $2s$, where s is the selective advantage of the heterozygote
(Section 7.8). Deleterious mutations can also become fixed by genetic drift, but
the probability is low except in very small populations. The mean time to fixation
of a neutral mutation that is destined for fixation is about $4N_e$ generations; for an
advantageous mutation it is shorter. During the substitution process, more than
one allele will be present in the population; a polymorphism will be observed.
Review Chapter 7 for a more extensive discussion of these ideas.

The neutral theory claims that most of the amino acid and nucleotide substitutions between groups are the result of random fixation of neutral mutations,
rather than selective fixation of advantageous mutations. The second claim is that
most of the molecular polymorphisms within groups are due to interactions between genetic drift and neutral mutations; that is, they are neutral polymorphisms, rather than balanced polymorphisms maintained by natural selection.

We must be clear on what the neutral theory does *not* claim. It does not claim
that natural selection is unimportant in evolution, or that most mutations are neutral. Most morphological adaptations are obviously the result of natural selection;
the neutral theory does not deny this. Nor does it deny that most mutations are
deleterious, if only slightly so. It claims only that most of the molecular variation

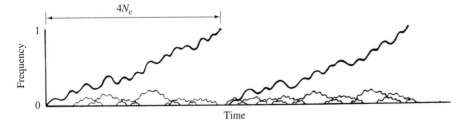

Figure 10.1 **Fate of neutral mutations in a finite population.**   Mutations fluctuate in frequency and most are lost, but a few are fixed by genetic drift (heavy lines). The
average time to fixation for a neutral mutation that is fixed is $4N_e$ generations. Advantageous mutations are rarer, but have a higher probability of fixation and go to fixation
faster. *Source:* Modified from Kimura (1983).

that we see is neutral. Deleterious mutations have been eliminated by natural selection, and (rare) advantageous mutations have been fixed. The molecular variants we see are those that have escaped the action of natural selection.

When comparing two groups, it is relatively easy to estimate the number of differences between them. For example, humans and rhesus monkeys differ at 4 of 141 amino acids in the α-hemoglobin protein; humans and cows have 17 differences. The number of *differences*, however, is not necessarily the same as the number of *substitutions*. The reason is that any given site may have changed more than once. Therefore, the observed number of differences is an underestimate of the number of substitutions expected to have occurred. Table 10.1 illustrates several examples. The longer two groups have been evolving independently, the more likely that a site has changed more than once, and the greater the disparity between the observed number of differences and the expected number of substitutions. There are various ways to estimate the expected number of substitutions from the observed number of differences; we will defer the details until Chapter 11.

When we talk about rates of molecular evolution, we mean substitution rates, that is, the expected number of amino acid or nucleotide substitutions per site per unit time. This is equivalent to the probability of a substitution per site per unit time. When comparing two sequences, we use the symbols p to indicate the

---

TABLE 10.1 Multiple changes in the nucleotide sequence of a gene (multiple hits).

The left column represents a partial sequence of an ancestral gene. The arrows represent different evolutionary lineages. Whenever more than one substitution occurs during evolution, the observed number of differences may not be the same as the actual number of substitutions.

Ancestor (unknown)	Current (observed)	Observed differences	Actual substitutions
AAAAA → AAAAT AAAAA → AAAAA	AAAAT AAAAA	1	1
AAAAA → AAAAC AAAAA → AAAAG	AAAAC AAAAG	1	2
AAAAA → AAAAT → AAAAC AAAAA → AAAAA	AAAAC AAAAA	1	2
AAAAA → AAAAC AAAAA → AAAAC	AAAAC AAAAC	0	2
AAAAA → AAAAA AAAAA → AAAAT → AAAAA	AAAAA AAAAA	0	2

observed proportion of differences (equivalent to the observed number of differences per site), K to indicate the expected number of substitutions per site, and k to indicate the substitution rate (expected number of substitutions per site per time period). We usually use subscripts to indicate the kinds of substitutions we are considering; for example, K_{AA} is the expected number of amino acid substitutions per site, and k_{AA} is the rate of amino acid substitutions.

Apparent Constancy of Amino Acid Substitution Rates

The earliest molecular data were amino acid sequences of homologous proteins from different organisms. For any pair of species, we can count the number of amino acid differences in a given protein and obtain the divergence time of those two species from the fossil record. For example, Table 10.2 compares the amino acid sequences of the α-hemoglobin protein of several vertebrates. Above the diagonal are the observed proportion of differences; below the diagonal are the expected number of substitutions per site, as estimated from the observed differences (see Section 11.1).

When the expected number of substitutions versus divergence time is plotted for various pairs of species, a remarkable result emerges. The points representing species pairs lie approximately on a straight line. The data in Table 10.2 are graphed in Figure 10.2. The slope of the line is an estimate of the rate of amino acid substitutions per year. Figure 10.3 shows other examples. *For a given protein, the rate of amino acid substitutions appears to be approximately constant over tens or hundreds of millions of years.* This was first pointed out by Zuckerkandl and Pauling (1962, 1965) with respect to vertebrate hemoglobins. They used the phrase "molecular evolutionary clock" to describe this constant rate of change.

Kimura (1968) noted the near constancy of amino acid substitution rates and argued that these rates are too constant and too high to be accounted for by natural selection favoring substitution of advantageous mutations. He argued as follows: If u_a is the mutation rate to advantageous mutations per generation and N is the population size, then the total number of advantageous mutations per generation is $2N_e u_a$. The substitution rate, designated by k, is then the total number of mutations per generation multiplied by the probability of fixation for any single mutation, $\Pr(fix)$:

TABLE 10.2   Proportions of amino acid differences in α-hemoglobin among eight species of vertebrates.
Above the diagonal: observed number of differences per site (p_{aa}). Below the diagonal: expected numbers of substitutions per site (K_{aa}), as calculated from equation (11.1) in Chapter 11. *Source:* From Kimura (1983).

	Shark	Carp	Newt	Chicken	Echidna	Kangaroo	Dog	Human
Shark	—	0.594	0.614	0.597	0.604	0.554	0.568	0.532
Carp	0.901	—	0.532	0.514	0.536	0.507	0.479	0.486
Newt	0.952	0.759	—	0.447	0.504	0.475	0.461	0.440
Chicken	0.909	0.722	0.592	—	0.340	0.291	0.312	0.248
Echidna	0.926	0.768	0.701	0.416	—	0.348	0.298	0.262
Kangaroo	0.807	0.707	0.644	0.344	0.428	—	0.234	0.191
Dog	0.839	0.652	0.618	0.374	0.354	0.267	—	0.163
Human	0.759	0.666	0.580	0.285	0.304	0.212	0.178	—

Figure 10.2 **Expected number of amino acid substitutions per site in α-hemoglobin among eight species, plotted against the time to common ancestor (millions of years).** Each dot represents one pairwise comparison from Table 10.2; the line is the best-fit linear regression line.

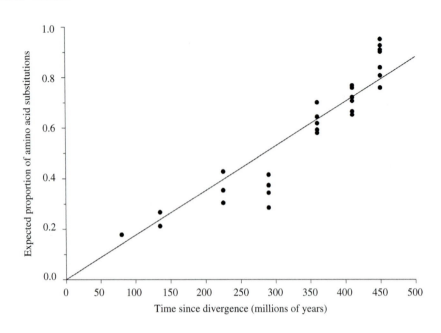

$$k = (2N_e u_a) \times \Pr(fix) \tag{10.1}$$

We saw in Section 7.8 that if fitnesses are $1, 1 + s$, and $1 + 2s$, the probability of fixation for a newly arisen advantageous allele is approximately $2s$ [equation (7.23)]. Therefore, the substitution rate for advantageous mutations is

$$k = 4N_e u_a s$$

If substitution rates are approximately constant among many different species, this implies that $N_e u_a s$ is approximately constant in different species. Kimura argued that it is very unlikely that the product of effective population size, advantageous mutation rate, and selection coefficient would be constant over many species and over long periods of time. There is no reason to expect that natural selection should act consistently, or that population sizes should be constant over millions of years.

On the other hand, the probability of fixation for a newly arisen neutral mutation is $1/2N_e$. Therefore, the substitution rate for neutral mutations is

$$k = (2N_e u) \times \left(\frac{1}{2N_e}\right) = u \tag{10.2}$$

where u (without a subscript) is the neutral mutation rate. The substitution rate is equal to the neutral mutation rate. This important result was first pointed out by Kimura (1968). Thus, assuming the neutral mutation rate is relatively constant for a given locus, the neutral theory predicts an approximately constant rate of molecular evolution. Kimura argued that this is a more parsimonious explanation of the observed constant substitution rate than natural selection.

Note that the neutral mutation rate is per generation. Therefore, the neutral theory predicts a constant substitution rate per generation, whereas the initial observations suggested a constant rate per year. We will return to this issue, and the controversy over the molecular clock, in Section 10.2.

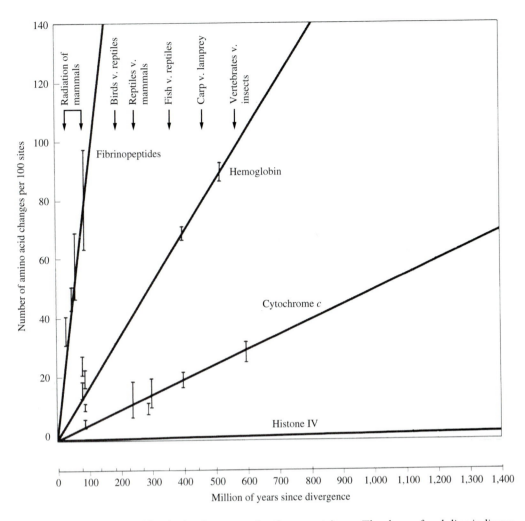

Figure 10.3 **Amino acid substitution rates for four proteins.** The slope of each line indicates the rate of evolution for that protein. *Source:* Modified from Dickerson (1972).

High Levels of Allozyme Variation and Segregational Load

The first allozyme surveys (Harris 1966; Lewontin and Hubby 1966) revealed surprisingly high levels of protein variation in humans and *Drosophila*. Subsequent studies have produced similar results for many species in a wide variety of taxa (Table 2.5). These allozyme polymorphisms were initially thought to be maintained by overdominance and, later, by some general form of balancing selection. Kimura (1968) claimed that such extensive balanced polymorphism would impose an intolerable genetic load on a population, and argued that allozyme polymorphisms are more easily explained by interactions between neutral mutation and genetic drift.

The genetic load is the amount by which the mean fitness of a population is reduced, compared to its maximum possible fitness (Section 6.6). Any process, for example, deleterious mutation or overdominance, that causes the actual population fitness to be less than its maximum possible value imposes a genetic load on the population. Single-locus overdominance creates a genetic load because heterozygous individuals can produce less fit homozygous offspring.

The potential problem of genetic load was mentioned by Lewontin and Hubby (1966) in the very first paper on allozyme heterozygosity in natural populations. Kimura (1968) used similar calculations to argue that the observed levels of allozyme polymorphisms cannot be maintained by natural selection because it would create an intolerable genetic load.

The neutral theory gets around the load problem by assuming that most allozyme variation is not maintained by natural selection, but is a transient phase in the random fixation (substitution) of neutral mutations. Most neutral mutations will be lost within a few generations, but a small proportion are destined for fixation. This fixation will take a long time (about $4N_e$ generations), and during this time, neutral alleles will be seen as molecular polymorphisms. This was first suggested by Kimura (1968) and explicitly described by Kimura and Ohta (1971a).

The argument from segregational load is based on assumptions that are now considered unrealistic. The most important is that overall fitness is determined by multiplicative interactions among individual loci (Section 6.6). More realistic models of multilocus fitness generally predict a lower genetic load than multiplicative interactions. Similarly, other models of balancing selection often predict lower genetic loads than single locus overdominance. (We consider some of these models in Chapter 12.) For these reasons, segregational load is no longer considered a compelling argument against selective maintenance of allozyme polymorphisms, even though it did play an important role in the initial development of the neutral theory.

The neutral theory was initially proposed to explain these two kinds of observations, the apparently constant substitution rates among species, and the high levels of allozyme polymorphisms within species. In this role, it has been controversial, although less so today than when it was first proposed. Its greatest use today is as an explicit null hypothesis against which many hypotheses about natural selection, gene flow, and so forth can be tested. We will examine these ideas throughout the rest of this chapter.

10.2 Predictions of the Neutral Theory

We now examine some of the predictions of the neutral theory and the experimental data bearing on these predictions. First, we need to consider a question that we have so far glossed over: What does "neutral" really mean?

How Weak Must Selection Be for an Allele to Be Neutral?

It is unlikely that a mutation will have *absolutely* no effect on fitness. Kimura recognized this; in his first paper on the neutral theory (1968), he used the phrases "almost neutral" and "neutral or nearly neutral." So our first question is, How small must the selection coefficient be for a mutation to behave as if it were neutral?

Let A_2 be a new mutant allele (with initial frequency $1/2N$), and let A_1 represent all other alleles in the population. We will scale the fitnesses as in Section 7.8:

$$
\begin{array}{ccc}
A_1A_1 & A_1A_2 & A_2A_2 \\
1 & 1+s & 1+2s
\end{array}
$$

Note that s is positive for an advantageous mutation and negative for a deleterious mutation. Note also that we have assumed that fitness of the heterozygote is midway between the two homozygotes. This assumption is justified by the experiments described in Section 6.4 which suggest that for weak selection, h is approximately 0.5.

From equation (7.22) the probability of fixation for a new mutation subject to weak selection (advantageous or deleterious) is approximately

$$\Pr(fix) \cong \frac{1 - e^{-2s}}{1 - e^{-4Ns}}$$

From Section 7.2, the probability of fixation for a newly arisen neutral mutation is $1/2N$. How small must s be for these two fixation probabilities to be approximately equal? One approach to this question is to consider the ratio of the fixation probabilities (Kimura 1983). Write this ratio as

$$r = \frac{(1 - e^{-2s})/(1 - e^{-4Ns})}{1/2N}$$

where the numerator is the fixation probability for a selected allele and the denominator is the fixation probability for a neutral allele. First, if s is small, then $1 - e^{-2s} \cong 2s$, and the ratio reduces to

$$r = \frac{4Ns}{1 - e^{-4Ns}} \qquad \textbf{(10.3)}$$

This approximation is valid as long as s is near zero but not equal to zero. Figure 10.4 plots this ratio against $4Ns$. For a strictly neutral allele, the ratio is one; for a nearly neutral allele, it is close to one. If we arbitrarily define an effectively neutral allele as one with a ratio between 0.9 and 1.1, then

$$0.9 < \frac{4Ns}{1 - e^{-4Ns}} < 1.1$$

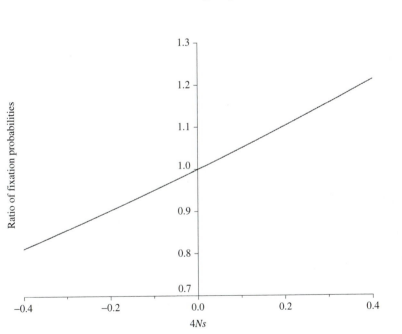

Figure 10.4 **Ratio of fixation probabilities for selected and neutral alleles, equation (10.3), plotted against 4Ns.** If an effectively neutral allele is defined as one with a ratio between 0.9 and 1.1, this corresponds to $4Ns$ between about -0.2 and $+0.2$, or $|s| \ll 1/2N$.

From the graph, this is true if $4Ns$ is between -0.2 and $+0.2$, approximately. Therefore, an allele is effectively neutral if

$$|4Ns| < 0.2$$

or,

$$|s| < \frac{0.1}{2N}$$

This is usually written as

$$|s| \ll \frac{1}{2N} \qquad\qquad \textbf{(10.4)}$$

where $\ll$ means less than by about a factor of 10. We have used the absolute value symbols to indicate that s can be either positive or negative. This was Kimura's (1983) definition of effectively neutral. More or less stringent definitions are possible.

Note that the definition of effectively neutral depends on population size. This is because genetic drift has a greater effect in small populations, and can overpower weak selection. We will discuss this idea in more detail in Section 10.3

Equation (10.4) has important implications. For an allele to behave as if it were neutral, selection for or against it must be very weak, or the population size must be very small. In a population of 50,000 an allele will be effectively neutral only if $|s| \ll 10^{-5}$. In other words, selection coefficients as small as 10^{-5} may be large enough to determine the fate of a new mutation. If the population size is 100, then effectively neutral means $|s| \ll 0.005$. The conclusion is that, unless the population is extremely small, there is a very narrow range of selection coefficients that will qualify as effectively neutral. Selection that is far too weak to detect in laboratory or field experiments may be strong enough to determine the fate of a new mutation.

The Molecular Clock

The neutral theory predicts that amino acid and nucleotide substitution rates should be approximately constant as long as the neutral mutation rate remains constant. First, you should realize that the molecular clock is not expected to tick at an absolutely constant rate. Fixation of neutral mutations is a random process. Even if the underlying events that control substitution are approximately constant, we should expect some variation in the substitution rate. Critics of the molecular clock claim that this variation is too high.

If the protein clock ticks at a relatively constant rate, there should be an approximately linear relationship between divergence time and number of amino acid substitutions, as suggested by Figure 10.2 and equation (10.2). Early studies suggested that this relationship was approximately true for several proteins, as discussed earlier. Since the 1970s, many more protein sequences have accumulated. Scherer (1990) conducted a comprehensive review of ten proteins, with over 500 sequences. He concluded that none of the ten proteins exhibits a constant molecular clock, including the hemoglobins, which were the initial stimulus for the molecular clock hypothesis. Scherer concluded that as more data have accumulated, the evidence for the molecular clock has become much weaker. For example, the initial studies of the cytochrome c clock compared only about 15 to 20

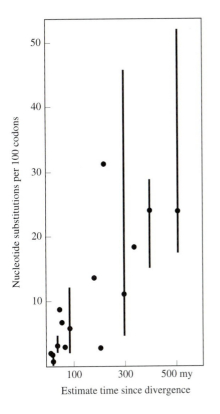

Figure 10.5 **Substitution rates at the cytochrome c locus among 87 species of animals.**   Horizontal axis is time since common ancestor. Vertical axis is inferred number of nucleotide substitutions. Vertical lines represent ranges. *Source:* From Scherer (1990).

species and produced a reasonably linear clock (Figure 10.3), whereas his study compared 87 species and produced a very "sloppy evolutionary clock" (Figure 10.5). Scherer concludes that the protein clock hypothesis should be rejected, and wonders "why the concept survived such a long period at all."

It is frequently assumed that the number of substitutions in a lineage is a random variable with a Poisson distribution. For a Poisson-distributed random variable, the mean and the variance are equal (Appendix A). Thus, if substitutions are Poisson distributed, the variance of the substitution rate should equal the mean substitution rate, and the ratio of variance to mean should be 1 (approximately).

The first test of this prediction was done by Ohta and Kimura (1971). They compared the amino acid sequences of α- and β-hemoglobins and cytochrome c in mammals, and concluded that the variance of both β-hemoglobin and cytochrome c was significantly higher than expected. Langley and Fitch (1974) conducted a much more extensive analysis and concluded that the data from four proteins (α- and β-hemoglobin, cytochrome c, and fibrinopeptide A) are inconsistent with the protein clock.

What do DNA sequences have to tell us about the protein molecular clock? First, recall the difference between silent (synonymous) substitutions, which do not change the amino acid sequence of the protein, and replacement (nonsynonymous) substitutions, which do change the amino acid sequence. Kimura (1986) reviewed some of the early data for several proteins and found that silent substitution rates differ by less than twofold, while replacement rates differ by nearly two orders of magnitude (Figure 10.6). More recent data extend the ranges of both types, but do not change the main conclusion.

Figure 10.6 **Silent (synonymous) and replacement (nonsynonymous) substitution rates for various genes.**   Synonymous rates are less variable than nonsynonymous rates. *Source:* From Kimura (1986).

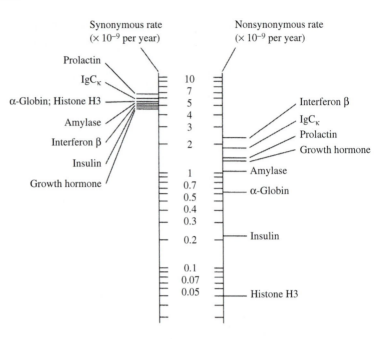

Gillespie (1989, 1991) compared DNA sequences of 20 proteins in three orders of mammals (rodents, artiodactyls, and primates). After correcting for the generation time effect (see below) and bias in his statistical procedures, he estimated that the average ratio of variance to mean was about 7.8 for replacement substitutions and 3.3 for silent substitutions. He concluded that "there is ample evidence that the Poisson clock is incompatible with protein evolution" (Gillespie 1991, p. 121). On the other hand, it appears that silent substitutions evolve at a more or less constant rate in mammals, consistent with the molecular clock. Given all the uncertainties of the estimations, one is hesitant to conclude that a ratio of 3.3 is much different from 1. Gillespie concluded that silent and replacement substitutions are controlled by different evolutionary processes.

Ohta (1995) conducted a similar study of 49 proteins in the same three orders of mammals. After correcting for generation time effect, she estimated the ratio of variance to mean to be 5.89 for silent and 5.60 for replacement substitutions, respectively. However, she did not correct for statistical bias. Bulmer (1989) showed that the bias is small for replacement substitutions, but much higher for silent substitutions. Ohta therefore concluded that her estimate for silent substitutions was probably inflated. If this is true, her results are more or less consistent with Gillespie's. The essential point here is that her estimate of the ratio of variance to mean was much greater than 1 for replacement substitutions, as was Gillespie's.

To complicate the issue, Zeng et al. (1998) obtained results for three species of *Drosophila* that were the opposite of Gillespie's and Ohta's. For 24 protein-coding genes, the average ratio of variance to mean was about 1.6 for replacement substitutions, and about 4.4 for silent substitutions. These results will be discussed further in Section 10.3.

We can conclude that the protein molecular clock is significantly overdispersed; that is, the variance of amino acid substitution rates is higher than predicted by the neutral theory. Most of the recent evidence suggests that substitution rates are more variable than initially thought. At the DNA level, things are more complicated: Several surveys suggest that silent and replacement substitutions are controlled by different evolutionary processes, but they lead to different conclusions. In mammals, replacement substitutions are overdispersed but silent substitutions are not. The opposite is true in *Drosophila*. We will discuss a possible resolution of this apparent contradiction in Section 10.3.

The Generation Time Effect

Some have argued that the neutral theory does not, in fact, predict the observed constant substitution rate, even if it is real. One of the initial observations leading to the neutral theory was the observed constancy *per year* of amino acid substitution rates (e.g., Figures 10.2 and 10.3). However, the neutral theory predicts a constant substitution rate *per generation*. From equation (10.2),

$$k = u$$

where k is the substitution rate per site and u is the neutral mutation rate per site per generation. (Here site can mean either a protein, an amino acid site within a protein, or a nucleotide site.) Since the neutral mutation rate is measured in generations, the substitution rate must be constant per generation.

We can calculate the substitution rate per year. If we define k_y as the substitution rate per year and g as the generation time in years, then

$$k_y = \frac{k}{g} = \frac{u}{g}$$

This equation predicts an inverse relationship between the substitution rate per year and generation time. This **generation time effect** is not seen in protein evolution.

Kimura and Ohta (1971a,b) initially argued that this means the mutation rate per year must be constant. But this is contrary to general biological understanding. In higher organisms, mutation rates are almost always measured per generation. The reason is that most mutations are thought to occur as copy errors during DNA replication associated with cell division. The number of cell divisions between zygote and gamete production is much more constant among different species than is generation time. Therefore, mutation rates should be more constant per generation than per year, contrary to the initial observations of protein evolution.

DNA sequences tell a different story. Silent substitutions do show a generation time effect, but replacement substitutions generally do not. The easiest way to see this is to use Gillespie's (1989) approach by comparing substitution rates with and without a correction for generation time effects. This is the same study mentioned above, comparing DNA sequences of 20 loci among three orders of mammals. His statistical procedures to correct for generation time effect are complex, and will not be discussed; we only summarize his results. For replacement substitutions, the uncorrected ratio of variance to mean was 8.26. After correcting for the generation time effect, the ratio was 6.95, only slightly less, implying that the generation time effect is minor for replacement substitutions. On the other hand, the uncorrected ratio for silent substitutions was 14.41. After correcting for generation

Figure 10.7   **Generation time effect for silent substitutions.**   Organisms with shorter generation times (e.g., *Drosophila*) have higher silent substitution rates than organisms with longer generation times (e.g., primates). *Source:* From Gillespie (1991).

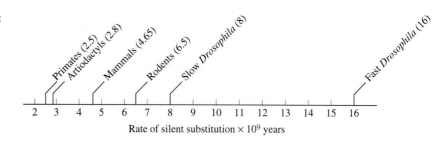

time effect, the ratio was 4.64. The great reduction suggests that generation time effects are much more important for silent substitutions than for replacement substitutions. Ohta (1995) came to essentially the same conclusion.

Figure 10.7 summarizes some of the data on silent substitution rates for three orders of mammals and various species of *Drosophila*. It is generally accepted that primates have a longer generation time than rodents, with artiodactyls intermediate, and that this relationship has held through evolutionary time. We see that, in fact, organisms with shorter generation times (e.g., *Drosophila*) have higher silent substitution rates. We must be cautious, but these data do suggest a generation time effect for silent substitutions.

Heterozygosity and Population Size

Recall that the predicted equilibrium heterozygosity under the infinite alleles model, from equation (7.19), is about

$$\widetilde{H} \cong \frac{4N_e u}{4N_e u + 1} \tag{10.5}$$

Thus, the neutral theory predicts an increasing relationship between heterozygosity and population size; see Figure 7.15. Nei and Graur (1984) tested this prediction by examining the relationship between allozyme heterozygosity and population size in 77 species for which they had reasonable estimates of population size. The correlation was 0.65, significant at the 0.001 level. However, even though the relationship between population size and heterozygosity is positive, it does not fit the relationship predicted by the neutral theory. In Figure 10.8, the solid line represents the predicted heterozygosity if $u = 10^{-7}$, a value commonly assumed for the neutral mutation rate to electrophoretically detectable alleles. Clearly, the data points do not fit the prediction; nearly all heterozygosities are less than predicted. No matter how we manipulate the mutation rate, the fit is terrible. We can conclude that the predicted relationship between heterozygosity and population size is weak at best.

How do we explain these observations? One possibility is that effective population sizes are much smaller and much more constant than actual population sizes. We saw in Section 7.4 that fluctuations in actual population size can have severe effects on effective population size. In particular, effective population size is very sensitive to small population sizes (bottlenecks). Nei and Graur (1984) argue that many species suffered severe bottlenecks during the last glaciation several thousand years ago. This would reduce heterozygosity to a low level, from which populations are still recovering. This is consistent with the theory of Nei et al. (1975) discussed in Section 7.6 and summarized in Figure 7.13. If heterozygosity is reduced by a bottleneck one or more generations long, it will take tens

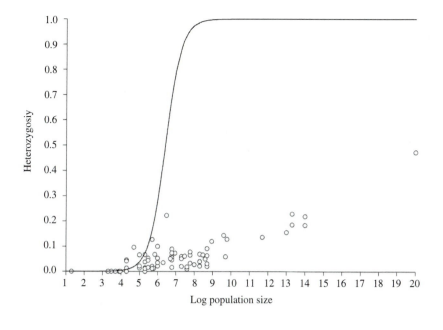

Figure 10.8 **Heterozygosity versus estimated population size for various species.** The line is the predicted heterozygosity based on equation (10.5), with $u = 10^{-7}$. *Source:* Data from Nei and Graur (1984).

or hundreds of thousands of generations to recover to the original heterozygosity. But recall that reduction of heterozygosity is minimal unless the bottleneck population size is very small (Figure 7.12) or the bottleneck lasts for a long time. Also, we must ask if it is reasonable that essentially *all* of the species plotted in Figure 10.8 experienced a severe bottleneck in the recent past.

A second possible explanation is that heterozygosity has been reduced by "hitchhiking." If an advantageous allele goes to fixation by natural selection, neutral alleles linked to it will be carried along passively, reducing heterozygosity. We will discuss hitchhiking in Section 10.4.

A third possible explanation is that molecular evolution and polymorphism are due mainly to fixation of slightly deleterious alleles. We will examine this "nearly neutral theory" in Section 10.3.

The consensus seems to be that the predicted relationship between population size and heterozygosity is not seen in natural populations, but the reasons are not well understood.

Substitution Rates and Selective Constraints

The neutral theory does not claim that natural selection is unimportant in molecular evolution. To the contrary, it argues that selective constraint is a powerful force. The basic idea is simple: The more important a molecule, or part of a molecule, the stronger will be natural selection to preserve its function. That molecule will evolve slowly, because most mutations will interfere with its function and natural selection will eliminate them (purifying selection). On the other hand, a molecule that has little or no function, for example a pseudogene, will have little or no selective constraint and will evolve more rapidly due to a higher neutral mutation rate.

We can quantify this idea to some extent by considering the fraction of all mutations that are neutral. Let u_T be the total mutation rate, and f be the fraction of all mutations that are neutral. Then the neutral mutation rate, u, is

$$u = fu_T$$

TABLE 10.3   Amino acid substitution rates for several proteins.

k_{aa} is the estimated number of substitutions per amino acid site per year. *Source:* From Kimura (1983a).

Protein	$k_{aa}(\times 10^{-9})$
Fibrinopeptides	8.3
Pancreatic ribonuclease	2.1
Lysozyme	2.0
α-Hemoglobin	1.2
Myoglobin	0.89
Insulin	0.44
Cytochrome c	0.3
Histone H4	0.01

The idea is that f, the fraction of mutations that are neutral, varies for different molecules, and for different parts of the same molecule. Under the neutral theory, the substitution rate is equal to the neutral mutation rate, so the prediction is that different molecules should evolve at different rates, depending on the degree of selective constraint.

Different molecules do, in fact, evolve at dramatically different rates. Table 10.3 summarizes amino acid substitution rates of several proteins. The rates vary by almost three orders of magnitude. Do these differences correspond to differences in selective constraint?

Consider histones, for example. Histones are intimately involved in chromosome structure, and are tightly bound to the DNA molecule (Figure 10.9). Most of the amino acids in histones 3 and 4 are directly involved in formation of the nucleosome. Nearly any change will have some effect, almost certainly deleterious. In other words, $f \cong 0$ for histones. Therefore, the neutral mutation rate is very low, and the neutral theory predicts that histones should evolve very slowly. The data confirm this prediction; histones are among the most slowly evolving proteins known (Table 10.3).

At the other extreme, consider fibrinopeptide molecules. They function as spacers to maintain the tertiary structure of fibrinogen, a molecule involved in blood clotting. During blood clotting, the fibrinopeptide is removed and fibrinogen folds

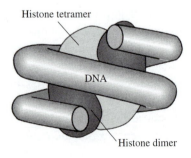

Figure 10.9 **Nucleosome structure, showing DNA molecule and histones.** The DNA coils around the histone core consisting of an H3-H4 tetramer (light shading) and two H2A-H2B dimers (dark shading). The histones are tightly bound to the DNA, and tolerate very little change in structure or amino acid sequence. *Source:* From Weaver and Hedrick (1997).

to form fibrin, which is actively involved in clotting. As long as the fibrinopeptide is the correct size, and can be spliced out at the appropriate time, there seems to be little other selective constraint. So f is near one, and the neutral mutation rate is near the total mutation rate. The neutral theory predicts that fibrinopeptide should evolve rapidly, and this is confirmed by sequence data (Table 10.3).

Pseudogenes provide the best example of rapid evolution due to minimal functional constraint. Pseudogenes are nonfunctional genes that have arisen as duplicate copies of functional genes, and then lost their function as a result of deleterious mutations. (For example, pseudogenes frequently have stop codons in the middle of the coding sequence.) Pseudogenes do not code for a functional protein, and are widely considered to be evolutionary junk, with no function. Given the lack of selective constraint, the neutral theory predicts that pseudogenes should evolve more rapidly than any functional gene. Again, DNA sequence data confirm this prediction. Li (1997) estimated the average substitution rate for pseudogenes to be about 3.9×10^{-9} substitutions per nucleotide site per year, much higher than the replacement substitution rates of most functional genes. Pseudogenes are frequently considered the neutral standard; that is, evolutionary rates in pseudogenes are considered representative of evolution under complete neutrality. Thus, substitution rates lower than those typical of pseudogenes probably indicate some selective constraint.

The idea of selective constraint also applies to different parts of a molecule. For example, amino acid positions 70 to 80 in cytochrome c are invariant in all species examined. These positions correspond to part of the active site of the enzyme, which apparently has essentially no tolerance for change. The amino acid substitution rate for this part of cytochrome c is zero. However, other parts of the molecule show much variation among different species (Dickerson 1971, 1972).

At the DNA level, we can predict that different parts of a gene should be subject to different selective constraints, and should evolve at different rates. First, introns are removed during RNA processing and have no effect on the final protein. They should evolve faster than functional parts of a gene. Second, it is reasonable to assume that fourfold degenerate sites (sites at which all nucleotide changes are silent) should evolve faster than twofold degenerate sites (sites at which two of the three possible nucleotide changes are silent), which in turn should evolve faster than nondegenerate sites (sites at which none of the possible changes are silent). These predictions are confirmed by DNA sequence data. Figure 10.10 compares average substitution rates for various parts of mammalian genes and pseudogenes. Pseudogenes evolve fastest (about 3.9×10^{-9} substitutions per nucleotide site per year), with introns and fourfold degenerate sites evolving almost as fast. As expected, nondegenerate sites evolve slowest (less than 1×10^{-9} substitutions per nucleotide site per year). Regions outside the coding sequence show intermediate rates, suggesting moderate selective constraints.

We shall describe one more example of different parts of a gene evolving at different rates due to different selective constraints. The insulin molecule consists of two polypeptide chains, A and B, linked by disulfide bonds. The precursor molecule, proinsulin, consists of a single chain with three sections, the A and B sections corresponding to the A and B chains of the insulin molecule, and a C section between them (Figure 10.11). When the proinsulin molecule is activated, it folds so that the A and B chains can form disulfide bonds with one another,

Figure 10.10 **Average substitution rates for parts of genes and pseudogenes.** *Source:* From Li (1997).

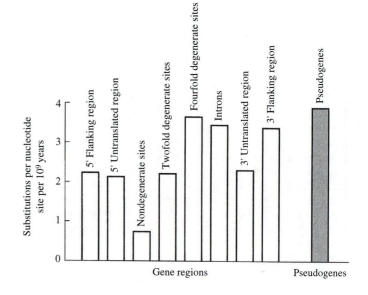

and the C chain is spliced out. The C chain plays no part in the functional insulin molecule. Replacement substitution rates in the A and B chain are about 0.20×10^{-9} substitutions per nucleotide site per year, whereas in the C chain the rate is about 1.07×10^{-9}, more than five times higher.

To summarize, molecules and parts of molecules with less selective constraint do evolve faster than molecules with more selective constraint. Pseudogenes, introns, and silent substitutions seem to evolve as if they are neutral or nearly so. Replacement substitutions evolve faster in molecules with fewer physiological restrictions than in highly constrained molecules.

Geographic and Environmental Correlations

If molecular variants are effectively neutral, there should be no correlations of allele frequencies among isolated populations, since those frequencies are controlled primarily by genetic drift, which acts independently in different populations. Moreover, the pattern of variation should be random, and uncorrelated with environmental variation.

Allozyme data are frequently inconsistent with these predictions. Many experimental studies have demonstrated that allozyme frequencies are sometimes similar in different populations of the same species, and even in different species. For example, Ayala et al. (1971, 1972) studied many populations of *Drosophila willistoni*, a widespread South American species. Figure 10.12 compares allele frequencies at four polymorphic loci in four widely separated populations. Each group of bars represents the alleles at a locus. The same allele is most common in all four populations, and alleles that are rare in one population are rare in others also.

Powell (1997) reviewed the data for several species of *Drosophila*, and concluded:"The most remarkable result from this and almost all other studies of *Drosophila* is the homogeneity of allele frequencies from population to population despite great distances between them. *The same alleles are found in all populations at about the*

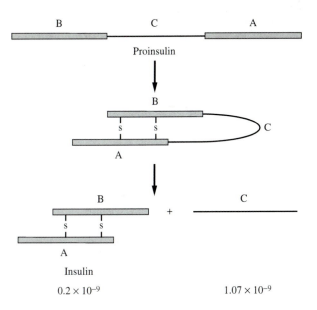

***Figure 10.11* Conversion of proinsulin to insulin.** The C peptide is not part of the functional insulin molecule. Numbers are replacement substitution rates in the functional insulin molecule (A and B peptides) and the C peptide.

same frequency" (Powell 1997, p. 39; his emphasis). Other groups show similar patterns. In general, the degree of differentiation among populations is inversely related to the perceived mobility of the species (Chapter 9).

Another prediction of the neutral theory is that allele frequencies should show no correlation with environmental variation. The most dramatic exception to this prediction is the alcohol dehydrogenase (*Adh*) locus in *Drosophila melanogaster*. We have met *Adh* before. Recall there are two electrophoretic alleles, *Fast* and *Slow*, and that the *Fast* allele is more efficient, but the *Slow* allele is more heat stable. Several studies have shown that there is a gradual increase in the frequency of the *Slow* allele as populations are closer to the equator. This pattern is seen on three different continents, North America, Australia, and Eurasia (Figure 10.13). The worldwide consistency of this pattern strongly suggests that it is due to natural selection. The physiological explanation is not known, but Oakeshott et al. (1982) showed that the best correlation is with rainfall and not temperature.

There are two possible explanations for these observations. First, the allele-frequency polymorphisms are maintained by natural selection. Some kind of balancing selection is acting to keep alleles in similar frequencies in different populations. Second, the alleles are indeed neutral, and the similarities are due to extensive gene flow among populations. We saw in Chapter 9 that for several models of gene flow, only a few migrants per generation are sufficient to prevent divergence of populations in the absence of natural selection.

These two alternatives were extensively debated in the 1970s (the so-called neutralist-selectionist debate). We saw in Section 5.6 that selection on allozymes can sometimes be convincingly demonstrated, but these results cannot be generalized because they are examples of unusually strong selection. We have seen that selection too weak to detect experimentally can still be strong enough to control allele frequencies in a population. In retrospect, it seems that the neutralist-selectionist debate was irresolvable with the tools available at the time. It is only

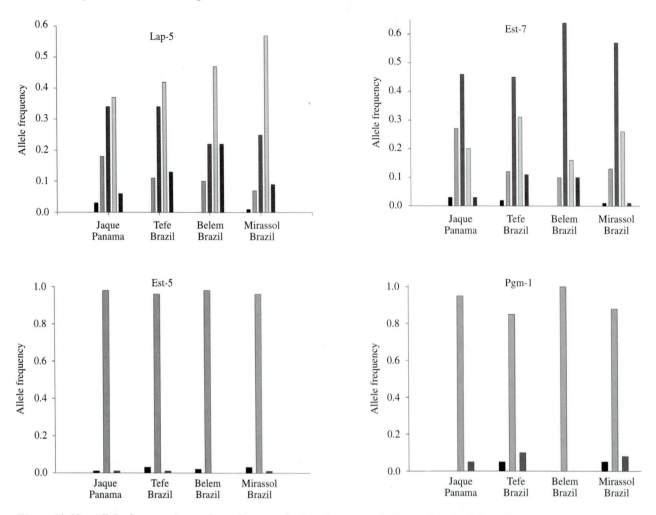

Figure 10.12 **Allele frequencies at four allozyme loci in four populations of *Drosophila willistoni*.**   Each group of bars represents the alleles at a locus. Alleles that are common in one population tend to be common in all. Populations are separated by hundreds to thousands of kilometers. *Source:* Data from Ayala (1972).

since extensive DNA sequences have became available that we have been able to convincingly detect weak selection at the molecular level. We will consider this topic in Section 10.4.

10.3 The Nearly Neutral Theory

As we saw in the previous section, several kinds of experimental observations are inconsistent with the predictions of the neutral theory. Three of these are (1) overdispersion of the protein molecular clock; (2) lack of a generation time effect in protein evolution; (3) heterozygosity much less than predicted from population size. The neutral theory can be modified to explain these observations. These modifications are primarily due to Tomoko Ohta, and the modified theory is commonly called the nearly neutral theory. (See review papers in Ohta 1974, 1992, 1996, as well as her original papers listed in the bibliography.)

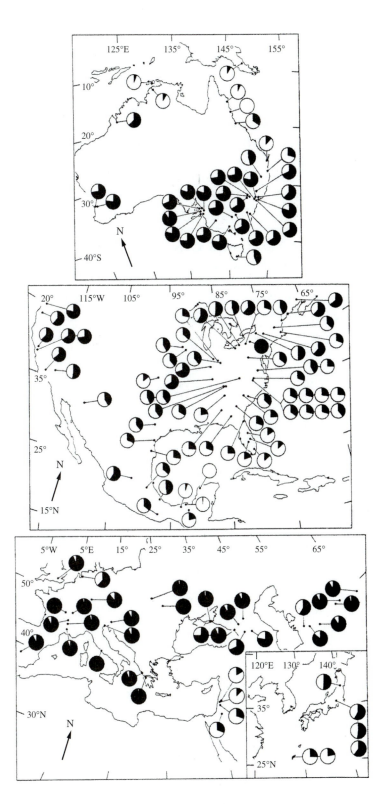

Figure 10.13 **Clines in frequencies of *Adh* alleles in Australia, North America, and Eurasia.**   Solid part of circles represents the proportion of the *Fast* allele. *Source:* From Oakeshott et al. (1982).

The neutral theory originally considered mutations to be either deleterious, neutral, or advantageous. This was based primarily on the biochemical properties of the mutation and is essentially independent of population size. But Ohta realized that selection is less effective in small populations; a slightly deleterious mutation has a greater chance of becoming fixed or drifting to a relatively high frequency. In other words, a slightly deleterious mutation might behave as if it were neutral in a small population, but be effectively selected against in a large population.

Ohta proposed that there is a large class of mutations with selection coefficients near zero that are effectively neutral in small populations but effectively deleterious in large populations. There are various ways to define effectively neutral. Let the fitnesses of heterozygotes and homozygotes for a mutant allele be $1 + s$ and $1 + 2s$, respectively, as in Section 10.2. Here, s is positive for an advantageous mutation and negative for a deleterious mutation. Ohta called a mutation effectively neutral if $|s| < 1/2N$. [Compare this to Kimura's definition of neutral, in equation (10.4)]. The idea is illustrated in Figure 10.14, which shows a hypothetical distribution of selection coefficients. The horizontal axis is s, and the vertical axis is the probability density function (Appendix A) of s. It can be interpreted, loosely, as a

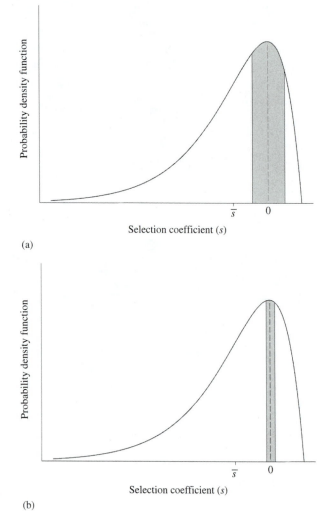

Figure 10.14 **Proportion of mutations that are effectively neutral in a small population (a), and a large population (b).** $\bar{s}$ is the mean selection coefficient (negative). The shaded region represents the proportion of mutations with $|s| < 1/2N$.

measure of the relative probability that a mutation will have a selection coefficient near s. Since most mutations are assumed to deleterious, the mean of s is negative. The shaded region represents values of s which are effectively neutral, that is, $-1/2N < s < 1/2N$. A greater proportion of mutations is effectively neutral in a small population (Figure 10.14a) than in a large population (Figure 10.14b). The mutation rate to effectively neutral mutations is smaller in large populations, since more mutations are "selectable" in large populations.

What is the probability that a slightly deleterious mutation will become fixed in a population? Ohta assumed that advantageous mutations are so rare they can be ignored in this context. For technical reasons, the notation is slightly different from our previous use. The fitness of the heterozygote is defined as $1 - s$; in other words, s is positive for a deleterious mutation. The selection coefficient is a random variable, assumed to have an exponential distribution with mean $\bar{s}$. The logic behind the assumption of an exponential distribution is that mutations with very small deleterious effects are more common than mutations with larger effects, and this distribution of effects (selection coefficients) is described by the exponential distribution (Appendix A). The equation for the exponential distribution is

$$f(s) = be^{-bs} \tag{10.6}$$

where $f(s)$ is the probability density function of s, and $b = 1/\bar{s}$. Make sure you understand that $f(s)$ is not a mutation rate. It can be interpreted as a measure of the relative probability that a mutant allele will have a selection coefficient near s, *given that a deleterious mutation has occurred.*

The general form of the exponential distribution is shown in Figure 10.15. Note that as s increases, the probability that a mutant allele will have a selection coefficient near s decreases. Note also that we are assuming that selection is weak, so most selection coefficients are near zero; therefore, $\bar{s}$ is very small.

It can be shown that, under certain assumptions, the probability of fixation for a deleterious mutation is approximately

$$\Pr(fix) \cong \frac{1}{8N^2\bar{s}} \tag{10.7}$$

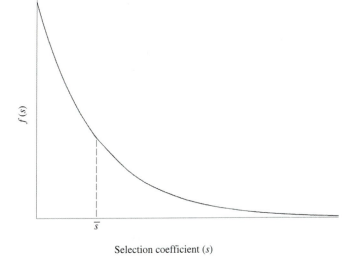

Figure 10.15 **Exponential distribution of selection coefficients of deleterious mutations.** $f(s)$ is the probability density function of s, as given by equation (10.6), and $\bar{s}$ is the mean of s.

$f(s)$

$\bar{s}$

Selection coefficient (s)

as long as $4N\bar{s}$ is large. The derivation of this equation is outlined in Box 10.1. The important thing is to see that, because of the N^2 in the denominator, the fixation probability decreases very rapidly as N increases.

We can now determine the substitution rate for deleterious mutations (k_d). Analogous to equation (10.1) we can write

$$k_d = 2Nu_d \times \Pr(fix)$$

where u_d is the mutation rate to deleterious alleles, and $2Nu_d$ is the total number of deleterious mutations that occur in a generation. Substituting $\Pr(fix)$ from equation (10.7), we get

$$k_d \cong 2Nu_d\left(\frac{1}{8N^2\bar{s}}\right) \cong \frac{u_d}{4N\bar{s}}$$

Box 10.1   Fixation probability for nearly neutral mutations

Ohta assumed that deleterious mutations have an exponential distribution, with mean $\bar{s}$. Here, s is the selection coefficient against the heterozygote, and a positive s indicates a deleterious mutation. Using the formula for the exponential distribution (Appendix A), we can write the probability that a deleterious mutation will have a selection coefficient of s as

$$f(s) = be^{-bs} \tag{1}$$

where $b = 1/\bar{s}$. Think of $f(s)$ as the relative probability that a deleterious mutation will have a selection coefficient near s.

Ohta defined a mutation as effectively neutral if $s < 1/2N$. The proportion of mutations that behave as if they were neutral is then the proportion of mutations with selection coefficients of $1/2N$ or less. To get this, we need to integrate equation (1) over the range 0 to $1/2N$

$$u_{nn} = \int_0^{1/2N} be^{-bs}ds \tag{2}$$

which is easy to integrate if you remember your calculus (Problem 10.3). The solution is

$$u_{nn} = 1 - e^{-b/2N} \tag{3}$$

The probability of fixation for a new mutation with selection coefficient s is, from equation (7.22)

$$\Pr(fix|s) = \frac{1 - e^{-2s}}{1 - e^{-4Ns}} \tag{4}$$

where the left side means the probability of fixation given a particular value of s.

The total probability of fixation for all deleterious mutations is just the probability of fixation for a given s multiplied by the probability of that s, summed over all possible values of s, or

$$\Pr(fix) \cong \int_{all\ s} [\Pr(fix|s)][f(s)]ds$$

where $\Pr(fix|s)$ is given by equation (4) and $f(s)$ by (1). Substituting and integrating from $s = 0$ to $s = 1$, we get

$$\Pr(fix) \cong \int_0^1 \left(\frac{1 - e^{-2s}}{1 - e^{-4Ns}}\right)(be^{-bs})ds$$

The right-hand side can be approximated by numerical integration on a computer. According to Gillespie (1991), if $4N\bar{s}$ is large, the result is approximately

$$\Pr(fix) \cong \frac{1}{8N^2\bar{s}}$$

This is the approximate fixation probability for a deleterious mutation. We can use this to determine substitution rates and heterozygosity, as described in the main text.

This equation says that, all else being equal, the substitution rate per generation should be lower in large populations.

On the other hand, the substitution rate per year is

$$k_y = \frac{k_d}{g} = \frac{u_d}{4N g \overline{s}}$$

where g is the generation time in years. Ohta initially assumed that there is a negative correlation between N and g; that is, animals with larger population sizes (e.g., rodents) tend to have shorter generation times than animals with smaller population sizes (e.g., primates). This has since been documented by Chao and Carr (1993); see Figure 10.16. If the product of N and g remains approximately constant, the substitution rate per year will be approximately constant, as seen in the amino acid substitution data. Thus, the nearly neutral theory can explain the approximately constant amino acid substitution rates per year.

Moreover, as population sizes fluctuate, the neutral mutation rate fluctuates, and therefore the substitution rate fluctuates. So variation in substitution rates can be explained by variation in population sizes. The nearly neutral theory can also explain overdispersion of the protein molecular clock.

Under the neutral theory, expected heterozygosity at equilibrium is

$$H \cong \frac{4Nu}{1 + 4Nu}$$

where u, remember, is the mutation rate to neutral mutations. This equation predicts that H should increase toward a maximum value of one as N increases (Figure 10.17, dashed line). Under the nearly neutral theory the same general equation holds, except that the neutral mutation rate decreases as N increases. The result is that H approaches a maximum value of less than one, and becomes essentially independent of N for large N. Figure 10.17 compares the predicted heterozygosities under the neutral and nearly neutral theories. For large populations, the predicted heterozygosity increases more slowly under the nearly neutral theory, and is more consistent with observed heterozygosities in natural populations.

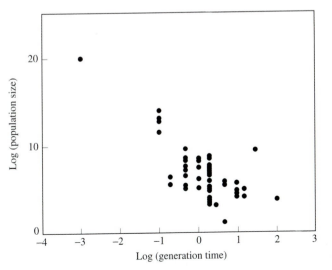

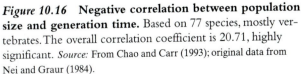

Figure 10.16 **Negative correlation between population size and generation time.** Based on 77 species, mostly vertebrates. The overall correlation coefficient is 20.71, highly significant. *Source:* From Chao and Carr (1993); original data from Nei and Graur (1984).

Figure 10.17 **Predicted heterozygosity under the neutral theory (dashed line) and nearly neutral theory (solid line).** Note log scale for population size.

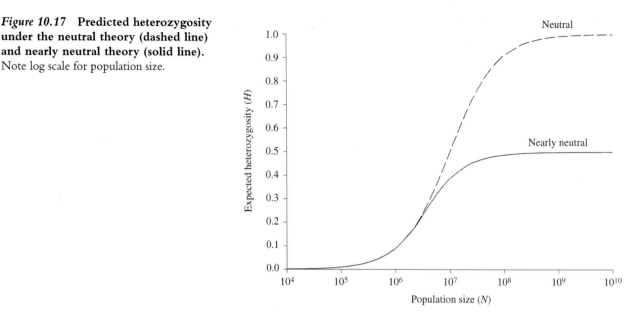

We have seen that the nearly neutral theory can explain three of the major inconsistencies between the neutral theory and the experimental data. We have only touched the surface of the nearly neutral theory, and have mostly ignored its various assumptions and technical details. For more on the nearly neutral theory, see Ohta's original papers and Kimura (1979). For a lucid, nonmathematical description and critique, see Gillespie (1995). For a recent debate, see Ohta (1996) and Kreitman (1996).

The nearly neutral theory has gained wide acceptance because slightly deleterious mutations are thought to be common (see Section 6.6), and because it can help explain a variety of observations and experimental results. The nearly neutral theory seems to bridge the neutral theory and the theory of weak selection at the molecular level. As Ohta (1996, p. 674) wrote, "Discrimination between the selection theory and the nearly neutral theory ... is very difficult. Indeed, the two theories are not quite distinguishable conceptually." However, the nearly neutral theory is difficult to test experimentally because some sequence of N can explain virtually any observation. Therefore, the neutral theory remains the null hypothesis against which to test for weak selection (or near neutrality).

Codon Bias

One possible example of nearly neutral evolution is the phenomenon of **codon bias** observed in some organisms. To understand codon bias, recall that, due to the redundancy of the genetic code, many amino acids are specified by more than one codon. For example, leucine is specified by six codons and glycine by four. It is often assumed that different codons specifying the same amino acid (synonymous codons) are selectively equivalent, since they do not change the amino acid sequence of the protein. However, this is not necessarily so. There is more than one tRNA molecule for some amino acids. Consider a mutation in a codon from CUA to CUC. (This is the mRNA sequence, which is directly derived from the DNA sequence.) Both codons specify leucine, but they require different tRNA molecules. If there are differences in the numbers of the two tRNA molecules, or their energy requirements for synthesis, or their affinity for leucine, or for the codon, they may result in different rates of protein synthesis. These different rates may translate into different fitnesses. Similarly, changes in the DNA sequence may alter the ability of regulatory proteins to bind to the DNA or the stability of the mRNA molecule, even if those changes do not change the amino acid sequence of the protein. In summary, it is possible that synonymous codons may not be selectively neutral with respect to one another.

If synonymous codons are selectively neutral, then codon usage should be random, and all synonymous codons for a given amino acid should occur with about equal frequency. On the other hand, if one codon is favored over the others, it should have a higher frequency than the others; a codon bias should be observed. In fact, several organisms show nonrandom codon usage. Figure 10.18 illustrates codon bias for three amino acids in four species.

There are patterns to the nonrandom use of codons. In *E. coli* and yeast, the preferred codon corresponds to the most common tRNA molecule for the amino acid. The evidence is less clear for *Drosophila*. Highly expressed genes show higher codon bias than genes that are expressed at lower levels. In *Drosophila*, genes with high codon bias show lower substitution rates than genes with less codon bias. Patterns in mammals are similar, though complicated by regions of high or low GC content.

These patterns of nonrandom codon usage suggest the action of natural selection. However, selection for codon bias cannot be very strong. Silent substitutions occur much more frequently than replacement substitutions, even in organisms with high codon bias. But the existence of codon bias argues that selection may sometimes favor one codon over another.

Zeng et al. (1998) compared DNA sequences of 24 protein-coding genes in three species of *Drosophila*. For each gene they compared the mean substitution rate to the variance among the three species. Recall that the ratio of variance to mean is expected to be one under the neutral theory. Averaged over all proteins, the ratio was 1.6 for replacement substitutions, and 4.4 for silent substitutions. The first value is not significantly different from one, but the second is. The conclusion is that replacement substitutions are evolving as predicted by the neutral theory, but silent substitutions are not. This is the opposite result from Gillespie's and Ohta's studies of mammalian genes described earlier.

Zeng et al. suggested that weak selection for codon bias may explain the high variance of silent substitutions in *Drosophila*, and the difference between *Drosophila* and mammals. The nearly neutral theory predicts that selection will be more effective in large populations than in small ones. If one codon is preferred over others, mutations to the preferred codon can be effectively selected for in large populations, resulting in high codon bias. Thereafter, most mutations will be from preferred codons to unpreferred, and can be effectively eliminated in very large *Drosophila* populations. But in mammals, with population sizes several orders of magnitude smaller than *Drosophila*, these mutations are effectively neutral. Thus, the dynamics of many silent substitutions may be driven by weak natural selection in *Drosophila* (large populations, effective selection, high codon bias) but by mutation and genetic drift in mammals (smaller populations, less effective selection, lower

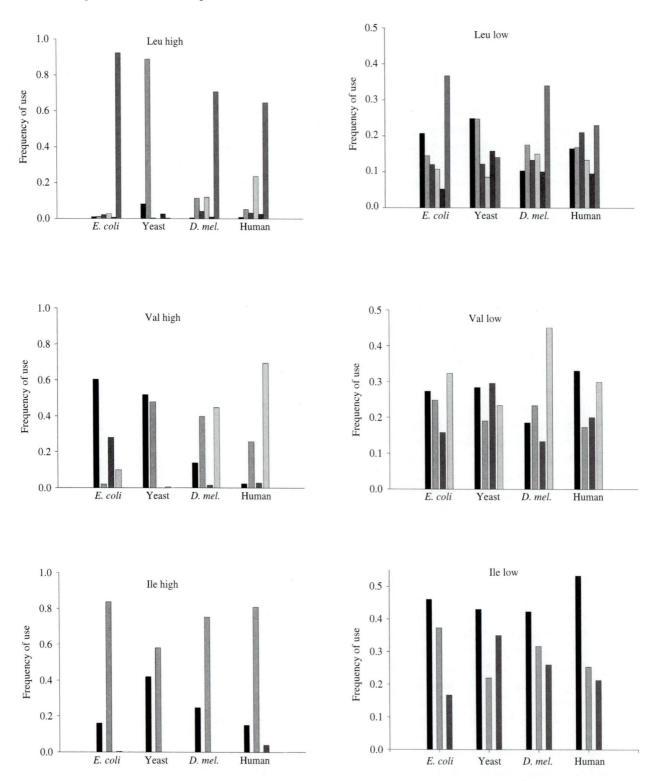

Figure 10.18 **Codon bias for three amino acids in four species. Each group of bars represents the different codons for an amino acid.**   High and low indicate genes expressed at high or low rates, except for mammals, where they represent regions of high or low GC content. Codon bias is generally higher for genes expressed at high rates. *Source:* Data from Sharp et al. (1988).

codon bias). The evolution of codon bias is still not well understood. For recent reviews, see Sharp et al. (1995), Kreitman (1996), Akashi (1997), and Hey (1999).

10.4 Detecting Natural Selection

It is sometimes possible to convincingly demonstrate that natural selection is acting on allozyme loci, as we saw with *Adh* in *Drosophila melanogaster* and *Pgi* in *Colias* butterflies (Section 5.6). These are examples of strong selection, and are probably not representative of selection on allozymes in general.

How strong is natural selection on protein or DNA variants likely to be? One approach to this question is to consider selection against null alleles. Null alleles are alleles that show no detectable activity by electrophoretic techniques. Voelker et al. (1980a) and Langley et al. (1981) estimated the frequencies of null alleles at 20 autosomal allozyme loci in *Drosophila melanogaster*. The average frequency of null alleles was $q \cong 0.0025$. Assuming these alleles are kept in a population by mutation-selection equilibrium (Section 6.5), we can estimate the strength of selection. From equation (6.3),

$$\widetilde{q} = \frac{u}{hs}$$

where $\widetilde{q}$ is the equilibrium frequency of a deleterious allele. Solving for hs, the strength of selection against heterozygotes, we get $hs = u/\widetilde{q}$. From an independent study, Voelker (1980b) estimated the mutation rate to null alleles to be about 3.86×10^{-6}. Using this estimate of u and the above estimate of $\widetilde{q}$ gives $hs = 0.0015$.

Gillespie (1991) argued that selection on allozyme variants cannot be stronger than selection on null alleles, and that selection coefficients must therefore be on the order of 0.001 or less for allozymes. There are obvious exceptions to this generalization, but it suggests that selection on segregating protein or DNA variation must usually be very weak. Selection coefficients on the order of 0.001 are almost certainly too weak to detect by experimental or field studies, but we have seen that such weak selection may still be strong enough to control the fate of an allele, depending on population size.

Clearly, if we are to detect weak selection, we must resort to indirect techniques. In this section, we briefly describe several statistical techniques to detect natural selection, based primarily on DNA sequence data. The general idea is to use predictions from the neutral theory about the amount and pattern of genetic variation as a null hypothesis, and to look for deviations from these predictions. Significant deviations may imply that natural selection is acting, although, as we shall see, interpretation is not always straightforward.

Kinds of Natural Selection

Before we consider patterns of variation due to natural selection, we must first review different kinds of natural selection. We can imagine three fundamental kinds of natural selection at the molecular (or any other) level. The most common is probably **purifying selection**. Deleterious mutations are eliminated to preserve the function of the protein or DNA sequence. Since many, if not most, mutations are probably deleterious, at least slightly so, purifying selection must be common.

The opposite of purifying selection is sometimes called **positive selection**, or Darwinian selection. Rare favorable mutations are selected for, resulting in the substitution of the new mutation for the previous best allele.

Finally, we must consider **balancing selection**, which acts to maintain two or more variants at a locus. Examples of balancing selection are overdominance, frequency dependent selection, and so forth (Chapters 5 and 12).

Each of these kinds of natural selection leaves its own footprint in the pattern of DNA variation. As we shall see, this footprint is sometimes detectable in a sample of DNA sequences.

Replacement Versus Silent Substitution Rates

We have seen that silent substitution rates are usually much higher than replacement rates (Table 7.1). This is evidence of purifying selection. If mutations occur randomly, most mutations in a coding sequence (about 75 percent) will change the amino acid sequence. The fact that we see far fewer replacement than silent substitutions means that most replacement mutations have been eliminated by natural selection. An extreme example is the histone proteins in Table 7.1, with replacement substitution rates of essentially zero. The same principle is true for silent and replacement polymorphisms within a species: Silent polymorphisms are usually more common because most replacement mutations have been eliminated by natural selection. A classic example is Kreitman's (1983) analysis of the *Adh* gene of *Drosophila melanogaster*. Within the coding sequence, 14 sites are polymorphic. All but one are silent, and the exception creates the *Fast/Slow* allozyme polymorphism. Other evidence suggests that this polymorphism is being maintained by balancing selection, but these data suggest purifying selection against replacement mutations at other sites.

We can imagine at least two situations in which replacement substitution rates might be higher than silent rates. First, if a protein is recruited for a new function, natural selection might temporarily favor replacement substitutions changing the structure and function of the protein. Second, if diversity itself is favored, then natural selection might favor replacement substitutions or polymorphisms. We will consider examples of each.

One example of a protein evolving a new function is the evolution of lysozyme in cows and langurs (colubine monkeys). In most mammals, the normal function of lysozyme is to destroy invading bacteria by breaking down their cell walls. It is found primarily in saliva, tears, white blood cells, and macrophages, and is usually expressed only at low levels in the pyloric region of the stomach. Ruminants (cows and relatives) and colubine monkeys have independently evolved foregut fermentation (by bacteria) to help break down otherwise indigestible plant tissues. In these animals, lysozyme is expressed at high levels in the anterior portion of the true stomach. Here, it apparently functions as a kind of digestive enzyme by breaking down the cell walls of bacteria coming from the foregut, so they can be digested by stomach enzymes. The stomach is a very different environment for lysozyme from other tissues; for example, it is much more acidic, and contains digestive enzymes such as pepsin. We would expect natural selection to favor mutations that make lysozyme more efficient in the stomach of ruminants and colubine monkeys. That is exactly what has occurred. The amino acid substitution rate along the line to langurs is about twice the rate in other primates. Lysozyme from cows and langurs is most active at lower pH, and is particularly resistant to the stomach enzyme pepsin.

Stewart et al. (1987) and Stewart and Wilson (1987) compared the amino acid sequence of lysozyme from several species of mammals. Langur lysozyme is most similar to that of other primates and is, in general, relatively distant from cow lysozyme (32 differences in 130 amino acids). However, langurs and cows share four unique substitutions (found in no other species). For comparison, langurs share zero unique substitutions with their closest relative, baboons. The pattern is one of general differentiation (random divergence) at most sites, combined with convergence at specific sites. Stewart et al. (1987) interpret this as indicating natural selection favoring convergent evolution of cow and langur lysozymes as they have adapted to their new function. This is an example of convergent evolution at the molecular level. Stewart and Wilson (1987) discuss the possible effect of the unique substitutions on the biochemical properties of lysozyme.

The lysozyme example was based on amino acid sequences of the proteins. Today, most analysis is done with DNA sequences directly. As we have seen, for protein-coding genes, silent substitution rates are typically several times higher than replacement rates; the averages from Table 7.1 are $K_S = 3.51$ and $K_A = 0.74$. (K_S and K_A are the expected numbers of silent and replacement substitutions per site, respectively; Section 10.1.) If selection favors replacement substitutions, we would expect K_S and K_A to be much more similar, and K_A might even be higher than K_S. Therefore, a ratio of K_A/K_S greater than one is strong evidence of positive natural selection. Numerous examples of this have been found. Yang and Bielawski (2000) list 55 genes or classes of genes in which positive selection has been detected by the K_A/K_S ratio.

Many marine invertebrates reproduce by releasing their gametes into the water. Fertilization occurs when a male gamete encounters a female gamete, and the sperm penetrates the egg. Specific biochemical interactions between the sperm and egg usually assure that fertilization occurs between members of the same species. This is a kind of premating reproductive isolation, discussed in Section 8.2.

In abalone, the sperm contains a protein called lysin, whose function is to penetrate the outer membrane of the egg so that the sperm can enter. It is reasonable to expect that this sperm-egg interaction should be very specific and as different from other species as possible, to prevent accidental interspecies fertilization. Vacquier and his colleagues have studied the evolution of the sperm lysin protein; it is one of the most rapidly evolving proteins known (Metz et al. 1998). This evolution is due to positive selection. Lee et al. (1995) compared the DNA sequences of the sperm lysin gene in 20 species of abalone. The pairwise K_A/K_S ratios for the entire sequence were as high as 4.10, among the highest ratios known for a complete sequence. Ratios for the most closely related species were always greater than one, suggesting positive selection for divergence.

If positive selection causes divergence in the sperm-egg interactions, we might expect that parts of the protein involved in these interactions would be more divergent than parts that are not involved. Hughes (1999) compared the

K_A/K_S ratios for different parts of the lysin gene in nine closely related species (Table 10.4). For parts of the gene coding for the signal peptide and the hydrophobic patch, the ratio was less than one. These two regions of the protein have functions unrelated to the penetration of the egg, and the ratios suggest purifying selection. The ratio for the rest of the protein, some of which must be involved in sperm penetration, was about 1.85, suggesting positive selection for diversification of sperm-egg interactions. The details of selection are not yet known, so this interpretation must be tentative. Other explanations are possible (Vacquier et al. 1997)

Selection for Diversity

Genetic diversity per se can sometimes be beneficial. Organisms must deal with a variety of parasites and pathogens, and have evolved complex defense mechanisms. Two well-known examples are the vertebrate immune system and the restriction/modification systems of bacteria. One can imagine that an organism that is able to defend itself against a variety of parasites and pathogens will have an advantage over one whose range of defenses is less. In other words, natural selection might favor a diversity of defense mechanisms.

The major histocompatibility complex (MHC) is a group of loci involved in the immune response of mammals. Class I MHC genes produce cell surface proteins that bind to foreign proteins, and allow recognition and destruction by T-lymphocytes. These cell surface proteins consist of two domains (called the α_1 and α_2 domains) that make up the antigen recognition site (ARS), and other domains not directly involved in recognition. There is great variation in the ARS, allowing cells to recognize a variety of foreign proteins; heterozygosity of MHC loci can be up to 80 percent or more. Hughes and Nei (1988) hypothesized that this variation is maintained by selection favoring variability at ARS sites. They predicted that sites within the ARS coding sequence should show more replacement variation than silent variation. They compared sequences of 12 alleles at three MHC loci in humans. Results confirmed their prediction: Within the ARS sequence there was a highly significant excess of replacement substitutions between individuals (66 comparisons; mean $K_A \cong .208$; mean $K_S \cong 0.068$), while in other parts of the gene there were significantly more silent substitutions. Similar comparisons in mice showed the same pattern, but the differences were not statistically significant due to small sample sizes. Hughes and Nei concluded that the ARS regions were subject to overdominant selection favoring ARS diversity, while other regions of the proteins were subject to purifying selection (elimination of deleterious mutations).

TABLE 10.4   Numbers of replacement (K_A) and silent (K_S) substitutions per site among sperm lysins of several species of abalone (*Haliotis*). *Source: From Hughes (1999); original data from Lee et al. (1995).*

Region of Protein	$K_A \pm$ s.e.	$K_S \pm$ s.e.	Ratio (K_A/K_S)
Signal peptide	0.047 ± 0.021	0.096 ± 0.095	0.49
Hydrophobic patch	0.023 ± 0.017	0.159 ± 0.128	0.14
Other	0.187 ± 0.015	0.101 ± 0.020	1.85

This pattern seems to be general for genes involved in the immune response. Hughes and Nei (1989) and Tanaka and Nei (1989) obtained similar results for class II MHC genes and for immunoglobulin variable-heavy genes. Moreover, similar processes occur in bacteria!

Colicins are toxic proteins produced by bacteria to kill other bacteria. Colicin genes are present on plasmids, and come paired with an "immunity gene," which confers immunity to the particular colicin produced by that plasmid. These genes appear to be important in bacterial competition and invasion into new habitats. Any bacterial cell that produces a colicin to which other cells are susceptible will have a competitive advantage. One can imagine a kind of diversifying selection similar to that in the immune response genes in mammals. In fact, the immunity region (the parts of the colicin and immunity genes that confer immunity) shows substitution rates several times higher than other parts of the colicin cluster. Riley (1993a,b) proposes a complex scenario involving plasmid transfer, recombination, and selection to explain the observed patterns. For our purposes, the essential point is that elevated substitution rates (in this case, both silent and replacement rates) are a result of natural selection for diversity. For reviews of the evolution of colicins and related proteins, see Riley (1993a,b; 1998).

Note that in these examples, as well as the abalone sperm lysin example above, we were able to detect natural selection by looking at the K_A/K_S ratio in different regions of the gene. Thus, local K_A/K_S ratios can indicate which parts of a protein are subject to natural selection, even if the ratio for the entire sequence is not significantly greater than one. In fact, it is rare for the K_A/K_S ratio to be greater than one for the entire gene. Endo et al. (1996) conducted an extensive survey and found that only 17 of 3595 gene groups (homologous genes in different species) had a K_A/K_S ratio of greater than one for the entire sequence. This is in dramatic contrast to the numerous cases in which the K_A/K_S ratio is greater than one for part of the gene (Yang and Bielawski 2000; Ford 2002).

The Expected Distribution of Allele Frequencies

We now consider several statistical tests designed to detect deviations from the predictions of the neutral theory. The goal will be to understand the principles of the tests, with less emphasis on the computational details, which are usually handled by computer programs.

The quantity $4Nu$ appears frequently in molecular population genetics; for example, see equation (7.19), which predicts equilibrium heterozygosity under the infinite alleles model. Often, we use the symbol θ to represent $4Nu$, where u is the mutation rate per locus.

The neutral theory predicts not only the equilibrium heterozygosity, but also the equilibrium distribution of allele frequencies under the infinite alleles model. Ewens (1972) showed that the expected number of different alleles, k, in a sample of size n is

$$E(k) = 1 + \frac{\theta}{\theta + 1} + \frac{\theta}{\theta + 2} + \cdots + \frac{\theta}{\theta + (n - 1)}$$

Ewens also worked out the expected distribution of allele frequencies at equilibrium, given n and k. Surprisingly, this distribution depends only on n and k. Variation around this expected distribution can be estimated by computer simulations. This means that an observed allele frequency distribution can be compared to the distribution expected under neutrality with the same n and k. An example comparing the observed and expected distributions at a human locus is shown in Figure 10.19.

We can imagine two main kinds of deviation from a neutral sample.[1] First, the most common allele may be at a higher frequency than expected, and uncommon alleles may be at lower frequencies than expected (Figure 10.20a). The allele frequency distribution of the sample is more uneven than predicted under neutrality; common alleles are too common and rare alleles are too rare. This is sometimes referred to as an excess of rare alleles (meaning many alleles whose frequency is lower than expected). Certain kinds of natural selection or demographic processes will produce allele frequency distributions that are too uneven compared to a neutral sample with the same n and k. Second, common alleles may have lower frequencies than expected, with intermediate frequency alleles at higher frequencies than expected (Figure 10.20b). The allele frequency of the sample is more even than predicted under neutrality. Again, some kinds of natural selection or demographic processes will produce allele frequency distributions that are too even compared to a neutral sample.

We next describe two tests designed to detect deviations from the allele frequency distribution expected under neutrality.

The Ewens–Watterson Test

Watterson (1978a,b) developed a statistical test for comparing the observed distribution of allele frequencies to the expected distribution derived by Ewens. Watterson showed that the expected homozygosity of the sample $(=\Sigma \hat{p}_i^2)$ is a good statistic for testing the null hypothesis that the observed allele frequency distribution is consistent with the neutral prediction. This expected homozygosity of the sample is compared to the expected homozygosity predicted by the neutral theory. The latter homozygosity and the significance of the difference are estimated by computer simulation. Watterson (1978b) summarizes these simulations with a table for various values of n and k. For example, with $n = 50$ and $k = 7$, an approximate 95 percent confidence interval for the sample homozygosity is from 0.18 to 0.57. A sample homozygosity outside this range is cause to reject the null hypothesis of neutrality.

The interpretation of the test, with respect to natural selection, is as follows: Homozygosity is determined mostly by common alleles; rare alleles contribute little to expected homozygosity. A sample homozygosity greater than predicted by the neutral theory means that the most common allele is more common than expected, and there are more rare alleles than expected (as in Figure 10.20a). An excess of rare alleles suggests that some alleles are kept at lower frequency than in a neutral sample because they are deleterious and their frequencies are being suppressed by purifying selection. A significant excess of rare alleles by the Watterson test is evidence of purifying selection; not all alleles in the sample are neutral.

---

[1] We will use the phrase neutral sample as shorthand for a sample with the same n and k from a population that is evolving according to the neutral theory.

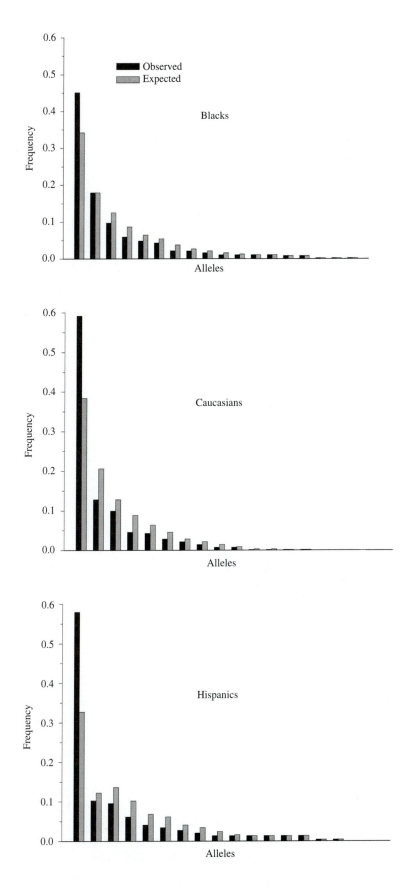

Alleles

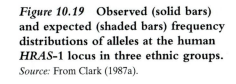

Figure 10.19   **Observed (solid bars) and expected (shaded bars) frequency distributions of alleles at the human *HRAS*-1 locus in three ethnic groups.** *Source:* From Clark (1987a).

Figure 10.20 **Allele frequency distributions that are (a) too uneven, or (b) too even, compared to the neutral expectation.**

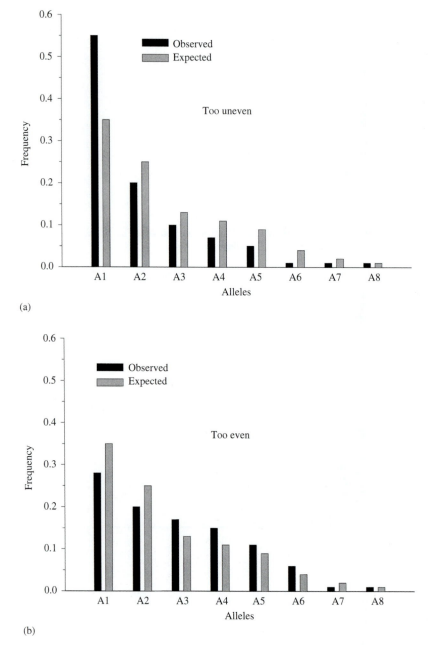

(a)

(b)

On the other hand, a sample homozygosity less than predicted (equivalently, heterozygosity greater than predicted) means that there are more alleles of intermediate frequency than expected, and fewer rare alleles than expected. The sample distribution is more even than predicted under neutrality (Figure 10.20b). This suggests that more alleles are being maintained at intermediate frequencies by some kind of balancing selection.

Unfortunately, demographic factors cloud the picture. A population that has recently been through a severe bottleneck will show excess homozygosity by the Watterson test. This is because rare alleles present in the sample after the bottleneck will be mostly newly arisen alleles, and at lower frequency than in a neutral sample at equilibrium. Without further information, one cannot distinguish between a recent bottleneck and purifying selection as explanations of excess homozygosity. Similarly, a recent population expansion can mimic balancing selection. We will discuss demographic factors in more detail below.

The example in Figure 10.19 shows RFLP variability at the human *HRAS*-1 locus in three ethnic groups sampled around New York City. Table 10.5 shows the observed and expected homozygosities for each group, and the result of Watterson's homozygosity test. In none of the three groups was the difference significant, suggesting that the distributions in Figure 10.19 are consistent with the neutral expectations. However, in each of the three groups the observed frequency of the most common allele was greater than expected and intermediate frequency alleles were less common than expected. Clark (1987a) performed a simple goodness of fit test (G-test), similar to a chi-square test, and showed that the test was highly significant for the Caucasian and Hispanic samples. This suggests that Watterson's homozygosity test is not very powerful; that is, it may not detect slight departures from neutrality. One reason is that it condenses the data into a single summary statistic, the expected homozygosity. A goodness of fit test uses the data more efficiently by explicitly considering allele frequencies. Clark (1987a) suggested that Watterson's test may be particularly weak for testing loci with many alleles, but also suggested that the assumptions of the G-test may be violated. We can conclude that these data suggest purifying selection at the *HRAS*-1 locus, but this is not a very strong conclusion.

Tajima's Test

Watterson's test assumes the infinite alleles model and, as suggested above, may be weak when applied to highly polymorphic loci. As DNA sequence data began to accumulate, it became necessary to devise alternative tests, based on the infinite sites model of DNA sequence evolution. One of the simplest tests is Tajima's (1989) test based on different estimates of $\theta\ (= 4Nu)$.

The quantity θ can be estimated in several ways, for example, by solving equation (7.19) to get

$$\theta = \frac{\widetilde{H}}{\widetilde{H} - 1}$$

TABLE 10.5   Observed and expected homozygosities at the *HRAS*-1 locus in three ethnic groups.

Numbers in parentheses are sample size and number of alleles observed in each group. By Watterson's test, the observed and expected homozygosities are not significantly different for any group, but a goodness of fit test is significant for the Caucasian and Hispanic samples (see text). *Source:* From Clark (1987a); original data from Baird et al. (1986).

Group	Observed	Expected	P-Value
Blacks (602, 18)	0.249	0.218	0.261
Caucasians (490, 14)	0.379	0.269	0.127
Hispanics (308, 16)	0.319	0.220	0.114

and using the observed heterozygosity to estimate θ. This is based on the infinite alleles model. Watterson (1975) showed that under the infinite sites model, θ can be estimated from the number of segregating sites, S in a group of DNA sequences:

$$\hat{\theta}_S = \frac{S}{a_1},$$

where a_1 is a constant that depends only on the sample size, n. Similarly, Tajima (1983) showed that θ can also be estimated from π, the nucleotide diversity of a group of sequences.

$$\hat{\theta}_\Pi = \pi L = \Pi$$

where L is the length of the sequences being examined.

Note that $\hat{\theta}_S$ depends only on the number of segregating sites in the sample, and does not consider their frequencies, but $\hat{\theta}_\Pi$ depends on the frequencies of the segregating sites. If natural selection is not acting on the sequences, then the expected value of both of these estimates is θ, the true value of $4N\mu$. However, natural selection affects the two estimates differently.

If deleterious mutations are occurring, there will be more segregating sites than in a neutral sample of the same size. In addition to sites in mutation-drift equilibrium, some sites will be polymorphic because of mutation-selection equilibrium. Therefore, S and $\hat{\theta}_S$ will be greater than expected under strict neutrality. On the other hand, π and $\hat{\theta}_\Pi$ will not be affected much because low frequency variants contribute little to π.

Under balancing selection, π and $\hat{\theta}_\Pi$ will be greater than under neutrality, because there will be more segregating sites at intermediate frequencies. On the other hand, S and $\hat{\theta}_S$ will be little affected.

To summarize, under neutrality we expect $\hat{\theta}_S = \hat{\theta}_\Pi$. Under purifying selection we expect $\hat{\theta}_S > \hat{\theta}_\Pi$, and under balancing selection we expect $\hat{\theta}_S < \hat{\theta}_\Pi$. Tajima (1989) proposed a statistical test to compare the two estimates of θ. His test statistic is

$$D = \frac{\hat{\theta}_\Pi - \hat{\theta}_S}{\sqrt{\hat{V}(\hat{\theta}_\Pi - \hat{\theta}_S)}}$$

Under the null hypothesis of neutrality, the expected value of D is zero. Tajima performed computer simulations and suggested that under neutrality, D has an approximate β-distribution. Based on these simulations, he provided a table of confidence intervals for D with various sample sizes. Under purifying selection (mutation-selection equilibrium) D will be negative, and under balancing selection it will be positive.

Tajima illustrated the test with an example based on the data of Miyashita and Langley (1988). They studied three kinds of polymorphisms: RFLPs, small indels, and large indels, in the *white* locus of *Drosophila melanogaster*. Their results, along with Tajima's D, are summarized in Table 10.6. D was significant only for large indels ($D = -2.0709$, $P < 0.05$), suggesting that large indels are being maintained

TABLE 10.6   Estimates of θ_S, θ_Π, and Tajima's D for three kinds of polymorphisms at the *white* locus of *Drosophila melanogaster*.

The difference is significant only for large indels. *Source:* Table based on Tajima (1989); data from Miyashita and Langley (1988).

Kind of Polymorphism	S	$\hat{\theta}_S$	$\hat{\theta}_\Pi$	D
RFLP	53	11.21	11.92	0.213
Small indels	40	8.46	10.02	0.607
Large indels	15	3.17	0.94	−2.071

by mutation-selection balance. This is consistent with our intuition that large indels are likely to be deleterious.

The HKA Test

The neutral theory makes predictions about both intraspecific polymorphism and interspecific divergence. Both depend on the neutral mutation rate, u, which will vary for different loci, but should be constant for the same locus in different species. A high neutral mutation rate at a locus will lead to high levels of polymorphism, and high levels of divergence. Conversely, a low neutral mutation rate will lead to low levels of polymorphism and divergence. If all polymorphism and divergence are neutral, then loci with high levels of intraspecific polymorphism should show high levels of interspecific divergence. The neutral theory therefore predicts a correlation between levels of intraspecific polymorphism and interspecific divergence. If two loci are both evolving neutrally, the ratio of polymorphism to divergence should be the same in both.

Hudson et al. (1987) devised a test of this hypothesis, commonly called the HKA test (after Hudson, Kreitman, and Aguadé, the authors of the paper). The test requires nucleotide polymorphism data from at least two loci in one species, and data on divergence from another species for the same loci. The polymorphism data are based on the number of segregating sites. Let S_{1A} and S_{2A} be the number of segregating sites at locus 1 and locus 2 in species A. Similarly, let S_{1B} and S_{2B} be the numbers of segregating sites at the same loci in species B. Let L_1 and L_2 be the length, in nucleotides, of locus 1 and locus 2; assume the lengths are the same in both species. Then the proportions of segregating sites at each locus in species A are S_{1A}/L_1 and S_{2A}/L_2. Similarly, the proportions of segregating sites at each locus in species B are S_{1B}/L_1 and S_{2B}/L_2.

Let D_1 and D_2 be the numbers of fixed differences between species A and B at locus 1 and locus 2, respectively. Then the proportions of different sites are D_1/L_1 and D_2/L_2. The HKA test assumes the infinite sites model, so no corrections for multiple hits are necessary.

The HKA test is computationally complex. It is essentially a chi-square test comparing the Ss and the Ds to their expectations. Finding these expectations and the variances requires solving systems of equations. We will not discuss the computational details, but will illustrate the test with an example.

Hudson et al. compared the sequences of the *Adh* locus and the 5′ flanking region (noncoding) upstream from the *Adh* locus in *Drosophila melanogaster* and

D. sechellia. Let locus 1 be the flanking locus and locus 2 be the *Adh* locus, and let species A be *D. melanogaster* and species B be *D. sechellia.* Hudson et al. considered only silent sites at the *Adh* locus. They had no polymorphism data for *D. sechillia,* so the test could be performed only on *D. melanogaster.* Their data are summarized in Table 10.7. (The lengths for polymorphism and divergence are different because introns were considered in the divergence estimates, but not in the polymorphism estimates.) The proportion of fixed differences was about the same at each locus (0.052 and 0.056), but the amount of polymorphism was more than four times higher at the *Adh* locus than at the flanking locus (0.101 vs. 0.022). The ratios (0.42 for the flanking locus vs. 1.82 for *Adh*) are significantly different according to the HKA test. Hudson et al. assume that the flanking locus is neutral since it is noncoding DNA, and interpret their results as indicating selection at the *Adh* locus. They hypothesize that excess polymorphism at the *Adh* locus (rather than a deficiency at the flanking locus) is due to balancing selection. If balancing selection occurs at a site, then polymorphism at closely linked sites will be higher than the neutral expectation because neutral mutations tightly linked to the selected site will accumulate as long as the polymorphism is maintained.

The HKA test makes many assumptions, the most important of which are no multiple hits (infinite sites model), no recombination within loci, free recombination among loci, and constant population sizes. The recombination assumptions are conservative, and make the HKA test less likely to reject neutrality. The assumption of constant population size is more important; increasing or decreasing population sizes can affect the amount of polymorphism in a population. We will discuss this problem below.

TABLE 10.7   Polymorphism at two loci in *Drosophila melanogaster* and divergence from *D. sechellia* at the same two loci.

Only silent sites were considered. See text for meanings of symbols. The lengths for polymorphism and divergence are different because introns were considered in the divergence estimates, but not in the polymorphism estimates. The ratios of polymorphism to divergence were significantly different by the HKA test, interpreted as an excess of polymorphism at the *Adh* locus due to balancing selection. No polymorphism data were available for *D. sechellia. Source:* Data from Hudson et al. (1987).

	Locus 1 (Flanking)	Locus 2 (*Adh*)
Polymorphism	$S_{1A} = 9$	$S_{2A} = 8$
	$L_1 = 414$	$L_2 = 79$
	poly $= 0.022$	poly $= 0.101$
Divergence	$D_1 = 210$	$D_2 = 18$
	$L_1 = 4052$	$L_2 = 324$
	div $= 0.052$	div $= 0.056$
Ratio	poly/div $= 0.419$	poly/div $= 1.823$

The McDonald-Kreitman Test

McDonald and Kreitman (1991) proposed a test that is conceptually similar to the HKA test, but compares silent versus replacement differences at a single locus. Like the HKA test, the McDonald-Kreitman (MK) test assumes that multiple hits do not occur, but it does not assume that population sizes are constant and at equilibrium. According to the neutral theory, intraspecific polymorphism is a transient phase of the substitution process. Therefore, if both polymorphisms and substitutions are neutral, the ratio of replacement to silent polymorphisms within a species should be the same as the ratio of replacement to silent substitutions between two species. This is the null hypothesis of the MK test. Let the numbers of substitutions and polymorphisms be as follows:

	Substitutions	Polymorphisms
Replacement	a	b
Silent	c	d

where a and c are the numbers of fixed differences between species, and b and d are the numbers of sites that are polymorphic within two species. The null hypothesis is

$$H_o: \quad \frac{a}{c} = \frac{b}{d}$$

McDonald and Kreitman suggested a G-test (conceptually similar to a chi-square test) be used to test the null hypothesis.

McDonald and Kreitman illustrated the test by comparing the DNA sequences of the coding region of the Adh locus in three species of $Drosophila$. Table 10.8 summarizes their data. The computational details of the G-test are somewhat complicated, and are explained in Box 10.2. The result is $G = 7.43$, with a P-value of about 0.006. McDonald and Kreitman conclude that there is a significant excess of replacement substitutions between the species. They attribute this to fixation of advantageous mutations. This is consistent with the experiments described in Section 5.6, which have demonstrated biochemical differences between Adh alleles, and that natural selection can act on these differences.

TABLE 10.8 · Numbers of replacement and silent substitutions between species and polymorphisms within species for the coding region of the Adh locus in three species of $Drosophila$.

Source: From McDonald and Kreitman (1991).

Kind of Difference	Substitutions	Polymorphisms	Row Totals
Replacement	7	2	9
Silent	17	42	59
Column totals	24	44	68

Box 10.2   The McDonald-Kreitman test

The McDonald-Kreitman test compares the ratio of replacement to silent substitutions between species to the ratio of replacement to silent polymorphisms within species. We illustrate with the *Adh* data for three *Drosophila* species in Table 10.8.

To perform the *G*-test, we first arrange the data in a table as follows:

	Substitutions	Polymorphisms	Row Totals
Replacement	a	b	R_1
Silent	c	d	R_2
Column totals	C_1	C_2	T

where a and c are the numbers of fixed differences between species, and b and d are the numbers of sites that are polymorphic within species. R_1 and R_2 are the row totals, and C_1 and C_2 are the column totals. T is the grand total $(= R_1 + R_2 = C_1 + C_2)$.

The null hypothesis, under neutrality, is

$$H_o: \quad \frac{a}{c} = \frac{b}{d}$$

We first calculate three preliminary quantities:

$$A = a \ln a + b \ln b + c \ln c + d \ln d$$

$$B = R_1 \ln R_1 + R_2 \ln R_2 + C_1 \ln C_1 + C_2 \ln C_2$$

$$C = T \ln T$$

From these we calculate the unadjusted value of *G*:

$$G_u = 2(A - B + C)$$

Sokal and Rohlf (1995) recommend a correction factor, which is calculated as

$$q = 1 + \frac{\left(\dfrac{T}{R_1} + \dfrac{T}{R_2} - 1\right)\left(\dfrac{T}{C_1} + \dfrac{T}{C_2} - 1\right)}{6T}$$

Finally, we calculate *G* as

$$G = \frac{G_u}{q}$$

This value of *G* is compared to the critical value of the chi-square distribution with one degree of freedom.

For the *Drosophila Adh* data in Table 10.8, we get

$$A = 7 \ln(7) + 2 \ln(2) + 17 \ln(17) + 42 \ln(42)$$
$$= 220.154$$

$$B = 9 \ln(9) + 59 \ln(59) + 24 \ln(24) + 44 \ln(44)$$
$$= 503.127$$

$$C = 68 \ln(68) = 286.927$$

$$G_u = 2(220.154 - 503.127 + 286.927) = 7.908$$

$$q = 1 + \frac{\left(\dfrac{68}{9} + \dfrac{68}{59} - 1\right)\left(\dfrac{68}{24} + \dfrac{68}{44} - 1\right)}{6(68)}$$
$$= 1.064$$

$$G = \frac{7.908}{1.064} = 7.43$$

The critical value at the 0.05 level is 3.84. Therefore, we reject the null hypothesis. It can also be rejected at the 0.01 level (critical value = 6.64).

The MK test has been applied to numerous loci whose sequence is available in more than one line in at least two species. Brookfield and Sharp (1994) reviewed several early applications of the test and concluded that the test rejected neutrality at three of six loci (Figure 10.21). In other words, there is evidence of natural selection at about half of the loci examined. It appears that this is a general pattern: The MK test detects natural selection at some loci, but not others.

Recombination, Hitchhiking, and Background Selection

Until now, we have ignored the fact that nucleotides within a locus are tightly linked to one another. This tight linkage and associated low level of recombination

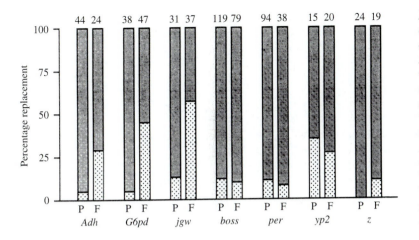

Figure 10.21 Proportions of replacement differences (stippled) and silent differences (shaded) for seven genes in *Drosophila*. P represents polymorphisms within species and F represents fixed differences between species (substitutions). Numbers at the top indicate numbers of variable sites. The McDonald-Kreitman test indicates a significant excess of replacement substitutions in *Adh* (alcohol dehydrogenase), *G6pd* (glucose-6-phosphate dehydrogenase, and *jgw* (jingwei). The test could not be performed on *z* (zeste) because there were no polymorphic replacement substitutions. *Source:* From Brookfield and Sharp (1994).

can affect levels and patterns of DNA variation, especially in the presence of natural selection.

Consider positive selection: If an advantageous mutation goes to fixation by natural selection, neutral sites linked to it on the chromosome may be carried along passively, reducing variation at tightly linked sites. This process is known as **genetic hitchhiking**, or simply hitchhiking. In extreme cases, all nucleotides tightly linked to the selected site are swept to fixation along with the selected site. This is known as a **selective sweep**, and the result is little or no variation at sites tightly linked to the selected site.

Maynard Smith and Haigh (1974) first suggested that hitchhiking can reduce variation at closely linked loci. Kaplan et al. (1989) studied the effect of hitchhiking on DNA variation, and concluded that positive selection at a nucleotide site should reduce neutral variation at tightly linked sites because of hitchhiking. The favored mutation occurs only once, in a single copy of DNA. As this favored mutation goes to fixation, the haplotype in which it initially occurs is carried along, and tightly linked sites will undergo little or no recombination to create variation around the selected site. The result is that, after fixation, nucleotide positions tightly linked to the selected site will have less variation than predicted by the neutral theory. There will be a region of reduced variability around the selected site, and the size of that region will depend on the recombination rate in the region. Regions with low recombination should show relatively large areas of reduced variation.

The first test of this prediction was done by Begun and Aquadro (1992). They compared nucleotide diversity (π) and recombination rates in 20 genes of *Drosophila melanogaster*. They found a highly significant positive relationship (Figure 10.22). They rejected the hypothesis of lower neutral mutation rates in regions of low recombination, and concluded that hitchhiking was the best explanation of their results.

Many studies have since confirmed that the pattern found by Begun and Aquadro occurs in other species, including humans (Nachman et al. 1998). For example, Yan et al. (1998) found decreased RFLP variation in regions surrounding the genes conferring resistance to organophosphate insecticides in populations of

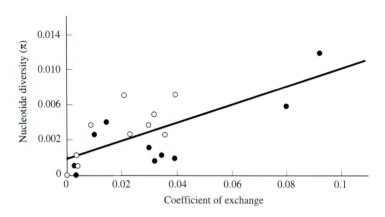

Figure 10.22 **Relationship between nucleotide diversity (π) and recombination rate.** Coefficient of exchange is a measure of recombination rate adjusted to account for differences in recombination rates between autosomal (open circles) and X-linked (solid circles) loci. The line is the best-fit linear regression. The probability that the true slope of the line is zero, given the data, is approximately 0.0007. *Source:* From Begun and Aquadro (1992).

the mosquito *Aedes aegypti* that had recently been subjected to organophosphate treatments. It is clear that, in many species, regions of low recombination frequently show reduced DNA variation. This reduced variation has frequently been attributed to the hitchhiking effect.

Charlesworth et al. (1993) proposed an alternative explanation. If deleterious mutations occur and are eliminated by natural selection, any neutral variation tightly linked to the deleterious site will also be eliminated. The magnitude of the effect should depend on the recombination rate in the region. Charlesworth et al. called this **background selection**, to distinguish it from hitchhiking. Both hitchhiking and background selection will cause reduced variation near a selected site, but the former depends on positive selection and the latter on purifying selection. Two very different processes might produce similar patterns of variation. How can we distinguish between them?

Fay and Wu (2000) proposed a test to distinguish between hitchhiking and background selection. The test depends on the fact that, immediately after a hitchhiking event, neutral variants near the selected site should be present in either high frequency, because they hitchhiked with the selected site, or low frequency, because they have arisen since the fixation. On the other hand, under background selection, neutral variants near the selected site should nearly always be at low frequencies, because they are newly arisen. Therefore, hitchhiking should produce a relative excess of high-frequency neutral variants compared to the frequency expected under neutrality or background selection. Fay and Wu's statistic compares the frequency of intermediate and high-frequency variants in a sample, compared to the prediction under neutrality. The power of the test (ability to reject the null hypothesis when it is false) depends primarily on the recombination rate, the strength of selection, and the time since the hitchhiking event. Fay and Wu apply the test to several published *Drosophila* sequences, and find significant evidence for hitchhiking in two cases. We do not yet know enough about the relative importance of hitchhiking versus background selection, but the initial consensus seems to be that both can contribute to the observed correlation between levels of variation and recombination, but that extreme reductions of variation are probably due mostly to hitchhiking.

Demographic Considerations

Most of the tests we have described assume that the population size is stable. Unfortunately, this assumption is seldom true, and changing population sizes can have dramatic effects on the results of these tests.

In what follows, we will be comparing a sample from a population that has recently changed in size to a sample of the same size and with the same number of alleles (same n and k) from a stable population at neutral equilibrium. It is important to understand that the comparison is between samples with the same n and k.

Consider a population that has recently been through a severe bottleneck. A bottleneck causes a loss of rare alleles and a slow recovery to mutation-drift equilibrium (Section 7.6). If a population has recently been through a bottleneck, a sample will contain relatively more rare alleles than a sample (with the same n and k) from an equilibrium population. This is because most alleles will have arisen since the bottleneck, and will still be at low frequency. There will be an excess of rare alleles, and a corresponding deficiency of intermediate-frequency alleles, compared to the neutral equilibrium expectation. The distribution will be more uneven than expected, a situation that mimics purifying selection. The Ewens-Watterson test can detect a frequency distribution that is more uneven than expected, but it cannot distinguish between purifying selection and a recent bottleneck as causes.

Similarly, $\hat{\pi}$ and $\hat{\theta}_\Pi$ will be reduced after a bottleneck, compared to a sample from a stable population with the same n and k, because rare alleles contribute little to π. However, $\hat{\theta}_S$ depends only on S, n, and k, and not on frequencies, so will not be affected. $\hat{\theta}_S$ will be greater than $\hat{\theta}_\Pi$, and Tajima's D will be negative, again mimicking purifying selection.

Now consider a population that is rapidly increasing in size. Loss of rare alleles will be slower than in a stable population (Section 7.8); therefore, S will be larger in a growing population than in a stable one. But $\hat{\theta}_\Pi$ will be little affected, and therefore D will be negative. Note that a negative D can be caused by either a population bottleneck, a recent expansion, or a recent hitchhiking event (see above). More information is needed to distinguish between the possible causes.

Tests based on comparisons of intraspecific polymorphism versus interspecific divergence can also be affected by changing population size. For example, McDonald and Kreitman (1991) found an excess of replacement substitutions compared to replacement polymorphisms at the *Adh* locus of *Drosophila*. They attributed this to positive selection leading to fixation of adaptive mutations. However, they suggested a possible alternative hypothesis. If past population sizes were small, slightly deleterious mutations may have been effectively neutral and could have become fixed in different populations. If these populations later rapidly expanded in size, then slightly deleterious mutations could be eliminated by purifying selection, and would not appear as polymorphisms. This would lead to an excess of (old) replacement substitutions compared to (new) replacement polymorphisms, the same pattern expected if adaptive mutations had been fixed by positive selection. McDonald and Kreitman were able to reject this hypothesis based on other considerations, but it remains a possibility that must be considered in other studies.

The magnitude, and sometimes the direction, of demographic effects depends on the severity of the bottleneck and how recently it occurred, and on how fast and for how long the population has been expanding or contracting. The essential conclusion is that natural selection and changing population sizes can have similar effects on the statistics used to detect natural selection, and without further information we usually cannot differentiate between these possible causes. This is an especially important problem when considering human variation, because human populations are known to have rapidly increased during the last 10,000 years or more. Kreitman (2000) reviews these issues.

One possible solution is to look at several different kinds of variation. Changing population size should affect all kinds of variation in the same way, but selection may act on one kind of variation but not another. For example, Miyashita and Langley (1988) examined variation in small indels, large indels, and RFLPs. Tajima (1989) applied the *D*-test and obtained a significant (negative) value only for large indels. Because *D* was not significantly different from zero for small indels and RFLPs, Tajima could conclude that the deviation from the neutral expectation in large indels was due to purifying selection, and not to a population bottleneck.

There are more sophisticated ways to differentiate between the effects of natural selection and changing population size on the patterns of DNA variation. These methods require an estimate of the evolutionary relationships among the haplotypes. We will discuss these ideas in Chapter 11. See also Emerson et al. (2001) for a recent review.

Cautions and Prospects

We have described several tests designed to detect natural selection on protein or DNA sequences. These tests take predictions of the neutral theory as the null hypothesis, and many applications of these tests have, in fact, rejected the null hypothesis. But rejection of the null hypothesis is not necessarily the same as demonstrating natural selection. This is because the null hypotheses of most of these tests include many other assumptions in addition to neutrality. Some of these assumptions are as follows:

1. All observed variation is neutral.
2. Populations are diploid and randomly mating with nonoverlapping generations.
3. The neutral mutation rate is constant.
4. Populations are at mutation-drift equilibrium.
5. Effective population size has been constant.
6. There is no gene flow among populations.
7. There is no recombination within a sequence.
8. There is free recombination between different loci.
9. There are no multiple hits (although some tests correct for multiple hits).

Most of the tests described here are based on most of these assumptions. When the null hypothesis is rejected, the tendency is to attribute the rejection to natural selection. However, violation of any of the above assumptions can lead to rejection of the null hypothesis. It can be difficult to differentiate among possible causes when the null hypothesis is rejected. In particular, we have seen that demographic processes (population bottlenecks, rapid population expansion, etc.) can affect levels and patterns of genetic variation, and these patterns can mimic those caused by various kinds of natural selection.

Another issue is that most of these tests are weak. That is, they have little power to detect deviations from the null hypothesis unless these deviations are large. One reason for this is that the tests do not use all of the information contained in the data. They usually are based on summary statistics such as the number of segregating sites, or estimated nucleotide diversity. However, the sequences themselves contain much useful information. For example, the phylogenetic relationships among the sequences can be estimated, and this can help us to understand whether deviation from the null hypothesis is due to selection or to demographic factors. We will discuss these ideas in Chapter 11.

Looking for evidence of selection in DNA sequences is currently one of the most active areas in molecular population genetics and evolution. New data and new statistical tests are published almost monthly. Kreitman (2000) provides a useful table summarizing several characteristics of many statistical tests. For recent reviews and discussions of methods designed to overcome some of the above problems, see Kreitman and Akashi (1995), Wayne and Simonsen (1998), Hey (1999), Otto (2000), Yang and Bielawski (2000), and Emerson et al. (2001). Hughes (1999) describes many examples of adaptive evolution at the DNA and protein level. Ford (2002) summarizes ecological applications of neutrality tests.

Summary

1. The neutral theory of molecular evolution was developed to explain the apparent constancy of amino acid substitution rates and the unexpectedly high levels of genetic variation revealed by allozyme surveys. It proposes that most substitutions among species and polymorphisms within species are the result of interactions between neutral mutations and genetic drift, and are not due to natural selection.

2. The number of observed differences between two amino acid or nucleotide sequences is an underestimate of the number of substitutions that have occurred, because more than one substitution may have occurred at a site. There are various ways to correct for this, making different assumptions about the evolution of amino acid and nucleotide sequences.

3. In DNA coding sequences, replacement substitution rates are lower and more variable than silent substitution rates. This suggests that purifying selection is acting on mutations that cause a change in the amino acid sequence of a protein.

4. Selection that is too weak to detect experimentally may nevertheless be the major factor affecting allele frequencies over evolutionary time.

5. The molecular clock hypothesis predicts that substitution rates for a given protein or DNA molecule should be approximately constant over long periods of time. It appears that rate variation in protein evolution is greater than predicted, but silent substitution rates often appear to be relatively constant.

6. The neutral theory predicts that organisms with shorter generation times should show higher substitution rates per year. This generation time effect is not seen in proteins, but is seen in silent substitutions.

7. The neutral theory predicts higher levels of heterozygosity than are observed in natural populations. The reasons for the inconsistency are not clear.

8. Different molecules, and different parts of a molecule, evolve at different rates, depending on the degree of selective constraint. Pseudogenes and four-fold degenerate sites, which are subject to little or no selection, show highest substitution rates, while nondegenerate sites show the lowest rates. This is as predicted by the neutral theory.

9. In many species, the distributions of allele frequencies are similar in different populations. This is contrary to the prediction of the neutral theory, but gene flow among populations can explain the homogeneity of allele frequencies.

10. The nearly neutral theory claims that there is a large fraction of mutations that are very slightly deleterious. These mutations behave as if they were neutral in small populations, but are selected against in large populations. The nearly neutral theory can explain several inconsistencies between predictions of the neutral theory and observations from natural populations.

11. Codon bias is one possible example of nearly neutral evolution. In several organisms, synonymous codons are used nonrandomly, suggesting that natural selection favors some codons over others. Selection for codon bias must be very weak.

12. Replacement substitution rates are usually lower than silent rates. Exceptions to this generalization often indicate the action of natural selection on protein sequences. Numerous examples are known.

13. The neutral theory predicts both the equilibrium heterozygosity and the equilibrium distribution of allele frequencies under the infinite alleles model or the infinite sites model. Several statistical tests have been developed to test whether the allele frequency distribution of a sample is consistent with the neutral prediction. Most tests are weak and are affected by changing population sizes.

14. The neutral theory predicts a positive correlation between levels of intraspecific polymorphism and rates of interspecific substitution. The HKA test and the MK test provide ways to test this prediction.

15. Neutral variation at sites tightly linked to a selected site may be indirectly affected by natural selection.

16. If a newly arisen advantageous mutation goes to fixation (positive selection), neutral variants linked to the selected site may be carried along passively. The result is reduced variation near the selected site; this is known as genetic hitchhiking.

17. As deleterious mutations are eliminated by natural selection (purifying selection), neutral variation tightly linked to the selected site will also be eliminated. This is known as background selection.

18. Changing population sizes can cause deviations from neutral expectations that are similar to deviations caused by positive or purifying selection, and it can be difficult to distinguish between selective or demographic explanations.

Problems

10.1. The following table gives estimated divergence times between humans and several other mammalian species:

Species	Estimated Divergence Time ($\times 10^6$ years)
Dog	80
Kangaroo	135
Echidna	225
Chicken	290
Newt	360
Carp	410
Shark	450

   a. Plot the observed proportion of amino acid differences (p_{aa}) for α-hemoglobin from the upper half of Table 10.2 versus estimated divergence time.

   b. On the same graph, plot the expected number of substitutions per site (K_{aa}) from the lower half of Table 10.2 versus estimated divergence time.

   c. Draw a linear regression line through each.

   d. Which, if either, looks like a constant rate? Discuss.

10.2. Estimate the range of selection coefficients that are effectively neutral (Kimura's definition) in a population of 100, 1000, 10,000 and 100,000. Which combination of N and s do you think is most reasonable for allozyme variants in humans? In insects? Based on these considerations alone, does it seem reasonable to think that most allozyme variants are effectively neutral in either group?

10.3. Integrate equation (2) in Box 10.1.

10.4. The replacement substitution rate for ribosomal protein S14 is about 0.02×10^{-9} substitutions per site per year (Table 7.1). What can you conclude about selective constraints on this protein?

10.5. In a sample of 50 sequences, five alleles were found, with frequencies of 0.025, 0.025, 0.025, 0.025, and 0.90. What is the expected homozygosity for this sample? According to Watterson's test, is this consistent with the predicted homozygosity under neutrality? (Watterson's estimate 95 percent confidence interval for $n = 50$ and $k = 5$ is about 0.25 to 0.78.)

   a. Interpret with respect to natural selection.

   b. Interpret with respect to changes in population size.

10.6. Perform a heterogeneity chi-square test (Section 9.2, Box 9.6) on the data in Table 10.8. Compare the P-values for the chi-square and the G-test in Box 10.2. (You can use the CHIDIST function of Excel to get the P-values.) Why are they different? Which test is more conservative? See Sokal and Rohlf (1995) for a good discussion of the G-test and other goodness of fit tests.

10.7. The following table gives the numbers of substitutions between *Arabidopsis thaliana* and *A. gemmifera* at the *PgiC* locus, and the numbers of polymorphisms within *A. thaliana* (Kawabe et al. 2000):

	Substitutions	Polymorphisms
Replacement	34	8
Silent	4	4

Does the McDonald-Kreitman test give evidence of natural selection at this locus? Discuss.

10.8. Long and Langley (1993) compared sequences of part of the *Adh* locus and *jgw*, a purported pseudogene derived from *Adh*, in *Drosophila yakuba* and *D. tessieri*. Their results for silent differences for can be summarized as follows:

	Adh	*jgw*
Polymorphic sites (*D. yak.*)	18	19
Substitutions (*yak.-tess.*)	11	23

Sequence lengths were the same in all comparisons. The HKA test was significant. Interpret these results with respect to possible natural selection at the *jgw* locus.

10.9. Long and Langley also compared silent and replacement differences at the *jgw* locus. Their results are

	Substitutions	Polymorphisms
Replacement	21	4
Silent	16	27

Are these results consistent with your answer to the previous problem? Do you think *jgw* is a pseudogene? Discuss.

10.10. The 5′ untranslated areas upstream from coding sequences often show moderate selective constraints (Figure 10.10). Discuss possible reasons for this.

CHAPTER 11

Molecular Evolution and Phylogenetics

The affinities of all the beings of the same class have sometimes been represented by a great tree. I believe this simile largely speaks the truth. The green and budding twigs may represent existing species; and those produced during each former year may represent the long succession of extinct species. ... The limbs divided into great branches, and these into lesser and lesser branches, were themselves once, when the tree was small, budding twigs; and this connexion of the former and present buds by ramifying branches may well represent the classification of all extinct and living species in groups subordinate to groups.

—*Charles Darwin (1859)*

Anyone who has studied evolution, even superficially, has seen numerous trees suggesting evolutionary relationships among species or other taxa. Historically, these trees have been based on morphological comparisons; however, modern molecular techniques allow us to infer evolutionary relationships based on comparative protein and DNA sequence data. These trees are often based on extensive sequence data and many evolutionary biologists think they are more accurate than trees based on morphological comparisons.

In the previous chapter, we saw how to use the predictions of the neutral theory to help us interpret patterns of protein and DNA sequence variation within and among species. Here, we will see how we can use the information in these sequences to infer evolutionary relationships among groups. First, we consider ways to estimate evolutionary distances between groups and how to estimate rates of molecular evolution. Then we introduce various ways to draw evolutionary trees based on molecular data, and how to evaluate them. Finally, we consider how to use the information contained within polymorphic sequences to make inferences about the evolutionary and demographic history of a population.

This chapter provides only a general introduction to molecular evolution. We will gloss over most of the technical and computational details. For more detailed treatments, the standard texts on molecular evolution are Li (1997) and Nei and Kumar (2000).

11.1 Amino Acid and Nucleotide Substitutions

Substitution numbers and rates are fundamental to many areas of molecular population genetics and evolution. By comparing substitution rates of homologous proteins in different organisms, we can test hypotheses about the constancy of substitution rates, and sometimes detect the action of natural selection. Comparison of substitution rates in different kinds of sequences can provide information about selective constraints or lack thereof. By comparing rates at different positions within a gene, we can sometimes determine which regions of the gene are subject to natural selection and which appear to be evolving neutrally. The estimated numbers of substitutions between a group of species, or populations, can be used to estimate the phylogenetic history of the group. Sometimes this information can be used to estimate divergence times between species or populations. All of these applications depend on estimates of the numbers of substitutions between groups.

We noted in the previous chapter that the observed number of differences between two amino acid or nucleotide sequences is an underestimate of the number of substitutions that have occurred since their common ancestor, because more than one substitution may have occurred at a given site (Table 10.1). In this section, we address the problem of how to estimate the expected number of substitutions from the observed number of differences, and how to estimate substitution rates, provided we have an independent estimate of divergence time.

Let p (with appropriate subscript) represent the observed proportion of differences between two sequences. It can be interpreted as the observed number of differences per site. Let K (with appropriate subscript) be the expected number of substitutions per site, based on some model of amino acid or nucleotide substitution. Note that both p and K are measured *per site*, to account for possible differences in lengths among sequences. In general, the expected number of substitution, per site, K, is some function of the observed number of differences per site, p,

$$K = f(p)$$

We wish to find the form of the function, which depends on whether we are considering amino acid sequences, noncoding DNA sequences, or coding DNA sequences, and on the model of amino acid or nucleotide substitution assumed.

Amino Acid Substitutions

First, consider amino acid substitutions. If n_{aa} is the number of amino acids in the sequence being compared, and d_{aa} is the number of different amino acids, then the observed proportion of differences (equivalently, the observed number of differences per site) is

$$p_{aa} = \frac{d_{aa}}{n_{aa}}$$

Let K_{aa} be the expected number of substitutions per site since the common ancestor of the two sequences. We wish to find the relationship between K_{aa} and p_{aa}. We assume that all amino acid substitutions are equally likely to occur, and that they are selectively neutral.

A substitution is a rare event with a low probability. The probability distribution of the number of substitutions since the common ancestor can be approximated by the Poisson distribution (Appendix A). If X is a random variable indicating the number of substitutions per site since the common ancestor, and x is its realized value, then, by the Poisson distribution

$$\Pr(X = x) = \frac{e^{-\lambda}\lambda^x}{x!}$$

where λ is the mean (expected value) of X. But if X is the number of substitutions, its expected value is K_{aa}, and we can substitute K_{aa} for λ to write

$$\Pr(X = x) = \frac{e^{-K_{aa}}K_{aa}^x}{x!}$$

The probability that zero substitutions will occur is

$$\Pr(X = 0) = \frac{e^{-K_{aa}}K_{aa}^0}{0!} = e^{-K_{aa}}$$

and the probability that one or more substitutions will occur is

$$\Pr(X \geq 1) = 1 - \Pr(X = 0) = 1 - e^{-K_{aa}}$$

We can estimate $\Pr(X \geq 1)$ by p_{aa}, the observed proportion of sites at which at least one substitution has occurred. So we have

$$p_{aa} = 1 - e^{-K_{aa}}$$

We can solve this equation for K_{aa} to get an estimate of K_{aa}

$$\hat{K}_{aa} = -\ln(1 - \hat{p}_{aa}) \tag{11.1}$$

where the $^\wedge$ indicates an estimate.

For example, humans and cows differ at 17 of 141 amino acids in the α-hemoglobin protein. Therefore, $p_{aa} = 17/141 = 0.121$ and $K_{aa} = -\ln(1 - 0.121) = 0.128$ substitutions per amino acid site since the common ancestor.

Figure 11.1 shows the relationship between K_{aa} and p_{aa}. For small values of p_{aa}, K_{aa} and p_{aa} are approximately equal; for larger values of p_{aa}, K_{aa} can be substantially larger than p_{aa}. This is what we expect if substitutions are random and the probability of multiple substitutions at a site increases with time.

We usually express substitution rates as the number of substitutions per site per year. Let k_{aa} be the amino acid substitution rate. Note that K_{aa} (uppercase) is the *number* of amino acid substitutions per site since divergence, and k_{aa} (lowercase) is the *rate* of substitutions per site per year. Let T be the number of years since the common ancestor. Then, since divergence, each line has undergone T years of evolution, for a total of $2T$ years of evolution. The substitution rate per site per year is then

$$\hat{k}_{aa} = \frac{\hat{K}_{aa}}{2T} \tag{11.2}$$

Figure 11.1 **Expected number of amino acid substitutions (K_{aa}) versus observed number of differences (p_{aa}).**   The expected number of substitutions is estimated from equation (11.1). The dashed line is the line $K_{aa} = p_{aa}$. K_{aa} is greater than p_{aa}, and the difference increases with increasing p_{aa}.

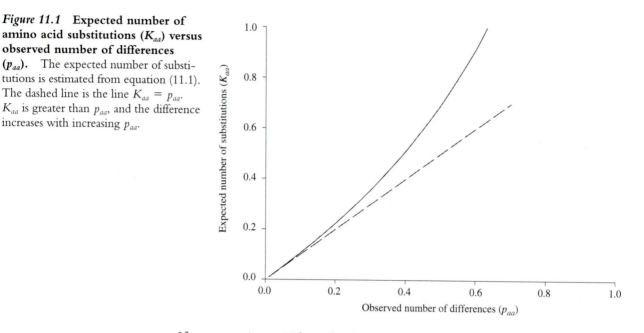

If we can estimate T from the fossil record (or some other independent source), we can estimate k_{aa}. In the above example, humans and cows diverged about 80 million years ago. Therefore,

$$\hat{k}_{aa} = \frac{0.128}{2(8 \times 10^7)} = 8 \times 10^{-10} \text{ substitutions per site per year}$$

For α-hemoglobin, estimates of k_{aa} are relatively constant for different pairs of species. For example, humans and kangaroos differ at 26 sites. This gives $p_{aa} = 0.184$, and $K_{aa} = 0.204$. The divergence time is about 135 million years ago, giving $k_{aa} \cong 7.5 \times 10^{-10}$ substitutions per site per year. Early observations of this constancy led to the molecular clock hypothesis (Section 10.2).

Estimated substitution rates vary greatly for different proteins. Several examples were given in Table 10.3.

Nucleotide Substitutions: Noncoding Sequences

Estimating the number of nucleotide substitutions is more complex than estimating amino acid substitutions for at least two reasons: First, back substitutions are more likely to occur. The probability that a second amino acid substitution at a site will restore the original amino acid is only 1/19, but the probability that a second nucleotide substitution will restore the original is 1/3, a fraction too high to ignore. Second, nucleotides within a protein-coding sequence are expected to evolve at different rates. Silent mutations are much more common at third codon positions than at first or second positions; therefore, assuming that most replacement mutations are eliminated by natural selection, we expect substitution rates per site to be higher at third codon positions. This means that we must consider noncoding and coding sequences separately, and within coding sequences we must consider silent and replacement substitutions separately. We consider the simpler case of noncoding sequences first.

Table 11.1a shows a generic substitution matrix. It shows the probability of any nucleotide changing to any other nucleotide. For example, α_{AG} is the probability

---

TABLE 11.1   Substitution matrices for nucleotide substitutions.
(a) Generic substitution matrix; each of the sixteen possible substitutions has its own rate. (b) Jukes-Cantor one parameter model; all substitutions are equally likely. (c) Kimura's two parameter model; the probability of a transition substitution is α and the probability of a transversion substitution is β.

a. Generic Model

	A	T	G	C
A	α_{AA}	α_{AT}	α_{AG}	α_{AC}
T	α_{TA}	α_{TT}	α_{TG}	α_{TC}
G	α_{GA}	α_{GT}	α_{GG}	α_{GC}
C	α_{CA}	α_{CT}	α_{CG}	α_{CC}

b. One-parameter Model

	A	T	G	C
A	$1 - 3\alpha$	α	α	α
T	α	$1 - 3\alpha$	α	α
G	α	α	$1 - 3\alpha$	α
C	α	α	α	$1 - 3\alpha$

c. Two-parameter Model

	A	T	G	C
A	$1 - \alpha - 2\beta$	β	α	β
T	β	$1 - \alpha - 2\beta$	β	α
G	α	β	$1 - \alpha - 2\beta$	β
C	β	α	β	$1 - \alpha - 2\beta$

---

that A will change to G in some time period, and α_{AA} is the probability that A will still be A. Note that $\alpha_{AA} = 1 - (\alpha_{AT} + \alpha_{AG} + \alpha_{AC})$; that is, the probability that a nucleotide will not change is one minus the probability that it will change to some other nucleotide.

We can make a number of assumptions about the structure of the substitution matrix. The simplest is that the probability of any nucleotide changing to any other nucleotide is the same for all nucleotides; that is, $\alpha_{AG} = \alpha_{AT} = \alpha_{GC} =$ etc. (Table 11.1b). A more realistic assumption incorporates the observation that transitions are more common than transversions. We can specify one rate for transitions (α_{AG}, α_{GA}, α_{TC}, and α_{CT}) and a different (lower) rate for transversions (all other changes) (Table 11.1c). More complex assumptions can also be made, including a different rate for each possible change, and different rates for different nucleotide positions within a sequence.

The simplest assumptions are that the substitution rate is the same at all sites and the probability of any nucleotide changing to any other nucleotide is the same for all nucleotides; this probability is designated by α. Table 11.1b shows the substitution matrix for these assumptions. This is the Jukes-Cantor model of nucleotide substitution (Jukes and Cantor 1969).

Our goal is to find an expression for the expected number of substitutions per site, K, as a function of the observed number of differences per site, p. We approach the problem by deriving a recursion equation for p and analyzing it. Let p_t be the observed proportion of nucleotide sites that are different between two sequences at time t, and let $q_t (=1 - p_t)$ be the observed proportion of sites that are the same. To simplify notation, let $r = 3\alpha$ be the total substitution rate, that is, the probability that a nucleotide will change to another in a given time period.

The probability that two nucleotides will be the same in generation $t + 1$, given that they are the same in generation t is equal to the probability that neither changes:

$$\Pr(\text{same in } t + 1 | \text{same in } t) = (1 - r)^2 = 1 - 2r + r^2$$

Since r is small, terms containing r^2 are very small, and can be neglected. This gives

$$\Pr(\text{same in } t + 1 | \text{same in } t) \cong 1 - 2r$$

Similarly, the probability that two nucleotides will be the same in generation $t + 1$ given that they are different in generation t is the probability that one does not change $(1 - r)$ times the probability that the other does change to the same nucleotide $(r/3)$. There are two ways this can happen (the first changes and the second does not, or vice versa). Assuming both ways are equally probable, this gives

$$\Pr(\text{same in } t + 1 | \text{different in } t) = 2\left(\frac{r}{3}\right)(1 - r)$$

which, ignoring terms with r^2, simplifies to

$$\Pr(\text{same in } t + 1 | \text{different in } t) \cong \frac{2r}{3}$$

Putting these together, and using the law of total probability, we get

$$\Pr(\text{same in } t + 1) = \Pr(\text{same in } t + 1 | \text{same in } t) \times \Pr(\text{same in } t)$$
$$+ \Pr(\text{same in } t + 1 | \text{different in } t) \times \Pr(\text{different in } t)$$

or,

$$q_{t+1} \cong (1 - 2r)q_t + \frac{2r}{3}(1 - q_t)$$

$$\cong \left(1 - \frac{8r}{3}\right)q_t + \frac{2r}{3}$$

Using the techniques in Box 9.7, this linear recursion equation can be solved to give

$$q_t \cong \frac{1}{4} + \left(1 - \frac{8r}{3}\right)^t\left(q_o - \frac{1}{4}\right)$$

But the two sequences are initially identical, so $q_o = 1$. This gives

$$q_t \cong \frac{1}{4} + \frac{3}{4}\left(1 - \frac{8r}{3}\right)^t$$

Now use the approximation $(1 - 8r/3) \cong e^{-8r/3}$ to get

$$q_t \cong \frac{1}{4} + \frac{3}{4}e^{-8rt/3} \qquad \textbf{(11.3)}$$

Note that if r is the substitution rate per site, and t is the number of generations since divergence, then $2rt$ is the expected number of substitutions per site since divergence, or K. We can rewrite equation (11.3) as

$$q_t \cong \frac{1}{4} + \frac{3}{4}e^{-4K/3} \qquad \textbf{(11.4)}$$

We want K in terms of p, not q. But, since $p = 1 - q$, we can rewrite equation (11.4) as (after some simplification)

$$p = \frac{3}{4}(1 - e^{-4K/3})$$

Note, we have dropped the subscript t. Finally, solve for K, to get

$$K = -\frac{3}{4}\ln\left(1 - \frac{4}{3}p\right) \qquad \textbf{(11.5)}$$

This gives the expected number of nucleotide substitutions, K, in terms of the observed number of nucleotide differences, p, for any time, t. We can estimate K by

$$\hat{K} = -\frac{3}{4}\ln\left(1 - \frac{4}{3}\hat{p}\right) \qquad \textbf{(11.6)}$$

where the hats indicate estimates from the observed sequence data.

As with amino acid substitutions, if we know the time since the common ancestor, we can express the *rate* of nucleotide substitution as

$$k = \frac{K}{2T}$$

where T is the number of years since divergence from the common ancestor.

This Jukes–Cantor model assumes that any nucleotide has the same probability of changing to any other nucleotide. This is inconsistent with known facts; we know that transitions occur more frequently than transversions. We can account for this by specifying one rate (α) for transitions, and a different rate (β) for transversions (Table 11.1c). This is commonly known as the Kimura two-parameter model, after Kimura (1980). More complex models can also be specified. See Li (1997) and Nei and Kumar (2000) for more information.

Nucleotide Substitutions: Coding Sequences

Any nucleotide in the coding sequence of a gene can change to any of three other nucleotides. Some of those will change the amino acid sequence of the protein and some will not. We have defined a silent (synonymous) change as one that does not change the amino acid sequence of the protein, and a replacement (nonsynonymous) change as one that does. We must consider silent and replacement substitutions separately, because their rates are likely to be different. Replacement mutations are more likely to be affected by natural selection (usually negatively), and thus substitution rates are likely to be different (usually lower).

We define a **synonymous site** as one at which any change is silent, and a **nonsynonymous site** as one at which any change is a replacement change. As we shall see below, some changes at a site may be silent and others may be replacement changes. When considering substitutions in coding sequences, we consider silent substitutions per synonymous site, and replacement substitutions per nonsynonymous site.

The observed proportion of silent differences among synonymous sites is

$$p_S = \frac{s}{n_S} \tag{11.7}$$

where s is the observed number of silent differences, and n_S is the estimated number of synonymous sites (to be explained below). Think of p_S as the observed number of silent differences per synonymous site. Similarly, the number of replacement differences per nonsynonymous site is found by

$$p_N = \frac{r}{n_N} \tag{11.8}$$

where r and n_N are the observed number of replacement differences and the estimated number of nonsynonymous sites, respectively. It is the observed number of replacement differences per nonsynonymous site.

First, consider the numerators of equations (11.3) and (11.4). It is fairly straightforward to determine whether a difference is silent or not. If there is only one difference within a codon, and that difference changes the amino acid, it is a replacement difference; if it does not change the amino acid, it is silent. If there are two (or three) differences within a codon, it is somewhat more complicated, because you do not know in what order the changes occurred. For example assume the homologous codons in two sequences are ACG (thr) and CGG (arg). Two changes have occurred. One possibility is

ACG (thr) → CCG (pro) → CGG (arg)

Here, both changes are replacement. The other possibility is

ACG (thr) → AGG (arg) → CGG (arg)

in which case the first change is replacement and the second is silent. One way to estimate the number of silent changes is to average the numbers from the two possibilities. In this example, the average of 0 and 1 is 0.5 silent changes (and therefore 1.5 replacement changes). This assumes that both pathways are equally likely; it is possible to weight the alternatives in various other ways. If all three positions are different in homologous codons, more possibilities must be considered, but the principle is the same. In this way, the number of silent differences, s, and the number of replacement differences, r, can be counted.

We must now consider how to estimate the numbers of synonymous and nonsynonymous sites in a sequence. The problem is that some changes at a site may be silent, while others may be replacement changes. We will consider a simple method proposed by Nei and Gojobori (1986). A **fourfold degenerate site** is one at which all three possible changes are silent. For example, CCU, CCC, CCA, and CCG all code for proline. The third position of these codons is a fourfold degenerate site. A **threefold degenerate site** is one at which two of the three possible changes are silent. AUU, AUC, and AUA all code for isoleucine. The third position of these codons is the only threefold degenerate site in the standard genetic code. It is usually considered twofold degenerate for convenience. A **twofold degenerate site** is one at which one of the three possible changes is silent. For example, AAA and AAG both code for lysine. The third position of these codons is a twofold degenerate site. A **nondegenerate site** is one at which any change will be a replacement change. For example, CCC codes for proline; the first position is nondegenerate, since any change will specify some other amino acid (UCC = serine, ACC = threonine, GCC = alanine).

Given the coding sequence of a gene and a table of the genetic code, it is possible to determine whether each site is twofold degenerate, fourfold degenerate, or nondegenerate. Table 11.2 makes this job a bit easier; various computer programs make it even easier. Let

n_o = the number of nondegenerate sites in a coding sequence

n_2 = the number of twofold degenerate sites

n_4 = the number of fourfold degenerate sites.

All of the possible changes at fourfold degenerate sites will be synonymous, and 1/3 of the possible changes at twofold degenerate sites will be synonymous. In other words, fourfold degenerate sites count as one synonymous site, and twofold degenerate sites count as 1/3 synonymous site. The number of synonymous sites, n_S, is then

$$n_S = n_4 + \frac{1}{3}n_2$$

Similarly, the number of nonsynonymous sites, n_N, is

$$n_N = n_o + \frac{2}{3}n_2 = n_T - n_S$$

where n_T is the total number of sites in the sequence. The fraction 2/3 is because 2/3 of all changes at twofold degenerate sites will be replacement changes.

When comparing two sequences, the number of synonymous sites is frequently different (and therefore, the number of nonsynonymous sites is also different). When this occurs, we consider the average number of synonymous and nonsynonymous sites.

We can now use equations (11.7) and (11.8) to estimate the number of silent differences per synonymous site and the number of replacement differences per nonsynonymous site. Note that p_S and p_N are the numbers of *differences* per site, and not the number of substitutions. In order to convert p_S and p_N to numbers of substitutions, we must use some substitution model. For example, using the Jukes-Cantor model, the expected number of silent substitutions per synonymous site is

$$K_S = -\frac{3}{4}\ln\left(1 - \frac{4}{3}p_S\right) \qquad \textbf{(11.9)}$$

and the expected number of replacement substitutions per nonsynonymous site is

$$K_N = K_A = -\frac{3}{4}\ln\left(1 - \frac{4}{3}p_N\right) \qquad \textbf{(11.10)}$$

The subscript A (presumably for amino acid substitutions) is usually used instead of the more consistent N. (Sometimes d_S and d_N are used instead of K_S and K_A;

TABLE 11.2   Degeneracy in the DNA genetic code.

Numbers above each codon indicate whether that position is nondegenerate (0), twofold degenerate (2), threefold degenerate (3), or fourfold degenerate (4). See text for details.

		T		C		A		G	
		002		004		002		002	
		TTT	phe	TCT	ser	TAT	tyr	TGT	cys	T
		002		004		002		002	
T		TTC	phe	TCC	ser	TAC	tyr	TGC	cys	C
		202		004		xxx		xxx	
		TTA	leu	TCA	ser	TAA	(STOP)	TGA	(STOP)	A
		202		004		xxx		000	
		TTG	leu	TCG	ser	TAG	(STOP)	TGG	trp	G
		004		004		002		004	
		CTT	leu	CCT	pro	CAT	his	CGT	arg	T
		004		004		002		004	
C		CTC	leu	CCC	pro	CAC	his	CGC	arg	C
		204		004		002		004	
		CTA	leu	CCA	pro	CAA	gln	CGA	arg	A
		204		004		002		004	
		CTG	leu	CCG	pro	CAG	gln	CGG	arg	G
		003		004		002		002	
		ATT	ile	ACT	thr	AAT	asn	AGT	ser	T
		003		004		002		002	
A		ATC	ile	ACC	thr	AAC	asn	AGC	ser	C
		003		004		002		002	
		ATA	ile	ACA	thr	AAA	lys	AGA	arg	A
		000		004		002		002	
		ATG	met	ACG	thr	AAG	lys	AGG	arg	G
		004		004		002		004	
		GTT	val	GCT	ala	GAT	asp	GGT	gly	T
		004		004		002		004	
G		GTC	val	GCC	ala	GAC	asp	GGC	gly	C
		004		004		002		004	
		GTA	val	GCA	ala	GAA	glu	GGA	gly	A
		004		004		002		004	
		GTG	val	GCG	ala	GAG	glu	GGG	gly	G

e.g. Nei and Kumar 2000.) Other, more realistic substitution models can be used. See Li (1997) for more information.

We illustrate with an example. The following are the first 30 nucleotides from two different alleles of a hypothetical gene. Silent differences are marked by S and replacement differences by R. The 0, 2, or 4 above each site indicates whether it is nondegenerate, twofold degenerate, or fourfold degenerate, as determined from Table 11.2.

Sequence 1:	004	002	202	004	002	004	004	003	002	004
	ACG	TAT	TTG	CGC	AAA	GGC	GCT	ATT	TGT	ACA
	thr	tyr	leu	arg	lys	gly	ala	ile	cys	thr
		S		R			R	S		S
Sequence 2:	004	002	202	002	002	004	004	003	002	004
	ACG	TAC	TTG	CAC	AAA	GGC	ACT	ATC	TGT	ACG
	thr	tyr	leu	his	lys	gly	thr	ile	cys	thr

For the first sequence, we have (counting the threefold degenerate site as twofold),

$$n_o = 19$$

$$n_2 = 6$$

$$n_4 = 5$$

The number of synonymous sites in the first sequence is

$$n_S = n_4 + \frac{1}{3}n_2 = 5 + \frac{6}{3} = 7.0$$

and the number of nonsynonymous sites is

$$n_N = n_o + \frac{2}{3}n_2 = 19 + \frac{2}{3}(6) = 23.0$$

For the second sequence, we count $n_o = 19$, $n_2 = 7$, and $n_4 = 4$, which gives $n_S = 6.33$ and $n_N = 23.67$. The averages for the two sequences are therefore

$$n_S = 6.67$$

$$n_N = 23.33$$

There are three silent differences and two replacement differences. Therefore,

$$p_S = \frac{3}{6.67} = 0.450$$

and

$$p_N = \frac{2}{23.33} = 0.086$$

To estimate the numbers of substitutions, we use equations (11.9) and (11.10) to get

$$K_S = -\frac{3}{4}\ln\left(1 - \frac{4}{3}(0.450)\right) = 0.687$$

and

$$K_A = -\frac{3}{4}\ln\left(1 - \frac{4}{3}(0.086)\right) = 0.091$$

As with noncoding sequences, if we have an estimate of T, the time since divergence of two sequences, we can estimate substitution rates. For silent substitutions,

$$k_S = \frac{K_S}{2T} \tag{11.11}$$

and, for replacement substitutions,

$$k_A = \frac{K_A}{2T} \tag{11.12}$$

We discussed silent and replacement substitution rates in Sections 7.5 and 10.4. Replacement mutations are likely to affect the function of a protein, usually deleteriously, and therefore most will probably be eliminated by natural selection. On the other hand, silent mutations are less likely to be affected by natural selection (but see Section 10.3) and are more likely to drift to fixation. Therefore, we expect silent substitution rates to be higher than replacement rates. That is exactly what we usually find. Table 7.1 summarizes rates for many proteins; the average silent substitution rate is about 3.51×10^{-9} substitutions per site per year, compared to the replacement rate of only 0.74×10^{-9}.

There are exceptions to this generalization. Sometimes we find a protein, or part of a protein, in which replacement rates are higher than silent rates. This typically indicates natural selection favoring some change in the protein, as discussed in Section 10.4.

11.2 Molecular Phylogenetics

Darwin was one of the first biologists to suggest that the evolutionary relationships between all living and extinct organisms can be visualized as an evolutionary tree. Phylogenetic analysis attempts to reconstruct the evolutionary history of a group of taxa—species, genera, families, and so forth. Results are typically summarized in a phylogenetic tree; any book on evolution is likely to contain many such trees. These trees have traditionally been based on morphological characteristics, and can be constructed using a variety of techniques and assumptions.

With the availability of comparative DNA sequences, it has become possible to reconstruct the evolutionary history of a group of homologous sequences. These sequences can come from the same species or from different species. As with traditional phylogenetic analysis, results are frequently summarized in tree form. Such trees are called **gene trees**, and represent the evolutionary relationships among the haplotypes (different sequences) studied.[1] These gene trees may or may not represent the evolutionary history of the groups from which the sequences came (see below).

Gene trees can be useful for studying a number of evolutionary questions concerning mutation rates, gene flow, natural selection, and so forth. In this section, we introduce some of the general principles of drawing and analyzing gene trees, and consider a few examples of their use. We will not go into the computational details, which are complex and handled by specialized computer programs. For more detailed references, see Li (1997), Nei and Kumar (2000), and Hall (2001). The latter is a particularly clear elementary introduction. Felsenstein (2003) provides a thorough treatment of phylogenetic analysis.

When constructing and analyzing gene trees, we must consider three basic questions:

---

[1]Trees can also be based on amino acid sequences of proteins, but we will consider only gene trees.

1. How do we use the sequence data to construct a tree?

2. What criteria do we use to determine which of various alternative trees is best?

3. How confident are we that our best tree is actually the true tree?

We will consider each of these questions, but first we introduce a few basic concepts and definitions.

Basic Concepts

Figure 11.2 illustrates a simple gene tree; for now, consider only the tree in Figure 11.2a. There are three basic parts to the tree: the end points, internal nodes, and the branch lengths.

The end points, h_1 through h_4, represent the haplotypes (different sequences) studied. End points are sometimes called **external nodes**. These sequences may come from different taxonomic groups (species, genera, etc.), different populations of the same species, or different individuals within the same population. Because phylogenetic trees can be constructed based on any of these, we call these external nodes **operational taxonomic units** (OTUs). **Internal nodes** are the

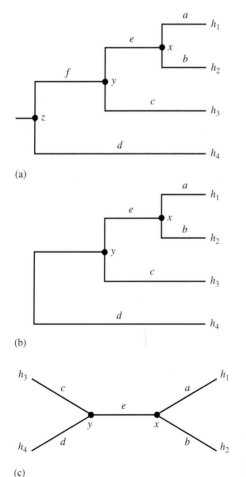

(a)

(b)

(c)

Figure 11.2 **Three ways to draw a gene tree.** (a) A rooted tree; (b) and (c) are identical unrooted trees. Sampled sequences (haplotypes) are designated $h_1 - h_4$; internal nodes are x, y, and z; branch lengths are designated $a-e$.

branch points in the tree. In Figure 11.2a, these nodes are indicated by x, y, and z. We usually do not know the sequences at the internal nodes, but sometimes they can be inferred. **Branch lengths** are some measure of the evolutionary distance between an internal node and an external node, or between two internal nodes. In Figure 11.2a, the quantities a through f represent branch lengths.

Figure 11.2a is a **rooted tree**. A rooted tree is one that shows relationships through time. Rooted trees are usually drawn with time progressing from left to right; traditional phylogenetic trees are usually rooted. In rooted trees, internal nodes represent the **most recent common ancestor** (MRCA) of all haplotypes connected to that node and to the right of it (assuming time goes from left to right). For example, x is the MRCA of h_1 and h_2; y is the MRCA of h_1, h_2, and h_3. The root of the tree (z in the figure) is the most recent common ancestor of all taxa in the tree.

An **unrooted tree**, on the other hand, shows only the relationships among the haplotypes, and does not imply direction to the evolutionary changes. Trees based on molecular data are commonly drawn as unrooted trees. However, unrooted trees are sometimes drawn to look as if they are rooted. Figure 11.2b and c show unrooted trees for the same haplotypes as Figure 11.2a. Figure 11.2b may look like a rooted tree, but there is not a node between h_4 and node y. The tree in Figure 11.2c is more obviously unrooted, and unrooted trees should generally be drawn this way.

There are several possible ways to root a tree. If, for example, we know that h_4 in Figure 11.2 is more distantly related to h_1, h_2, and h_3 than any of the latter are to each other, then we can establish the root somewhere along the line from h_4 to node y. Figure 11.2a shows this rooted version of the tree. Gene trees are often rooted by including a taxon or sequence that is known to be distantly related to the others. Such a group is called an **outgroup**. For example, a tree of several species of primates can be rooted by including a rodent as an outgroup.

Drawing a gene tree consists of determining the branching order, or topology, of the tree, and estimating the branch lengths. There are many ways to do this, making various assumptions about the evolutionary processes leading to the haplotypes examined.

The fundamental assumption behind gene trees is that haplotypes that are very similar in sequence share a more recent common ancestor than haplotypes that are less similar. Figure 11.3 illustrates a tree that violates this assumption. In the line leading to h_1, two substitutions have occurred at the fifth site, $A \rightarrow T$ and $T \rightarrow A$, so that the original nucleotide is restored. These are marked F (forward) and B (back) in the figure. In addition, the same substitution has occurred at the first site in two different lineages. These are parallel substitutions, marked P in the figure. Because of back and parallel substitutions, h_1 and h_5 have identical sequences, even though they are separated by four substitutions and are the most distantly related sequences in the tree. Similarity due to causes other than common ancestry, for example, parallel or back substitutions, is called **homoplasy**. The likelihood of homoplasy increases with large data sets and for distantly related groups, in which the number of substitutions is large. If homoplasy is common, it may be impossible to infer the true tree with any confidence.

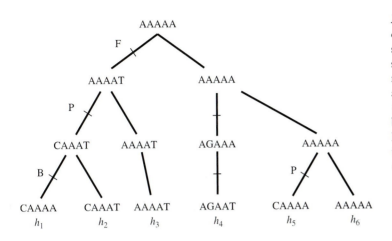

Figure 11.3 **An example of how homoplasy can cause distantly related haplotypes to have similar sequences.** Substitutions are marked by short bars across the lines of descent. Substitutions marked *F* and *B* are forward and back substitutions at the same site. Substitutions marked *P* are parallel (identical) substitutions at the same site. Because of back and parallel substitutions, haplotypes h_1 and h_5 have identical sequences, even though four substitutions have occurred in the lines of descent from their common ancestor.

Gene Trees Versus Species Trees

We can think of two different kinds of phylogenetic trees. Species trees show the evolutionary history of a group of related species; for example, a tree might show the relationships among all the species in a genus. We can also draw trees showing relationships among families, populations, or other groups. The essential idea is that we are showing the evolutionary relationships among (formal or informal) taxonomic groups.

Gene trees, on the other hand, show only the evolutionary relationships among different DNA sequences. If we are comparing sequences from different species, the gene tree may or may not correspond to the species tree. The main reason for lack of correspondence is DNA polymorphism at the time of speciation. If a population is polymorphic at the time speciation begins, subsequent fixations may cause some haplotypes in different species to be more closely related than haplotypes in the same species. Figure 11.4 shows an example. The shaded area represents the topology of a species tree, with two speciation events. The

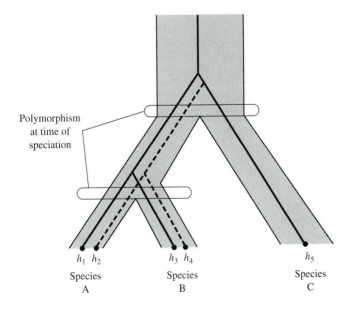

Figure 11.4 **An example of how polymorphism at the time of speciation can cause a gene tree to differ from a species tree.** The broad shaded bars are the species tree; solid and dashed lines are the gene tree. Haplotypes h_1 and h_2 are both in species A, even though h_1 is more closely related to h_3 in species B than it is to h_2.

lines within the species tree represent a hypothetical gene tree. Sequences h_1 and h_2 are both in species A, but by the gene tree, h_2 is more closely related to h_4, in species B, and h_5, in species C, than it is to h_1. Similarly, h_3 and h_4 are both in species B, but h_3 is more closely related to h_1 in species A than it is to h_4.

Topological differences between gene trees and species trees are more likely if polymorphism is high at the time of speciation and the intervals between speciation events are short. If we are to use gene trees to estimate species trees, we must independently examine many loci. Relationships common to independent gene trees are more likely to indicate true relationships than any single gene tree.

Gene duplications and multigene families complicate the situation further. If gene duplication occurred before speciation, then genes in different species may be more closely related than genes in the same species. Consider the vertebrate hemoglobins, for example. There are four main hemoglobins, designated α, β, γ, and δ. The α and β genes diverged about 450 to 500 million years ago (Figure 6.2b). Because of this ancient divergence, the α genes in all mammals are all more closely related to each another than any α gene is to the β gene in the same species (Figure 11.5). The point is that, when we construct a gene tree, we must be certain we are examining the same gene in each OTU.

Sequence Alignment

The first step in constructing a tree is to align the sequences in the sample. By alignment, we mean arranging the sequences so that each sequence is in a row, and homologous sites are aligned in the same column. Keep in mind that homologous means that two sites are derived from a common ancestor, and not that they are identical. Alignment is a straightforward process if the sequences are the same length, and are very similar with only a few substitutions. Frequently, the sequences are different lengths because some contain small insertions or deletions. When aligning sequences of different lengths, we must insert one or more gaps to keep homologous sites in the same column. The basic idea is that because substitutions are rare events, there should be more similarities than differences between two homologous sequences. Therefore, the alignment with the most similarities is probably the correct alignment.

We illustrate the process with an example. Consider two sequences to be aligned:

Sequence 1	A	T	G	C	A	T	G	C	A	T	G	C
Sequence 2	A	T	C	C	G	C	T	T	G	C		

The sequences are different lengths, so one or more gaps have to be inserted in sequence 2. If you play around with the sequences, you will find that the alignment that maximizes the number of sites with the same nucleotide is

Sequence 1	A	T	G	C	A	T	G	C	A	T	G	C
Sequence 2	A	T	**C**	C	–	–	G	C	**T**	T	G	C

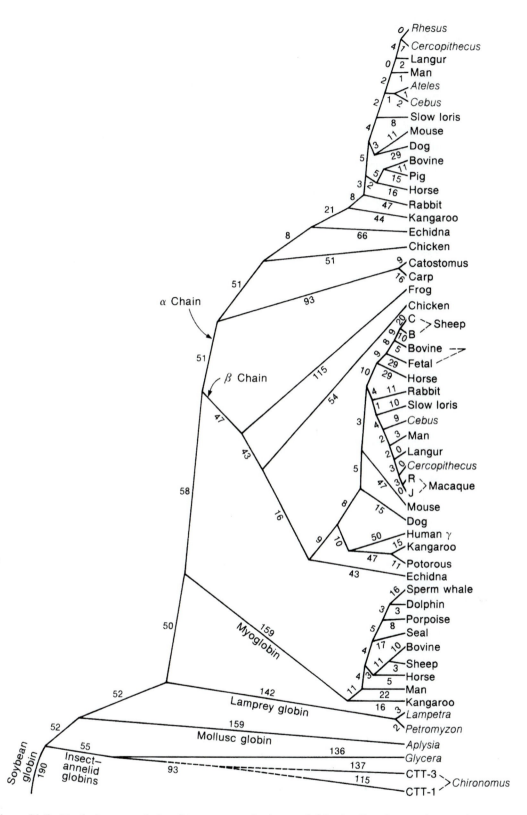

Figure 11.5 **Evolutionary relationships among the hemoglobin family of genes in vertebrates (based on amino acid sequences).** The α-hemoglobins of distantly related species are more closely related than the α- and β-hemoglobin of the same species. Numbers on branches are the estimated numbers of nucleotide substitutions. *Source:* From Dobzhansky et al. (1977); modified from Goodman et al. (1975).

By inserting a gap of two sites in sequence 2, all but two sites (boldface in the lower sequence) can be made identical in the two sequences. The evolutionary inference is that the sequences differ by (at least) two substitutions and an indel two nucleotides in length. This does not imply that one sequence is ancestral and the other is descendent. If both are from existing organisms, they are descended from a common ancestor whose sequence was probably different from either of them. We are only comparing one to another, so that the nucleotides in each column are homologous.

Alignment programs typically consider each site and give a score of 1 if the sequences are identical or 0 if they are different. The alignment with the highest total score is considered the best alignment. Obviously, by inserting enough gaps in the appropriate places, we can align two sequences so that every site is identical between them. For example, we could align the above sequences as

| Sequence 1 | A | T | – | G | C | A | T | G | C | A | – | T | G | C |
| Sequence 2 | A | T | C | – | C | – | – | G | C | – | T | T | G | C |

Every column that does not contain a gap contains the same nucleotides. However, this requires inserting five gaps, and we should consider this an unlikely possibility. We usually wish to find the best compromise between the maximum number of similarities and the minimum number of gaps. Alignment programs typically evaluate alternative alignments by using a gap penalty, which depends on the number and length of the gaps in the alignment. For example, the first alignment above has eight similarities and one gap of two nucleotides. If we subtract one for each missing nucleotide, the score is 6. The second alignment also has eight similarities, but six missing nucleotides, for a score of only 2. Alignment programs usually allow you to specify the gap penalty.

Alignment of two closely related sequences is usually fairly straightforward. Several computer programs are available, and alignment can often be done by eye. Alignment of multiple sequences is a much more complex task, especially if the sequences are different lengths, and contain many substitutions. The most popular program for creating multiple alignments is ClustalX, a free program available on the Internet (see Appendix B).

Once sequences are properly aligned, a tree is constructed by comparing homologous sites. The assumption is that sequences with the most similarities at homologous sites are more closely related than sequences with fewer similarities at homologous sites (but see discussion of homoplasy above). Proper alignment is the critical first step in determining which sites are homologous.

How Many Trees?

The basic idea of phylogenetic reconstruction is to examine different trees for a group of OTUs and use some criterion to decide which of the many possible trees is best. We first consider the problem of searching for possible trees. Criteria for evaluating those trees will be discussed below, even though the processes of searching and evaluating are often coupled in practice.

The number of possible trees that can be drawn for a group of OTUs is extraordinarily high. Any unrooted tree can be rooted by placing the root along any branch; therefore, the number of possible rooted trees is always larger than the number of unrooted trees.

If n is the number of OTUs, then the number of unrooted trees is

$$T_U = \frac{(2n - 5)!}{2^{n-3}(n - 3)!}$$ **(11.13)**

and the number of rooted trees is

$$T_R = \frac{(2n - 3)!}{2^{n-2}(n - 2)!}$$ **(11.14)**

Table 11.3 gives the numbers of unrooted and rooted trees for various numbers of OTUs. Notice how quickly the number of trees becomes unmanageably large. For $n = 20$, the number of unrooted trees is more than 10^{20}! Many studies include more than 20 OTUs.

The obvious conclusion is that, for even relatively small studies, you cannot examine all possible trees. Some strategy for searching through the tree space is necessary.

Tree Searching Strategies

Finding the best tree involves two steps: searching among possible trees, and evaluating each tree by some criterion of "best." We will discuss criteria for evaluating alternative trees below. For small data sets, computer programs can evaluate all possible trees; such a search is called an **exhaustive search**. Obviously, if all possible trees are evaluated, the program will find the best tree based on the specified criteria.

Frequently, however, an exhaustive search is impossible, and some subset of all possible trees must be chosen for evaluation. One method is called the **branch and bound** method. Figure 11.6 illustrates the procedure for five OTUs. The first step is to generate a random tree and calculate its score based on the specified criteria. This is the tree for initial comparisons; call its score S. Next, three OTUs are chosen and the only possible unrooted tree is constructed (tree a). A fourth OTU is added to each of the three possible branches, creating three trees, $b1$, $b2$, and $b3$. Tree $b1$ is evaluated. If its score is worse than S, then all larger trees

TABLE 11.3   Numbers of unrooted trees (T_U) and rooted trees (T_R) for different numbers of OTUs, as calculated from equations (11.13) and (11.14).

Number of OTUs n	Number of Unrooted Trees T_U	Number of Rooted Trees T_R
3	1	3
4	3	15
5	15	105
6	105	945
8	10,395	135,135
10	2,027,025	34,459,425
15	7.9×10^{12}	2.1×10^{14}
20	2.2×10^{20}	8.2×10^{21}

Figure 11.6 **The branch and bound method of tree searching.**   Tree *a* is the only possible unrooted tree with three OTUs (A, B, and C). Trees *b1*, *b2*, and *b3* are the three possible trees made by adding a fourth OTU (D) to one of the branches of tree *a*. The five trees below *b1*, *b2*, and *b3* are the five possible trees made by adding a fifth OTU (E) to the different branches of b_1, b_2, or b_3. At each step each tree is evaluated and trees with scores worse than the current lowest score are discarded. See text for details.

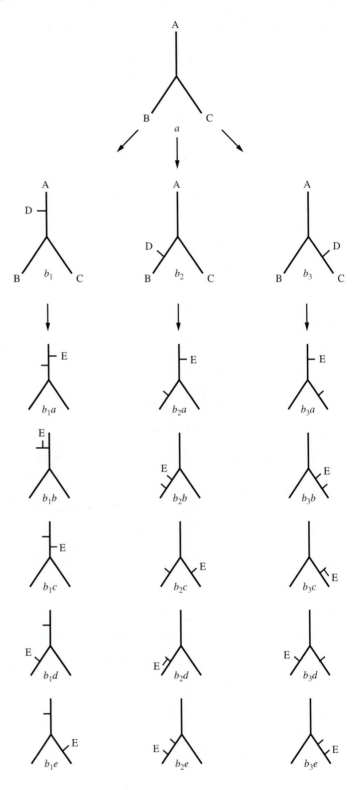

derived from it will have worse scores. Tree *b1* and all trees derived from it (e.g., trees *b1a* through *b1e*) can be eliminated. But if tree *b1* has a score better than *S*, then a fourth OTU is added, creating five larger trees (trees *b1a* through *b1e*). Trees with scores worse than *S* are eliminated. If any trees *b1a* through *b1e* have scores better than *S*, then a sixth OTU is added, creating seven possible trees with six OTUs (not shown in the figure). Each tree is evaluated and kept or enlarged based on its score. The procedure is repeated until all trees derived from *b1* are eliminated, or a tree with all *n* OTUs is found with a score better than *S*. If a better tree is found, it is the new comparison tree and its score becomes the new *S*. If a better tree is not found, the original tree and its score remain the comparison tree. Trees *b2* and *b3* and their derivative trees are then evaluated in the same way. The final best tree is the one with the best score for all *n* OTUs.

The advantage of the branch and bound method is that it can eliminate large groups of trees without actually evaluating them. Like an exhaustive search, it will find the best tree, but will usually do so much more efficiently.

Even the branch and bound method is impractical for large data sets. In such cases, a **heuristic search** must be done. Heuristic searches examine only a subset of all possible trees, and are not guaranteed to find the best tree. The general approach is to sequentially build a series of trees, keeping only the best tree at each step and building upon it. At each step, the best tree is best, given the history of tree building, but it is not necessarily the best tree globally. It is analogous to the example of adaptive landscapes in Section 5.9; natural selection will sometimes cause population fitness to move toward a local maximum, which is not necessarily the global maximum.

There are a variety of approaches to heuristic searches. We will briefly describe only one, the **stepwise addition** method. Initially three OTUs are chosen, and the single unrooted tree constructed. A fourth OTU is added, creating three possible trees. The best of these trees is chosen based on the specified criteria, and a fifth OTU is added on each possible branch, resulting in five trees. The best of these is kept, and a sixth OTU added. The best of the resulting trees is kept, and so on. The result is the best tree, given the initial three OTUs and the order in which the other OTUs are added. It is not necessarily the best tree overall. Adding OTUs in a different order may give a better tree. Various methods exist for choosing which three OTUs to start with, and the order in which others are added.

How do we know whether a heuristic search has found the globally best tree? That is a very difficult question; no one method will always find the best tree. One approach is to enter the OTUs in many different orders. If the different orders all yield the same tree, then you can be reasonably confident that it is the globally best tree. If different orders yield different trees, they may have certain clusters in common. If so, you can be reasonably confident that these clusters are real.

Other methods exist for conducting heuristic searches and for evaluating their results. See Swofford et al. (1996), Nei and Kumar (2001), or Felsenstein (2003) for more on heuristic searches.

Tree Construction: Distance Methods Versus Character State Methods

There are two basic methods of constructing phylogenetic trees. **Distance methods** use some measure of evolutionary distance between pairs of OTUs and construct a tree based on those distances. Examples of evolutionary distance are K_{AA}, K_S, or K_A of Section 11.1; many other measures are possible. Distance methods do not use the sequence data directly, but construct trees based on distance measures derived from the sequence data. **Character state methods** use the actual sequence data. Each site in the sequence is considered a character, and each character can have one of four possible states, for example, A, T, G, or C for DNA sequence data. Character state methods are more efficient than distance methods because they use all of the information in the sequences; however they are computationally demanding.

Constructing phylogenetic trees is not something you want to do without a computer. Calculations are extensive, even for the simplest methods, and for the more complex methods they are impossible without a computer. The two most popular software packages for constructing and analyzing phylogenetic trees are PHYLIP (Felsenstein 1993) and PAUP* (Swofford 2000). Several other programs are available. See Appendix B for information on these programs.

Distance Methods

The first step when using distance methods is to construct a matrix of observed differences between pairs of OTUs. These are the *p*s from Section 11.1, and can be based on differences in amino acid sequences or DNA sequences. Let p_{ij} be the observed proportion of differences between sequence i and sequence j, as calculated in Section 11.1. We then use some model of evolutionary change (e.g., Jukes-Cantor model) to convert the observed differences into some measure of evolutionary distance; call that distance d_{ij}. These d_{ij} can be, for example, K_{aa} for amino acid sequences, or K_S or K_A for DNA coding sequences. There are other possibilities.

Once we have a matrix of evolutionary distances between OTUs, we use some kind of clustering algorithm to group OTUs. The two main distance methods are **UPGMA** (unweighted pair group method with arithmetic means) and **neighbor joining**.

UPGMA begins with an $n \times n$ distance matrix (where n is the number of OTUs). It finds the two OTUs with the smallest distance and combines them into a single OTU. For example, let A, B, C, D, and E represent the OTUs. Assume A and C have the smallest distance. Combine these into (AC). These represent the first two OTUs to be clustered; the branch lengths along each lineage from their common ancestor are each half the distance between A and C. In other words, equal rates of evolution are assumed in the two lineages. UPGMA then constructs an $(n - 1) \times (n - 1)$ distance matrix with (AC) as a single OTU. Distance between B and (AC), for example, is calculated as $d_{B(AC)} = (d_{BA} + d_{BC})/2$, that is, the average of the distances between B and A and between B and C. The process is repeated, each time clustering two OTUs or combined units and reducing the size of the distance matrix by one, until the matrix is reduced to a single entry.

UPGMA is conceptually simple and easy to implement. It simultaneously determines tree topology and branch lengths, and builds a single tree based on the clustering algorithm. The problem is that it assumes equal rates of evolution along all branches of the tree. We often have evidence that this is not true, thus making UPGMA inappropriate in many cases. Moreover, the end result is a single tree, and it is often difficult to judge how much confidence to place in that tree. For these reasons, UPGMA is seldom used any more.

The Neighbor Joining method is similar to UPGMA in that it clusters OTUs by constructing a series of distance matrices. The basic principle of the neighbor joining method is to minimize the total evolutionary distance in the tree. Unlike UPGMA, it does not assume equal rates along all branches of the tree. For this reason, it is more widely used than UPGMA. Unfortunately, neither method is guaranteed to produce the true tree, or even the best tree given the data.

There are several other distance matrix methods; Li (1997) and Nei and Kumar (2000) discuss them in detail. All distance methods suffer from the same fundamental weakness: They are based on some estimate of evolutionary distance derived from the sequence data, and do not use the sequence data directly. Information contained within the sequences is lost by converting the sequence data to a distance matrix (see Problem 11.5). Thus, distance matrix methods are inefficient because they do not use all of the data available.

Distance methods are fast and efficient, and can easily be implemented on most computers. Until fairly recently, they were the only practical way to construct phylogenetic trees, because of limited computer power and speed. Methods that use all of the sequence data are computationally intensive, and only in the last few years have readily available computers had sufficient speed and power to make these newer methods practical. This increased computer power has led to much interest in character state methods and ways to evaluate the trees produced by them.

Character State Methods: Parsimony

One of the most common methods of reconstructing gene trees uses the **principle of parsimony**. According to this principle, the best tree is the one that requires the fewest evolutionary changes to account for the observed haplotypes.

Programs that implement parsimony examine each site in the sequence individually. They first determine whether a site is informative or not. An **informative site** is a site at which at least two different nucleotides occur in the sample of haplotypes, and each variant occurs at least twice. An **uninformative site** is an invariant site, or a site at which a variant nucleotide occurs in only a single haplotype in the sample. Informative sites allow us to differentiate between alternative trees based on the parsimony principle.

Consider the following small example of variable sites in a sample of four haplotypes:

Site	1	2	3	4	5	6
h_1	G	A	G	C	T	C
h_2	A	C	G	A	C	C
h_3	G	A	A	A	T	T
h_4	A	A	A	A	T	T

Only variable sites are shown. Note that sites 2, 4, and 5 contain only a single variant nucleotide, and are thus uninformative. Sites 1, 3, and 6 are informative sites.

Parsimony techniques consider different trees and examine each informative site in each tree. For each site, the computer program finds the tree that requires the fewest changes, and assigns a score (minimum number of changes required) to that tree. It repeats the procedure for each informative site and adds the scores. The tree with the lowest total score (fewest total changes required) is the best tree.

We will use the above sequences to illustrate the principle. For four haplotypes, there are only three possible unrooted trees. These are shown in Figure 11.7a. We will consider all of the possibilities for site 1 in each tree.

Figure 11.7b shows the possibilities for site 1 in tree 1. We assume that both internal nodes are either A or G, as any other assumption would obviously require more changes. Thus, there are four possibilities shown in Figure 11.7b. If we assume that both internal nodes are A, then two changes are required to produce the four haplotypes. These changes are indicated by lines on the branches. Similarly, if we assume that both internal nodes are G, then two changes are also required (not the same changes). Finally, if we assume one internal node is A and one is G, then three changes are required. The score (minimum number of changes) for tree 1 at site 1 is therefore 2.

Figure 11.7c shows the possibilities for site 1 in tree 2. If we assume that both internal nodes are either A or G, then two changes are required. But if we assume the left internal node is G and the right is A, then only one change is required. The opposite requires five changes. Therefore, the minimum number of changes is one and the score for tree 2 at site 1 is 1.

Figure 11.7d shows the possibilities for site 1 in tree 3. If we assume that both internal nodes are the same, then two changes are required. If we assume that they are different, then three changes are required. Therefore, the score for tree 3 at site 1 is 2.

At site 1, tree 2 is the most parsimonious tree, with a score of one. We can repeat the procedure for each informative site, and obtain a score for each tree for each site. The tree with the lowest total score is the best tree. For this example, the total scores are 4, 5, and 6 for trees 1, 2, and 3, respectively (Problem 11.6). Therefore, the best tree by the parsimony criterion is tree 1.

It is possible that more than one tree will have the lowest score, or that the lowest score is insignificantly different from the next lowest. In that case, we must use some other criteria to choose among the competing trees.

Note that parsimony produces unrooted trees. These trees can be rooted by using an outgroup as described above, or by other methods. Most programs that implement parsimony estimate branch lengths by estimating the average number of substitutions required along each branch. See Nei and Kumar (2000) for details.

There are numerous variations of parsimony routines. For example, some weight transitions and transversions differently. Some consider the probability of change in one direction (e.g., A → T) to be different from the probability of change in the opposite direction (e.g., T → A). This is a consideration when drawing trees based on RFLPs; it is easier to lose a restriction site than to regain it. See Swofford et al. (1996) and Nei and Kumar (2000) for more information.

If homoplasy occurs, the true tree may not be the tree with the minimum number of changes. If the differences among haplotypes are small, homoplasy is probably a minor concern, but if differences are great, then homoplasy becomes more likely, and parsimony techniques become less reliable. Parsimony is also less reliable if the number of haplotypes is large, or if there are large differences in substitution rates along different branch lengths. See Felsenstein (1978, 2003) for a discussion of these problems.

If the number of haplotypes is small, parsimony programs can examine every possible tree. But for even moderate numbers of haplotypes, the number of possible trees is prohibitively large for an exhaustive search; for only 20 haplotypes, the

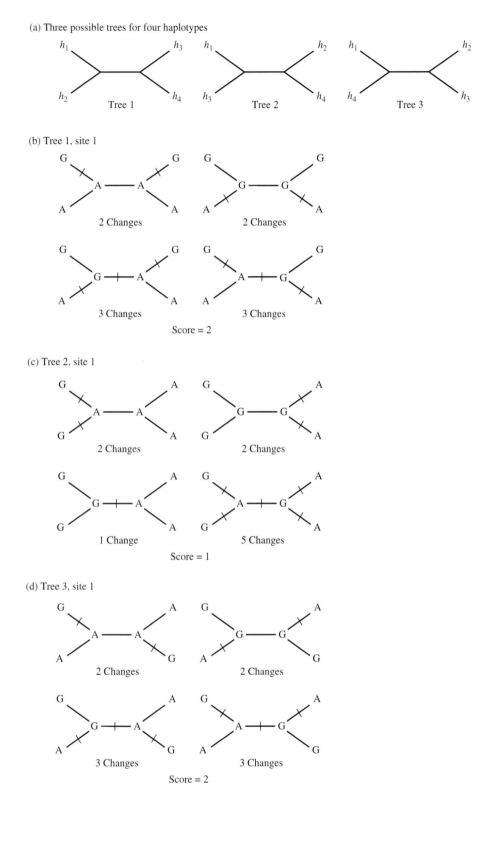

(a) Three possible trees for four haplotypes

Tree 1   Tree 2   Tree 3

(b) Tree 1, site 1

2 Changes   2 Changes

3 Changes   3 Changes

Score = 2

(c) Tree 2, site 1

2 Changes   2 Changes

1 Change   5 Changes

Score = 1

(d) Tree 3, site 1

2 Changes   2 Changes

3 Changes   3 Changes

Score = 2

Figure 11.7 **The parsimony principle of tree construction.** (a) Three possible unrooted trees. (b) Possible evolutionary scenarios for tree 1, site 1. Letters at external nodes indicate observed nucleotides at site 1 of each haplotype. Letters at internal nodes indicate possible nucleotides of ancestral sequences; only four of the 16 possibilities are shown. Changes are indicated by short bars. (c), (d) Possible evolutionary scenarios at site 1 for trees 2 and 3. See text for details.

number of unrooted trees is greater than 10^{20} (Table 11.3). Therefore, for even moderately sized studies, parsimony programs cannot perform an exhaustive search, and heuristic searches must be employed. This increases the likelihood that the most parsimonious tree found is not the correct tree.

Character State Methods: Maximum Likelihood

A second method for searching through the tree space and evaluating the competing trees is based on the principle of maximum likelihood. Here, we hypothesize a particular model of sequence evolution and search for the tree that maximizes the probability of obtaining the observed sequences in a sample, given that model. We can write

$$L = \Pr(\text{data}|\text{hypothesis})$$

where L is the likelihood (probability) of observing the data, given a particular model and its parameters. Here, data means the observed sequences in the sample, and hypothesis is a model of sequence evolution (e.g., Jukes-Cantor model) along with its estimated parameters, and a hypothesized tree. The goal is to find the tree that maximizes L. Note that maximum likelihood requires that we specify an explicit model of sequence evolution, whereas parsimony does not.

We illustrate the method using the same example we used for parsimony. For a model of sequence evolution, we will assume that the probability of any nucleotide changing to any other nucleotide in one time period is α, and that the a priori probability of observing any nucleotide at a specific site is 1/4; that is, all nucleotides are equally frequent. This is the simplest possible model of sequence evolution. For the four sequences in the sample, there are three possible trees (Figure 11.8a). First, consider site 1 in tree 1 (Figure 11.8b). Let X and Y be the nucleotides at the two internal nodes. Each can be either A, T, G, or C, so there are 16 possibilities for site 1 in tree 1. Under the model specified, the probability of change in one direction is the same as the probability of change in the other direction, so we can arbitrarily root the tree anywhere that is convenient. Let X be the root node. Then, the probability of observing the data at site 1, given tree 1 is (starting from X)

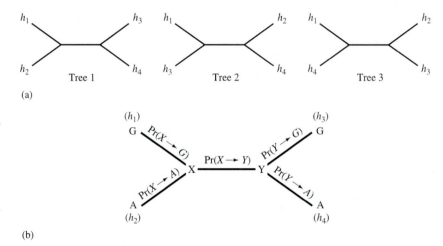

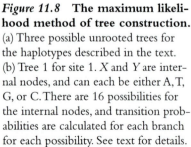

Figure 11.8 **The maximum likelihood method of tree construction.** (a) Three possible unrooted trees for the haplotypes described in the text. (b) Tree 1 for site 1. X and Y are internal nodes, and can each be either A, T, G, or C. There are 16 possibilities for the internal nodes, and transition probabilities are calculated for each branch for each possibility. See text for details.

$$\text{Pr}(\text{data}|\text{tree 1}) = \sum_{X}\sum_{Y} \text{Pr}(X) \times \text{Pr}(X \to G) \times \text{Pr}(X \to A)$$

$$\times \text{Pr}(X \to Y) \times \text{Pr}(Y \to G) \times \text{Pr}(Y \to A)$$

where $\text{Pr}(X)$ is the probability that X is A, T, G, or C, and the other probabilities on the right are the transition probabilities along each branch of the tree. The double summation means we have to consider all possibilities for X and Y (the 16 possibilities mentioned above). Each of these probabilities can be calculated from the model, although you wouldn't want to do it by hand.

Designate this probability by L_{11}, where the subscripts mean tree 1, site 1. We now repeat the calculation for every site in the sample, given tree 1. Note, we must consider every site, not just informative sites as we did with parsimony. If we define L_{1j} as the likelihood for tree 1, site j, given, then the total likelihood for tree 1 is the product of the likelihoods for the individual sites,

$$L_1 = \prod_{j=1}^{s} L_{1j}$$

where s is the number of sites in the sequence. Because the likelihoods are extremely small numbers, we work with their logarithms

$$\ln(L_1) = \ln(L_{11}) + \cdots + \ln(L_{1s}) = \sum_{j=1}^{s} L_{1j}$$

(Remember, the logarithm of a product is the sum of the individual logarithms.)

We now repeat the entire procedure for tree 2, and all other possible trees. If L_i is the likelihood of the ith tree, then

$$L_i = \prod_{j=1}^{s} L_{ij}$$

and

$$\ln(L_i) = \sum_{j=1}^{s} \ln(L_{ij})$$

The tree with the highest log-likelihood is the best tree, according to the maximum likelihood criterion.

The maximum likelihood method obviously requires extensive calculations. It has become practical only with the widespread availability of fast computers with extensive memory. Computer programs usually have ways to eliminate obviously unlikely trees; even so, we are likely to run into the problem of too many trees to evaluate. The above was written under the assumption that all possible trees could be evaluated. If this is not possible, a branch and bound or heuristic search is necessary. For reviews of likelihood methods, see Huelsenbeck and Crandall (1997) and Lewis (2001).

Character State Methods: Bayesian Methods

We met Bayesian analysis in Section 4.5, where we tried to estimate the probability that a person carrying a known marker allele actually carries an associated disease allele. The basic idea of Bayesian analysis is that, before obtaining any data, we have a **prior probability** that a hypothesis is true, based on some kind of model. After we have some data, we can modify this probability to obtain a **posterior probability** that the hypothesis is true. For example, before flipping a coin, you might believe the probability of getting heads is 0.5. If you flip the coin 100 times and get 80 heads, you might modify your belief and think that the probability of heads is greater than 0.5.

Recall that the maximum likelihood method tries to maximize the probability of obtaining the observed data, given the evolutionary model and the hypothesized tree. It searches for the tree that maximizes

$$\Pr(\text{data}|\text{hypothesis})$$

where hypothesis means a specified evolutionary model and a hypothesized tree. This is a useful concept from a statistical point of view, but is not particularly intuitive to interpret. The question we are really interested in is, What is the probability that a hypothesis is true, given the data? Bayesian methods attempt to estimate this probability,

$$\Pr(\text{hypothesis}|\text{data})$$

which is a more intuitive concept to most biologists. According to Bayes' Theorem (Appendix A) this probability can be calculated by the formula

$$\Pr(\text{hypothesis}|\text{data}) = \frac{\Pr(\text{hypothesis}) \times \Pr(\text{data}|\text{hypothesis})}{\Pr(\text{data})}$$

Pr(hypothesis) is the prior probability that the hypothesis is correct. Typically, we assume that all hypotheses have the same prior probability, although some restrictions can be made. Pr(hypothesis|data) is the posterior probability that the hypothesis is correct, given the data. This is what we are interested in. For phylogenetic inference, the above equation is deceptively complex, because "hypothesis" includes a tree topology, along with branch lengths and specific parameters of the model. There is an essentially infinite number of hypotheses. Similarly, Pr(data) is the probability of obtaining the data, considering all possible hypotheses. From the law of total probability,

$$\Pr(\text{data}) = \sum_{\substack{all \\ hypotheses}} \Pr(\text{data}|\text{hypothesis}) \times \Pr(\text{hypothesis})$$

These complexities make it impossible to calculate the posterior probability, except for the simplest problems. However, it can be estimated by computer simulation, not for all possible hypotheses, but for the most probable hypotheses.

The simulation procedure works approximately as follows: (1) Start with a specified or random tree. (2) Make a random change by changing either the tree topology or branch length, or by randomly changing a parameter value. (3) Calculate the likelihood [Pr(data|hypothesis)] of the new tree. (4) If the new tree has a higher likelihood than the old tree, accept the new tree. If the new tree has a lower likelihood than the old tree, accept or reject the new tree according to specified probabilistic rules. (5) Repeat steps 2 through 4 over and over again. (6) Every specified number of iterations, the current tree is saved. This kind of "random walk" from tree to tree is called a Markov chain, and the simulation process is called Markov chain Monte Carlo (MCMC) simulation. In general, the walk will proceed toward trees with higher likelihood. As the simulation proceeds, some trees will be "visited" over and over. It turns out that the number of times a particular tree is visited is proportional to its posterior probability. Thus, by running the simulation long enough, the most probable trees, given the data, can be found and their probabilities estimated.

The main advantage of this Bayesian approach over maximum likelihood is that not all trees need be evaluated; trees are sampled in proportion to their posterior probabilities, which is what we are interested in. Also, the Bayesian approach gives you a group of trees, all with high posterior probabilities. You can then examine these trees and, for example, determine which clusters are common to all or most of them. You can have a high degree of confidence that these clusters are real.

Bayesian techniques in phylogenetic analysis are relatively new, and much new work is being published. They will probably become more sophisticated and widely used over the next few years. For reviews, see Yang and Rannala (1997), Larget and Simon (1999), Huelsenbeck et al. (2001), and Lewis (2001). Hall (2001) describes how to use a relatively simple program to construct Bayesian trees.

Assessing Confidence and Reliability of Gene Trees

Assume we have constructed a gene tree by one of the above methods. How confident can we be that we have found the true tree? Usually, we are mainly interested in finding the correct relationships among the OTUs; that is, we are mainly interested in the topology of the tree, and less concerned about the branch lengths. We wish to find some way to judge how confident we can be that the topology of our tree reflects the true evolutionary relationships among the OTUs. It may be that the tree accurately indicates the relationships of some clusters, but not others.

A **clade** is a monophyletic group; all OTUs within a clade have a common ancestor, and are more closely related to each other than to any OTU outside the clade. When evaluating a gene tree (or any other kind of phylogenetic tree) we want to know how confident we can be that the various clades in the tree represent actual relationships.

A common method of evaluating a gene tree uses the statistical technique of the **bootstrap**. This is a general technique of estimating the confidence in some statistical conclusion. For example, assume we have a particular tree obtained by parsimony technique. The null hypothesis is that this is the true tree. How confident can we be in accepting that null hypothesis? As mentioned above, this can be refined to ask how confident we can be in accepting each clade in the tree as real.

In general, bootstrapping involves resampling the data multiple times, with replacement. The statistical or phylogenetic analysis is done on each pseudosample, and the distribution of the test statistic or of the phylogenetic trees is estimated from the pseudosamples. It is widely applied to estimate the P-values of statistics whose distribution is unknown. Here, we explain how it is applied to analysis of gene trees.

Consider a simple example of four sequences with five polymorphic sites:

Site	1	2	3	4	5
Haplotype 1	A	T	A	A	A
Haplotype 2	T	A	A	C	A
Haplotype 3	A	A	G	A	T
Haplotype 4	T	A	G	A	A

We construct a tree based on, for example, parsimony. How confident can we be in this tree? Let s be the number of sites in the sequence. Bootstrapping involves randomly picking a site s times, with replacement, and drawing a tree based on that pseudosample. Sampling with replacement means that some sites may be represented more than once and some not at all in the pseudosample. For example, the first pseudosample might be

Site	2	4	1	4	3
Haplotype 1	T	A	A	A	A
Haplotype 2	A	C	T	C	A
Haplotype 3	A	A	A	A	G
Haplotype 4	A	A	T	A	G

A tree is drawn from this pseudosample, using the same method used for the original tree. The sampling process is then repeated; for example, the second pseudosample might be

Site	5	3	1	2	1
Haplotype 1	A	A	A	T	A
Haplotype 2	A	A	T	A	T
Haplotype 3	T	G	A	A	A
Haplotype 4	A	G	T	A	T

A tree is drawn from this sample. The process of sampling and tree construction is repeated many times; the recommended number is 1000. We end up with, say, 1000 trees based on random samples of the data. The computer program that does the resampling counts the number of times each clade appears, and writes this number, as a percentage, on the original tree. Figure 11.9 shows an example; it shows a bootstrap tree for the celF protein and related proteins in several species of bacteria. The clade containing *Bacillus subtilis* α-galactosidase and *E. coli* α-galactosidase appeared in 100 percent of the bootstrap trees. We can be quite confident that this clade is real. However, the clade clustering these two OTUs with *Bacillus subtilis* hydrolase appeared in only 50 percent of the bootstrap trees. We should be skeptical about the reality of this clade.

The bootstrap is widely used to estimate the reliability of gene trees, and most phylogenetic programs implement it. Other methods exist; see Swofford et al. (1996) for a good discussion.

Examples

In addition to the obvious applications to understanding the relationships among different groups, molecular phylogenetics has applications in many other fields, such as epidemiology, biotechnology, gene duplications, conservation biology, animal behavior, and others. Here, we describe several examples, to give a sense of the diversity of potential applications.

The human immunodeficiency virus (HIV), the cause of AIDS, has a very high mutation rate, and therefore evolves very rapidly (Section 6.3). This rapid

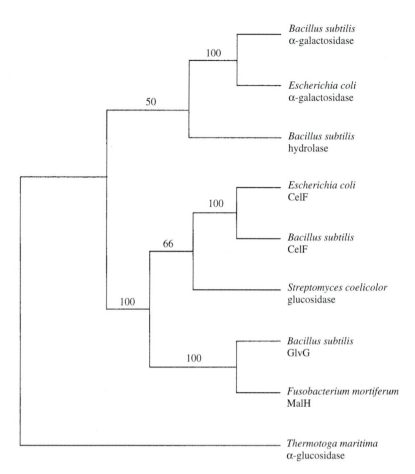

Figure 11.9 **A bootstrap tree for the celF and related proteins in several species of bacteria.** Numbers on branches indicate the percentage of times each clade occurred during the bootstrap procedure. *Source:* Modified from Hall (2001).

evolution allows us to trace the evolutionary history of different strains of HIV. From 1990 to 1992, several patients of a Florida dentist with AIDS were found to be HIV positive. Most had no other risk factors, and it was hypothesized that they were infected by the dentist. This hypothesis was confirmed by a phylogenetic analysis of part of the *env* gene of HIV (Ou et al. 1992). The gene tree based on DNA sequences from the dentist, several of his patients, and several randomly chosen HIV-positive individuals is shown in Figure 11.10. (Two samples were included from the dentist and from patient A.) The dentist and five of his patients cluster together in a single clade, the so-called dental clade, indicated by the dashed box in Figure 11.10. This tree was constructed using parsimony techniques. Several equally parsimonious trees were found; they all contained the dental clade. Bootstrap analysis showed that 79 of 100 bootstrap trees contained the dental clade as shown in Figure 11.10. None of the five patients in the dental clade had any other risk factors, and it was concluded that they were infected by the dentist. The two patients outside the dental clade had other risk factors, and probably were infected by some other means.

Hillis et al. (1994) reanalyzed the above data along with some additional data. They showed that substitution rates varied significantly for different kinds of substitutions, and incorporated these differences into their evolutionary model. They constructed trees using several variations of parsimony, UPGMA, and neighbor

Figure 11.10 **A gene tree for part of the *env* gene of HIV isolated from a dentist and several of his patients.** The dashed box outlines the dental clade; all patients in the box are inferred to have received HIV from the dentist. Two samples were taken from the dentist and from Patient A. *Source:* From Page and Holmes (1998); modified from Ou et al. (1992).

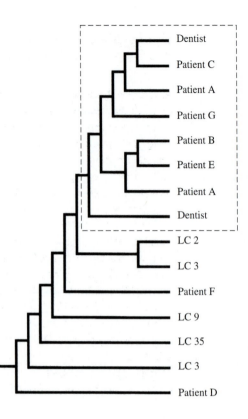

joining. All methods except UPGMA recovered the dental clade 100 percent of the time. Recall that UPGMA is sensitive to variations in substitution rate, which were documented by Hillis et al. This study reinforces the conclusion of Ou et al. (1992) that several of the patients were infected by the dentist. It also illustrates the importance of considering differences in substitution rates.

An understanding of molecular phylogenetics, or lack thereof, can affect ecological and management decisions in conservation biology. The dusky seaside sparrow (*Ammodramus maritimus nigrescens*) was a dark plumaged form of the seaside sparrow. It was originally considered a separate species, but later reclassified as one of several subspecies of *A. maritimus*. It was found only in coastal salt marshes along the Atlantic Coast of Brevard County, Florida (Figure 11.11a). The subspecies was declared endangered in 1966, but continued to decline in numbers due to habitat destruction. In 1980 only six birds were left, all males. Obviously, the subspecies was doomed to extinction, but wildlife managers thought they could save some of the *nigrescens* genome by crossing the males to females of a closely related subspecies, and then backcrossing the female offspring back to the *nigrescens* males. By repeating this backcross for several generations, as long as the *nigrescens* males lived, they hoped to reconstitute a population that was mostly *nigrescens* genetically. Based on morphological similarity, they crossed the *nigrescens* males to females of the subspecies *peninsulae*, on the Gulf Coast of Florida (Figure 11.11a). Backcrosses produced individuals with as much as 87.5 percent *nigrescens* genes. The last pure *nigrescens* bird died in 1987. The endangered species act does not recognize hybrids and the dusky seaside sparrow was officially declared extinct in 1990.

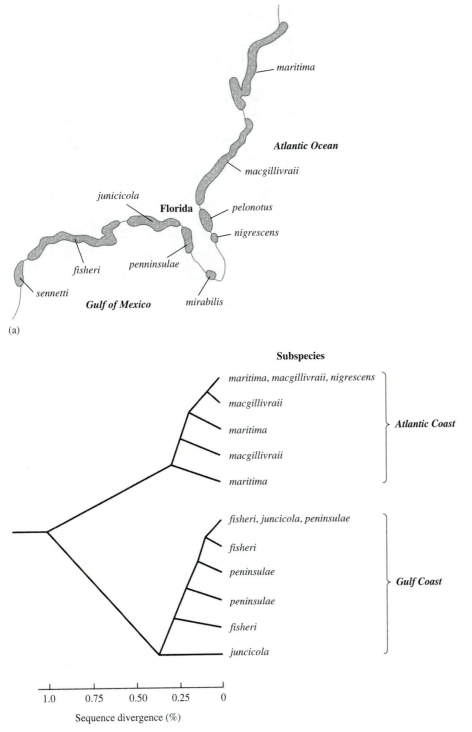

(a)

(b)

Figure 11.11   Relationship of the dusky seaside sparrow (*Ammodramus maritimus nigrescens*) to other subspecies.   (a) Distribution of subspecies. (b) Gene tree based on RFLP variation in mitochondrial DNA. *Source:* Both from Graur and Li (2000); originals from Avise and Nelson (1989).

Avise and Nelson (1989) sought to answer two questions: First, was the dusky seaside sparrow really a genetically unique population? Second, was the subspecies *peninsulae* really the subspecies most closely related to *nigrescens*? They analyzed mtDNA from several subspecies along the Atlantic and Gulf coasts of Florida. Mitochondrial DNA is useful for these kinds of studies because, at least in vertebrates, it evolves several times faster than nuclear DNA, making it useful for differentiating among closely related groups. Avise and Nelson did not sequence the mtDNA molecule, but estimated variation and differentiation by looking at RFLP variation. The analysis of RFLP variation is technically more complicated than analysis of sequence variation, but is conceptually similar. Gene trees can be drawn using modifications of the techniques described above.

Avise and Nelson's results are summarized in Figure 11.11b. It turns out that, at least at the level of RFLP variation, the dusky seaside sparrow is indistinguishable from nearby Atlantic Coast populations. The tree shown is a UPGMA tree. Parsimony produced the same tree, along with several equally parsimonious trees. All trees showed the Atlantic Coast subspecies as a single clade, and the Gulf Coast subspecies as a separate clade. There was a relatively deep division between the two clades that was unrecognized by morphological analysis. We must be cautious about these results, because the mtDNA molecule is effectively a single locus, and analysis based on RFLPs is weaker than analysis based on actual sequences. If taken at face value, these results imply two things: First, the dusky seaside sparrow may not have been a genetically distinct unit after all. Second, the captive breeding program used females from the wrong population.

Unfortunately, the dusky seaside sparrow is extinct, and molecular phylogenetic analysis could not have saved it. However, this example provides a cautionary tale for wildlife managers attempting to save an endangered taxon by a captive breeding program. Management decisions must be based on accurate evolutionary relationships among the groups of interest. Molecular phylogenetic analysis, especially with modern DNA sequencing techniques, can provide important information for making biologically sound decisions.

Finally, we consider how molecular data can contribute to understanding of the recent evolution of our own species. Paleontologists agree that until about 1 to 2 million years ago, the ancestors of *Homo sapiens* evolved in Africa. How modern humans have evolved and spread throughout the rest of the world is still controversial. The traditional view is that *Homo erectus* (an alternative name is *Homo ergaster*) left Africa about 1 million years ago and migrated into Europe and Asia. Populations in Africa, Europe, and Asia then more or less simultaneously evolved into modern *Homo sapiens*. Extensive gene flow prevented these populations from evolving into different species. This has been called the multiregional evolution model. The fossil evidence is mixed, but most paleontologists seem to accept this view.

An alternative view is that there were two (or more) migrations out of Africa. The first occurred about 1 million years ago and established populations of *Homo erectus*, *H. neanderthalensis*, and possibly primitive *H. sapiens* in Europe and Asia. Modern *Homo sapiens* evolved in Africa and migrated into Europe and Asia about 100,000 to 200,000 years ago, replacing the existing populations without interbreeding with them. This hypothesis has been called the African replacement model, or Out of Africa model, and remains controversial. It is supported by most of the molecular evidence.

The first molecular evidence in support of the African replacement model was presented by Cann et al. (1987). They sampled 147 individuals, and constructed a parsimony tree based on RFLP variation in mitochondrial DNA (mtDNA). Mitochondrial DNA is good for studies of recently diverged populations because it evolves faster than nuclear DNA. Moreover, it is inherited maternally; thus, gene trees based on mtDNA can be readily interpreted as maternal phylogenies. The root of the tree represents one (not the only) maternal ancestor of all individuals in the tree. The popular press promptly dubbed this maternal common ancestor "Eve." The gene tree of Cann et al. showed deep divisions separating African branches from a mixture of African and non-African branches. Also, more variation occurred in Africans than in non-Africans. Both of these observations support the hypothesis that the most recent common ancestor lived in Africa, and African populations moved into other regions. Based on estimated rates of mtDNA substitutions, Cann et al. estimated that the most recent common ancestor lived about 140,000 to 290,000 years ago, and that the migration out of Africa occurred about 90,000 to 180,000 years ago.

The study of Cann et al. was criticized on several grounds. The most important were, first, that RFLP analysis is weak, and therefore the conclusions of Cann et al. were statistically weak. Second, Cann et al. did not obtain any samples from actual Africans; they used African-Americans as surrogates for Africans. Third, Cann et al. arbitrarily placed the root of the tree at the midpoint of the branch separating the two deepest divisions of the tree.

In a follow-up study, Vigilant et al. (1991) addressed these criticisms. They studied 189 individuals, including 121 native Africans. They sequenced 610 nucleotides of the mtDNA molecule, and rooted their tree using a chimpanzee sequence as an outgroup. Their results were essentially the same as those of Cann et al. The parsimony tree that Vigilant et al. published is shown in Figure 11.12a. The deepest branches of their tree separated African from non-African sequences, and more variation occurred in Africa than elsewhere. They estimated that the most recent common ancestor lived in Africa about 166,000 to 249,000 years ago.

The tree of Vigilant et al. was immediately criticized. The number of possible trees for 189 individuals is essentially infinite [try substituting substitute $n = 189$ into equation (11.14)]. Vigilant et al. were able to evaluate only a very small sample of the possible trees. Templeton (1991) and Hedges et al. (1991) found hundreds of trees more parsimonious than the one published by Vigilant et al. Some showed African origins and some did not. Figure 11.12b shows a tree found by Templeton; it is more parsimonious than the tree of Vigilant et al., and the deepest branches lead to non-Africans, not Africans. Hedges et al. found trees more parsimonious than Templeton's. These results cast doubt on the African origin postulated by Cann et al. (1987) and Vigilant et al. (1991).

But recall the difference between gene trees and species trees. A gene tree may or may not reflect the evolutionary history of the species (or, in this case, populations) from which the genes were taken. If we examine gene trees based on different genes, we may find commonalities that indicate true relationships among the populations sampled. Several researchers have examined DNA sequences at other loci and constructed gene trees from them. For example, Hammer (1995, Hammer et al. 1997) sequenced part of the Y-chromosome and also concluded

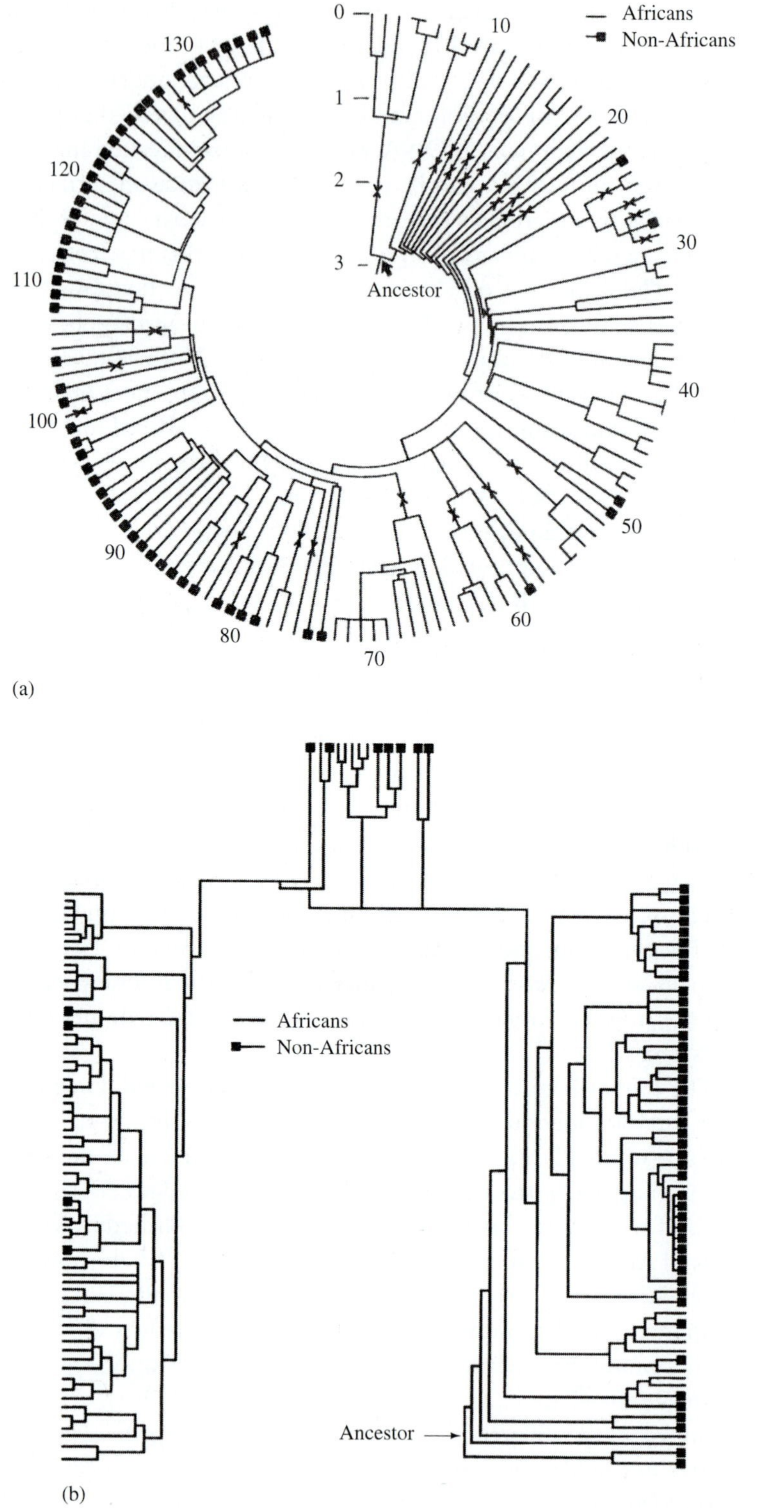

Figure 11.12 **Gene trees based on human mtDNA sequences.** (a) Tree of Vigilant et al. (1991) showing an African ancestor. (b) Tree of Templeton (1992) based on the same data. This tree is more parsimonious than the tree in (a) and does not show an African ancestor. *Source:* From Vigilant et al. (1991) and Templeton (1992).

(a)

(b)

that the most recent common ancestor lived in Africa about 188,000 years ago, consistent with the mtDNA studies. Analyses of autosomal and X-linked loci generally support this conclusion, as do microsatellites (Goldstein et al. 1995), and linkage disequilibrium studies (Tishkoff et al. 1996). Nearly all loci show more variation in African populations than in other populations, implying that African populations are older and have had more time to accumulate variation. This part of the evidence for an African origin is not disputed.

After much debate, the consensus seems to be in favor of the African replacement model, although not in its purist form. There is some evidence for more than one migration out of Africa, and for some gene flow back into Africa from other parts of the world (Hammer et al. 1997, 1998; Templeton 2002). Also, unequal gene flow in males and females, and changing population sizes (expansions and/or bottlenecks) complicate the story.

This example illustrates the potential problems caused by incomplete searching of the tree space, and the necessity of analyzing several loci before making definite conclusions about population or species trees. Also, as when trying to detect natural selection (Section 10.4), demographic issues must be considered.

11.3 Gene Genealogies and Coalescence

In the previous section, we saw how to use the information contained in a sample of DNA sequences to estimate the evolutionary relationships among them. We used this information to draw gene trees that summarize our best estimate of the relationships among the sequences, given a particular evolutionary model.

In this section, we ask what kinds of patterns we might expect to see in samples from populations that have been evolving under different evolutionary processes. The general idea is to simulate gene trees based on some model of evolution. This model might consider mutation, natural selection, changing population size, gene flow, recombination, and so forth. We then ask what characteristics we can expect to see in a sample from a population that has been evolving under this model. By characteristics of the sample, we mean the topology of the gene tree based on the sample sequences, the number of segregating sites, nucleotide diversity, estimated time to most recent common ancestor, and so forth. We can then compare the gene trees derived from an actual sample to the simulated gene trees, to see if the hypothesized evolutionary model does a good job of describing the patterns seen in the sample. We begin with a simple neutral model based on genetic drift.

Introduction to Coalescence

We saw in Section 7.2 that genetic drift will cause alleles to be lost from a population, and eventually all alleles but one will be lost. The probability that a newly arisen allele will eventually become fixed is equal to its initial frequency, $1/2N$, and the expected time to fixation is about $4N$ generations [equation (7.8), Problem 7.4]. This simple idea can be extended to provide powerful methods for inferring information about populations.

Now consider a sample of n alleles (sequences) from a population of size N. In this context, "allele" means a copy of a gene, and alleles are not necessarily different in sequence. We can follow these alleles back in time to their most recent common ancestor. At some point in the past, two alleles will be derived from the same allele in the previous generation. This leaves $n - 1$ independent alleles. At some more distant time, two of these alleles will come from the same allele in the previous generation. This process continues until all alleles are derived from the same ancestral allele. Each time two alleles come from the same allele in the previous generation, we say a **coalescence** occurs. For a sample of *n* alleles, $n - 1$ coalescent events will lead back to their most recent common ancestor.

The lines of ancestry running from the alleles in the sample back to their most recent common ancestor trace out a **gene genealogy** for the sample alleles (Figure 11.13). A gene genealogy is like a gene tree in that it shows the relationships among alleles; we usually use the phrase gene genealogy when considering the relationships among alleles from a single population.

Initially, we assume a diploid locus; constant population size with random mating; and no mutation, natural selection, subdivision, or recombination within the gene. We wish to estimate the times between coalescent events, and the total time back to the common ancestor (the **coalescence time**). This information will allow us to infer the general structure of gene trees based on sequences evolving under genetic drift alone. Later, we will consider more complex models including mutation, varying population size, natural selection, subdivision, and so forth, and compare the expected structure of trees under these models to the expected structure of trees under genetic drift only.

In all of what follows, we will assume a diploid locus. When considering a haploid species, mtDNA, or Y-chromosome DNA, we must replace $2N$ with N, N_f (number of females), or N_m (number of males), respectively. Think of $2N$ as the number of copies of the locus in the population. Also, remember that N is really the effective population size, and may differ from the actual population size (Section 7.4).

We will consider a process of sampling with replacement, which means that an allele can leave more than one copy of itself in the next generation. Consider some arbitrary generation. Pick two alleles, with replacement. The probability these are identical by descent from the previous generation is $1/2N$ (Section 7.2). The probability that these two alleles are not identical by descent is therefore $1 - 1/2N$, or $(2N - 1)/2N$. Now pick a third allele. The probability that it is identical by descent to one of the first two is $2/2N$, and the probability that it is

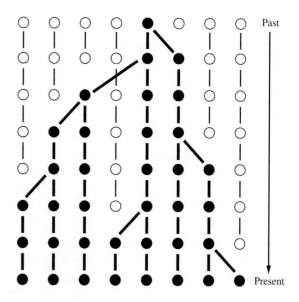

Figure 11.13   The principle of coalescence.   Each circle represents one copy of a gene; time flows from top to bottom. The eight genes at the bottom of the figure are all descended from the single shaded gene at the top. As we move from bottom (present) to top (past) each time a gene fails to leave a copy of itself, a coalescence occurs. Ultimately, all genes coalesce to a single common ancestor. Distances between horizontal rows represent more than one generation, and are not drawn to scale.

not is $1 - 2/2N$, or $(2N - 2)/2N$. Similarly, the probability that the fourth allele is different from the first three is $(2N - 3)/2N$, and probability that the nth allele in the sample is different from the previous $n - 1$ alleles is then $[2N - (n - 1)]/2N$. The probability that all n alleles in the sample are different is the product of these probabilities. This is the probability that there is no coalescence in the first generation back, or

$$\Pr\begin{pmatrix} \text{no coalescence} \\ \text{in first gen} \end{pmatrix} = \frac{2N - 1}{2N} \times \frac{2N - 2}{2N} \times \cdots \times \frac{2N - (n - 1)}{2N}$$

Rewrite this as

$$\Pr\begin{pmatrix} \text{no coalescence} \\ \text{in first gen} \end{pmatrix} = \left(1 - \frac{1}{2N}\right) \times \left(1 - \frac{2}{2N}\right) \times \cdots \times \left(1 - \frac{n - 1}{2N}\right)$$

If we multiply this out and ignore terms with N^2 in the denominator, we can approximate it as

$$\Pr\begin{pmatrix} \text{no coalescence} \\ \text{in first gen} \end{pmatrix} \cong 1 - \left(\frac{1}{2N} + \frac{2}{2N} + \cdots + \frac{n - 1}{2N}\right)$$

or

$$\Pr\begin{pmatrix} \text{no coalescence} \\ \text{in first gen} \end{pmatrix} \cong 1 - \left(\frac{1 + 2 + \cdots + (n - 1)}{2N}\right) \quad \textbf{(11.15)}$$

The numerator of the fraction can be simplified to $n(n - 1)/2$, which gives

$$\Pr\begin{pmatrix} \text{no coalescence} \\ \text{in first gen} \end{pmatrix} \cong 1 - \frac{n(n - 1)}{4N}$$

The probability that a coalescence will occur in the first generation is one minus the probability that it will not occur, or

$$\Pr\begin{pmatrix} \text{coalescence} \\ \text{in first gen} \end{pmatrix} \cong \frac{n(n - 1)}{4N} \quad \textbf{(11.16)}$$

If a coalescence does not occur in the first generation, the probability that it will occur in the second generation is the same. The probability that a coalescence will occur t generations back is the probability that it does not occur in the first $(t - 1)$ generations times the probability that it does occur in the tth generation. If p is the probability that a coalescence will occur in any generation, as given by equation (11.16), and $1 - p$ is the probability that it will not occur, as given by equation (11.15), then

$$\Pr(T_n = t) = (1 - p)^{t-1}p \quad \textbf{(11.17)}$$

where T_n is the time, in generations, until coalescence from n to $n - 1$ alleles, and $\Pr(T_n = t)$ is the probability that the first coalescence occurs in generation t. This is a geometric distribution (Appendix A). The mean of a geometric distribution is the inverse of the probability of success. Therefore, the expected time until the first coalescence is, from equation (11.16),

$$E(T_n) = \frac{4N}{n(n - 1)} \quad \textbf{(11.18)}$$

This is the expected time to coalescence from n to $n - 1$ alleles. If we let i be any number of alleles between 2 and n, and T_i be the coalescence time from m to $i - 1$ alleles, we can generalize equation (11.18) to

$$E(T_i) = \frac{4N}{i(i - 1)} \tag{11.19}$$

Note that as we go back in time, i gets smaller, so that the expected time until the next coalescence is longer. Figure 11.14 shows an example. The expected time required to coalesce from n alleles to a single allele is then the sum of the coalescence times,

$$E(T) = E(T_n) + E(T_{n-1}) + E(T_{n-2}) + \cdots + E(T_2)$$

$$= \sum_{i=2}^{n} E(T_i) = 4N \sum_{i=2}^{n} \frac{1}{i(i - 1)}$$

It can be shown (see any calculus text) that the summation is $(n - 1)/n$. Therefore,

$$E(T) = 4N\left(\frac{n - 1}{n}\right) \tag{11.20}$$

The expected coalescence time depends only on the population size and the number of alleles in the sample.

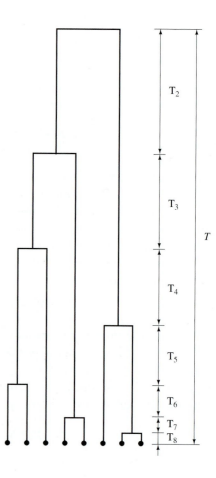

Figure 11.14   **The expected time to the next coalescence ($T_8 \ldots T_2$) increases as we move from the present (bottom) back in time (top) to the common ancestor.**

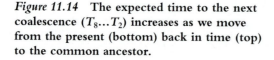

The second important thing to note is that all of the T_i are random variables, and equation (11.19) only gives their expectations. Independent realizations of the coalescent process (for example, evolution of different loci under the same conditions) will produce different genealogies of the alleles. Figure 11.15 shows six independent simulations of the process with a sample of $n = 5$ alleles. You can think of these as genealogies of six independent neutral loci.

Mutation and Coalescence

Until now we have considered only whether alleles are identical by descent; we have paid no attention to whether they are identical in state. As time proceeds forward from the allele destined for fixation, mutations will occur so that some of the alleles in the sample may have different sequences, even though they are identical by descent from the original sequence. We can superimpose mutation on the gene genealogy and predict the amount and pattern of variation expected in the sample.

How many mutations can we expect? If u is the mutation rate per sequence per generation, and T_C is the total number of generations in the coalescent genealogy (*not* the time to the most recent common ancestor), then the total number of mutations is uT_C. We can easily get T_C. In the period between n and $n - 1$ alleles, there are n lineages, each T_n generations long; in the period between $n - 1$ and $n - 2$ alleles, there are $n - 1$ lineages, each T_{n-1} generations long; and so forth. So, the total number of generations in the entire genealogy is

$$T_C = \sum_{i=2}^{n} iT_i \qquad \textbf{(11.21)}$$

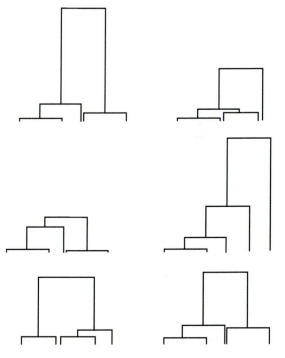

Figure 11.15 **Six independent simulations of the coalescent process for a sample of five alleles.** *Source:* From Donelly and Tavaré (1995).

Under the infinite alleles model, each mutation creates a segregating site. Therefore, the expected number of segregating sites in the sample of n alleles is

$$E(S_n) = uE(T_C) = uE\left(\sum_{i=2}^{n} iT_i\right)$$

Recall that the expectation of a sum is the sum of the expectations. Applying this, we get

$$E(S_n) = u\left(\sum_{i=2}^{n} iE(T_i)\right)$$

But from equation (11.19), the expectation of T_i is $4N/[i(i-1)]$. Therefore,

$$E(S_n) = u\left(\sum_{i=2}^{n} i\frac{4N}{i(i-1)}\right) = 4Nu\sum_{i=2}^{n}\frac{1}{i-1}$$

Change the index of summation, by letting $j = i - 1$, and recall that $4Nu = \theta$:

$$E(S_n) = \theta\sum_{j=1}^{n-1}\frac{1}{j} \tag{11.22}$$

This is the expected number of segregating sites in a sample of n sequences. From this equation, we can estimate θ:

$$\hat{\theta} = \frac{S_n}{\displaystyle\sum_{j=1}^{n-1}\frac{1}{j}} \tag{11.23}$$

where S_n is the observed number of segregating sites in a sample of n sequences. This is Watterson's estimate of θ, discussed in Section 10.4.

Of what use is this theory? We can simulate genealogies with mutation and compare the simulated trees to the observed tree. The simulated tree (under assumptions of constant population size, no selection, no gene flow, no recombination, and so forth) depends only on the sample size and the observed number of segregating sites. The observed tree may be the result of many other evolutionary processes. Significant differences may indicate that one or more of these processes is acting or has acted in the past. In what follows, we will briefly describe the expected genealogies based on several kinds of evolutionary models, usually skipping over the mathematical derivations. Keep in mind that coalescence, under any model, is a highly stochastic process, so that obtaining the expected genealogies is highly dependent on computer simulations.

Changing Population Size

Consider first a population that has been growing. Population size in recent generations will be large compared to more distant ones, meaning coalescence is less likely in recent generations than in distant past ones. The terminal branches will be relatively longer, and branches near the root will be relatively shorter, than in a population of constant size.

In a population that has recently been declining, the opposite effect occurs. Compared to a population of constant size, the tree will be compacted near the tips (small population, short coalescence times) and stretched near the root (large population, long coalescence times).

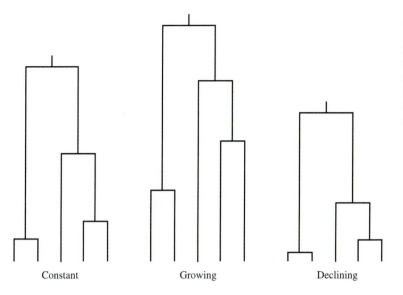

Figure 11.16 **Representative gene genealogies for constant, growing, and declining population sizes.** In a growing population, the terminal branches are elongated compared to a population of constant size. The opposite is true for a declining population. Total time to coalescence is also affected by changing population size.

Constant   Growing   Declining

Figure 11.16 shows representative genealogies for growing and declining populations, compared to a population of constant size. It is possible to test whether a tree differs significantly from the null hypothesis of a constant population size; for example, see Emerson et al. (2001) and Templeton (1998).

Natural Selection

Both positive selection and purifying selection tend to reduce the amount of genetic variation in a sample, compared to the neutral expectation. Part of the reason is that coalescence occurs faster.

First consider positive selection: Favored alleles are more likely to be transmitted than other alleles. The result is that coalescence is more likely to occur because two sequences are likely to be descended from the favored sequence. The expected time to coalescence from n alleles to one is less than under neutrality (Figure 11.17a,b). With less time in the genealogy for mutations to collect, the number of segregating sites will be less than under neutrality.

The process is similar for purifying selection. Some sequences are unlikely to be transmitted, increasing the likelihood of coalescence. As under purifying selection, the expected time to coalescence and number of segregating sites will be less than under neutrality (Figure 11.17c).

Balancing selection has quite a different effect. If two sequences have been maintained by balancing selection, both will have accumulated neutral mutations since their divergence. If balancing selection has acted for a long time, the time to the most recent common ancestor will be longer than the neutral expectation. The result is that the genealogy will have two deep branches representing the favored sequences, and many shallow branches representing neutral variation within each sequence (Figure 11.17d).

Population Subdivision

In a randomly mating population, there should be no correlation between alleles and geography. That is, alleles sampled from nearby locations should not cluster

Figure 11.17   **Representative gene genealogies under different kinds of natural selection.**   Under both positive selection and purifying selection, the total time to coalescence is less than under neutrality. Under balancing selection, it is longer, and branches near the root are elongated compared to neutrality. *Source:* Modified from Wayne and Simonson (1998).

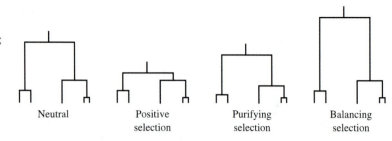

Neutral   Positive   Purifying   Balancing
                  selection   selection   selection

together more often than alleles sampled from distant locations. But if the population is subdivided, with limited gene flow between subpopulations, alleles sampled from the same subpopulation will be more similar than alleles sampled from different subpopulations. If there are two subpopulations, the resulting tree will show two deep branches, corresponding to the two subpopulations, with alleles sampled from the same subpopulation tending to cluster together in shallower branches (Figure 11.18). With more extensive gene flow, the tree will take some intermediate form.

Slatkin and Maddison (1989) showed how to estimate *Nm*, the number of immigrants per generation, from a sample genealogy. See also Hudson (1990), Donnelly and Tavaré (1995), and Fu and Li (1999) for more details and references to the original literature.

Recombination

Until now, we have ignored the possibility of recombination within a sequence. For mitochondrial DNA or Y-chromosome DNA, this is reasonable, but for diploid autosomal loci we cannot ignore recombination over evolutionary time, even among tightly linked sites within a sequence.

If there is no recombination between parts of a sequence, then the entire sequence has a single genealogical history, as described above. If there is free

Figure 11.18   **A representative gene genealogy in a subdivided population with little gene flow.** *A* and *B* indicate whether genes were sampled from subpopulation *A* or *B*. The two major clades mostly represent the two subpopulations, with some gene flow between them. Sequence A_5 apparently originated in subpopulation *B* but was sampled in subpopulation *A*, and vice versa for sequence B_1.

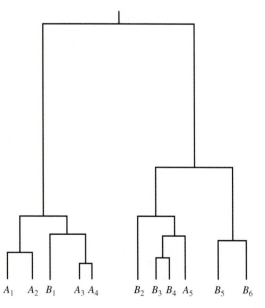

A_1 A_2 B_1 A_3 A_4 B_2 B_3 B_4 A_5 B_5 B_6

recombination between parts of a sequence, then each part evolves independent-ly of the others, and each part will represent an independent realization of the coalescent process, as shown in Figure 11.15. If there is limited recombination, each part will be a different, but not independent, process. The genealogies of the parts will be correlated. This means that no single genealogy can represent the history of a sample of recombining sequences.

Figure 11.19 illustrates a simple example with two alleles in the sample, desig-nated a_1b_1 and a_2b_2. Let a_ob_o represent the ancestral sequence. The letters repre-sent regions of the sequence, and not individual sites. It is probably easiest to first work forward from the ancestral sequence. The first event is a mutation that con-verts a_ob_o to a_1b_o. Three more mutations convert a_ob_o to a_ob_1, a_ob_o to a_2b_o, and a_ob_o to a_ob_2. Two recombination events occur. The first is between a_1b_o and a_ob_1 to produce haplotype a_1b_1, the first sample sequence. The second recombination is between a_2b_o and a_ob_2 to create haplotype a_2b_2, the second sample sequence. If we trace back to the most recent common ancestor for the a regions (heavy solid lines), we reach the node labeled CA_a. If we trace back to the most recent com-mon ancestor for the b regions (dashed lines), we reach the node labeled CA_b, which is more recent than CA_a. A more recent common ancestor means fewer segregating sites (per site) in region b than in region a.

A second effect of recombination is to stretch the tree near the tips and to in-crease the total time to coalescence (Schierup and Hein 2000). A tree will look similar to one of a growing population without recombination (Figure 11.16b).

Recombination can throw a monkey wrench into coalescent theory. If re-combination is relatively common, different parts of a sequence will have differ-ent genealogies, even if they are tightly linked. This can severely limit our ability to make inferences about other evolutionary processes. Much recent work has been done to address this problem (e.g., Schierup and Hein 2000; Brown et al. 2001; Posada and Crandall 2001; Posada et al. 2002); see also Fu and Li (1999).

An Example

We describe one example of how coalescent theory can be useful in interpreting the variation seen in a sample of DNA sequences. Hudson et al. (1994, 1997) se-quenced 41 copies of the superoxide dismutase (*Sod*) locus in *Drosophila*

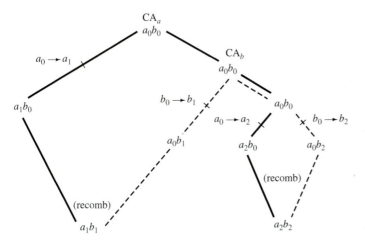

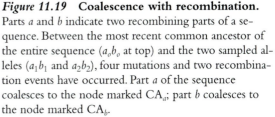

Figure 11.19 **Coalescence with recombination.** Parts *a* and *b* indicate two recombining parts of a se-quence. Between the most recent common ancestor of the entire sequence (a_ob_o at top) and the two sampled al-leles (a_1b_1 and a_2b_2), four mutations and two recombina-tion events have occurred. Part *a* of the sequence coalesces to the node marked CA_a; part *b* coalesces to the node marked CA_b.

melanogaster. Samples came from California and Barcelona, Spain. Allozyme studies had previously shown that two forms of *Sod*, designated *F* (*Fast*) and *S* (*Slow*), are found in most populations. The frequency of the *S* allele varied from about 0.05 to 0.18 in the California and Barcelona populations. Hudson et al. sequenced 19 *S* and 22 *F* alleles. They chose to sample approximately equal numbers of the *F* and *S* alleles; thus, their sample was not a random sample from the populations, although the 19 *S* and 22 *F* alleles were random samples of *S* and *F* alleles, respectively. All 19 *S* sequences were identical. Of the 22 *F* sequences, 9 were identical and were designated *Fast* A; they differed from the *S* sequence at a single site.

This unusual pattern suggests that the *S* sequence has recently arisen from the *Fast* A sequence, and has not had time to accumulate variation. Hudson et al. suggested that some kind of natural selection was responsible for this pattern, but both Tajima's *D* test and the HKA test (Section 10.4) failed to reject neutrality. Hudson et al. developed a more powerful test based on coalescent theory. The basic idea is this: Consider a random sample of n sequences, with S_n segregating sites. What is the probability that a subset of size m ($m < n$) of this sample will have only one segregating site?

Because their sequences were not a random sample of population sequences, Hudson et al. had to construct an artificial random sample from their data set. This sample was constructed to reflect the population frequencies of the *F* and *S* alleles; it contained all 22 *F* sequences and 3 *S* sequences (all alike). The number of segregating sites in this constructed random sample was 63; a subset of 12 sequences had only one segregating site.

Hudson et al. then asked what is the probability under neutrality that, of a sample of 25 sequences, a subset of 12 will have only one segregating site. To answer this question, they simulated 10,000 genealogies based on the neutral coalescent model with $n = 25$ and $S_n = 63$, and asked what proportion of the simulated genealogies contained subset of 12 sequences with only one segregating site. The answer was 0.0081, assuming no recombination within the sequences. If low levels of recombination were allowed (as suggested by detailed analysis of the sequences), the proportion was lower. Hudson et al. concluded that the observed pattern in their data was very unlikely under the neutral coalescent model.

Hudson et al. hypothesized a scenario to explain the observed pattern. First, a beneficial mutation occurred in or near the *Sod* locus. The *Fast* A haplotype increased to a relatively high frequency by hitchhiking with the favored mutation. Then a second mutation occurred creating the *Slow* allele from the *Fast* A haplotype. Either further hitchhiking or selection favoring the *Slow* allele caused it to increase to the observed frequency of 5 to 18 percent. This explains the relatively high frequencies of both the *Fast* A and *Slow* haplotypes. From the observed data, it is impossible to tell whether the observed variation is a transient polymorphism in which a favored allele is on the way to fixation, or a recent balanced polymorphism.

The Importance of Coalescent Theory

Coalescent theory provides a powerful way of looking at the evolution of DNA sequences. It is relatively straightforward to simulate a coalescent process for some evolutionary model and compare the simulated trees to the trees based on sample sequences. Keep in mind that any gene genealogy represents a single realization of the coalescent process. Computer simulations must produce many replicate genealogies and search for their common features. One reason coalescent methods are so useful is that we only have to keep track of individuals that actually leave ancestors in the genealogy. For large samples and populations, this is much more efficient than keeping track of all $2N$ individuals in the population.

Coalescent theory provides a rigorous statistical framework for hypothesis testing and inference about populations from samples of DNA sequences. In this section, we have tried to develop a qualitative understanding of the differences in gene genealogies due to different evolutionary processes. Much current work is aimed at developing rigorous statistical procedures to compare these models to the neutral model (e.g., Templeton 1998).

Coalescent theory is the most important development in theoretical population genetics since the neutral theory, and much new work is being published frequently. We have only barely introduced the subject. For more details, Hudson (1990) provides the standard introduction to coalescent theory for a variety of models. Fu and Li (1999) provide a more recent review, and a good nonmathematical introduction. As computer power increases and simulation methods improve, more progress will be made.

Summary

1. The relationship between the observed number of amino acid differences per site, p_{aa}, and the expected number of substitutions per site, K_{aa}, is given by equation (11.1), which assumes that the number of substitutions is a random variable with a Poisson distribution.

2. Estimating the number of substitutions in DNA sequences is more complex than estimating the number of substitutions in protein sequences. Coding and noncoding sequences must be considered separately, and within coding sequences, synonymous and nonsynonymous sites must be considered separately.

3. For noncoding DNA sequences, the simplest substitution model is the Jukes-Cantor model, which assumes that the substitution rate is the same at all sites and that all substitutions are equally likely. The relationship between the expected number of substitutions per site and the observed number of differences per site is given by equation (11.6).

4. Within protein-coding DNA sequences, a synonymous site is one at which any change is silent, and a nonsynonymous site is one at which any change is a replacement change. There are different ways to estimate the numbers of synonymous and nonsynonymous sites in a sequence.

5. Synonymous or nonsynonymous substitution rates are always expressed as the numbers of synonymous or nonsynonymous substitutions per synonymous or nonsynonymous site. The relationships between the observed numbers of differences and the expected numbers of substitutions under the Jukes-Cantor model are given by equations (11.9) and (11.10). More complex substitution models are possible.

6. It is possible to draw gene trees showing the evolutionary relationships among a group of DNA sequences. These trees may or may not correspond to population trees or species trees.

7. The two main methods of constructing gene trees are distance methods, which utilize pairwise estimates of some measure of evolutionary distance, and character state methods, which use the sequence information directly. The latter are more powerful, but computationally more demanding.

8. The most widely used character state method is parsimony, which attempts to find the tree with the fewest evolutionary changes.

9. Two newer methods of tree construction are maximum likelihood and Bayesian. Both require an explicit evolutionary model and are computationally intensive.

10. Confidence in a gene tree may be assessed by the bootstrap procedure, a method of resampling data to construct numerous pseudosamples from which statistical inferences can be made.

11. Coalescent theory allows us to predict the general shape of a gene tree and various characteristics of a sample of DNA sequences, based on different evolutionary models.

12. Coalescent theory sometimes allows us to make inferences about mutation rates, present and past population sizes, natural selection, gene flow, or recombination, based on a sample of DNA sequences.

13. Coalescent theory provides a rigorous statistical framework for testing hypotheses about the above-mentioned evolutionary processes. As computer power increases and simulation techniques improve, coalescent theory will provide more powerful ways to differentiate between neutral evolution and other evolutionary models.

Problems

11.1. The following are partial amino acid sequences (single-letter code) near the beginning of the sperm lysin proteins of two species of abalone (Lee et al. 1995):

| *Haliotis rufescens* | SWHYVEPKFL | NKAFEVALKV | QIIAGFDRGL |
| *Haliotis cracherodii* | RYQFVQHQYI | RKAFEVALKV | EIIAGFDRTL |

Estimate p_{aa} and K_{aa}. What might be the function of this part of the protein? Explain your reasoning.

11.2. Consider the two sequences discussed under the section "Sequence Alignment," pages 434–436. Assume they are noncoding DNA sequences and estimate p and K.

11.3. The following are the first 30 nucleotides of the coding sequence of *Adh* from *Drosophila melanogaster* (Kreitman 1983) and *D. erecta* (Jeffs et al. 1994):

| *D. melanogaster* | ATG | TCG | TTT | ACT | TTG | ACC | AAC | AAG | AAC | GTG |
| *D. erecta* | ATG | GCA | TTC | ACC | TTG | ACC | AAC | AAG | AAC | GTC |

Estimate p_S, p_N, K_S, and K_A.

11.4. Plot the expected number of substitutions per site (K) versus the observed number of differences per site (p) for amino acid substitutions and for noncoding DNA substitutions on the same graph. Also plot $K = p$. Which substitution rate diverges faster from the line of equality? Why?

11.5. Consider the following five DNA haplotypes. The sequence is 500 nucleotides in length, and only variable sites are shown.

h_1	A	...	A	...	A	...	A	...	A
h_2	A	...	A	...	A	...	C	...	T
h_3	A	...	G	...	A	...	A	...	A
h_4	A	...	A	...	A	...	A	...	T
h_5	G	...	G	...	A	...	A	...	A

a. Assume the number of substitutions is equal to the number of differences. Under what circumstances would this be a reasonable assumption?

b. Construct a table showing the pairwise differences between the haplotypes.

c. Try to draw a UPGMA gene tree based on the difference table. Describe the problems you encounter.

d. Try to draw an unrooted tree, based on the relationships among the sequences. Does considering the sequences themselves help you to draw a tree with greater confidence than considering only the distance table? Discuss.

11.6. For the four sequences in the parsimony example (p. 441) determine the score for each tree at sites 3 and 6. Which tree has the lowest total score based on all informative sites?

11.7. Consider a sample of 6 DNA sequences from a population of $N = 100$. Calculate $E(T_i)$ for $i = 2\ldots6$. What is the expected time for all 6 sequences to coalesce?

CHAPTER 12

Natural Selection II: Balancing Selection and Advanced Models

It is my personal view that no serious and capable investigator has believed the values of genotypic fitnesses to be constant in an absolute sense.

—*Kojima (1971)*

Genetic variation may be maintained by a balance of selective forces acting in different ways, so that no single allele is favored under all circumstances. **Balancing selection** is a general term that applies to any kind of natural selection that tends to maintain genetic variation in a population. Single-locus over-dominance, discussed in Chapter 5, is one example. More complex kinds of natural selection frequently produce balancing selection and genetic polymorphism, for example, the self-incompatibility systems discussed in Section 8.3.

In the basic model of natural selection of Chapter 5, we assumed that there are only two alleles at a locus and that the fitness of a genotype is constant. That model has been a useful guide for predicting and interpreting the course of natural selection, but it is unrealistic in many ways. For example,

Many loci show more than two alleles, sometimes many more.
The fitness of a genotype can vary spatially.
The fitness of a genotype can vary temporally.
The fitness of a genotype can be different in males and females.
The fitness of a genotype can depend on the frequency of that genotype.
The fitness of a genotype can depend on the population size.
The fitness of a genotype can depend on the genotype at other loci.

In this chapter, we introduce these and other kinds of natural selection. One goal is to survey the diversity of ways in which natural selection can operate in the real world. A second goal is to assess the importance of these kinds of selection in natural populations, and to evaluate their potential for maintaining genetic variation.

12.1 Natural Selection with Multiple Alleles at a Locus

In Section 5.2 we concluded that, for natural selection acting at a single locus with two alleles, genetic variation will be maintained only if the heterozygote has a higher fitness than either homozygote. Many allozyme, microsatellite, and other loci have more than two alleles. An extreme case is the major histocompatibility complex (MHC) in vertebrates, with as many as 100 alleles at some loci. In this section, we ask whether single-locus overdominance, or natural selection in general, is a reasonable explanation for these multiple allele polymorphisms.

The condition for a stable polymorphism with two alleles is simple: The heterozygote must have a fitness higher than either homozygote. What are the conditions for a stable polymorphism with three or more alleles? We first need to phrase the question more carefully. Consider a locus with k alleles, designated $A_1, \ldots, A_k$. The genotype A_iA_j has fitness w_{ij}. Under what conditions will a stable polymorphism be maintained, with all k alleles kept in the population?

By analogy with the two-allele case, we might guess that, for any two alleles, the heterozygote A_iA_j should have a higher fitness than either A_iA_i or A_jA_j. In our usual notation,

$$w_{ij} > w_{ii} \quad \text{and} \quad w_{ij} > w_{jj}$$

for all pairs of alleles. It turns out that this is neither necessary nor sufficient to maintain a k-allele polymorphism.

Similarly, we might guess that if all heterozygotes have higher fitness than all homozygotes, then all alleles will be maintained in the population. Again, this is neither necessary nor sufficient.

If these rather intuitive conditions are neither necessary nor sufficient, what *are* the conditions for maintenance of a k-allele polymorphism? We will describe a model for k-allele selection theory, and see what conclusions we can draw from it.

Assume there are k alleles at a locus. Let p_i be the frequency of the ith allele, and w_{ij} be the fitness of A_iA_j. By analogy with equation (5.23), the marginal fitness (average fitness) of the ith allele is

$$w_i = \sum_{j=1}^{k} p_j w_{ij} \tag{12.1}$$

In other words, the marginal fitness of an allele is the weighted average of the fitnesses of all genotypes containing that allele.

The mean fitness of the population is the weighted average of the marginal fitnesses:

$$\overline{w} = \sum_{i=1}^{k} p_i w_i \tag{12.2}$$

By analogy with equation (5.27) the change in frequency of the ith allele is

$$\Delta p_i = p_i \left(\frac{w_i - \overline{w}}{\overline{w}} \right) \qquad \textbf{(12.3)}$$

Note that this simple looking equation represents k equations, one for each allele.

These are simple extensions of the two-allele case, analogous to equations (5.23), (5.25), and (5.27). However, analysis of these equations is much more complicated than for two alleles.

Let $\mathbf{W}$ be the matrix of fitness values. For three alleles, $\mathbf{W}$ will be

$$\mathbf{W} = \begin{bmatrix} w_{11} & w_{12} & w_{13} \\ w_{21} & w_{22} & w_{23} \\ w_{31} & w_{32} & w_{33} \end{bmatrix}$$

where each entry is a fitness of one of the genotypes. The extension to more than three alleles is obvious. We wish to know the conditions on the fitness matrix that will allow all alleles to be maintained in a stable polymorphic equilibrium.

An equilibrium is any set of frequencies for which $\Delta p_i = 0$ for all alleles. Mathematically, the ps can be any value, but we must place some biological restrictions on them. The ps represent allele frequencies, so each p must be between zero and one, and they must sum to one. If these conditions are met, we say an equilibrium exists; sometimes it is called a feasible equilibrium, meaning it is biologically feasible. A stable equilibrium is one that is actually approached with time.[1] A locally stable equilibrium is approached, provided the initial conditions are sufficiently close to it. A globally stable equilibrium is approached irrespective of the initial conditions. We are interested in feasible stable equilibria.

Analysis of the k-allele selection model requires knowledge of linear algebra. Boxes 12.1, 12.2, and 12.3 outline the procedure and an example. Here we summarize the main results:

1. For k alleles, there may be as many as $2^k - 1$ possible equilibria.
2. There is only one possible equilibrium with all k alleles present.
3. That equilibrium is stable if and only if the matrix $\mathbf{W}$ has one positive eigenvalue and $k - 1$ negative eigenvalues. (See Box 12.2 for an introduction to eigenvalues.)
4. If the internal equilibrium is locally stable, it is also globally stable, and the population will move toward it.
5. The mean fitness will increase as the population approaches equilibrium.

Roughgarden (1979) gives a good summary of the mathematics leading to these results; see also Ewens (1969), Mandel (1959), and Kingman (1961) for details.

For two alleles, result 3 above is equivalent to the condition that the heterozygote must have a fitness higher than either homozygote. But for more than two alleles, no such straightforward biological interpretation is possible. By analogy with the two-allele case, one might assume that if all heterozygotes have higher fitness than all homozygotes, a stable equilibrium would exist. But this is not necessarily true; Table 12.1a gives a set of fitnesses where all heterozygotes have higher fitness than all homozygotes, but a stable equilibrium does not exist.

---

[1]Technically, this is an asymptotically stable equilibrium, but is usually called just a stable equilibrium.

Box 12.1   Analysis of the *k* allele selection model

We summarize the theory for three alleles. Using techniques of linear algebra, extension to any number of alleles is straightforward.

Let $\mathbf{W}$ be the matrix of fitness values. For three alleles,

$$\mathbf{W} = \begin{bmatrix} w_{11} & w_{12} & w_{13} \\ w_{21} & w_{22} & w_{23} \\ w_{31} & w_{32} & w_{33} \end{bmatrix}$$

Now, let $\mathbf{W}_i$ be the matrix $\mathbf{W}$, with the *i*th column replaced by ones. For example,

$$\mathbf{W_1} = \begin{bmatrix} 1 & w_{12} & w_{13} \\ 1 & w_{22} & w_{23} \\ 1 & w_{32} & w_{33} \end{bmatrix}$$

Finally, let Δ_i be the determinant of $\mathbf{W}_i$. For example,

$$\Delta_1 = \begin{vmatrix} 1 & w_{12} & w_{13} \\ 1 & w_{22} & w_{23} \\ 1 & w_{32} & w_{33} \end{vmatrix}$$

where the absolute value bars indicate the determinant.

Using this notation, it can be shown that the equilibrium frequency of the *i*th allele is

$$\widetilde{p}_i = \frac{\Delta_i}{\displaystyle\sum_{i=1}^{k} \Delta_i} \tag{1}$$

Roughgarden (1979) describes a relatively clear way to obtain equation (1).

Under what conditions will a stable equilibrium exist with all *k* alleles present? The answer is simple to state, but difficult to prove. Kingman (1961) showed that an equilibrium is stable if and only if the matrix $\mathbf{W}$ has one positive eigenvalue and all other eigenvalues are negative. Box 12.2 gives an introduction to eigenvalues; Box 12.3 gives a numerical example of a stable equilibrium with three alleles.

Box 12.2   Eigenvalues of a matrix

Consider an $n \times n$ matrix $\mathbf{M}$

$$\mathbf{M} = \begin{bmatrix} m_{11} & \cdots & m_{1n} \\ \vdots & \ddots & \vdots \\ m_{n1} & \cdots & m_{nn} \end{bmatrix}$$

where m_{ij} is the entry in the *i*th row and *j*th column.

The characteristic equation of $\mathbf{M}$ is

$$|\mathbf{M} - \lambda\mathbf{I}| = 0$$

where the absolute value bars indicate the determinant. $\mathbf{I}$ is the identity matrix, in which the diagonal entries are all ones and all other entries are zeros. λ is an eigenvalue, a number that makes the above equation true. An $n \times n$ matrix has n eigenvalues. They may be real or complex, and all may not be unique. To illustrate, consider a 2×2 matrix:

$$\mathbf{M} = \begin{bmatrix} a & b \\ c & d \end{bmatrix}$$

Then,

$$\mathbf{M} - \lambda\mathbf{I} = \begin{bmatrix} a - \lambda & b \\ c & d - \lambda \end{bmatrix}$$

and the characteristic equation is

$$(a - \lambda)(d - \lambda) - bc = 0$$

This equation has two roots (eigenvalues), usually designated λ_1 and λ_2. They can be found using the quadratic formula.

For matrices larger than 2×2, the solution to the characteristic equation is usually obtained numerically, using computer programs such as Mathematica, MathCad, or Matlab. The value and nature of eigenvalues is the key to formal stability analysis of differential and difference equations. See any textbook on linear algebra for more details.

Box 12.3   A stable three allele polymorphism

Consider the following fitness matrix, from Lewontin et al. (1978):

$$\mathbf{W} = \begin{bmatrix} 0.236 & 0.846 & 0.748 \\ 0.846 & 0.184 & 0.393 \\ 0.748 & 0.393 & 0.396 \end{bmatrix}$$

Note that one homozygote fitness (w_{33}) is greater than one heterozygote fitness (w_{23}).

As in Box 12.1, define $\mathbf{W}_i$ as the fitness matrix $\mathbf{W}$, with the ith column replaced by ones. Then, Δ_i is the determinant of $\mathbf{W}_i$. This leads to

$$\Delta_1 = 0.072$$
$$\Delta_2 = 0.032$$
$$\Delta_3 = 0.063$$

Next, using equation (1) of Box 12.1, the equilibrium allele frequencies are

$$\tilde{p}_1 = 0.430$$
$$\tilde{p}_2 = 0.194$$
$$\tilde{p}_3 = 0.376$$

The allele frequencies are all between zero and one, and they sum to one. Therefore, the equilibrium is biologically feasible. The equilibrium is stable if the fitness matrix $\mathbf{W}$ has one positive and two negative eigenvalues. The eigenvalues of $\mathbf{W}$ turn out to be

$$\lambda_1 = -0.714$$
$$\lambda_2 = -0.080$$
$$\lambda_3 = 1.609$$

Therefore, the fitness matrix $\mathbf{W}$ will lead to a stable polymorphic equilibrium with all three alleles present.

TABLE 12.1   Two examples of three allele fitness sets.

The second example is explained in more detail in Box 12.3 *Source*: From Lewontin et al. (1978).

a.   All heterozygotes have higher fitness than all homozygotes.

$$\mathbf{W} = \begin{bmatrix} 0.6563 & 0.7462 & 0.8861 \\ 0.7462 & 0.2817 & 0.7654 \\ 0.8861 & 0.7654 & 0.6121 \end{bmatrix}$$

Using the techniques in Box 12.1, the equilibrium allele frequencies are

$$\tilde{p}_1 = 0.552 \qquad \tilde{p}_2 = -0.013 \qquad \tilde{p}_3 = 0.462$$

Since $\tilde{p}_2$ is negative, a biologically feasible, three-allele equilibrium does not exist.

---

b.   One homozygote (A_3A_3) has higher fitness than one heterozygote (A_2A_3).

$$\mathbf{W} = \begin{bmatrix} 0.236 & 0.846 & 0.748 \\ 0.846 & 0.184 & 0.393 \\ 0.748 & 0.393 & 0.396 \end{bmatrix}$$

The equilibrium allele frequencies are

$$\tilde{p}_1 = 0.430 \qquad \tilde{p}_2 = 0.194 \qquad \tilde{p}_3 = 0.376$$

The equilibrium is biologically feasible. The eigenvalues of $\mathbf{W}$ are (see Box 12.3)

$$\lambda_1 = -0.714 \qquad \lambda_2 = -0.080 \qquad \lambda_3 = 1.609$$

Therefore, the equilibrium is stable by Kingman's (1961) criterion.

Table 12.1b gives an example where one homozygote (A_3A_3) has higher fitness than one of its heterozygotes (A_2A_3), but a stable equilibrium does exist. Overdominance is neither a necessary nor a sufficient condition to maintain a stable polymorphism with three alleles. For more than three alleles, interpretation is even more difficult.

Result 3 above gives the *mathematical* condition for a stable k-allele polymorphism, but it provides little *biological* insight. We might try another approach, and ask, What proportion of all possible fitness matrices, **W**, will satisfy this condition? If the proportion is relatively large, we might conclude that multiple allele polymorphisms should be relatively common; but if the proportion is very small, we might want to consider other possible mechanisms.

Lewontin et al. (1978) took this approach. They generated fitness sets at random, and asked what proportion of these produced a stable polymorphic equilibrium with all k alleles present. They considered three kinds of fitness sets: (1) completely random, with no restrictions on any of the fitnesses; (2) pairwise heterosis, where each heterozygote has higher fitness than either of its constituent homozygotes; and (3) total heterosis, where all heterozygotes have higher fitness than all homozygotes. Their results were surprising (Table 12.2). Even if all heterozygotes are more fit than all homozygotes (the most favorable case for polymorphism), the proportion of fitness sets leading to a stable polymorphic equilibrium was very small for more than a few alleles. For five alleles, the proportion of fitness sets that produced a stable polymorphic equilibrium was only about 10 percent; for seven alleles it was only about 0.1 percent; for eight or more alleles, it was zero. The obvious conclusion from these simulations is that a small proportion of random fitness sets will allow a stable multiple allele polymorphism, and that as the number of alleles increases, the proportion decreases.

Lewontin et al. generated random k-allele fitness sets. An alternative approach is to ask whether we can build a multiple allele polymorphism, one allele at a

TABLE 12.2 Proportion of biologically feasible stable equilibria in randomly generated fitness sets for different numbers of alleles.

Completely random means all fitnesses were generated randomly, with no restrictions. Pairwise heterosis means each heterozygote has higher fitness than either of its constituent homozygotes. Total heterosis means all heterozygotes have higher fitness than all homozygotes. Numbers in parentheses are numbers of fitness sets generated. *Source:* From Lewontin et al. (1978).

Number of Alleles	Completely Random (100,000)	Pairwise Heterosis (10,000)	Total Heterosis (10,000)
2	0.33466	1	1
3	0.04237	0.5524	0.7120
4	0.00240	0.1259	0.3433
5	0.00006	0.0116	0.1041
6	0	0.0003	0.0137
7		0	0.0011
8		0	0
9		0	0

time. For example, start with one allele, and add a new allele by mutation. What is the probability that it can "invade" and produce a stable polymorphism? Then add a third allele and see if it can invade to produce a three-allele polymorphism, and so forth. This approach is closer to the way mutation and natural selection interact to produce polymorphisms than the random fitness set approach of Lewontin et al.

The criterion for an allele to invade when rare is straightforward. Assume we have a stable k-allele polymorphism at equilibrium. A new mutation arises. If the new allele is initially rare (e.g., a new mutation) it will be present only in heterozygotes. The marginal fitness of the new allele will be, by equation (12.1),

$$w_{k+1} = \sum_{i=1}^{k} p_i w_{i,k+1}$$

The new allele will survive if its marginal fitness is greater than mean fitness of the population with k alleles, as given by equation (12.2):

$$w_{k+1} > \overline{w} \qquad \textbf{(12.4)}$$

In other words, the average fitness of heterozygotes containing the new allele must be greater than the mean fitness of the existing population. One important implication of this conclusion is that the fate of an allele depends on alleles already in the population, and not solely on its own characteristics.

Spencer and Marks (1988, 1992; Marks and Spencer 1991) assumed a k-allele polymorphism, and generated new mutant alleles with random fitnesses. They asked how many alleles could be maintained in a stable polymorphism by this process. Their results were very different from those of Lewontin et al. They found that under a variety of assumptions, stable polymorphisms with a moderate number of alleles can easily be constructed by adding one allele at a time. In other words, the small proportion of fitness space that can maintain multiple allele polymorphisms can easily be "found" by natural selection. In their most complex simulations, Spencer and Marks (1992) found that most mutant alleles were quickly lost, but that a few invaded the existing population. Their simulations generated polymorphisms with a mean of 7.8 alleles, a number sufficient to explain most observed allozyme polymorphisms (but not highly polymorphic loci, such as the MHC loci). Two caveats must be added to their conclusions: First, we do not know whether mutation rates and effects in nature are sufficient to generate these results. Second, as shown in Section 7.8, most beneficial mutations are lost by genetic drift before they can become established. Spencer and Marks did not include genetic drift in their simulations.

Both the analytical results and the computer simulations suggest that natural selection is not likely to be the sole cause of most multiple allele polymorphisms. This should come as no surprise; we have seen that interactions between neutral mutations and genetic drift seem to explain some of these polymorphisms, especially at the DNA level.

But extensive multiple allele polymorphisms do exist, and natural selection is sometimes suggested as a cause. For example, we saw in Section 10.4 that selection for a diversity of defense mechanisms probably explains the extensive polymorphisms seen at MHC loci. Similarly, self-incompatibility systems in plants require many alleles in order to avoid reduced fertility and low population fitness (Section 8.3).

To summarize, natural selection at a single locus sometimes helps to maintain multiple allele polymorphisms, but the theory suggests that it is unlikely to be the sole cause of the extensive genetic variation seen in most populations. We must consider alternative explanations for this variation. Two things that we have so far assumed are that the locus we are studying does not interact with other loci, and that fitnesses are constant over space and time. We address these issues in the next few sections.

12.2 Two-Locus Models of Selection

Genes do not evolve in isolation; they interact with other genes, both linked and unlinked. Enzymes controlling steps in metabolic pathways must work together to coordinate production of the end product. Genes controlling development must be carefully coordinated in their expression and effects. Thus, gene interaction must be common and, at least sometimes, subject to natural selection. In this section, we examine simple models of interaction between two loci, with the hope that we can gain general insights about the effects of gene interaction on genetic variation and evolution.

Review of Two-Locus Models Without Selection

We begin by reviewing two-locus models without selection. Consider two loci, with two alleles each. Alleles at locus 1 are A and a; at locus 2, B and b. (We designate alleles with upper- and lowercase letters only to minimize subscripts; there is no implication of dominance.) Let p_A and p_a be the allele frequencies at the first locus, and p_B and p_b the allele frequencies at the second locus. The frequency of recombination between the loci is r. There are four gamete types, $AB, Ab, aB,$ and ab, with frequencies $g_1, g_2, g_3,$ and g_4. D is the coefficient of gametic disequilibrium,

$$D = g_1 g_4 - g_2 g_3$$

The recursion equations for the gamete frequencies are given by equations (4.18) through (4.21):

$$g_{1,t+1} = g_1 - rD$$
$$g_{2,t+1} = g_2 + rD$$
$$g_{3,t+1} = g_3 + rD$$
$$g_{4,t+1} = g_4 - rD$$

The recursion equation for the disequilibrium coefficient is

$$D_{t+1} = (1 - r)D$$

These equations are derived and explained in Section 4.2. The essential conclusions are that, without selection, D will decay to zero at a rate that depends on the recombination rate, and that each locus will approach Hardy–Weinberg equilibrium. We now investigate how selection changes these conclusions.

A General Two-Locus Model of Viability Selection

Assume that fitness differences are due to viability differences only; also assume random union of gametes. Under these assumptions, the genotypic fitnesses in a two-locus system can be written as

	AB	*Ab*	*aB*	*ab*
AB	w_{11}	w_{12}	w_{13}	w_{14}
Ab	w_{21}	w_{22}	w_{23}	w_{24}
aB	w_{31}	w_{32}	w_{33}	w_{34}
ab	w_{41}	w_{42}	w_{43}	w_{44}

where the subscripts indicate gamete types. We can simplify the notation somewhat by assuming that it does not matter which parent a gamete comes from; that is, $w_{12} = w_{21}$, and so forth. Also assume that double heterozygotes have the same fitness, whether *AB/ab* or *Ab/aB*; that is, assume that $w_{14} = w_{23}$. Making these simplifications, and rewriting so that the rows and columns of the fitness matrix represent the genotypes at locus 1 and 2, respectively, we get

	BB	*Bb*	*bb*
AA	w_{11}	w_{12}	w_{22}
Aa	w_{13}	w_{14}	w_{24}
aa	w_{33}	w_{34}	w_{44}

Keep in mind that the subscripts still represent gamete types, and not genotypes.

Now, define the marginal fitnesses of the gametes as

$$w_i = \sum_{j=1}^{4} g_j w_{ij}$$

This is just the weighted average of the fitnesses of all genotypes containing gamete type i, weighted by the frequency of the other gamete in the genotype. The w_is are analogous to marginal fitnesses of the one-locus model, equations (5.23) and (12.1). Next, define the mean fitness of the population in the usual way:

$$\overline{w} = \sum_{i=1}^{4} \sum_{j=1}^{4} g_i g_j w_{ij}$$

We can now find the recursion equations for the gamete frequencies. The procedure for finding them is fairly straightforward; Table 12.3 shows the general procedure. The results are

$$g_{1,t+1} = (g_1 w_1 - w_{14} r D)/\overline{w} \tag{12.5}$$

$$g_{2,t+1} = (g_2 w_2 + w_{14} r D)/\overline{w} \tag{12.6}$$

$$g_{3,t+1} = (g_3 w_3 + w_{14} r D)/\overline{w} \tag{12.7}$$

$$g_{4,t+1} = (g_4 w_4 - w_{14} r D)/\overline{w} \tag{12.8}$$

TABLE 12.3   Derivation of the recursion equations for a two-locus, two-allele model.
Fitnesses are as described in the text; r is the recombination rate between the two loci. Only the recursion for g_1 is shown; the others are obtained in the same way.

Genotype	Frequency	Fitness	Gamete Types Produced			
			AB	Ab	aB	ab
AB / AB	g_1^2	w_{11}	$g_1^2 w_{11}$			
AB / Ab	$2g_1 g_2$	w_{12}	$g_1 g_2 w_{12}$	$g_1 g_2 w_{12}$		
AB / aB	$2g_1 g_3$	w_{13}	$g_1 g_3 w_{13}$	$g_2 g_4 w_{24}$	$g_1 g_3 w_{13}$	
AB / ab	$2g_1 g_4$	w_{14}	$(1 - r)g_1 g_4 w_{14}$	$r g_1 g_4 w_{14}$	$r g_1 g_4 w_{14}$	$(1 - r)g_1 g_4 w_{14}$
Ab / Ab	g_2^2	w_{22}		$g_2^2 w_{22}$		
Ab / aB	$2g_2 g_3$	w_{23}	$r g_2 g_3 w_{23}$	$(1 - r)g_2 g_3 w_{23}$	$(1 - r)g_2 g_3 w_{23}$	$r g_2 g_3 w_{23}$
Ab / ab	$2g_2 g_4$	w_{24}		$g_2 g_4 w_{24}$		$g_2 g_4 w_{24}$
aB / aB	g_3^2	w_{33}			$g_3^2 w_{33}$	
aB / ab	$2g_3 g_4$	w_{34}			$g_3 g_4 w_{34}$	$g_3 g_4 w_{34}$
ab / ab	g_4^2	w_{44}				$g_4^2 w_{44}$
Sum	1		$g_{1,t+1}$	$g_{2,t+1}$	$g_{3,t+1}$	$g_{4,t+1}$

$$g_{1,t+1} = \frac{g_1^2 w_{11} + g_1 g_2 w_{12} + g_1 g_3 w_{13} + (1 - r)g_1 g_4 w_{14} + r g_2 g_3 w_{23}}{\overline{w}}$$

$$= \frac{g_1 w_1 - w_{14} r D}{\overline{w}}$$

These equations describe how gamete frequencies change over time. We wish to answer three questions: (1) Under what conditions will there be a stable equilibrium with both loci polymorphic? (2) Will gametic disequilibrium be maintained indefinitely? (3) Will mean population fitness increase with time and reach a maximum at equilibrium?

Equations (12.5) through (12.8) are deceptively difficult equations. (Recall that both $\overline{w}$ and D contain the products of gamete frequencies.) No general solution is known. However, we can analyze special cases and gain some insights from them.

The easiest special case to consider is when $r = 0$. If there is no recombination, then the four gamete types act as four alleles at a single locus, and we can apply the results of Section 12.1. For any given set of fitnesses, we can determine whether a polymorphic equilibrium exists, and whether it is stable.

For $r > 0$, the problem is much more difficult. In general, multiple equilibria are possible, and more than one may be stable. The existence and stability of these equilibria sometimes depend on r. Sometimes gametic disequilibrium decays to zero at equilibrium and sometimes it does not, again depending on r.

Why do we care whether gametic disequilibrium decays to zero? If $D = 0$, each locus evolves independently; one locus selection theory is sufficient to describe the evolution of the population. However, if $D \neq 0$, different loci do not evolve independently; selection at one locus affects allele frequencies at the other. We cannot describe the evolution of the population based on single-locus fitnesses alone.

Natural selection can create gametic disequilibrium by favoring certain gamete types (e.g., *AB* and *ab*) and certain combinations of gametes (e.g., *AB/ab*). But during sexual reproduction, recombination will break up gamete types. Whether alleles at different loci will associate nonrandomly at equilibrium depends on the degree of interaction and the amount of recombination. In general, strong interaction and low recombination rates favor nonrandom associations ($D \neq 0$).

We now examine several special cases in which the fitness matrix takes on a specific structure.

Additive Viability Model

One of the simplest two-locus models is to assume that fitness is additive over loci. This might be the case for two duplicated genes that both produce the same product (enzyme or RNA molecule). Fitness depends on the total amount of the product produced; therefore, the two-locus fitness is the sum of the single-locus fitnesses. Note, this involves no interaction between loci. We can write the fitnesses of an additive viability model as

	BB	**Bb**	**bb**
AA	$\alpha_1 + \beta_1$	$\alpha_1 + \beta_2$	$\alpha_1 + \beta_3$
Aa	$\alpha_2 + \beta_1$	$\alpha_2 + \beta_2$	$\alpha_2 + \beta_3$
aa	$\alpha_3 + \beta_1$	$\alpha_3 + \beta_2$	$\alpha_3 + \beta_3$

This model is mathematically tractable. The general results are:

1. The condition for a stable two-locus polymorphic equilibrium is that there is overdominance at each locus.
2. If this equilibrium exists, it is the only internal equilibrium, and is globally stable.
3. At this equilibrium, $\widetilde{D} = 0$, and the gamete frequencies are equal to the products of the appropriate allele frequencies.
4. Mean fitness of the population is maximized at this polymorphic equilibrium.
5. These results are independent of *r*.

To summarize, in the two-locus additive viability model, each locus behaves independently of the other at equilibrium. (However, during the approach to equilibrium, there will be interaction between the loci affecting allele frequencies.) For the mathematical details, see Moran (1967), Bodmer and Felsenstein (1967), Ewens (1969), and Karlin and Feldman (1970).

Multiplicative Viability Model

Another simple two-locus model assumes that fitness is multiplicative across loci. This could happen if different genes affect viability at different stages in the life cycle. For example, if viability of *AA* is 0.9 and viability of *Bb* is 0.7, then the viability of *AABb* would be 0.63. We can write the fitnesses of such a model as follows:

	BB	Bb	bb
AA	$\alpha_1\beta_1$	$\alpha_1\beta_2$	$\alpha_1\beta_3$
Aa	$\alpha_2\beta_1$	$\alpha_2\beta_2$	$\alpha_2\beta_3$
aa	$\alpha_3\beta_1$	$\alpha_3\beta_2$	$\alpha_3\beta_3$

Some general results are:

1. The condition for a stable two-locus polymorphic equilibrium is that there is overdominance at each locus.

2. If condition 1 holds, then, for sufficiently high recombination, $r > r^*$ (where r^* is a critical value that depends on the fitnesses), there is a single polymorphic equilibrium with $\widetilde{D} = 0$.

3. If condition 1 holds, for sufficiently low recombination, $r < r^*$, there are two stable polymorphic equilibria, one with $\widetilde{D} > 0$ and one with $\widetilde{D} < 0$. Which one is approached depends on whether the initial D is positive or negative.

4. The mean fitness of the population does not necessarily increase each generation.

Note that, unlike the additive viability model, the outcome depends on r, and on the initial conditions.

Symmetric Viability Models

More complex (but still not completely general) models assume various kinds of symmetry in the fitness matrix. The most general form of these symmetric viability models is

	BB	Bb	bb
AA	$1 - \alpha$	$1 - \beta$	$1 - \delta$
Aa	$1 - \gamma$	1	$1 - \gamma$
aa	$1 - \delta$	$1 - \beta$	$1 - \alpha$

Symmetric viability models are very complex. There can be many equilibria, including several stable polymorphic equilibria. The outcome of selection depends on initial gamete frequencies, on initial D, and on r. For example, consider the harmless-looking fitness matrix

	BB	Bb	bb
AA	.9	.8	.9
Aa	.6	1	.6
aa	.9	.8	.9

(Here, $\alpha = \delta = 0.1$, $\beta = 0.2$, and $\gamma = 0.4$.) Hastings (1985) discovered that this fitness set can have *fifteen* equilibria! There are four corner equilibria, a which one gamete type is fixed, and four edge equilibria, at which fixation occurs at one locus and polymorphism is maintained at the other. All eight of these equilibria are unstable. There are three central equilibria, all with $\widetilde{p}_A = \widetilde{p}_B = 0.5$, but with $\widetilde{D} = 0$, $\widetilde{D} > 0$, or $\widetilde{D} < 0$. (The $\sim$ indicates equilibrium, as usual.) These are also unstable.

Depending on the recombination rate, there can be as many as four stable polymorphic equilibria. These are listed in Table 12.4, for $r = 0.09$. All have the same magnitude for $\widetilde{D}$, but two have $\widetilde{D} > 0$ and two have $\widetilde{D} < 0$. These polymorphic equilibria are all locally stable. Which one the population approaches depends on the initial gamete frequencies and the initial D. Figure 12.1a shows several examples. Moreover, depending on initial conditions, natural selection may cause the mean fitness of the population to decrease. Figure 12.1b shows the change in mean fitness for these same four examples. In one case, the mean fitness decreases sharply before beginning to increase again.

No general solution to symmetric viability models is known. As the example above suggests, the mathematics are very complex. Much has been discovered by computer simulations; much remains unknown. See Franklin and Feldman (2000) for a recent review and references to the original literature.

General Two-Locus Models

Even the symmetric viability models are unrealistically restrictive due to the assumed symmetries. How much more complex might general two-locus models be?

Little is known about the behavior of general two-locus models. One approach is to first assume that $r = 0$ and use the results of one-locus multiple allele theory. In this case, the maximum number of possible equilibria is $2^k - 1 = 15$ (four gametes equal four alleles if $r = 0$). It has been argued that the same result holds if r is sufficiently small, and, in fact, we have seen that 15 feasible equilibria can exist for versions of the symmetric viability model. It is not known how many of these equilibria can be stable in a general viability model. For r greater than some critical value, even less is known. The following summarizes some generalizations about general two-locus viability models, based on analytic and simulation studies:

1. If $r = 0$, the two-locus model is equivalent to a one-locus, four-allele model. This implies that there is at most one stable equilibrium with both loci polymorphic.

TABLE 12.4   Stable equilibria for a symmetric viability model with $\alpha = \delta = 0.1$, $\beta = 0.2$, and $\gamma = 0.4$.
The recombination rate is $r = 0.09$. There are also 11 feasible unstable equilibria. *Source:* From Hastings (1985).

Equilibrium	p_A	p_B	D
1	0.125	0.1692811	0.0744118
2	0.125	0.8307189	−0.0744118
3	0.875	0.1692811	−0.0744118
4	0.875	0.8307189	0.0744118

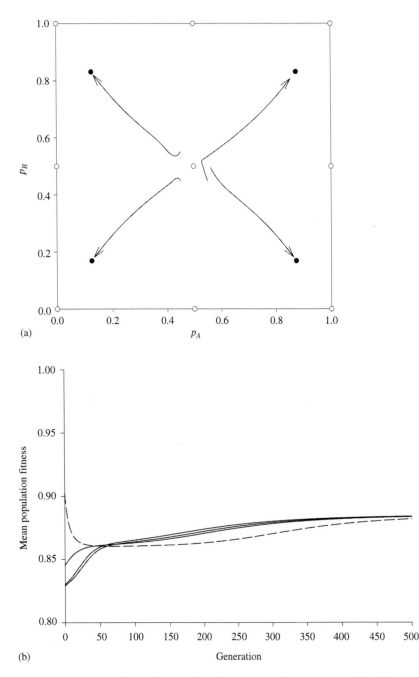

(a)

(b)

Figure 12.1   **Multiple stable equilibria of a two-locus symmetric viability model.**   Horizontal and vertical axes represent the frequency of the A or B allele, respectively. (a) Change in allele frequencies. Each generation, the population moves from one (p_A, p_B) pair to another, according to the recursion equations. Each line represents the movement through time of a population starting with different initial frequencies of A and B, with the arrow indicating the direction of movement. Solid circles indicate stable equilibria; open circles are unstable equilibria. A graph such as this is called a **phase plane**, and the movement of a population through time is called a **trajectory**. Four different trajectories are shown; the population moves toward different stable equilibria, depending on initial allele frequencies and initial gametic disequilibrium. (b) Change in $\overline{w}$ for each of the trajectories in (a). In one case (dashed line), the mean population fitness decreases sharply before gradually rising again; at equilibrium, $\overline{w}$ is less than the initial $\overline{w}$. This case corresponds to the trajectory moving toward the upper right stable equilibrium in (a).
Source: Numerical example from Hastings (1985).

2. Tight linkage alone will not lead to equilibria at which $D \neq 0$. Some degree of gene interaction is required. For higher r, stronger interaction is required.

3. At some stable polymorphic equilibria, $D = 0$, and at some $D \neq 0$. Smaller r and greater gene interaction are more likely to lead to equilibria with $D \neq 0$.

4. If r is less than some critical value (which depends on fitnesses), there can be at most two stable polymorphic equilibria (one with $D > 0$ and one with $D < 0$).

5. If r is greater than some critical value, the number of possible stable polymorphic equilibria is unknown. At least four are possible.

6. As r increases, the population mean fitness at equilibrium *usually* decreases, but not always. This is because high recombination breaks up the best gametes more frequently.

7. The population mean fitness does not always increase with time, and is not necessarily maximized at equilibrium. However, if gene interaction is relatively weak and recombination is not too rare, mean fitness usually increases.

See Hedrick et al. (1978) and Franklin and Feldman (2000) for reviews, references to the original literature, and speculations.

If fitnesses are additive across loci (the additive viability model), one-locus models are adequate. It is only when we have significant fitness interactions among loci that two-locus models become necessary. If gene interaction is rare and weak, then the additive and multiplicative models are reasonable approximations, and the relatively simple conclusions derived for them are approximately true. Maybe we can hold out some hope that the theoretical complexities described above are only theoretical.

How Common Is Gene Interaction?

The earliest evidence for gene interaction came from experiments on chromosomal inversions in *Drosophila*. For example, Dobzhansky and Pavlovsky (1953) established population cages of *D. pseudoobscura* with various combinations of ST (standard) and CH (Chiricahua) inversions. They found that ST and CH chromosomes from the same local population exhibited chromosomal heterosis; a stable and repeatable equilibrium was reached. However, when the ST and CH chromosomes came from different populations, the results were much more variable and less repeatable. They concluded that loci within ST and CH inversions from the same population contained alleles that worked well with each other (gene interaction), and with alleles in the other chromosome type (chromosomal heterosis). This coadaptation had not evolved when ST and CH came from different populations.

Spassky et al. (1965) used a modification of chromosome extraction techniques (Section 2.7) to study interactions between second and third chromosomes in *Drosophila pseudoobscura*. Excluding lethal and semilethal chromosomes, the average viability of flies homozygous for second or third chromosomes was 0.854 or 0.897, respectively, compared to outbred controls. If there was no interaction between the second and third chromosomes, then flies homozygous for both would be expected to have a viability equal to the product of the individual viabilities (multiplicative viabilities). Spassky et al. found that flies homozygous for both chromosomes had an average viability less than the product, indicating interactions between loci on the second and third chromosomes. They found much variation among different chromosome pairs; sometimes, flies homozygous for both chromosomes had higher viability than predicted by the multiplicative model. Seager and Ayala (1982) and Seager et al. (1982) performed similar experiments with *Drosophila melanogaster*. They also found that flies homozygous for both chromosomes had significantly higher fitness than predicted by the multiplicative model, especially when total fitness was considered (Seager et al. 1982). Overall, these experiments indicate extensive and variable interaction between loci on different chromosomes, although it is not known how many loci are involved.

Elena and Lenski (1997b) estimated the extent of gene interactions in *E. coli*. They used transposon mutagenesis to construct mutant strains containing one, two, or three mutations in random genes, and estimated fitnesses of the mutant strains by comparing growth rates to a standard genotype. Their *overall* results were consistent with multiplicative fitnesses and gave no evidence for gene interaction, but this conclusion is misleading. In another series of experiments, they constructed recombinant double-mutant genotypes from pairs of single-mutant strains. This allowed them to compare double-mutant genotypes to each of the single mutants. In 14 out of 27 comparisons, the fitness of the double-mutant genotype differed significantly from the fitness predicted from multiplicative fitnesses. However, seven were higher than predicted and seven were lower. Elena and Lenski concluded that in *E. coli*, gene interaction is common and that positive interactions are about as common as negative interactions.

As we have seen, *D* will decay to zero unless there is interaction between the two loci (ignoring other forces). So, a stable and consistent nonzero *D* can be evidence of gene interaction, if other factors can be eliminated.

One might predict that loci in the same developmental or biochemical pathway might be more likely to interact than pairs of unrelated loci, and therefore to show higher values of *D*. This is frequently true. For example, Zouros and Johnson (1976) studied four allozyme loci on the second chromosome of *Drosophila mojavensis*. They estimated *D* for each pair of loci. The only pair of loci for which *D* was significantly different from zero was *XDH* (xanthine dehydrogenase) and *AO* (acetaldehyde oxidase). These two enzymes are under the same regulatory control, and probably share a cofactor or subunit. Moreover, mutations at other loci simultaneously affect the activity of both enzymes. Thus, there is strong evidence that these loci interact at the biochemical level, and the prediction of significant disequilibrium between them is confirmed.

Gametic disequilibrium between tightly linked nucleotide sites is usually assumed to be transient due to tight linkage and minimal recombination, but in some cases it may indicate interaction between individual nucleotide sites. Schaeffer and Miller (1993) obtained 99 DNA sequences of the *Adh* region of *Drosophila pseudoobscura*. They found 359 segregating sites in a 3.5 kb region. Most pairs of sites, even those tightly linked, showed no significant disequilibrium, but two clusters of sites showed significant disequilibrium. Schaeffer and Miller eliminated population subdivision and lack of recombination as causes, and concluded that disequilibrium was probably maintained by interactions among sites in these clusters, possibly due to selective constraints on the secondary structure of the pre-mRNA molecule.

Kirby et al. (1995) tested this hypothesis. They inferred the secondary structure of the *Adh* pre-mRNA molecules. Both clusters of disequilibrium found by Schaeffer and Miller were involved in stem-loop structures of the pre-mRNA molecule. Figure 12.2 shows the inferred stem-loop structure of one part (the adult intron) from the two most common haplotypes. Asterisks indicate polymorphic sites, and significant disequilibrium exists between almost all polymorphic sites in this region of the molecule. Kirby et al. attribute this disequilibrium to intramolecular interactions that stabilize the secondary structure of the pre-mRNA molecule, as proposed by Schaeffer and Miller.

Keep in mind that other forces, for example, genetic drift, mutation, population subdivision, population bottlenecks, or directional selection with hitchhiking, can generate gametic disequilibrium. Tight linkage or restricted recombination retards the decay of gametic disequilibrium. Thus, nonzero D does not necessarily indicate gene interaction. However, the evidence is strong that gene interaction is common in natural populations.

Evolution of Recombination Rates

Natural selection can favor nonrandom associations between alleles at different loci. But recombination will break up these associations. If selection strongly favors nonrandom associations, we might expect selection to also favor reduced recombination. This idea is known as the **reduction principle**, and was first suggested by Fisher (1930).

For example, a chromosomal rearrangement that brings interacting genes into tight linkage might be favored. This is thought to be how supergenes have evolved. We discussed supergenes briefly in Section 4.2; they are clusters of tightly linked genes that effectively act as a single gene because there is very little recombination among them. The genes controlling shell color and banding patterns in the snail *Cepaea nemoralis* are a well-known example. Natural selection

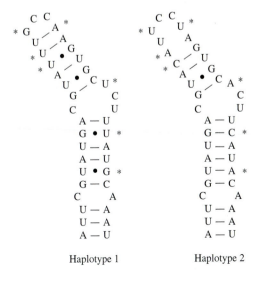

Haplotype 1   Haplotype 2

Figure 12.2 **Inferred secondary structure of part of the pre-mRNA molecule coded for by the *Adh* gene in *Drosophila pseudoobscura*.** The two most common haplotypes are shown. Asterisks indicate polymorphic sites, and disequilibrium among these sites is nearly complete. This disequilibrium is interpreted as the result of natural selection to preserve the stem-loop structure of this part of the molecule. *Source:* From Kirby et al. (1995); asterisks added.

favors certain combinations of shell color and banding pattern, and Ford (1975) has suggested that chromosomal rearrangements that reduce recombination between these genes have been favored, resulting in a supergene. See Section 4.2 for more on these genes. The genes controlling heterostyly in the primrose are another well-known example (Section 8.3). For reviews of supergenes, see Ford (1975), Jones et al. (1977), and Hedrick et al. (1978).

Selection can also affect modifier loci that control recombination rates at other loci. Consider two loci with alleles A and a or B and b, as we have done so far. Assume random mating and that the fitnesses are such that a stable two-locus polymorphism exists with $D \neq 0$, and that the population is at or near this equilibrium. Now, consider a third locus that itself does not affect fitness (viability), but affects the recombination rate between the first two loci. Introduce a new mutation at this locus that changes the recombination rate. Under what conditions will this new mutation be favored? Feldman et al. (1980) and Liberman and Feldman (1986) showed that modifier alleles will invade the population if they decrease the recombination rate. In other words, under the conditions specified, natural selection will always favor reduced recombination.

This conclusion is based on several important assumptions: The population mates randomly and is at or near the polymorphic equilibrium; fitnesses are constant over space and time; and genetic drift can be ignored. If these conditions are not met, natural selection may favor increased recombination. The conditions under which increased recombination is favored are complex, and sensitive to the form of the fitness matrix and the way in which fitnesses change. For example, we saw in Section 6.6 that under some circumstances, recombination may aid in the elimination of deleterious mutations. Also, since recombination creates genetic variation (new combinations of alleles) it may be favored in variable environments.

Consider directional selection in which alleles A and B are favored, irrespective of the gamete type or genotype they are in. If disequilibrium is positive, A and B are associated more often than expected. Selection will favor reduced recombination to keep them together. Conversely, if disequilibrium is negative, A and B are together less often than expected. Selection will initially favor increased recombination to bring them together. This can occur, for example, if A and B are new advantageous mutations that arise in different individuals. The conclusion is that under directional selection, either increased or decreased recombination can be favored, depending on initial conditions.

For selection on recombination to be effective, a population must have genetic variation for recombination rates. Several studies have shown that such variation does exist. For example, Brooks and Marks (1986) demonstrated substantial variation in recombination rates among loci on three chromosomes of *Drosophila melanogaster*. Otto and Barton (2001) list several laboratory experiments in which recombination rates changed substantially (usually increasing) as a result of selection for other characteristics.

As this summary suggests, the evolution of recombination rates is a complex topic, and the mathematical modeling represents a "formidable challenge" (Otto and Michalakis 1998). For good reviews and an introduction to the original literature, see Feldman et al. (1996) and Otto and Michalakis (1998).

Beyond Two Loci

We have seen that going from one-locus to two-locus selection models introduces an enormous complexity, and that the proper units of study are no longer allele frequencies, but gamete frequencies. Most organisms have hundreds or thousands of polymorphic loci, both linked and unlinked. It is reasonable to assume, based on biochemistry and developmental biology, that a large proportion of these loci potentially interact with one another, and these interactions may involve many loci, not just pairs, as we have considered so far. We have little theoretical understanding of the consequences of these higher-order gene interactions.

If we cannot comprehend the potential complexity of two-locus systems, how can we hope to understand the interactions of hundreds or thousands of loci? To illustrate the point, consider some results of simple three-locus models studied by Feldman et al. (1974) and Karlin and Liberman (1976). They found that as many as 255 equilibria can exist, of which 193 can be internal (all three-loci polymorphic). Simulation results typically showed from four to ten stable interior equilibria, but more may exist. Little is known about multilocus models, and one wonders whether they are likely to provide useful biological insight.

Why does it matter? If multilocus gene interaction is minimal, then evolution can be described by changes in allele frequencies. However, if gene interaction is extensive, then allele frequencies are insufficient, and we must consider larger units of selection, gametes, chromosomes, or even entire genomes. Thus, it is important to know something about the frequency and strength of multilocus interactions.

Langley et al. (1974) found few three-locus interactions among allozymes in *Drosophila melanogaster*. This is probably due to high recombination and weak selection, as mentioned above. In another experiment, Clark (1987b) found few and weak higher-order interactions among three or more chromosomes in *D. melanogaster*, but offered several cautions about generalizing his results. Both of these results are consistent with Hastings' (1986) suggestion that with weak selection and weak interaction, higher-order disequilibrium should be rare.

One approach to multilocus population genetics is to (temporarily) abandon the explicit genetic approach and to study the statistics of phenotypic variation. This is the subject of quantitative genetics, to be discussed in Chapter 13.

12.3 Selection in Spatially Varying Environments

It seems obvious that environments vary spatially, and that the fitness of a genotype may depend on the environment in which it is found. In Section 9.4, we concluded that if selection is strong compared to gene flow, equilibrium allele frequencies in different subpopulations will depend primarily on the local selection coefficients, but increasing gene flow limits the effectiveness of local selection. This assumed that random mating occurred within each subpopulation. In this section, we consider models in which selection differs in each subpopulation, but mating occurs more or less randomly among all individuals in all subpopulations. Think of a single randomly mating population inhabiting an environment divided into different ecological niches in which selection acts differently. We are interested in whether a stable polymorphism is more likely in a spatially varying environment, compared to a uniform environment.

Consider a simple model. Let the environment be divided into two niches. Let the A_1 allele be favored in niche 1, and the A_2 allele in niche 2, with the heterozygote having intermediate fitness in both niches. Then, if there is no gene flow between niches, the A_1 allele will become fixed in niche 1 and the A_2 allele will become fixed in niche 2. The overall frequency of the A_1 allele will depend on the proportion of the environment that is niche 1. A polymorphism will persist as long as both niches are available, but each niche will be monomorphic.

What happens if gene flow occurs between niches? The conditions for polymorphism become much more complicated, as we shall see.

The Levene Model

Levene (1953) was the first to address the question of polymorphism in a multiple-niche model. Consider a population which inhabits n different niches. We will standardize fitnesses to the heterozygote, and designate the fitnesses in the ith niche as

genotype	A_1A_1	A_1A_2	A_2A_2
fitness	W_i	1	V_i

The relative proportion of individuals in the ith niche is c_i. Assume the entire population mates at random, and then individuals disperse randomly into the different niches. If we let p be the overall frequency of the A_1 allele before selection, and p_i be the frequency of the A_1 allele in niche i after selection, then we can use equation (5.7) to determine the frequency of A_1 after selection in the ith niche

$$p_i = \frac{p^2 W_i + p(1 - p)}{\overline{w}_i}$$

This comes from equation (5.7) by letting $w_{11} = W_i$, and $w_{12} = 1$. The denominator, $\overline{w}_i$, is the mean fitness in the ith niche, calculated in the usual way.

If we let $\Delta p_i = p_i - p$ be the change in frequency due to selection in the ith niche, then a lot of algebra will give

$$\Delta p_i = \frac{p(1 - p)[1 - V_i + p(W_i + V_i - 2)]}{p^2(W_i + V_i - 2) + 2p(1 - V_i) + V_i}$$

Finally, the *overall* change in allele frequency is the weighted average of the individual changes

$$\Delta p = \sum_{i=1}^{n} c_i \Delta p_i = \sum_{i=1}^{n} c_i \frac{p(1 - p)[1 - V_i + p(W_i + V_i - 2)]}{p^2(W_i + V_i - 2) + 2p(1 - V_i) + V_i} \tag{12.9}$$

Remember, our goal is to find the conditions under which a polymorphism will be maintained in the population. A *sufficient* condition for this is that Δp be positive for p near zero (i.e., A_1 will increase when rare), and Δp be negative for p near one (A_2 will increase when rare). This ensures that neither allele will be lost. We are not concerned with what happens between $p = 0$ and $p = 1$, as long as neither of these end points can be reached. This is what Prout (1968) has called a **protected polymorphism**.

By analogy to the single-niche case, we might speculate that a protected polymorphism will exist if the arithmetic mean of the heterozygote fitnesses is higher than the arithmetic mean of the homozygote fitnesses; that is,

$$\overline{W} < 1 \quad \text{and} \quad \overline{V} < 1$$

where $\overline{W}$ and $\overline{V}$ are the arithmetic means of the Ws and Vs, respectively. (Remember that the fitness of the heterozygote is 1 in all niches.) This is known as **marginal overdominance**; the average fitness of heterozygotes is higher than the average fitness of either homozygote. In a single niche, marginal overdominance is the necessary and sufficient condition for a stable polymorphism. Is this true for multiple niches also? Levene showed that marginal overdominance is not necessary for a multiple-niche polymorphism. He showed that a sufficient condition for a protected polymorphism is

$$\sum_{i=1}^{n} \frac{c_i}{W_i} > 1 \quad \text{and} \quad \sum_{i=1}^{n} \frac{c_i}{V_i} > 1$$

We can rewrite these inequalities as

$$\frac{1}{\sum_{i=1}^{n} \frac{c_i}{W_i}} < 1 \quad \text{and} \quad \frac{1}{\sum_{i=1}^{n} \frac{c_i}{V_i}} < 1 \tag{12.10}$$

The left-hand sides of these inequalities are the **harmonic means** of the Ws and Vs, respectively (see Box 12.4). It can be shown that the harmonic mean is always less than the arithmetic mean. Therefore, a sufficient condition for a multiple-niche polymorphism is less restrictive than that for a single-niche model. Multiple niches make a polymorphism easier to maintain.

How much easier? Not much, as it turns out. Figure 12.3 illustrates a two-niche example, where if $c_1 = c_2 = 0.5$, a protected polymorphism exists. But, for the same fitnesses, if $c_1 = 0.53$ and $c_2 = 0.47$, the polymorphism disappears. The sizes of the niches and the fitnesses must be within a narrow range of values for both inequalities in equation (12.10) to hold. Furthermore, the weaker the selection, the narrower the permissible range of the niche sizes.

Variations on the Levene Model

The Levene model makes many unrealistic assumptions. Numerous alternative models have been proposed, making different assumptions about population regulation (Dempster 1955), dominance (Prout 1968), and so forth. Hedrick (1986) reviews many of them.

Perhaps the most unrealistic assumptions of the Levene model are that mating is random in the entire population, and that individuals settle into niches at random. We can imagine models of nonrandom mating, in which individuals are more likely to mate with individuals from the same niche, or of habitat selection, in which individuals mate at random but are more likely to lay eggs in the niche in which they were raised. These models have been discussed by Maynard Smith

---

BOX 12.4   Arithmetic, geometric, and harmonic means

You are familiar with the **arithmetic mean**; it is just the average of a collection of numbers. The algebraic formula is

$$\overline{X}_a = \frac{1}{n}\sum_{i=1}^{n} X_i$$

where the X_i are the individual items and n is the number of items being averaged.

The **geometric mean** is defined as

$$\overline{X}_g = \left(\prod_{i=1}^{n} X_i\right)^{1/n} = \sqrt[n]{X_1 X_2 \cdots X_n}$$

The **harmonic mean** is defined as

$$\overline{X}_h = \frac{n}{\sum_{i=1}^{n}\frac{1}{X_i}}$$

For example, consider the five numbers 1, 3, 5, 7, and 9. The arithmetic mean is

$$\overline{X}_a = \frac{1}{5}(1 + 3 + 5 + 7 + 9) = 5.0$$

The geometric mean is

$$\overline{X}_g = \sqrt[5]{1 \times 3 \times 5 \times 7 \times 9} = 3.94$$

The harmonic mean is

$$\overline{X}_h = \frac{5}{\frac{1}{1} + \frac{1}{3} + \frac{1}{5} + \frac{1}{7} + \frac{1}{9}} = 2.80$$

You will often need to calculate means based on frequencies of different items. Let X_i be the value of one item and f_i be its relative frequency ($\Sigma f_i = 1$); n is now the number of *different* items. The equivalent formulas for the three kinds of means are

$$\overline{X}_a = \sum_{i=1}^{n} f_i X_i$$

$$\overline{X}_g = \prod_{i=1}^{n} (X_i)^{f_i}$$

$$\overline{X}_h = \frac{1}{\sum_{i=1}^{n}\frac{f_i}{X_i}}$$

The last equation is the form seen in equation (12.10).

For example,

X	frequency
3	0.3
5	0.5
6	0.2

The arithmetic mean is

$$\overline{X}_a = 0.3 \times 3 + 0.5 \times 5 + 0.2 \times 6 = 4.60$$

The geometric mean is

$$3^{0.3} \times 5^{0.5} \times 6^{0.2} = 4.45$$

The harmonic mean is

$$\overline{X}_h = \frac{1}{\frac{0.3}{3} + \frac{0.5}{5} + \frac{0.2}{6}} = 4.29$$

An important relationship among the means is

$$\overline{X}_h \leq \overline{X}_g \leq \overline{X}_a$$

where equality holds only if all the items are the same. This relationship is used in Sections 12.3 and 12.4 in judging the likelihood of polymorphism in spatially or temporally varying environments compared to constant environments.

(1966, 1970), Karlin (1982), and others. Figure 12.4 compares the likelihood of polymorphism among these models. The solid lines represent the upper and lower limits of c_1 that will permit polymorphism in a two-niche Levene model (random mating and no habitat selection). Clearly, for weak selection, the sizes of the two niches must be very near 0.5 each. The dashed lines represent the limits

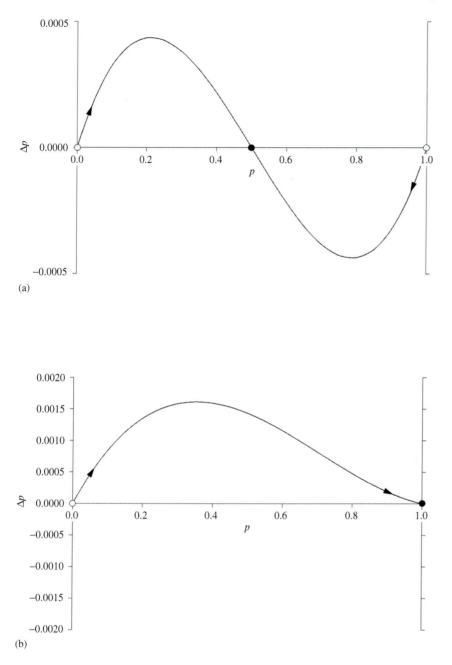

(a)

(b)

Figure 12.3 **Polymorphism in a two-niche Levene model.**   Solid circles are stable equilibria; open circles are unstable equilibria. The fitnesses are $W_1 = V_2 = 1.1$, and $W_2 = V_1 = 0.91$. (a) The niches are equal size, $c_1 = c_2 = 0.5$, and a stable polymorphism is maintained. (b) $c_1 = 0.53$ and $c_2 = 0.47$; no polymorphism is possible.

under a model of habitat selection, but with random mating. The range of permissible niche sizes is only slightly greater than under the pure Levene model. The dotted lines represent the limits under a model of both habitat selection and nonrandom mating. The range of permissible niche sizes is substantially greater than in the other models.

The general conclusion from these models is that spatial variation in fitnesses can increase the likelihood of a stable polymorphism, but the conditions are fairly restrictive, especially if selection is weak and niche sizes are very different.

Figure 12.4   Comparison of stability requirements for a Levene model and variations.   Fitnesses are $1, 1 - .5s$, and $1 - s$ in niche 1; and $1 - s, 1 - .5s$, and 1 in niche 2. The vertical axis is c_1, with $c_2 = 1 - c_1$. The region between the solid lines allows polymorphism under a standard Levene model with random mating within and among niches, and no habitat selection (preference). For weak selection, the sizes of the two niches must be very nearly equal. The region between the dashed lines allows polymorphism under a model of random mating and partial habitat selection (individuals lay a greater than random proportion of their eggs in the niche in which they were born). The region between the dotted lines allows polymorphism under a model of partial habitat selection and mating within niches only. *Source:* From Hedrick (1986).

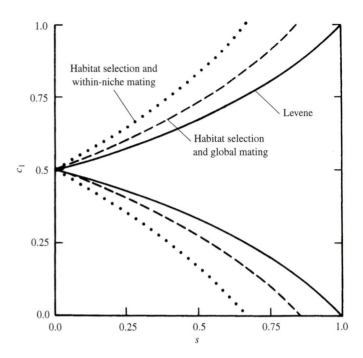

Nonrandom mating and habitat selection both can increase the likelihood of a stable polymorphism. For further discussion of models of spatial variation, see Maynard Smith and Hoekstra (1980), Hoekstra et al. (1985), Hedrick (1986), Gillespie (1991), and references therein. For an interesting example of a stable two-niche polymorphism with heterozygote *inferiority*, see Wilson and Turelli (1986).

Spatially Varying Selection in Natural Populations

Multiple-niche polymorphisms are common in nature. For most, the genetic basis of the polymorphism is either unknown or the character under selection is a quantitative trait (Chapter 13). Only a few cases have been documented in which the character is based primarily on a single gene, and selection has been demonstrated in different ecological niches.

Clarke et al. (1963) studied larval color patterns in the butterfly *Papilio demodocus*. Two main patterns were found, which they called the citrus pattern and the umbellifer pattern, based on alternative host plants. Breeding studies indicated that the alternative patterns were controlled mainly by a single gene. The umbellifer pattern was partially dominant, with some variation in the degree of dominance, due to modifier genes. By comparing frequencies of the two patterns among early and late instar larvae, they demonstrated viability selection favoring the citrus pattern on citrus plants and the umbellifer pattern on umbellifers. They concluded that selection was due to predation by birds, and that some degree of habitat choice aided maintenance of the polymorphism.

Schmidt and Rand (2001) studied a population of barnacles that appears to fit the assumptions of the Levene model. Acorn barnacles (*Semibalanus balanoides*) were collected at five stages of the life cycle: planktonic larvae, recently settled larvae, metamorphs, early juveniles, and midseason juveniles. The latter four stages were collected from four different habitats:

XH   exposed (without an algal canopy), high intertidal zone
AH   algal canopy, high intertidal zone
XL   exposed, low intertidal zone
AL   algal canopy, low intertidal zone

These habitats differ in temperature and desiccation induced stress, with XH being the most stressful and AL the least. Three genetic markers were examined for each individual: *Gpi* (glucose-6-phosphate isomerase), *Mpi* (mannose-6-phosphate isomerase), and a restriction site polymorphism in the mitochondrial DNA control region. Only the *Mpi* locus showed significant spatial and temporal heterogeneity. Schmidt and Rand found that, between the metamorph and early juvenile stages, the frequency of the *Mpi FF* genotype increased in the XH environment and decreased in the AH environment. The *Mpi SS* genotype showed the opposite pattern. From changes in genotype frequencies, Schmidt and Rand were able to estimate relative viabilities of the three genotypes in each environment. For the XH and AH environments, their results were (standardized to the heterozygote):

Genotype	*SS*	*SF*	*FF*
Viability in XH	0.696	1	1.424
Viability in AH	1.519	1	0.880

The conditions of (12.10) are not met. However, recall that these are *sufficient* but not *necessary* conditions; it is possible that a stable polymorphism will occur even if these conditions are not met. Schmidt and Rand found that for these fitnesses, a stable polymorphism does exist for some range of niche sizes (see Problem 12.5). They concluded that their data are consistent with the Levene model and that selection in different habitats is the primary factor maintaining the *Mpi* polymorphism.

Schmidt and Rand discuss reasons to believe that selection is acting directly on the *Mpi* locus. For example, laboratory experiments demonstrated that *FF* genotypes grow better in stressful environments when their diets are supplemented with mannose. Regardless of the mechanism of selection, this study is a good example of a single-locus polymorphism apparently maintained by environmental heterogeneity as in a Levene-type model.

12.4 Selection in Temporally Varying Environments

Just as environments vary in space, they vary in time. Any long-lived organism lives through seasonal changes: Winter and summer are very different in temperate regions; many areas alternate wet and dry seasons. Environments also vary randomly and unpredictably over time. Droughts and floods, hot spells and cold spells are facts of life. Organisms must deal with these changing conditions if they are to survive and reproduce. Different conditions may exert very different selection pressures on these organisms at different times.

Intuition might suggest that temporally varying environments would favor polymorphism in a population. If a population must experience different environments at different times, it would seem advantageous to have some individuals adapted to each environment.

We can imagine two kinds of temporally varying environments. Selection can vary among generations, where different generations experience different environments, but any individual experiences only one. Selection is constant within a generation, but varies among generations. Second, selection can vary within a generation. Long-lived organisms experience seasonal and annual variation, and an individual will experience different selection pressures at different times in its life.

In this section, we examine these ideas. Again, the main question of interest is; Under what conditions will temporally varying environments promote the maintenance of a stable genetic polymorphism? We consider only a few simple models and examples.

Temporal Variation Among Generations

Imagine an insect that has one generation per year, and overwinters in, say, the pupal stage. Assume that most winters are "typical," but occasionally a particularly severe winter occurs. Assume also that a genotype exists that is particularly resistant to severe winters, but develops more slowly during normal years. This genotype has an advantage during severe winters, but is at a disadvantage during normal years. Under what conditions will the population remain polymorphic for both the normal form and the cold resistant form?

Haldane and Jayakar (1963) were the first to study this question. They considered a simple model in which different generations experience different environments, and genotypes have different fitness in different environments. Using similar notation as in the previous section, we will designate the fitnesses in the ith generation as

Genotype	A_1A_1	A_1A_2	A_2A_2
Fitness in ith generation	W_i	1	V_i

We have again scaled the fitnesses to the heterozygote. We wish to know under what conditions both alleles will be maintained in the population, when the fitnesses vary as described above. Haldane and Jayakar (1963) showed that a protected polymorphism will exist for T generations if

$$\left(\prod_{i=1}^{T} W_i\right)^{1/T} < 1 \quad \text{and} \quad \left(\prod_{i=1}^{T} V_i\right)^{1/T} < 1 \qquad \textbf{(12.11)}$$

where T is the number of generations being considered. The left-hand sides of these two inequalities are the **geometric means** of the Ws and Vs, respectively (see Box 12.4). So, under the assumptions of the model, a sufficient condition for a protected polymorphism in a temporally varying environment is that the geometric means of both homozygote fitnesses be less than one (the fitness of the heterozygote). If we let $\overline{W}_g$ and $\overline{V}_g$ represent the geometric means of the

Ws and Vs, respectively, then we can write the condition for polymorphism as

$$\overline{W}_g < 1 \quad \text{and} \quad \overline{V}_g < 1$$

Consider a simple three generation example. Assume the fitnesses are as follows:

Genotype	A_1A_1	A_1A_2	A_2A_2
Fitness in generation 1	1.1	1	0.80
Fitness in generation 2	1.1	1	0.80
Fitness in generation 3	0.80	1	1.1

In other words, the direction of selection reverses every third generation. Will a polymorphism be maintained? The geometric means of the Ws and Vs are 0.989 and 0.890, respectively. Therefore, a polymorphism will be maintained.

However, consider the following, slightly different fitnesses:

Genotype	A_1A_1	A_1A_2	A_2A_2
Fitness in generation 1	1.1	1	0.85
Fitness in generation 2	1.1	1	0.85
Fitness in generation 3	0.85	1	1.1

The geometric means of the Ws and of the Vs are 1.01 and 0.926. A polymorphism will *not* be maintained. This suggests that the conditions for polymorphism are quite restrictive. This is the same conclusion we reached about weak selection in spatially varying environments.

Equation (12.11) implies that a polymorphism will be maintained if directional selection usually favors the A_1A_1 genotype, but occasionally favors the A_2A_2 genotype, provided the occasional selection favoring A_2A_2 is strong enough, and is not too rare. For example, if in five out of six years, the fitnesses are $W_i = 1.1$ and $V_i = 0.9$, but in the sixth year, $W_i = 0.55$ and $V_i = 1.1$, then $\overline{W}_g = 0.98$ and $\overline{V}_g = 0.93$, and both alleles will be maintained in the population. On the other hand, if seven of eight years favor A_1A_1, then $\overline{W}_g = 1.009$, and the A_2 allele will be lost. The direction of selection must change relatively frequently if a polymorphism is to be preserved.

The classic example of temporally varying selection is seasonal variation of inversion frequencies in *Drosophila pseudoobscura*. Dobzhansky (1943), in the first of many papers on the subject, documented seasonal changes in the frequencies of the ST and CH inversions at two locations in California. In general, the frequency of ST increased in warm weather and decreased in colder weather. The changes were cyclic, and repeated over several years. Figure 12.5 shows the results for one location. In subsequent experiments, Wright and Dobzhansky (1946) experimentally reproduced some of these changes. The ST inversion consistently increased in frequency in population cages kept at 25 °C, but not in cages kept at 16.5 °C. Clearly, the ST inversion is better adapted to higher temperatures than CH. These were the first of many studies on temporal variation of inversion frequencies in this and related species; for a good summary, see Powell (1997).

Peter and Rosemary Grant and their colleagues have documented reversal of selection on body size and beak shape in one of Darwin's finches, the medium ground finch *Geospiza fortis* (Grant and Grant 1995, and papers cited therein). In

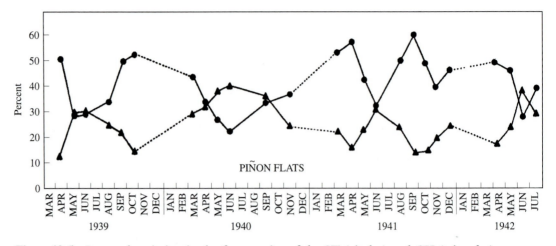

Figure 12.5 **Seasonal variation in the frequencies of the ST (circles) and CH (triangles) gene arrangements of *Drosophila pseudoobscura* at Piñon Flats, CA.** *Source:* From Dobzhansky (1943).

dry years with plants producing few small seeds, selection favors larger birds with deeper beaks. In wet years, selection favors smaller birds with shallower beaks. These are quantitative traits, not single-locus traits, but the principle of varying selection in different years holds, and clearly contributes to genetic variation for both traits. We will consider quantitative traits, and this example, in more detail in Chapter 13.

In Section 12.3 we found that the condition for polymorphism in a spatially varying environment was that the harmonic means of the Ws and Vs must be less than one. In Box 12.4, it is shown that

$$\bar{X}_h < \bar{X}_g < \bar{X}_a$$

where the subscripts indicate the harmonic, geometric, and arithmetic means, respectively. It follows that the conditions for polymorphism in a spatially varying environmentt ($\overline{W}_h < 1$ and $\overline{V}_h < 1$) are less restrictive than the conditions for polymorphism in a temporally varying environment ($\overline{W}_g < 1$ and $\overline{V}_g < 1$). Both are less restrictive than the conditions for polymorphism in a constant environment ($\overline{W}_a < 1$ and $\overline{V}_a < 1$).

What can we conclude from these models of temporal variation? They suggest that polymorphism is only slightly more likely in a temporally varying environment than in a constant environment. More complex models often lead to the same general conclusion; however, see Section 12.5.

Temporal Variation Within Generations

A second kind of temporally varying environment occurs when the environment changes within the time span of a single generation; individuals experience an environment that changes over time. For example, insects that develop in decaying fruit or animal dung experience a change in the environment as they mature. The composition of organic compounds changes; metabolic waste products build up; and the fruit or dung dries out. Individuals must be able to withstand a variety of conditions as they develop. Similarly, long-lived organisms experience cyclic or randomly varying environments from year to year.

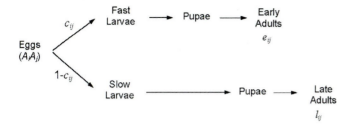

Figure 12.6 **Model of a temporally changing environment within a single generation.**   For genotype A_iA_j, a fraction c_{ij} of eggs mature early, with egg to adult viability of e_{ij}. A fraction $(1 - c_{ij})$ of eggs mature late, and have egg to adult viability of l_{ij}. *Source:* Based on model of Borash et al. (1998).

Borash et al. (1998) investigated a model of development in a deteriorating environment in which some individuals emerge early and others emerge late. Figure 12.6 illustrates the model in terms of the *Drosophila* life cycle, which motivated it. Fast developing larvae become adults early in the environmental progression. Slow developing larvae emerge later and experience a different larval environment, with different selective pressures and different viability. Let c_{ij} be the proportion of genotype A_iA_j that emerge early, and $1 - c_{ij}$ be the proportion that emerge late. Let e_{ij} and l_{ij} be the viability of early and late emerging individuals of genotype A_iA_j. Borash et al. showed that, for two alleles, the conditions for a protected polymorphism are

$$c_{12}e_{12} + (1 - c_{12})l_{12} > c_{11}e_{11} + (1 - c_{11})l_{11}$$

$$\text{and} \tag{12.12}$$

$$c_{12}e_{12} + (1 - c_{12})l_{12} > c_{22}e_{22} + (1 - c_{22})l_{22}$$

Look at the first equation. The left-hand side is the weighted average of viabilities of the early and late emerging heterozygotes. The right-hand side is the weighted average of the early and late emerging A_1A_1 homozygotes. The equation says that A_2 will increase when rare if the weighted average of heterozygotes is greater than the weighted average of A_1A_1 homozygotes. Similar logic applies to the second equation, which shows that A_1 will increase when rare if heterozygotes have a higher average viability than the A_2A_2 homozygotes. This should sound familiar and make intuitive sense. Once again, we find that the condition for polymorphism is a kind of "average overdominance" of the heterozygote.

How often will equation (12.12) be satisfied? According to Borash et al., this model has a broad range of parameter values that will lead to a protected polymorphism. They suggest that this kind of temporally variable selection may be important in many kinds of organisms.

Borash et al. conducted experiments with *Drosophila* to test their hypothesis. *Drosophila* medium undergoes complex changes as development occurs: Larval crowding increases, ammonia levels increase, acetic acid levels increase, ethanol levels decrease, and so forth. To summarize a complex series of experiments, late emerging larvae had higher viability in crowded, but not in uncrowded, conditions, and had significantly higher viabilities in high ammonia environments. Early emerging larvae had lower viability under crowded conditions and high ammonia concentration. Borash et al. conclude that genetic variation for developmental time, competitive ability at high densities, and ammonia tolerance is maintained by selection in a temporally varying environment that favors two alternative phenotypes, early and late emerging, with characteristics favored by the two different environments. They think that this kind of balancing selection may be common.

This model is somewhat different from another commonly encountered situation. It is possible that individuals may experience conflicting selection pressures at different stages in the life cycle; for example, A_1A_1 may be favored in larvae, and A_2A_2 favored in adults. The difference is that all individuals must encounter both kinds of selection. This is known as antagonistic pleiotropy, and will be discussed in Section 12.7.

12.5 Selection in Randomly Varying Environments

Until now, we have assumed that the environment varies in some regular or repeatable way. Obviously, that is not always true. Temperature, rainfall, food availability, predator density, and so forth can all vary in random and unpredictable ways. In this section, we consider the effect of randomly varying environments on the maintenance of polymorphisms.

John Gillespie has studied a class of models involving both spatial and temporal variation in fitnesses. He calls these models SAS–CFF models, and they were originally developed to explain extensive enzyme polymorphism revealed by electrophoresis in the 1970s. SAS–CFF stands for stochastic additive scale—concave fitness function. The meaning of this cryptic phrase will become clearer as we describe the models.

Gillespie assumes that some biological characteristic, for example, enzyme activity, is additive with respect to genotype; that is, heterozygotes are halfway between the two homozygotes. Gillespie considers multiple alleles, but we will consider only a two-allele model, and symbolize this assumption as

Genotype	A_1A_1	A_1A_2	A_2A_2
Enzyme activity	$1 + x$	1	$1 - x$
Fitness	1	$1 - hs$	$1 - s$

(Fitnesses are also given for reference below.) Here, enzyme activity is scaled so that the heterozygote has an activity of one. See Figure 12.7a. The assumption of additivity is based on the idea that total enzyme activity is the sum of the activities of the individual alleles, so that the activity of the heterozygote is the average of the two homozygotes. Heterozygote intermediacy with respect to enzyme activity is well documented.

Next, it is assumed that x is a random variable, which can be positive (A_1A_1 has highest activity) or negative (A_2A_2 has highest activity), but is always near zero. This explains the "stochastic additive scale" part of the model.

The third assumption is that enzyme activity affects fitness in such a way that fitness is a concave function of enzyme activity (concave fitness function; Figure 12.7b). The concavity of the fitness function assures that the fitness of the heterozygote is always closer to the fitness of the best homozygote ($h < 0.5$). The idea of a concave fitness function was proposed by Wright (1934), and extended by Gillespie and Langley (1974) and by Kacser and Burns (1981). We saw in Section 6.4 that for weak selection, $s \cong 0$ and $h \cong 0.5$; therefore, the fitness of the heterozygote will be approximately halfway between the two homozygotes, but always closer to the best homozygote. The degree of concavity of the fitness

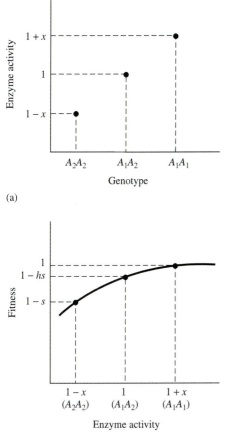

Figure 12.7   **A two-allele model of random variation in fitness.**
(a) Enzyme activity is assumed to be additive, with genotypes A_2A_2, A_1A_2, and A_1A_1 having activity $1 - x$, 1, and $1 + x$, respectively; x is a random variable, shown here as positive for convenience only. (b) Fitness is then assumed to be a concave function of enzyme activity (genotype). The concavity assures that for weak selection, fitness of the heterozygote is closer to fitness of the best homozygote.

function depends on the distribution of x, on the strength of selection, the degree of dominance, and other factors.

These models are very complex, requiring advanced probability theory to analyze. Here, we only summarize some of the main conclusions. The main factors considered in various versions of the models are (1) the degree of dominance; (2) spatial variation in fitness; (3) temporal variation in fitness; (4) the degree of spatial and temporal autocorrelation among environments; (5) genetic drift; and (6) mutation.

Some of the main predictions are

1. Under a variety of biologically reasonable assumptions, these models predict large amounts of genetic variation, especially if both spatial and temporal variation in fitnesses are considered.

2. Spatial subdivision does not necessarily lead to increased polymorphism, but it does so under some selection schemes.

3. Temporal variation increases the amount of polymorphism, but temporal autocorrelation (the degree to which consecutive time periods resemble one another) decreases it.

4. Under some circumstances, the predicted distribution of allele frequencies is identical to, or very similar to, the predicted distribution under the neutral theory (Section 10.4).

One of the potential weaknesses of these models is that they require a reversal of dominance. To see this, recall that fitness of the homozygotes depends on the random variable x. If x is positive, A_1A_1 has the highest fitness; if x is negative, A_2A_2 is best. But in both cases, the assumption of a concave fitness function requires that the heterozygote be closer in fitness to the best homozygote. It is not known how common this reversal of dominance may be.

In general, models with randomly varying selection coefficients sometimes predict large amounts of genetic variation, and thus are serious contenders in our search for the causes of the extensive genetic variation seen in many populations. However, because they postulate random variation in very small selection coefficients, they are nearly impossible to test directly. For more details and references, see Gillespie (1991, 1994, and the papers cited therein).

12.6 Frequency Dependent Selection

Frequency dependent natural selection occurs when the fitness of a genotype depends on its frequency, and on the frequencies of other genotypes in the population. It almost always refers to the situation in which the fitness of a genotype increases when it is rare. Levin (1988) called this stabilizing frequency dependent selection, as opposed to disruptive frequency dependent selection, in which fitness of a genotype declines as that genotype becomes rarer. The former can lead to a protected polymorphism, where the latter always results in the rare genotype being lost from the population. We will consider only the former and, following common usage, call it simply frequency dependent selection.

It is easy to imagine situations in which frequency dependent selection might occur. For example, predators frequently form a "search image" of what their prey is expected to look like. Rare prey, which deviate from this search image, will have higher survival rates. But as this rare form increases in frequency, predators will learn to recognize it, and its survival will begin to decrease.

We first examine a simple model of frequency dependent selection to get some idea of the theoretical possibilities that it presents. We will then look at a few examples and consider the significance of frequency dependent selection in natural populations.

A Model of Frequency Dependent Selection

There are various ways in which frequency dependent selection can be modeled. We will consider a simple case in which the fitness of a genotype is inversely related to its frequency. Assume two alleles, A_1 and A_2 with frequencies p and q, and random mating. The fitnesses of the three genotypes can be written as

$$w_{11}(p) = 1 - p^2 a \tag{12.13}$$

$$w_{12}(p) = 1 - 2pqa \tag{12.14}$$

$$w_{22}(p) = 1 - q^2 a \tag{12.15}$$

We have written the fitnesses as a function of p to emphasize the frequency dependence; the parameter a is a measure of the strength of frequency dependence $(0 \leq a \leq 1)$.

The recursion equation for p can be derived in the usual way, to get

$$p_{t+1} = \frac{p^2 w_{11}(p) + pq w_{12}(p)}{\overline{w}(p)} \qquad (12.16)$$

where

$$\overline{w}(p) = p^2 w_{11}(p) + 2pq w_{12}(p) + q^2 w_{22}(p) \qquad (12.17)$$

These are the same equations as (5.7) and (5.6), except for the frequency dependence.

   To find the equilibria of this model, first substitute equations (12.13) through (12.15) into (12.16). Then set $p_{t+1} = p_t = \widetilde{p}$ and solve for $\widetilde{p}$. After a lot of algebra, the result is

$$a[p^5 - 4p^4 + 6p^3 - 4p^2 + p] = 0 \qquad (12.18)$$

There are two things to notice about equation (12.18). First, we can divide out a. This means the equilibria are independent of the strength of frequency dependence (although the approach to equilibrium and the population fitness are not). Second, the factor in brackets can be factored to give

$$p(1 - p)(2p - 1)(3p^2 - 3p + 1) = 0$$

The first three factors give equilibria of $\widetilde{p} = 0$, $\widetilde{p} = 1$, and $\widetilde{p} = 0.5$. The fourth factor gives complex roots that are irrelevant. Thus, there are three possible equilibria. Intuition should suggest that $\widetilde{p} = 0$ and $\widetilde{p} = 1$ should be unstable. It can be shown formally that this is true, and that $\widetilde{p} = 0.5$ is locally and globally stable.

   There are several interesting things about this model. Figure 12.8a shows the fitnesses as functions of p (with $a = 1$). At equilibrium, the fitnesses are $w_{11} = w_{22} = 0.75$ and $w_{12} = 0.5$. We have a stable equilibrium with the heterozygote having the lowest fitness! Figure 12.8b shows mean fitness as a function of p. We see that $\overline{w}$ is maximized at equilibrium, and that as the population approaches equilibrium, $\overline{w}$ will always increase. The same is true for $a < 1$ (weaker frequency dependence). The mean fitness at equilibrium increases as a decreases. [Look at equations (12.13), (12.14), and (12.15) to see this.]

   In some other models of frequency dependent selection, all genotypes have the same fitness at equilibrium. This means there is no segregational load (Sections 6.6 and 10.1), a desirable characteristic from an evolutionary perspective. Other models do not always maximize $\overline{w}$ at equilibrium. Mean fitness may remain constant, or even decrease to a minimum at equilibrium. Once again, we see that natural selection will not always direct a population toward its best possible state.

   The dynamic behavior of frequency dependent selection models can be very complex, including cycles, chaos, and intermittency. See Altenberg (1991) and Gavrilets and Hastings (1995) for discussions of these mathematical complexities.

Frequency Dependent Selection in Natural Populations

Frequency dependent selection is attractive as a means of maintaining genetic variation, and is widely thought to be important in natural populations (e.g., Clarke 1979). In Chapter 9, we discussed one important kind of frequency

Figure 12.8 The frequency dependent selection model described by equations (12.13)–(12.15).   (a) Genotypic fitnesses as functions of p. There is a globally stable equilibrium at $p = 0.5$. At equilibrium, the fitnesses are 0.75, 0.50, and 0.75. (b) Mean population fitness ($\overline{w}$) as a function of p. $\overline{w}$ is maximized at the stable equilibrium.

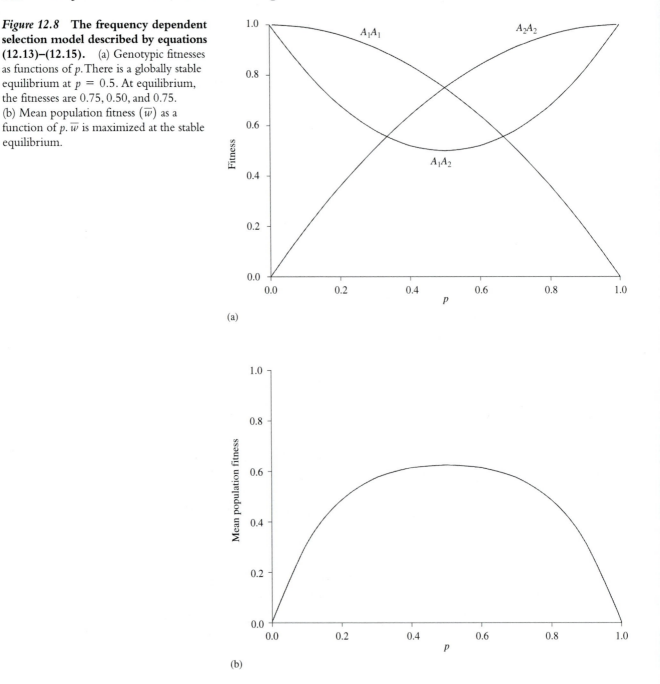

(a)

(b)

dependent selection, the phenomenon of self-incompatibility systems in many species of plants. A rare allele at a self-incompatibility locus will have fewer aborted pollinations or fertilizations than a common allele, and thus have a rare allele advantage. This kind of frequency dependent selection sometimes maintains dozens of alleles at a locus.

A similar phenomenon has been observed in *Drosophila* and some other insects. Many experiments have demonstrated that males with rare genotypes often

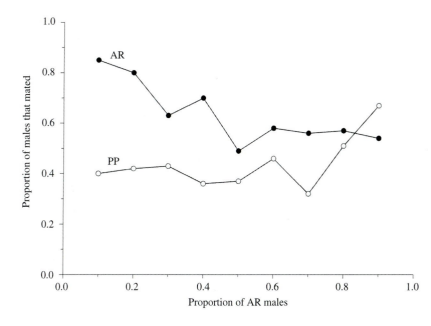

Figure 12.9 **Rare male mating advantage for the AR and PP inversions of *Drosophila pseudoobscura***. *Source:* Data from Spiess (1968).

have a mating advantage. Figure 12.9 shows one example. Spiess (1968) varied the proportion of *D. pseudoobscura* males carrying the third chromosome inversions Arrowhead (AR) and Pikes Peak (PP) and observed the number of males of each type that mated successfully. Figure 12.9 shows that each type mated more successfully when rare than when common. This kind of rare male advantage seems to be a common phenomenon. In *Drosophila*, it has been demonstrated for a variety of characteristics, including chromosomal variants, visible mutations, and allozyme variants. It has also been demonstrated in a variety of other insects, and some vertebrates. See Petit and Ehrman (1969) and Partridge (1988) for reviews and different perspectives on the evolutionary importance of rare male advantage.

Rare male mating advantage is only one component of fitness that may be subject to frequency dependent selection. In several species of insects, larval viability has been shown to be subject to frequency dependent selection. For example, Kojima and Yarbrough (1967) established cultures of *Drosophila melanogaster* that were homozygous for the *F* or *S* allele at the *Est*-6 locus. They then took mated females from all four mating types and put them in culture vials in proportions that resulted in eggs in Hardy-Weinberg expected genotype frequencies for various frequencies of the *F* allele. They then determined the genotypes of surviving offspring (males only, for technical reasons). They estimated relative viability as the ratio of the observed proportion of a genotype to the expected proportion. The results are summarized in Figure 12.10. Both the *FF* and *SS* genotypes had higher viability when rare. In addition, Kojima and Yarbrough estimated the equilibrium frequency for this locus to be about 0.3. At that frequency, the viabilities of the three genotypes were not significantly different. This is a prediction of some frequency dependent models (but not of the one considered above). Note that selection is not necessarily acting directly on the *Est*-6 locus, but may be acting on a tightly linked locus that remained in linkage disequilibrium during the course of the experiment. Ayala and Campbell (1974) give other examples of frequency dependent viabilities.

Figure 12.10 **Frequency dependent viabilities of the *FF* and *SS* genotypes at the *Est-6* locus of *Drosophila melanogaster*.** Each genotype has higher viability when rare in larvae. Shown is the average of two replicates. *Source:* Data from Kojima and Yarbrough (1967).

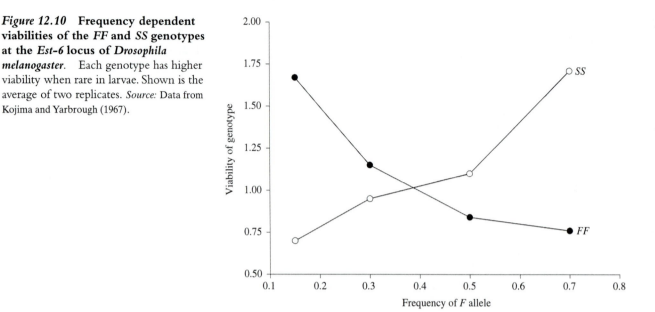

The usual explanation for these observations is differential resource utilization. If a rare genotype can effectively use some resource that common genotypes cannot, it will have a selective advantage. Possible examples are allozymes with slightly different substrate specificities, the ability to feed on rare foods, or different resistance to environmental toxins. Direct evidence for this hypothesis is rare. See Antonovics and Kareiva (1988) for a review.

Another important and very widespread example of frequency dependent selection is the 1:1 sex ratio seen in most sexual organisms. Fisher (1930) was the first to propose an explanation for this. In any sexually reproducing diploid population, males and females must contribute equally to the gamete pool (each zygote results from one male gamete and one female gamete). If one sex is less frequent than the other, the relative contribution from each individual of that sex will be higher than from the other. For example, if there are twice as many males as females, then, all else being equal, each female will contribute twice as many gametes to the next generation as a male. Individuals that produce the rarer sex will have a reproductive advantage. Any genetic tendency to produce the rarer sex will be favored by selection, and the sex ratio will evolve toward 1:1. This argument assumes that parental investment in each sex is equal, and that prereproductive mortality is equal in both sexes.

Fisher's argument has been widely accepted (with some modifications and extensions) as the explanation of the 1:1 sex ratio. However, it has been difficult to test, because there is essentially no genetic variation for sex ratio in most organisms. Conover and Van Voorhees (1990) took advantage of an unusual sex determination mechanism in the Atlantic silverside (*Menida menida*), a fish found in bays and estuaries along the east coast of North America. Sex is determined jointly by major (variable) genes and by environmental temperature at a particular stage of larval development. Larvae experiencing low temperatures usually develop into females, and most larvae experiencing high temperatures become males. Conover and Van Voorhees established several populations at either high

(28 °C) or low (17 °C) temperatures. For example, in the South Carolina high (SC-H) population, the initial sex ratio (proportion of females) was 0.18. In the South Carolina low (SC-L) population, the initial sex ratio was 0.70. They maintained these populations at constant temperature for six to eight generations, and estimated the sex ratio each generation. Their results are summarized in Figure 12.11. In all experimental populations, the sex ratio moved toward 0.5. There was some variation among lines, and some oscillation, but of 19 generations (5 lines) where the sex ratio differed significantly from 0.5, 18 moved toward 0.5 in the next generation. These results are as predicted by Fisher's hypothesis. Conover and Van Voorhees suggest that their results are also relevant to debates over global warming. It has been argued that rapid climate change might significantly distort the sex ratio in species with environmentally determined sexual development. The results of Conover and Van Voorhees suggest that such species can evolve rapidly enough to maintain a balanced sex ratio. This assumes that these species possess genetic variation for sex determination.

We have already mentioned the idea that predators may form a search image of their prey, and prey that deviate from this image may have a higher survival. This may lead to color or pattern polymorphisms, in which rare forms suffer less predation than common forms. Numerous experiments have shown that many kinds of predators take mostly common forms of prey and overlook rare forms. Clarke (1979) and Allen (1988) review the role of predation in frequency dependent selection.

The opposite can also occur. If prey density is high, as in flocks of birds or schools of fish, then predators may take prey that stand out from the crowd. Selection favors uniformity instead of polymorphism. This is what Levin (1988) called disruptive frequency dependent selection.

Host-parasite coevolution can lead to frequency dependent selection. Haldane (1949) first suggested that parasites would evolve in response to the most common genotypes in their hosts. The basic idea is similar to that of predator-prey interactions: Rare host genotypes will be attacked less effectively by the parasite, until they become common enough for the parasite to evolve effective means of attacking them.

The major histocompatibility complex (MHC) in vertebrates is a series of tightly linked genes involved in the immune response against parasites or pathogens. The genes are highly variable, with over 100 alleles at some loci, and much molecular and biomedical evidence indicates that this variation is maintained by some form of balancing selection. The main function of the MHC proteins is to display foreign peptides on the surface of infected cells. Killer T-cells then recognize these foreign peptides and kill the infected cell, and the parasite within. This process leads to strong selection for diversity of the MHC proteins, resulting in the extraordinarily high levels of genetic variation in the MHC system (Section 10.4). Parasites will evolve to recognize and evade the common MHC proteins. A rare MHC protein may be able to recognize a parasite, whereas the common proteins may not; thus, frequency dependent selection will favor rare MHC alleles. This is not the entire story, however. Other kinds of balancing selection, for example, heterozygote advantage, are probably involved. At any rate it is easy to imagine an evolutionary arms race in which both host and parasite are evolving constantly and rapidly in response to one another. See Hedrick (1994)

Figure 12.11 **Evolution of sex ratio in experimental populations of the Atlantic silverside.** SC, NY, and NS indicate South Carolina, New York, and Nova Scotia. H and L indicate experimental populations raised at high (28 °C) or low (17 °C) temperatures. Initial sex ratio was manipulated experimentally. In most populations, the sex ratio moved toward 0.5. See text for details. *Source:* Modified from Conover and Van Voorhees (1990); error bars omitted.

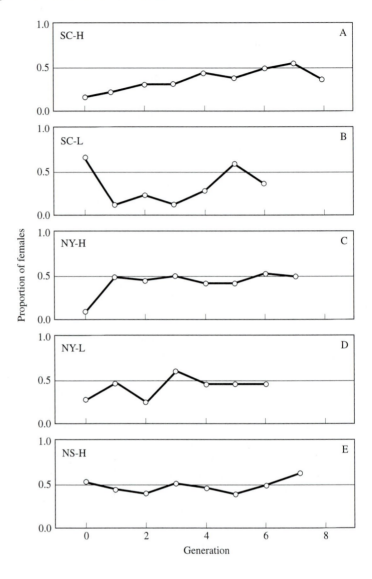

and Hedrick and Kim (2000) for reviews of the genetics and evolution of the MHC genes.

An analogous process can occur in bacteria. Many species of bacteria possess restriction enzymes that recognize and destroy foreign DNA, for example, from bacteriophage. They also possess modification enzymes to modify their own DNA so that it is not destroyed. Bacteriophage can sometimes evolve to evade these restriction enzymes, perhaps by modifying their DNA so that the bacterial cell does not recognize it as foreign. Such phage can then successfully infect normally resistant bacterial cells. A bacterial cell that has an unusual restriction/modification system will be less likely to be infected by bacteriophage, which evolve in response to the common systems. Ultimately, this advantage will be lost, as the bacteriophage learn to recognize and evade the formerly rare type. Again, we expect an evolutionary arms race between bacteria and phage. Levin (1986, 1988) discusses theoretical models and experimental results of this and other aspects of frequency dependent selection in bacteria.

In Batesian mimicry, a palatable mimic resembles an unpalatable model (of another species) and is protected because predators recognize the pattern as unpalatable. In order for this to work, the mimic must be rare compared to the model; otherwise, predators would learn to associate the pattern with the palatable species. Batesian mimicry occurs in several species of butterflies. For example, *Papilio dardanus* in East Africa is polymorphic for wing patterns that mimic three different model species. The wing patterns are controlled by several tightly linked genes in gametic disequilibrium (coadapted gene complexes, or supergenes), which produce the different patterns. The mimicry of three different models instead of one is apparently an adaptation to keep the frequency of each mimic low compared to the model, a classic case of frequency dependent selection.

An essential point of these coevolving systems is that the advantage of a rare genotype is lost as that genotype becomes more common. Predators or parasites will evolve the ability to recognize and attack that genotype as it becomes common. Without this loss of advantage as frequency increases, we have not frequency dependent selection, but directional selection. It is not always clear which kind of selection is occurring.

From these examples, it is tempting to conclude that frequency dependent selection is common, and may be an important factor in maintaining genetic variation. However, keep in mind that it is sometimes difficult to distinguish frequency dependent selection from heterozygote superiority or simple directional selection favoring a new allele. In all three cases, a rare allele will initially increase in frequency.

For early reviews of the classic studies of frequency dependent selection, see Kojima (1971), Ayala and Campbell (1974), and Clarke (1979). For more recent reviews, see the special issue on frequency dependent selection in *Philosophical Transactions of the Royal Society, London*, volume B319, July 1988. Heino et al. (1998) discuss the relationship between frequency dependent selection and density dependent selection.

12.7 Antagonistic Pleiotropy

Selection can occur during any stage of the life cycle. If genes have pleiotropic effects affecting different stages, then a particular genotype may be favored in one stage and selected against in another. A genotype may have positive effects on one component of fitness (e.g., viability) and negative effects on another (e.g., fertility). Such conflicting selection pressures on different components of fitness are called **antagonistic pleiotropy**.

The classic example comes (as usual) from *Drosophila*. Natural selection favors fast development during the larval stage. However, adults that develop rapidly are smaller and have lower fecundity than adults that develop slowly. So we have conflicting selection pressures; selection for development time favors rapid development, but selection for fecundity favors slower development. Curtsinger et al. (1994) cite several other examples of conflicting selective pressures in *Drosophila*. Antagonistic pleiotropy is thought to be very common.

Rose (1982, 1985) suggested that antagonistic pleiotropy would frequently lead to genetic polymorphism. However, although evidence for the existence of antagonistic pleiotropy is strong, there seems to be little evidence that it is a major contributor to polymorphism in natural populations (Curtsinger et al. 1994; Hedrick 1997).

Consider a model in which one genotype has the highest viability and another has the highest fertility. For example, plants can be either small, with many flowers (low survival, high fertility) or larger, with fewer flowers (higher survival, lower fertility). For a single locus with two alleles, let v_{ij} be the relative viability of genotype A_iA_j, and f_{ij} be the relative fertility. Assume directional selection in opposite directions, with heterozygote intermediacy for both components. Then we have

Genotype	A_1A_1		A_1A_2		A_2A_2
Viability	v_{11}	$>$	v_{12}	$>$	v_{22}
Fertility	f_{11}	$<$	f_{12}	$<$	f_{22}

In this example, the two fitness components are multiplicative, and can be combined into a single fitness value; that is, if $v_{ij}N_{ij}$ is the number of genotype A_iA_j that survive to reproductive age, and each adult produces f_{ij} offspring, then the reproductive rate (absolute fitness) of A_iA_j is $v_{ij}f_{ij}N_{ij}$. We have the familiar one-locus model of Section 5.2. The condition for a stable polymorphism is then

$$v_{11}f_{11} < v_{12}f_{12} > v_{22}f_{22}$$

That is, the relative fitness of the heterozygote must be greater than either homozygote.

Another example of multiplicative fitness components would be survival at different stages of the life cycle; overall survival is the product of survival at each stage. However, different fitness components are not always multiplicative. For example, fertilities at two different stages of the life cycle are additive; total fertility is the sum of fertilities at different stages. Other components can interact in complex ways that are neither multiplicative nor additive, and require complex models.

Hedrick (1999b) and Prout (2000) reviewed a variety of one- and two-locus models and concluded that, under a variety of conditions, antagonistic pleiotropy is more likely to result in fixation than in polymorphism. This does not mean that antagonistic pleiotropy is rare, but only that it probably is not a major factor in maintaining genetic variation in natural populations. Factors that increase the likelihood of polymorphism are strong selection, similar strength of selection in different fitness components (e.g., $v_{11} \cong f_{22}$ and $v_{22} \cong f_{11}$ in the above model), and reversal of dominance, so that the heterozygote is closer to the better homozygote for both fitness components. The latter is especially favorable to polymorphism, but is considered unlikely. For theoretical analyses leading to these conclusions, see Curtsinger et al. (1994), Hedrick (1997), and Prout (2000).

Pemberton et al. (1991) described an interesting example of antagonistic pleiotropy in the red deer (*Cervus elaphus*) on the Isle of Rhum, Scotland. They monitored juvenile survival (0–2 years) and several components of female reproductive fitness associated with genotypes at two allozyme loci, *Mpi* (mannose phosphate isomerase) and *Idh-2* (isocitrate dehydrogenase). Each locus had two

TABLE 12.5   Effect of two allozyme loci on fitness components in red deer.

At the *Mpi* locus, genotypes *ss* and *ff* were combined due to small sample sizes. *Source:* From Pemberton et al. (1991).

	Fitness Component			
	Juvenile Survival	**Adult Survival**	**Age at First Calving**	**Fecundity (calves per year)**
Mpi				
ss	0.62	0.53	4.59	0.53
sf and *ff*	0.25	0.35	4.25	0.67
Significance	P < 0.001	NS	P < 0.02	P < 0.10
Idh-2				
ss	0.41	0.80	4.36	0.61
sf	0.60	0.31	4.63	0.57
ff	0.46	0.59	4.43	0.46
Significance	P < 0.10	P < 0.02	P < 0.05	P < 0.10

alleles, *f* (fast) and *s* (slow). Some of their results are summarized in Table 12.5. At the *Mpi* locus, *ss* homozygotes had higher survival rates than did *sf* and *ff* (combined because of small sample sizes), but calved later and less frequently. At the *Idh*-2 locus, heterozygotes had the highest juvenile survival, but lowest adult survival, and calved latest of the three genotypes. These two allozyme loci clearly have opposing effects on different fitness components.

No one believes that the fitness components studied by Pemberton et al. are controlled by two loci, let alone random allozyme loci. The allozyme loci are almost certainly just markers for chromosomal regions containing one or more genes affecting the fitness components. In fact, the different fitness components may actually be affected by different genes, each in strong disequilibrium with the allozyme loci, and the apparent antagonistic pleiotropy may be spurious. Nevertheless, these data illustrate negative correlations between fitness components in nature.

This example raises another issue. Fitness components are quantitative traits, controlled not by a single gene, but by many genes. The theory summarized above suggests that single-locus antagonistic pleiotropy is unlikely to result in a stable polymorphism, but the situation is much less clear for quantitative traits. We will discuss antagonistic pleiotropy in quantitative traits in Section 13.4.

12.8 Opposing Selection in Males and Females

It is easy to imagine situations in which natural selection acts differently in males and females. A classic example is plumage color in birds. During the breeding season, males of many species develop brightly colored plumage to attract females. This also makes them more conspicuous to predators, but the advantages of attracting a mate outweigh the disadvantages. On the other hand, selection favors dull coloration in females, allowing them to be inconspicuous when incubating their eggs. Genes that have opposing effects on fitness in males and females are called **sexually antagonistic genes** (Rice 1992).

We can develop a model to investigate how opposing selection in the sexes can affect polymorphism. Consider a single (autosomal) locus with two alleles. Define the fitnesses as follows:

Genotype	A_1A_1	A_1A_2	A_2A_2
Males	w_{11m}	w_{12m}	w_{22m}
Females	w_{11f}	w_{12f}	w_{22f}

Similarly, let p_m and q_m $(= 1 - p_m)$ be the allele frequencies in males, and p_f and $q_f (= 1 - p_f)$ be the allele frequencies in females. Assume random mating, so that the genotype frequencies are $p_m p_f$, $(p_m q_f + q_m p_f)$, and $q_m q_f$. Define the mean fitnesses in males and females separately, taking into account the allele frequency differences between males and females:

$$\overline{w}_m = p_m p_f w_{11m} + (p_m q_f + q_m p_f)w_{12m} + q_m q_f w_{22m}$$

$$\overline{w}_f = p_m p_f w_{11f} + (p_m q_f + q_m p_f)w_{12f} + q_m q_f w_{22f}$$

Following the procedure described in Section 5.2, we can obtain the recursion equations for allele frequencies in males and females:

$$p_{m,t+1} = \frac{p_m p_f w_{11m} + (1/2)(p_m q_f + q_m p_f)w_{12m}}{\overline{w}_m} \tag{12.19}$$

$$p_{f,t+1} = \frac{p_m p_f w_{11f} + (1/2)(p_m q_f + q_m p_f)w_{12f}}{\overline{w}_f} \tag{12.20}$$

Here, we have simply applied the one-locus, two-allele model to males and females separately.

Equations (12.19) and (12.20) are coupled, nonlinear equations. They can be solved for equilibrium values of p_m and p_f (Kidwell et al. 1977), but these solutions yield little biological insight. Alternatively, we can determine the conditions for a protected polymorphism. The condition for A_2 to increase when rare is

$$\frac{w_{12m}}{w_{11m}} + \frac{w_{12f}}{w_{11f}} > 2$$

and the condition for A_1 to increase when rare is

$$\frac{w_{12m}}{w_{22m}} + \frac{w_{12f}}{w_{22f}} > 2$$

Derivation of these conditions requires some knowledge of stability analysis and linear algebra. See Kidwell et al. (1977) and Prout (2000) for the mathematical details.

If we scale the fitnesses so that the heterozygote has fitness of 1 in each sex, these equations can be rewritten as

$$\frac{1}{\frac{1}{2}\left(\frac{1}{w_{11m}} + \frac{1}{w_{11f}}\right)} < 1 \qquad \textbf{(12.21)}$$

and

$$\frac{1}{\frac{1}{2}\left(\frac{1}{w_{22m}} + \frac{1}{w_{22f}}\right)} < 1 \qquad \textbf{(12.22)}$$

The left-hand sides of these two equations are the harmonic means of the homozygote fitnesses. (The 1/2 in the denominators is because we assume equal numbers of males and females.) So, sufficient conditions for a protected polymorphism are that the harmonic mean of w_{11m} and w_{11f} and the harmonic mean of w_{22m} and w_{22f} both be less than one.

These are the same conditions that exist for a two-niche polymorphism under the Levene model (Section 12.3). In fact, we could consider the males inhabiting one niche and females the other, with a special kind of gene flow between the niches (disassortative mating between males and females). How likely are these conditions to be met? Since the mathematical conditions are the same as for a two-niche model, the conclusion is the same: Especially if selection is weak, the fitnesses differences between males and females must be within a narrow range for conditions (12.21) and (12.22) to hold. We conclude that differential selection in males and females *by itself* is unlikely to maintain a stable polymorphism.

These are the conditions under which neither allele will be lost. We would also like to know what happens between the two unstable fixation points. The answer, of course, depends on the fitnesses, but some interesting examples exist. For example, there can be two simultaneously stable polymorphic equilibria, separated by a third unstable equilibrium. Figure 12.12 illustrates an example.

How common are sexually antagonistic genes? Rice (1992) suggests that they may be common. Such genes might cause an evolutionary arms race between the sexes. A characteristic that increases male fitness but decreases female fitness will increase if the disadvantage to females is not too great. There will be selection favoring female defense, producing selection for males to counter the defense, leading to selection for further female defense, and so forth. Genes controlling these characteristics might frequently exhibit transient polymorphisms as one allele replaces another due to changing selection pressures.

The best understood sexually antagonistic genes are those coding for the seminal proteins in *Drosophila melanogaster*. Seminal fluid in *Drosophila* is a complex substance consisting of, in addition to sperm, many proteins, called accessory gland proteins, because they are produced by the male accessory gland. Both males and females mate more than once. Thus, one might expect natural selection to favor males that can prevent their sperm from being displaced by a sperm from a subsequent mating (defense), and selection favoring males with the ability to displace sperm from a prior mating (offense). Seminal fluid is involved in both of these functions. After mating, it helps produce a plug that prevents sperm from a subsequent mating from entering the reproductive tract of the female. It is also

Figure 12.12 **Multiple equilibria under a model of sexually antagonistic selection.**   Fitnesses for males are 0.5, 1.0, and 0.6. Fitnesses for females are 1.0, 0.4, and 0.9. There are three unstable equilibria (open circles) and two stable equilibria (solid circles). Arrows show general direction of movement away from unstable equilibria and toward stable equilibria. *Source:* Modified from Hartl and Clark (1989).

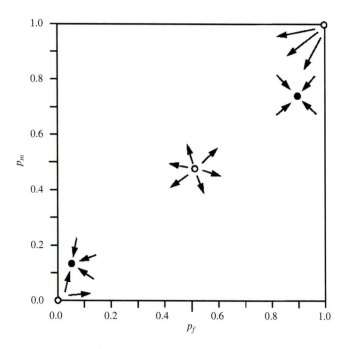

toxic to sperm from a prior mating. In addition, it decreases the female's propensity to remate and increases her fecundity. There is genetic variation for all of these characteristics. Thus, we can view the female reproductive tract as a battleground where chemical warfare is waged between different males. The females are the innocent civilians caught in the middle, as the seminal fluid is toxic to them.

William Rice and his colleagues have conducted experiments that suggest such an evolutionary arms race is actually occurring. In one experiment (Rice 1996), *Drosophila melanogaster* females were prevented from evolving by a complex breeding scheme involving attached X chromosomes and artificial selection. Males, on the other hand, were allowed to evolve in response to the females, who could not counterevolve. Experimental males rapidly increased their fitness. They showed increased ability to successfully remate a previously mated female (offense) and increased resistance to sperm displacement by a subsequent male (defense). On the other hand, female mortality increased, and the higher the frequency of remating, the higher the mortality. Rice attributed this to increased quantity and/or toxicity of seminal fluid.

In another experiment, Holland and Rice (1999) studied the effects of enforced monogamy on sexually antagonistic genes. The prediction is that if the arms race is fueled by competition among males, then removal of that competition should eliminate the conflicting selection pressures so that sexual antagonism should decrease. Holland and Rice maintained single-pair matings for 47 generations; pairs were chosen randomly, so that there was no opportunity for selection by either male or female choice. The results were as predicted: Survival of monogamous females was greater when they were paired with monogamous males than with control males. Monogamous males courted and remated less frequently than did control males. Monogamous pairs had a higher reproductive rate than did control pairs. Holland and Rice conclude that multiple mating imposes a sexually antagonistic load on the population, resulting in antagonistic coevolution between the sexes.

If the antagonistic coevolution hypothesis is correct, we might expect that sexually antagonistic genes would evolve faster than other kinds of genes. Begun et al. (2000) compared the sequences of several genes encoding *Drosophila* accessory gland proteins (*Acp* genes). Replacement substitution rates between *D. melanogaster* and *D. simulans* averaged 0.0497 for *Acp* genes, compared to 0.0107 for non-*Acp* genes. Intraspecific nucleotide polymorphism was also higher in *Acp* genes. A McDonald-Kreitman test (Section 10.4) for all *Acp* genes combined suggested a highly significant excess of replacement substitutions, although the significance was due to only two genes. Begun et al. concluded that the evidence that a large fraction of the amino acid substitutions is due to directional selection was weak, but that this did not preclude the possibility of directional selection at relatively few amino acid positions.

In a similar study, Swanson et al. (2001) used expressed sequence tags (ESTs) to identify putative *Acp* genes. The average rate of replacement substitutions (again between *Drosophila melanogaster* and *D. simulans*) was 0.052 for *Acp* genes, compared to 0.024 for other genes. Many of the putative *Acp* genes had K_A/K_S ratios greater than 1, strong evidence of positive selection (Section 10.4). Swanson et al. conclude that the rapid divergence of these genes is due to positive selection.

The picture that emerges from these studies is that different selection pressures in males and females may be quite common, and the result is an evolutionary arms race leading to antagonistic coevolution. Sexually antagonistic genes are constantly and rapidly evolving, and in the process transient polymorphisms will be seen as one allele replaces another.

12.9 Sexual Selection

Sexual antagonism is one aspect of a more general phenomenon known as **sexual selection**. Darwin (1871) was the first to suggest that natural selection might act differently on males and females. He asked why males of so many species have morphological or behavioral traits that probably reduce their survival. His answer was that these traits exist because they give their bearer an advantage over other males in competition for mates. Darwin defined sexual selection as competition among individuals for mates.

Several components of fitness were discussed in Section 5.1 (see Figure 5.1). Two of these involve sexual selection: mating ability and gamete competition. We will discuss each of these, and examine how selection acts differently in males and females.

Once an individual reaches reproductive maturity, it is critical that he or she successfully mate and produce offspring; otherwise, from an evolutionary point of view, that individual may as well have died as a zygote. Reproductive behavior is usually different in males and females. Males usually compete for females, while females tend to be selective with respect to their mates.

These behavioral differences are due to differences in parental investment. In most species, the male invests much less time and energy in reproduction and rearing of young than does the female. In some species, the male simply provides sperm and does not participate at all in the rearing of young. On the other hand, females typically invest much more in reproduction. For example, the physiology of pregnancy and nursing is very energy demanding and stressful. It is also dangerous; pregnant or nursing females are much more susceptible to predation and disease than are males or other females. Females also tend to invest more time and energy in parental care than do males.

These differences suggest that the limiting factor in male reproductive success is the number of females a male can mate with; the more, the better. Males should therefore compete for as many females as possible. On the other hand, the limiting factor for females is the ability to produce and rear offspring. They should therefore be more careful about whom they mate with, choosing males who will maximize their reproductive success. These are the main predictions of sexual selection theory.

Bateman (1948) was the first to test these predictions. He compared reproductive success (number of offspring produced) versus the number of matings in *Drosophila melanogaster*. The prediction is that reproductive success should increase with multiple matings for males, but not for females. Bateman's results were consistent with this prediction (Figure 12.13). This was one of the first experiments on sexual selection, and the general results have been confirmed many times since.

Competition Among Males for Mates

Males compete for females in two ways. Direct competition involves fighting for rights to a female. Many species have developed weapons for such fighting, for example, horns in bighorn sheep and beetles, antlers in deer. These structures are usually absent or reduced in females. Sexual selection favors such structures in males because they give their possessors a competitive advantage in obtaining females with which to mate. The enormous size of male elephant seals compared

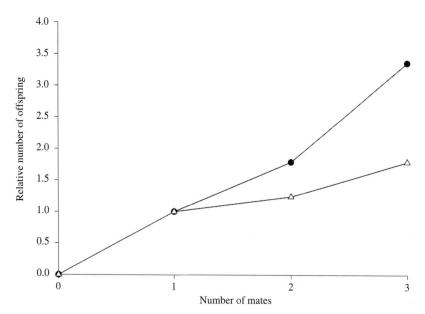

Figure 12.13 Relationship between number of mates and number of offspring produced for males (circles) and females (triangles). Data points represent the average of six replicates combined. The number of offspring produced increases with number of mates much more rapidly for males than for females. *Source:* Data from Bateman (1948).

to females is another example. Large size gives the males an advantage in fighting to defend their harems.

Direct competition among males can occur after mating. Since in many species, females will mate with more than one male, it is to a male's advantage to make sure his sperm take precedence over other sperm that the female may have in her body. Such sperm competition is best known in insects. It can be very direct. For example, the male damselfly has a special structure on its penis to scoop out pre-existing sperm from a female's reproductive tract (Waage 1984). This results in almost all of the female's eggs being fertilized by the most recent male she has mated with.

Sperm competition has been most thoroughly studied in *Drosophila*. As discussed in Section 12.8, females mate more than once, and store sperm in the seminal receptacle. Sperm competition among males takes two forms: offense, in which a later male's sperm displaces that of an earlier male, and defense, in which an earlier male's sperm resists displacement by that of a later male. Clark et al. (1995) have documented extensive genetic variation for both kinds of sperm competition. Thus, we have sexual selection favoring both offensive and defensive capabilities in sperm competition, with the associated trade-offs. Moreover, as discussed in Section 12.8, the seminal fluid of males is toxic to females, so we have sexual antagonism as well. Females are not passive sperm recipients in this battle; female choice and resistance are discussed below.

An analogous process of pollen competition occurs in plants. Many plants receive more pollen grains on the stigma than are necessary for fertilization of the ova. A race ensues, as different pollen tubes grow to reach the ova. Snow and Spira (1996) used allozyme markers to demonstrate genetic variation for pollen competitive ability in the rose mallow (*Hibiscus moscheutos*). This form of gamete competition may be common in plants.

Direct competition can occur even after birth of the offspring. Male lions sometimes kill cubs sired by other males. Females are not receptive to mating again until after their cubs are weaned. Thus, killing unweaned cubs not only removes offspring of competing males, but brings the females into breeding condition months earlier than otherwise (Packer et al. 1988).

Indirect competition occurs when males call attention to themselves with showy displays. Bright plumage in the males of many species of birds and bright coloration in males of some fish species are examples. Typically, these bright color patterns develop only just prior to the breeding season, and are much less obvious at other times of the year. Males of many species of frogs make loud and distinctive mating calls to attract females. Some birds build complex bowers to attract females for mating, or develop elaborate structures and displays to attract mates. The tails of male peacocks and the tails and puffed-up chests of male sage grouse are familiar examples.

Competition for mates is dangerous for males. Fighting involves risk of injury; bright color patterns, elaborate displays, or loud mating calls make the males more obvious to potential predators. Thus, there is a trade-off between sexual selection and natural selection. This explains why the males' colors are brightest just prior to mating, and why females usually are duller than males.

Endler (1978, 1980) has confirmed the trade-off between sexual and natural selection. Guppies (*Poecilia reticulata*) are native to clear streams in northeastern

South America and the island of Trinidad. The fish vary in the number, size, and coloration of spots. Populations with high risk of predation by other fish have fewer, smaller, and less colorful spots, making them less conspicuous against the gravel stream beds (Endler 1978). Endler hypothesized that the pattern of spots reflects a balance between natural and sexual selection. In predator-free populations, sexual selection should favor brightly colored and conspicuous males, but in environments with predators, natural selection should favor cryptic coloration. He tested this hypothesis by raising guppies in predator-free tanks. After several months, the fish had larger and more colorful spots than they did initially. He then introduced a dangerous predator to some of the tanks. The guppies evolved back to having smaller spots and duller colors. These results are exactly as predicted. Endler (1980) also performed similar experiments by transplanting guppies from a stream containing a dangerous predator to a stream containing no guppies and only a minor predator. After about 15 generations (about two years) the introduced guppies had larger and more colorful spots. These experiments confirm the opposing selection between competition for mates (sexual selection) and predator avoidance (natural selection).

Female Choice

The fact that males develop elaborate displays and calls to attract mates implies that the females choose among different males. Again, Darwin suggested this, but until fairly recently it has been a controversial idea. As suggested earlier, females might choose males that maximize their own reproductive success. There are three main hypotheses to explain the evolution of female choice.

Females may choose males that provide resources to them. Male birds with large or high-quality territories may be able to provide more food for their offspring. In many insects the male offers the female a nuptial gift, which she eats while mating. If the gift is too small or of low quality, it, and the male, are rejected.

A second hypothesis is that females will choose males with good genes. Brightly colored males or males that can call or display longer than others indicate that they are healthy, parasite-free, and vigorous. Females, by choosing such males, acquire good genes for their offspring. This hypothesis can explain the bright colors and elaborate displays produced by males of many species: Bright colors are indicators of high fitness, and are reinforced by female choice.

Barn swallow males have long tails, which they display to attract females. Møller (1988) showed by direct observation of banded individuals and experimental manipulation of tail length, that females prefer males with longer tails, and that longer tailed males fledge more young in a season. Barn swallow adults and nestlings are parasitized by mites. Møller (1994) showed that these mites reduce both male tail length and nestling growth rate. Males with longer tails have fewer parasites. Both they and their offspring have a longer lifespan. So, females that choose males with longer tails get mates that have greater parasite resistance and higher reproductive rates. Both of these traits are passed to their offspring.

A third hypothesis to explain female choice is that it is based on preexisting sensory biases in the female. Females prefer traits that elicit the greatest amount of sensory stimulation (Ryan and Keddy-Hector 1992). In other words, a female is predisposed to respond to a particular stimulus, and male behavior evolves to provide that stimulus during courtship and mating. For example, Ryan and Keddy-Hector

(1992) summarize studies on three species of frogs. Females prefer male calls of lower frequency. In all three species, the preferred frequency was that to which the auditory system was most sensitive, and not the average call frequency of males. The reason for the mismatch between male call frequency and female auditory sensitivity is not known.

These three hypotheses are not mutually exclusive. The question is not which one is appropriate, but their relative importance and interactions in any particular species.

Fisher (1915) suggested that the interaction between sexual selection for male ornaments and female preference for those ornaments can lead to a process of runaway selection. If there is genetic variation for both male ornamentation and for female preference, then males with greater ornaments will have greater mating success. Offspring that receive the alleles for greater ornamentation from their fathers will receive alleles for preference from their mothers. Nonrandom mating will lead to gametic disequilibrium in which alleles for greater ornaments are associated with alleles for female preference. Any mutation leading to greater ornaments in males will be favored by females, and any mutation increasing preference in females will lead to selection for greater ornaments in males. Female preference and male ornamentation reinforce one another. The result is a positive feedback system in which male ornaments become more and more exaggerated. Eventually, the disadvantage of greater male ornamentation may outweigh the advantage and the process may stabilize. Alternatively, female preferences may be so strong that the process leads to extinction. The extinct Irish elk, with its enormous antlers, is thought to be an example of the positive feedback between female choice and exaggerated male characteristics.

This positive feedback hypothesis predicts a genetic correlation, due to gametic disequilibrium, between the degree of male ornamentation and the strength of female preference. Several experiments have confirmed this prediction. For example, female stickleback fish prefer males with intense red coloration on their abdomens. Bakker (1993) bred intense red and pale red males to several females and examined coloration in the sons and preference in the daughters. He found a strong positive correlation between the intensity of coloration in the sons and the strength of preference in the daughters (Figure 12.14). From the design of the breeding experiment, he was able to estimate the genetic correlation at about 0.75, a high correlation consistent with the positive feedback hypothesis.

Female choice can go beyond deciding whom to mate with. Birkhead and Møller (1993) suggested that females can control paternity at four stages of the breeding cycle: (1) before copulation; (2) during copulation; (3) after copulation but before fertilization; and (4) after fertilization. Birkhead and Møller review evidence for each of these.

There has been much recent interest in sperm selection by females. We have already seen an analogous process of pollen incompatibility in plants. The pollen tube does not grow if the pollen grain contains the same allele as either allele in the stigma (Section 8.3). This is a mechanism to avoid inbreeding and consequent inbreeding depression in the offspring. Does a similar process occur in animals? That is, can multiply mated females choose which sperm to fertilize their eggs with? Olsson et al. (1996) suggested that female Swedish sand lizards (*Lacerta agilis*) do this. Females mate with many different males, including full siblings. Olsson et al. used

Figure 12.14 **Correlation between intensity of red coloration in male three-spined sticklebacks and preference for red in their female siblings.** Each open circle represents the average of sons' intensity and average of daughters' preference among the offspring of one male. Altogether, 103 sons and 61 daughters were measured. Horizontal and vertical error bars indicate ±1 standard deviation for sons' intensity and daughters' preference within each family. *Source:* From Bakker (1993).

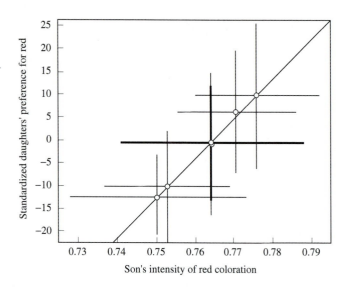

the proportion of shared DNA fingerprint bands as an index of relationship between mating males and females. They found that more distantly related males sired a higher proportion of offspring than more closely related males. This was true for both natural matings and laboratory matings. They concluded that females actively choose sperm from more distantly related males. However, as Wirtz (1997) pointed out, it is difficult to distinguish female sperm choice from male sperm competition or early embryonic mortality due to inbreeding depression. The frequency and significance of sperm choice by females is still controversial.

Reversal of Sex Roles

We have assumed that females invest more in reproduction and offspring care than do males, and as a consequence males compete for mates and females are choosy. This is true for most species that care for their offspring, but in some species males rear the offspring and have greater parental investment. If the theory discussed in this section is true, we would expect females to compete for mates and males to be choosy in these species. That prediction is generally confirmed. For example, female spotted sandpipers fight and compete for males. A female may mate with several different males; she mates, lays her eggs, and leaves them for the male to care for while she goes off to mate with another male. Access to males is the limiting factor for female reproductive rate; therefore, females compete for males. One aspect of this competition is that the sexual dimorphism associated with sexual selection is reversed. Females of these species are often more brightly colored than the males.

Sexual selection is currently an active field of research, and we have only introduced the subject here. For more information, see Andersson (1994), Arnold and Duvall (1994), Andersson and Iwasa (1996), Stockley (1997), Ryan (1998), and Holland and Rice (1998).

12.10 A Case Study: Soay Sheep

If there is one lesson to be learned from this chapter, it is that natural selection is not a consistent force. It varies in space, in time, and between sexes. It acts on different characters simultaneously, and can act in different directions at different stages of the life cycle. If we are to understand the effects of selection on natural populations, we must consider those populations in their physical and biological environments, as well as the complex interactions between different genes and different phenotypes.

In this section, we summarize the main results of one long-term study to illustrate how natural selection acts in the real world. This study should serve as a caution not to think exclusively about one mode of natural selection. The main papers on which the following summary is based are Clutton-Brock et al. (1991, 1997), Moorcroft et al. (1996), Coltman et al. (1999), and the references therein.

Soay sheep (*Ovis aries*) were introduced in 1932 to the island of Hirta in the St. Kilda archipelago of Scotland. The population is unmanaged (natural), with a long-term effective population size of about 260. The population has been intensively studied since the 1960s. Mating occurs during the fall, most mortality occurs in winter, and lambing occurs in the spring. The population undergoes an approximate three-year cycle due to overcompensating density-dependent mortality. At high densities, winter mortality is high and the population crashes to a low level. It takes about two years to build up to high density again, at which time high mortality causes the population to decline, and the cycle is repeated. In crash years, up to 70 percent of the population may die. Survival during crashes appears to be nonrandom with respect to several allozyme and microsatellite loci, as well as for horn type and coat color (see below).

The population is polymorphic for horn morphology. Three phenotypes are seen: polled (no horns), horned (normal horns), and scurred (small vestigial horns). The presence or absence of horns is controlled by a single gene, and whether horns are normal or scurred is controlled by one or two genes. Phenotype frequencies differ in males and females. In males, the polled phenotype is absent, and the frequency of the horned phenotype is about 0.88. In females, the frequencies of polled, scurred, and horned phenotypes are about 0.41, 0.24, and 0.35, respectively. Selection operates differently in the sexes. Scurred males have lower mating success than horned males due to inferior fighting abilities, resulting in less access to females. However, scurred females have higher breeding success, as measured by conception rate and weaning rate. Thus, sexual selection favors horned males and fertility selection favors scurred females. In addition, scurred males and females both have higher survival rates, especially at high densities.

A second morphological polymorphism also exists. Coat color is either light or dark, and is controlled by a single gene, with dark dominant over light. The frequency of the light form is relatively stable at about 25 percent. Coat color appears to be independent of horn morphology.

Selective pressures vary with population size. At low population densities, overwinter survival is high and approximately equal for horned versus scurred and light versus dark phenotypes. As population density increases, survival rates decrease for all phenotypes, but at high densities, light coat color and scurred horns have significantly higher survival rates than do dark color or normal horns.

Selection also varies with age. In adults (age > 36 months), scurred females have higher conception rates at all densities, but in juveniles (age 3–13 months), scurred females have higher conception rates at low densities but lower rates at high densities. Similarly, juvenile scurred females have lighter offspring at low densities, but heavier offspring at high densities. This is true for younger age classes, but not for adults. Not only the strength, but also the direction of selection, depends on population density.

Paternity of lambs was estimated from microsatellite data. The lifetime breeding success (number of lambs sired) was lowest for rams born in high-density years. This was because juveniles had low mating success in their first year (sexual selection), and low survival rates through their first winter (viability selection).

This example illustrates the real-life complexity of natural selection. Selection pressure varies from year to year. Antagonistic selection between males and females affects horn phenotype and mating success. Selection is simultaneously acting on several related traits, and is sometimes density dependent and age dependent. Sexual selection occurs in males as competition for mates. This complexity is probably typical of the way natural selection acts in nature. The models we have discussed in this chapter suggest possibilities for different kinds of selection, but we must not fall into the trap of thinking that selection (natural or sexual) acts uniformly and consistently.

Recall Wright's metaphor of an adaptive landscape to describe the operation of natural selection (Section 5.9). Imagine the fitness possibilities as a multidimensional landscape, with mountain peaks being fitness maxima, and valleys and sinks representing minima. The population will (usually) seek to climb to the summit of the nearest peak (local fitness maximum). However, due to conflicting and changing selection pressures, the landscape is not so much like a mountain range as like a waterbed in motion, with the location of peaks, valleys, and ridges constantly shifting. A population may adapt to a local peak, but find itself in a valley a few generations later.

Summary

1. Simple models of natural selection that assume constant fitnesses are unrealistic. Fitness varies in space and time, and sometimes depends on genotypic frequencies and/or population size. Genes may have pleiotropic effects that oppose each other in males and females, or at different times in the life cycle.

2. Single-locus overdominance is unlikely to be the main explanation of extensive genetic variation observed in protein and DNA sequences. Theory and computer simulations suggest that as more alleles are considered, the probability of a stable polymorphism decreases rapidly.

3. Gene interaction among different loci is common, and may lead to permanent gametic disequilibrium. If this occurs, the evolutionary fates of alleles at one locus are tied to alleles at other loci, and single-locus fitness estimates and predictions can be misleading. A general principle is that permanent gametic

disequilibrium is more likely if there is strong interaction and/or tight linkage between loci.

4. Two-locus selection theory is very complex. Multiple stable equilibria may exist, and the direction of evolution of a population may depend on initial allele frequencies and initial gametic disequilibrium.

5. The reduction principle suggests that natural selection will favor reduced recombination whenever it strongly favors nonrandom associations among loci. It is not a universal principle; under some conditions, natural selection will favor increased recombination.

6. Various models of spatially varying selection suggest that conditions for a stable polymorphism are somewhat less restrictive than in a uniform environment, but not dramatically so. The probability of polymorphism increases with habitat preference and nonrandom mating. Spatial variation in natural selection is commonly observed in natural populations, but only a few studies have tried to relate it directly to theoretical models.

7. Models of temporally varying selection also suggest that conditions for a stable polymorphism are only somewhat less restrictive than in a uniform environment. Temporally varying allele (or inversion) frequencies are commonly observed, but temporally varying selection is only one of several possible explanations.

8. Models of weak selection that varies randomly in space and time sometimes predict large amounts of genetic variation.

9. Under frequency dependent selection, the fitness of a genotype depends on its frequency. If fitness increases as a genotype becomes rarer, the inevitable result is a protected polymorphism.

10. Frequency dependent selection is thought to be common. Examples are the maintenance of a 1:1 sex ratio, self-incompatibility systems of some plants, and evolutionary response to parasites or predators.

11. Antagonistic pleiotropy occurs when natural selection favors different genotypes at different stages of the life cycle, or favors different genotypes for different components of fitness. It is very common, but single-locus models suggest that it is not likely to be a major contributor to polymorphism.

12. Natural selection can act in opposite directions in males and females. Such sexually antagonistic genes may be common, and may cause an evolutionary arms race between males and females. These genes evolve rapidly and as one allele replaces another, transient polymorphisms will be seen.

13. Differences in parental investment lead to competition among males for multiple mates, and discrimination by females to choose the best mates. This leads to the evolution of male ornaments and displays to attract females, and to female preferences for certain male appearances or behavior.

14. A long-term study of Soay sheep illustrates the diverse ways in which natural and sexual selection act in natural populations.

15. The essential lesson of this chapter is that natural selection is not a consistent force. The many conflicting modes and directions of natural selection collectively maintain much of the genetic variation observed in most natural populations.

Problems

12.1. In addition to the well-known A and S alleles of the β-hemoglobin gene, a third allele, designated C, exists. Following are estimated fitnesses for the different genotypes in West African populations (Cavalli-Sforza and Bodmer 1971):

	A	S	C
A	0.89	1.00	0.89
S	—	0.20	0.70
C	—	—	1.31

  a. Assume the C allele is absent. Estimate the equilibrium frequencies of the A and S alleles.

  b. In some populations, the A and S alleles are near equilibrium, with a low frequency of the C allele. Why doesn't the C allele go to fixation?

12.2. Verify the recursion equations for g_2, g_3, and g_4, using Table 12.3.

12.3. Show that, if natural selection is not acting, equations (12.5) through (12.8) reduce to equations (4.18) through (4.21).

12.4. Consider a two-niche Levene model. Let $W_1 = V_2 = 1.05$ and $W_2 = V_1 = 0.95$. Assume $c_1 = c_2 = 0.5$.

  a. Will a multiple-niche polymorphism be maintained under these conditions?

  b. Now, let $W_1 = 1.06$, with all of the other fitnesses unchanged. Will a stable polymorphism be maintained?

  c. Assume the original fitnesses, but let $c_1 = 0.53$. Will a stable polymorphism be maintained?

  d. Is the Levene model useful in explaining observed polymorphisms in natural populations? If it is not useful as an explanatory tool, is it useful for anything?

12.5. Schmidt and Rand (2001) estimated fitnesses of acorn barnacles in different habitats based on *Mpi* genotypes.

  a. Show that the estimated fitnesses on page 493 do not satisfy the sufficient conditions for a stable polymorphism under the Levene model.

  b. Using these fitnesses and $c_1 = 0.45$, plot Δp versus p in equation (12.9). Label all equilibria as either stable or unstable. Under what conditions will p approach a stable equilibrium?

  c. Repeat for $c_1 = 0.40$ and $c_1 = 0.50$.

  d. Do these calculations convince you that selection in different habitats is the primary factor maintaining the *Mpi* polymorphism? What other factors might increase your confidence that this explanation is or is not correct?

12.6. Consider the frequency dependent selection model of Section 12.6; let $a = 1$.

  a. Plot Δp versus p for p from 0 to 1. Label all equilibria as either stable or unstable.

  b. Plot $\bar{w}(p)$ versus p for p from 0 to 1. Plot for $a = 1$ and $a = .5$ on the same graph.

12.7. What is the difference between a stable polymorphism and a transient polymorphism? How might you tell the difference?

12.8. Following are partial data for two *Acp* genes studied by Begun et al. (2000):

Gene	Kind of Difference	Substitutions	Polymorphisms
Acp36DE	Replacement	72	10
	Silent	52	24
Acp53Ea	Replacement	7	5
	Silent	11	6

Perform a McDonald–Kreitman test on each gene. What can you conclude about natural selection at each locus?

12.9. Consider the following fitnesses in males and females:

	A_1A_1	A_1A_2	A_2A_2
Males	.7	1.0	1.2
Females	1.2	1.0	0.9

Will a stable polymorphism be maintained? If yes, which allele will have the higher frequency at equilibrium? If no, which allele will go to fixation?

13

Quantitative Genetics

*Those who are not specifically concerned with understanding
quantitative genetics, but are interested in a character from some
other point of view, may well be forgiven if they often seem to regard
the demonstration that the character is inherited "polygenically" as
an end to investigation rather than a beginning.*

—*J.M. Thoday (1961)*

Until now, we have generally considered characters that show discrete phenotypes and are controlled by a single gene with no significant environmental
effects. Many characters are not so simple. For example, body weight shows a
continuous distribution, is controlled by many genes, and is affected by environmental factors (e.g., diet and exercise).

A **quantitative trait** is one that shows a continuous range of phenotypes.
There are two kinds of quantitative traits: (1) True quantitative traits show a continuous distribution, in which the phenotype can take on any value. Body weight
and height are examples. (2) Meristic traits take on an integer value. Examples
are fingerprint ridge counts or the number of scales along the lateral line of a
fish. Both are analyzed the same way.

Closely related are **threshold traits**, in which any individual is classified as either
having the trait or not, but for which the underlying basis is quantitative. Many
human diseases are threshold traits, for example, high blood pressure or obesity. If
these characteristics are affected by many genes and environmental factors, the
quantitative genetic methods discussed in this chapter can be used to analyze them.

Quantitative traits have two distinguishing features: (1) They are usually controlled by many genes acting together. (2) Environmental influences usually have
significant effects on the phenotype. Consider height in humans. People are not
simply tall or short; they come in all sizes. It is common knowledge that tall
people tend to have tall children and short people tend to have short children.
Therefore, there is a genetic component to height. The number of genes controlling height is unknown, but probably large. Environmental influences also
affect height. For example, the average height of the Japanese population is

several inches taller than it was two generations ago, and this is due mainly to improved diet.

In this chapter, we will consider several questions about quantitative traits:

1. What is the genetic basis of quantitative traits; that is, are they subject to the rules of Mendelian inheritance?
2. How do we separate genetic effects from environmental effects on a quantitative trait?
3. How many loci affect quantitative traits, and how large are their effects?
4. How much genetic variation for quantitative traits is there in natural populations?
5. How is this genetic variation maintained?
6. How important are mutation, linkage, dominance, epistasis, and pleiotropy in the evolution of quantitative traits?

Several of these questions are the quantitative trait analogs of questions about molecular variation that we have investigated in previous chapters.

Much of quantitative genetics theory originated as a tool for animal and plant breeding. For that reason, the terminology and symbolism are sometimes different from the terminology of conventional population and evolutionary genetics. We will initially follow the breeding approach and later apply the principles to natural populations. If the initial theory seems abstract and overly theoretical, keep in mind that if you have ever eaten a food that has been artificially selected for increased yield or nutritional value (for example, corn, wheat, tomatoes, milk, eggs, pork, chicken, or beef), you have benefited from quantitative genetics theory.

Perhaps a caution is in order here. Quantitative genetics relies heavily on theoretical and applied statistics. In order to fully understand much of what follows, you must be comfortable with the ideas of variance, covariance, and correlation. Analysis of variance (ANOVA) is a fundamental tool in quantitative genetics, but we will not consider it in detail. Appendix A summarizes much of what you need to know, but a thorough knowledge of probability and statistics is essential for complete understanding.

Quantitative genetics is currently a very active and exciting field. This chapter only scratches the surface of a very complex subject. Those interested in further study should refer to one of the standard texts on quantitative genetics (Falconer and Mackay 1996; Roff 1997; Lynch and Walsh 1998), upon which much of this chapter is heavily based.

13.1 Genetic and Environmental Effects on Quantitative Traits

In the early 1900s, soon after the rediscovery of Mendel's work, there was some doubt about the mechanism of inheritance of quantitative traits. It was initially believed that such traits do not follow the laws of Mendelian inheritance (see Section 1.3). However, within a few years, several important experiments demonstrated that the inheritance of quantitative traits is consistent with Mendel's laws, and in 1918 Fisher demonstrated mathematically that the patterns of inheritance seen for quantitative traits can be explained by Mendelian inheritance at multiple loci. This landmark paper established the theoretical foundations of quantitative genetics.

The two essential characteristics of quantitative traits are that they are affected by both genetic and environmental factors, and that they are controlled by several loci, each of which follow the laws of Mendelian inheritance. In this section, we review two classic experiments that demonstrated these facts.

One of the first experiments to show that quantitative traits are affected by both genetic and environmental factors was performed by Johannsen (1903). He studied seed weight in inbred lines of beans. Inbred lines are almost completely homozygous. Every individual within a line is homozygous for the same alleles, but different inbred lines are homozygous for different alleles. Johannsen studied several different inbred lines with different seed weights. He planted seeds from each line and compared the seed weight of the offspring to the seed weight of the parent. Johannsen showed that phenotypic differences *among* lines were inherited, but that phenotypic variation *within* a line was not. Lines with heavier seeds produced offspring with heavier seeds, indicating that the differences among lines were due to genetic differences (Figure 13.1a). However, within any single line, there was no relationship between parent seed weight and offspring seed weight (Figure 13.1b). The variation in seed weight within a line was not heritable; that is, it was due to environmental effects. Thus, Johannsen showed that seed weight in beans is affected by both genetic and environmental factors.

Johannsen did not determine how many genes affected seed weight, but it is almost certainly more than one. Nilsson-Ehle (1909) was one of the first to demonstrate that a nearly continuous distribution of phenotypes can be produced by several genes acting together. Nilsson-Ehle studied kernel color in wheat, which varies from dark red to white. Nilsson-Ehle crossed different red lines to the white line, and examined kernel color in the F_1 and F_2 generations. The F_1 always showed a shade of red intermediate between the two parents, and the F_2 showed a range of phenotypes varying from the redness of the red parent to white. Some lines segregated approximately 3:1 (red:white) in the F_2, whereas some segregated 15:1 and some 63:1. Here, red means any shade of red. Three kinds of crosses and the results are described in Table 13.1.

Nilsson-Ehle interpreted his results as follows: Kernel color is controlled by three independently assorting genes. Each gene has an allele that produces a red pigment (call the alleles A_1, B_1, or C_1 for the three loci) and an allele that produces no pigment (A_2, B_2, or C_2). These three genes act additively and independently; in other words, the kernel color (degree of redness) is determined solely by the total number of A_1, B_1, or C_1 alleles. The relationship between the phenotypes and genotypes is summarized in Table 13.2.

Under this hypothesis, the results of the three crosses are explained in Figure 13.2. The degree of redness in the F_2 depends only on the number of A_1, B_1, or C_1 alleles received from the F_1 parents, and this number is determined by the Mendelian laws of segregation and independent assortment. The F_2 phenotypic ratios come from the 2×2, 4×4, and 8×8 Punnett squares for crosses 1, 2, and 3, respectively.

Much subsequent work has confirmed Nilsson-Ehle's interpretation. In this case, we have a nearly continuous distribution controlled by three genes acting additively and independently.

Figure 13.1 **Variation among and within inbred lines of beans.** (a) Mean parental seed weight (cg) versus mean off-spring seed weight for different lines. Each data point represents parent and offspring weights for a different line. The slope of the line is significantly positive, indicating that lines with heavier seeds produce off-spring with heavier seeds. (b) Variation within lines. Each line represents a different inbred line; data points represent different individuals within the same line and their offspring seed weights. If variation within a line is inherited, we expect a significantly positive slope as in (a). None of the slopes is significantly different from zero, indicating that variation within a line is not inherited. *Source:* Data from Johannsen (1903).

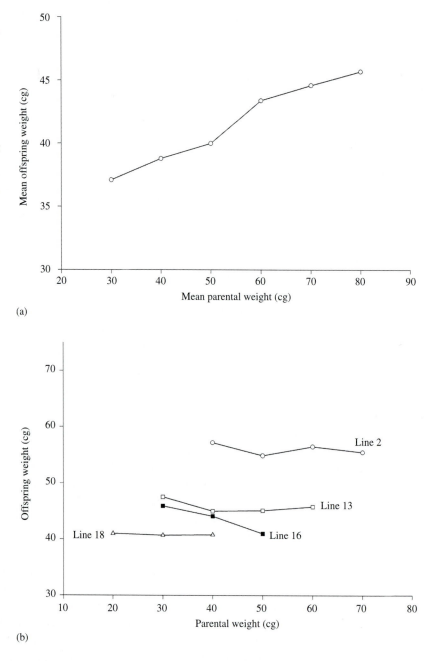

(a)

(b)

TABLE 13.1 **Results of crosses among inbred lines of wheat that differed in kernel color.**

Kernel color ranges from white through increasingly dark shades of red. *Source:* From Nilsson-Ehle (1909).

Cross	F_1 Phenotype	Approximate Proportion of White Kernels in F_2
Light red × white	very light red	1/4
Dark red × white	light red	1/16
Darkest red × white	intermediate red	1/64

TABLE 13.2   Three-locus explanation of Nilsson-Ehle's results.

The kernel color is determined by the number of A_1, B_1, or C_1 alleles.

Color	Number of Redness Alleles	Genotypes
White	0	$A_2A_2B_2B_2C_2C_2$
Very light red	1	$A_1A_2B_2B_2C_2C_2,$ $A_2A_2B_1B_2C_2C_2,$ $A_2A_2B_2B_2C_1C_2$
Light red	2	$A_1A_1B_2B_2C_2C_2,$ $A_1A_2B_1B_2C_2C_2,$ $A_2A_2B_2B_2C_1C_1,$ etc.
Intermediate red	3	$A_1A_2B_1B_2C_1C_2,$ $A_1A_1B_1B_2C_2C_2,$ $A_2A_2B_1B_2C_1C_1,$ etc.
Dark red	4	$A_1A_1B_1B_1C_2C_2,$ $A_1A_1B_1B_2C_1C_2,$ $A_2A_2B_1B_1C_1C_1,$ etc.
Very dark red	5	$A_1A_1B_1B_1C_1C_2,$ $A_1A_1B_1B_2C_1C_1,$ $A_1A_2B_1B_1B_1C_1C_1$
Darkest red	6	$A_1A_1B_1B_1C_1C_1$

Note in Figure 13.2 that, as the number of segregating loci increases, the distribution of the F_2 phenotypes begins to resemble a normal distribution. For even a moderately large number of loci, the phenotypic distribution would approximate a normal distribution very closely. In fact, many quantitative traits show an approximately normal distribution; several examples are shown in Figure 13.3. We shall see later why this is expected, and why much quantitative genetics theory assumes a normal distribution. In the following sections, we will sometimes invoke results from normal distribution theory without explaining why they are true.

The interpretation illustrated in Figure 13.2 assumes that any kernel can be unambiguously assigned to one of seven phenotypic classes. That is usually not possible, which is why Nilsson-Ehle expressed his results in terms of the ratio of red (any shade) to white. His approach avoided the necessity of distinguishing between various shades of red. The difficulty is due not only to the subtleties of judging shades of red, but also to environmental effects that can modify kernel color. Consider what would happen if environmental effects caused some variation around the typical redness value for each genotype: Individuals with the same

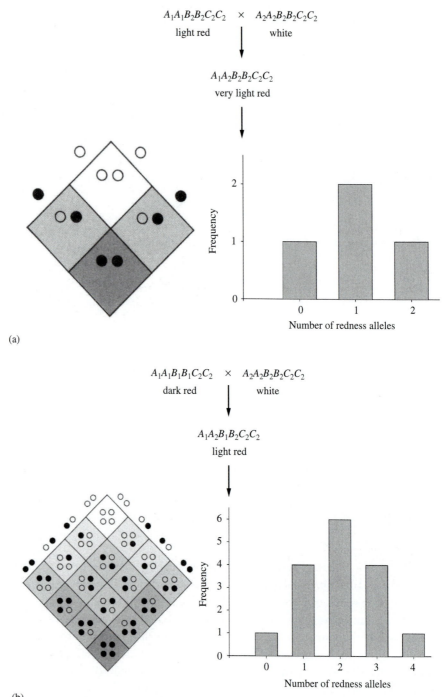

(a)

(b)

Figure 13.2 **Explanation of Nilsson-Ehle's experimental results with kernel color in wheat. See text for details.** (a) Cross between inbred lines differing at one locus affecting kernel color. (b) Cross between inbred lines differing at two loci affecting kernel color. (c) Cross between inbred lines differing at three loci affecting kernel color. Dark circles represent redness-producing alleles; white circles, nonredness alleles. The histograms represent expected frequency distributions of the number of redness-producing alleles in the F_2. As the number of segregating loci increases, the histograms more closely approximate a normal distribution. *Source:* Punnett squares from Ayala (1982).

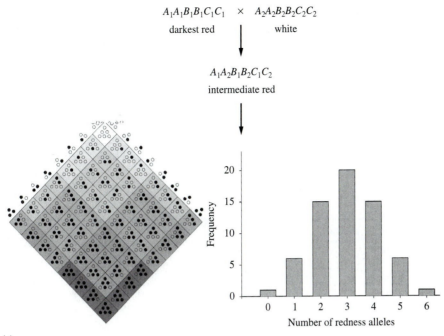

$A_1A_1B_1B_1C_1C_1$ $\times$ $A_2A_2B_2B_2C_2C_2$

darkest red   white

$A_1A_2B_1B_2C_1C_2$

intermediate red

(c)

Figure 13.2   **(Continued)**

genotype would exhibit a range of phenotypes (redness values) centered on the value determined by the genotype (Figure 13.4). If these phenotypic ranges overlap, there would be a continuous distribution of redness values, and any given redness value could be caused by more than one genotype. For example, the degree of redness indicated by the dashed line in Figure 13.4 could be caused by either two, three, or four redness-producing alleles. The result of superimposing environmental variation onto genetic variation is that we now have a truly continuous distribution of phenotypes and there is no longer a unique association between genotype and phenotype. This is the situation for most quantitative traits.

In this example, there was no dominance among alleles affecting kernel color; at any locus, the heterozygote was halfway between the two homozygotes in redness. Clearly, this need not always be the case. Similarly, there was no interaction among the genes affecting kernel color, but we cannot ignore the possibility of gene interaction for other quantitative traits. Many experiments have demonstrated that gene interaction can be important in determining a phenotype affected by more than one gene. For example, coat color in mice is controlled by at least five interacting genes that affect various aspects of pigment type and distribution (Griffiths et al. 2000).

To summarize, we have shown that an individual's phenotype for a quantitative trait is due to both genetic effects and environmental effects. We can symbolize this as

$$P = G + E \qquad \textbf{(13.1)}$$

where P is the phenotypic value of an individual, and G and E are the genetic and environmental effects, respectively. We can further partition the genetic effect

Figure 13.3 **Examples of quantitative characters that show an approximately normal distribution.**
(a) Growth rate in mice; (b) litter size in mice; (c) abdominal bristle number in *Drosophila melanogaster*; (d) facet number in eyes of *Drosophila melanogaster* with Bar eye mutation. Histograms represent actual data; curved lines are the best fitting normal curve. *Source:* From Falconer and Mackay (1996).

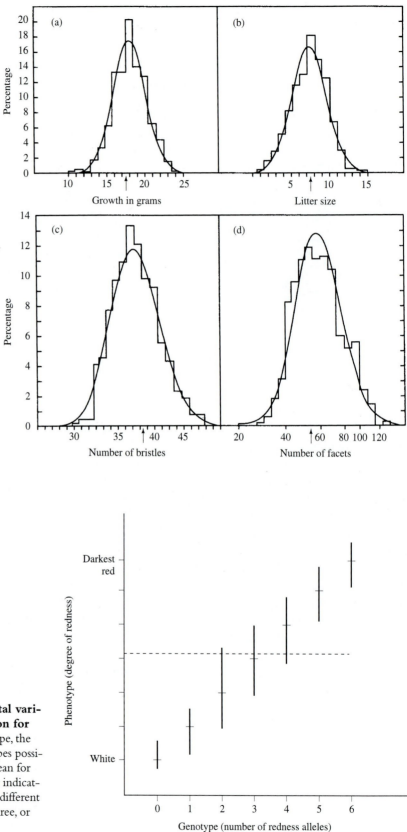

Figure 13.4 **Hypothetical environmental variation superimposed on genetic variation for kernel color in wheat.** For each genotype, the vertical bar represents the range of phenotypes possible, and the horizontal bar represents the mean for that genotype. Individuals with kernel color indicated by the dashed line can have any of three different genotypes (i.e., they may have either two, three, or four redness-producing alleles).

into an additive effect, as illustrated by the kernel color example, a possible effect due to dominance interaction between alleles at the same locus, and a possible effect due to interaction among alleles at different loci. We will consider each of these components in more detail in the next section.

13.2 The Genetics of Quantitative Characters

We begin by considering a single locus with two alleles. Assume random mating, no environmental effects, and no natural selection, mutation, or other evolutionary processes affecting the locus. Therefore, the genotypes will be in the Hardy-Weinberg frequencies. Scale the phenotypic values so that the A_1A_1 genotype has a phenotypic value of a and A_2A_2 has a value of $-a$. The value of A_1A_2 is designated by d, which is a measure of dominance. Consider an example from mice (Falconer and Mackay 1996). The pygmy gene affects body weight as follows (letting A_1 be the wild type allele and A_2 the pg allele)

Genotype	Mean Weight
A_1A_1	14g
A_1A_2	12g
A_2A_2	6g

Figure 13.5 shows the genotypes with their phenotypic values and their rescaled values. The zero point for the rescaled values is halfway between the two homozygotes, or 10. Therefore, $a = 4$ and $d = 2$, as shown in the figure.

We can now construct another table, remembering that the genotypes are in Hardy-Weinberg frequencies.

Genotype	Value	Frequency
A_1A_1	a	p^2
A_1A_2	d	$2pq$
A_2A_2	$-a$	q^2

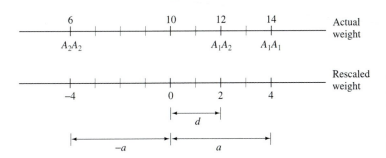

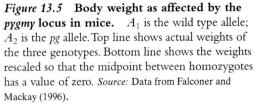

Figure 13.5 **Body weight as affected by the pygmy locus in mice.** A_1 is the wild type allele; A_2 is the pg allele. Top line shows actual weights of the three genotypes. Bottom line shows the weights rescaled so that the midpoint between homozygotes has a value of zero. *Source:* Data from Falconer and Mackay (1996).

From this, we can find the mean value of the population. It is just the weighted average of the values

$$\mu = p^2 a + 2pqd + q^2(-a)$$

which simplifies to

$$\mu = a(p - q) + 2pqd \qquad (13.2)$$

Next, we express the value of each genotype as a deviation from the population mean. We call these the **genotypic values**, and symbolize them by G with the appropriate subscripts. In other words,

$$G_{11} = a - \mu$$

$$G_{12} = d - \mu$$

$$G_{22} = -a - \mu$$

Substituting the right side of equation (13.2) and simplifying gives

$$G_{11} = 2qa - 2pqd \qquad (13.3)$$

$$G_{12} = a(q - p) + d(1 - 2pq) \qquad (13.4)$$

$$G_{22} = -2pa - 2pqd \qquad (13.5)$$

The genotypic values are a measure of how superior (or inferior) each genotype is compared to the population mean, and thus these values depend on the allele frequencies. However, as we shall see, an individual's genotypic value is not actually passed to its offspring.

Decomposition of the Phenotype

Suppose you wish to conduct a selective breeding experiment in which you want to increase the average body weight in a population of mice. Which individuals in the population should you choose to breed? The obvious first response is to breed those individuals with the highest weights. However, those individuals may not have the best genotypes. They may have experienced especially favorable environments and thus their good phenotypes may be due to favorable environmental effects, and not to good genotypes. These environmental effects will usually not be passed on to their offspring. Before you begin a selective breeding program, it is wise to have some idea of how much of the phenotypic variation in a population is actually due to genetic variation among individuals.

It gets worse. Ignoring for now environmental effects, the phenotype of an individual is determined by its genotype. However, in sexual reproduction, genotypes are not transmitted to the offspring. The genotypes of the parents are broken up during Mendelian segregation, and each parent transmits only one allele to the offspring. So the value of an individual in a breeding program does not depend on its genotype; it depends only on the allele it passes to its offspring. We need some measure of the average value of an allele when combined at random with other alleles in the population.

First consider allele A_1: Under random union of gametes (random mating), it will combine with another A_1 allele with a frequency of p and it will combine with an A_2 allele with a frequency of q. The resultant genotypes are A_1A_1 and

A_1A_2, with values a and d, respectively. Thus, the average value of genotypes containing A_1 is

$$\mu_1 = pa + qd$$

Express this as a deviation from the population mean:

$$\alpha_1 = \mu_1 - \mu = [pa + qd] - [a(p - q) + 2pqd]$$

which simplifies to

$$\alpha_1 = q[a + d(q - p)] \qquad \textbf{(13.6)}$$

α_1 is called the **average effect** of A_1.

Similarly, we can calculate the average effect of A_2. First,

$$\mu_2 = pd - qa$$

from which we get

$$\alpha_2 = -p[a + d(q - p)] \qquad \textbf{(13.7)}$$

α_1 and α_2 are the average effects[1] of alleles A_1 and A_2, expressed as deviations from the population mean. Note that the quantity in brackets is the same in equations (13.6) and (13.7). We will call this quantity α (without a subscript)

$$\alpha = a + d(q - p) \qquad \textbf{(13.8)}$$

So, we can express α_1 and α_2 as

$$\alpha_1 = q\alpha$$

and

$$\alpha_2 = -p\alpha$$

We now define the **breeding value** of an individual as the sum of the average effects of the two alleles it carries. In symbols,

$$BV_{11} = \alpha_1 + \alpha_1 = 2q\alpha$$

$$BV_{12} = \alpha_1 + \alpha_2 = \alpha(q - p)$$

$$BV_{22} = \alpha_2 + \alpha_2 = -2p\alpha$$

Because the breeding value is the sum of the average effects of the alleles, it is frequently called the **additive effect** and symbolized by A, with subscripts to indicate the genotype. Thus, the additive effects of the three genotypes are

$$A_{11} = 2q\alpha \qquad \textbf{(13.9)}$$

$$A_{12} = \alpha(q - p) \qquad \textbf{(13.10)}$$

$$A_{22} = -2p\alpha \qquad \textbf{(13.11)}$$

The breeding values are *not* the same as the genotypic values unless $d = 0$. To see the relationship between them, express the genotypic values in terms of α. First, consider the genotypic value G_{11} as described in equation (13.3). Solve

---

[1]Technically, α_1 and α_2 are the *average excess* of alleles A_1 and A_2, but under random mating average effect and average excess are the same. We will follow Falconer and Mackay (1996) and call them the average effects.

equation (13.8) for *a* and substitute the result into (13.3). After some algebraic manipulation, we get

$$G_{11} = 2q\alpha - 2q^2d \tag{13.12}$$

Similarly,

$$G_{12} = \alpha(q - p) + 2pqd \tag{13.13}$$

and

$$G_{22} = -2p\alpha - 2p^2d \tag{13.14}$$

The first term on the right-hand side of these three equations is the breeding value, or additive value. The second term is due to interactions between the alleles of a genotype, that is, due to dominance effects. These are called the **dominance deviations**:

$$D_{11} = -2q^2d \tag{13.15}$$

$$D_{12} = 2pqd \tag{13.16}$$

$$D_{22} = -2p^2d \tag{13.17}$$

So, we can write the genotypic values as

$$G_{11} = A_{11} + D_{11}$$

$$G_{12} = A_{12} + D_{12}$$

$$G_{22} = A_{22} + D_{22}$$

In biological terms, the genotypic value of an individual is partitioned into the additive effects of the two alleles, plus the effect due to interaction between the alleles (dominance effect). Remember, all these are expressed as deviations from the population mean. The only part that matters in a breeding program is the additive effect, because during sexual reproduction the dominance effects in the parents are destroyed as the parental genotypes are broken up and new genotypes are formed in the offspring. An exception is found in some plant breeding programs in which an entire genotype can be reproduced by asexual propagation.

So far, we have considered only a single locus. This is unrealistic because many loci typically affect a quantitative trait, and the effects of a single locus cannot usually be measured. We can observe only the overall effects of all loci combined. Let us write the genotypic value of an individual as

$$G = A + D \tag{13.18}$$

where G, A, and D (without subscripts) are the genotypic value, additive effects, and dominance effects, summed over all loci affecting the trait.

There is an additional complication when more than one locus contributes to a quantitative trait. That is the possibility of epistasis, or gene interaction among different loci. So we must add another term to equation (13.18).

$$G = A + D + I \tag{13.19}$$

where *I* is the sum of all gene interactions among different loci.

As an example of gene interactions, consider two genes controlling pigmentation in the coat of mice (Russell 1949, cited in Falconer and Mackay 1996). The degree of pigmentation depends on two loci that control the number of pigment granules deposited in a hair and on the size of the granules. The mean sizes of pigment granules for the nine two-locus genotypes are shown in Table 13.3. Consider the B locus: If the genotype at the C locus is CC or Cc^e, the difference between BB and bb is $2a = 0.67$; therefore, $a = 0.335$. But if the genotype at the C locus is $c^e c^e$, the difference between BB and bb is 0.17, giving $a = 0.085$. The contribution of a B allele depends on the genotype at the C locus.

We will consider gene interaction in more detail in Section 13.6.

If we allow environmental variation to affect each phenotype, then the phenotype of an individual is due to a genetic component, as described above, and an environmental component, as seen in equation (13.1). Combining equations (13.1) and (13.19) we get

$$P = A + D + I + E \tag{13.20}$$

In practice, this equation is not very useful because we usually cannot separate the genetic from the environmental effects on an individual. In order to make progress, we must study variation within the population (see below).

At this point we must consider a critical assumption. We assume that each trait is controlled by a large number of unlinked loci, each of which has a small effect on the phenotype. We can now invoke the central limit theorem which says, roughly, that the sum of a large number of random variables approaches a normal distribution as the number of variables becomes very large. Therefore, under the above assumptions, the genotypic value of a quantitative trait is approximately normally distributed.

Furthermore, we assume that environmental effects are normally distributed, with a mean environmental effect of zero. Therefore, the phenotype will be approximately normally distributed with the phenotypic mean equal to the genotypic mean.

Much of quantitative genetics relies on the assumption that the phenotypic values and the genotypic values are normally distributed. This assumption allows us to analyze quantitative characters with a purely statistical approach and ignore the underlying genetic complexities. This is valuable because we usually have little or no understanding of the actual genetic control of quantitative traits. The observation that quantitative traits frequently show an approximately normal distribution (Figure 13.3) gives reason to think that this assumption is not unreasonable, but we must not forget that much of what follows depends on it.

TABLE 13.3   Size of pigment granules in hairs of mice (microns) *Source:* From Russell (1949); cited in Falconer and Mackay (1996).

Genotype at C Locus	Genotype at B Locus		
	BB	Bb	bb
CC	1.44	1.44	0.77
Cc^e	1.44	1.44	0.77
$c^e c^e$	0.94	0.94	0.77

Variance Components for Quantitative Traits

If we can describe the phenotype of an individual by $P = G + E$, we can describe the variance in the population by

$$V_P = V_G + V_E + 2\,\text{cov}(G, E)$$

We usually assume that there is no genotype $\times$ environment correlation, so the covariance term is zero, and the phenotypic variance is partitioned into genetic and environmental components. We can easily estimate the phenotypic variance; we need to find some way to estimate how much of this phenotypic variation is due to genetic variation within the population. This is important in breeding programs. For example, if you wish to breed for improved milk production in a herd of dairy cattle, you will not make much progress if most of the variation among different cows is environmental and not genetic.

Initially, we will ignore environmental variation, and concentrate on genetic variation. We will also consider only a single locus, so that interaction terms are not present. In this simplified case, we have

$$V_G = V_A + V_D + 2\,\text{cov}(A, D)$$

It can be shown that the covariance term is zero (Box 13.1), so we have

$$V_G = V_A + V_D$$

Recall that the variance of a random variable can be calculated as

$$\text{Var}(X) = \sum_{\text{all } X} f_i(X_i - \mu)^2$$

where X is a random variable, μ is its mean, and f_i is the frequency of the ith value of X. Applying this to the genotypic values gives (assuming Hardy-Weinberg equilibrium genotype frequencies)

$$V_G = p^2 G_{11}^2 + 2pq G_{12}^2 + q^2 G_{22}^2$$

Box 13.1   The covariance between A and D

The covariance between A and D is defined as

$$\text{cov}(A, D) = E[(A - EA)(D - ED)]$$

We must first find the expectations of A and D. The following table will be helpful:

Genotype	Frequency	A	D
A_1A_1	p^2	$2q\alpha$	$-2q^2d$
A_1A_2	$2pq$	$\alpha(q - p)$	$2pqd$
A_2A_2	q^2	$-2p\alpha$	$-2p^2d$

From this, we can calculate EA:

$$EA = p^2(2q\alpha) + 2pq\alpha(q - p) + q^2(-2p\alpha)$$
$$= 2pq\alpha(p + q - p - q)$$
$$= 0$$

Similarly,

$$ED = p^2(-2q^2d) + 2pq(2pqd) + q^2(-2p^2d)$$
$$= -2p^2q^2d + 4p^2q^2d - 2p^2q^2d$$
$$= 0$$

Finally, using the computational formula for covariance from equation (13.28), with $EA = ED = 0$,

$$\text{cov}(A, D) = p^2(2q\alpha)(-2q^2d)$$
$$+ 2pq\alpha(q - p)(2pqd)$$
$$+ q^2(-2p\alpha)(-2p^2d)$$
$$= 4p^2q^2\alpha d(-q + q - p + p)$$
$$= 0$$

(Remember, the Gs are defined as deviations from the population mean.) We have already calculated the Gs; they are given by equations (13.12), (13.13), and (13.14). Substituting these values into the above equation, we get, after much algebraic simplification,

$$V_G = 2pq\alpha^2 + (2pqd)^2$$

Following our earlier convention of considering terms with α as additive effects and terms with d as dominance deviation, we can define

$$V_A = 2pq\alpha^2 \qquad \textbf{(13.21)}$$

and

$$V_D = (2pqd)^2 \qquad \textbf{(13.22)}$$

V_A and V_D are called the **additive genetic variance** and **dominance variance**, respectively. When we consider more than one locus, we must also consider the variation due to gene interactions, designated by V_I:

$$V_G = V_A + V_D + V_I \qquad \textbf{(13.23)}$$

This assumes that all the covariances are zero.

Finally, assuming no genotype $\times$ environment correlation, the phenotypic variance can be decomposed into its components

$$V_P = V_A + V_D + V_I + V_E \qquad \textbf{(13.24)}$$

Again, this assumes that all the covariances are zero. In principle, we can partition the phenotypic variance still further to account for nonrandom mating, consistent environmental effects within families, and so forth, but we will stop here.

Heritability

We have partitioned the phenotypic variance into a genetic component and an environmental component. We can now determine what proportion of the phenotypic variation is due to genetic variation. We define the **broad sense heritability** as the ratio of the total genetic variance to the phenotypic variance

$$H^2 = \frac{V_G}{V_P} \qquad \textbf{(13.25)}$$

Similarly, we can define the **narrow sense heritability** as the ratio of the additive genetic variance to the phenotypic variance

$$h^2 = \frac{V_A}{V_P} \qquad \textbf{(13.26)}$$

The symbols H^2 and h^2 are historical artifacts, and refer to the heritabilities and not to the squares of the heritabilities. (Sometimes H is used for the broad sense heritability instead of H^2.) Clearly, H^2 is greater than h^2 unless V_D and V_I are both zero, in which case $H^2 = h^2$.

The narrow sense heritability is most important in determining how a quantitative character evolves under artificial or natural selection, as we shall see in Sections 13.3 and 13.4. Whenever we mention heritability, it is assumed we mean the narrow sense heritability unless we say otherwise. Table 13.4 gives estimates of narrow sense heritability for several quantitative traits in different organisms. These are mostly based on laboratory or field experiments; we will consider heritability in natural populations in Section 13.6.

TABLE 13.4   Heritabilities of various traits in different organisms.

Many references are secondary sources.

Organism	Trait	Heritability (h^2)	Reference
Dairy cattle	Milk yield	0.35	Falconer and Mackay (1996)
Pigs	Litter size	0.05	Falconer and Mackay (1996)
Chickens	Body weight	0.55	Falconer and Mackay (1996)
Mice	Tail length	0.40	Falconer and Mackay (1996)
	Litter size	0.20	Falconer and Mackay (1996)
Humans	Height	0.65	Falconer and Mackay (1996)
	Fingerprint ridge count	0.92	Jorde et al. (1995)
	Schizophrenia	0.70	Jorde et al. (1995)
	Blood pressure (systolic)	0.60	Jorde et al. (1995)
	Score on IQ tests	0.50	Jorde et al. (1995)
Drosophila melanogaster	Abdominal bristle number	0.53	Clayton et al. (1957)
	Body size	0.40	Falconer and Mackay (1996)
	Egg production	0.20	Falconer and Mackay (1996)
Tribolium castaneum	Fecundity	0.36	Mousseau and Roff (1987)
	Pupa weight	0.48	Halliburton and Gall (1981)
Eurytemora affinis (/)	Temperature tolerance	0.14	Lynch and Walsh (1998)
Eurytemora affinis (?)	Temperature tolerance	0.72	Lynch and Walsh (1998)
Medium ground finch	Body weight	0.91	Boag (1983)
	Wing length	0.84	Boag (1983)
	Bill length	0.65	Boag (1983)
Song sparrow	Body weight	0.038	Smith and Zach (1979)
	Wing length	0.135	Smith and Zach (1979)
	Bill length	0.333	Smith and Zach (1979)
Collared flycatcher	Lifetime reproductive success	0.21	Merila and Sheldon (2000)
	Clutch size	0.35	Merila and Sheldon (2000)
	Wing length	0.47	Merila and Sheldon (2000)
Red deer	Number of offspring	0.00	Kruuk et al. (2000)

Note from equation (13.26) that heritability is a ratio. If *either* the additive variance or the environmental variance changes, then the heritability will change. Thus, we must be very cautious about extrapolating heritability estimates from one population or experiment to another.

The broad sense heritability is useful in plant breeding programs, because entire genotypes, including dominance and interaction effects, can be replicated by asexual propagation.

Covariance Among Relatives

A parent shares half its alleles with each of its offspring. Full siblings share half of their alleles with one another. Similarly, a grandparent shares one-fourth of its alleles with its grandoffspring, and half siblings share one-fourth of their alleles. So if a quantitative character is to some extent heritable, then close relatives should have more similar genotypes, and therefore more similar phenotypes, than unrelated individuals or distant relatives. This basis of similarity among relatives is one of the most important ideas in quantitative genetics. We will express this similarity as a covariance. The definition of a covariance between two random variables, X and Y, is (see Appendix A)

$$\text{cov}(X, Y) = E[(X - EX)(Y - EY)] \tag{13.27}$$

An equivalent formula, which we will use later, is

$$\text{cov}(X, Y) = \sum_{\substack{\text{all} \\ XY \text{ pairs}}} f_i(X_i - EX)(Y_i - EY) \tag{13.28}$$

where f_i is the frequency of the ith XY pair. A positive covariance means that X and Y increase or decrease together, and a negative covariance means that as one increases, the other decreases.

We must distinguish between the phenotypic and genetic covariance. Let X_P and Y_P be random variables representing the phenotypic values of two individuals. Then $X_P = X_G + X_E$ and $Y_P = Y_G + Y_E$, where the subscripts G and E indicate genetic and environmental components. Then the covariance of X_P and Y_P is, using the rules for covariances from Appendix A,

$$\text{cov}(X_P, Y_P) = \text{cov}(X_G + X_E, Y_G + Y_E)$$
$$= \text{cov}(X_G, Y_G) + \text{cov}(X_G, Y_E) + \text{cov}(X_E, Y_G) + \text{cov}(X_E, Y_E)$$

We now make some critical assumptions. If we assume that environmental effects vary randomly among individuals, and are independent of genotype, then the last three covariances on the right are zero, and we have

$$\text{cov}(X_P, Y_P) = \text{cov}(X_G, Y_G)$$

We could further decompose the genotypic value into additive, dominance, and interaction components. Under assumptions that these effects are independent, the new covariances will be zero and the above equation is still valid. It must be stressed that these are rather restrictive assumptions, and what follows depends on them. They are sometimes reasonable in laboratory experiments or breeding programs, but are probably not valid in natural populations. See Lynch and Walsh (1998) for details of relaxing these assumptions.

Two related individuals may share one or more alleles that are identical by descent from their common ancestor(s). Because of this, related individuals will, on the average, have more similar genotypes, and therefore more similar phenotypes, than unrelated individuals; both the phenotypic and genetic covariance will be positive. The size of the genetic covariance should depend on the proportion of alleles that are identical by descent.

Let X and Y be the genotypic values of two relatives, for example, one parent and its offspring, or two full siblings. Then, $\text{cov}(X, Y)$ is the genotypic covariance. At a single locus, these relatives can share either 0, 1, or 2 alleles that are identical by descent from their common ancestor(s).

If they have no alleles that are identical by descent, then variation in X will be independent of variation in Y, and $\text{cov}(X, Y) = 0$.

If they share one allele that is identical by descent, the covariance is half the additive variance because only one allele is contributing to variation, the other being identical in the two individuals; that is, $\text{cov}(X, Y) = V_A/2$. Dominance variance cannot contribute to the covariance because the relatives share only one allele.

If X and Y have two alleles that are identical by descent, they are genetically identical, and $\text{cov}(X, Y) = \text{cov}(X, X) = \text{var}(X) = V_A + V_D$, ignoring the interaction variance and assuming the covariance between genetic and environmental effects is zero.

Let r_0, r_1, and r_2 be the probability that X and Y share 0, 1, or 2 alleles identical by descent. Then the covariance between X and Y is

$$\text{cov}(X, Y) = r_0(0) + r_1(V_A/2) + r_2(V_A + V_D) \qquad \textbf{(13.29)}$$

If P and O are parent and offspring, then $r_0 = 0$, $r_1 = 1$, and $r_2 = 0$. From equation (13.29) we get

$$\text{cov}(P, O) = 0(0) + (1)(V_A/2) + 0(V_A + V_D)$$

$$= \frac{1}{2}V_A$$

Similarly, if S_1 and S_2 are full siblings, then $r_0 = 1/4$, $r_1 = 1/2$, and $r_2 = 1/4$, and

$$\text{cov}(S_1, S_2) = (1/4)(0) + (1/2)(V_A/2) + (1/4)(V_A + V_D)$$

$$= \frac{1}{2}V_A + \frac{1}{4}V_D$$

It is possible to calculate the covariances between other kinds of relatives. A few examples are summarized in Table 13.5. Notice that the covariances are simple fractions of V_A and V_D. In pairs where the coefficient of V_D is zero, we can obtain an estimate of the additive genetic variance by estimating the covariance between the relatives.

We can use the degree of phenotypic resemblance between relatives to estimate the heritability of a trait. To illustrate, let P and O be random variables representing the phenotypic values of one parent and its offspring. Then the

TABLE 13.5   Covariances among relatives expressed as coefficients of V_A and V_D

Relationship	Covariance
Midparent-Offspring	$(1/2)V_A$
One parent-Offspring	$(1/2)V_A$
Full siblings	$(1/2)V_A + (1/4)V_D$
Half siblings	$(1/4)V_A$
Grandparent-Grandoffspring	$(1/4)V_A$
First cousins	$(1/8)V_A$
Identical twins	$V_A + V_D$

regression of O on P is

$$b_{OP} = \frac{\text{cov}(P, O)}{\text{var}(P)} \qquad \textbf{(13.30)}$$

Under the assumptions stated above, the numerator is the genetic covariance, which for parent-offspring pairs is $V_A/2$. Also, if the parents are chosen at random, then $\text{var}(P) = V_P$. Substituting these into the above equation gives

$$b_{OP} = \frac{1}{2}\left(\frac{V_A}{V_P}\right)$$

or

$$b_{OP} = \frac{h^2}{2} \qquad \textbf{(13.31)}$$

This gives a convenient way to estimate the heritability of a quantitative trait. We measure the phenotypes of parents and their offspring in the laboratory, or sometimes in natural populations, and estimate the regression coefficient of offspring phenotype on parent phenotype. Then

$$\hat{h}^2 = 2\hat{b}_{OP} \qquad \textbf{(13.32)}$$

is an estimate of the narrow sense heritability. It is easy to show (Problem 13.7) that if $\overline{P}$ is the midparent value (the average of the two parents) the regression of offspring on midparent value is

$$b_{O\overline{P}} = h^2 \qquad \textbf{(13.33)}$$

Figure 13.6 shows the results of a typical experiment comparing midparent and offspring values for a quantitative trait, in this case tarsus length in one of Darwin's finches. The estimated heritability was 0.82.

It must be emphasized again that this method depends on several assumptions about genetic and environmental covariances. These assumptions are suspect in

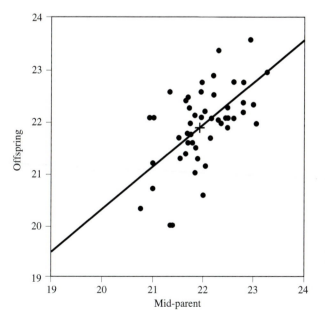

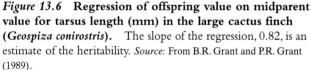

Figure 13.6 **Regression of offspring value on midparent value for tarsus length (mm) in the large cactus finch (Geospiza conirostris).**   The slope of the regression, 0.82, is an estimate of the heritability. *Source:* From B.R. Grant and P.R. Grant (1989).

natural populations. For example, the heritability estimate above may be inflated by environmental covariances.

13.3 Artificial Selection on a Quantitative Trait

Plant and animal breeders typically practice a kind of artificial selection called **truncation selection**. Individuals with a phenotypic value greater than some minimum value (called the truncation point) are selected to be the parents of the next generation. The hope is that individuals with high phenotypic values have good genotypes and will pass those good genes to their offspring. We have seen that the degree to which this happens depends, in part, on the heritability of the trait. Let

$\overline{X}_p$ = phenotypic mean of all individuals in the parent population

$\overline{X}_s$ = phenotypic mean of individuals selected to breed

$\overline{Y}_p$ = phenotypic mean of offspring if all individuals in parent generation were allowed to breed

$\overline{Y}_s$ = phenotypic mean of the offspring of selected individuals

Typically, the mean of the offspring is greater than the mean of all individuals in the parental generation, but less than the mean of the selected individuals.

We define the **selection differential** as the difference between the mean of selected individuals and the mean of all individuals in the parental generation

$$S = \overline{X}_s - \overline{X}_p \qquad \textbf{(13.34)}$$

Similarly, we define the **response to selection** as the difference between the mean of the offspring of the selected individuals and the mean of all individuals in the parental generation

$$R = \overline{Y}_s - \overline{X}_p \qquad \textbf{(13.35)}$$

Figure 13.7 illustrates these ideas.

Now look at Figure 13.8. Each circle represents one parent-offspring pair; the solid circles represent the selected parent-offspring pairs. $\overline{X}_p$ and $\overline{Y}_p$ are the midparent mean of *all* parents, and the (hypothetical) mean of their offspring. The line is the best-fit linear regression line of Y_p on X_p, assuming all parents are allowed to breed. As mentioned previously, the slope of this line is the heritability. A useful fact from regression theory is that the point $(\overline{X}_p, \overline{Y}_p)$ is on the regression line. $\overline{X}_s$ is the midparent mean of selected parents, and $\overline{Y}_s$ is the predicted mean of their offspring. The point $(\overline{X}_s, \overline{Y}_s)$ is also on the regression line, since the selected group is just a subset of the entire group.

Now, from the two-point definition of the slope of a line, we have

$$b = \frac{\overline{Y}_s - \overline{Y}_p}{\overline{X}_s - \overline{X}_p} \qquad \textbf{(13.36)}$$

If all parents are allowed to breed, and natural selection is not acting, then the mean will not change in the next generation. In other words, $\overline{Y}_p = \overline{X}_p$, and

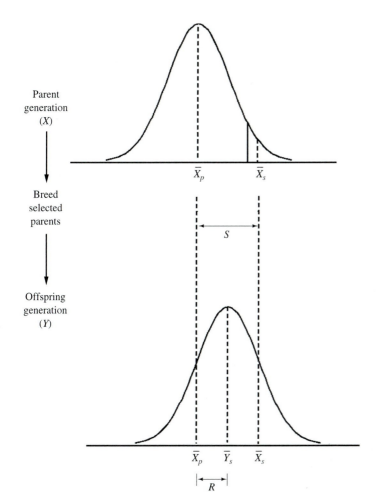

Figure 13.7   Response to truncation selection. $\overline{X}_p$ is the mean of the parental generation before selection, and $\overline{X}_s$ is the mean of selected parents. $\overline{Y}_s$ is the mean of the offspring of selected parents. S is the selection differential and R is the response to selection.

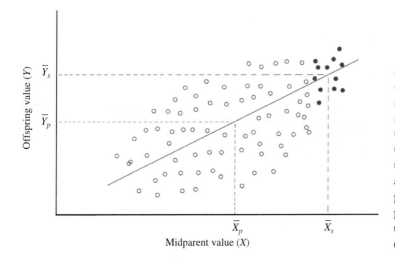

Figure 13.8   Regression of offspring value (Y) on midparent value (X). $\overline{X}_p$ is the mean of all individuals in the parental generation, and $\overline{Y}_p$ is the mean of their offspring (assuming all individuals in the parental generation are allowed to breed). $\overline{X}_s$ is the mean of selected parents and $\overline{Y}_s$ is the mean of their offspring (assuming only selected parents are allowed to breed). Solid circles represent selected parents. The response to selection $(\overline{Y}_s - \overline{Y}_p)$ depends on the slope of the regression line (the heritability). *Source:* Modified from Falconer and Mackay (1996).

the numerator in equation (13.36) is R, the response to selection. Next, note that the denominator is the selection differential, S. Finally, recall that the slope of the midparent-offspring regression line is h^2. Putting all of these facts together, we get

$$h^2 = \frac{R}{S}$$

Rearranging, we can see how the response to selection depends on the heritability

$$R = h^2 S \tag{13.37}$$

This important equation is known as the **breeder's equation** because it can be used to predict the response to directional selection on a quantitative trait. The response depends on both the heritability of the trait being selected, which must be estimated by some independent method, and on the strength of selection, as measured by the selection differential, S.

We illustrate the use of equation (13.37) with a classic example of response to selection (Clayton et al. 1957). The character selected was abdominal bristle number in *Drosophila melanogaster*. The mean of the parental population was 35.3 and the mean of the selected individuals was 40.6, making $S = 5.3$. Heritability had been previously estimated to be about 0.52. Therefore, the predicted response was $R = h^2 S = 0.52(5.3) = 2.8$. The mean of the offspring of the selected individuals was 37.9, giving an actual response of 2.6.

We can also use equation (13.37) to estimate heritability from a selection experiment. If we know R and S, we can estimate the heritability as

$$\hat{h}^2 = \frac{R}{S}$$

Heritability, as estimated in this way, is called the **realized heritability**.

Sometimes it is convenient to standardize the selection differential. Define the standardized **intensity of selection** as

$$i = \frac{S}{\sigma_P} \tag{13.38}$$

where σ_p is the square root of the phenotypic variance of all individuals in the parental generation. Using the intensity of selection instead of the selection differential, equation (13.37) becomes

$$R = h^2 i \sigma_P \tag{13.39}$$

which can be rewritten as

$$R = i h \sigma_A \tag{13.40}$$

where σ_A is the square root of the additive genetic variance.

The standardized intensity of selection is sometimes useful when comparing selection on different traits, as we shall see when we consider correlated response to selection. Another advantage is that, assuming that phenotype follows a normal distribution, the intensity of selection can be calculated from the proportion of

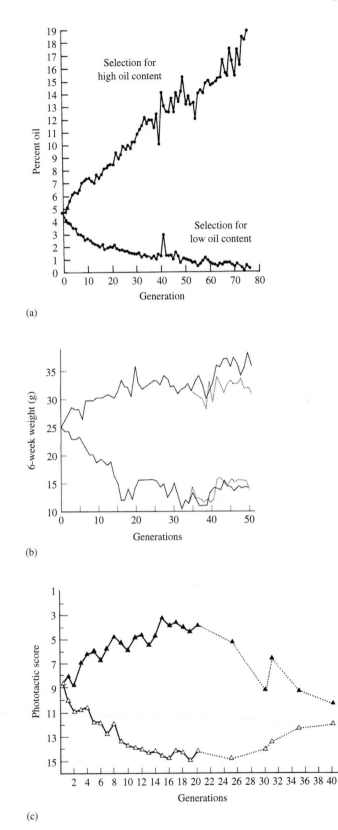

(a)

(b)

(c)

Figure 13.9 **Response to repeated selection.**
(a) Oil content of corn kernels (Dudley 1977); (b) Six-week body weight in mice; (Roberts 1966); (c) Photo-tactic response in *Drosophila pseudoobscura* (Dobzhansky and Spassky 1969). Dotted lines in (b) and (c) indicate selection in reverse direction. *Source:* (a) and (b) modified from Falconer and Mackay (1996); (c) modified from Spiess (1977).

the population selected under truncation selection. Thus, one can control the intensity of selection by specifying what proportion of the population is saved. For example, a selection intensity of 1 corresponds to saving about 38 percent of the population. Falconer and Mackay (1996) and Gillespie (1998) give useful tables and graphs relating the intensity of selection to proportion of the population saved.

Continued Selection on a Quantitative Trait: Short–Term Response

Typically, selection experiments continue for many generations, showing a continued improvement in the trait selected for. Figure 13.9 shows several examples. Several generalities have emerged from hundreds of laboratory selection experiments:

1. Selection is usually effective, and the response is relatively constant over the first few generations.
2. There is usually a plateau beyond which selection is ineffective.
3. Upon relaxation of selection, the population sometimes tends to regress back toward the original mean value (but not all the way).
4. Selection response is sometimes asymmetrical; that is, selection in the high direction sometimes proceeds at a rate different from selection in the low direction.
5. Viability and/or fertility often decrease as selection proceeds.

We wish to understand the reasons for these generalities.

Equation (13.37) predicts the response to a single generation of selection. What about the predicted response to selection over many generations? We need to consider short-term response and long-term response separately.

In the short term, we can assume that heritability does not change significantly. The reason for this is that we have assumed that many loci are affecting the trait and that each locus has a small effect. If this is true, then the effect of selection will be spread over many loci, and allele frequencies at any single locus will change very slowly. Therefore, V_A and hence heritability will change only very slowly.

Under this assumption, it is possible to estimate the realized heritability over several generations of a selection experiment. Table 13.6 shows the result of selection for increased plasma cholesterol level in mice. The last two columns show the cumulative response and the cumulative selection differential, obtained by adding the current value to all the values of all previous generations. Figure 13.10a shows how the population mean changed each generation. Figure 13.10b plots the cumulative response against the cumulative selection differential. The slope of the regression line is an estimate of the realized heritability, in this case, 0.63.

To summarize, the short-term response to selection can, in theory, be predicted from the heritability of the trait. Cumulative response should be an approximately linear function of the cumulative selection differential, as long as heritability remains constant. Many experiments have confirmed this prediction.

TABLE 13.6   Results of selection for increased plasma cholesterol level in mice.
The selection differential in each generation is the amount of selection applied to produce that generation. Thus, there is no selection differential in generation 0, because those individuals had not yet been subject to selection. The cumulative selection differential is the sum of the selection differentials for the current and all preceeding generations. *Source:* From Falconer and Mackay (1996).

Generation	$\overline{X}_p$	$\overline{X}_s$	Selection Differential (S)	Cumulative Selection Differential	Cumulative Response
0	2.16	2.32	0	0	0
1	2.26	2.34	0.16	0.16	0.10
2	2.26	2.37	0.08	0.24	0.10
3	2.33	2.41	0.11	0.35	0.17
4	2.45	2.47	0.08	0.43	0.29
5	2.44		0.02	0.45	0.28

Continued Selection on a Quantitative Trait: Long-Term Response

A selection plateau is frequently reached in long-term selection experiments (e.g., Figure 13.9c before the direction was reversed). One possible cause of a selection plateau is the depletion of additive variance. In the long term, additive genetic variance is expected to decrease for two reasons. First is the effect of genetic drift. Recall from Section 7.2 that, in the absence of selection, genetic drift reduces heterozygosity by

$$H_t = H_0\left(1 - \frac{1}{2N}\right)^t$$

An analogous formula for the decay of additive genetic variance is

$$V_A(t) = V_A(0)\left(1 - \frac{1}{2N}\right)^t \tag{13.41}$$

where $V_A(0)$ is the initial additive variance, and $V_A(t)$ is the additive variance after t generations. Therefore, as V_A decays, heritability decays, and response to selection will eventually cease because there is no additive variance left in the population. The rate at which this occurs depends on the (effective) population size, and is probably not important for most stable natural populations. However it may be significant in small endangered populations. Storfer (1996) discusses this issue.

The second reason for the long-term decay of additive variance is the effect of selection. Selection changes allele frequencies. As selection proceeds, favorable alleles will increase toward fixation. As this occurs, the additive genetic variance will decrease. Again, the response to selection will decrease or cease as additive variance is eliminated. We will discuss the effect of selection on genetic variance in more detail in Section 13.4.

Thus, two factors predict that additive variance will eventually be eliminated and response to selection will cease. These predictions have been partially confirmed in experiments. Selected populations often reach a plateau, beyond which

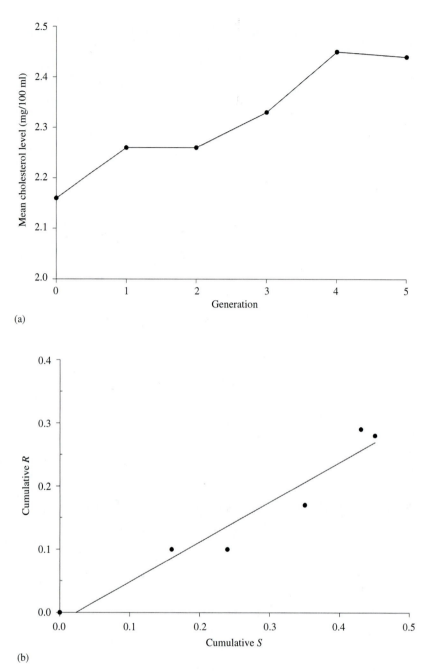

Figure 13.10 **Response to selection for plasma cholesterol level in mice.** (a) Change in mean cholesterol level over five generations. (b) Cumulative response plotted against the cumulative selection differential. The slope of the regression line is 0.63, and is an estimate of the heritability. *Source:* Data from Weibust (1973).

artificial selection is ineffective. However, selection in the opposite direction is sometimes effective, as illustrated in Figure 13.10c. This indicates that additive variance is still present in the population, even though response to selection has ceased. A common reason for the lack of response is that at a certain point, natural selection opposes artificial selection, so that further artificial selection is ineffective, even though additive genetic variance is still present.

Studies of natural populations also indicate that there is frequently much additive genetic variation present. We will return to this issue in Section 13.6.

Correlated Response to Selection

We have seen how we can use equation (13.37) to predict the response to selection on a quantitative trait. Frequently, this selection can affect another quantitative trait. If two traits are genetically correlated, selection on one will affect the other also.

Genetic correlation can be due to pleiotropy or to gametic disequilibrium. Pleiotropy is when a gene affects more than one character. Whenever this occurs, selection on one character will obviously affect the other character also. Similarly, if alleles at a locus affecting one trait are in gametic disequilibrium with alleles at a locus affecting a second trait, selection on the first trait will affect the second trait. Genetic correlations are very common. Table 13.7 shows a few examples.

We need to find a way to predict how selection on one quantitative trait, X, will affect a correlated trait, Y. First, review the definition of a correlation: If X and Y are two random variables, the correlation between them, usually designated by r, is

$$r = \frac{\text{cov}(X, Y)}{\sigma_X \sigma_Y} \tag{13.42}$$

where σ_X and σ_Y are the standard deviations of X and Y. The correlation coefficient can range from -1 to $+1$.

As with variances and covariances, the phenotypic correlation between two traits consists of two components, a genetic correlation and an environmental correlation. Genetic correlation is due to pleiotropy or gametic disequilibrium; environmental correlation is due to environmental factors affecting both traits simultaneously. The relationship between the genetic and environmental correlations is derived in Box 13.2. It is

$$r_P = h_X h_Y r_A + e_X e_Y r_E \tag{13.43}$$

TABLE 13.7   Genetic (additive) correlations between different traits in various organisms.

Organism	Traits	r_A	Reference
Humans	IgG/IgM levels	0.07	Falconer and Mackay (1996)
Cattle	Milk yield/butterfat percent	−0.38	Falconer and Mackay (1996)
Pigs	Weight gain/backfat thickness	0.13	Falconer and Mackay (1996)
Chickens	Body weight/egg weight	0.42	Falconer and Mackay (1996)
	Egg weight/egg production	−0.31	Falconer and Mackay (1996)
Mice	Body weight/tail length	0.29	Falconer and Mackay (1996)
Drosophila melanogaster	Abdominal/sternopleural bristles	0.41	Falconer and Mackay (1996)
	Wing length/thorax length	0.80	Roff (1997)
	Body weight/fecundity	0.14	Roff (1997)
	Body weight/egg size	0.01	Roff (1997)
	Body weight/adult emergence	−0.10	Roff (1997)
Milkweed bug	Wing length/head width	0.68	Roff (1997)
Medium ground finch	Body weight/wing length	0.88	Grant (1986)
	Bill length/tarsus length	0.71	Grant (1986)

Box 13.2   The phenotypic, genetic, and environmental correlations

The phenotypic covariance between traits X and Y can be written as

$$\text{cov}(X_P, Y_P) = \text{cov}(X_G + X_E, Y_G + Y_E)$$
$$= \text{cov}(X_G, Y_G) + \text{cov}(X_E, Y_E)$$
$$+ \text{cov}(X_G, Y_E) + \text{cov}(X_E, Y_G)$$

Assuming no genetic $\times$ environmental covariance, the last two terms on the right are zero and this simplifies to

$$\text{cov}(X_P, Y_P) = \text{cov}(X_G, Y_G) + \text{cov}(X_E, Y_E) \quad (1)$$

Ignoring dominance and interaction effects, $\text{cov}(X_G, Y_G) = \text{cov}(X_A, Y_A)$, and we can rewrite equation (1) as

$$\text{cov}(X_P, Y_P) = \text{cov}(X_A, Y_A) + \text{cov}(X_E, Y_E) \quad (2)$$

From the definition of the correlation coefficient given in equation (13.42), we can rewrite this as

$$r_P \sigma_{PX} \sigma_{PY} = r_A \sigma_{AX} \sigma_{AY} + r_E \sigma_{EX} \sigma_{EY} \quad (3)$$

Next, since $h^2 = \sigma_A^2 / \sigma_P^2$, we can write

$$\sigma_A = h \sigma_P \quad (4)$$

Similarly if we define $e^2 = \sigma_E^2 / \sigma_P^2$, we can write

$$\sigma_E = e \sigma_P \quad (5)$$

Substituting equations (4) and (5) into equation (3), we get

$$r_P \sigma_{PX} \sigma_{PY} = r_A (h_X \sigma_{PX})(h_Y \sigma_{PY}) + r_E (e_X \sigma_{PX})(e_Y \sigma_{PY})$$

which simplifies to

$$r_P = h_X h_Y r_A + e_X e_Y r_E$$

This is equation (13.43) in the main text.

where the subscripts A, G, and E, indicate phenotypic, additive, and environmental correlations, h_X^2 and h_Y^2 are the heritabilities of traits X and Y, $e_X^2 = 1 - h_X^2$, and $e_Y^2 = 1 - h_Y^2$. We have ignored dominance and interaction effects and assumed the genetic correlation is equal to the additive correlation. It is important to note that the genetic and environmental correlations may or may not have the same sign.

We wish to predict the correlated response in trait Y when selection is on trait X. Note that X and Y are two different traits in an individual, and not two different individuals, as they were previously. It should be intuitive that the correlated response in Y should depend in some way on the heritabilities of both X and Y, on the genetic correlation between them, and on the selection differential. Just as the direct response to selection depends on the additive genetic variance, the indirect response depends on the additive correlation, r_A. The correlated response in Y to selection on X can be predicted by

$$R_Y = \frac{r_A h_X h_Y S \sigma_{PY}}{\sigma_{PX}} \quad (13.44)$$

See Falconer and Mackay (1996) for a derivation of this equation. An equivalent form, using the selection intensity, $i = S/\sigma_{PX}$, is

$$R_Y = r_A h_X h_Y i \sigma_{PY} \quad (13.45)$$

Equations (13.44) or (13.45) describe how character Y changes when selection is performed on character X. Compare equation (13.45) with (13.39). You can think of $r_A h_X h_Y$ as a kind of "modified heritability" that converts the selection on X to response in Y. Falconer and Mackay (1996) call it the **coheritability**.

Note that in order to predict the correlated response, we need an estimate of the additive correlation. The phenotypic correlation between two quantitative traits is easily estimated; estimating the additive correlation is much more difficult. [See Lynch and Walsh (1998) for discussion of some methods.] This raises the question of whether we can use r_P, which is easily estimated, as a reasonable estimate of r_A. Roff (1997) and Lynch and Walsh (1998) reviewed the literature and concluded that r_A and r_P usually have the same sign, and are of similar magnitude. However, this is by no means universal, and this generalization must be treated very cautiously (e.g., Willis et al. 1991). See also Steppan et al. (2002) for more on this issue.

Alternatively, one can turn the problem on its head and estimate r_A from the observed correlated response, just as one can estimate heritability from the observed direct response to selection. Correlation estimated this way is called the **realized correlation** (Roff 1997). Such estimates are generally less reliable than estimates of heritability based on direct response.

Decrease in Fitness Associated with Artificial Selection

Correlated response is often a problem in breeding programs or selection experiments. Major components of fitness often decrease as a consequence of artificial selection. For example, directional selection for bristle number in *Drosophila* usually results in decreased viability in selected lines. Equation (13.45) shows that if the additive correlation between two traits is negative, then selection to increase one will cause a correlated decrease in the other. For example, if X is body weight, and Y is a component of fitness such as litter size or survival, and r_A is negative, then fitness will decrease as selection is performed for increased body weight.

There are reasons to expect negative genetic correlations between a quantitative trait and a component of fitness, at least sometimes. Consider a large, relatively stable population with no recent history of direct selection on a quantitative trait, X. Assume that Y is some component of fitness, and is genetically correlated with X, either by pleiotropy or gametic disequilibrium. Natural selection will usually act on Y so that fitness will approach a local maximum: Allele frequencies will be at the optimum to maximize fitness, and coadapted gene complexes will have been stabilized to maximize the desirable effects of gene interaction. Now begin artificial selection on X. Selection will change allele frequencies, moving them away from their previously established optima, and will favor new gene complexes different from those favored by natural selection. The result is that the correlated component of fitness, Y, which had been previously maximized, will decrease. The only time this will not occur is if X is determined entirely by genes that have no effect on fitness.

13.4 Natural Selection on Quantitative Traits

In our theoretical development, we have assumed that an individual's phenotypic value for a quantitative trait has no effect on that individual's ability to survive or reproduce. As hinted at the end of the previous section, that frequently is not the case, especially in natural populations. Traits such as body weight, growth rate, and so forth are almost certainly subject to natural selection.

The main components of fitness, for example, viability, fecundity, or mating ability, are themselves quantitative traits with large effects on net fitness, and therefore under strong natural selection. Other characters, morphological traits such as bristle number in fruit flies or body weight in many organisms, are less directly related to fitness and probably less subject to natural selection, but not entirely immune from it. In this section, we examine the effect of natural selection on quantitative traits.

Kinds of Natural Selection

Remember that natural selection acts on the phenotype, which is determined to a greater or lesser extent by the genotype. The relationship between the phenotypic value of a quantitative trait and fitness can take several different forms, as shown in Figure 13.11. For comparison, the first graph shows a quantitative trait that is not subject to natural selection.

We have already discussed truncation selection (Figure 13.11b). It is an extreme form of **directional selection**, in which high (or low) values of the trait are favored. Figure 13.11c shows less extreme forms of directional selection. Major components of fitness are probably subject to directional selection. For example, Mackay (1985) showed that viability and fertility were under strong directional selection in *Drosophila melanogaster* (Figure 13.12).

Morphological characters can also be subject to directional selection. Peter and Rosemary Grant have studied the ecology and evolution of Darwin's finches in the Galapagos islands for many years [summarized in Grant (1986) and Grant and Grant (1989, 1995)]. These birds are mainly seed eaters, and the size and shape of their bills are correlated with the kinds of seeds they eat. Grant and Grant (1995) summarize several aspects of a long-term study of the medium ground finch (*Geospiza fortis*). Birds with larger bills can more easily handle larger and harder seeds. Moreover, variation in bill size and shape is highly heritable. Grant and Grant have shown that body size and beak depth are subject to natural selection. In 1976–1977, a severe drought occurred and more than 80 percent of the *Geospiza fortis* population on the island of Daphne Major died, mostly from starvation. As the drought persisted, the seeds these birds normally eat became fewer and fewer, and birds had to subsist on the large, hard seeds of *Opuntia* and *Tribulus*, which they normally ignore. Table 13.8 shows the means of several quantitative characters in 1976, before the drought, and in 1978, after the drought. The difference between years is statistically significant for all characters. Directional selection during the drought favored larger birds with larger bills. All of the traits in Table 13.8 are positively correlated (e.g., larger birds have deeper beaks), but, using multivariate statistical techniques, it is possible to partially differentiate between direct selection and a correlated response (see below). Grant and Grant concluded that both beak depth and body size (weight) were under direct selection, the former for feeding efficiency, and the latter because larger birds were metabolically more efficient and perhaps behaviorally more dominant in competing with smaller birds.

This hypothesis has been confirmed. The years 1980 and 1982 were dry years, although not as severe as 1976–1977, and in both years natural selection favored larger birds with deeper beaks (Price et al. 1984). Moreover, in 1983, El Niño created a very wet year, resulting in plentiful small, soft seeds. As a result, body size

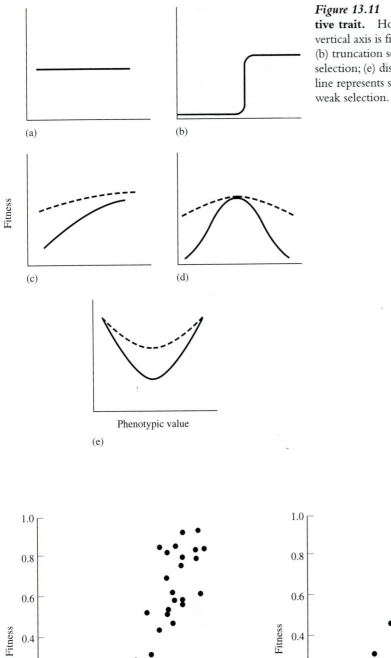

Figure 13.11 **Kinds of natural selection on a quantitative trait.**   Horizontal axis is the value of the quantitative trait; vertical axis is fitness, or a major component: (a) no selection; (b) truncation selection; (c) directional selection; (d) stabilizing selection; (e) disruptive selection. In (c) through (e), the solid line represents strong selection and the dashed line represents weak selection.

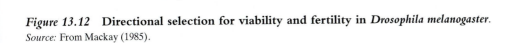

Figure 13.12 **Directional selection for viability and fertility in *Drosophila melanogaster*.**
Source: From Mackay (1985).

TABLE 13.8 Change in quantitative traits in the medium ground finch (*Geospiza fortis*) before and after a severe drought.
All differences between years are statistically significant.
Source: From Grant and Grant (1995).

Character	Mean (1976)	Mean (1978)
Body weight (g)	16.06	17.13
Wing length (mm)	67.88	68.87
Tarsus length (mm)	19.08	19.29
Bill length (mm)	10.63	10.95
Bill depth (mm)	9.21	9.70
Bill width (mm)	8.58	8.83
Sample size	634	135

decreased, exactly as would be expected (Gibbs and Grant 1987). Note that directional selection changed direction in different years.

Stabilizing selection occurs when natural selection favors phenotypes with intermediate values (Figure 13.11d). A classic example is birth weight in humans. Babies with exceptionally high or low birth weights have lower survival rates than babies with birth weights close to the mean (Figure 13.13).

If natural selection eliminates extreme phenotypes, the phenotypic variance should decrease. For example, Rendel (1943) weighed 960 duck eggs, of which 619 hatched. The mean and variance of all 960 eggs were 73.9 grams and 52.7. The mean and variance of the eggs that hatched were 73.8 and 43.9. The difference between the means is not significant, but the difference between the variances is highly significant. Presumably at least some of the mortality was due to genetic causes.

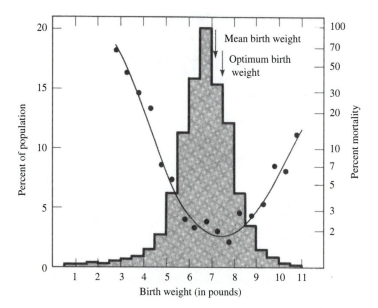

Figure 13.13 **Stabilizing selection on birth weight in human babies.** The histogram represents the percentage of babies in each weight class (left vertical axis). The solid circles indicate percent mortality in each weight class (right vertical axis). The solid line is a best-fit curve to the mortality data.
Source: From Cavalli-Sforza and Bodmer (1971); original data from Karn and Penrose (1951).

Stabilizing selection is often assumed to be common in natural populations (but see below). For example, Schluter and Smith (1986) found that song sparrows with intermediate body size had higher winter survival and reproductive rates than individuals with higher or lower weights. Endler (1986) and Roff (1997) cite many other examples.

Disruptive selection occurs when high and low values of the trait are both favored and intermediate values are selected against (Figure 13.11e). Numerous laboratory experiments have examined the consequences of disruptive selection, but its importance in natural populations is unclear. Endler (1986), in his review of selection in natural populations, found that disruptive selection was less common than either stabilizing or directional selection, but Kingsolver et al. (2001) found that it was about as common as stabilizing selection.

Bill size in African finches, *Pyrenestes ostrinus*, is subject to disruptive selection (Smith 1993). Large-billed birds feed on hard seeds, and small-billed birds feed on soft seeds. Birds with intermediate bill sizes have lower survival rates than either large- or small-billed birds (Figure 13.14). This example is unusual because the bill size difference seems to be due mainly to a single gene.

Antagonistic Pleiotropy

We saw in the previous section that the additive correlation (r_A) between a quantitative trait under directional selection and a component of fitness is likely to be negative in a population in approximate equilibrium. The result is that artificial selection on a quantitative trait is expected to reduce fitness.

It is also true that the additive covariance between two components of fitness is likely to be negative. For example, consider developmental time and fecundity in *Drosophila*. Fast development and high fecundity are favored by natural selection. However, flies that develop rapidly are smaller, and smaller flies lay fewer eggs. Conversely, more slowly developing flies are larger and lay more eggs. Therefore, selection favoring one of these fitness components is selection against the other. The result is a kind of stabilizing selection on body weight.

Negative correlations between fitness components are sometimes due to **antagonistic pleiotropy** (see Section 12.7). Alleles that favor both components of

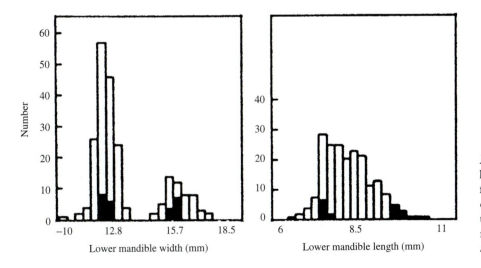

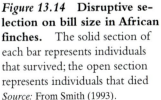

Figure 13.14 **Disruptive selection on bill size in African finches.** The solid section of each bar represents individuals that survived; the open section represents individuals that died
Source: From Smith (1993).

fitness will be selected for and become fixed relatively rapidly. Those that favor one component and depress the other will be subject to conflicting selection pressures and will remain segregating in the population for a longer time. As a result, most of the genetic covariance between the two components will be due to this latter group of alleles.

Schluter and Smith (1986) found that short tarsus length and long beak length favored juvenile overwinter survival in song sparrows. However, these same characteristics reduced adult female reproductive success. Thus, there was a trade-off between juvenile survival and adult reproductive success. Based on tarsus length and bill length, they estimated the genetic correlation between juvenile overwinter survival and adult female reproductive success to be about -0.99. (This is a partial correlation based on these two morphological traits only; the overall correlation is probably weaker.) A similar trade-off was found between female survival and breeding time in one of Darwin's finches (Price 1984b).

Figure 13.15 illustrates another example of antagonistic pleiotropy, in the water strider *Aquarius remigis*. The quantitative character is body length, and the two components of fitness are daily fecundity and longevity. Fecundity increases with total length, while longevity decreases (top two graphs in Figure 13.15). As a result, lifetime fecundity, a measure of overall fitness, is highest at intermediate body lengths (bottom graph). This is apparent stabilizing selection on body length, but is actually the result of conflicting selection pressures on two components of fitness correlated with body length.

Natural Selection on Correlated Traits

Natural selection acts on the overall phenotype of the individual. This overall phenotype consists of many different aspects; for example, in birds it includes body weight, wing length, bill size, and so forth, and all of these have some effect on net fitness. Many of these traits are genetically correlated. Thus, natural selection can cause simultaneous changes in several morphological characters. Is there a way to determine which of these characters is under *direct* natural selection, as opposed to being correlated with a character under direct selection?

Lande and Arnold (1983) extended a statistical method first proposed by Pearson (1903) for separating the effects of direct selection on a character from indirect effects. The principle is straightforward; here, we only outline the main ideas. See Lande and Arnold (1983) for details.

Assume you have measurements before and after selection on several correlated traits, and estimates of their phenotypic variances and covariances. Now, direct natural selection on one trait will cause a correlated response in correlated traits. Following Pearson, Lande and Arnold showed that the partial regression coefficient of fitness (or a major component) on any character value is a measure of the strength of *direct* selection on that character. Using the standard terminology of multiple regression, we assume that the fitness of an individual can be approximated by a **linear model**

$$w = \alpha + \beta_1 P_1 + \beta_2 P_2 + \cdots + \beta_n P_n + \varepsilon$$

This is analogous to the standard linear regression equation ($Y = a + bX + \varepsilon$). The dependent variable, analogous to Y, is w, the fitness of an individual. The independent

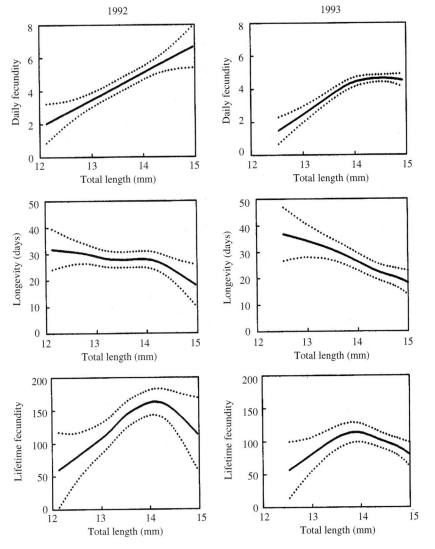

Figure 13.15 Antagonistic pleiotropy in a water strider.   Daily fecundity increases with body length (top two graphs) but longevity decreases with body length (middle two graphs). Consequently, lifetime fecundity (a measure of overall fitness) is highest at intermediate body lengths. Dashed lines are 95 percent confidence intervals. *Source:* From Roff (1997); original data from Preziosi and Fairbairn (1997).

variables, analogous to X, are the P's, the phenotypic values of n quantitative characters. The βs are partial regression coefficients, analogous to b in the two-variable model. α is a constant analogous to the intercept, a, and ε is a residual term that accounts for variation unexplained by the P's. Lande and Arnold showed that the βs are measures of *direct* natural selection on the corresponding quantitative traits. Thus, the methods of multivariate regression analysis can disentangle the effects of direct and indirect selection. One way to interpret the βs is that a one-unit change in the value of P_i, with all the other P's unchanged, will result in a change in relative fitness of β_i.

We will outline Lande and Arnold's results for directional selection, using two quantitative traits to illustrate. We will use matrix algebra, which allows compact representation of the process, and easily extends to any number of correlated characters. See Lande and Arnold (1983) for details.

Recall that, for a single character, the response to directional selection is given by equation (13.37)

$$R = h^2 S$$

Note for future reference, we can rewrite this as

$$R = V_A(V_P)^{-1}S \qquad (13.46)$$

For two characters, X and Y, we define the vector of selection differentials as

$$\mathbf{S} = \begin{bmatrix} S_X \\ S_Y \end{bmatrix} = \begin{bmatrix} \overline{X}_s - \overline{X}_p \\ \overline{Y}_s - \overline{Y}_p \end{bmatrix}$$

where the subscripts s and p indicate the selected group and the entire parental population. The vector of selection responses is

$$\mathbf{R} = \begin{bmatrix} R_X \\ R_Y \end{bmatrix}$$

We wish to predict the selection responses due to directional natural selection acting simultaneously on both traits, and to separate direct responses from correlated responses.

We now define the additive and phenotypic covariance matrices:

$$\mathbf{V_A} = \begin{bmatrix} V_{AX} & \mathrm{cov}_A \\ \mathrm{cov}_A & V_{AY} \end{bmatrix}$$

$$\mathbf{V_P} = \begin{bmatrix} V_{PX} & \mathrm{cov}_P \\ \mathrm{cov}_P & V_{PY} \end{bmatrix}$$

where cov_A and cov_P are shorthand for the additive and phenotypic covariances between X and Y.[2]

Lande and Arnold made use of the fact that, under certain assumptions,

$$\mathbf{R} = \mathbf{V_A}\mathbf{V_P^{-1}}\mathbf{S}$$

where $\mathbf{V_P^{-1}}$ is the inverse matrix of $\mathbf{V_P}$. This is the multivariate analog of equation (13.46) above. Lande and Arnold next defined a new vector, β as

$$\beta = \begin{bmatrix} \beta_X \\ \beta_Y \end{bmatrix} = \mathbf{V_P^{-1}}\mathbf{S}$$

So the response vector can now be written as

$$\mathbf{R} = \mathbf{V_A}\beta \qquad (13.47)$$

The vector β is called the vector of **selection gradients**. β_X is a measure of the strength of direct selection on X, and β_Y is a measure of the strength of direct selection on Y. Writing equation (13.47) out in longhand will help to understand the meaning of β:

$$R_X = V_{AX}\beta_X + \mathrm{cov}_A\beta_Y$$
$$R_Y = \mathrm{cov}_A\beta_X + V_{AY}\beta_Y$$

In each of these equations, the term on the right side containing the variance is the direct response and the term containing the covariance is the correlated response.

---

[2] $\mathbf{V_A}$ and $\mathbf{V_P}$ are often symbolized by $\mathbf{G}$ and $\mathbf{P}$ in the literature, but we will use the more intuitive $\mathbf{V_A}$ and $\mathbf{V_P}$.

It can be shown, using matrix algebra, that

$$\beta_X = \frac{V_{PY} S_X - \text{cov}_P S_Y}{V_{PX} V_{PY} - (\text{cov}_P)^2}$$

$$\beta_Y = \frac{-\text{cov}_P S_X + V_{PX} S_Y}{V_{PX} V_{PY} - (\text{cov}_P)^2}$$

Note that the right-hand sides of these two equations contain only phenotypic variances and covariances, along with the selection differentials. All of these quantities can be estimated from a sample of individuals before and after selection. Thus, we can estimate the strength of direct selection on a character by observing the phenotypes before and after selection, and the phenotypic covariances.

The matrix equations above easily extend to any number of quantitative characters. All that is needed are estimates of the selection differential on each trait, and the phenotypic variance/covariance matrix. With that information it is possible to estimate the direct effect of (directional) natural selection on each trait separately.

Table 13.9 summarizes an example used by Lande and Arnold (1983). After a storm, they collected 94 pentatomid bugs (*Euschistus variolarius*), of which 39 survived and 55 died. They measured four morphological characters on all bugs, and calculated the selection differential for each character as the mean of the surviving bugs minus the mean of all 94 bugs. Table 13.9 gives the phenotypic correlations, selection differentials (S and i), and selection gradients (β) for each character. The selection gradients indicate strong direct selection for increased thorax width and decreased wing length. However, thorax width did not change significantly, as indicated by the selection differential. Because these two characters are highly correlated ($r \cong 0.71$), direct selection for increased thorax width and indirect selection for decreased thorax width due to direct selection for decreased wing length balanced one another, and thorax width changed very little. If we had looked only at the selection differentials, we would not have detected selection on thorax width. Conversely, scutellum length changed significantly, as indicated by S, but the selection gradient was not significant, indicating that the

TABLE 13.9   Phenotypic correlations, selection differentials (*S*), selection intensities (*i*), and selection gradients (*β*) for four morphological characteristics in a population of bugs, *Euschistus variolarius*.
All analysis was done on natural logarithms of the actual measurements. *Source:* From Lande and Arnold (1983)

Character	Head	Thorax	Scutellum	Wing	*S*	*i*	*β* ± standard error
			Phenotypic Correlations				
Head width	—				−0.004	−0.11	−0.7 ± 0.17
Thorax width	0.72	—			−0.003	−0.06	11.6 ± 3.9**
Scutellum length	0.50	0.59	—		−0.016*	−0.28*	−2.8 ± 2.7
Wing length	0.60	0.71	0.62	—	−0.019**	−0.43**	−16.6 ± 4.0**

*P < 0.05
**P < 0.01

change in scutellum length was a correlated response to direct selection on thorax width and/or wing length.

This method has been widely used to study the effects of natural selection on quantitative characters. For example, recall the previous example of selection in Darwin's finches. Price et al. (1984) showed that body weight and beak depth were under strong direct selection compared to the other (correlated) morphological traits.

Lande and Arnold (1983) also derived a way to estimate the direct effects of stabilizing or disruptive selection. Analogous to the (linear) selection gradient β, which is a measure of direct directional selection on a character, they defined the quadratic selection gradient, γ, as a measure of direct stabilizing or disruptive selection on the character. For stabilizing selection $\gamma < 0$, and for disruptive selection $\gamma > 0$. See their original paper for details.

The Strength of Natural Selection

It is frequently assumed that natural selection is usually weak in nature. There is some evidence that this is not the case. Natural selection in nature is sometimes as strong as artificial selection in laboratory experiments. For example, 83 percent of the ground finches discussed above died in the 1976–1977 drought, mostly due to starvation (Price et al. 1984).

Endler (1986) reviewed the literature of directional selection on quantitative traits. He concluded that selection intensities in nature extensively overlap the intensities of artificial selection experiments. Selection intensities of 1 or greater were not uncommon. (Recall that a selection intensity of 1 corresponds to a selective mortality of about 62 percent under truncation selection.) Keep in mind that if quantitative traits are affected by many loci, then selection on any given locus is much weaker than on the character as a whole.

Surprisingly, Endler (1986) found that stabilizing selection was less common than directional selection. However, the general conclusion was similar for both: Selection can take a wide range of values, and is not necessarily weak. Endler also found that disruptive selection was less common than either directional or stabilizing selection.

Endler's review covered the literature through the end of 1983, about the time that Lande and Arnold's (1983) paper was published. Since that landmark paper, there have been dozens, perhaps hundreds, of studies estimating the strength of selection on quantitative traits in natural populations. Kingsolver et al. (2001) summarized the main literature from 1984 through 1997. They looked at over 2500 estimates of selection which satisfied their rather rigorous criteria for inclusion (e.g., field estimates of natural populations). These included invertebrates, vertebrates, and plants, a total of 62 different species. Their results suggest that some of Endler's conclusions need to be revised.

For directional selection, Kingsolver et al. considered 993 estimates of β. Since the direction of selection (up or down) was irrelevant, they considered only the absolute value of β. The median absolute value of β was 0.16, but only 25 percent of the estimates were significantly greater than zero. Directional selection was significantly stronger for morphological traits than for life history traits (Table 13.10). Overall, selection was weaker than estimated by Endler (1986), but a few

TABLE 13.10   Summary of estimates of directional selection gradients (β) and quadratic selection gradients (γ).

For stabilizing selection, $\gamma < 0$, and for directional selection, $\gamma > 0$. Only absolute values are shown. Note, most estimates of β or γ are not significantly different from zero due to low statistical power. See also Figure 12.19. *Source:* From Kingsolver et al. (2001).

	Directional Selection (β)	Stabilizing or Disruptive Selection (γ)
Number of estimates	993	574
Overall median estimate $\lvert\beta\rvert$ or $\lvert\gamma\rvert$	0.16	0.10
Median estimate by kind of trait		
Morphological	0.17	not given
Life history	0.08	not given
Median estimate by kind of fitness component		
Survival	0.09	0.02
Fecundity	0.16	0.14
Mating success	0.18	0.16
Percentage of estimates significantly different from zero	25%	16%

studies estimated fairly strong selection (Figure 13.16a). Perhaps surprisingly, sexual selection (selection for mating success) and fecundity selection were both stronger than viability selection (Table 13.10).

Kingsolver et al. found 574 estimates of γ, Lande and Arnold's (1983) measure of the strength of stabilizing or disruptive selection. Recall from above that $\gamma < 0$ for stabilizing selection and $\gamma > 0$ for disruptive selection. Kingsolver et al. found that stabilizing and disruptive selection were typically weak; the overall median absolute value of γ was 0.10, with only 16 percent of the estimates significantly different from zero (Table 13.10). The frequency distribution of γ was symmetrical about zero, indicating that stabilizing and disruptive selection were about equally common, and of about the same strength (Figure 13.16b). As with directional selection, they found that selection for mating success and fecundity selection were both stronger than viability selection (Table 13.10).

To summarize, Kingsolver et al. found that natural selection on quantitative traits, as estimated by β or γ, was typically rather weak, with occasional exceptions. Directional selection was stronger than stabilizing or disruptive selection, and the latter two were about equal in frequency and strength. Sexual selection was stronger than viability selection, and directional selection for morphological traits was stronger than for life history traits. Unfortunately, most estimates of β or γ were not significantly different from zero, due to small sample size and consequent low statistical power. Kingsolver et al. suggest (among other things) that future studies, especially of stabilizing or disruptive selection, use much larger sample sizes in order to be able to detect weak selection.

Figure 13.16 **Frequency distributions of selection gradients based on estimates from natural populations.** (a) Absolute value of linear selection gradients (β) indicating strength of directional selection. Based on 993 estimates. (b) Quadratic selection gradients (γ) indicating strength of stabilizing ($\gamma < 0$) or disruptive ($\gamma > 0$) selection. Based on 465 estimates. In both cases, darkly shaded areas represent estimates that are significantly different from zero. *Source:* From Kingsolver et al. (2001).

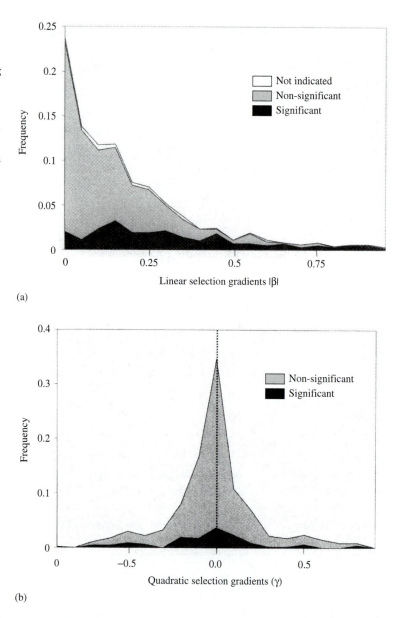

(a)

(b)

13.5 Quantitative Trait Loci

Until now, we have been vague about the number and nature of the genes controlling quantitative traits. Initially we considered a single locus, and then *assumed* a large number of loci with small effects. This allowed us to use a statistical approach to analyze quantitative traits, without knowing the genetic basis of the trait. However, for a thorough understanding of quantitative characters, we must know something about the actual number of loci affecting a trait, and the magnitude of their effects. This is important for understanding the long-term evolution of quantitative traits, and how genetic variation is maintained in nature. Moreover, many human diseases are influenced by many genes. Identification and study of these genes have already proved useful in diagnosis and treatment of complex human diseases, and will certainly become more so with completion of the Human Genome Project.

Loci that affect a quantitative trait are called **quantitative trait loci**, or **QTLs**. A QTL is a relatively small region of a chromosome, and does not necessarily correspond to a single gene. Current techniques usually do not allow us to know whether a QTL consists of a single gene or of several tightly linked genes.

In this section, we examine methods to estimate the number of QTLs, and the magnitude of their effects. We will also see how it is sometimes possible to map the genes controlling a quantitative trait to specific regions of a chromosome. This is a very complex and rapidly advancing field; we only introduce the general principles.

How Many Loci Affect a Quantitative Trait?

The earliest method to estimate the number of loci affecting a quantitative trait was proposed by Castle (1921) and extended by Wright (1968). It is therefore called the Castle-Wright estimator. It is based on the phenotypic difference between two inbred lines. Let M_1 and M_2 be the means for two different inbred lines. It is assumed that these lines are fixed for alternative alleles at loci affecting the quantitative trait. These two lines are crossed, and the F_1 are then crossed with one another to produce an F_2. The variance in the F_2 will be higher than the variance in either inbred line or the F_1, because the alternative alleles are now segregating. We call this excess variance the segregational variance, V_{seg}:

$$V_{seg} = V_{F2} - V_{F1}$$

Castle showed that an estimate of the minimum number of loci affecting the trait is

$$n_e = \frac{(M_1 - M_2)^2}{8V_{seg}} \qquad \textbf{(13.48)}$$

This assumes each locus has an equal effect on the trait, and that there is no dominance, epistasis, or linkage. Under these assumptions, n_e is an estimate of the number of loci that affect the trait. Violation of any of these assumptions makes n_e an underestimate of the true number of loci; therefore, n_e is a minimum estimate of the number of loci. It is called the effective number of loci affecting the trait. Lynch and Walsh (1998) discuss the Castle-Wright estimator in detail, and more recent modifications of it.

Table 13.11 gives estimates of n_e based on the Castle-Wright estimator, or variations of it. In general, n_e is usually 20 or fewer. Given the assumptions, we must interpret these numbers very cautiously.

Today, individual quantitative trait loci are usually located and mapped using molecular markers (see below). The numbers of QTLs detected are generally not inconsistent with estimates based on the Castle-Wright estimator.

Mapping Quantitative Trait Loci

The procedure for mapping genes associated with quantitative traits is similar to the procedure for mapping genes associated with human diseases, discussed in Section 4.4. Detecting a QTL depends on the existence of linkage disequilibrium between the QTL and a marker locus whose position is known. We first describe the simpler situation of mapping QTLs in organisms whose breeding can be controlled.

TABLE 13.11   Estimates of the effective number of loci (n_e) affecting quantitative traits in various organisms. *Source:* From Roff (1997) and Lynch and Walsh (1998), who give original references.

Organism	Trait	n_e (± s.e.)
Corn	log (oil percent)	18 ± 2
	ln (ear length)	13 ± 3
	ln (seed weight)	13 ± 3
Lima beans	Seed size	17 ± 2
Red peppers	Fruit shape	3 ± 1
	Fruit weight	13 ± 1
Tomato	log (fruit weight)	12 ± 1
Drosophila melanogaster	ln (longevity)	1 ± 1
	Female head shape	6–9
	Abdominal bristles	98
	Sternopleural bristles	5
	Sternopleural bristles	18
Tribolium castaneum	Pupa weight	157–485
Cave fish	Eye diameter	6 ± 1
Chickens	Body weight	5 ± 1
Mice	log (body weight)	12 ± 1
	Litter size	2
	Litter size	164
Rats	Coat color	5–9
Rabbits	Body weight	14
	Ear length	19
Humans	Skin color	5

Sax (1923) was the first to demonstrate an association between a specific gene and a quantitative trait. He crossed two inbred lines of beans that differed in pigmentation and seed weight. The pigmented line, genotype *PP*, had an average seed weight of about 48 cg; the unpigmented line, genotype *pp*, had an average seed weight of about 21 cg. Sax crossed the two lines to obtain an F_1 and then crossed the F_1 to one another to obtain an F_2 generation. The mean seed weights of the F_2 genotypes were as follows:

Genotype	*PP*	*Pp*	*pp*
Mean weight	30.7	28.3	26.4

The difference between homozygotes is statistically significant. Genotypic differences at the pigmentation locus account for $(30.7 - 26.4)/(48 - 21)$, or about 16 percent of the difference between the inbred lines. This does not necessarily mean that the pigmentation gene itself has an effect on seed weight; the effect may be due to one or more genes tightly linked to the pigmentation gene.

Thoday (1961) suggested that this principle could be used to map loci affecting a quantitative trait. If we have many marker loci scattered throughout the genome, we can search for associations between marker genotypes and the quantitative trait. Significant associations, as illustrated by the bean data, would indicate that a gene affecting the quantitative trait is located near the marker locus.

The procedure is fairly straightforward for organisms whose breeding can be experimentally controlled. The first step is to create two inbred lines that differ for loci controlling the trait of interest (QTLs) and a number of marker loci. This is frequently done by performing artificial selection in opposite directions for many generations. Let Q_1 and Q_2 be the alleles at a locus affecting the quantitative trait, and M_1 and M_2 be forms of a marker whose chromosomal position is known. Assuming Q_1 is associated with M_1, we have the following:

Line	Genotype	Genotypic Value
L_1	Q_1M_1/Q_1M_1	a
L_2	Q_2M_2/Q_2M_2	$-a$

The two inbred lines are then crossed to produce an F_1. The F_1 are heterozygous for all QTLs and marker loci that differ in the two lines; in the above example, the F_1 are genotype Q_1M_1/Q_2M_2. The essential point is that gametic disequilibrium is complete in the F_1. Now cross the F_1 to each other. If there is no recombination between the QTL and the marker locus, the F_2 will be as follows:

Genotype	Frequency	Marker Genotype	Genotypic Value
Q_1M_1/Q_1M_1	0.25	M_1M_1	a
Q_1M_1/Q_2M_2	0.50	M_1M_2	d
Q_2M_2/Q_2M_2	0.25	M_2M_2	$-a$

In this case, the marker genotypes differ for the quantitative trait, indicating that either the marker locus itself or a locus tightly linked to it is a QTL.

On the other hand, if the QTL and the marker locus are unlinked, then alleles at the Q locus will be randomly associated with alleles at the M locus, and the marker genotypes will all have the same value for the quantitative trait.

Finally, if the QTL and the marker locus are linked, but with some recombination between them, then the marker genotypes will be different for the quantitative trait, but the differences will be weakened to the extent that recombination breaks up the initial disequilibrium between Q_1 and M_1. The difference between the homozygous marker genotypes will be

$$E(M_1M_1) - E(M_2M_2) = 2a(1 - 2r) \qquad \textbf{(13.49)}$$

where $E(M_1M_1)$ and $E(M_2M_2)$ are the expected genotypic values of the marker genotypes and r is the recombination rate between the marker locus and the QTL. The derivation of this equation is outlined in Problem 13.13. The left-hand side is the difference between the expected genotypic values of the homozygous marker genotypes, and can be estimated from the F_2 data. If $r = 0$, then the QTL and the marker locus are effectively the same, and the difference between M_1M_1 and M_2M_2 is $2a$, the same as the difference between the QTL homozygotes. If $r = 0.5$ (unlinked loci), then the right-hand side of equation (13.49) is zero, and the marker genotypes do not differ for the quantitative trait. That is what we expect if marker alleles are randomly associated with QTL alleles. For r between 0 and 0.5, the difference is less than $2a$, by an amount that depends on r.

Applying equation (13.49) to the bean data of Sax, the difference between the two marker classes was about 4.3 cg. If the QTL is absolutely linked to the pigmentation locus $(r = 0)$, this corresponds to $a \cong 2.15$ for that locus. If there is some recombination between the pigmentation gene and the QTL, then a is greater than 2.15 by an amount that depends on r. For example, if $r = 0.1$, then $a \cong 2.7$.

Equation (13.49) reveals one of the weaknesses of this approach. The expected difference between the two marker homozygotes will be small unless linkage is tight $(r \cong 0)$. This requires a high density of markers in order to detect linkage. Until recently, this has been possible in only a few organisms with well-known genetic maps, such as *Drosophila*. Furthermore, if the QTL has a small effect (a is small), the expected difference between the homozygous marker classes is small, even with tight linkage. The conclusion is that large sample sizes are required for detection of QTLs and the QTLs detected are likely to be those with relatively large effects.

A more sophisticated approach uses markers in pairs. This procedure is called **interval mapping**. Assume we have two marker loci, M and N, whose positions are known. Let r be the known recombination rate between the two marker loci; assume that r is small enough that double crossovers can be ignored. Now assume a QTL is located between M and N. Let r_1 be the recombination rate between the M locus and the QTL, and r_2 the recombination rate between the N locus and the QTL. Figure 13.17 illustrates the situation.

As before, we have two inbred lines as follows:

Line	Genotype	Genotypic Value
L_1	$M_1Q_1N_1/M_1Q_1N_1$	a
L_2	$M_2Q_2N_2/M_2Q_2N_2$	$-a$

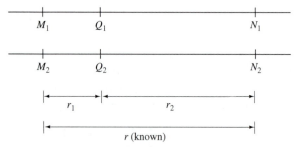

Figure 13.17 **Interval mapping of quantitative trait loci.** M and N are two marker loci, with known recombination rate, r. Q is a QTL located between them. The figure shows a heterozygote in which M_1, N_1, and Q_1 are in coupling phase, as would occur in an F_1 of a cross between two lines selected in opposite directions.

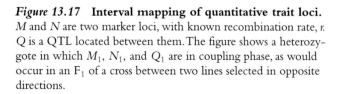

Cross L_1 to L_2, and backcross the F_1 to L_1. The possible F_2 genotypes, their marker types, and genotypic values are shown in Table 13.12. Note, this is just a standard three-point cross, with double crossovers assumed to be absent.

From Table 13.12, it is possible to derive the following two equations:

$$E(M_1M_1N_1N_1) - E(M_1M_2N_1N_2) = a - d \qquad \textbf{(13.50)}$$

$$E(M_1M_1N_1N_2) - E(M_1M_2N_1N_1) = \frac{(a - d)(r_2 - r_1)}{r} \qquad \textbf{(13.51)}$$

These are analogous to equation (13.49) from the single marker mapping. The left-hand sides are the expected genotypic values of the marker genotypes. They can be estimated by the observed values in the F_2 generation. Remembering that $r_1 + r_2 = r$, with r known, we have three equations in three unknowns, r_1, r_2, and $(a - d)$. The additive and dominance effects are confounded, and cannot be separated. The solution to these equations is straightforward, but the notation is cumbersome. To simplify, let $\hat{A}$ and $\hat{B}$ be the estimates of the left-hand sides of equations (13.50) and (13.51), respectively. Remember, these can be estimated from the F_2 data. Solving for r_1 and r_2, we get

$$r_1 = \frac{r(\hat{A} - \hat{B})}{2\hat{A}}$$

$$r_2 = r - r_1$$

Thus, we can estimate the position of the QTL between the two marker loci. Interval mapping allows a more precise estimate of the location of a QTL, and gives an estimate of the effect of the QTL, as $(a - d)$. However, it does not allow us to separate the additive and dominance effects.

Repeating this procedure for pairs of marker loci throughout the genome, we can find QTLs, map their positions, and estimate their effects. Until recently, only a few organisms, such as *Drosophila* and mice, had enough marker loci to make this approach practical. Most of the early work (before molecular markers) was done on bristle number in *Drosophila melanogaster*, because of the large number of mapped genes in that organism. See Mackay (1995) for a review. However, recent molecular techniques have made it possible to map hundreds or thousands of molecular markers (RFLPs, microsatellites, single nucleotide polymorphisms, etc.)

TABLE 13.12   Interval mapping of a QTL.
It is assumed that the marker loci are close enough that double crossovers can be ignored. See text for details.

F_2 Genotype	Marker Type	Genotypic Value
$M_1Q_1N_1/M_1Q_1N_1$	$M_1M_1N_1N_1$	a
$M_2Q_2N_2/M_1Q_1N_1$	$M_1M_2N_1N_2$	d
$M_1Q_2N_2/M_1Q_1N_1$	$M_1M_1N_1N_2$	d
$M_2Q_1N_1/M_1Q_1N_1$	$M_1M_2N_1N_1$	a
$M_1Q_1N_2/M_1Q_1N_1$	$M_1M_1N_1N_2$	a
$M_2Q_2N_1/M_1Q_1N_1$	$M_1M_2N_1N_1$	d

in almost any organism. This has resulted in many studies attempting to map individual QTLs in various species.

One of the early studies using molecular markers to map QTLs was done by Paterson et al. (1988), who studied three quantitative characters in tomatoes: fruit mass, concentration of soluble solids, and pH. Instead of inbred lines, they crossed two different species and then backcrossed the F_1 to one of the parental species. They studied 70 marker loci: 63 RFLPs, 5 allozymes, and 2 morphological traits. These markers covered all 12 chromosomes, with an average distance between adjacent markers of about 14.3 centimorgans.

Paterson et al. used interval mapping to map QTLs affecting each of these three quantitative traits. For each potential QTL, they calculated LOD scores, as described in Section 4.4. The LOD score is an indication of the probability that a QTL is linked to the marker locus. They used a cutoff of 2.4 for significant linkage. This corresponds to an overall probability of one or more false positives of about 0.05, the conventional level of significance (see Section 4.4). The LOD score map for chromosome 10 is shown in Figure 13.18. It indicates a QTL for pH at position *CD34A*.

For fruit mass, concentration of soluble solids, and pH, Paterson et al. found six, four, and five QTLs, respectively. These QTLs accounted for about 47 percent of the difference in fruit weight between the two species, and about 91 percent of the difference in soluble solids. (Data were not given for pH.)

The quantitative characters studied by Paterson et al. are important economic characters, affecting yield of tomato paste and preservation characteristics of the fruits. Mapping (some of) the genes affecting these characters will allow tomato breeders to more easily and accurately plan and monitor their breeding programs. Similarly, mapping QTLs affecting characters such as drought, heat, or cold resistance, and disease and pest resistance might allow breeders to maximize these desirable traits in many agriculturally and economically important organisms, either by conventional breeding programs or by using recombinant DNA technology.

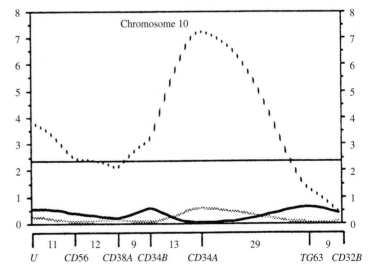

Figure 13.18 **LOD score map for chromosome 10 in tomatoes.** Solid line is fruit mass; wavy line is concentration of soluble liquids; dashed (upper) line is pH. Letters and numbers on the horizontal axis indicate marker loci and distances between them; vertical axis is LOD score, a measure of the probability that a QTL is located at a particular chromosomal position. The horizontal line at 2.4 is the critical LOD score for detection of a QTL; corresponding to an overall probability of a false positive of about 0.05. The upper curve indicates a QTL affecting fruit pH near position *CD34A*. *Source:* From Paterson et al. (1988).

Table 13.13 summarizes QTL mapping studies for a variety of organisms, excluding *Drosophila*. For a review of *Drosophila* studies, see Mackay (1995). That same journal issue (*Trends in Genetics*, vol. 11, no. 12, December 1995) has several other reviews of QTL studies. In general, it seems that about 5 to 20 QTLs are detected by these studies, and that these account for a large proportion of the phenotypic variance for the character. A single QTL typically accounts for about 10 percent of the phenotypic variance. This suggests that variation in many quantitative traits is controlled by a relatively small number of loci with moderately large effects. However, most of these studies are unable to detect QTLs with small effects. A large number of QTLs with small effects could have gone undetected.

Mapping QTLs is expensive, labor intensive, and fraught with statistical problems. Most studies are unable to detect QTLs of small effect. Statistical analysis is complicated by the fact that sometimes hundreds of tests are done, requiring huge sample sizes and very rigid criteria to avoid false positives. We have only touched the surface of this very complex and active field. See Lynch and Walsh (1998) for a thorough discussion.

Quantitative Trait Loci in Humans and Natural Populations

The QTL mapping techniques described above require the ability to construct homozygous inbred lines and to breed them in controlled ways. This is obviously impossible with humans and natural populations, so other approaches have to be used.

Recall the first step was to create an F_1 generation in which gametic (linkage) disequilibrium between QTLs and marker loci was complete. From there, it is easy enough in principle to detect associations between marker loci and the trait value. For humans we must rely on naturally existing disequilibrium. This creates two serious problems: First, we must be able to *infer* gametic phase from pedigree data. Not all families will provide enough information to do this. Second, sample sizes are typically small, and statistical power is very low. See Section 4.4 for a discussion of these problems with respect to mapping genes associated with human diseases.

TABLE 13.13   Estimated numbers of QTLs for various traits in several organisms. *Source:* Summarized from Tanksley (1993) and Roff (1997).

Organism	Trait	No. QTL Detected	Percent of V_P Explained
Tomato	Soluble liquids	4	44
	Fruit mass	6	58
	Fruit pH	5	48
Corn	Height	3–7	34–73
	Yield	3–13	10–87
Potato	Tuber shape	1	60
Cowpea	Seed weight	2	53
Aedes aegypti	*Plasmodium* resistance	2	67
Mouse	Epilepsy	2	50
Rat	Blood pressure	2	18–30

Sophisticated techniques have been developed to map QTLs in humans and other outbred populations. Again, see Lynch and Walsh (1998) for more information. Thomson and Esposito (2000) summarize recent progress in mapping genes associated with human complex diseases. See also the special issue of *Trends in Genetics* mentioned previously.

13.6 Evolutionary Quantitative Genetics

So far in this chapter, we have been mainly concerned with descriptions and predictions over the short term, primarily as they apply to artificial selection programs or laboratory experiments. When we consider natural populations, we must consider long-term evolutionary processes. Forces such as mutation and genetic drift, which can be ignored when considering only a few generations, must be taken into account when considering evolutionary time.

Initially, we made a number of simplifying assumptions about the nature of quantitative traits. We assumed a large number of loci, each having a small effect. Dominance, epistasis, pleiotropy, and mutation have almost always been ignored. We have seen that, under these assumptions, we can make reasonable predictions about the short-term effects of artificial and natural selection by using a purely statistical approach, ignoring the underlying genetic basis of the trait. However, long-term effects depend on the genetic basis of the trait, and on mutation rates, population sizes, etc. As Barton and Turelli (1989) pointed out, minor violations of the assumptions of the model will have little effect in the short term, but may dominate the evolutionary process over the long term. So, inevitably we must consider the topics of dominance, epistasis, pleiotropy, mutation, linkage, and genetic drift. This creates serious problems, since we know almost nothing about the actual genetic basis of most quantitative traits in natural populations. Also, the theoretical models get very complex and difficult. In this section, we will only summarize some of the major ideas in the very active field of evolutionary quantitative genetics. For good reviews, see Barton and Turelli (1989), Roff (1997), and Lynch and Walsh (1998).

Our organizing question will be: How much genetic variation for quantitative traits is there in natural populations? This is important for at least two reasons (Houle 1992). First, the ability of a population to evolve in response to natural or artificial selection depends on the amount of genetic variation in the population. Houle called this "evolvability." Second, as with molecular variation, the amount of quantitative genetic variation is relevant to understanding the effects and relative strengths of various evolutionary processes that affect natural populations.

We will examine the predicted effects of natural selection, mutation, and genetic drift on the expected amount of genetic variation, and compare these predictions to what we actually see in natural populations. As with molecular variation, we will find that no single theory can adequately explain all of our observations.

The Effect of Natural Selection on Genetic Variation

It should be obvious that stabilizing selection will decrease the *phenotypic* variance in a population. If extreme phenotypes are eliminated, the phenotypic variance must decrease. But it is not obvious whether this means a decrease in the genetic variance. To illustrate, consider a trait controlled by two loci, with alleles A_1 and B_1 increasing the value of the trait, and A_2 and B_2 decreasing the value. If stabilizing selection is operating, the best genotypes will have two "increasing" alleles. Double heterozygotes of the form A_1B_1/A_2B_2 (coupling) and A_1B_2/A_1B_2 (repulsion) will be favored. In the first, genetic variation will be preserved at each locus; in the latter, it will not. Which is more likely to evolve?

Roff (1997) reviews several theoretical models of stabilizing selection. They differ in genetic assumptions and details, but most predict that additive genetic variation will decrease in response to stabilizing selection. This prediction has been confirmed in laboratory experiments. For example, Kaufman et al. (1977) found that both heritability and genetic variance decreased in a long-term selection experiment on pupa weight in the flour beetle, *Tribolium castaneum*.

Directional selection will also decrease additive variance (Falconer and Mackay 1996; Roff 1997). As selection proceeds, alleles favored by selection will approach fixation and genetic variation will therefore decrease in the long term. This prediction has also been experimentally confirmed; see Roff (1997) for discussion and examples.

The case of disruptive selection is less certain. Both theoretical models and laboratory experiments suggest that genetic variance will increase over the short term, but the long-term effect is unclear. Figure 13.19 shows the results of disruptive selection for pupa weight in *Tribolium castaneum*. After 15 generations, the genetic variance increased as much as 50-fold. Keep in mind that this was a short-term result based on artificial selection. Whether both groups would be maintained under natural conditions is unclear. Theoretical considerations suggest that it is unlikely. For example, recall from Chapter 12 that several models of

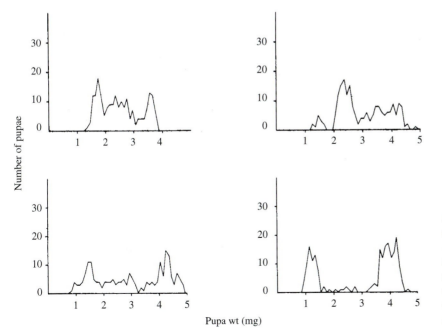

Pupa wt (mg)

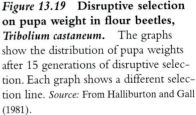

Figure 13.19 **Disruptive selection on pupa weight in flour beetles, *Tribolium castaneum*.** The graphs show the distribution of pupa weights after 15 generations of disruptive selection. Each graph shows a different selection line. *Source:* From Halliburton and Gall (1981).

spatially and temporally varying selection and of antagonistic pleiotropy all predict that polymorphism would be maintained only under fairly restrictive conditions. The same general result is true of most quantitative genetic models of disruptive selection.

To summarize, the consensus is that most forms of natural selection should cause a long-term decrease in genetic variance. This suggests that natural populations should have relatively little genetic variation for quantitative traits, and that traits subject to strong natural selection should have less genetic variation than traits less directly related to fitness, for example, morphological traits. Is this what we see?

Heritabilities in Natural Populations

It is widely assumed that life history traits (viability, fertility, etc.) are under stronger directional selection than morphological traits (but see Section 13.4 for some evidence to the contrary). As mentioned above, this leads to the prediction that in natural populations, life history characters should have less additive genetic variation than morphological characters.

Heritability is frequently assumed to be a good estimate of the amount of genetic variation in a population (but see below). If this is true, then life history traits should have lower heritabilities than morphological traits. This prediction has been confirmed. Mousseau and Roff (1987) reviewed more than 1000 heritability estimates for various kinds of traits in outbred animal populations, excluding *Drosophila*. Their results are summarized in Table 13.14. Mean heritability estimates for life history characters are significantly lower than for morphological characters (0.26 vs. 0.46), with physiological and behavioral characters intermediate. Similar results were obtained for heritability estimates of *Drosophila* characters (Roff and Mosseau 1987).

These results have been interpreted to mean that life history traits have less genetic variation than morphological traits. Note, however, that heritabilities are not zero, indicating that significant genetic variation is present for life history characters.

The studies summarized by Mosseau and Roff (1987) were of populations derived from wild, outbred populations; however, most of the heritabilities were actually estimated in the laboratory. This raises the question of whether laboratory estimates of heritability are good estimates of heritability in natural populations. There are at least two reasons to think they may not be. First, it is frequently assumed that environmental variation is less in laboratory environments than in nature; therefore, laboratory estimates of heritability are probably overestimates of

TABLE 13.14   Heritabilities of different kinds of quantitative traits.

Source: From Mosseau and Roff (1987).

Kind of Trait	Sample Size	Mean Heritability	Standard Error
Life history	341	0.262	0.012
Physiology	104	0.330	0.027
Behavior	105	0.302	0.023
Morphology	570	0.461	0.004

heritability in nature. Second, if there is any kind of genotype $\times$ environment interaction, such that genotypes respond differently to laboratory environments than to natural environments, then additive variance may be different in the laboratory.

If we assume that environmental variance is greater in natural environments than in the laboratory, and that additive variance is not different (no genotype $\times$ environment interaction), then some simple algebra will show that

$$\frac{h_L^2}{h_N^2} = \frac{V_{PN}}{V_{PL}} \tag{13.52}$$

where h_L^2 and V_{PL} are the heritability and phenotypic variance in the laboratory, and h_N^2 and V_{PN} are the heritability and phenotypic variance in nature. Roff (1997) suggested that, excluding *Drosophila*, the ratio of V_{PN} to V_{PL} was about 1.3. If this is true, then laboratory estimates of heritability are not greatly biased. Moreover, he reviews cases where heritability has been estimated in both laboratory and natural conditions, and concludes that the estimates are not significantly different.

Drosophila data challenge this conclusion. For example, Coyne and Beecham (1987) and Prout and Barker (1989) devised experimental methods to compare heritability in seminatural environments (where the parents matured in nature, but their offspring were reared in the laboratory) to heritability in laboratory environments (where both parents and offspring are reared in the laboratory). They found that natural heritabilities were sometimes substantially lower than laboratory heritabilities, but that natural heritabilities were significantly greater than zero (Table 13.15).

With some organisms, it is possible to estimate heritability under truly natural conditions. You need information on relatives (e.g., measurements on both parents and offspring) and large sample sizes (many families) to reduce standard errors to reasonable levels. Until recently, this has most frequently been done with relatively isolated populations of birds, where individuals can be banded and followed through the reproductive cycle. Table 13.16 compares heritability estimates of several morphological characters in two species of birds. For all characters, the heritability estimates were higher for the finches, but most were significantly greater than zero in both species.

Heritability estimates for fitness-related traits under truly natural conditions are even rarer than for morphological traits. Merila and Sheldon (2000) estimated the heritability of lifetime reproductive success in collared flycatchers (*Ficedula*

TABLE 13.15   Estimates of heritability under pseudonatural and laboratory conditions in two species of *Drosophila*. *Source:* From Coyne and Beecham (1987) and Prout and Barker (1989).

Species	Character	h^2 Nature	h^2 Lab
Drosophila melanogaster	Wing length	0.22 ± 0.08	0.58 ± 0.08
	Abdominal bristle number	0.47 ± 0.14	0.57 ± 0.12
Drosophila buzzatii	Thorax length	0.06 ± 0.01	0.38 ± 0.02
		0.09 ± 0.008	

TABLE 13.16 Heritabilities (± standard errors) of morphological characteristics of medium ground finches (*Geospiza fortis*) and song sparrows (*Melospiza melodia*).

Based on midparent-offspring regression. N is the number of families. *Source:* Ground finch data from Boag (1983); song sparrow data from Smith and Zach (1979).

Character	Medium Ground Finch ($N = 39$)	Song Sparrow ($N = 64$)
Body weight	0.91 ± 0.09	0.04 ± 0.16
Wing length	0.84 ± 0.14	0.13 ± 0.10
Tarsus length	0.71 ± 0.10	0.32 ± 0.11
Bill length	0.65 ± 0.15	0.33 ± 0.09
Bill depth	0.79 ± 0.09	0.51 ± 0.13
Bill width	0.90 ± 0.10	0.50 ± 0.14

albicollis) to be about 0.07 in males and 0.21 in females; only the latter is significantly different from zero. Kruuk et al. (2000) estimated the heritability of total fitness (number of offspring produced in a lifetime) in red deer to be about 0.02 in males and 0.00 in females; neither value is significantly different from zero. Similarly, Grant and Grant (2000) failed to find significant heritability for lifetime number of recruits in Darwin's finches. All of these studies are consistent with the findings of Mosseau and Roff (1987) that heritabilities for fitness traits are lower than for morphological traits. All are plagued by low statistical power.

To summarize, in natural populations fitness-related traits usually have lower heritabilities than morphological traits. This is usually assumed to be because fitness related traits have lower additive genetic variance. Next, we will consider an alternative interpretation.

Heritability Versus Additive Genetic Variance

What we are really interested in is the amount of additive genetic variance in a population. So far, we have used heritability as a surrogate estimate of V_A, but this may not be appropriate. Recall the definition of heritability:

$$h^2 = \frac{V_A}{V_P}$$

where V_A is the additive genetic variance and V_P is the total phenotypic variance. It has frequently been assumed that heritability is a good indicator of the amount of (additive) genetic variance for a trait, but this is not necessarily true. To see this, we can decompose the phenotypic variance as in equation (13.20), to get

$$h^2 = \frac{V_A}{V_A + V_D + V_I + V_E} \tag{13.53}$$

From equation (13.53) it is obvious that heritability depends not only on additive variance, but on three other kinds of variance. These last three terms in the denominator are sometimes collectively called the residual variance, V_R. If any component of the residual variance changes, the heritability will change.

In the studies summarized by Mosseau and Roff (1987), the lower heritability of life history traits and traits strongly associated with fitness was attributed to lower additive variance due to stronger natural selection. However, equation (13.53) suggests an alternative explanation. If fitness traits have higher residual variance than morphological traits, their heritabilities will be lower, even though the amount of additive variance may be the same, or even higher.

Houle (1992) suggested that heritability is frequently not a good indicator of additive genetic variance, and that the coefficient of additive genetic variation is a better measure to compare variation among traits. Recall that the coefficient of variation for a random variable is the standard deviation divided by the mean, expressed as a percent. Therefore, the coefficient of additive genetic variation is

$$CV_A = \frac{\sqrt{V_A}}{\mu} \times 100$$

where μ is the population mean. If we define the residual variation as all of the nonadditive components of phenotypic variance $(V_D + V_I + V_E)$, then we can similarly define the coefficient of residual variation as

$$CV_R = \frac{\sqrt{V_P - V_A}}{\mu} \times 100$$

The point is that low heritability of fitness traits can be due to either low additive genetic variation or to high residual variation (or both), and these quantities can be compared among traits by comparing their coefficients of variation.

Houle (1992) surveyed the literature for studies from which CV_A and CV_R could be determined. He found that life history and fitness-related traits had both higher CV_A and higher CV_R than morphological traits. He found a highly significant negative correlation between h^2 and CV_R, indicating that low heritabilities of fitness-related traits are likely due to high amounts of residual variation. This is the opposite of the usual interpretation that low heritabilities of fitness-related traits are due to low amounts of additive variance.

Several recent studies have confirmed Houle's results. For example, Merila and Sheldon (2000) found that lifetime reproductive success and lifetime fledgling production in collared flycatchers had lower heritabilities than morphological traits. However, these fitness traits sometimes had higher additive genetic variances (expressed as CV_A). Kruuk et al. (2000) obtained similar results for several components of fitness in the red deer. In both species, fitness traits had higher residual variance than morphological traits. These studies confirm that low heritabilities of fitness traits can be due to low additive variance or high residual variance, or both. They also suggest that additive variance can sometimes be higher for fitness traits.

These results raise two questions: First, why is residual variance for fitness traits higher than for morphological traits? Second, why is additive variance sometimes higher for fitness traits than for morphological traits, in apparent contradiction to the theory summarized above?

We can imagine at least two possible reasons for higher residual variance in fitness traits. One possibility is that the environmental variance is greater for fitness traits. This was first suggested by Price and Schluter (1991). They argued that morphological traits affect fitness. Therefore, all environmental variation affecting morphological traits also affects fitness. But additional environmental variation may affect fitness traits but not morphological traits. For example, longevity and associated breeding success depend on body size, but also on many factors not related to body size. So the environmental variance for longevity is expected to be higher than for body size.

Merila and Sheldon (2000) suggested a second possibility: Fitness traits are probably affected by more genes than are morphological traits. If so, the potential for gene interaction is greater, increasing the interaction variance without affecting the additive variance.

Price and Schluter (1991) also discussed why additive variance is sometimes higher for fitness traits than for morphological traits. They argued that quantitative genetics theory does not necessarily predict low genetic variation for fitness traits. They developed a model in which additive variance can actually increase under natural selection, and showed that the amount of additive variance for fitness traits can be either greater than or less than for morphological traits.

Another possibility is that mutation can create significant additive variance. If fitness traits are affected by more loci than morphological traits, as hypothesized above, then the variance created by mutation will be greater for fitness traits (Houle et al. 1996). We discuss the effect of mutation on genetic variance below.

To summarize, natural populations appear to harbor much genetic variation for quantitative characters. Even traits closely associated with fitness typically show some genetic variation, sometimes more than for morphological traits. This is inconsistent with the predictions of most theory. We have a new version of a familiar question: Why is there so much quantitative genetic variation in natural populations? We will examine several models that seek to answer this question.

Mutation Rates for Quantitative Characters

Clayton and Robertson (1955) were the first to suggest that mutation might be a significant source of genetic variation for quantitative traits. If, in fact, quantitative traits are affected by many loci, each of which can mutate, we might expect new mutations to be a significant source of genetic variation in natural populations.

We got our first hint of the effect of mutation on quantitative characters when we discussed mutation accumulation experiments in Section 6.6. In *Drosophila*, estimates for the total number of new mutations affecting viability averaged about 0.6 per zygote.[3] Viability is a quantitative character, presumably affected by many loci. One might expect that morphological characters would be affected by fewer loci, thus the total mutation rate might be lower. The best data come from experiments on bristle number in *Drosophila melanogaster*. Lynch and Walsh (1998) reviewed these studies and estimated that the total mutation rate affecting bristle numbers was about 0.1 per zygote. Other studies of *Drosophila*, mice, and corn have produced estimates of similar magnitude. These estimates are surprisingly

---

[3]Here we are using the classical results; see Section 6.6 for doubts about these estimates.

high. If U is the per-character mutation rate, u is the per-locus mutation rate, and n is the number of loci affecting a character, then

$$U = 2nu \qquad (13.54)$$

(The 2 comes from diploidy; there are two copies of each locus that can mutate.) If we take 0.1 as an estimate of U, and 10^{-5} as a typical per-locus mutation rate (a high estimate), then we can estimate the number of loci that mutate as

$$n = \frac{U}{2u} \cong 5000$$

It is hard to believe that 5000 loci affect bristle number in *Drosophila*. If the per-locus mutation rate is 10^{-6}, then $n \cong 5 \times 10^4$, more loci than exist in the *Drosophila* genome. Moreover, the QTL studies summarized in Section 13.5 suggest that the number of loci affecting quantitative traits is on the order of 100 or fewer, with only a few of these causing most of the observed variation. If we assume $n = 100$, we get $u \cong 5 \times 10^{-4}$, which seems very high. Barton and Turelli (1989) called this contradiction the paradox of per-character versus per-locus mutation rates. Several possible explanations have been proposed: (1) Quantitative traits may really be affected by thousands of loci; (2) per-locus mutation rates may be higher for quantitative trait loci than for other loci; (3) the estimates of per-character mutation rates may be much too high. So far, we have little information to help us choose among these alternatives.

Effects of Mutation and Genetic Drift on Genetic Variance

Mutation will create new alleles each generation. This should increase the genetic variance. The amount of variance created by new mutations each generation is called the **mutational variance**, symbolized by V_M. The (additive) genetic variance will increase by V_M each generation:

$$V_A(t + 1) = V_A(t) + V_M \qquad (13.55)$$

The value of V_M depends on the number of loci mutating, the mutation rate per locus, and the phenotypic effect of a mutation. Consider a single locus affecting a quantitative trait. Let the mutation rate be u and the average effect of a mutation on the phenotype be zero. In other words, a mutation is as likely to increase the phenotype as it is to decrease it. Different mutations will have different effects on the phenotype, and the variance of effects is designated m^2.

Now consider n loci affecting the trait. For simplicity, we will assume that all loci have the same mutation rate and variance of effects, and that there is no interaction among loci. Then the total mutation rate per character is $2nu$, as explained above, and the total mutational variance per character is

$$V_M = 2num^2$$

How large is V_M? Clayton and Robertson (1955) were the first to estimate it. They monitored the change in genetic variance for abdominal bristle number in inbred lines of *Drosophila melanogaster*, and suggested that the additive variance would increase by about 0.1 percent per generation due to new mutations. Lynch (1988) reviewed several studies on *Drosophila* traits, mouse body size, and others, and concluded that for a variety of quantitative traits, $V_M \cong 10^{-3}V_E$.

(We will see below why it is convenient to express V_M as a proportion of the environmental variance.)

Equation (13.55) predicts that, ignoring drift, the genetic variance will increase indefinitely. Of course, we cannot ignore drift. Because genetic drift causes alleles to be lost, intuition might suggest that it would cause a decrease in genetic variance of quantitative traits. That intuition would be correct; by analogy to the loss of heterozygosity described by equation (7.5), the loss of additive genetic variance due to drift is

$$V_A(t + 1) = \left(1 - \frac{1}{2N_e}\right)V_A(t) \tag{13.56}$$

N_e is the effective population size, as discussed in Section 7.4. Thus, under genetic drift alone, the genetic variance should eventually decay to zero, but the process will be slow except in very small populations.

Drift decreases the amount of genetic variation each generation, while mutation increases it. Just as we did for molecular variation, we can ask at what point do these two processes reach an equilibrium. The recursion equation for the combined effects of drift and mutation is, combining equations (13.55) and (13.56),

$$V_A(t + 1) = \left(1 - \frac{1}{2N_e}\right)V_A(t) + V_M$$

where the first term on the right gives the decrease due to genetic drift and the second term is the increase due to mutation. At equilibrium, $V_A(t + 1) = V_A(t) = \tilde{V}_A$. Some algebraic rearrangement then gives the equilibrium value of the genetic variance

$$\tilde{V}_A = 2N_e V_M$$

The equilibrium phenotypic variance is then

$$\tilde{V}_P = \tilde{V}_A + V_E$$

(This assumes that V_E is constant and there is no dominance or epistasis.) We can now find the heritability at equilibrium

$$\tilde{h}^2 = \frac{\tilde{V}_A}{\tilde{V}_P} = \frac{2N_e V_M}{2N_e V_M + V_E}$$

If we assume $V_M \cong 10^{-3}V_E$, as estimated above, this gives

$$\tilde{h}^2 = \frac{2N_e \times 10^{-3}}{2N_e \times 10^{-3} + 1} \tag{13.57}$$

(The V_Es cancel.) Figure 13.20 plots $\tilde{h}^2$ versus N_e. Clearly, heritabilities in the range observed in nature can be maintained by mutation-drift equilibrium for moderately sized populations, on the order of 100 or more. For very large populations, heritability should approach one. We conclude that mutation-drift balance can explain the relatively high heritabilities commonly found in natural populations.

What's wrong with this conclusion? We have left out what may be the most important evolutionary process acting on quantitative characters. We have seen much evidence that natural selection acts on many, if not most, quantitative characters,

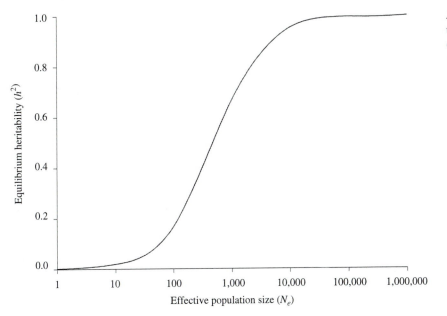

Figure 13.20 **Equilibrium heritability, as predicted by equation (13.57).** This assumes $V_M \cong 0.001 V_E$.

and that its expected effect is to decrease genetic variance. Can mutation create enough variation to counteract the depleting effects of genetic drift and natural selection, and thus explain the large amounts of quantitative genetic variation found in natural populations?

Joint Effects of Mutation and Natural Selection on Genetic Variance

Several theoretical models have investigated the joint effects of selection and mutation on (additive) genetic variance (e.g., Latter 1960; Kimura 1965; Bulmer 1972; Lande 1975; Turelli 1984; Caballero and Keightley 1994). We will only summarize some of the main assumptions and conclusions.

Most models assume a life cycle as follows:

$$zygotes \xrightarrow{\text{selection}} adults \xrightarrow{\text{mutation}} gametes \xrightarrow{\text{random mating}} zygotes$$

Now, selection will decrease the genetic variance, and mutation will increase it. We can write a general recursion equation for genetic variance

$$V_G(t + 1) = V_G(t) - V_S(t) + V_M(t)$$

where $V_S(t)$ is the amount of variance lost by selection in generation t, and $V_M(t)$ is the amount gained by mutation. These two processes will reach an equilibrium at which $V_G(t + 1) = V_G(t) = \widetilde{V}_G$, where $\widetilde{V}_G$ is the equilibrium genetic variance. In order to make further progress, we have to make assumptions about the nature of selection and mutation.

The usual assumption about selection is that it is stabilizing; a commonly assumed fitness function is

$$w(P) = \exp\left(\frac{-P^2}{2w}\right) \tag{13.58}$$

where P is the phenotypic value, scaled so that the best phenotype has a value of zero, w is a measure of the strength of selection, and $w(P)$ is the fitness. If w is small, selection is strong; if w is large, selection is weak. This fitness function was first suggested by Haldane (1954), and is illustrated in Figure 13.21 for weak and strong selection. We assume that natural selection keeps the population near a mean phenotypic value of zero.

Now let mutation act as follows: Let x be the average effect of an allele on the quantitative trait before mutation and x' be its effect after mutation. Then,

$$x' = x + M$$

where M is a random variable with mean zero and variance m^2. So mutation is equally likely to increase or decrease the effect of an allele. If most mutations have small effects, as discussed in Chapter 6, then m^2 is small. Note that, although a mutation is equally likely to increase or decrease the average effect of an allele, its effect on fitness depends on the phenotype of its carrier (Figure 13.21). A mutation that moves the phenotype closer to the mean will increase fitness of the carrier.

Given these assumptions, the problem is to find the amount of genetic variation in the population at mutation/selection equilibrium. In its most general form, the answer will depend on the form of the fitness function, the strength of natural selection, the mutation rate per locus, the distribution of mutational effects, the number of loci affecting the trait, the linkage relationships among the loci, the degree of dominance at a locus, the degree of gene interaction, and the environmental variance. The general problem has not been solved. Several workers have made various assumptions and approximations to get some results. One of the early influential models was that of Lande (1975). He modeled a trait subject to stabilizing selection, with each locus having additive effects on the quantitative trait. Using estimates of mutation rates from several organisms and

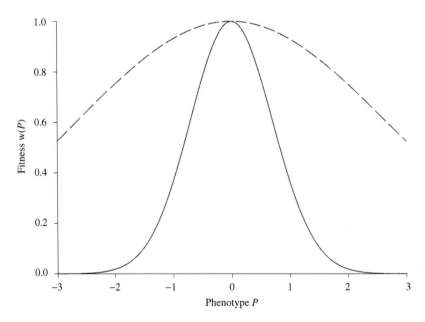

Figure 13.21 **The fitness function described by equation (13.58).** The solid line indicates strong selection, and the dashed line indicates weak selection.

assuming moderate numbers of loci controlling a trait, he concluded that large amounts of genetic variation ($h^2 \cong 0.3$ to 0.5) could be maintained in natural populations, even under strong natural selection. This suggests that mutation/selection balance can explain much of the quantitative genetic variation observed in natural populations.

However, Turelli (1984) showed that Lande's results depended on a mathematical approximation that is valid only if mutation rates are very high or selection is very weak. He claimed that the data from laboratory experiments are inconsistent with this assumption. He then reanalyzed the model using different mathematical approximations, and concluded that the equilibrium amount of genetic variation should be much less than Lande predicted. However, Turelli pointed out that neither the effects of various mathematical assumptions nor the relevant mutation rates and variances are very well understood, so neither the theory nor the experimental data allow us to make any robust conclusions about the joint effects of mutation and selection on the genetic variance of quantitative traits.

Lande's (1975) model and Turelli's (1984) extension both assumed additive effects and direct stabilizing selection on the quantitative trait, and ignored genetic drift. Caballero and Keightley (1994) analyzed more complex models, which included dominance, indirect (pleiotropic) effects of mutation on viability, and various distributions of the effects of mutation on the quantitative trait and on fitness. The main variables they included in their models were

1. The distribution of mutational effects on the quantitative trait
2. The distribution of mutational effects on viability
3. The correlation between mutational effect on the trait and on viability
4. The degree of dominance with respect to the trait
5. The degree of dominance with respect to viability
6. The strength of (indirect, directional) selection on viability
7. The per-character mutation rate
8. Population size

The main differences between Caballero and Keightley's models and earlier models were that they allowed dominance, and they assumed *indirect* selection on viability through pleiotropic effects of mutations.

Caballero and Keightley estimated the above parameters from recent experiments on bristle number and viability in *Drosophila melanogaster* (Mackay et al. 1992) and from various other sources. Using these estimates, their main conclusions were

1. At equilibrium, heritabilities for the trait will be moderate (about 0.4 to 0.6) for population sizes from 10^4 to 10^6.
2. The main factor affecting heritability of the trait is the shape of the distribution of mutational effects on fitness. Distributions with higher kurtosis (more peaked than a normal distribution) produce higher heritabilities.
3. The equilibrium amount of genetic variance for the trait is nearly independent of the dominance of new mutations.
4. At equilibrium, only about 10 percent of the genetic variance for the trait is dominance variance.

5. Heritability of viability at equilibrium was about 0.09 to 0.15, depending on the distribution of mutational effects.

6. Dominance variance for viability was generally very low at equilibrium.

These conclusions seem to agree reasonably well with experimental results from *Drosophila* (upon which the parameter estimates were based). Taken at face value, they suggest that pleiotropic effects on fitness are important, but dominance effects are not. However, we still know little about how valid Caballero and Keightley's assumptions and parameter estimates are for other traits and other organisms.

How Important Are Dominance, Linkage, Epistasis, and Pleiotropy?

The simplest model of quantitative traits assumes a large number of loci affecting the trait, that all loci have small effects, and that dominance, epistasis, and linkage are absent. As mentioned at the beginning of this section, this allows us to make reasonably accurate short-term predictions about the evolution of quantitative traits without knowing the detailed genetic basis of the trait. However, in order to understand the long-term evolution of quantitative traits and the maintenance of quantitative genetic variation, we must know more about their genetic basis. Here, we summarize what is known about the genetic architecture of quantitative traits, and how violations of the above assumptions affect their evolution.

The first assumption is that the trait is determined by many loci, all loci having small effects. As discussed in Section 13.5, many recent studies of QTLs have shown that quantitative traits are typically affected by a few loci with relatively large effects, with an unknown but possibly large number of loci with small effects (Falconer and Mackay 1996; Lynch and Walsh 1998). This suggests that models of selection and variation that assume many loci with small effects should be reevaluated.

We next consider the issue of dominance. We saw in Section 6.4 that mutations with large effect on viability tend to be recessive, while mutations with small effect tend to be nearly additive. Is the same true with QTLs? In other words, do QTLs with small effects act more or less additively, while QTLs with large effects show dominance? That seems to be the case for the best-studied quantitative trait, bristle number in *Drosophila*. Mackay (1995) reviewed several studies and concluded that mutations at QTLs affecting bristle number have varying degrees of dominance, with mutations with large effects tending to be recessive. These induced or spontaneous mutations are thought to be similar in effect to the alleles existing in natural populations.

However, the degree of dominance does seem to be quite variable in general. For example, deVicente and Tanksley (1993) detected 74 QTLs affecting several quantitative traits in tomatoes. Effects ranged from completely recessive to additive to completely dominant. Overdominance and underdominance were also detected. Dominance of alleles that increased the value of the trait was as common as dominance of alleles that decreased the value. Lynch and Walsh (1998) review many QTL studies and conclude that some degree of dominance is common. If this is true, estimates of V_G may seriously overestimate the amount of additive variance in a population.

Crnokrak and Roff (1995) surveyed many studies and concluded that morphological traits show little or no dominance, whereas life history traits show relatively greater dominance effects. They suggest that this is because life history traits are expected to be subject to stronger natural selection, which will reduce additive variance more than dominance variance.

We saw in Chapters 4 and 12 that gametic disequilibrium is expected to decay to zero unless some process (e.g., genetic drift, natural selection, or gene flow) acts to maintain it. Therefore, we might expect the long-term effect of linkage to be minor in relatively large, isolated, outbreeding populations. A variety of theoretical models considering stabilizing selection and mutation, with additive effects and no epistasis for the quantitative trait, suggest that linkage among QTLs will have minor effects on the amount of genetic variance maintained in a population (Lande 1975; Turelli 1985; Lynch and Hill 1986; Bürger et al. 1989).

Thus, the issue is not whether QTLs are linked, but whether significant gametic disequilibrium develops among them. (Recall that gametic disequilibrium can develop among unlinked loci.) This in turn depends on how the character is affected by natural selection and gene interaction. Lynch and Walsh (1998) give equations for the additive and dominance variance with gametic disequilibrium. In our notation, they are:

$$V_A = \sum_{i=1}^{n} 2p_i q_i \alpha_i^2 + 2\sum_{i=1}^{n}\sum_{j \neq i}^{n} \alpha_i \alpha_j D_{ij}$$

$$V_D = \sum_{i=1}^{n} (2p_i q_i d_j)^2 + 4\sum_{i=1}^{n}\sum_{j \neq i}^{n} d_i d_j D_{ij}^2$$

where i and j represent the individual loci, and there are n loci. You need not understand the details of these equations, but note that the right-hand side of each consists of two terms. The first term of each equation is the variance assuming gametic equilibrium. Lynch and Walsh call these the equilibrium genetic variances; they are the multilocus analogs of equations (13.21) and (13.22). The second term of each equation is a deviation due to gametic disequilibrium, and includes the parameter D_{ij}, the disequilibrium coefficient between the ith and jth locus. If all of the D_{ij} decay to zero, then obviously gametic disequilibrium has no long-term effect on the additive or dominance variance. But if gene interaction (or some other ongoing force) acts to maintain one or more of the D_{ij} at some nonzero value, then the actual (expressed) variance can be either greater than or less than the variance under gametic equilibrium, depending on the signs of the D_{ij}.

Several theoretical models (e.g., Bulmer 1971; see also Lynch and Walsh 1998) suggest that both stabilizing selection and directional selection will generate negative disequilibrium. This will decrease, sometimes substantially, the amount of expressed additive variance compared to the amount expected under gametic equilibrium. Conversely, disruptive selection is expected to generate positive disequilibrium, thereby increasing the expressed genetic variance (Bulmer 1971).

To summarize, the theory suggests that linkage *alone* will not have much effect on the amount of genetic variation maintained in a population. The degree and kind of gene interaction is much more important. So we must ask what the experimental evidence for epistasis is.

The word epistasis has several meanings (see Phillips 1998 and Wade et al. 2001 for discussions). We shall use it in the general sense of gene interaction: The effect of a genotype at one locus depends on the genotype at another locus.

Epistasis is difficult to quantify because gene interactions are expected to be weak relative to main effects, and large sample sizes are required to detect them. One approach is to perform a two-way analysis of variance (ANOVA), with the effects of two QTLs as the main effects. Significant interaction effects indicate interaction between the two QTLs. See any statistics textbook for details of ANOVA.

Tanksley (1993) reviewed several QTL studies using molecular markers. He concluded that the number of significant interactions was about the same as expected by chance, and most interactions were weak. He tentatively concluded that strong epistatic interactions "are the exception and not the rule for naturally occurring polygenes" (Tanksley 1993, p. 231).

There are important exceptions. For example, Long et al. (1995) detected seven QTLs affecting abdominal bristle number in *Drosophila melanogaster*. Four of these showed strong interactions, so that two QTLs from one line had a greater effect than expected from their individual effects. Surprisingly, the interaction effects were about the same magnitude as the main effects. Lynch and Walsh (1998) and Mackay (2001) cite other examples of significant epistasis, and discuss reasons why it may be underdetected in experimental studies.

In addition to the maintenance of quantitative genetic variation, epistasis is also important to the understanding of numerous other evolutionary topics. Examples are population differentiation and speciation, mating systems, and conservation of endangered populations. Fenster et al. (1997) review these topics and their relationship to epistasis.

Finally, we must address the issue of pleiotropy. As Barton and Turelli (1989) wrote, "It is widely believed that pleiotropy is ubiquitous." One reason for this belief is the observation that different quantitative characters are frequently correlated (Section 13.3), although, as we have seen, this can be due to either pleiotropy or stable gametic disequilibrium. It seems reasonable to assume that at least some of these correlations are due to pleiotropy. Certainly, there are many known examples. Loci affecting sternopleural bristle number also affect abdominal bristle number and larval viability (e.g., Long et al. 1995). QTLs in corn can affect many characters; for example, Edwards et al. (1987) found that some QTLs affected as many as 78 of the 82 characters they examined.

Many, perhaps most, quantitative traits have pleiotropic effects on one or more components of fitness. Many experiments have shown that loci affecting bristle number also affect larval viability. Figure 13.22 shows one example (Kearsey and Barnes 1970). Figure 13.22a shows the distribution of bristles on adult flies raised in uncrowded (solid line) or crowded (dotted line) larval conditions. Competition under crowded conditions eliminated larvae destined to have extreme bristle numbers. Figure 13.22b shows the survival rate for flies with different bristle numbers. Clearly, flies with intermediate bristle numbers had higher survival rates. Note that selection is acting on larvae that do not have sternopleural bristles; therefore, selection must be acting on some pleiotropic effect of genes that affect bristle number.

The concept of antagonistic pleiotropy, discussed in Sections 12.7 and 13.4, also suggests that loci affecting one fitness component frequently have pleiotropic effects on other components.

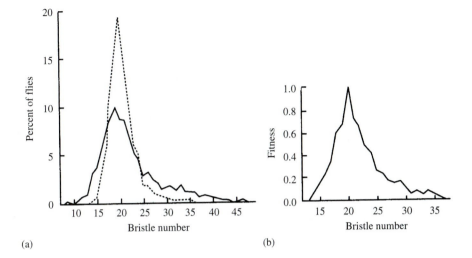

Figure 13.22 **Pleiotropy and natural selection on sternopleural bristle number in *Drosophila melanogaster.*** (a) Distribution of bristle number in adults that experienced uncrowded (solid line) and crowded (dotted line) conditions during larval development. Development under crowded conditions eliminates larvae destined to have very high or very low bristle numbers. (b) Relative viability of files with different bristle numbers. Stabilizing selection occurs at loci that affect both larval survival and adult bristle number. *Source:* From Falconer and Mackay (1996); original from Kearsey and Barnes (1970).

Pleiotropy has complex and variable effects on the level of quantitative genetic variation in a population. Turelli (1985) explored theoretical models of pleiotropy and concluded that the effect on additive variance depended critically on the details of the model, and on mutation rates and selection intensities. Caballero and Keightley (1994) concluded that pleiotropic effects on viability were important, but equilibrium levels of genetic variance also depended on the parameter estimates and details of their models.

To summarize, recent molecular and statistical techniques have allowed us to examine the fundamental assumptions of the basic quantitative genetics model. In short, *all* of these assumptions are violated to a greater or lesser degree. This fact may have little effect on short-term predictions, but it demands a reexamination of long-term predictions. Theoretical quantitative geneticists are still working out the ramifications of this new understanding. For an overview of recent advances in quantitative genetics, see Mackay (2001).

What Maintains Genetic Variation for Quantitative Traits?

We have seen that, contrary to most theoretical predictions, genetic variation for quantitative traits is common in natural populations. As with molecular variation, we are confronted with a long list of possible explanations: mutation, selection varying in space and time, antagonistic pleiotropy, epistasis, overdominance, gene flow, frequency dependent selection, and so forth. We still do not know which of these forces are most important. There is much work yet to be done, but recent approaches using molecular marker loci and powerful statistical techniques suggest that progress will be rapid.

Summary

1. A quantitative trait is one that shows a continuous range of phenotypes. Quantitative traits are controlled by many genes and are affected by environmental factors.

2. The phenotypic value of a quantitative trait can be partitioned into a genetic component and an environmental component. The genetic component can be further partitioned into an additive component, due to effects of individual alleles, a dominance component, due to interaction between alleles at the same locus, and an interaction component, due to interactions between alleles at different loci.

3. The phenotypic variance of a population can be similarly partitioned into additive genetic variance, dominance variance, interaction variance, and environmental variance. Only the additive variance is important in most artificial selection programs.

4. The standard quantitative genetic model assumes that a quantitative trait is controlled by a large number of unlinked loci, each with small and additive effects on the trait. This allows us to analyze quantitative traits with statistical methods, without having to know the genetic details of the trait.

5. The narrow sense heritability is the ratio of the additive variance of a population to the phenotypic variance. It is useful in predicting how a quantitative trait will evolve in response to artificial or natural selection.

6. Relatives are more similar phenotypically than unrelated individuals, because they may share alleles that are identical by descent. The degree of similarity depends on the degree of relatedness, and is measured by the covariance between relatives. This provides a method of estimating heritability, for example, equation (13.32) for parent-offspring regression.

7. Truncation selection is a form of artificial selection in which only individuals with phenotypes greater than a specified value are allowed to breed. The response to truncation selection depends on the proportion of the population allowed to breed, and on the heritability of the character being selected for [equations (13.37) and (13.39)].

8. Artificial or natural selection on one character will cause a correlated response on other characters that are genetically correlated to the one under selection. The correlated response depends on the heritabilities of the two characters and on their additive genetic correlation [equations (13.44) and (13.45)].

9. Natural selection on a quantitative character can be either directional, stabilizing, or disruptive (Figure 13.11).

10. Major components of fitness are sometimes negatively correlated with quantitative traits, so that selection acting on a quantitative trait often reduces fitness.

11. Natural selection acts on net fitness, which is affected by many quantitative characters. Using multiple regression analysis, it is possible to separate the effects of direct selection on a character from the effects of indirect selection.

12. Natural selection on quantitative traits is typically weak, with occasional exceptions. Directional selection is stronger than stabilizing or disruptive selection, and the latter two are about equal in frequency and strength.

13. Using molecular markers, it is possible to map loci affecting a trait (quantitative trait loci) and to estimate the magnitude of their effects. Most traits studied so far are controlled by a small number of loci having relatively large effects, and an unknown, but possibly large, number of loci with small effects.

14. Natural selection on a quantitative trait is generally expected to reduce the additive genetic variance for that trait.

15. Natural populations show surprisingly large amounts of genetic variation for quantitative traits, even those closely related to fitness. This raises the question of why there is so much quantitative genetic variation, since natural selection is generally expected to reduce it.

16. Traits closely associated with fitness have lower heritabilities than morphological traits, but sometimes have high levels of additive genetic variance. The reason for the apparent contradiction is that fitness traits frequently have higher levels of residual (nonadditive) variance than morphological traits.

17. Mutation probably contributes significant genetic variation for quantitative traits. Whether it is sufficient to maintain large amounts of variation in the face of natural selection and genetic drift is still not well understood.

18. The degree of dominance at quantitative trait loci is variable. Alleles with large effects tend to be recessive, and alleles with small effects tend to be partially dominant. There are exceptions to this generalization.

19. Gene interaction is common, and its effects can be relatively strong compared to single-locus effects. It can either increase or decrease the amount of quantitative genetic variation in a population, with the latter being more common.

20. Pleiotropy is also common, and quantitative traits often have pleiotropic effects on one or more components of fitness.

21. We still do not have a clear understanding of how the different evolutionary processes (mutation, genetic drift, natural selection, epistasis, gene flow, and so forth) interact to determine the amount of quantitative genetic variation in natural populations.

Problems

13.1. Consider the pygmy gene discussed in Section 13.2. Let $p = 0.8$ be the frequency of the wild type allele and $q = 0.2$ be the frequency of the pg allele.

    a. Calculate the population mean, the genotypic values, additive effects, and dominance deviations of the three genotypes.

    b. Calculate V_A and V_D.

13.2. Plot V_A and V_D versus p for the *pygmy* gene of Problem 13.1.

13.3. Body weight in lambs has a phenotypic mean of 54 kg and standard deviation of 12.1 kg. The heritability is $h^2 = 0.45$.

    a. Estimate V_A.

    b. Assume $V_E = 39.2$. Estimate $V_D + V_I$.

13.4. Two inbred lines of beans have mean weights of 35 and 65 cg. When crossed, the phenotypic variance of the F_1 was 8. When the F_1 were crossed to each other, the phenotypic variance of the F_2 was 41. Estimate the minimum number of loci affecting bean weight.

13.5. Estimate the realized heritability from the following parent-offspring data on height (cm).

Father's Height	Son's Height
180	175
170	172.5
167.5	165
177.5	170
182.5	180
165	172.5
172.5	167.5
177.5	172.5
170	175

13.6. Why might estimates of realized heritability differ from heritability estimates based on other methods?

13.7. Show that if $\overline{P}$ is the midparent value, the regression of offspring on midparent value is given by equation (13.33).

13.8. Show that the regression of cumulative response on cumulative selection differential is the heritability. (*Hint*: Use the two point definition of the slope of a line.)

13.9. Derive equation (13.52).

13.10. Body weight in *Geospiza fortis* before the 1977 drought was 15.79 grams. Mean body weight of survivors was 16.85 grams (Boag and Grant 1981). Predict the weight of the offspring of the survivors. Use the heritability estimates in Table 13.16. What other factors might complicate this prediction?

13.11. Long et al. (1995) crossed lines of *Drosophila melanogaster* that had been selected in opposite directions for abdominal bristle number. The means of the selected lines differed by 26.74. Heritability was estimated to be 0.248, and the phenotypic variance in the F_2 was 32.49. Estimate the minimum number of loci affecting abdominal bristle number in these lines.

13.12. Two inbred lines, with genotypes $M_1M_1Q_1Q_1$ and $M_2M_2Q_2Q_2$, had mean phenotypic values of 50 and 30. When these two lines were crossed and the F_1 crossed to one another, the mean phenotypic values for the marker genotypes in the F_2 were

Marker Genotype	Value
M_1M_1	44
M_1M_2	42
M_2M_2	36

a. What proportion of variation in the F_2 does this QTL account for?

b. Previous experiments estimated $a \cong 12$ for the QTL. Estimate the recombination rate between the marker locus and the QTL.

13.13. Equation (13.49) gives the expected value of the difference between the marker genotypes in the F_2 generation of a cross between two lines differing at both the

QTL and the marker locus. The following steps will show how to derive this equation:

a. Note that the expected genotypic value of (M_1M_1) is given by

$$E(M_1M_1) = \Pr(Q_1Q_1|M_1M_1)(a) + \Pr(Q_1Q_2|M_1M_1)(d)$$
$$+ \Pr(Q_2Q_2|M_1M_1)(-a)$$

This is the average of the genotypic values of all QTL genotypes that are also M_1M_1.

b. Using the rules for conditional probability, find the probabilities on the right side.

c. Combine these and simplify to get an expression for $E(M_1M_1)$.

d. Write an equation analogous to that in part (a) for $E(M_2M_2)$.

e. Find the conditional probabilities on the right side of the equation in part (d).

f. Combine these to get an expression for $E(M_2M_2)$.

g. Show that the difference between parts (c) and (d) is given by equation (13.49).

Probability and Random Variables

Any formal reasoning regarding genetic transmission has to allow for the role played by chance; and is necessarily expressed in terms of probabilistic concepts.

—*A. Jacquard (1975)*

Probability is one of the most important concepts in population genetics; much of this book requires an understanding of the basic ideas of probability theory. This appendix summarizes most of what you need to know. It is intended as a summary and reference only; ideas will be explained and illustrated, but proofs will not be given. For details and more information, see any textbook on probability theory.

A.1 Populations and Samples

We begin by reviewing several ideas discussed in Box 2.2. A **population** consists of every possible object in the study, or every possible outcome of an experiment. These may be individuals in a biological population, genes in a genome, automobiles in a city, or any other group. The essential idea is that a population is a *complete collection* of objects, or outcomes. A **sample** is a subset of the population. We usually cannot study the population in its entirety, so must make inferences about it based on a sample.

A population can be described by one or more parameters. A **parameter** is a characteristic of a population, for example, the allele frequency or heterozygosity at a particular locus, or the average body weight. We cannot know the true values of these parameters unless we examine every individual in the population, which is usually impossible. We must estimate them from a sample. For example, we estimate the allele frequency in a population by examining a sample of individuals and calculating the allele frequency in that sample.

Assume you want to estimate the allele frequency at a certain locus in a population of birds. You collect a sample and calculate an allele frequency for that sample. Now imagine that another person collects a different sample from the same population. His or her sample allele frequency will probably be different from yours. Thus, the sample allele frequency varies from one sample to the next, due to random factors. This is an example of what we call a random variable (see Sections A.3 and A.4). Two important characteristics of a random variable are its mean, or average value, and its variance. These are unknown and must be estimated from the sample. We hope that different estimates are tightly clustered around the true parameter value.

In what follows, we will first define the ideas of probability, mean, variance, covariance, and so forth in terms of population parameters. At the end we will discuss how to estimate them from a sample. It is not always straightforward to obtain a good estimate of a parameter. Much of statistics is concerned with techniques of finding reliable estimates of population parameters.

A comment on notation: In statistics, parameters are often designated by Greek letters and their estimates by corresponding Latin letters; for example, σ^2 is the population variance and s^2 is the sample variance that estimates σ^2. Alternatively, we sometimes use the same symbol for both the parameter and the estimate, and put a circumflex ("hat") over the symbol to indicate the estimate. For example, if p is the allele frequency in a population, then $\hat{p}$ is its estimate. This convention is common in population genetics, and is used throughout this book.

A.2 Probability

There is usually some uncertainty, or randomness, associated with an experiment. However, if the experiment is repeated many times, there is often some predictability about the overall results. For example, if you flip a coin once there is uncertainty whether the outcome will be heads or tails. However, if you flip the coin 100 times, you can be fairly certain that the number of heads will be about 50.

In probability theory, an **event** is defined as the outcome of an experiment, or the result of an observation. If we repeat the experiment many times, the **frequency** of an event is the number of times it occurs. The **relative frequency** is the number of times it occurs divided by the number of times the experiment was performed. For example, if you flip a coin 100 times and get 48 heads, the frequency of heads is 48, and the relative frequency is 0.48. Unfortunately, frequency is often used when relative frequency is meant; for example, when we say allele frequency we really mean relative frequency. The meaning is usually clear from the context.

If an experiment is repeated many times, the relative frequencies of the possible outcomes often seem to approach some limit. The **probability** of an event is the relative frequency of that event when the experiment is repeated a very large number of times. Let N be the number of replications of the experiment, and $n(E)$ the number of times event E occurs. Then, if N is very large, we can define the probability of event E as

$$\Pr(E) = \frac{n(E)}{N} \tag{A.1}$$

In other words, the probability of an event is its relative frequency in the long run. An alternative interpretation is that $n(E)$ is the number of ways that event E can occur, and N is the number of possible outcomes of the experiment. For example, if you roll two dice of different colors, there are 36 possible outcomes (counting, for example, $4 + 3$ as different from $3 + 4$). There are six ways to get a sum of seven. Therefore, assuming both dice are fair, the probability of rolling a seven is 6/36.

From the definition of probability, three facts follow immediately:

1. The probability of an event is between zero and one. A probability of zero means the event cannot occur, and a probability of one means the event is certain.

2. The sum of the probabilities of all possible outcomes of an experiment must equal one. If A, B, or C must happen, then

$$Pr(A) + Pr(B) + Pr(C) = 1$$

3. The probability that something will not occur is one minus the probability that it will occur.

$$Pr(\text{not } E) = 1 - Pr(E)$$

This is useful because sometimes it is easier to calculate the probability that something will not occur than the probability that it will occur.

Probability is a theoretical concept. We never know the exact probability of an event, because we can never repeat an experiment an infinite number of times. Even in the dice example above, the probability of rolling a seven was an estimate based on the assumption that both dice are fair; that is, we assumed that the probabilities of the 36 different outcomes were all the same.

We can estimate probabilities in two ways: First, we can repeat the experiment as often as practical and estimate the probability from the observed relative frequency. Second, we can estimate the probability from some theoretical model. For example, under the rules of Mendelian segregation, the probability that an Aa heterozygote will pass the A allele to its offspring is 0.5.

In the examples that follow, we will assume that coins and dice are fair. This is consistent with the results of many experiments in which the observed relative frequencies of heads and tails are very near 0.5 each, and the observed relative frequencies of each of the six numbers on a die are very near 1/6 each.

Law of Addition

The probability that *either* of two events will occur is the *sum* of their individual probabilities minus the probability that both will occur simultaneously:

$$Pr(A \text{ or } B) = Pr(A) + Pr(B) - Pr(A \text{ and } B) \tag{A.2}$$

The last term on the right is the overlap between A and B. It is counted in both $Pr(A)$ and in $Pr(B)$, and must be subtracted out to avoid counting it twice; see Figure A.1. For example, flip a penny and a nickel simultaneously. The probability that at least one will turn up heads is

$$Pr(\text{at least one heads}) = Pr(P = H) + Pr(N = H) - Pr(P = H \text{ and } N = H)$$
$$= 0.5 + 0.5 - 0.25$$
$$= 0.75$$

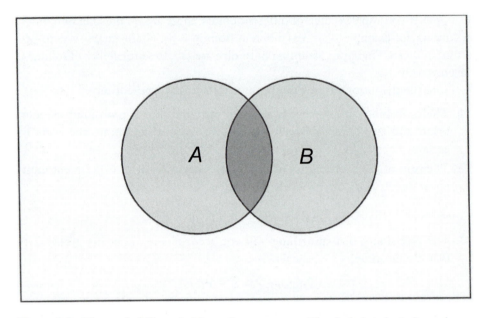

Figure A.1 **The probability of either of two events.** The shaded circles indicate the probabilities of events, A and B. The darker shaded area is part of both A and B and is counted twice in $\Pr(A) + \Pr(B)$. In order to get the probability of A or B, the overlap must be subtracted from the sum, that is, $\Pr(A \text{ or } B) = \Pr(A) + \Pr(B) - \Pr(A \text{ and } B)$.

To confirm this result, note that the four possibilities are *HH, HT, TH,* and *TT* (where the first letter represents the outcome for the penny and the second letter the outcome for the nickel). Three of the four possibilities have at least one head.

If A and B are mutually exclusive, that is, both cannot happen simultaneously, then $\Pr(A \text{ and } B) = 0$, and

$$\Pr(A \text{ or } B) = \Pr(A) + \Pr(B)$$

For example, roll a single die. The probability that it will turn up 3 or 4 is

$$\Pr(D = 3 \text{ or } 4) = \Pr(D = 3) + \Pr(D = 4)$$

$$= \frac{1}{6} + \frac{1}{6} = \frac{1}{3}$$

Conditional Probability

Let A be a particular outcome of some experiment. Then the probability that A will occur is $\Pr(A)$. Let B be some event that affects whether or not A will occur. Then we define the conditional probability of A, assuming that B has already occurred as $\Pr(A|B)$, read as the probability of A given B. It can be shown from probability theory that

$$\Pr(A|B) = \frac{\Pr(A \text{ and } B)}{\Pr(B)} \tag{A.3}$$

where $\Pr(A \text{ and } B)$ is the probability that both A and B occur.

For example, assume you have two different dice, designated D_1 and D_2 (assume they are different colors so you can tell them apart). Let S be the sum of the dots on the two dice. What is the probability that the sum will be 4? There are 36 possible outcomes, of which only three sum to four $(1 + 3, 2 + 2,$ or $3 + 1)$. So, the probability of rolling a 4 is

$$\Pr(S = 4) = 3/36$$

Now, what is the probability that the sum is 4, given that the first die has already been rolled and shows a 1? In this case, the only possibility is that the second die shows a 3, and the probability of that is $1/6$. We can write

$$\Pr(S = 4 | D_1 = 1) = 1/6$$

which is not the same as $\Pr(S = 4)$.

Total Probability

If two events, B and C, both affect whether A will occur, and either B or C must occur, then the law of total probability says

$$\Pr(A) = \Pr(A | B) \times \Pr(B) + \Pr(A | C) \times \Pr(C) \qquad \textbf{(A.4)}$$

This is the weighted average of the two conditional probabilities.

In the above example,

$$
\begin{aligned}
\Pr(S &= 4) \\
&= \Pr(S = 4 | D_1 = 1) \times \Pr(D_1 = 1) + \Pr(S = 4 | D_1 \neq 1) \times \Pr(D_1 \neq 1) \\
&= (1/6)(1/6) + (2/30)(5/6) \\
&= 3/36
\end{aligned}
$$

as shown above. The only part of this that is not obvious is $\Pr(S = 4 | D_1 \neq 1)$. You should convince yourself that this is indeed $2/30$ (see Figure A.2).

Independence

Two events are **independent** if the occurrence or nonoccurrence of one has no effect on the occurrence or nonoccurrence of the other. For example, if you roll a pair of dice, the number that shows on the first is independent of the number that shows on the second. In the language of conditional probability, A and B are independent if

$$\Pr(A | B) = \Pr(A) \qquad \textbf{(A.5)}$$

In other words, knowing that B has occurred gives us no new information about the probability that A will occur.

Consider a double homozygote of the form $AaBb$. Under independent assortment, the probabilities of the four gamete types are $\Pr(AB) = \Pr(Ab) = \Pr(aB) = \Pr(ab) = 1/4$. Then, from equation (A.3),

$$\Pr(A | B) = \frac{\Pr(A \text{ and } B)}{\Pr(B)} = \frac{\Pr(AB)}{\Pr(B)} = \frac{1/4}{1/2} = \frac{1}{2} = \Pr(A)$$

Knowing that a gamete receives the B allele tells us nothing about whether it receives the A allele. On the other hand, if the genes are linked and in coupling (AB/ab), with recombination rate r, the probabilities of the four gamete types are $\Pr(AB) = \Pr(ab) = (1 - r)/2$ and $\Pr(Ab) = \Pr(aB) = r/2$. Then,

Figure A.2 The possible outcomes of rolling a pair of dice. If the two dice are distinguishable (e.g., different colors), there are 36 possible joint outcomes. The numbers in the squares are the sum for each outcome. There are six ways to get a sum of 7; therefore, the probability of getting a 7 is $6/36 \cong 0.167$, assuming that any number has an equal probability of showing on either die (i.e., assuming both dice are fair). The probabilities of getting the other sums can be similarly determined.

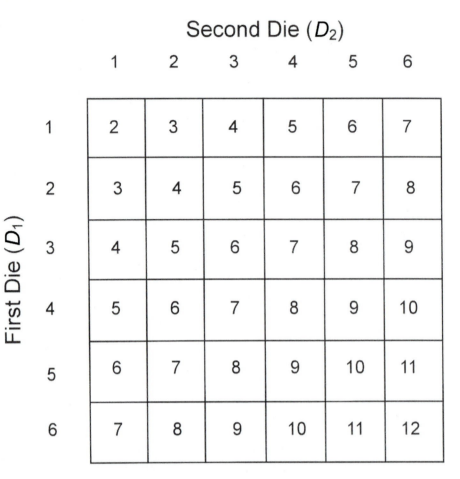

$$\Pr(A|B) = \frac{\Pr(A \text{ and } B)}{\Pr(B)} = \frac{\Pr(AB)}{\Pr(B)} = \frac{(1-r)/2}{1/2} = 1 - r \neq \Pr(A)$$

For example, if $r = 0.1$, then $\Pr(A|B) = 0.9$. Knowing that a gamete has received the B allele provides information about the probability that it also received the A allele.

Law of Multiplication

Rewrite equation (A.3) as

$$\Pr(A \text{ and } B) = \Pr(A|B) \times \Pr(B) \tag{A.6}$$

If A and B are independent, then $\Pr(A|B) = \Pr(A)$ and equation (A.6) becomes

$$\Pr(A \text{ and } B) = \Pr(A) \times \Pr(B) \tag{A.7}$$

This is known as the law of multiplication. The probability that *both* of two independent events will occur is the *product* of their individual probabilities.

For example, roll two dice simultaneously. The probability that the first will show a 2 and the second will show a 5 is

$$\Pr(D_1 = 2 \text{ and } D_2 = 5) = \frac{1}{6} \times \frac{1}{6} = \frac{1}{36}$$

Bayes' Theorem

In Section 4.5 we saw how it is possible to estimate the probability that a person carries a disease causing allele, given that he or she has a particular marker allele at a different locus. Bayes' theorem allows us to estimate this probability if we know something about the association between the marker allele and the disease allele.

Let A and B be two events, for example, genotypes at different loci. Then, from the definition of conditional probability,

$$\Pr(A|B) = \frac{\Pr(A \text{ and } B)}{\Pr(B)} \qquad \textbf{(A.8)}$$

Similarly,

$$\Pr(B|A) = \frac{\Pr(A \text{ and } B)}{\Pr(A)} \qquad \textbf{(A.9)}$$

Rewrite equation (A.8) as

$$\Pr(A \text{ and } B) = \Pr(A|B) \times \Pr(B)$$

Now substitute this into the numerator of equation (A.9) to get

$$\Pr(B|A) = \frac{\Pr(A|B) \times \Pr(B)}{\Pr(A)} \qquad \textbf{(A.10)}$$

Equation (A.10) is known as **Bayes' theorem**.

We illustrate with a simple example. Let D_1 and D_2 be the number of dots on two dice, and S be their sum. What is the probability that $D_1 = 5$, given that $S = 8$? Using Bayes' theorem,

$$\Pr(D_1 = 5|S = 8) = \frac{\Pr(S = 8|D_1 = 5) \times \Pr(D_1 = 5)}{\Pr(S = 8)}$$

We know that $\Pr(D_1 = 5) = 1/6$. There are five ways to get a sum of 8 on the two dice, so $\Pr(S = 8) = 5/36$. Finally, if $D_1 = 5$, then there is only one way the sum can be 8, that is, $D_2 = 3$, with probability 1/6. Therefore, $\Pr(S = 8|D_1 = 5) = 1/6$. Substituting these into the above equation, we get

$$\Pr(D_1 = 5|S = 8) = \frac{(1/6) \times (1/6)}{5/36} = \frac{1}{5}$$

You can verify this by examining Figure A.2.

Bayes' theorem is useful in gene mapping and disease diagnosis (Section 4.5) and in phylogenetic reconstruction (Section 11.2).

A.3 Discrete Random Variables

Some experiments have a limited number of possible outcomes. For example, if you flip a coin it can come up either heads or tails. If you roll a single die, the only possible outcomes are 1, 2, 3, 4, 5, or 6. A double heterozygote *AaBb* can produce only four gamete types, *AB, Ab, aB,* or *ab*. In an experiment like this, each possible outcome has an associated probability, and the probabilities sum to one.

A **discrete random variable** is a function that takes on particular values, each with a specified probability. For example, if X is a random variable indicating the number of dots that show when you roll a single die, the possible values of X are 1, 2, 3, 4, 5, or 6. The probability of each is 1/6. In this example, the probabilities are the same for each possible value of the random variable, but this need not be the case.

We usually indicate random variables by uppercase letters and specific values of that random variable by lowercase letters. In the above example, X is a random variable, and x is a particular value, sometimes called its realized value. The value of x can be 1, 2, 3, 4, 5, or 6. The probability that the random variable X takes on the specified value x is written as

$$\Pr(X = x)$$

The **probability distribution** of a discrete random variable describes the possible values of the random variable, along with the associated probability for each value. For example, if X is the number of dots that shows on a single die, the probability distribution of X can be described as

$$\Pr(X = 1) = 1/6$$
$$\Pr(X = 2) = 1/6$$
$$\Pr(X = 3) = 1/6$$
$$\Pr(X = 4) = 1/6$$
$$\Pr(X = 5) = 1/6$$
$$\Pr(X = 6) = 1/6$$

The probabilities associated with the different values of x can often be written as a function of x. The **probability distribution function** of X, often symbolized by $p(x)$ is defined as

$$p(x) = \Pr(X = x)$$

For example, we can write the probability distribution function of X above as

$$p(x) = 1/6 \quad \text{for } x = 1, 2, 3, 4, 5, 6$$
$$= 0 \quad \text{otherwise}$$

Expected Value of a Discrete Random Variable

The expected value, or expectation, of a discrete random variable is symbolized by $E(X)$, and is defined as

$$E(X) = \sum_{\text{all } x} x p(x) \tag{A.11}$$

We sometimes write EX instead of $E(X)$ if no ambiguity is created. The expected value is the weighted average of all possible values of X, weighted by their probabilities. It is also called the mean, and is often symbolized by the Greek letter μ.

In the die example,

$$E(X) = 1\left(\frac{1}{6}\right) + 2\left(\frac{1}{6}\right) + 3\left(\frac{1}{6}\right) + 4\left(\frac{1}{6}\right) + 5\left(\frac{1}{6}\right) + 6\left(\frac{1}{6}\right)$$
$$= 3.5$$

This illustrates the idea that the expected value is not necessarily "expected" in the nontechnical sense of the word. We would never expect 3.5 dots to show on a die.

Variance of a Discrete Random Variable

The variance of a random variable is defined as the expected value of the squared deviation from the mean. If X is a random variable and EX is its expected value, then

$$V(X) = E(X - EX)^2 \qquad \text{(A.12)}$$

Expanding the square and noting that the expected value of a sum is equal to the sum of the expected values (Section A.5), we obtain a useful equivalent formula

$$V(X) = E(X^2) - (EX)^2 \qquad \text{(A.13)}$$

There are several common ways to symbolize the variance, for example, $V(X)$ as above, var(X), or σ^2. Sometimes a subscript is necessary to distinguish among different random variables, for example, σ^2_X.

The variance is a measure of how much the individual values are spread around the mean. A low variance means that nearly all individual values are close to the mean; a high variance means that some individual values are far from the mean.

Using the definition of expected value and representing it by μ, we can write the variance of a discrete random variable as

$$V(X) = \sum_{\text{all } x} (x - \mu)^2 p(x) \qquad \text{(A.14)}$$

In the die example, we can calculate the variance as

$$V(X) = (1 - 3.5)^2 \left(\frac{1}{6}\right) + (2 - 3.5)^2 \left(\frac{1}{6}\right) + (3 - 3.5)^2 \left(\frac{1}{6}\right)$$

$$+ (4 - 3.5)^2 \left(\frac{1}{6}\right) + (5 - 3.5)^2 \left(\frac{1}{6}\right) + (6 - 3.5)^2 \left(\frac{1}{6}\right)$$

$$= 2.92$$

The **standard deviation** of a random variable is the positive square root of the variance, and is usually indicated by σ, sometimes with a subscript.

We now describe several common discrete probability distributions, and give their expected value and variance.

The Binomial Distribution

Imagine an experiment with only two possible outcomes. Call them success and failure, and let p be the probability of success and $1 - p$ the probability of failure. Such an experiment is called a Bernoulli experiment. Repeat the experiment n times; each repetition is independent of the others. Let X be a random variable indicating the number of successes in n independent replications of the experiment. Then X can take on integer values 1, 2, 3, $\ldots$, n. The probability distribution function of X is

$$p(x) = \Pr(X = x) = \binom{n}{x} p^x (1 - p)^{n-x} \qquad \text{(A.15)}$$

where

$$\binom{n}{x} = \frac{n!}{x!(n - x)!}$$

and $n!$ is the factorial of n; that is, $n! = 1 \times 2 \times \cdots \times (n - 1) \times n$.

This is the binomial distribution. The expected value and variance of X are

$$E(X) = np \tag{A.16}$$

$$V(X) = np(1 - p) \tag{A.17}$$

The binomial distribution appears frequently in many areas of population genetics, for example, gene mapping (Section 4.4) and the variance of allele frequency change under genetic drift (Box 7.1).

The Poisson Distribution

The Poisson distribution describes the random occurrence of events that are individually rare, but for which there are many opportunities for the event to occur. For example, the number of new mutations in a gamete is often assumed to follow a Poisson distribution (Box 6.1).

Again, imagine an experiment with two possible outcomes, success or failure. Let p be the probability of success in any single experiment, and n be the number of times the experiment is repeated, with each replication independent of the others. Then, np is the expected number of successes. Now assume that p is small and n is large, but that the product $np = \lambda$ is constant as n increases (and therefore p decreases). Let X be the number of successes in n independent replications. Then, in the limit as n gets larger and p gets smaller, the probability distribution function of X is

$$p(x) = \frac{e^{-\lambda}\lambda^x}{x!} \tag{A.18}$$

where $\lambda = np$. The mean and variance of a Poisson distributed random variable are

$$E(X) = \lambda \tag{A.19}$$

$$V(X) = \lambda \tag{A.20}$$

If n is large and p is small, the binomial distribution can be approximated by a Poisson distribution. In Section 6.7, we used the Poisson distribution to estimate the distribution of the number of new mutations in a gamete, and the probability that a new mutation will be lost.

Geometric Distribution

Once again, imagine a series of n identical and independent experiments with p being the probability of success for each experiment. How many experiments must be performed until the first success? Let X be the number of the trial in which the first success occurs. Then X has a geometric distribution, and the probability distribution function of X is

$$p(x) = p(1 - p)^{x-1} \tag{A.21}$$

The mean and variance of X are

$$E(X) = \frac{1}{p} \tag{A.22}$$

$$V(X) = \frac{1 - p}{p^2} \tag{A.23}$$

The geometric distribution was used in the discussion of coalescent theory in Section 11.3.

TABLE A.1   **Common discrete probability distributions**

Distribution	Probability Distribution Function	Possible Values for X	Expected Value	Variance
Bernoulli	$p(x) = p^x(1 - p)^{1-x}$	0,1	p	$p(1 - p)$
Binomial	$p(x) = \binom{n}{x}p^x(1 - p)^{n-x}$	$0, 1, 2, \ldots, n$	np	$np(1 - p)$
Poisson	$p(x) = \dfrac{e^{-\lambda}\lambda^x}{x!}$	$0, 1, 2, \ldots$	$\lambda = np$	$\lambda = np$
Geometric	$p(x) = p(1 - p)^{x-1}$	$1, 2, 3, \ldots$	$\dfrac{1}{p}$	$\dfrac{1 - p}{p^2}$
Multinomial	$p(x_1, x_2, \cdots x_k)$ $= \dfrac{n!}{x_1!x_2!\cdots x_k!}p_1^{x_1}p_2^{x_2}\cdots p_k^{x_k}$	$x_i = 0, 1, 2, 3, \cdots, n$	$E(X_i) = np_i$	$V(X_i) = np_i(1 - p_i)$

Table A.1 lists these and other common probability distributions, along with their means and variances.

A.4 Continuous Random Variables

Not all random variables are limited to a specified set of possible values. For example, height in humans is a continuous variable. Allele frequency can be any value between zero and one. Random variables that have a continuous distribution of possible values are called **continuous random variables**.

Let X be a continuous random variable that can take on any value in the range from a to b. Then there is an infinite number of possible values for X. From the definition of probability,

$$\Pr(X = x) = \frac{\text{number of times that } x \text{ occurs}}{\text{number of possible values for } X}$$

If the denominator is infinitely large, $\Pr(X = x) = 0$ for any particular x. But we know that *some* result must occur, so $\Pr(X = x)$ is a meaningless concept for a continuous random variable.

We cannot think about the probability distribution of a continuous random variable in the same way that we think about the probability distribution of a discrete random variable. Instead, we define a **probability density function** for continuous random variables. The probability density function describes the probability that a continuous random variable will take on a value within a specified range.

Let $f(x)$ be the probability density function of the random variable X. Probability density functions are scaled so that the total area under the curve is one. Then the probability that X will take on a value between a and b is

$$\Pr(a < x < b) = \int_a^b f(x)dx \qquad \textbf{(A.24)}$$

Make sure you understand that $f(x)$ is *not* the probability that $X = x$. However, $f(x)$ is, in a general sense, proportional to the probability that X will take on a value *near* x.

We illustrate with an example. Let

$$f(x) = \lambda e^{-\lambda x} \qquad x \geq 0 \tag{A.25}$$

This is the exponential distribution, and one example is graphed in Figure A.3. First, verify that the total area under the curve is one:

$$\Pr(0 < x < \infty) = \int_0^\infty \lambda e^{-\lambda x} dx = -e^{-\lambda x} \Big|_0^\infty = -\left[e^{-\infty} - e^0\right] = 1$$

So $f(x)$ is indeed a probability density function. The probability that X will be between 0 and 1 is

$$\Pr(0 < x < 1) = \int_0^1 \lambda e^{-\lambda x} dx = -e^{-\lambda x} \Big|_0^1 = -\left[e^{-\lambda} - e^0\right] = 1 - e^{-\lambda}$$

If $\lambda = 1$, as in Figure A.3, this becomes approximately 0.63.

Expected Value and Variance of a Continuous Random Variable

The expected value of a continuous random variable is defined as

$$E(X) = \int_{-\infty}^\infty x f(x) dx \tag{A.26}$$

Similarly, the variance of a continuous random variable is

$$V(X) = E(X - EX)^2 = \int_{-\infty}^\infty (x - \mu)^2 f(x) \tag{A.27}$$

These look like equations (A.11) and (A.14), with integration replacing summation, but they are not quite equivalent because $f(x)$ is not a probability.

Some continuous random variables are defined only for a particular range of values; for example, the exponential distribution described by equation (A.25) is defined only for positive values of X. In this case, the limits of integration are replaced by the range of possible values for X.

For example, the expected value for the exponential random variable described by equation (A.25) is

$$E(X) = \int_0^\infty x \lambda e^{-\lambda x} dx$$

Integrating by parts gives

$$E(X) = -x e^{-\lambda x} \Big|_0^\infty + \int_0^\infty e^{-\lambda x} dx$$

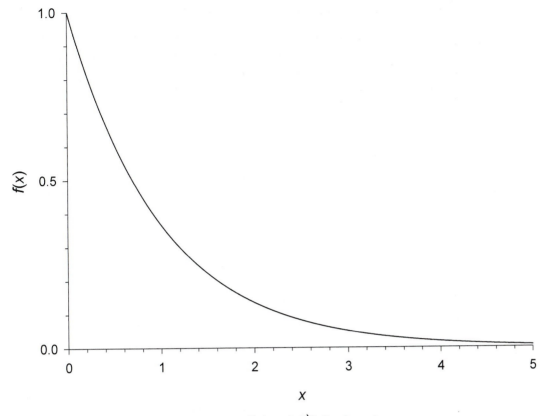

Figure A.3   **The exponential distribution, $f(x) = \lambda e^{-\lambda x}$, for $\lambda = 1$.**

which simplifies to

$$E(X) = \frac{1}{\lambda} \qquad \textbf{(A.28)}$$

Similarly, it can be shown that the variance is

$$V(X) = \frac{1}{\lambda^2} \qquad \textbf{(A.29)}$$

We now describe several common continuous random variables.

The Exponential Distribution

The exponential distribution describes the waiting time until an event occurs. It is the continuous analog of the geometric distribution. A common example is radioactive decay. If X is the time until a particular radioactive nucleus decays, then X has an exponential distribution. Similarly, the failure time of many electronic parts has an approximate exponential distribution.

We introduced the exponential distribution in the example above. The probability density function is

$$f(x) = \lambda e^{-\lambda x} \qquad x \geq 0 \qquad \textbf{(A.30)}$$

The mean and variance of an exponential random variable are given by equations (A.28) and (A.29).

The Normal Distribution

The most important continuous distribution in probability theory is the normal distribution. Many random variables follow an approximately normal distribution, and it is used extensively in quantitative genetics (Chapter 13).

The probability density function of a normal random variable is

$$f(x) = \frac{1}{\sigma\sqrt{2\pi}} \exp\left(\frac{-(x - \mu)^2}{2\sigma^2}\right) \tag{A.31}$$

The mean and variance are

$$E(X) = \mu \tag{A.32}$$

$$V(X) = \sigma^2 \tag{A.33}$$

The normal distribution is very important in statistical analysis and hypothesis testing. Many statistical procedures assume that the variable under study is approximately normally distributed.

The Standard Normal Distribution

Let X be a normally distributed random variable with probability density function given by equation (A.31). Define Z as

$$Z = \frac{X - \mu}{\sigma} \tag{A.34}$$

Then Z has a normal distribution with an expected value of zero and a variance of one. The probability density function of Z is

$$f(z) = \frac{1}{\sqrt{2\pi}} \exp\left(\frac{-z^2}{2}\right) \tag{A.35}$$

This is the standardized normal distribution. It is useful in comparing different random variables, and many statistical tables are based on it.

Table A.2 summarizes several common continuous distributions.

A.5 Functions of Random Variables

It is often necessary to find the expected value or variance of some function of a random variable. Here, we present without proof several useful rules for working with simple functions of random variables. They apply to either discrete or continuous random variables.

Rules for Expected Values

The following are rules for finding the expected value of a function of a random variable. If X and Y are two random variables and a and b are constants, then:

1. The expected value of a constant is that constant:

$$E(a) = a \tag{A.36}$$

2. The expected value of a constant times a random variable is the constant times the expected value of the random variable:

$$E(bX) = bE(X) \tag{A.37}$$

TABLE A.2   Common continuous probability densities

Distribution	Probability Density Function	Possible Values for X	Expected Value	Variance
Uniform	$f(x) = \dfrac{1}{b-a}$	$a < x < b$	$\dfrac{a+b}{2}$	$\dfrac{(b-a)^2}{12}$
Exponential	$f(x) = \lambda e^{-\lambda x}$	$x \geq 0$	$\dfrac{1}{\lambda}$	$\dfrac{1}{\lambda^2}$
Normal	$f(x) = \dfrac{1}{\sigma\sqrt{2\pi}} \exp\left(\dfrac{-(x-\mu)^2}{2\sigma^2}\right)$	$-\infty < x < \infty$	μ	σ^2
Standard normal	$f(z) = \dfrac{1}{\sqrt{2\pi}} \exp\left(\dfrac{-z^2}{2}\right)$	$-\infty < z < \infty$	0	1

3. The expected value of the sum of random variables is the sum of the expected values:

$$E(X + Y) = E(X) + E(Y) \qquad \textbf{(A.38)}$$

4. Combining equations (A.36), (A.37), and (A.38), the expected value of a linear function of a random variable is

$$E(a + bX) = a + bE(X) \qquad \textbf{(A.39)}$$

Note two other facts. In general,

$$E(XY) \neq (EX)(EY) \qquad \textbf{(A.40)}$$

and

$$E(X^2) \neq (EX)^2 \qquad \textbf{(A.41)}$$

Rules for Variances

If X and Y are two random variables and c is a constant, then:

1. The variance of a constant is zero:

$$V(c) = 0 \qquad \textbf{(A.42)}$$

2. The variance of a constant times a random variable is

$$V(cX) = c^2 V(X) \qquad \textbf{(A.43)}$$

3. If X and Y are independent, then

$$V(X + Y) = V(X) + V(Y) \qquad \textbf{(A.44)}$$

4. If X and Y are not independent, then

$$V(X + Y) = V(X) + V(Y) + 2\mathrm{cov}(X, Y) \qquad \textbf{(A.45)}$$

where $\mathrm{cov}(X, Y)$ is the covariance of X and Y, to be explained in the next section.

A.6 Pairs of Random Variables

We often encounter two or more random variables that are related in some way, for example, allele frequencies at two loci, or body weight in parents and offspring. In this section we summarize ways to describe the relationship between two random variables.

Covariance

If X and Y are two random variables, then the covariance between them is defined as

$$\text{cov}(X, Y) = \sigma_{XY} = E[(X - EX)(Y - EY)] \qquad \textbf{(A.46)}$$

Multiplying out the quantities in brackets and simplifying gives an equivalent formula:

$$\text{cov}(X, Y) = E(XY) - (EX)(EY) \qquad \textbf{(A.47)}$$

The covariance is sometimes symbolized by σ_{XY}, as in equation (A.46).

If X and Y increase together, the covariance between them is positive; if they decrease together, the covariance is negative. A covariance of zero does not necessarily mean that there is no relationship between X and Y; it means only that there is no linear relationship. For example, Figure A.4 shows a situation in which the covariance is essentially zero, but there is clearly a strong relationship between X and Y.

Two random variables are independent if

$$\Pr(Y = y | X = x) = \Pr(Y = y)$$

In other words, variation in X has no effect on variation in Y. If X and Y are independent, their covariance is zero. However, the converse is not true, as illustrated in Figure A.4.

We state without proof several useful rules for working with covariances. If X, Y, Z, and W are random variables, and a and b are constants, then

$$\text{cov}(a, X) = 0 \qquad \textbf{(A.48)}$$

$$\text{cov}(a + X, Y) = \text{cov}(X, Y) \qquad \textbf{(A.49)}$$

$$\text{cov}(aX, bY) = ab\,\text{cov}(X, Y) \qquad \textbf{(A.50)}$$

$$\text{cov}(X, X) = V(X) \qquad \textbf{(A.51)}$$

$$\text{cov}(X + Y, W + Z) = \text{cov}(X, W) + \text{cov}(X, Z) + \text{cov}(Y, W) + \text{cov}(Y, Z)$$
$$\textbf{(A.52)}$$

We can now relate the variance of a sum to the sum of the variances. Let $Z = X + Y$. Then

$$V(Z) = E(Z^2) - (EZ)^2$$
$$= E\left[(X + Y)^2\right] - \left[E(X + Y)\right]^2$$

A bit of algebraic manipulation will produce

$$V(X + Y) = E(X^2) - (EX)^2 + E(Y^2) - (EY)^2 + 2[E(XY) - (EX)(EY)]$$

The first two terms on the right are the variance of X, the second two are the variance of Y, and the terms in the square brackets are the covariance. Therefore,

$$V(X + Y) = V(X) + V(Y) + 2\,\text{cov}(X, Y)$$

If X and Y are independent, then $\text{cov}(X, Y) = 0$, and

$$V(X + Y) = V(X) + V(Y)$$

Both of these were given in the rules for variances in Section A.5.

Correlation

The covariance of two random variables can be any value. Sometimes it is useful to have a standardized covariance. Such a standardized covariance is called the

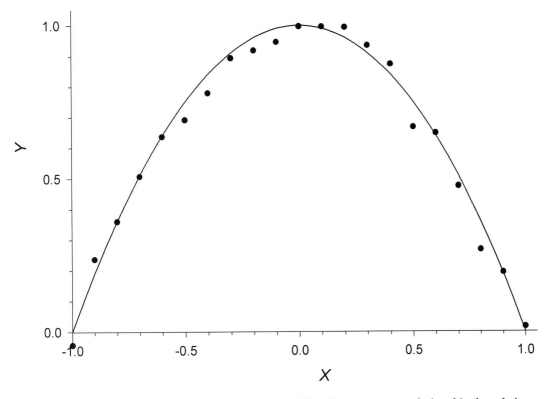

Figure A.4   **An example in which two random variables show a strong relationship, but their covariance is zero.**   The curve is $Y = 1 - X^2$. The actual values of X and Y are very close to the curve, indicating a strong dependency of Y on X, with some random effect. The actual covariance for the data points is -0.00054.

correlation coefficient, and is defined as

$$\rho = \frac{\text{cov}(X, Y)}{\sqrt{V(X)V(Y)}} = \frac{\sigma_{XY}}{\sigma_X \sigma_Y} \qquad \textbf{(A.53)}$$

(The symbol on the left is the Greek letter rho.)

The correlation coefficient can range from -1 to $+1$. Like the covariance, it is a measure of the degree to which the two variables are linearly related. If ρ is near $+1$ (or -1), Y increases (or decreases) as X increases, and the relationship is very nearly linear. If ρ is near zero, there is essentially no linear relationship between X and Y.

Linear Regression

Two variables can have a functional relationship. For example,

$$y = a + bx$$

describes a linear relationship between the variables x and y. If you plot y versus x, every point will be on the line (Figure A.5a). Random variables can also show a linear relationship, but the relationship is never perfect. Figure A.5b shows an example. The random variable Y is linearly related to X, but the points do not fall exactly on the line due to random variation in Y. The value of Y can be partitioned into a part due to the linear relationship between Y and X, and a random part that is unexplained by the linear relationship. We can write

$$Y = \alpha + \beta X + \varepsilon \tag{A.54}$$

where $\alpha + \beta X$ describes the linear relationship, and ε is a random variable with mean zero, describing the deviation from perfect linearity. β is the slope of the line, and is called the regression coefficient of Y on X. β is defined as

$$\beta = \frac{\text{cov}(X, Y)}{V(X)} = \frac{\sigma_{XY}}{\sigma_X^2} \tag{A.55}$$

The other parameter, α, is the Y-intercept, and is defined as

$$\alpha = \mu_Y - \beta\mu_X \tag{A.56}$$

where μ_X and μ_Y are EX and EY.

Linear regression is used frequently in population genetics. For examples, see Figures 6.9, 9.6b, 10.2, and 12.14. It is used extensively in quantitative genetics (Chapter 13).

It is possible that Y can be a linear function of several independent variables. This is known as multiple linear regression, and the general model is

$$Y = \alpha + \beta_1 X_1 + \beta_2 X_2 + \cdots + \beta_k X_k + \varepsilon$$

where the Xs are different random variables and the βs are the regression coefficients for each X. Multiple linear regression was used in Section 13.4, in studying natural selection on correlated quantitative traits.

A.7 Estimates of Population Parameters

We have defined the mean, variance, and so forth as population parameters. As mentioned in Section A.1, we can never know the true values of these parameters, and must estimate them from a sample. There are standard estimators for the parameters we have discussed here. We follow the convention of using Greek letters for the parameters and Latin letters for their estimators. (An exception is the common use of $\bar{x}$ for the estimate of the population mean.)

These estimators are random variables. For example, if σ^2 is the variance, then s^2 is a random variable that estimates the variance. It is desirable that the expected value of the estimator be equal to the true parameter value. If so, the estimator is **unbiased**. Natural estimators, obtained by replacing the population values by their sample values, are sometimes biased, and must be adjusted to create an unbiased estimator. For example, a natural estimator of the population variance is

$$s_n^2 = \frac{1}{n}\sum_{i=1}^{n}(x_i - \bar{x})^2$$

It can be shown that the expected value of s_n^2 is not σ^2. An adjustment has to be made to eliminate bias. An unbiased estimator of the population variance is

$$s^2 = \frac{1}{n-1}\sum_{i=1}^{n}(x_i - \bar{x})^2 \tag{A.57}$$

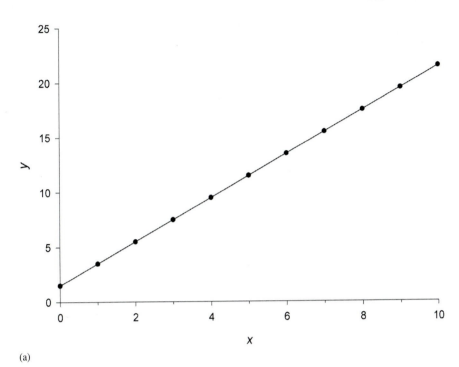

(a)

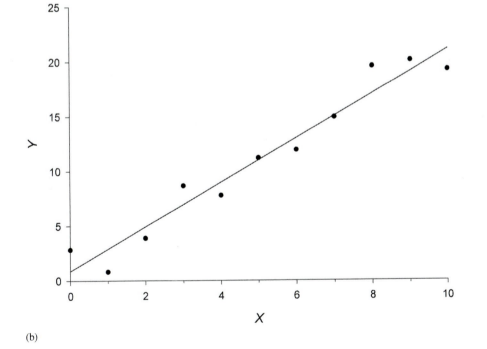

(b)

Figure A.5 **Functional and statistical relationships between variables.** (a) A linear functional relationship in which the (nonrandom) variable y is determined exactly by x. In this example, $y = 1.5 + 2x$. Every (x, y) pair falls exactly on the line. (b) A statistical relationship in which the random variable Y is linearly related to X, but with some random variation. The line is the best-fit regression line, as explained in the text.

Another example is the estimator for expected heterozygosity from equation (3.17)

$$\hat{H}_{\text{exp}} = \frac{2N}{2N-1}\left(1 - \sum \hat{p}_i^2\right)$$

Here we have used the $^\wedge$ over a symbol to indicate an estimate of the population parameter, the convention used throughout most of this book.

We will not discuss in detail how to estimate the parameters we have discussed, as the calculations are almost always handled by computers. Table A.3 summarizes the definitions of these parameters and formulas for their standard estimators. Computer algorithms use algebraically equivalent formulas that are computationally efficient and minimize round-off error. For example, the formula for sample variance is given by equation (A.57) above. An equivalent formula is

$$s^2 = \frac{1}{n-1}\left[\sum x^2 - \frac{\left(\sum x\right)^2}{n}\right] \tag{A.58}$$

Similar computational formulas exist for other estimators.

TABLE A.3   Population parameters and the sample statistics used to estimate them.

The formulas for the sample statistics are written to emphasize the relationship to the parameters. Computer algorithms use algebraically equivalent formulas that are more computationally efficient and minimize round-off error.

Parameter	Definition	Sample Statistic
Expected value (Discrete random variable)	$\mu = \sum\limits_{\text{all } x} xp(x)$	$\bar{x} = \frac{1}{n}\sum\limits_{i=1}^{n} x_i$
Expected value (Continuous random variable)	$\mu = \int\limits_{-\infty}^{\infty} xf(x)dx$	$\bar{x} = \frac{1}{n}\sum\limits_{i=1}^{n} x_i$
Variance (Discrete random variable)	$\sigma_X^2 = E(X - EX)^2 = \sum\limits_{\text{all } x}(x - \mu)^2 p(x)$	$s_x^2 = \frac{1}{n-1}\sum\limits_{i=1}^{n}(x_i - \bar{x})^2$
Variance (Continuous random variable)	$\sigma_X^2 = E(X - EX)^2 = \int\limits_{-\infty}^{\infty}(x - \mu)^2 f(x)$	$s_x^2 = \frac{1}{n-1}\sum\limits_{i=1}^{n}(x_i - \bar{x})^2$
Covariance	$\sigma_{XY} = E[(X - EX)(Y - EY)]$	$s_{xy} = \frac{1}{n}\sum\limits_{i=1}^{n}(x_i - \bar{x})(y_i - \bar{y})$
Correlation	$\rho = \frac{\sigma_{XY}}{\sigma_X \sigma_Y}$	$r = \frac{s_{xy}}{s_x s_y}$
Linear regression coefficient	$\beta = \frac{\sigma_{XY}}{\sigma_X^2}$	$b = \frac{s_{xy}}{s_x^2}$
Linear regression intercept	$\alpha = \mu_Y - \beta\mu_X$	$a = \bar{y} - b\bar{x}$

Computer Software for Population Genetics

I think there is a world market for about five computers.

—*Thomas J. Watson (1947)*

Modern population genetics is heavily dependent on computers. This appendix lists some of the common software packages used by population geneticists. It is intended as an introductory guide, not as a complete list of useful programs. Many specialized programs exist for specific tasks; they are not listed here.

Most of these programs are for personal computers and many are available for both Macintosh and Windows. A few run under Unix. Both commercial and free programs are listed; many of the population genetics data analysis programs are free.

The Web site for information about each program is listed. These addresses were accurate as of May 2003; however, Web sites change frequently. If an address has changed, a search with one of the World Wide Web search engines will usually find the new site.

Spreadsheets

Excel
http://www.microsoft.com
A good spreadsheet is essential for simulating population genetics models, and for basic data recording and analysis. Most of the calculations and simulations in *Introduction to Population Genetics* were done with Microsoft Excel. This spreadsheet has extensive mathematical and statistical capabilities, and one can also write complex macros for specialized tasks. The add-on Analysis ToolPak adds significantly to Excel's capabilities; however, Excel's graphing capabilities are somewhat limited. They are acceptable for student projects, but not for publication-quality graphs (see below).

Other spreadsheet programs are available, but none seems to be as popular as Excel.

Equation Editor

MathType
http://www.mathtype.com
If you write anything that contains equations, you will need an equation editor. The equations in this book were written with MathType, an easy-to-use and trouble-free equation editor that integrates seamlessly into Word and other word processors.

Graphing

The graphing capabilities of spreadsheets are limited. You will need a specialized graphing program to produce publication-quality graphs.

SigmaPlot
http://www.spss.com
SigmaPlot is the standard for creating scientific graphs. It integrates with Excel and allows manipulation of most parts of a graph.

DeltaGraph
http://www.redrocksw.com/deltagraph/
A cheaper alternative to SigmaPlot.

Most statistics and mathematics programs also have graphing abilities.

Population Genetics: Simulations

Populus
http://www.cbs.umn.edu/populus/
Populus is a program that simulates a variety of population genetics and ecological models. Parameter values of the models are entered in an input window, and results are shown in one or more graphs. This is one of the best ways to learn about the effects of different parameter values in a model. Free and highly recommended!

Excel
http://www.microsoft.com
One of the best ways to learn population genetics is to construct and simulate different models with a spreadsheet. This complements Populus by requiring you to understand the calculations and assumptions of the model. It allows you to change initial conditions, parameter values, and assumptions, and to see the results. Most of the models in this book can be explored with any spreadsheet; Excel is available on most personal computers. For a good introduction to spreadsheet modeling in ecology, evolution, and (basic) population genetics, see Donovan and Welden (2002).

Coalescent Simulation
http://www.birc.dk/Software/index.html
Web-based simulations of coalescent models.

Population Genetics: Data Analysis

These programs perform a variety of tests on population genetics data. They test for Hardy-Weinberg equilibrium and gametic disequilibrium, estimate F-coefficients and their analogs, calculate various kinds of genetic distances, test for evidence of natural selection, and so forth.

ARLEQUIN
http://lgb.unige.ch/arlequin/
Arlequin is a versatile program that can analyze many kinds of data, both conventional and molecular. It does analysis of molecular variance (AMOVA; see Section 9.5). It is free and easy to use, and comes with a useful manual.

Genetic Data Analysis
http://lewis.eeb.uconn.edu/lewishome/software.html
This is a collection of programs designed to accompany the book of the same title (Weir 1996). It tests for Hardy-Weinberg equilibrium and gametic disequilibrium, calculates *F*-coefficients, as well as containing other capabilities.

GENEPOP
http://wbiomed.curtin.edu.au/genepop/index.html
Another general purpose data analysis program.

MEGA (Molecular Evolutionary Genetic Analysis)
http://evolgen.biol.metro-u.ac.jp/MEGA/
Analyzes nucleotide or amino acid sequences, estimates evolutionary distances, and performs related functions.

FSTAT
http://www.unil.ch/izea/softwares/fstat.html
Estimates *F*-coefficients by several methods.

RSTCALC
http://helios.bto.ed.ac.uk/evolgen/rst/rst.html
Calculates R_{ST} and related statistics based on microsatellite data.

Migrate
http://evolution.genetics.washington.edu/lamarc/migrate.html
Estimates migration (gene flow) rates and effective population sizes based on DNA sequence, microsatellite, or allozyme data. Part of a larger package called Lamarc.

Microsat
http://hpgl.stanford.edu/projects/microsat/
Calculates distance measures from microsatellite data. No longer supported.

DnaSP (DNA Sequence Polymorphism)
http://www.ub.es/dnasp/
Calculates various estimates of polymorphism from DNA data, tests for gametic disequilibrium, estimates recombination and gene flow rates, and tests for natural selection.

Assignment Calculator (Doh)
http://www2.biology.ualberta.ca/jbrzusto/Doh.php#AboutTest
Analyzes multiple-locus genotypes and determines which of several populations an individual is most likely to have come from. Web-based.

Gene Mapping

These programs perform linkage analysis, gene mapping, and related functions.

FASTLINK

http://www.ncbi.nlm.nih.gov/CBBresearch/Schaffer/fastlink.html

GeneHunter

http://www-genome.wi.mit.edu/resources.html

MapMaker

http://www-genome.wi.mit.edu/resources.html

MapQTL

http://www.plant.dlo.nl/default.asp?section=products&page=/products/ mapping/mapqtl/mqintro.htm
Maps quantitative trait loci based on data from molecular markers.

QTL Cartographer

http://statgen.ncsu.edu/qtlcart/cartographer.html
Maps quantitative trait loci based on data from molecular markers.

A comprehensive list of linkage analysis software is available at
http://linkage.rockefeller.edu/soft/list1.html

Phylogenetic Analysis

These programs construct and analyze phylogenetic trees based on molecular or morphological data.

ClustalX

http://inn-prot.weizmann.ac.il/software/ClustalX.html
This is the standard program for multiple sequence alignment.

PAUP* (Phylogenetic Analysis Using Parsimony, and other methods)

http://paup.csit.fsu.edu/
PAUP* is one of the two major programs for inferring phylogenies. It originally used only parsimony techniques, but the current (4.0 beta) version now includes maximum likelihood and distance methods.

PHYLIP (Phylogeny Inference Package)

http://evolution.genetics.washington.edu/phylip.html
PHYLIP is the other major program for inferring phylogenies. This is actually a comprehensive set of programs to perform all aspects of phylogenetic analysis.

MacClade

http://macclade.org/macclade.html

WinClada/Nona

http://www.cladistics.com/index.html
These two programs work together to construct and analyze phylogenetic trees based on parsimony.

PAML (Phylogenetic Analysis by Maximum Likelihood)

http://abacus.gene.ucl.ac.uk/software/paml.html
Constructs and analyzes trees based on DNA or protein sequences, using maximum likelihood techniques.

MrBayes
http://morphbank.ebc.uu.se/mrbayes/
Constructs and analyzes phylogenetic trees using Bayesian techniques.

TreeView
http://taxonomy.zoology.gla.ac.uk/rod/treeview.html
A program for displaying trees constructed by other phylogeny programs.

A comprehensive list of software for phylogenetic analysis is available at
http://evolution.genetics.washington.edu/phylip/software.html

Statistical Analysis

Following are some of the more common commercial statistical packages. All are complex and expensive. Student or Academic editions are usually available.

JMP
http://www.jmp.com

MINITAB
http://www.minitab.com/

SigmaStat
http://www.spss.com

S-PLUS
www.insightful.com

SPSS
http://www.spss.com

STATISTICA
http://www.statsoftinc.com

Mathematical Analysis

The following programs perform a variety of mathematical functions. They can perform complex calculations and solve equations either symbolically or numerically. They can manipulate matrices and perform numerical integration and differentiation. They have their own programming languages, and their graphing capabilities are good. These are powerful (and expensive) programs for sophisticated mathematical analysis.

Mathematica
http://www.wolfram.com/

Mathcad
http://www.mathcad.com/

MATLAB
http://www.mathworks.com/

Answers to Problems

Chapter 1

1.3. $\tilde{p} = \dfrac{v}{u + v}$

1.4. $\tilde{p} \cong 0.091$

Chapter 2

2.1.

Locus	Strawberry Canyon	Flagstaff
pt-8	$p_s = 0.55$	$p_s = 0.25$
	$p_f = 0.45$	$p_f = 0.75$
	$V(s) = 0.0124$	$V(s) = 0.0117$
	$V(f) = 0.0124$	$V(f) = 0.0117$
	$H_{obs} = 0.70$	$H_{obs} = 0.25$
est-5	$p_a = 0.0909$	$p_a = 0.1111$
	$p_b = 0.3636$	$p_b = 0.4444$
	$p_c = 0.0455$	$p_c = 0$
	$p_d = 0.3182$	$p_d = 0.4444$
	$p_e = 0.1818$	$p_e = 0$
	$V(a) = 0.0038$	$V(a) = 0.0055$
	$V(b) = 0.0105$	$V(b) = 0.0137$
	$V(c) = 0.0020$	$V(c) = 0$
	$V(d) = 0.0099$	$V(d) = 0.0137$
	$V(e) = 0.0068$	$V(e) = 0$
	$H_{obs} = 1.00$	$H_{obs} = 0$
combined	$H_{obs} = 0.85$	$H_{obs} = 0.125$

2.4. Fast $\hat{\pi} = 0.0029$

Slow $\hat{\pi} = 0.0055$

Chapter 3

3.1a. Three alleles

  Homozygous genotypes $= 3$
  Heterozygous genotypes $= 3$
  Degrees of freedom $= 3$

Four alleles

  Homozygous genotypes $= 4$
  Heterozygous genotypes $= 6$
  Degrees of freedom $= 6$

3.2.

Locus	Caucasian	African	Asian
ABO	0.48	0.45	0.52
Lutheran	0.068	0.053	0
Duffy	0.52	0.11	0.18

3.3.

Genotype	Exp. Freq.
A_1A_1	0.154
A_2A_2	0.093
A_3A_3	0.007
A_4A_4	0.048
A_1A_2	0.239
A_1A_3	0.065
A_1A_4	0.173
A_2A_3	0.050
A_2A_4	0.134
A_3A_4	0.036

$\chi^2 = 19.03$
$df = 6$
$H_{obs} = 0.715$
$H_{exp} = 0.698$
$F = -0.025$

3.4. For Strawberry Canyon:

Locus pt-8		Locus est-5	
Genotype	**Exp. Freq.**	**Genotype**	**Exp. Freq.**
ss	0.303	aa	0.008
sf	0.495	bb	0.132
ff	0.203	cc	0.002
		dd	0.101
		ee	0.033
		ab	0.066
		ac	0.008
		ad	0.058
		ae	0.033
		bc	0.033
		bd	0.231
		be	0.132
		cd	0.029
		ce	0.017
		de	0.116

$\chi^2 = 1.72$ $\chi^2 = 8.35$

$df = 1$ $df = 10$

$H_{obs} = 0.70$ $H_{obs} = 1.0$

$H_{exp} = 0.52$ $H_{exp} = 0.76$

$F = -0.34$ $F = -0.32$

Note: The sample sizes are so small that these estimates are essentially meaningless statistically.

3.5.

Genotype	Obs. Freq.	Exp. Freq.	p
Zygotes:			
AA	0.25	0.25	0.50
Aa	0.50	0.50	
aa	0.25	0.25	
Adults:			
AA	0.28	0.28	0.53
Aa	0.50	0.50	
aa	0.22	0.22	

3.6. $\hat{q} \cong 0.025$

  $H_{exp} \cong 0.048$

3.9.

Population	H_{obs}	H_{exp}	F
Miami, FL	0.140	0.206	0.321
Miami, FL	0.195	0.247	0.210
Portland, ME	0.567	0.490	−0.156
Raleigh, NC	0.394	0.445	0.113

Chapter 4

4.1. Above the diagonal = D; below the diagonal = χ^2.

	BaimHI	**HindIII (1)**	**HindIII (2)**	**XhoI**
BamHI		−0.076	0.062	−0.125
HindIII (1)	4.16		−0.021	0.042
HindIII (2)	1.99	0.49		−0.055
XhoI	5.06	1.24	1.57	

4.6. $p_M = 0.774$

  $p_N = 0.226$

  $p_S = 0.464$

  $p_s = 0.536$

  χ^2 (MN locus) = 0.14; accept Ho

  χ^2 (Ss locus) = 4.74; reject Ho

  $\hat{D} = 0.234$

  $D' = 0.223$

  χ^2 (disequilibrium) = 13.52; reject Ho

4.8. There is one recombinant offspring (the sixth) in generation III.

The LOD score peaks at about $r \cong 0.12$. LOD score at this point is about 1.1; no conclusion can be made about linkage.

4.9. $\Pr(D|M) \cong 2.2 \times 10^{-3}$

Chapter 5

5.5.

	Generation 1		Generation 50	
	w_1	w_2	w_1	w_2
a.	0.901	0.801	0.976	0.876
b.	0.801	0.901	0.876	0.976
c.	0.998	0.604	0.867	0.867
d.	0.802	0.996	0.867	0.867
e.	0.720	0.740	0.600	0.800
f.	0.740	0.730	1.000	0.600

5.10. The fitnesses must be rescaled so that $w_{11} = 1$.

    a. $s = 0.2$; $h = 0.5$
    b. $s = -0.25$; $h = 0.5$
    c. $s = 0.25$; $h = -1$
    d. $s = 0.25$; $h = -1$
    e. $s = 0.2$; $h = 2$
    f. $s = 0.2$; $h = 2$

5.13. Males: v_{11} (eyeless) $= 0.84$; v_{12} (wild type) $= 1$; v_{22} (shaven) $= 0.77$

Females: $v_{11} = 0.87$; $v_{12} = 1$; $v_2 = 0.90$

5.15. $t = \dfrac{\ln(x_t) - \ln(x_o)}{\ln(w)}$

5.16. $t \cong 1388$ generations

5.17. There is a consistent increase in the frequency of A_1, but in each generation the observed and expected genotype frequencies match.

5.20. $\widetilde{p} = 0.40$ for all fitness sets

5.21. Minimum sample size is 236

Chapter 6

6.1. 6.5×10^9 mutations/nucleotide/day

6.3.

	mut/gene/cell div	mut/nuc/cell div
E. coli	3.95×10^{-8}	3.7×10^{-11}
Drosophila	6.13×10^{-7}	4.6×10^{-11}
Humans	1.56×10^{-6}	2.1×10^{-11}

6.5. Pr(0 mutations) $= 0.996$
Pr(1 mutation) $= 3.98 \times 10^{-3}$
Pr(≥ 2 mutations) $= 2 \times 10^{-5}$

6.8. If $u = 100v$, then $\widetilde{q} \cong 0.99$

6.9. For a lethal mutation $hs = 0.03$
For a mildly deleterious mutation $hs = 0.02$

6.11. a. $\widetilde{q} = 0.01$
    b. $\widetilde{q} = 2 \times 10^{-4}$
    c. $\widetilde{q} = 1 \times 10^{-4}$
    d. $\widetilde{q} = 0.022$
    e. $\widetilde{q} = 0.001$

6.13. Drosophila: $\overline{w}_n \cong 0.73$, $L_n \cong 0.27$
Humans: $\overline{w}_n \cong 0.20$, $L_n \cong 0.80$

6.14. Relative fitness of individuals with Huntington's disease is about 0.997.

6.15. a. $\hat{u} \cong 4 \times 10^{-4}$
    b. $s_1 \cong 0.02$

Chapter 7

7.1. $H_{10} = 0.38$

7.3. Expected times to fixation:

Initial p	$N = 10$	$N = 100$
0.1	38	379
0.5	28	277
0.9	10	102

7.8. Pr(at least one A_1 allele) $= 1 - (1 - p)^{2N}$

7.10. Population sizes are 100, 130, 169, 220, 286, and 371
$N_e \cong 175$

7.12. $t = \dfrac{-0.693}{\ln\left[1 - \left(\dfrac{1}{2N}\right)\right]}$

$t \cong 1.39\text{N}$

7.13. $H_{10} = 0.056 H_o$

7.15.

	$N = 500$	$N = 50$
a.	0.001	0.01
b.	0.005	0.016
c.	0.020	0.023
d.	0.040	0.040
e.	4.16×10^{-11}	0.003

7.16. $N \geq 25$

7.17.

	Large Populations		*Small Populations*	
	5 Months	**18 Months**	**5 Months**	**18 Months**
Mean	0.378	0.274	0.333	0.327
Variance	0.0017	0.0030	0.0030	0.0132

7.18. For four offspring, Pr(extinction) $= 0.125$

Chapter 8

8.1. About 32.5 generations, or 650 years

8.3. Figure 8.2a; $g_{XY} = 1/2$
Figure 8.2b; $g_{XY} = 1/4$
Figure 8.2c; $g_{XY} = 1/4$
Figure 8.2d; $g_{XY} = 1/8$
Figure 8.2e; $g_{XY} = 1/16$
Figure 8.2f; $g_{XY} = 1/64$

8.4. Figure 8.16a; $f = 0.0625$
Figure 8.16b; $f = 0.1758$

8.5.

Locus	Lake Population	Pond Population
32-b	0.43	0.75
61	0.77	0.55
19	0.56	1
83	1	0.87
27	0.80	0.75
59-b	0.66	0.60
9	1	1

8.6. About 105 times more likely

8.9. $f = 0.20$

  $\delta = 0.002$

8.10. $H_{10} = 0.028$

8.13. $I = 0.33$

  $\sigma_I = 0.03$

8.15. $\chi^2 = 32.45$

  df = 33

  Accept H_o

Chapter 9

9.1. $F_{IS} = 0$; $F_{IT} = 0.19$; $F_{ST} = 0.19$

9.2. $H_{obs} = 0.34$; $H_{exp} = 0.42$; $F = 0.19$

9.3. Lewis $F_{ST} = 0.21$

  Duffy $F_{ST} = 0.57$

9.4. $F_{ST} = 0.01$

9.5. pt-8 locus $F_{ST} = 0.09$

  est-5 locus $F_{ST} = 0.02$

9.7. $F_{ST} = 0.11$

9.9. $F_{ST} = 0.045$

9.13. $\chi^2 = 11.74$

  df = 4

  Reject Ho at 0.05 level of significance, but not at 0.01 (P-value $\cong 0.02$)

9.15. The second and third individuals in population X and the third in population Y appear to be immigrants. The probabilities obtained by multiplying together the expected genotype frequencies do not account for sampling bias. Computer programs that perform assignment tests make adjustments.

9.19. $s = 0.031$; $h = -0.667$

  $\tilde{p} = 0.714$ with no gene flow

  $\tilde{p} \cong 0.75$ with $m = 0.01$

  $\tilde{p} \cong 0.80$ with $m = 0.10$

9.21. $\hat{N}_{ST} = 0.052$

Chapter 10

10.2.

N	$|s|$
10^2	5×10^{-3}
10^3	5×10^{-4}
10^4	5×10^{-5}
10^5	5×10^{-6}

For humans, $N = 10^4$ seems reasonable; for insects, $N = 10^5$. (These are effective population sizes.) By Kimura's definition, for an allozyme variant to be neutral, the absolute value of the selection coefficient must be much less than the values given above. It seems reasonable that any change large enough to affect the mobility of a protein is likely to affect the function of the protein more than this. This suggests that most allozyme variants may not be neutral, but this is a judgment call. Less strict definitions of neutral are possible.

10.5. Expected homozygosity = 0.81. This is outside Watterson's confidence limits, but barely. It (weakly) suggests either purifying selection or a recent bottleneck.

10.6. $\chi^2 = 8.7$; P-value = 0.0032

  $G = 7.43$; P-value = 0.0064

  The P-value is the probability of getting the data under the null hypothesis. The G-test is more conservative. In this example, both tests clearly reject the null hypothesis.

10.7. $G = 2.84$; do not reject the null hypothesis.

10.8. There is an excess of substitutions in *jgw*; this implies selection for a new function.

10.9. $G = 14.6$; reject the null hypothesis. There is a significant excess of replacement substitutions, again implying selection for a new function.

Chapter 11

11.1. $p_{aa} = 0.40$; $K_{aa} = 0.51$

11.2. $p = 0.20$; $K = 0.23$

11.3. $p_S = 0.50$; $p_N = 0.42$

  $K_S = 0.82$; $K_A = 0.62$

11.6. Scores:

	Site 1	Site 3	Site 6	Total
Tree 1	2	1	1	4
Tree 2	1	2	2	5
Tree 3	2	2	2	6

Tree 1 is the best tree.

11.7.

m	$E(T_m)$
2	200
3	66.7
4	33.3
5	20
6	13.3
$E(T) = 333.3$	

Chapter 12

12.1. $\tilde{p}_A = 0.88$, $\tilde{p}_S = 0.12$; $\bar{w} = 0.90$

The marginal fitness of the C allele is 0.87, which is less than $\bar{w}$. Therefore, C cannot invade.

12.4.

	Harmonic Mean of Ws	Harmonic Mean of Vs	Stable Polymorphism?
a.	0.9975	0.9975	Yes
b.	1.002	0.9975	No
c.	1.0005	0.9945	No

12.5. a. Harmonic mean of Ws $= 0.934$
Harmonic mean of Vs $= 1.103$
Sufficient conditions are not met.

b. There is a stable polymorphic equilibrium at $p \cong 0.85$, and an unstable polymorphic equilibrium at $p \cong 0.47$.

c. For $c_1 = 0.4$, $p = 0$ and $p = 1$ are stable equilibria. There is an unstable equilibrium at $p = 0.2$. For $c_1 = 0.5$, $p = 0$ is stable and $p = 1$ is unstable.

12.8. *Acp36DE* Locus: $G = 8.81$. Reject neutrality; there appears to be an excess of replacement substitutions.

Acp53Ea Locus: $G = 0.12$. No evidence for selection.

12.9. Harmonic mean of $w_{11} = 0.88$; harmonic mean of $w_{22} = 1.03$. The sufficient conditions for a protected polymorphism are not met. This does not guarantee that no polymorphism will exist, but if you graph p_m and p_f versus t you will see that both always decrease to zero, regardless of initial frequencies. The A_1 allele will be lost.

Chapter 13

13.1. $\mu = 3.04$, $\alpha = 2.8$ (*Note*: the mean is expressed as a deviation from the rescaled zero point. The actual mean is 13.04.)
$G_{11} = 0.96$, $G_{12} = -1.04$, $G_{22} = -7.04$
$A_{11} = 1.12$, $A_{12} = -1.68$, $A_{22} = -4.48$
$D_{11} = -0.16$, $D_{12} = 0.64$, $D_{22} = -2.56$
$V_A = 2.51$, $V_D = 0.41$

13.3. $V_A = 65.9$, $V_D + V_I = 41.3$

13.4. $n_e = 3.4$

13.5. $h^2 = 0.80$

13.10. $\bar{Y}_S = 16.75$

13.11. $n_e = 11.1$

13.12. The QTL accounts for about 40 percent of the observed variation.
$r = 0.33$

Literature Cited

Abbott, R.J. and M.F. Gomes. 1988. Population genetic structure and outcrossing rate of *Arabidopsis thaliana* (L.) Heynh. *Heredity* 62:411–418.

Ainsworth, P.J., L.C. Surh, and M.B. Coulter-Mackie. 1991. Diagnostic single strand conformational polymorphism, (SSCP): a simplified non-radioisotopic method as applied to a Tay-Sachs B1 variant. *Nucleic Acids Research* 19: 405–406.

Akashi, H. 1997. Codon bias evolution in *Drosophila*. Population genetics of mutation-selection drift. *Gene* 205: 269–278.

Allard, R.W., G.R. Babbel, M.T. Clegg, and A.L. Kahler. 1972. Evidence for coadaptation in *Avena barbata*. *Proc. Natl. Acad. Sci. USA* 69:3043–3048.

Allen, J.A. 1988. Frequency-dependent selection by predators. *Phil. Trans. Royal. Soc. Lond. B* 319:485–503.

Allendorf, F.W. 1997. Genetically effective sizes of grizzly bear populations. In G.K. Meffe and C.R. Carrol (eds.), *Principles of Conservation Biology*, 2nd ed. Sunderland, MA: Sinauer Associates Inc., 174–175.

Allendorf, F.W. and R.F. Leary. 1986. Heterozygosity and fitness in natural populations of animals. In M.E. Soule (ed.), *Conservation Biology: The Science of Scarcity and Diversity*. Sunderland, MA: Sinauer Associates Inc., 57–76.

Allendorf, F.W. and C. Servheen. 1986. Genetics and the conservation of grizzly bears. *Trends Ecol. Evol.* 1:88–89.

Allison, A.C. 1954a. The distribution of the sickle-cell trait in East Africa and elsewhere, and its apparent relationship to the incidence of subtertian malaria. *Trans. Royal Soc. Trop. Med. Hyg.* 48:312–318.

Allison, A.C. 1954b. Protection afforded by sickle-cell trait against subtertian malarial infection. *Brit. Med. J.* 1:290–294.

Altenberg, L. 1991. Chaos from linear frequency-dependent selection. *Amer. Natur.* 138:51–68.

Anderson, W., Th. Dobzhansky, O. Pavlovsky, J.R. Powell, and D. Yardley. 1975. Genetics of natural populations. XLII. Three decades of genetic change in *Drosophila pseudoobscura*. *Evolution* 29:24–36.

Anderson, W.W. et al. 1991. Four decades of inversion polymorphism in *Drosophila pseudoobscura*. *Proc. Natl. Acad. Sci. USA* 88:10367–10371.

Andersson, M. 1994. *Sexual Selection*. Princeton: Princeton University Press.

Andersson, M. and Y. Iwasa. 1996. Sexual selection. *Trends Ecol. Evol.* 11:53–58.

Antonovics, J. and P. Kareiva. 1988. Frequency-dependent selection and competition: empirical approaches. *Phil. Trans. Royal Soc. Lond. B* 319:601–613.

Arcese, P. and J.N.M. Smith. 1988. Effects of population density and supplemental food on reproduction in the song sparrow. *J. Anim. Ecol.* 57:119–136.

Arnold, S.J. and D. Duvall. 1994. Animal mating systems: A synthesis based on selection theory. *Amer. Natur.* 143: 317–348.

Avise, J.C. 1994. *Molecular Markers, Natural History, and Evolution*. New York: Chapman and Hall.

Avise, J.C. and W.S. Nelson. 1989. Molecular genetic relationships of the extinct dusky seaside sparrow. *Science* 243:646–648.

Ayala, F.J. 1972. Darwinian versus non-Darwinian evolution in natural populations of *Drosophila*. *Proc. 6th Berkeley symposium on mathematical statistics and probability V: Darwinian, neo-Darwinian, and non-Darwinian evolution*, 211–236.

Ayala, F.J. 1977. "Nothing in biology makes sense except in the light of evolution." *J. Hered.* 68:3–10.

Ayala, F.J. 1982. *Population and Evolutionary Genetics: A Primer*. Menlo Park, CA: Benjamin Cummings.

Ayala, F.J. and C.A. Campbell. 1974. Frequency-dependent selection. *Ann. Rev. Ecol. Syst.* 5:115–138.

Ayala, F.J., J.R. Powell, and Th. Dobzhansky. 1971. Polymorphisms in continental and island populations of *Drosophila* willistoni. *Proc. Natl. Acad. Sci. USA* 68:2480–2483.

Ayala, F.J., M.L. Tracey, C.A. Mourao, and S. Perez-Salas. 1972. Enzyme variability in the *Drosophila willistoni* group IV. Genetic variation in natural populations of *Drosophila willistoni*. *Genetics* 70:113–139.

Baird, M., I. Balazs, A. Giusti, L. Miyazake, L. Nicholas, K. Wexler, E. Kanter, J. Glassbert, F. Allen, P. Rubinstein, and L. Sussman. 1986. Allele frequency distribution of two highly polymorphic DNA sequences in three ethnic groups and its application to the determination of paternity. *Amer. J. Human Genet.* 39:489–501.

Baker, A.J. and A. Moeed. 1987. Rapid genetic differentiation and founder effect in colonizing populations of common mynas (*Acridotheres tristis*). *Evolution* 41:525–538.

Bakker, T.C.M. 1993. Positive genetic correlation between female preference and preferred male ornament in sticklebacks. *Nature* 363:255–257.

Ballou, J.D. 1997. Ancestral inbreeding only minimally affects inbreeding depression in mammalian populations. *J. Hered.* 88:169–178.

Baquero, F. and J. Blazquez. 1997. Evolution of antibiotic resistance. *Trends Ecol. Evol.* 12:482–487.

Barker, J.S.F. 1982. Population genetics of *Opuntia* breeding *Drosophila* in Australia. In J.S.F. Barker, and W.T. Starmer (eds.), *Ecological Genetics and Evolution*. Sydney: Academic Press, 209–224.

Barker, J.S.F., F.M. Sene, P.D. East, and M.A.Q.R. Pereira. 1985. Allozyme and chromosomal polymorphism in *Drosophila buzzatii* in Brazil and Argentina. *Genetica* 67:161–170.

Barton, N.H. and M. Turelli. 1989. Evolutionary quantitative genetics: How little do we know? *Ann. Rev. Genet.* 23:337–370.

Bateman, A.J. 1948. Intra-sexual selection in *Drosophila*. *Heredity* 2:349–368.

Bateman, A.J. 1959. The viability of near-normal irradiated chromosomes. *Intl. J. Rad. Biol.* 1:170–180.

Beerli, P. and J. Felsenstein. 1999. Maximum likelihood estimation of migration rates and effective population numbers in two populations. *Genetics* 152:763–773.

Beerli, P. and J. Felsenstein. 2001. Maximum likelihood estimation of a migration matrix and effective population sizes in *n* subpopulations by using a coalescent approach. *Proc. Natl. Acad. Sci.* 98:4563–4568.

Begun, D.J. and C.F. Aquadro. 1992. Levels of naturally occurring DNA polymorphism correlate with recombination rates in *D. melanogaster*. *Nature* 356:519–520.

Begun, D.J., P. Whitley, B.L. Todd, H.M. Waldrip-Dail, and A.G. Clark. 2000. Molecular population genetics of male accessory gland proteins in *Drosophila*. *Genetics* 156:1879–1888.

Bell, R.G. and J.T. Matschiner. 1972. Warfarin and the inhibition of vitamin K activity by an oxide metabolite. *Nature* 237:32–33.

Berger, J. 1990. Persistence of different-sized populations: An empirical assessment of rapid extinctions in bighorn sheep. *Conserv. Biol.* 4(1):91–98.

Berger, J. 1999. Intervention and persistence in small populations of bighorn sheep. *Conserv. Biol.* 13:432–435.

Birch, L.C. 1960. The genetic factor in population ecology. *Amer. Natur.* 94:5–24.

Birkhead, T. and A. Møller. 1993. Female control of paternity. *Trends Ecol. Evol.* 8:100–104.

Bishop, J.A. 1981. A NeoDarwinian approach to resistance: Examples from mammals. In J.A. Bishop and L.M. Cook (eds.), *Genetic Consequences of Man Made Change*. New York and London: Academic Press, 37–51.

Boag, P.T. 1983. The heritability of external morphology in Darwin's ground finches (*Geospiza*) on Isla Daphne Major, Galapagos. *Evolution* 37:877–894.

Boag, P.T. and P.R. Grant. 1978. Heritability of external morphology in Darwin's finches. *Nature* 274:793–794.

Boag, P.T. and B.R. Grant. 1981. Intense natural selection in a population of Darwin's finches (Geospizinae) in the Galapagos. *Science* 214:82–85.

Bodmer, W.F. and J. Felsenstein. 1967. Linkage and selection: Theoretical analysis of the deterministic two locus random mating model. *Genetics* 57:237–265.

Bohonak, A.J., N. Davies, G.K. Roderick, and F.X. Villablanca. 1998. Is population genetics mired in the past? *Trends Ecol. Evol.* 13:360.

Bonnell, M.L. and R.K. Selander. 1974. Elephant seals: Genetic variation and near-extinction. *Science* 184:908–909.

Borash, D.J., A.G. Gibbs, A. Joshi, and L.D. Mueller. 1998. A genetic polymorphism maintained by natural selection in a temporally varying environment. *Amer. Natur.* 151:148–156.

Bossart, J.L. and D.P. Prowell. 1998. Genetic estimates of population structure and gene flow: limitations, lessons, and new directions. *Trends Ecol. Evol.* 13: 202–206.

Bowen, B., J.C. Avise, J.I. Richardson, A.B. Meylan, D. Margaritoulis, and S.R. Hopkins-Murphy. 1993. Population structure of loggerhead turtles (*Caretta caretta*) in the northwestern Atlantic Ocean and Mediterranean Sea. *Conserv. Biol.* 7:834–844.

Bradshaw Jr., H.D., K.G. Otto, B.E. Frewen, J.K. McKay, and D.W. Schemske. 1998. Quantitative trait loci affecting differences in floral morphology between two species of monkeyflower (*Mimulus*). *Genetics* 149:367–382.

Brakefield, P.M. 1987. Industrial melanism: Do we have the answers? *Trends Ecol. Evol.* 2:117–122.

Bromham, L. and P.H. Harvey. 1996. Naked mole-rats on the move. *Curr. Biol.* 6:1082–1083.

Brookfield, J.F.Y. and P.M. Sharp. 1994. Neutralism and selectionism face up to DNA data. *Trends in Genet.* 10:109–111.

Brooks, L.D. and R.W. Marks. 1986. The organization of genetic variation for recombination in *Drosophila melanogaster*. *Genetics* 114:525–547.

Brown, C.L., E.C. Garner, A.K. Dunker, and P. Joyce. 2001. The power to detect recombination using the coalescent. *Mol. Biol. Evol.* 18:1421–1424.

Brown, W.L. and E.O. Wilson. 1956. Character displacement. *Syst. Zool.* 5:49–64.

Brzustowski, J. 2002. Doh assignment test calculator. Available at: http://www2.biology.ualberta.ca/jbrzusto/Doh.php.

Bull, J.J., M.R. Badgett, H.A. Wichman, J.P. Huelsenbech, D.M. Hillis, A. Gulati, C. Ho, and I.J. Molineaux. 1997. Exceptional convergent evolution in a virus. *Genetics* 147:1497–1507.

Bulmer, M.G. 1971. The effect of selection on genetic variability. *Amer. Natur.* 105:201–211.

Bulmer, M.G. 1972. The genetic variability of polygenic characters under optimizing selection, mutation, and drift. *Genet. Res.* 19:17–25.

Bulmer, M.G. 1989. Estimating the variability of substitution rates. *Genetics* 123:615–619.

Bürger, R., G.P. Wagner, and F. Stettinger. 1989. How much heritable variation can be maintained in finite populations by mutation-selection balance? *Evolution* 43:1748–1766.

Buri, P. 1956. Gene frequency in small populations of mutant *Drosophila*. *Evolution* 10:367–402.

Burke, T. 1989. DNA fingerprinting and other methods for the study of mating success. *Trends Ecol. Evol.* 4:139–144.

Burke, T., N.B. Davies, M.W. Bruford, and B.J. Hatchwell. 1989. Parental care and mating behaviour of polyandrous dunnocks *Prunella modularis* related to paternity by DNA fingerprinting. *Nature* 338:249–251.

Byers, D.L. and D.M. Waller. 1999. Do plant populations purge their genetic load? Effects of population size and mating history on inbreeding depression. *Ann. Rev. Ecol. Syst.* 30:479–513.

Caballero, A and P.D. Keightley. 1994. A pleiotropic nonadditive model of variation in quantitative traits. *Genetics* 138:883–900.

Cabe, P.R. 1998. The effects of founding bottlenecks on genetic variation in the European starling (*Sturnus vulgaris*) in North America. *Heredity* 80:519–525.

Cain, A.J. and P.M. Sheppard. 1950. Selection in the polymorphic land snail *Cepaea nemoralis*. *Heredity* 4:275–294.

Cain, A.J. and P.M. Sheppard. 1954a. Natural selection in *Cepaea*. *Genetics* 39:89–116.

Cain, A.J. and P.M. Sheppard. 1954b. The theory of adaptive polymorphism. *Amer. Natur.* 88:321–326.

Camin, J.H. and P.R. Ehrlich. 1958. Natural selection in water snakes (*Natrix sipedon* L.) on islands in Lake Erie. *Evolution* 12:504–511.

Cann, R.L., M. Stoneking, and A.C. Wilson. 1987. Mitochondrial DNA and human evolution. *Nature* 325:31–36.

Cardon, L.R. and J.I. Bell. 2001. Association study designs for complex diseases. *Natur. Rev. Genet.* 2:91–99.

Carter, P.A. and W.B. Watt. 1988. Adaptation at specific loci. V. Metabolically adjacent enzyme loci may have very distince experiences of selective pressures. *Genetics* 119:913–924.

Case, T.J. 2000. *An Illustrated Guide to Theoretical Ecology*. New York and Oxford: Oxford University Press.

Castle, W.E. 1903. The laws of heredity of Galton and Mendel and some laws governing race improvement by selection. *Proc. Amer. Acad. Arts Sci.* 39:223–242.

Castle, W.E. 1921. An improved method of estimating the number of genetic factors concerned in cases of blending inheritance. *Proc. Natl. Acad. Sci. USA* 81:6904–6907.

Cavalli-Sforza, L.L. 1998. The DNA revolution in population genetics. *Trends Genet.* 14:60–65.

Cavalli-Sforza, L.L. and W.F. Bodmer. 1971. *The Genetics of Human Populations*. New York: W.H. Freeman Co.

Cavener, D.R. and M.T. Clegg. 1981. Multigenic response to ethanol in *Drosophila melanogaster*. *Evolution* 35:1–10.

Chakraborty, R., M. Kimmel, D.N. Stivers, L.J. Davison, and R. Deka. 1997. Relative mutation rates at di-, tri-, and tetranucleotide microsatellite loci. *Proc. Natl. Acad. Sci. USA* 94:1041–1046.

Chambers, G.K. 1988. The *Drosophila* alcohol dehydrogenase gene-enzyme system. *Advances in Genetics* 25:39–107.

Chao, L. 1990. Fitness of RNA virus decreased by Muller's ratchet. *Nature* 348:454–455.

Chao, L and D.E. Carr. 1993. The molecular clock and the relationship between population size and generation time. *Evolution* 47:688–690.

Chao, L., T.T. Tran, and T.T. Tran. 1997. The advantage of sex in the RNA virus ϕ6. *Genetics* 147:953–959.

Charlesworth, D., B. Charlesworth, and G.A.T. McVean. 2001. Genome sequences and evolutionary biology, a two-way interaction. *Trends Ecol. Evol.* 16:235–242.

Charlesworth, B., M.T. Morgan, and D. Charlesworth. 1993. The effect of deleterious mutations on neutral molecular evolution. *Genetics* 134:1289–1303.

Chetverikov, S.S. [1926] 1961. On certain aspects of the evolutionary process from the standpoint of modern genetics. Originally published in Russian in *Zhurnal Eksperimental' noi Biologii* A2:3–54. Translated and reprinted in *Proc. Amer. Phil. Soc.* 105:167–195.

Christiansen, F.B., O. Frydenberg, and V. Simonsen. 1977. Genetics of *Zoarces* populations. X. Selection component analysis of the EstIII polymorphism using samples of successive cohorts. *Hereditas* 87:129–150.

Clark, A.G. 1987a. Neutrality tests of highly polymorphic restriction-fragment-length polymorphisms. *Amer. J. Human Genet.* 41:948–956.

Clark, A.G. 1987b. A test of multilocus interaction in *Drosophila melanogaster*. *Amer. Natur.* 130:283–299.

Clark, A.G., M. Aguadé, T. Prout, L.G. Harshman, and C. H. Langley. 1995. Variation in sperm displacement and its association with accessory gland protein loci in *Drosophila melanogaster*. *Genetics* 139:189–201.

Clark, R.W. 1969. *JBS: The Life and Work of J.B.S. Haldane*. New York: Coward-McCann.

Clarke, B.C. 1979. The evolution of genetic diversity. *Proc. Royal Soc. Lond.* B 205:453–474.

Clarke, C.A., C.G.C. Dickson, and P.M. Sheppard. 1963. Larval color pattern in *Papilio demodocus*. *Evolution* 17:130–137.

Clarke, C.A. and P.M. Sheppard. 1966. A local survey on the distribution of the industrial melanic forms in the moth *Biston betularia*

and estimates of the selective values of these in an industrial environment. *Proc. R. Soc. Lond. B.* 165:424–439.

Clausen J., D.D. Keck, and W.M. Hiesey. 1941. Regional differentiation in plant species. *Amer. Natur.* 75:231–250.

Clayton, G. and A. Robertson. 1955. Mutation and quantitative variation. *Amer. Natur.* 89:151–158.

Clayton, G.A., J.A. Morris, and A. Robertson. 1957. An experimental check on quantitative genetical theory. I. Short-term responses to selection. *J. Genetics* 55:131–151.

Clutton-Brock, T.H., A.W. Illius, K. Wilson, B.T. Grenfell, A.D.C. MacColl, and S.D. Albon. 1997. Stability and instability in ungulate populations: An empirical analysis. *Amer. Natur.* 149:195–219.

Clutton-Brock, T.H., K. Wilson, and I.R. Stevenson. 1997. Density-dependent selection on horn phenotype in Soay sheep. *Phil. Trans. Royal Soc. Lond.* 352:839–850.

Clutton-Brock, T.H., O.F. Price, S.D. Albon, and P.A. Jewell. 1991. Persistent instability and population regulation in Soay sheep. *J. Anim. Ecol.* 60:593–608.

Cockerham, C.C. 1969. Variance of gene frequencies. *Evolution* 23:72–84.

Cockerham, C.C. 1973. Analyses of gene frequencies. *Genetics* 74:679–700.

Cohan, F.M. 2000. Genetic structure of prokaryotic populations. In R.S. Singh and C.B. Krimbas (eds.), *Evolutionary genetics: From Molecules to Morphology*. Cambridge: Cambridge Univ. Press, 475–489.

Coltman, D.W., J.A. Smith, D.R. Bancroft, J. Pilkington, A.D.C. MacColl, T.H. Clutton-Brock, and J.M. Pemberton. 1999. Density-dependent variation in lifetime breeding success and natural and sexual selection in Soay rams. *Amer. Natur.* 154:730–746.

Conant, R. and W. Clay. 1937. A new subspecies of watersnake from the islands of Lake Erie. *Occasional papers of the Museum of Zoology, University of Michigan* 346:1–9.

Conover, D.O. and D.A. Van Voorhees. 1990. Evolution of a balanced sex ratio by frequency-dependent selection in a fish. *Science* 250:1556–1558.

Constantino, R.F., et al. 1997. Chaotic dynamics in an insect population. *Science* 275:389–391.

Cooch, F.G. and J.A. Beardmore. 1959. Assortative mating and reciprocal difference in the blue-snow goose complex. *Nature* 183:1833–1834.

Coyne, J.A. 1976. Lack of genic similarity between two sibling species of *Drosophila* as revealed by varied techniques. *Genetics* 84:593–607.

Coyne, J.A. 1998. Not black and white. *Nature* 396:35–36.

Coyne, J.A. and E. Beecham. 1987. Heritability of two morphological characters within and among natural populations of *Drosophila melanogaster*. *Genetics* 117:727–737.

Coyne, J.A. and H.A. Orr. 1989. Patterns of speciation in *Drosophila*. *Evolution* 43:362–381.

Coyne, J.A. and H.A. Orr. 1997. "Patterns of speciation in *Drosophila*" Revisited. *Evolution* 51:295–303.

Coyne, J.A. and H.A. Orr. 1998. The evolutionary genetics of speciation. *Phil. Trans. Royal Soc. Lond.* B 353:287–305.

Coyne, J.A., N.H. Barton, and M. Turelli. 1997. A critique of Sewall Wright's shifting balance theory of evolution. *Evolution* 51:643–671.

Coyne, J.A., N.H. Barton, and M. Turelli. 2000. Is Wright's shifting balance process important in evolution? *Evolution* 54:306–318.

Coyne, J.A., A.A. Felton, and R.C. Lewontin. 1978. Extent of genetic variation at a highly polymorphic locus in *Drosophila pseudoobscura*. *Proc. Natl. Acad. Sci. USA* 75:5090–5093.

Crandall, K.A., D. Posada, and D.Vasco. 1999. Effective population sizes: missing measures and missing concepts. *Animal Conserv.* 2:317–319.

Creed, E.R., D.R. Lees, and J.G. Duckett. 1973. Biological method of estimating smoke and sulphur dioxide pollution. *Nature* 244:278–280.

Creed, E.R., D.R. Lees, and M.G. Bulmer. 1980. Pre-adult viability differences of melanic *Biston betularia* (L.) (Lepidoptera). *Biol. J. Linn. Soc.* 13:251–262.

Crnokrak, P. and D.A. Roff. 1999. Inbreeding depression in the wild. *Heredity* 83:260–270.

Crnokrak, P. and D.A. Roff. 1995. Dominance variance: associations with selection and fitness. *Heredity* 75:530–540.

Crow, J.F. 1986. *Basic Concepts in Population, Quantitative, and Evolutionary Genetics.* New York:W.H. Freeman and Co.

Crow, J.F. 1990. R.A. Fisher, A centennial view. *Genetics* 124:207–211.

Crow, J.F. 1992. Centennial: J.B.S. Haldane, 1892–1964. *Genetics* 130:1–6.

Crow, J.F. 1997. The high spontaneous mutation rate: Is it a health risk? *Proc. Natl. Acad. Sci. USA* 94:8380–8386.

Crow, J.F. 1999. The odds of losing at genetic roulette. *Nature* 397:293–294.

Crow, J.F. 2002. Perspective: Here's to Fisher, additive genetic variance and the fundamental theorem of natural selection. *Evolution* 56:1313–1316.

Crow, J.F. and C. Denniston. 1988. Inbreeding and variance effective population numbers. *Evolution* 42:482–495.

Crow, J.F. and M. Kimura. 1970. *An Introduction to Population Genetics Theory.* New York: Harper & Row.

Crow, J.F. and M.J. Simmons. 1983. The mutational load in *Drosophila.* In M. Ashburner, H.L. Carson, and J.N. Thompson Jr. (eds.), *The Genetics and Biology of Drosophila.* Vol. 3c. New York: Academic Press,1-35.

Curtis, C.F., L.M. Cook, and R.J. Wood. 1978. Selection for and against insecticide resistance and possible methods of inhibiting the evolution of resistance in mosquitoes. *Ecol. Entom.* 3:273–287.

Curtsinger, J.W., P.M. Service, and T. Prout. 1994. Antagonistic pleiotropy, reversal of dominance, and genetic polymorphism. *Amer. Natur.* 144:210–228.

Darwin, C. [1859] 1964. *On the Origin of Species by Means of Natural Selection, or the Preservation of Favoured Races in the Struggle for Life.* Reprinted by Harvard Univ. Press.

Darwin, C. 1871. *The Descent of Man and Selection in Relation to Sex.* London: John Murray.

Darwin, C. 1872. *On the Origin of Species by Means of Natural Selection, or the Preservation of Favoured Races in the Struggle for Life,* 6th ed. London: John Murray.

Darwin, C. 1875. *The Variation of Animals and Plants under Domestication,* 2nd ed. Vol. II. London: John Murray.

Darwin, C. 1876. *The Effects of Cross and Self Fertilization in the Vegetable Kingdom.* London: John Murray.

David, P. 1998. Heterozygosity-fitness correlations: new perspectives on old problems. *Heredity* 80: 531–537.

Davidson, J. 1938a. On the ecology of the growth of the sheep population in South Australia. *Trans. Royal Soc. South Australia* 62:141–148.

Davidson, J. 1938b. On the growth of the sheep population in Tasmania. *Trans. Royal Soc. South Australia* 62:342–346.

Davies, N., F.X.Villablanca, and G.K. Roderick. 1999. Determining the source of individuals: multilocus genotyping in non-equilibrium population genetics. *Trends Ecol. Evol.* 14:17–21.

Dawson, P.S. 1970. Linkage and the elimination of deleterious mutant genes from experimental populations. *Genetica* 41:147–169.

Dempster, E.R. 1955. Maintenance of genetic heterogeneity. *Cold Spring Harbor Symp. Quant. Biol.* 70:25–32.

Dennis, B., P.L. Munholland, and J.M. Scott. 1991. Estimation of growth and extinction parameters for endangered species. *Ecol. Monogr.* 61(2):115–143.

Denver, D., K. Morris, M. Lynch, L.L. Vassilieva, and W.K. Thomas. 2000. High direct estimate of the mutation rate in the mitochondrial genome of *Caenorhabditis elegans. Science* 289:2342–2344.

Desgeorges, M., P. Boulot, P. Kjellberg, G. Lefort, M. Rolland, J. Demaille, and M. Claustres. 1993. Prenatal diagnosis of cystic fibrosis using SSCP analysis. *Prenat. Diagn.* 13:147–148.

deVicente, M.C. and S.D. Tanksley. 1993. QTL analysis of transgressive segregation in an interspecific tomato cross. *Genetics* 134:585–596.

Dickerson, R.E. 1971. The structure of cytochrome c and the rates of molecular evolution. *J. Mol. Evol.* 1:26–45.

Dickerson, R.E. 1972. The structure and history of an ancient protein. *Sci. Amer.* 226(4):58–72.

Doak, D. 1989. Spotted Owls and old growth logging in the Pacific northwest. *Conserv. Biol.* 3:389–396.

Dobzhansky, Th. 1937. *Genetics and the Origin of Species.* New York: Columbia University Press.

Dobzhansky, Th. 1943. Genetics of natural populations. IX. Temporal changes in the composition of populations of *Drosophila pseudoobscura. Genetics* 28:162–186.

Dobzhansky, Th. 1955. A review of some fundamental concepts and problems of population genetics. *Cold Spring Harbor Symp. Quant. Biol.* 20:1–15.

Dobzhansky, Th. 1970. *Genetics of the Evolutionary Process.* New York: Columbia University Press.

Dobzhansky, Th. 1973. Nothing in biology makes sense except in the light of evolution. *Amer. Biol. Teacher* 35:125–129.

Dobzhansky, Th. and H. Levene. 1951. Development of heterosis through natural selection in experimental populations of *Drosophila pseudoobscura. Amer. Natur.* 85:247–264.

Dobzhansky, Th. and O. Pavlovsky. 1953. Indeterminate outcome of certain experiments on *Drosophila* populations. *Evolution* 7:198–210.

Dobzhansky, Th. and O. Pavlovsky. 1957. An experimental study of interaction between genetic drift and natural selection. *Evolution* 11:311–319.

Dobzhansky, Th. and M.L. Queal. 1938. Chromosomal variation in populations of *Drosophila pseudoobscura* inhabiting isolated mountain ranges. *Genetics* 23:239–251.

Dobzhansky, Th. and B. Spassky. 1954. Genetics of natural populations XXII: A comparison of the concealed variability in *Drosophila prosaltans* with that in other species. *Genetics* 39:472–487.

Dobzhansky, Th. and B. Spassky. 1969. Artificial and natural selection for two behavioral traits in *Drosophila pseudoobscura. Proc. Natl. Acad. Sci. USA* 62:75–80.

Dobzhansky, Th., F.J. Ayala, G.L. Stebbins, and J.W.Valentine. 1977. *Evolution.* San Francisco:W.H. Freeman & Co.

Dobzhansky, Th., A.S. Hunter, O. Pavlovsky, B. Spassky, and B.Wallace. 1963. Genetics of natural populations. XXXI. Genetics of an isolated marginal population of *Drosophila pseudoobscura. Genetics* 48:91–103.

Dockhorn-Dworniczak, B., B. Dworniczak, L. Brommelkamp, J. Bulles, J. Horst, and W.W. Bocker. 1991. Non-isotopic detection of single strand conformation polymorphism (PCR-SSCP): a rapid and sensitive technique for the diagnosis of phenylketonuria. *Nucleic Acids Research* 19:2500.

Dodd, A.P. 1940. *The Biological Campaign against Prickly Pear.* Brisbane: A.H. Tucker, Government Printer.

Dodd, D.M.B. 1989. Reproductive isolation as a consequence of adaptive divergence in *Drosophila pseudoobscura*. *Evolution* 43:1308–1311.

Donnelly, P. and S. Tavaré. 1995. Coalescents and genealogical structure under neutrality. *Ann. Rev. Genet.* 29:401–421.

Donovan, T.M. and C.W. Welden. 2002. *Spreadsheet Exercises in Ecology and Evolution.* Sunderland, MA: Sinauer Associates.

Drake, J.W. 1991. A constant rate of spontaneous mutation in DNA-based microbes. *Proc. Natl. Acad. Sci. USA* 88:7160–7164.

Dronamraju, K.R. 1990. *Selected Genetic Papers of J.B.S. Haldane.* New York: Garland Publishing Inc.

Dubrova, Y.E., R.I. Bersimbaev, L.B. Djansugurova, M.K. Tankimanova, Z.Z. Mamyrbaeva, R. Mustonen, C. Lindholm, M. Hulten, and S. Salomaa. 2002. Nuclear weapons tests and the human germline mutation rate. *Science* 295:1037.

Dubrova, Y.E., V.N. Nesterov, N.G. Krouchinsky, V.A. Ostapenko, R. Neumann, D.L. Nell, and A. J. Jeffreys. 1996. Human minisatellite mutation rate after the Chernobyl accident. *Nature* 380:683–686.

Dubrova, Y.E., V.N. Nesterov, N.G. Krouchinsky, V.A. Ostapenko, G. Vergnaud, F. Giraudeau, J. Buard, and A.J. Jeffreys. 1997. Further evidence for elevated human minisatellite mutation rate in Belarus eight years after the Chernobyl accident. *Mutation Res.* 381:267–278.

Dubunin, N.P., D.D. Romashov, M.A. Heptner, and Z.A. Demidova. 1937. Aberrant polymorphism in *Drosophila fasciata* Meig. [in Russian]. *Biol. Zh.* 6:311–354.

Dudley, J.W. 1977. 76 generations of selection for oil and protein percentage in maize. In E. Pollak, O. Kempthorne, and T. Bailey (eds.), *International Congress of Quantitative Genetics.* Ames, IA: Iowa State Univ. Press, 459–473.

Dudley, J.W. and J.R. Lambert, 1992. Ninety generations of selection for oil and protein content in maize. *Maydica* 37:1–7.

Eanes, W.F. 1999. Analysis of selection on enzyme polymorphisms. *Ann. Rev. Ecol. Syst.* 30:301–326.

Eckert, C.G. and S.C.H. Barrett. 1992. Stochastic loss of style morphs from populations of tristylous *Lythrum salicaria* and *Decodon verticullatus* (Lythraceae). *Evolution* 46:1014–1029.

Eckert, C.G., D. Manicacci, and S.C.H. Barrett. 1996. Genetic drift and founder effect in native versus introduced populations of an invading plant, *Lythrum salicaria* (Lythraceae). *Evolution* 50:1512–1519.

Edwards, A.W.F. 1994. The fundamental theorem of natural selection. *Biol. Rev.* 69:443–474.

Edwards, M.D., C.W. Stuber, and J.F. Wendel. 1987. Molecular-marker-facilitated investigations of quantitative-trait loci in Maize; I. Numbers, genomic distribution and types of gene action. *Genetics* 116:113–125.

Ehrlich, P.R. and P.H. Raven. 1969. Differentiation of populations. *Science* 165:1228–1232.

Ehrlich, P.R., R.W. Holm, and P.H. Raven (eds.). 1969. *Papers on Evolution.* Boston: Little Brown and Co.

Ehrlich, P.R., D.S. Dobkin, and D. Wheye. 1988. *The Birder's Handbook.* New York: Simon & Schuster, Inc.

Ehrman, L. 1965. Direct observation of sexual isolation between allopatric and between sympatric strains of the different *Drosophila* paulistorum races. *Evolution* 19:459–464.

Eisen, J.A. 2000. Horizontal gene transfer among microbial genomes: new insights from complete genome analysis. *Current Opinion Genet. Devel.* 10:606–611.

Eldridge, M.D.B., J.M. King, A.K. Loupis, P.B.S. Spencer, A.C. Taylor, L.C. Pope, and G.P. Hall. 1999. Unprecedented low levels of genetic variation and inbreeding depression in an island population of the black-footed rock-wallaby. *Conserv. Biol.* 13:531–541.

Elena, S.F. and R.E. Lenski. 1997a. Long-term experimental evolution in *Escherichia coli*. VII. Mechanisms maintaining genetic variability within populations. *Evolution* 51:1058–1067.

Elena, S.F. and R.E. Lenski. 1997b. Test of synergistic interactions among deleterious mutations in bacteria. *Nature* 390:395–398.

Elena, S.F., V.S. Cooper, and R.E. Lenski. 1996. Punctuated equilibrium caused by selection of rare beneficial mutations. *Science* 272:1802–1804.

Ellegren, H., G. Lindgren, C.R. Primmer, and A.P. Møller. 1997. Fitness loss and germline mutations in barn swallows breeding in Chernobyl. *Nature* 389:593–596.

Ellstrand, N.C., H.C. Prentice, and J.F. Hancock. 1999. Gene flow and introgression from domesticated plants into their wild relatives. *Ann. Rev. Ecol. Syst.* 30:539–563.

Elton, C.S. 1958. *The Ecology of Invasions by Animals and Plants.* London: Chapman and Hall.

Emerson, B.C., E. Paradis, and C. Thebaud. 2001. Revealing the demographic histories of species using DNA sequences. *Trends Ecol. Evol.* 16:707–716.

Emerson, S. 1939. A preliminary survey of the *Oenothera organensis* population. *Genetics* 24:524–537.

Endler, J.A. 1977. *Geographic Variation, Speciation, and Clines.* Princeton: Princeton University Press.

Endler, J.A. 1978. A predator's view of animal color patterns. *Evol. Biol.* 11:319–364.

Endler, J.A. 1980. Natural selection on color patterns in *Poecilia reticulata*. *Evolution* 34:76–91.

Endler, J.A. 1986. *Natural Selection in the Wild.* Princeton, NJ: Princeton University Press.

Endo, T., K. Ikeo, and T. Gojobori. 1996. Large-scale search for genes on which positive selection may operate. *Mol. Biol. Evol.* 13:685–690.

Estoup, A., L. Garnery, M. Solignac, and J-M Cornuet. 1995. Microsatellite variation in honey bee (*Apis mellifera* L.) populations: Hierarchical genetic structure and test of the infinite allele and stepwise mutation models. *Genetics* 140:679–695.

Ewens, W.J. 1969. Mean fitness increases when fitnesses are additive. *Nature* 221:1076.

Ewens, W.J. 1972. The sampling theory of selectively neutral alleles. *Theor. Pop. Biol.* 3:87–112.

Ewens, W.J. 1989. An interpretation and proof of the fundamental theorem of natural selection. *Theor. Pop. Biol.* 36:167–180.

Excoffier, L., P.E. Smouse, and J.M. Quattro. 1992. Analysis of molecular variance inferred from metric distances among DNA haplotypes: Application to human mitochondrial DNA restriction data. *Genetics* 131:479–491.

Eyre-Walker, A. and P.D. Keightley. 1999. High genomic deleterious mutation rates in hominids. *Nature* 397:344–347.

Eyre-Walker, A., P.D. Keightley, N.G.C. Smith, and D. Gaffney. 2002. Quantifying the slightly deleterious mutation model of molecular evolution. *Mol. Biol. Evol.* 19:2142–2149.

Falconer, D.S. and T.F.C. Mackay. 1996. *Introduction to Quantitative Genetics.* Essex: Longman.

Fay, J.C. and C-I Wu. 2000. Hitchhiking under positive Darwinian selection. *Genetics* 155:1405–1413.

Fay, J.C., G.J. Wyckoff, and C-I Wu. 2001. Positive and negative selection on the human genome. *Genetics* 158:1227–1234.

Fay, J.C., G.J. Wyckoff, and C-I Wu. 2002. Testing the neutral theory of molecular evolution with genomic data from *Drosophila*. *Nature* 415:1024–1026.

Feldman, M.W., F.B. Christiansen, and L.D. Brooks. 1980. Evolution of recombination in a constant environment. *Proc. Natl. Acad. Sci. USA* 77:4838–4841.

Feldman, M.W., I.R. Franklin, and G.J. Thomson. 1974. Selection in complex genetic systems. I. The symmetric equilibria of the three-locus symmetric viability model. *Genetics* 76:135–162.

Feldman, M.W., S.P. Otto, and F.B. Christiansen. 1996. Population genetic perspectives on the evolution of recombination. *Ann. Rev. Genetics* 30:261–295.

Felsenstein, J. 1978. Cases in which parsimony or compatibility methods will be positively misleading. *Syst. Zool.* 27:401–410.

Felsenstein, J. 1993. *PHYLIP (Phylogeny Inference Package) version 3.5c.* Distributed by the author. Department of Genetics, University of Washington, Seattle.

Felsenstein, J. 2003. *Inferring Phylogenies.* Sunderland, MA: Sinauer Associates. (In press).

Fenster, C.B., L.F. Galloway, and L. Chao. 1997. Epistasis and its consequences for the evolution of natural populations. *Trends Ecol. Evol.* 12:282–286.

Fisher, R.A. 1915. The evolution of sexual preference. *Eugenics Rev.* 7:184–192.

Fisher, R.A. 1918. The correlation between relatives on the supposition of Mendelian inheritance. *Trans. Royal Soc. Edinburgh* 52:399–433.

Fisher, R.A. 1930a. *The Genetical Theory of Natural Selection.* Oxford: Clarendon Press.

Fisher, R.A. 1930b. The distribution of gene ratios for rare mutations. *Proc. Royal Soc. Edinburgh* 50:204–219.

Fisher, R.A. 1958. *The Genetical Theory of Natural Selection*, 2nd ed. New York: Dover Publications, Inc.

Foley, P. 1994. Predicting extinction times from environmental stochasticity and carrying capacity. *Conserv. Biol.* 8(1):124–137.

Forbes, S.H. and D.K. Boyd. 1997. Genetic structure and migration in native and reintroduced Rocky Mountain wolf populations. *Conserv. Biol.* 11:1226–1234.

Ford, E.B. 1975. *Ecological Genetics*, 4th ed. London: Chapman and Hall.

Ford, M.J. 2002. Applications of selective neutrality tests to molecular ecology. *Mol. Evol.* 11:1245–1262.

Frankham, R. 1995. Inbreeding and extinction: A threshold effect. *Conserv. Biol.* 9:792–799.

Frankham, R. 1997. Do island populations have less genetic variation than mainland populations? *Heredity* 78:311–327.

Frankham, R. 1998. Inbreeding and extinction: Island populations. *Conserv. Biol.* 12:665–675.

Franklin, I.R. and M.W. Feldman. 2000. The equilibrium theory of one- and two-locus systems. In R.S. Singh and C.B. Krimbas (eds.), *Evolutionary Genetics: From Molecules to Morphology.* Cambridge: Cambridge University Press, 258–283.

Freeman, S. and J.C. Herron. 2001. *Evolutionary Analysis*, 2nd ed. Upper Saddle River, NJ: Prentice Hall.

Fry, A.J. and R.M. Zink. 1998. Geographic analysis of nucleotide diversity and song sparrow (Aves: Emberizidae) population history. *Mol. Ecol.* 7:1303–1313.

Fry, J.D., P.D. Keightley, S.L. Heinsohn, and S.V. Nuzhdin. 1999. New estimates of the rates and effects of mildly deleterious mutation in *Drosophila melanogaster*. *Proc. Natl. Acad. Sci.* 96:574–579.

Fu, Y-X and W-H Li. 1999. Coalescing into the 21st century: An overview and prospects of coalescent theory. *Theor. Pop. Biol.* 56:1–10.

Futuyma, D. 1998. *Evolutionary Biology*, 3rd ed. Sunderland, MA: Sinauer Associates, Inc.

Gabriel, W., M. Lynch, and R. Burger. 1993. Muller's ratchet and mutational meltdowns. *Evolution* 47:1744–1757.

Gaggiotti, O.E., O. Lange, K. Rassmann, and C. Gliddon. 1999. A comparison of two indirect methods for estimating average levels of gene flow using microsatellite data. *Mol. Ecol.* 8:1513–1520.

García-Dorado, A. 1997. The rate and effects distribution of viability mutation in *Drosophila*: Minimum distance estimation. *Evolution* 51:1130–1139.

García-Dorado, A. and A. Caballero. 2000. On the average coefficient of dominance of deleterious spontaneous mutations. *Genetics* 155:1991–2001.

García-Dorado, A., C. López-Fanjul, and A. Caballero. 1999. Properties of spontaneous mutations affecting quantitative traits. *Genet. Research* 74:341–350.

Garza, J.C., M. Slatkin, and N.B. Freimer. 1995. Microsatellite allele frequencies in humans and chimpanzees, with implications for constraints on allele size. *Mol. Biol. Evol.* 12:594–603.

Gause, G.F. 1934 [1971]. *The Struggle for Existence.* Baltimore: Williams and Wilkins. Reprinted in 1971 by Dover Publications.

Gavrilets, S. and A. Hastings. 1995. Intermittency and transient chaos from simple frequency-dependent selection. *Proc. Royal Soc. Lond. B* 261:233–238.

Gershowitz, H., P.C. Junqueira, F.M. Salzano, and J.V. Neel. 1967. Further studies on the Xavante Indians. III. Blood groups and the *ABH-Le^a* secretor types in the Simoes Lopes and Sao Marcos Xavantes. *Amer. J. Human Genet.* 19:502–513.

Gibbs, H.L. and P.R. Grant. 1987. Oscillating selection on Darwin's finches. *Nature* 327:511–513.

Gibbs, H.L., P.J. Weatherhead, P.T. Boag, B.N. White, L.M. Tabak, and D.J. Hoysak. 1990. Realized reproductive success of polygynous red-winged blackbirds revealed by DNA markers. *Science* 250:1394–1397.

Gillespie, J.H. 1989. Lineage effects and the index of dispersion of molecular evolution. *Mol. Biol. Evol.* 6:636–647.

Gillespie, J.H. 1991. *The Causes of Molecular Evolution.* New York and Oxford: Oxford University Press.

Gillespie, J.H. 1994. Alternatives to the neutral theory. In B. Golding (ed.), *Non Neutral Evolution: Theories and Molecular Data.* New York: Chapman and Hall, 1–17.

Gillespie, J.H. 1995. On Ohta's hypothesis: Most amino acid substitutions are deleterious. *J. Mol. Evol.* 40:64–69.

Gillespie, J.H. 1998. *Population Genetics: A Concise Guide.* Baltimore: Johns Hopkins University Press.

Gillespie, J.H. and C.H. Langley. 1974. A general model to account for enzyme variation in natural populations. *Genetics* 76:837–848.

Gilman, M. and R. Hails. 1997. *An Introduction to Ecological Modeling: Putting Practice into Theory.* Oxford: Blackwell Science.

Gilpin, M.E. and M.E. Soule. 1986. Minimum viable populations: Processes of species extinction. In M.E. Soule (ed.), *Conservation Biology: The Science of Scarcity and Diversity.* Sunderland, MA: Sinauer Associates, Inc., 19–34.

Ginzburg, L.R. 1992. Evolutionary consequences of basic growth equations. *Trends Ecol. Evol.* 7:133.

Girman, D.J. 1996. The use of PCR-based single-stranded conformation polymorphism analysis (PCR-SSCP) in population genetics. In T.B. Smith and R.K. Wayne (eds.), *Molecular Genetic Approaches to Conservation*. New York: Oxford University Press, 167–182.

Gold, H.J. 1977. *Mathematical Modeling of Biological Systems*. New York: John Wiley & Sons.

Goldstein, D.B. and D.D. Pollock. 1997. Launching microsatellites: A review of mutation processes and methods of phylogenetic inference. *J. Heredity*. 88:335–342.

Goldstein, D.B., A. R. Linares, L.L. Cavalli-Sforza, and M.W. Feldman. 1995. Genetic absolute dating based on microsatellites and the origin of modern humans. *Proc. Natl. Acad. Sci. USA* 92:6723–6727.

Goodman, M., W. Moore, and G. Matsuda. 1975. Darwinian evolution in the genealogy of haemoglobin. *Nature* 253:603–608.

Goodman, S.J. 1997. R_{ST} Calc: A collection of computer programs for calculating estimates of genetic differentiation from microsatellite data and determining their significance. *Mol. Ecol.* 6:881–885.

Gotelli, N.J. 2001. *A Primer of Ecology*, 3rd ed. Sunderland, MA: Sinauer Associates Inc.

Gottlieb, L.D. and N.F. Weeden. 1979. Gene duplication and phylogeny in *Clarkia*. *Evolution* 33:1024–1039.

Gottlieb, L.D. and V.S. Ford. 1996. Phylogenetic relationships among the sections of *Clarkia* inferred from its nucleotide sequences of PgiC. *System. Botany* 21:45–62.

Gottlieb, L.D. and V.S. Ford. 1997. A recently silenced duplicate PgiC locus in *Clarkia*. *Mol. Biol. Evol.* 14:125–132.

Goudet, J. 1995. FSTAT (Version 1.2): A computer program to calculate *F*-statistics. *J. Hered.* 86:485–486.

Goudet, J. 2000. FSTAT: A program to estimate and test gene diversities and fixation indices (version 2.9.1). Available from http://www.unil.ch/izea/softwares/fstat.html. Updated from Goudet (1995).

Gould, J.L. and W.T. Keeton. 1996. *Biological Science*, 6th ed. New York: W.W. Norton and Company.

Gould, S.J. 1989. *Wonderful Life*. New York: W.W. Norton Co.

Grant, B.R. and P.R. Grant. 1989. *Evolutionary Dynamics of a Natural Population: The Large Cactus Finch of the Galapagos*. Chicago: Univ. of Chicago Press.

Grant, B.S. 1999. Fine tuning the peppered moth paradigm. *Evolution* 53:980–984.

Grant, B.S., D.F. Owen, and C.A. Clarke. 1996. Parallel rise and fall of melanic peppered moths in America and Britain. *J. Hered.* 87:351–357.

Grant, P.R. 1986. *Ecology and Evolution of Darwin's Finches*. Princeton, NJ: Princeton Univ. Press.

Grant, P.R. and B.R. Grant. 1995. Predicting microevolutionary responses to directional selection on heritable variation. *Evolution* 49:241–251.

Grant, P.R. and B.R. Grant. 2000. Non-random fitness variation in two populations of Darwin's finches. *Proc. Royal Soc. Lond. B* 267:131–138.

Graur, D. and W-H Li. 2000. *Fundamentals of Molecular Evolution*, 2nd ed. Sunderland, MA: Sinauer Associates.

Greaves, J.H., R. Redfern, P.B. Ayres, and J.E. Gill. 1977. Warfarin resistance: a balanced polymorphism in the Norway rat. *Genet. Res.* 30:257–263.

Greenberg, R. and J.F. Crow. 1960. A comparison of the effect of lethal and detrimental chromosomes from *Drosophila* populations. *Genetics* 45:1153–1168.

Greenway, Jr, J.C. 1967. *Extinct and Vanishing Birds of the World*. New York: Dover Publications, Inc.

Greig, J.C. 1979. Principles of genetic conservation in relation to wildlife management in Southern Africa. *S. Afr. Tydskr. Naturnav* 9:57–78.

Grenfell, B. and J. Harwood. 1997. (Meta)population dynamics of infectious diseases. *Trends Ecol. Evol.* 12:395–399.

Griffiths, A.J.F., J.H. Milleer, D.T. Suzuki, R.C. Lewontin, and W.M. Gelbart. 1996. *An Introduction to Genetic Analysis*, 6th ed. New York: W.H. Freeman.

Griffiths, A.J.F., J.H. Milleer, D.T. Suzuki, R.C. Lewontin, and W.M. Gelbart. 2000. *An Introduction to Genetic Analysis*, 7th ed. New York: W.H. Freeman.

Guilfoile, P. 2000. *A Photographic Atlas for the Molecular Biology Laboratory*. Englewood, CO: Morton Publishing Co.

Gusella, J.F. et al. 1983. A polymorphic DNA marker genetically linked to Huntington's disease. *Nature* 306:234–238.

Guttman, D.S. 1997. Recombination and clonality in natural populations of *Escherichia coli*. *Trends Ecol. Evol.* 12:16–22.

Halbach, U. 1979. Introductory remarks: strategies in population research exemplified by rotifer population dynamics. *Fortschritte der Zoologie* 25:1–27.

Haldane, J.B.S. 1924a. A mathematical theory of natural and artificial selection. I. *Trans. Camb. Phil. Soc.* 23:19–41.

Haldane, J.B.S. 1924b. A mathematical theory of natural and artificial selection. II. *Biol. Proc. Camb. Phil. Soc.* 1:158–163.

Haldane, J.B.S. 1926. A mathematical theory of natural and artificial selection. III. *Proc. Camb. Phil. Soc.* 23:363–372.

Haldane, J.B.S. 1927a. A mathematical theory of natural and artificial selection. IV. *Proc. Camb. Phil. Soc.* 23:607–615.

Haldane, J.B.S. 1927b. A mathematical theory of natural and artificial selection. V. Selection and mutation. *Proc. Camb. Phil. Soc.* 23:838–844.

Haldane, J.B.S. 1930. A mathematical theory of natural and artificial selection. VI. Isolation. *Proc. Camb. Phil. Soc.* 26:220–230.

Haldane, J.B.S. 1931a. A mathematical theory of natural and artificial selection. VII. Selection intensity as a function of mortality rate. *Proc. Camb. Phil. Soc.* 27:131–136.

Haldane, J.B.S. 1931b. A mathematical theory of natural and artificial selection. VIII. Metastable populations. *Proc. Camb. Phil. Soc.* 27:137–142.

Haldane, J.B.S. 1932a. A mathematical theory of natural and artificial selection. IX. Rapid selection. *Proc. Camb. Phil. Soc.* 28:244–248.

Haldane, J.B.S. 1932b [1966]. *The Causes of Evolution*. New York and London: Harper & Brothers; Reprint, Ithaca, NY: Cornell University Press.

Haldane, J.B.S. 1937. The effect of variation on fitness. *Amer. Natur.* 71:337–349.

Haldane, J.B.S. 1949. Disease and evolution. *Ricerca Scient. Supp.* 19:68–76.

Haldane, J.B.S. 1954. The measurement of natural selection. *Proc. 9th Intl. Congr. Genet.* (Caryologia, Supplement to Vol 6) 1:480–487.

Haldane, J.B.S. and S.D. Jayakar. 1963. Polymorphism due to selection of varying direction. *J. Genet.* 58:237–242.

Hall, B.G. 2001. *Phylogenetic Trees Made Easy: A How-To Manual for Molecular Biologists*. Sunderland, MA: Sinauer Associates.

Halliburton, R. 1993. The great Australian blight: Ecological invasion and biological control in Australia. *Conn. Review* 15(2):1–10.

Halliburton, R. and G.A.E. Gall. 1981. Disruptive selection and assortative mating in *Tribolium castaneum*. *Evolution* 35:829–843.

Halliburton, R. and J.S.F. Barker. 1993. Lack of mitochondrial DNA variation in Australian *Drosophila buzzatii*. *Mol. Biol. Evol.* 10:484–487.

Hammer, M.F. 1995. A recent common ancestry for human Y chromsomes. *Nature* 378:376–378.

Hammer, M.F., A.P. Spurdle, T. Karafet, M.R. Bonner, E.T. Wood, A. Novelletto, P. Malaspina, R.J. Mitchell, S. Horai, T. Jenkins, and S.L. Zegura. 1997. The geographic distribution of human Y chromosome variation. *Genetics* 145:787–805.

Hammer, M.F., T. Karafet, A. Rasanayagam, E.T. Wood, T.K. Altheide, T. Jenkins, R.C. Griffiths, A.R. Templeton, and S.L. Zegura. 1998. Out of Africa and back again: Nested cladistic analysis of human Y chromosome variation. *Mol. Biol. Evol.* 15:427–441.

Hansen, M.M., R.A. Hynes, V. Loeschcke, and G. Rasmussen. 1995. Assessment of the stocked or wild origin of anadromous brown trout (*Salmo trutta* L.) in a Danish river system, using mitochondrial DNA RFLP analysis. *Mol. Ecol.* 4:189–198.

Hardy, G.H. 1908. Mendelian proportions in a mixed population. *Science* 28:49–50.

Harris, H. 1966. Enzyme polymorphism in man. *Proc. Royal Soc. Lond. B.* 164:298–310.

Harris, H. and D.A. Hopkinson. 1976. *Handbook of Enzyme Electrophoresis*. New York: American Elsevier.

Harris, R.B. and F.W. Allendorf. 1989. Genetically effective population size of large mammals: Assessment of estimators. *Cons. Biol.* 3:181–191.

Hartl, D.L. and A.G. Clark. 1997. *Principles of Population Genetics*, 3rd ed. Sunderland, MA: Sinauer Associates.

Hastings, A. 1985. Four simultaneously stable polymorphic equilibria in two locus, two allele models. *Genetics* 109:255–261.

Hastings, A. 1986. Multilocus population genetics with weak epistasis. II. Equilibrium properties of multilocus models: What is the unit of selection? *Genetics* 112:157–171.

He, M. and D.S. Haymer. 1999. Genetic relationships of populations and the origins of new infestations of the Mediterranean fruit fly. *Mol. Ecol.* 8:1247–1257.

Hedges, S.B., S. Kumar, K. Tamura, and M. Stoneking. 1991. Human origins and analysis of mitochondrial DNA sequences. *Science* 255:737–739.

Hedrick, P.W. 1986. Genetic polymorphism in heterogeneous environments: A decade later. *Ann. Rev. Ecol. Syst.* 17:535–566.

Hedrick, P.W. 1987. Gametic disequilibrium measures: proceed with caution. *Genetics* 117:331–341.

Hedrick, P.W. 1994. Evolutionary genetics of the major histocompatibility complex. *Amer. Natur.* 143:945–964.

Hedrick, P.W. 1999a. Perspective: Highly variable genetic loci and their interpretation in evolution and conservation. *Evolution* 53:313–318.

Hedrick, P.W. 1999b. Antagonistic pleiotropy and genetic polymorphism: a perspective. *Heredity* 82:126–133.

Hedrick, P.W. and S.T. Kalinowski. 2000. Inbreeding depression in conservation biology. *Ann. Rev. Ecol. Syst.* 31:139–162.

Hedrick, P.W. and T.J. Kim. 2000. Genetics of complex polymorphisms: Parasites and maintenance of the major histocompatibility complex variation. In R.S. Singh and C.B. Costas (eds.), *Evolutionary genetics: From molecules to morphology*, Cambridge: Cambridge University Press, 204–234.

Hedrick, P.W. and E. Murray. 1978. Average heterozygosity revisited. *Amer. J. Hum. Genet.* 30:377–382.

Hedrick, P., S. Jain, and L. Holden. 1978. Multilocus systems in evolution. *Evol. Biol.* 11:101–184.

Heino, M., J.A.J. Metz, and V. Kaitala. 1998. The enigma of frequency-dependent selection. *Trends Ecol. Evol.* 13:367–370.

Hey, J. 1999. The neutralist, the fly and the selectionist. *Trends Ecol. Evol.* 14:35–38.

Hill, W.G. 1974. Estimation of linkage disequilibrium in randomly mating populations. *Heredity* 33:229–239.

Hillis, D.M., J.P. Huelsenbeck, and C.W. Cunningham. 1994. Application and accuracy of molecular phylogenies. *Science* 264:671–677.

Hochachka, P.W. and G.N. Somero. 1984. *Biochemical Adaptation*. Princeton, NJ: Princeton University Press.

Hoekstra, R.F., R. Bijlsma, and A.J. Dolman. 1985. Polymorphism from environmental heterogeneity: models are only robust if the heterozygote is close in fitness to the favoured homozygote in each environment. *Genet. Res.* 45:299–314.

Hoffman, A.A. and P.A. Parsons. 1991. *Evolutionary Genetics and Environmental Stress*. Oxford: Oxford University Press.

Holland, B. and W.R. Rice. 1998. Perspective: Chase-away sexual selection: antagonistic seduction versus resistance. *Evolution* 52:1–7.

Holland, B. and W.R. Rice. 1999. Experimental removal of sexual selection reverses intersexual antagonistic coevolution and removes a reproductive load. *Proc. Natl. Acad. Sci. USA* 96:5083–5088.

Houle, D. 1992. Comparing evolvability and variability of quantitative traits. *Genetics* 130:195–204.

Houle, D., D.K. Hoffmaster, S. Assimacopoulos, and B. Charlesworth. 1992. The genomic mutation rate for fitness in *Drosophila*. *Nature* 359:58–60.

Houle, D., B. Morikawa, and M. Lynch. 1996. Comparing mutational variabilities. *Genetics* 143:1467–1483.

Hubby, J.L. and R.C. Lewontin. 1966. A molecular approach to the study of genic heterozygosity in natural populations. I. The number of alleles at different loci in *Drosophila pseudoobscura*. *Genetics* 54:577–594.

Hudson, R.R. 1990. Gene genealogies and the coalescent process. *Oxford Surv. Evol. Biol.* 7:1–44.

Hudson, R.R., K. Bailey, D. Skarecky, J. Kwiatowski, and F.J. Ayala. 1994. Evidence for positive selection in the superoxide dismutase (*Sod*) region of *Drosophila melanogaster*. *Genetics* 136:1329–1340.

Hudson, R.R., M. Kreitman, and M. Aguadé. 1987. A test of neutral molecular evolution based on nucleotide data. *Genetics* 116:153–159.

Hudson, R.R., A.G. Sáez, and F.J. Ayala. 1997. DNA variation at the *Sod* locus of *Drosophila melanogaster*: An unfolding story of natural selection. *Proc. Natl. Acad. Sci. USA* 94:7725–7729.

Hudson, R.R., M. Slatkin, and W.P. Maddison. 1992. Estimation of levels of gene flow from DNA sequence data. *Genetics* 132:583–589.

Huelsenbeck, J.P. and K.A. Crandall. 1997. Phylogeny estimation and hypothesis testing using maximum likelihood. *Ann. Rev. Ecol. Syst.* 28:437–466.

Huelsenbeck, J.P., F. Ronquist, R. Nielsen, and J.P. Bollback. 2001. Bayesian inference of phylogeny and its impact on evolutionary biology. *Science* 294:2310–2314.

Hughes, A.L. 1999. *Adaptive Evolution of Genes and Genomes*. Oxford: Oxford University Press.

Hughes, A.L. and M. Nei. 1988. Pattern of nucleotide substitution at major histocompatibility complex class I loci reveals overdominant selection. *Nature* 335:167–170.

Hughes, A.L. and M. Nei. 1989. Nucleotide substitution at major histocompatibility complex class II loci: Evidence for overdominant selection. *Proc. Natl. Acad. Sci. USA* 86:958–962.

Hughes, C.R. and D.C. Queller. 1993. Detection of highly polymorphic microsatellite loci in a species with little allozyme polymorphism. *Mol. Ecol.* 2:131–137.

Husband, B.C. and S.C.H. Barrett. 1992. Genetic drift and the maintenance of the style length polymorphism in tristylous populations of *Eichhornia paniculata* (Pontederiaceae). *Heredity* 69:440–449.

Ioerger, T.R., A.G. Clark, and T. Kao. 1990. Polymorphism at the self-incompatibility locus in Solanaceae predates speciation. *Proc. Natl. Acad. Sci. USA* 87:9732–9735.

Irwin, S.D., K.A. Wetterstrand, C.M. Hutter, and C.F. Aquadro. 1998. Genetic variation and differentiation at microsatellite loci in *Drosophila simulans*: Evidence for founder effects in New World populations. *Genetics* 150:777–790.

Istock, C.A. and R.S. Hoffman (eds.). 1995. *Storm Over a Mountain Island*. Tucson: Univ. of Arizona Press.

Istock, C.A., K.E. Duncan, N. Ferguson, and X. Zhou. 1992. Sexuality in a natural population of bacteria—*Bacillus subtilis* challenges the clonal paradigm. *Mol. Ecol.* 1:95–103.

Jacquard, A. 1975. Inbreeding: One word, several meanings. *Theor. Pop. Biol.* 7:338–363.

Jarne, P. and P.J.L. Lagoda. 1996. Microsatellites, from molecules to populations and back. *Trends Ecol. Evol.* 11:424–429.

Jeffs, P.S., E.C. Holmes, and M. Ashburner. 1994. The molecular evolution of the alcohol dehydrogenase and alcohol dehydrogenase-related genes in the *Drosophila melanogaster* species subgroup. *Mol. Biol. Evol.* 11:287–304.

Johannsen, W.L. 1903. Elemente der exakten erblichkeitslehre. Gustav Fischer, Hena, Germany. Translated and reprinted in R. Milkman (ed.), *Benchmark papers in Genetics*. Vol.13. Stroudsberg, PA: Hutchinson Ross Publishing Co.

Johnson, M.S. and R. Black. 1998. Increased genetic divergence and reduced genetic variation in populations of the snail *Bembicium vittatum* in isolated tidal ponds. *Heredity* 80:163–172.

Johnson, T.C., C.A. Scholz, M.R. Talbot, K. Kelts, R.D. Rickets, G. Ngobi, K Beuning, I. Ssemmandra, and J.W. McGill. 1996. Late pleistocene dessication of Lake Victoria and rapid evolution of cichlid fishes. *Science* 273:1091–1093.

Johnston R.F. and R.K. Selander. 1964. House sparrows: Rapid evolution of races in North America. *Science* 144:548–550.

Johnson, W.E. and R.K. Selander. 1971. Protein variation and systematics in kangaroo rats (genus *Dipodomys*). *Syst. Zool.* 20:377–405.

Jones, H.L. and J.M. Diamond. 1976. Short-time-base studies of turnover in breeding bird populations on the California Channel Islands. *Condor* 78:526–549.

Jones, J.S., B.H. Leith, and P. Rawlings. 1977. Polymorphism in *Cepaea*: A problem with too many solutions? *Ann. Rev. Ecol. Syst.* 8:109–143.

Jorde, L.B., J.C. Carey, and R.L. White. 1995. *Medical Genetics*. St. Louis: Mosby.

Jukes, T.H. and C.R. Cantor. 1969. Evolution of protein molecules. In H.N. Munro (ed.), *Mammalian Protein Metabolism*. New York: Academic Press, 21–132.

Kacser, H. and J.A. Burns. 1981. The molecular basis of dominance. *Genetics* 97:639–666.

Kaplan, N., R.R. Hudson, and M. Izuka. 1991. The coalescent process in models with selection, recombination and geographic subdivision. *Genet. Res.* 57:83–91.

Kaplan, N.L., R.R. Hudson, and C.H. Langley. 1989. The "hitchhiking effect" revisited. *Genetics* 123:887–899.

Kaplinsky, N., D. Braun, D. Lisch, A. Hay, S. Hake, and M. Freeling. 2002. Maize transgene results in Mexico are artefacts. *Nature* 416:601.

Karlin, S. 1982. Classifications of selection-migration structures and conditions for a protected polymorphism. *Evol. Biol.* 14:61–204.

Karlin, S. and M.W. Feldman. 1970. Linkage and selection: Two locus symmetric viability model. *Theor. Pop. Biol.* 1:37–71.

Karlin, S. and U. Liberman. 1976. A phenotypic symmetric selection model for three loci, two alleles: the case of tight linkage. *Theor. Pop. Biol.* 10:334–364.

Karn, M.N. and L.S. Penrose. 1951. Birth weight and gestation time in relation to maternal age, parity, and infant survival. *Ann. Eugen.* 15:206–233.

Kaufman, P.K., F.D. Enfield, and R.E. Comstock. 1977. Stabilizing selection for pupa weight in *Tribolium castaneum*. *Genetics* 87:327–341.

Kawabe, A., K. Yamane, and N.T. Miyashita. 2000. DNA polymorphism at the cytosolic phosphoglucose isomerase (PgiC) locus of the wild plant *Arabidopsis thaliana*. *Genetics* 156:1339–1347.

Kearsey, M.J. and B.W. Barnes. 1970. Variation for metrical characters in *Drosophila* populations. *Heredity* 25:11–21.

Keightley, P.D. 1996. Nature of deleterious mutation load in *Drosophila*. *Genetics* 144: 1993–1999.

Keightley, P.D. and A. Caballero. 1997. Genomic mutation rates for lifetime reproductive output and lifespan in *Caenorhabditis elegans*. *Proc. Natl. Acad. Sci. USA* 94: 3823–3827.

Keightley, P.D. and A. Eyre-Walker. 1999. Terumi Mukai and the riddle of deleterious mutation rates. *Genetics* 153:515–523.

Keightley, P.D. and A. Eyre-Walker. 2000. Deleterious mutations and the evolution of sex. *Science* 290:331–333.

Keller, L.F. 1998. Inbreeding and its fitness effects in an insular population of song sparrows (*Melospiza melodia*). *Evolution* 52:240–250.

Keller, L.F. and P. Arcese. 1998. No evidence for inbreeding avoidance in a natural population of song sparrows (*Melospiza melodia*). *Amer. Natur.* 152:380–392.

Keller, L.F. and D.M. Waller. 2002. Inbreeding effects in wild populations. *Trends Ecol. Evol.* 17:230–241.

Keller, L.F., P. Arcese, J.N.M. Smith, W.M. Hochachka, and S.C. Stearns. 1994. Selection against inbred song sparrows during a natural population bottleneck. *Nature* 372:356–357.

Kennedy, P.K., M.L. Kennedy, P.L. Clarkson, and I.S. Liepins. 1991. Genetic variability in natural populations of the gray wolf, *Canis lupis*. *Canad. J. Zool.* 69:1183–1188.

Kere, J. 2001. Human population genetics: Lessons from Finland. *Ann. Rev. Genomics Hum. Genet.* 2:103–128.

Kettlewell, H.B.D. 1973. *The Evolution of Melanism*. Oxford: Clarendon Press.

Kibota, T.T. and M. Lynch. 1996. Estimate of the genomic mutation rate deleterious to overall fitness in *E. coli*. *Nature* 381:694–696.

Kidwell, J.F., M.T. Clegg, F.M. Stewart, and T. Prout. 1977. Regions of stable equilibria for models of differential selection in the two sexes under random mating. *Genetics* 85:171–183.

Kimura, M. 1957. Some problems of stochastic processes in genetics. *Annals Math. Stat.* 28:882–901.

Kimura, M. 1962. On the probability of fixation of mutant genes in a population. *Genetics* 47:713–719.

Kimura, M. 1965. A stochastic model concerning the maintenance of genetic variability in quantitative characters. *Proc. Natl. Acad. Sci. USA* 54:731–736.

Kimura, M. 1968. Evolutionary rate at the molecular level. *Nature* 217:624–626.

Kimura, M. 1969. The number of heterozygous nucleotide sites maintained in a finite population due to steady flux of mutations. *Genetics* 61:893–903.

Kimura, M. 1980. A simple method for estimating evolutionary rates of base substitutions through comparative studies of nucleotide sequences. *J. Mol. Evol.* 16:111–120.

Kimura, M. 1983a. *The Neutral Theory of Molecular Evolution*. Cambridge: Cambridge Univ. Press.

Kimura, M. 1983b. Rare variant alleles in the light of the neutral theory. *Mol. Biol. Evol.* 1:84–93.

Kimura, M. 1986. DNA and the neutral theory. *Phil. Trans. Royal Soc. Lond. B.* 312:343–354.

Kimura, M. and J.F. Crow. 1964. The number of alleles that can be maintained in a finite population. *Genetics* 49:725–738.

Kimura, M. and T. Maruyama. 1966. The mutational load with epistatic gene interactions in fitness. *Genetics* 54:1337–1351.

Kimura, M. and T. Ohta. 1971a. Protein polymorphism as a phase of molecular evolution. *Nature* 229:467–469.

Kimura, M. and T. Ohta. 1971b. On the rate of molecular evolution. *J. Mol. Evol.* 1:1–17.

Kimura, M. and T. Ohta. 1971c. *Theoretical Aspects of Population Genetics*. Princeton: Princeton University Press.

King, J.L. and T.H. Jukes. 1969. Non-Darwinian evolution. *Science* 164:788–798.

King, R.B. 1987. Color pattern polymorphism in the Lake Erie water snake, *Nerodia sipedon insularum*. *Evolution* 41:241–255.

King, R.B. 1993a. Color pattern variation in Lake Erie water snakes: Inheritance. *Canad. J. Zool.* 71:1985–1990.

King, R.B. 1993b. Color-pattern variation in Lake Erie water snakes: Prediction and measurement of natural selection. *Evolution* 47:1819–1833.

King, R.B. and R. Lawson. 1995. Color-pattern variation in Lake Erie water snakes: The role of gene flow. *Evolution* 49:885–896.

Kingman, J.F.C. 1961. A mathematical problem in population genetics. *Proc. Camb. Phil. Soc.* 57:574–582.

Kingsolver, J.G., H.E. Hoekstra, J.M. Hoekstra, D. Berrigan, S.N. Vignieri, C.E. Hill, A. Hoang, P. Gilbert, and P. Beerli. 2001. The strength of phenotypic selection. *Amer. Natur.* 157:245–261.

Kirby, D.A., S.V. Muse, and W. Stephan. 1995. Maintenance of pre-mRNA secondary structure by epistatic selection. *Proc. Natl. Acad. Sci. USA* 92:9047–9051.

Knibb, W.R., P.D. East, and J.S.F. Barker. 1987. Polymorphic inversion and esterase loci complex on chromosome 2 of *Drosophila buzzatii*. I. Linkage disequilibria. *Aust. J. Biol. Sci.* 40:257–269.

Kojima, K-I. 1971. Is there a constant fitness value for a given genotype? No! *Evolution* 25:281–258.

Kojima, K-I. and K.M. Yarbrough. 1967. Frequency-dependent selection at the esterase 6 locus in *Drosophila melanogaster*. *Proc. Natl. Acad. Sci.* 57:645–649.

Kondrashov, A.S. 1988. Deleterious mutations and the evolution of sexual reproduction. *Nature* 336:435–440.

Kondrashov, A.S. 1995. Contamination of the genome by very slightly deleterious mutations: Why have we not died 100 times over? *J. Theor. Biol.* 175:583–594.

Kondrashov, A.S. 2001. Sex and U. *Trends in Genet.* 17:75–77.

Koonin, E.V., L. Aravind, and A.S. Kondrashov. 2000. The impact of comparative genomics on our understanding of evolution. *Cell* 101:573–576.

Krebs, C.J. 1994. *Ecology: The Experimental Analysis of Distribution and Abundance*, 4th ed. New York: Harper Collins.

Krebs, R.A. and J.S.F. Barker. 1993. Coexistence of ecologically similar colonising species. II. Population differentiation in *Drosophila aldrichi* and *D. buzzatii* for competitive effects and responses at different temperatures and allozyme variation in *D. aldrichi*. *J. Evol. Biol.* 6:281–298.

Kreitman, M. 1983. Nucleotide polymorphism at the alcohol dehydrogenase locus of *Drosophila melanogaster*. *Nature* 304:412–417.

Kreitman, M. 1996. The neutral theory is dead. Long live the neutral theory. *Bioessays* 18:678–683.

Kreitman, M. 2000. Methods to detect selection in populations with applications to the human. *Ann. Rev. Genomics Human Genet.* 1:539–559.

Kreitman, M. and H. Akashi. 1995. Molecular evidence for natural selection. *Ann. Rev. Ecol. Syst.* 26:403–422.

Kruuk, L.E.B., T.H. Clutton-Brock, J. Slate, J.M. Pemberton, and S. Brotherstone. 2000. Heritability of fitness in a wild mammal population. *Proc. Natl. Acad. Sci. USA* 97:698–703.

Laake, K., M. N. Telatar, G.A. Geitvik, R.O. Hansen, A. Heilberg, A.M. Andresen, R. Gatti, and A.L. Borresendale. 1998. Identical mutation in 55% of the ATM alleles in 11 Norwegian AT families: Evidence for a founder effect. *Eur. J. Human Genet.* 6:235–244.

Lacy, R.C. 1987. Loss of genetic diversity from managed populations: Interacting effects of drift, mutation, immigration, selection, and population subdivision. *Conserv. Biol.* 1:143–158.

Lamotte, M. 1959. Polymorphism of natural populations of *Cepaea nemoralis*. *Cold Spring Harbor Symp. Quant. Biol.* 24:65–86.

Lande, R. 1975. The maintenance of genetic variability by mutation in a polygenic character with linked loci. *Genet. Res.* 26:221–235.

Lande, R. 1988. Genetics and demography in biological conservation. *Science* 241:1455–1460

Lande, R. 1994. Mutation and conservation. *Conserv. Biol.* 9:782–791.

Lande, R. and S.J. Arnold. 1983. The measurement of selection on correlated characters. *Evolution* 37:1210–1226.

Lander, E.S. and D. Botstein. 1989. Mapping Mendelian factors underlying quantitative traits using RFLP linkage maps. *Genetics* 121:185–199.

Lander, E.S. and N.J. Schork. 1994. Genetic dissection of complex traits. *Science* 265:2037–2048.

Langley, C.H. and W.M. Fitch. 1974. An examination of the constancy of the rate of molecular evolution. *J. Mol. Evol.* 3:161–177.

Langley, C.H., E. Montgomery, and W.F. Quattlebaum. 1982. Restriction map variation in the Adh region of *Drosophila*. *Proc. Natl. Acad. Sci. USA* 79:5631–5635.

Langley, C.H., Y.N. Tobari, and K. Kojima. 1974. Linkage disequilibrium in natural populatins of *Drosophila melanogaster*. *Genetics* 78:921–936.

Langley, C.F., R.A. Voelker, A.J.L. Brown, S. Ohnishi, B. Dickson, and E. Montgomery. 1981. Null allele frequencies at allozyme loci in natural populations of *Drosophila melanogaster*. *Genetics* 99:151–156.

Larder, B.A. and S.D. Kemp. 1989. Multiple mutations in HIV-1 reverse transcriptase confer high-level resistance to zidovudine (AZT). *Science* 246:1155–1158.

Larget, B. and D.L. Simon. 1999. Markov Chain Monte Carlo algorithms for the Bayesian analysis of phylogenetic trees. *Mol. Biol. Evol.* 16:750–759.

Latter, B.D.H. 1960. Natural selection for an intermediate optimum. *Aust. J. Biol Sci.* 13:30–35.

Lawrence, J.G. and J.R. Roth. 1996. Selfish Operons: Horizontal transfer may drive the evolution of gene clusters. *Genetics* 143:1843–1860.

Lebda, L. 1999. Testing the *r* and *K* selection trade-off in populations of *Escherichia coli*. MA Thesis, Dept. of Biology, Western Connecticut State University, Danbury, CT.

Ledig, F.T. 1986. Heterozygosity, heterosis and fitness in outbreeding plants. In M.E. Soule (ed.), *Conservation Biology: The Science of Scarcity and Diversity*. Sunderland, MA: Sinauer Associates Inc., 77–104.

Ledig, F.T. 2000. Founder effects and the genetic structure of Coulter pine. *J. Hered.* 91:307–315.

Lee, Y-H, T. Ota, and V.D. Vacquier. 1995. Positive selection is a general phenomenon in the evolution of abalone sperm lysin. *Mol. Biol. Evol.* 12:231–238.

Lenski, R.E. and M. Travisano. 1994. Dynamics of adaptation and diversification: A 10,000-generation experiment with bacterial populations. *Proc. Natl. Acad. Sci. USA* 91:6808–6814.

Lenski, R.E., C.L. Winkworth, and M.A. Riley. 2003. Rates of DNA evolution in experimental populations of *Escherichia coli* during 20,000 generations. *J. Mol. Evol.* 56:498–508.

Les, D.H., J.A. Reinartz, and E.J. Esselman. 1991. Genetic consequences of rarity in *Aster furcatus* (Asteraceae), a threatened, self-incompatible plant. *Evolution* 45:1641–1650.

Levene, H. 1953. Genetic equilibrium when more than one ecological niche is available. *Amer. Natur.* 87:311–313.

Levin, B.R. 1986. Restriction-modification and the maintenance of genetic diversity in bacterial populations. In E. Nevo and S. Karlin (eds.), *Proceedings of a Conference on Evolutionary Processes and Theory*. New York: Academic Press.

Levin, B.R. 1988. Frequency-dependent selection in bacterial populations. *Phil. Trans. Royal Soc. Lond. B* 319:459–472.

Levin, B.R. 2000. The population biology of antibiotic resistance. In R.S. Singh and C.B. Krimbas (eds.), *Evolutionary Genetics: From Molecules to Morphology*. Cambridge: Cambridge University Press, 235–253.

Levin, D.A. and W.L. Crepet. 1973. Genetic variation in *Lycopodium lucidulum*: A phylogenetic relic. *Evolution* 27:622–632.

Levy, S.B. 1998. The challenge of antibiotic resistance. *Sci. Amer.* 278(3):46–53.

Lewin, B. 2000. *Genes VII*. Oxford: Oxford Univ. Press.

Lewis, P.O. 2001. Phylogenetic systematics turns over a new leaf. *Trends Ecol. Evol.* 16:30–37.

Lewontin, R.C. 1974. *The Genetic Basis of Evolutionary Change*. New York: Columbia University Press.

Lewontin, R.C. 1988. On measures of gametic disequilibrium. *Genetics* 120:849–852.

Lewontin, R.C. 1997. Dobzhansky's Genetics and the Origin of Species: Is it still Relevent? *Genetics* 147:351–355.

Lewontin, R.C. and C.C. Cockerham. 1959. The goodness-of-fit test for detecting natural selection in random mating populations. *Evolution* 13:561–564.

Lewontin, R.C. and J.L. Hubby. 1966. A molecular approach to the study of genic heterozygosity in natural populations. II. Amount of variation and degree of heterozygosity in natural populations of *Drosophila pseudoobscura*. *Genetics* 54:595–609.

Lewontin, R.C., J.A. Moore, W.B. Provine, and B. Wallace (eds.). 1981. *Dobzhansky's Genetics of Natural Populations I-XLIII*. New York: Columbia Univ. Press.

Lewontin, R.C., L.R. Ginzburg, and S.D. Tuljapurkar. 1978. Heterosis as an explanation for large amounts of genic polymorphism. *Genetics* 88:149–170.

Li, W-H. 1997. *Molecular Evolution*. Sunderland, MA: Sinauer Associates.

Liberman, U. and M.W. Feldman. 1986. A general reduction principle for genetic modifiers of recombination. *Theor. Pop. Biol.* 30:341–371.

Litt, M. and J.A. Luty. 1989. A hypervariable microsatellite revealed by in vitro amplification of a dinucleotide repeat within the cardiac muscle actin gene. *Amer. J. Hum. Genet.* 44:397–401.

Long, A.D., R.F. Lyman, A.H. Morgan, C.H. Langley, and T.F.C. Mackay. 2000. Both naturally occurring insertions of transposable elements and intermediate frequency polymorphisms at the achaete-scute complex are associated with variation in bristle number in *Drosophila melanogaster*. *Genetics* 154:1255–1269.

Long, A.D., S.L. Mullaney, L.A. Reid, J.D. Fry, C.H. Langley, and T.F.C. Mackay. 1995. High resolution mapping of genetic factors affecting abdominal bristle number in *Drosophila melanogaster*. *Genetics* 139:1273–1291.

Long, M. and C.H. Langley. 1993. Natural selection and the origin of *jingwei*, a chimeric processed functional gene in *Drosophila*. *Science* 260:91–95.

Lugon-Moulin, N., H. Brünner, A. Wyttenbach, J. Hauser, and J. Goudet. 1999. Hierarchical analyses of genetic differentiation in a hybrid zone of *Sorex araneus* (Insectivora: Soricidae). *Mol. Ecol.* 8:419–431.

Luikart, G. and P.R. England. 1999. Statistical analysis of microsatellite DNA data. *Trends Ecol. Evol.* 14:253–256.

Lynch, M. 1988. The rate of polygenic mutation. *Genet. Res.* 51:137–148.

Lynch, M. and J.S. Conery. 2000. The evolutionary fate and consequences of duplicate genes. *Science* 290:1151–1155.

Lynch, M. and T.J. Crease. 1990. The analysis of population survey data on DNA sequence variation. *Mol. Biol. Evol.* 7:377–394.

Lynch, M. and W. Gabriel. 1990. Mutation load and the survival of small populations. *Evolution* 44:1725–1737.

Lynch, M. and W.G. Hill. 1986. Phenotypic evolution by neutral mutation. *Evolution* 40:915–935

Lynch, M. and B. Walsh. 1998. *Genetics and Analysis of Quantitative Traits*. Sunderland, MA: Sinauer Associates, Inc.

Lynch, M., J. Conery, and R. Burger. 1995. Mutational meltdowns in sexual populations. *Evolution* 49:1067–1080.

Lynch, M., L. Latta, J. Hicks, and M. Giorgianni. 1998. Mutation, selection, and the maintenance of life-history variation in a natural population. *Evolution* 53:727–733.

Lynch, M., J. Blanchard, D. Houle, T. Kibota, S. Schultz, L. Vassilieva, and J. Willis. 1999a. Perspective: Spontaneous deleterious mutation. *Evolution* 53:645–663.

Lynch, M., M. Pfrender, K. Spitze, N. Lehman, D. Allen, J. Hicks, L. Latta, M. Ottene, F. Bogue, and J. Colbourne. 1999b. The quantitative and molecular genetic architecture of subdivided species. *Evolution* 53: 100–110.

MacArthur, R.H. and E.O. Wilson. 1967. *The Theory of Island Biogeography*. Princeton: Princeton Univ. Press.

Mace, G.M. and R. Lande. 1990. Assessing extinction threats: Toward a reevaluation of IUCN threatened species categories. *Conserv. Biol.* 5(2):148–157.

Mackay, T.F.C. 1985. A quantitative genetic analysis of fitness and its components in *Drosophila melanogaster*. *Genet. Res.* 47:59–70.

Mackay, T.F.C. 1995. The genetic basis of quantitative variation: numbers of sensory bristles of *Drosophila melanogaster* as a model system. *Trends Genet.* 11:464–470.

Mackay, T.F.C. 1996. The nature of quantitative genetic variation revisited: Lessons from *Drosophila* bristles. *Bioessays* 18:113–121.

Mackay, T.F.C. 2001. The genetic architecture of quantitative traits. *Ann. Rev. Genet.* 35:303–339.

Mackay, T.F.C., R.F. Lyman, and M.S. Jackson. 1992. Effects of P element insertions on quantitative traits in *Drosophila melanogaster*. *Genetics* 130:315–332.

Majerus, M.E.N. 1998. *Melanism: Evolution in Action*. New York and Oxford: Oxford University Press.

Malogolowkin-Cohen, Ch., A.S. Simmons, and H. Levene. 1965. A study of sexual isolation between certain strains of *Drosophila paulistorum*. *Evolution* 19:95–103.

Mandel, S.P.H. 1959. The stability of a multiple allelic system. *Heredity* 13:289–302.

Mann, C.C. 2002. Has GM corn 'invaded' Mexico? *Science* 295:1617–1618.

Marks, R.W. and H.G. Spencer. 1991. The maintenance of single-locus polymorphism. II. The evolution of fitnesses and allele frequencies. *Amer. Natur.* 138:1354–1371.

Maroni, G. 2001. *Molecular and Genetic Analysis of Human Traits.* Malden, MA: Blackwell Science Inc.

Marshall, D.R. and R.W. Allard. 1970. Maintenance of isozyme polymorphisms in natural populations of *Avena barbata*. *Genetics* 66:393–399.

Marshall, A.J. and S.K. Jain. 1969. Interference in pure and mixed populations of *Avena fatua* and *A. barbata*. *J. Ecol.* 57:251–270.

Martinez, D.E. and J. Levinton. 1996. Adaptation to heavy metals in the aquatic oligochaete *Limnodrilus hoffmeisteri*: Evidence for control by one gene. *Evolution* 50:1339–1343.

May, R.M. 1985. Evolution of pesticide resistance. *Nature* 315:12–13.

Maynard Smith, J. 1966. Sympatric speciation. *Amer. Natur.* 100:637–650.

Maynard Smith, J. 1970. Genetic polymorphism in a varied environment. *Amer. Natur.* 104:487–490.

Maynard Smith, J. 1974. *Models in Ecology.* Cambridge: Cambridge Univ. Press.

Maynard Smith, J. 1998. *Evolutionary Genetics*, 2nd ed. Oxford: Oxford University Press.

Maynard Smith, J. and J. Haigh. 1974. The hitch-hiking effect of a favourable gene. *Genet. Res.* 23:23–35.

Maynard Smith, J. and R. Hoekstra. 1980. Polymorphism in a varied environment: how robust are the models? *Genet. Res.* 35:45–57.

Maynard Smith, J., N.H. Smith, M. O'Rourke, and B.G. Spratt. 1993. How clonal are bacteria? *Proc. Natl. Acad. Sci. USA* 90:4384–4388.

Mayr, E. 1940. Speciation phenomena in birds. *Amer. Natur.* 74:249–278.

Mayr, E. 1963. *Animal Species and Evolution.* Cambridge, MA: Harvard Univ. Press.

McCommas, S.A. and E.H. Bryant. 1990. Loss of electrophoretic variation in serially bottlenecked populations. *Heredity* 64:315–321.

McDonald, J.H. and M. Kreitman. 1991. Adaptive protein evolution at the Adh locus in *Drosophila*. *Nature* 351:652–654.

McKechnie, S.W. and B.W. Geer. 1993. Micro-evolution in a wine cellar population: an historical perspective. *Genetica* 90:201–215.

McKenzie, J.A. and P. Batterham. 1994. The genetic, molecular and phenotypic consequences of selection for insecticide resistance. *Trends Ecol. Evol.* 9:166–169.

McKusick, V.A., J.A. Hostetler, J.A. Egeland, and R. Eldridge. 1964. The distribution of certain genes in the Old Order Amish. *Cold Spring Harbor Symposium on Quantitative Biology* 29:99–114.

Meffe, G.K., C.R. Carrol, and Contributors. 1997. *Principles of Conservation Biology.* Sunderland, MA: Sinauer Associates Inc.

Menges, E.S. 1990. Population viability analysis for an endangered plant. *Conserv. Biol.* 4(1):52–62.

Menges, E.S. 1991. The application of minimum viable population theory to plants. In D.A. Falk and K. Holsinger (eds.), *Genetics and Conservation of Rare Plants.* New York: Oxford University Press.

Merilä, J. and B.C. Sheldon. 2000. Lifetime reproductive success and heritability in nature. *Amer. Natur.* 155:301–310.

Merola, M. 1994. A reassment of homozygosity and the case for inbreeding depression in the cheetah, *Acinonyx jubatus*: Implications for conservation. *Conserv. Biol.* 8:961–971.

Mettler, L.E. and T.G. Gregg. 1969. *Population Genetics and Evolution.* Englewood Cliffs, NJ: Prentice Hall.

Mettler, L.E., T.G. Gregg, and H.E. Schaffer. 1988. *Population Genetics and Evolution*, 2nd ed. Englewood Cliffs, NJ: Prentice Hall.

Metz, E.C., R. Robles-Sikisaka, and V.D. Vacquier. 1998. Nonsynonymous substitution in the ablaone sperm fertilization genes exceeds substitution in introns and mitochondrial DNA. *Proc. Natl. Acad. Sci. USA* 95:10676–10681.

Metz, M. and J. Futterer. 2002. Suspect evidence of transgenic contamination. *Nature* 416:600–601.

Michalakis, Y. and L. Excoffier. 1996. A generic estimation of population subdivision using distances between alleles with special reference to microsatellite loci. *Genetics* 142:1061–1064.

Mitton, J.B. 1994. Molecular approaches to population biology. *Ann. Rev. Ecol. Syst.* 25:45–69.

Mitton, J.B. 1997. *Selection in Natural Populations.* New York and Oxford: Oxford University Press.

Miyashita, N. and C.H. Langley. 1988. Molecular and phenotypic variation of the white locus region in *Drosophila melanogaster*. *Genetics* 120:199–212.

Mohri, H., M.K. Singh, W.T.W. Ching, and D.D. Ho. 1993. Quantitation of zidovudine-resistant humaan immunodeficiency virus type 1 in the blood of treated and untreated patients. *Proc. Natl. Acad. Sci. USA* 90:25–29.

Møller, A.P. 1988. Female choice selects for male sexual tail ornaments in the monogamous swallow. *Nature* 332:640–642.

Møller, A.P. 1994. *Sexual Selection and the Barn Swallow.* Oxford: Oxford University Press.

Møller, A.P. and T.A. Mosseau. 2001. Albinism and phenotype of barn swallows *(Hirundo rustica)* from Chernobyl. *Evolution* 55:2097–2104.

Monsutti, A. and N. Perrin. 1999. Dinucleotide microsatellite loci reveal a high selfing rate in the freshwater snail *Physa acuta*. *Mol. Ecol.* 8:1076–1078.

Moorcroft, P.R., S.D. Albon, J.M. Pemberton, I.R. Stevenson, and T.H. Clutton-Brock. 1996. Density-dependent selection in a fluctuating ungulate population. *Proc. Royal Soc. Lond. B* 263:31–38.

Moran, P.A.P. 1967. Unsolved problems in evolutionary biology. *Proc. Fifth Berkeley Symp. Math. Stat. Prob.* 4:457–480.

Morgan, T.H., A.H. Sturdevant, H.J. Muller, and C.B. Bridges. 1915. *The Mechanism of Mendelian Heredity.* New York: Henry Holt and Co.

Moriyama, E.N. and J.R. Powell. 1996. Intraspecific nuclear DNA variation in *Drosophila*. *Mol. Biol. Evol.* 13:261–277.

Motulsky, H. 1995. *Intuitive Biostatistics.* Oxford: Oxford University Press.

Mourant, A.E. 1954. *The Distribution of the Human Blood Groups.* Oxford: Blackwell Scientific Publ.

Mourant, A.E. 1983. *Blood Relations: Blood Groups and Anthropology.* Oxford: Oxford University Press.

Mourant, A.E., A.C. Kopec, and K. Domaniewska-Sobczek. 1976. *The Distribution of the Human Blood Groups and Other Polymorphisms.* Oxford: Oxford Univ. Press.

Mousseau, T.A. and D.A. Roff. 1987. Natural selection and the heritability of fitness components. *Heredity* 59:181–197.

Mueller, U.G. and L. Wolfenbarger. 1999. AFLP genotyping and fingerprinting. *Trends Ecol. Evol.* 14:389–394.

Mukai, T. 1964. The genetic structure of natural populations of *Drosophila melanogaster*. I Spontaneous mutation rate of polygenes controlling viability. *Genetics* 50:1–19.

Mukai, T, S.I. Chigusa, L.E. Mettler, and J.F. Crow. 1972. Mutation rate and dominance of genes affecting viability in *Drosophila melanogaster*. *Genetics* 72:335–355.

Muller, H.J. 1950. Our load of mutations. *Amer. J. Hum. Genet.* 2:111–176.

Muller, H.J. 1964. The relation of recombination to mutational advance. *Mutation Research* 1:2–9.

Myers, K. 1986. Introduced vertebrates of Australia, with emphasis on the mammals. In R.H. Groves and J.J. Burdon (eds.), *Ecology of Biological Invasions*. Cambridge Univ. Press.

Nachman, M.W. and S.L. Crowell. 2000. Estimate of the mutation rate per nucleotide in humans. *Genetics* 156:297–304.

Nachman, M.W., V.L. Bauer, S.L. Crowell, and C.F. Aquadro. 1998. DNA variability and recombination rates at X-linked loci in humans. *Genetics* 150:1133–1141.

National Park Service. 1998. Parks with endangered species: Some special cases. (http://www.nature.nps.gov/wv/espark.htm#griz) Accessed 23 October 1998.

Neel, J.V., C. Satoh, K. Goriki, M. Fujita, N. Takahashi, J. Asakawa, and R. Hazama. 1986. The rate with which spontaneous mutation alters the electrophoretic mobility of polypeptides. *Proc. Natl. Acad. Sci.* 83:389–393.

Nei, M. 1972. Genetic distance between populations. *Amer. Natur.* 106:283–291.

Nei, M. 1973. Analysis of gene diversity in subdivided populations. *Proc. Natl. Acad. Sci. USA* 70:3321–3323.

Nei, M. 1977. F-statistics and analysis of gene diversity in subdivided populations. *Ann. Hum. Genet.* 41:225–233.

Nei, M. 1986. Definition and estimation of fixation indices. *Evolution* 40:643–645.

Nei, M. 1987. *Molecular Evolutionary Genetics*. New York: Columbia University Press.

Nei, M. and R.K. Chesser. 1983. Estimation of fixation indices and gene diversities. *Ann. Hum. Genet.* 47:253–259.

Nei, M. and T. Gojobori. 1986. Simple methods for estimating the numbers of synonymous and nonsynonymous nucleotide substitutions. *Mol. Biol. Evol.* 3:418–426.

Nei, M. and D. Graur. 1984. Extent of protein polymorphism and the neutral mutation theory. *Evol. Biol.* 17:73–118.

Nei, M. and S. Kumar. 2000. *Molecular Evolution and Phylogenetics*. Oxford: Oxford University Press.

Nei, M. and A.K. Roychoudhury. 1982. Genetic relationship and evolution of human races. *Evol. Biol.* 14:1–59.

Nei, M., T. Maruyama, and R. Chakraborty. 1975. The bottleneck effect and genetic variability in populations. *Evolution* 29:1–10.

Neigel, J.E. 1996. Estimation of effective population size and migration parameters from genetic data. In T.B. Smith and R.K. Wayne (eds.), *Molecular Genetic Approaches to Conservation*. New York and Oxford: Oxford University Press, 329–346.

Neigel, J.E. 1997. A comparison of alternative strategies for estimating gene flow from genetic markers. *Ann. Rev. Ecol. Syst.* 28:105–128.

Nevo, E. 1988. Genetic diversity in nature: Patterns and Theory. *Evol. Biol.* 23:217–246.

Nevo, E., A. Beiles, and R. Ben-Shlomo. 1984. The evolutionary significance of genetic diversity: Ecological, demographic and life history correlates. In G.S. Mani (ed.), *Evolutionary Dynamics of Genetic Diversity*. New York: Springer-Verlag.

Newmark, W.D. 1985. Legal and biotic boundaries of western North American national parks: A problem of congruence. *Biol. Conserv.* 33:197–208.

Nielsen, E.E., M.M. Hansen, and V. Loeschcke. 1997. Analysis of microsatellite DNA from old scale samples of Atlantic salmon Salmo salar: a comparison of genetic composition over 60 years. *Mol. Ecol.* 6:487–492.

Nikali, K., A. Suomalainen, J. Terwilliger, T. Koskinen, J. Weissenbach, and L. Peltonen. 1995. Random search for shared chromosomal regions in four affected individuals: The assignment of a new hereditary ataxia locus. *Amer. J. Hum. Genet.* 56:1088–1095.

Nilsson-Ehle, H. 1909. Kreuzungsuntersuchungen an Hafer und Weizen. *Lunds Univ. Arsskrift*, n.s. series 2, 5(2):1–122.

Noor, M.A.F. 1999. Reinforcement and other consequences of sympatry. *Heredity* 83:503–508.

Nowak, M. 1990. HIV mutation rate. *Nature* 347:522.

Nuzhdin, S.V. and T.F.C. Mackay. 1995. The genomic rate of transposable element movement in *Drosophila melanogaster*. *Mol. Biol. Evol.* 12:180–181.

Oakeshott, J.G., J.B. Gibson, P.R. Anderson, W.R. Knibb, D.G. Anderson, and G.K. Chambers. 1982. Alcohol dehydrogenase and glycerol-3-phosphate dehydrogenase clines in *Drosophila melanogaster* on different continents. *Evolution* 36:86–96.

O'Brien, S.J. 1994. The cheeta's conservation controversy. *Conserv. Biol.* 8:1153–1155.

O'Brien, S.J., M.E. Roelke, L. Marker, A. Newman, C.A. Winkler, D. Meltzer, L. Colley, J.F. Evermann, M. Bush, and D.E. Wildt. 1985. Genetic basis for species vulnerability in the cheetah. *Science* 227:1428–1434.

O'Brien, S.J., D.E. Wildt, D. Goldman, C.R. Merril, and M. Bush. 1983. The cheetah is depauperate in genetic variation. *Science* 221:1459–1462.

O'Donald, P. 1959. Possibility of assortative mating in the Arctic Skua. *Nature* 183:1210–1211.

Ohnishi, O. 1977a. Spontaneous and ethyl methane-sulfonate induced mutations controlling viability in *Drosophila melanogaster*. I. Recessive lethal mutations. *Genetics* 87:519–527.

Ohnishi, O. 1977b. Spontaneous and ethyl methane-sulfonate induced mutations controlling viability in *Drosophila melanogaster*. II. Homozygous effect of polygenic mutations. *Genetics* 87:529–545.

Ohnishi, O. 1977c. Spontaneous and ethyl methane-sulfonate induced mutations controlling viability in *Drosophila melanogaster*. III. Heterozygous effect of polygenic mutations. *Genetics* 87:547–556.

Ohno, S. 1970. *Evolution by Gene Duplication*. Berlin: Springer-Verlag.

Ohta, T. 1974. Mutational pressure as the main cause of molecular evolution and polymorphism. *Nature* 252:351–354.

Ohta, T. 1992. The nearly neutral theory of molecular evolution. *Ann. Rev. Ecol. Syst.* 23:263–286.

Ohta, T. 1995. Synonymous and nonsynonymous substitutions in mammalian genes and the nearly neutral theory. *J. Mol. Evol.* 40:56–63.

Ohta, T. 1996. The current significance and standing of neutral and nearly neutral theories. *BioEssays* 18:673–677.

Ohta, T. 1998. Evolution by nearly-neutral mutations. *Genetica* 102/103:83–90.

Ohta, T. and M. Kimura. 1971. On the constancy of the evolutionary rate of cistrons. *J. Mol. Evol.* 1:18–25.

Ohta, T. and M. Kimura. 1973. A model of mutation appropriate to estimate the number of electrophoretically detectable alleles in a finite population. *Genet. Research* 22:201–204.

Olsson, M., R. Shine, and T. Madsen. 1996. Sperm selection by females. *Nature* 383:585.

Orr, H.A. 1990. "Why polyploidy is rarer in animals than in plants" revisited. *Amer. Natur.* 136:759–770.

Otto, S.P. 2000. Detecting the form of selection from DNA sequence data. *Trends in Genet.* 16:526–5529.

Otto, S.P. and N.H. Barton. 2001. Selection for recombination in small populations. *Evolution* 55:1921–1931.

Otto, S.P. and Y. Michalakis. 1998. The evolution of recombination in changing environments. *Trends Ecol. Evol.* 13:145–151.

Otto, S.P. and M.C. Whitlock. 1997. The probability of fixation in populations of changing size. *Genetics* 146:723–733.

Ou, C-Y and 15 others. 1992. Molecular epidemiology of HIV transmission in a dental practice. *Science* 256:1165–1171.

Packer, C., D.A. Gilbert, A.E. Pusey, and S.J. O'Brien. 1991. A molecular genetic analysis of kinship and cooperation in African lions. *Nature* 351:562–565.

Packer, C., L. Herbst, A.E. Pusey, J.D. Bygott, J.P. Hanby, S.J. Cairns, and M. Borgerhoff Mulder. 1988. Reproductive success of lions. In T.H. Clutton-Brock (ed.), *Reproductive success: Studies of Individual Variation in Contrasting Breeding Systems*. Chicago: University of Chicago Press, 263–283.

Paetkau, D., W. Calvert, I. Stirling, and C. Strobeck. 1995. Microsatellite analysis of population structure in Canadian polar bears. *Mol. Ecol.* 4:347–354.

Paetkau, D., G.F. Shields, and C. Strobeck. 1998. Gene flow between insular, coastal and interior populations of brown bears in Alaska. *Mol. Ecol.* 7:1283–1292.

Paetkau, D., L.P. Waits, P.L. Clarkson, L. Craighead, and C. Strobeck. 1997. An empirical evaluation of genetic distance statistics using microsatellite data from bear (Ursidae) populations. *Genetics* 147:1943–1957.

Page, R.D.M. and E.C. Holmes. 1998. *Molecular Evolution: A Phylogenetic Approach*. Oxford: Blackwell Science.

Palmblad, I.G. 1968. Competition in experimental populations of weeds with emphasis on the regulation of population size. *Ecology* 49:26–34.

Papadopoulos, D., D. Schneider, J. Meier-Eiss, W. Arber, R.E. Lenski, and M. Blot. 1999. Genomic evolution during a 10,000-generation experiment with bacteria. *Proc. Natl. Acad. Sci. USA* 96:3807–3812.

Partridge, G.G. 1979. Relative fitness of genotypes in a population of *Rattus norvegicus* polymorphic for warfarin resistance. *Heredity* 43:239–246.

Partridge, L. 1988. The rare-male effect: what is its evolutionary significance? *Phil. Trans. Royal. Soc. Lond.* B 319:525–539.

Paterson, A.H., E.S. Lander, J.D. Hewitt, S. Peterson, S.E. Lincoln, and S.D. Tanksley. 1988. Resolution of quantitative traits into Mendelian factors by using a complete linkage map of restriction fragment length polymorphisms. *Nature* 335: 721–726.

Pearl, R. 1927. The growth of populations. *Quart. Rev. Biol.* 2:532–548.

Pearson, K. 1903. Mathematical contributions to the theory of evolution. XI. On the influence of natural selection on the variability and correlation of organs. *Phil. Trans. Royal Soc. Lond.* A 200:1–66.

Pearson, K. 1904. On a generalized theory of alternative inheritance, with special reference to Mendel's laws. *Phil. Trans. Royal Soc. A* 203:53–86

Peck, S.L., S.P. Ellner, and F. Gould. 1998. A spatially explicit stochastic model demonstrates the feasibility of Wright's shifting balance theory. *Evolution* 52:1834–1839.

Pemberton, J.M., S.D. Albon, F.E. Guinness, and T.H. Clutton-Brock. 1991. Countervailing selection in different fitness components in female red deer. *Evolution* 45:93–103.

Petit, C. and L. Ehrman. 1969. Sexual selection in *Drosophila*. *Evol. Biol.* 3:177–223.

Petrie, M. and B. Kempenaers. 1998. Extra-pair paternity in birds: explaining variation between species and populations. *Trends Ecol. Evol.* 13:52–58.

Phillips, P.C. 1998. The language of gene interaction. *Genetics* 149:1167–1171.

Pimm, S.L. 1991. *The Balance of Nature: Ecological Issues in the Conservation of Species and Communities*. Chicago: Univ. of Chicago Press.

Pimm, S.L., J.L. Jones, and J. Diamond. 1988. On the risk of extinction. *Amer. Natur.* 132:757–785.

Pogson, G.H., C.T. Taggart, K.A. Mesa, and R.G. Boutilier. 2001. Isolation by distance in the Atlantic cod, *Gadus morhua*, at large and small geographic scales. *Evolution* 55:131–146.

Posada, D. and K.A. Crandall. 2001. Evaluation of methods for detecting recombination from DNA sequences: Computer simulations. *Proc. Natl. Acad. Sci. USA* 98:13757–13762.

Posada, D., K.A. Crandall, and E.C. Holmes. 2002. Recombination in evolutionary genomics. *Ann. Rev. Genet.* 36:75–97.

Postgate, J. 1994. *The Outer Reaches of Life*. Cambridge: Cambridge Univ. Press.

Powell, J.R. 1987. "In the air"—Theodosius Dobzhansky's Genetics and the Origin of Species. *Genetics* 117:363–366.

Powell, J.R. 1997. *Progress and Prospects in Evolutionary Biology: The Drosophila Model*. New York and Oxford: Oxford Univ. Press.

Prakash, S., R.C. Lewontin, and J.L. Hubby. 1969. A molecular approach to the study of genic heterozygosity in natural populations. IV. Patterns of genic variation in central, marginal, and isolated populations of *Drosophila pseudoobscura*. *Genetics* 61:841–858.

Pratt, D.M. 1943. Analysis of population development in *Daphnia* at different temperatures. *Biol. Bull.* 85:116–140.

Preziosi, R.F. and D.J. Fairbairn. 1997. Sexual size dimorphism and selection in the wild in the waterstrider *Aquarius remigis*: Lifetime fecundity selection on female total length and its components. *Evolution* 51:467–474.

Price, G.R. 1972. Fisher's "fundamental theorem" made clear. *Ann. Hum. Genet.* 36:129–140.

Price, T. and D. Schluter. 1991. On the low heritability of life-history traits. *Evolution* 45:853–861.

Price, T.D. 1984b. The evolution of sexual size dimorphism in Darwin's finches. *Amer. Natur.* 123:500–518.

Price, T.D., P.R. Grant, H.L. Gibbs, and P.T. Boag. 1984. Recurrent patterns of natural selection in a population of Darwin's finches. *Nature* 309: 787–789.

Primack, R.B. 1993. *Essentials of Conservation Biology*. Sunderland, MA: Sinauer Associates Inc.

Prout, T. 1965. The estimation of fitnesses from genotypic frequencies. *Evolution* 19:546–551.

Prout, T. 1968. Sufficient conditions for multiple niche polymorphism. *Amer. Natur.* 102:493–496.

Prout, T. 1969. The estimation of fitnesses from population data. *Genetics* 63:949–967.

Prout, T. 1971a. The relation between fitness components and population prediction in *Drosophila*. I: The estimation of fitness components. *Genetics* 68:127–149.

Prout, T. 1971b. The relation between fitness components and population prediction in *Drosophila*. II: Population prediction. *Genetics* 68:151–167.

Prout, T. 2000. How well does opposing selection maintain variation? In R.S. Singh and C.B. Costas (eds.), *Evolutionary Genetics:*

From Molecules to Morphology. Cambridge: Cambridge Univ. Press, 157–181.

Prout, T. and J.S.F. Barker. 1989. Ecological aspects of the heritability of body size in *Drosophila* buzzatii. *Genetics* 123:803–813.

Prout, T. and F. McChesney. 1985. Competition among immatures affects their adult fecundity: Population dynamics. *Amer. Natur.* 126:521–558.

Provine, W.B. 1971. *The Origins of Theoretical Population Genetics*. Chicago: University of Chicago Press.

Provine, W.B. 1986. *Sewall Wright and Evolutionary Biology*. Chicago: University of Chicago Press.

Przeworski, M., R.R. Hudson, and A. Di Rienzo. 2000. Adjusting the focus on human variation. *Trends Genet.* 16:296–302.

Quammen, D. 1996. *The Song of the Dodo*. New York: Scribner.

Quist, D. and I.H. Chapela. 2001. Transgenic DNA introgressed into traditional maize landraces in Oaxaca, Mexico. *Nature* 414:541–543.

Quist, D. and I.H. Chapela. 2002. Reply to Metz and Futterer (2002) and Kaplinsky et al. (2002). *Nature* 416:601.

Ralls, K., J.D. Ballou, and A. Templeton. 1988. Estimates of lethal equivalents and the cost of inbreeding in mammals. *Conserv. Biol.* 2:185–193.

Ramshaw, J.A.M., J.A. Coyne, and R.C. Lewontin. 1979. The sensitivity of gel electrophoresis as a detector of genetic variation. *Genetics* 93:1019–1037.

Rannala, B. and J.L. Mountain. 1997. Detecting immigration by using multilocus genotypes. *Proc. Natl. Acad. Sci. USA* 94:9197–9201.

Raymond, M., A. Callaghan, P. Forte, and N. Pasteur. 1991. Worldwide migration of amplified insecticide resistance genes in mosquitoes. *Nature* 350:151–153.

Reed, J.Z., D.J. Tollit, P.M. Thompson, and W. Amos. 1997. Molecular scatology: the use of molecular genetic analysis to assign species, sex, and individual identity to seal faeces. *Mol. Ecol.* 6:225–234.

Reich, D.E., M. Cargill, S. Bolk, J. Ireland, P.C. Sabeti, D.J. Richter, T. Lavery, R. Kouyoumjian, S.F. Farhadian, R. Ward, and E.S. Lander. 2001. Linkage disequilibrium in the human genome. *Nature* 411:199–204.

Reichow, D. and M.J. Smith. 2001. Microsatellites reveal high levels of gene flow among populations of California squid *Loligo opalescens*. *Mol. Ecol.* 10:1101–1109.

Reinartz, J.A. and D.H. Les. 1994. Bottleneck-induced dissolution of self-incompatibility and breeding system consequences in *Aster furcatus* (Asteraceae). *Amer. J. Botany* 81:446–455.

Rendel, J.M. 1943. Variations in the weights of hatched and un-hatched duck's eggs. *Biometrika* 33:48–58.

Renshaw, E. 1991. *Modelling Biological Populations in Space and Time*. Cambridge: Cambridge Univ. Press.

Rice, W.R. 1992. Sexually antagonistic genes: Experimental evidence. *Science* 256:1436–1439.

Rice, W.R. 1996. Sexually antagonistic male adaptation triggered by experimental arrest of female evolution. *Nature* 381:232–234.

Rich, S.S., A.E. Bell, and S.P. Wilson. 1979. Genetic drift in small populations of *Tribolium*. *Evolution* 33:579–584.

Ricklefs, R.E. 1990. *Ecology*. New York: WH Freeman & Co.

Riley, M.A. 1993a. Positive selection for colicin diversity in bacteria. *Mol. Biol. Evol.* 10:1048–1059.

Riley, M.A. 1993b. Molecular mechanisms of colicin evolution. *Mol. Biol. Evol.* 10:1380–1395.

Riley, M.A. 1998. Molecular mechanisms of bacteriocin evolution. *Ann. Rev. Genet.* 32:255–278.

Roberts, R.C. 1966. The limits to artificial selection for body weight in the mouse. II. The genetic nature of the limits. *Genet. Res.* 8:361–375.

Roff, D.A. 1997. *Evolutionary Quantitative Genetics*. New York: Chapman & Hall.

Roff, D.A. and T.A. Mousseau. 1987. Quantitative genetics and fitness: lessons from *Drosophila*. *Heredity* 58:103–118.

Rose, M.R. 1982. Antagonistic pleiotropy, dominance, and genetic variation. *Heredity* 48:63–78.

Rose, M.R. 1985. Life history evolution with antagonistic pleiotropy and overlapping generations. *Theor. Pop. Biol.* 28:342–358.

Ross, K.G. 1997. Multilocus evolution in fire ants: Effects of selection, gene flow, and recombination. *Genetics* 145:961–974.

Rossi, M.S., E. Barrio, A. Latorre, J.E. Quezada-Díaz, E. Hasson, A. Moya, and A. Fontdevila. 1996. The evolutionary history of *Drosophila buzzatii*. XXX. Mitochondrial DNA polymorphism in original and colonizing populations. *Mol. Biol. Evol.* 13:314–323.

Roughgarden, J. 1979. *Theory of Population Genetics and Evolutionary Ecology: An Introduction*. New York: MacMillan Pub. Co.

Roughgarden, J. 1998. *Primer of Ecological Theory*. Upper Saddle River, NJ: Prentice Hall.

Roush, R.T. and J.A. McKenzie. 1987. Ecological genetics of insecticide and acaricide resistance. *Ann. Rev. Entom.* 32:361–380.

Rousset, F. 1997. Genetic differentiation and estimation of gene flow from F-statistics under isolation by distance. *Genetics* 145:1219–1228.

Rousset, F. and M. Raymond. 1995. Testing heterozygote excess and deficiency. *Genetics* 140:1413–1419.

Rousset, F. and M. Raymond. 1997. Statistical analyses of population genetic data: new tools, old concepts. *Trends Ecol. Evol.* 12:313–317.

Russell, E.S. 1949. A quantitative histological study of the pigment found in the coat-color mutants of the house mouse. IV. The nature of the effects of genic substitution in five major allelic series. *Genetics* 34:146–166.

Ryan, M.J. 1998. Sexual selection, receiver biases, and the evolution of sex differences. *Science* 281:1999–2003.

Ryan, M.J. and A. Keddy-Hector. 1992. Directional patterns of female mate choice and the role of sensory biases. *Amer. Natur.* 139:s4–s35.

Ryman, N. 1983. Patterns of distribution of biochemical genetic variation in salmonids: differences between species. *Aquaculture* 33:1–21.

Saccheri, I., M. Kuussaari, M. Kankare, P. Vikman, W. Fortelius, and I. Hanski. 1998. Inbreeding and extinction in a butterfly metapopulation. *Nature* 392: 491–494.

Saccheri, I.J., I.J. Wilson, R.A. Nichols, M.W. Bruford, and P.M. Brakefield. 1999. Inbreeding of bottlenecked butterfly populations: Estimation using the likelihood of changes in marker allele frequencies. *Genetics* 151:1053–1063.

Saether, B.-E. 1997. Environmental stochasticity and population dynamics of large herbivores: a search for mechanisms. *Trends Ecol. Evol.* 12:143–149.

Sager, R. and F.J. Ryan. 1961. *Cell Heredity*. New York; John Wiley & Sons.

Sargent, T.D., C.D. Millar, and D.M. Lambert. 1998. The "classical" explanation of industrial melanism: Assessing the evidence. *Evol. Biol.* 30:299–322.

Sax, K. 1923. The association of size differences with seed-coat pattern and pigmentation in Phaseolus vulgaris. *Genetics* 8:552–560.

Schaeffer, S.W. and E.L. Miller. 1993. Estimates of linkage disequilibrium and the recombination parameter determined from segregating nucleotide sites in the alcohol dehydrogenase region of *Drosophila pseudoobscura*. *Genetics* 135:541–552.

Scherer, S. 1990. The protein molecular clock: Time for a reevaluation. *Evol. Biol.* 24:83–106.

Schierup, M.H. and J. Hein. 2000. Consequences of recombination on traditional phylogenetic analysis. *Genetics* 156:879–891.

Schlager, G. and M.M. Dickie. 1971. Natural mutation rates in the house mouse: Estimates for five specific loci and dominant mutations. *Mutation Research* 11:89–96.

Schluter, D. and J.N.M. Smith. 1986. Natural selection on beak and body size in the song sparrow. *Evolution* 40:221–231.

Schmidt, P.S. and D.M. Rand. 2001. Adaptive maintenance of genetic polymorphism in an intertidal barnacle: Habitat- and life-stage specific survivorship of MPI genotypes. *Evolution* 55:1336–1344.

Schug, M.D., C.M. Hunter, K.A. Wetterstrand, M.S. Gaudette, T.F.C. Mackay, and C.F. Aquadro. 1998. The mutation rates of di-, tri-, and tetranucleotide repeats in *Drosophila melanogaster*. *Mol. Biol. Evol.* 15:1751–1760.

Schwartz, M.K., L.S. Mills, K.S. McKelvey, L.F. Ruggiero, and F.W. Allendorf. 2002. DNA reveals high dispersal synchronizing the population dynamics of Canada lynx. *Nature* 415:520–522.

Seager, R.D. and F.J. Ayala. 1982. Chromosome interactions in *Drosophila melanogaster*. I. Viability studies. *Genetics* 102:467–483.

Seager, R.D., F.J. Ayala, and R.W. Marks. 1982. Chromosome interactions in *Drosophila melanogaster*. II. Total fitness. *Genetics* 102:485–502.

Selander, R.K. and B.R. Levin. 1980. Genetic diversity and structure in *Escherichia coli* populations. *Science* 210:545–547.

Selander, R.K., W.G. Hung, and S.Y. Yang. 1969. Protein polymorphism and genic heterozygosity in two European subspecies of the house mouse. *Evolution* 23:379–390.

Selander, R.K., S.Y. Yang, R.C. Lewontin, and W.E. Johnson. 1970. Genetic variation in the horseshoe crab (*Limulus polyphemus*), a phylogenetic "relic". *Evolution* 24:402–414.

Shaffer, M.L. 1981. Minimum population sizes for species conservation. *Bioscience* 31(2):131–134.

Shaffer, M.L. 1983. Determining minimum viable population sizes for the grizzly bear. *Intl. Conf. Bear Res. Manage.* 5:133–139.

Shaffer, M.L. and F.B. Samson. 1985. Population size and extinction: A note on determining critical population sizes. *Amer. Natur.* 125:144–152.

Sharp, P.M. and W.-H. Li. 1989. On the rate of DNA sequence evolution in *Drosophila*. *J. Mol. Evol.* 28:398–402.

Sharp, P.M., E. Cowe, D.G. Higgins, D.C. Shields, K.H. Wolfe, and F. Wright. 1988. Codon usage patterns in *Escherichia coli*, *Bacillus subtilis*, *Saccharomyces cerevisiae*, *Schizosaccharomyces pombe*, *Drosophila melanogaster* and *Homo sapiens*; a review of the considerable within-species diversity. *Nucleic Acids Res.* 16:8207–8211.

Sharp, P.M., M. Averof, A.T. Lloyd, G. Matassi, and J.F. Peden. 1995. DNA sequence evolution: the sounds of silence. *Phil. Trans. Royal Soc. Lond. B* 349:241–247.

Sheppard, P.M. 1951. Fluctuations in the selective value of certain phenotypes in the polymorphic land snail *Cepaea nemoralis* (L.). *Heredity* 5:125–134.

Sherwin, W.B., N.D. Murray, J.A. Marshall Graves, and P.R. Brown. 1991. Measurement of genetic variation in endangered populations: Bandicoots (Marsupalia: Peramelidae) as an example. *Conserv. Biol.* 5:103–108.

Simberloff, D. 1987. The spotted owl fracas: Mixing academic, applied and political ecology. *Ecology* 68:766–772.

Simmons, M.J. and J.F. Crow. 1977. Mutations affecting fitness in *Drosophila* populations. *Ann. Rev. Genet.* 11:49–78.

Singh, R. 1979. Genic heterogeneity within electrophoretic "alleles" and the pattern of variation among loci in *Drosophila pseudoobscura*. *Genetics* 93:997–1018.

Singh, R.S. and L.R. Rhomberg. 1987a. A comprehensive study of genic variation in natural populations of *Drosophila melanogaster*. I. Estimates of gene flow from rare alleles. *Genetics* 115:313–322.

Singh, R.S. and L.R. Rhomberg. 1987b. A comprehensive study of genic variation in natural populations of *Drosophila melanogaster*. II. Estimeate of heterozygosity and patterns of geographic differentiation. *Genetics* 117:255–271.

Singh, R.S., R.C. Lewontin, and A.A. Felton. 1976. Genetic heterogeneity within electrophoretic "alleles" of xanthine dehydrogenase in *Drosophila pseudoobscura*. *Genetics* 84:609–629.

Slate, J., L.E.B. Kruuk, T.C. Marshall, J.M. Pemberton, and T.H. Clutton-Brock. 2000. Inbreeding depression influences lifetime breeding success in a wild population of red deer (*Cervus elaphus*). *Proc. Royal Soc. Lond. B* 267:1657–1662.

Slatkin, M. 1985a. Rare alleles as indicators of gene flow. *Evolution* 39:53–65.

Slatkin, M. 1985b. Gene flow in natural populations. *Ann. Rev. Ecol. Syst.* 16:393–430.

Slatkin, M. 1993. Isolation by distance in equilibrium and non-equilibrium populations. *Evolution* 47:264–279.

Slatkin, M. 1995. A measure of population subdivision based on microsatellite allele frequencies. *Genetics* 139:457–462.

Slatkin, M. and N.H. Barton. 1989. A comparison of three indirect methods for estimating average levels of gene flow. *Evolution* 43:1349–1368.

Slatkin, M. and W.P. Maddison. 1989. A cladistic measure of gene flow inferred from the phylogenies of alleles. *Genetics* 123:603–613.

Slatkin, M. and W.P. Maddison. 1990. Detecting isolation by distance using phylogenies of genes. *Genetics* 126:249–260.

Smith, C.A.B. 1970. A note on testing the Hardy-Weinberg law. *Ann. Hum. Genet.* 33:377–383.

Smith, J.N.M. and R. Zach. 1979. Heritability of some morphological characters in a song sparrow population. *Evolution* 33:460–467.

Smith, J.N.M., P. Arcese, and W.M. Hochachka. 1991. Social behaviour and population regulation in insular bird populations: implications for conservation. In C.M. Perrins, J.-D. Lebreton, and G.J.M. Hirons (eds.), *Bird population studies: Relevance to conservation*. Oxford: Oxford University Press, 148–167.

Smith, N.G.C. and A. Eyre-Walker. 2002. Adaptive protein evolution in *Drosophila*. *Nature* 415:1022–1024.

Smith, T.B. 1993. Disruptive selection and the genetic basis of bill size polymprphism in the African finch *Pyrenestes*. *Nature* 363:618–620.

Snow, A.A. and T.P. Spira. 1991. Pollen vigour and the potential for sexual selection in plants. *Nature* 352:796–797.

Sokal, R.R. and F.J. Rohlf. 1995. *Biometry*. 3rd ed. New York: W.H. Freeman and Co.

Sorsa, V. 1988. *Chromosome Maps of Drosophila*. Vol. 1. Boca Raton, FL: CRC Press.

Spassky, B., Th. Dobzhansky, and W.W. Anderson. 1965. Genetics of natural populations. XXXVI. Epistatic interactions of the components of the genetic load in *Drosophila pseudoobscura*. *Genetics* 52:653–664.

Spencer, H.G. and R.W. Marks. 1988. The maintenance of single-locus polymorphism. I. Numerical studies of a viability selection model. *Genetics* 120:605–613.

Spencer, H.G. and R.W. Marks. 1992. The maintenance of single-locus polymorphism. IV. Models with mutation from existing alleles. *Genetics* 130:211–221.

Spencer, W.P. 1947. Mutations in wild populations of *Drosophila*. *Adv. Genetics* 1:359–402.

Spencer, W.P. 1957. Genetic studies on *Drosophila mulleri*. I. Genetic analysis of a population. *Univ. Texas Publ.* 5721:186–205.

Spiess, E.B. 1968. Low frequency advantage in mating od *Drosophila pseudoobscura* karyotypes. *Amer. Natur.* 102: 363–379.

Spiess, E.B. 1977. *Genes in Populations*. New York: John Wiley & Sons.

Spiess, E.B. (ed). 1962. *Papers on Animal Population Genetics*. Boston: Little Brown and Co.

Spratt, B.G. 1996. Antibiotic resistance: Counting the cost. *Current Biol.* 6:1219–1221.

Stadler, L.J. 1942. Some observations on gene variability and spontaneous mutation. *Spragg Mem. Lect.* 3:3–15.

Stebbins, G.L. and R.C. Lewontin. 1971. Comparative evolution at the levels of molecules, organisms, and populations. *Proceedings of the Sixth Berkeley Symposium on Mathematical Statistics and Probability* 5:23–42.

Stephens, J.C. et al. 2001. Haplotype variation and linkage disequilibrium in 313 human genes. *Science* 293:489–494.

Steppan, S.J., P.C. Phillips, and D. Houle. 2002. Comparative quantitative genetics: evolution of the G matrix. *Trends Ecol. Evol.* 17:320–327.

Stern, C. 1973. *Principles of Human Genetics*. San Francisco: W.H. Freeman.

Stevens, M.T., M.G. Turner, G.A. Tuskan, W.H. Romme, L.E. Gunter, and D.M. Waller. 1999. Genetic variation in postfire aspen seedlings in Yellowstone National Park. *Mol. Ecol.* 8:1769–1780.

Stewart, C.-B. and A.C. Wilson. 1987. Sequence convergence and functional adaptation of stomach lysozymes from foregut fermenters. *Cold Spring Harbor Symposia on Quantitative Biology* 52:891–899.

Stewart, C.-B., J.W. Schilling, and A.C. Wilson. 1987. Adaptive evolution in the stomach lysozymes of foregut fermenters. *Nature* 330:401–404.

Stibůrková, B., J. Majewski, I. Šebesta, W. Zhang, J. Ott, and S. Kmoch. 2000. Familial juvenile hyperuricemic nephropathy: Localization of the gene on chromosome 16p11.2-and evidence for genetic heterogeneity. *Amer. J. Hum. Genet.* 66:1989–1994.

Stockley, P. 1997. Sexual conflict resulting from adaptations to sperm competition. *Trends Ecol. Evol.* 12:154–159.

Storfer, A. 1996. Quantitative genetics: a promising approach for the assessment of genetic variation in endangered species. *Trends Ecol. Evol.* 8:343–348.

Strachan, T. and A.P. Read. 1999. *Human Molecular Genetics*, 2nd ed. New York: Wiley-Liss.

Strickberger, M. 2000. *Evolution*, 3rd ed. Sudbury, MA: Jones and Bartlett Publishers.

Sunnocks, P. 2000. Efficient genetic markers for population biology. *Trends Ecol. Evol.* 15:199–203.

Sutherland, G.R. and R.I. Richards. 1995. Simple tandem DNA repeats and human genetic disease. *Proc. Natl. Acad. Sci.* 92:3636–3641.

Swanson, W.J., A.G. Clark, H.M. Waldrip-Dail, M.F. Wolfner, and C.F. Aquadro. 2001. Evolutionary EST analysis identifies rapidly evolving male reproductive proteins in *Drosophila*. *Proc. Natl. Acad. Sci. USA* 98:7375–7379.

Swofford, D.L. 2000. PAUP*. Phylogenetic Analysis Using Parsimony (*and other methods). Version 4. Sunderland, MA: Sinauer Associates.

Swofford, D.L., G.J. Olsen, P.J. Waddell, and D.M. Hillis. 1996. Phylogenetic inference. In D.M. Hillis, C. Moritz, and B.K. Mable (eds.), *Molecular Systematics*, 2nd ed. Sunderland, MA: Sinauer Associates, 407–514.

Tajima, F. 1983. Evolutionary relationship of DNA sequences in finite populations. *Genetics* 105:437–460.

Tajima, F. 1989. The effect of change in population size on DNA polymorphism. *Genetics* 123:597–601.

Tan, C.C. 1946. Mosaic dominance in the inheritance of color patterns in the lady-bird beetle, Harmonia axyridis. *Genetics* 31:195–210.

Tanaka, T. and M. Nei. 1989. Positive Darwinian selection observed at the variable-region genes of immunoglobulins. *Mol. Biol. Evol.* 6:447–459.

Tanksley, S.D. 1993. Mapping polygenes. *Ann. Rev. Genetics* 27:205–233.

Taylor, C.E. and C. Condra. 1980. r- and K-selection in *Drosophila pseudoobscura*. *Evolution* 34: 1183–1193.

Templeton, A.R. 1991. Human origins and analysis of mitochondrial DNA sequences. *Science* 255:737.

Templeton, A.R. 1997. Coadaptation, local adaptation, and outbreeding depression. In G.K. Meffe, C.R. Carroll, et al. (eds.), *Principles of Conservation Biology*, 2nd ed. Sunderland, MA: Sinauer Associates, Inc., 171–172.

Templeton, A.R. 1998. Nested clade analyses of phylogeographic data: testing hypotheses about gene flow and population history. *Mol. Ecol.* 7:381–397.

Templeton, A.R. 2002. Out of Africa again and again. *Nature* 416:45–51.

Templeton, A.R. and B. Read. 1983. The elimination of inbreeding depression in a captive herd of Speke's gazelle. In C.M. Schonewald-Cox, S.M. Chambers, B. MacBryde, and W.L. Thomas (eds.), *Genetics and Conservation: A Reference for Managing Wild Animal and Plant Populations*. Menlo Park, CA: Benjamin/Cummings, 241–261.

Templeton, A. and B. Read. 1984. Factors eliminating inbreeding depression in a captive herd of Speke's gazelle (*Gazella spekei*). *Zoo Biology* 3:177–199.

Thoday, J.M. 1961. Location of polygenes. *Nature* 191:368–370.

Thomson, G. and M.S. Esposito. 2000. The genetics of complex diseases. *Trends in Genetics* 15(12):M17–M20.

Thuillet, A-C, D. Bru, J. David, P. Roumet, S. Santoni, and P. Sourdille. 2002. Direct estimation of mutation rate for 10 microsatellite loci in Durum wheat, *Triticum turgidum* (L.) Thell. ssp *durum* desf. *Mol. Biol. Evol.* 19:122–125.

Tishkoff, S.A. et al. 1996. Global patterns of linkage disequilibrium at the CD4 locus and modern human origins. *Science* 271:1380–1387.

Turelli, M. 1984. Heritable genetic variation via mutation-selection balance: Lerch's zeta meets the abdominal bristle. *Theor. Pop. Biol.* 25:138–193.

Turelli, M. 1985. Effects of pleiotropy on predictions concerning mutation-selection balance for polygenic traits. *Genetics* 111:165–195.

Turelli, M., N.H. Barton, and J.A. Coyne. 2001. Theory and speciation. *Trends Ecol. Evol.* 16:330–343.

Vacquier, V.D., W.J. Swanson, and Y-H Lee. 1997. Positive Darwinian selection on two homologous fertilization proteins: What is the selective pressure driving their divergence? *J. Mol. Evol.* 44(Suppl):S15–S22.

Vanschothorst, E.M., J.C. Jansen, E. Grooters, D.E.M. Prins, L.J. Wiersinga, A.G.L. Vandermey, G.J.B. Vanommen, P. Devilee, and C.J. Cornelisse. 1998. Founder effect at PGL1 in hereditary head and neck paraganglioma families from The Netherlands. *Amer. J. Human Genet.* 63:468–473.

Vassilieva, L.L., A.M. Hook, and M. Lynch. 2000. The fitness effects of spontaneous mutations in *Caenorhabditis elegans. Evolution* 54:1234–1246.

Via, S. 2001. Sympatric speciation in animals: The ugly duckling grows up. *Trends Ecol. Evol.* 16:381–390.

Viard, F., F. Justy, and P. Jarne. 1997. The influence of self-fertilization and population dynamics on the genetic structure of subdivided populations: A case study using microsatellite markers in the freshwater snail *Bulinus truncatus. Evolution* 51:1518–1528.

Vigilant, L., M. Hofreiter, H. Siedel, and C. Boesch. 2001. Paternity and relatedness in wild chimpanzee communities. *Proc. Natl. Acad. Sci. USA* 98:12890–12895.

Vigilant, L., M. Stoneking, H. Harpending, K. Hawkes, and A.C. Wilson. 1991. African populations and the evolution of human mitochondrial DNA. *Science* 253:1503–1507.

Vigue, C.L. and F.M. Johnson. 1973. Isozyme variability of the genus *Drosophila.* VI. Frequency-property-environment relationships of allelic alcohol dehydrogenases in *D. melanogaster. Biochem. Genet.* 9:213–227.

Voelker, R.A., C.H. Langley, A.J.L. Brown, S. Ohnishi, B. Dickson, E. Montgomery, and S.C. Smith. 1980a. Enzyme null alleles in natural populations of *Drosophila melanogaster.* Frequencies in a North Carolina population. *Proc. Natl. Acad. Sci. USA* 77:1091–1095.

Voelker, R.A., H.E. Schaffer, and T. Mukai. 1980b. Spontaneous allozyme mutations in *Drosophila melanogaster.* Rate of occurrence and nature of the mutants. *Genetics* 94:961–968.

Waage, J.K. 1984. Sperm competition and the evolution of Odonate mating systems. In R.L. Smith (ed.), *Sperm Competition and the Evolution of Animal Mating Systems.* New York: Academic Press, 251–290.

Wade, M.J. and C.J. Goodnight. 1998. The theories of Fisher and Wright in the context of metapopulations: when nature does many small experiments. *Evolution* 52:1537–1553.

Wade, M.J., R.G. Winther, A.F. Agrawal, and C.J. Goodnight. 2001. Alternative definitions of epistasis: dependence and interaction. *Trends Ecol. Evol.* 16:498–504.

Wahlund, S. 1928. Zusammensetzung von populationen und korrelationserscheinungen vom standpunkt der vererbungslehre aus betrachtet. *Hereditas* 11:65–106.

Wallace, B. 1958a. The role of heterozygosity in *Drosophila* populations. *Proc. 10th Intl. Congr. Genet.* 1:408–419.

Wallace, B. 1958b. The comparison of observed and calculated zygotic distributions. *Evolution* 12:113–115.

Wallace, B. 1963. The elimination of an autosomal lethal from an experimental population of *Drosophila melanogaster. Amer. Natur.* 97:65–66.

Wallace, B. 1968. Polymorphism, population size, and genetic load. In R.C. Lewontin (ed.), *Population Biology and Evolution.* Syracuse, NY: Syracuse University Press, 87–108.

Wallace, B. 1968. *Topics in Population Genetics.* New York: W.W. Norton and Co. Inc.

Wallace, B. 1981. *Basic Population Genetics.* New York: Columbia University Press.

Waples, R.S. 1998. Separating the wheat from the chaff: patterns of genetic differentiation in high gene flow species. *J. Hered.* 89:438–450.

Waser, P.M. and C. Strobeck. 1998. Genetic signatures of interpopulation dispersal. *Trends Ecol. Evol.* 13:43–44.

Watt, W.B. 1977. Adaptation at specific loci. I. Natural selection on phosphoglucose isomerase of *Colias* butterflies: Biochemical and population aspects. *Genetics* 87:177–194.

Watt, W.B. 1983. Adaptation at specific loci. II. Demographic and biochemical elements in the maintenance of PGI polymorphism. *Genetics* 103:691–724.

Watt, W.B., R.C. Cassin, and M.S. Swan. 1983. Adaptation at specific loci. III. Field behavior and survivorship differences among *Colias* PGI genotypes are predictable from in vitro biochemistry. *Genetics* 103:725–739.

Watt, W.B., P.A. Carter, and S.M. Blower. 1985. Adaptation at specific loci. IV. Differential mating success among glycolytic allozyme genotypes of *Colias* butterflies. *Genetics* 109:157–175.

Watt, W.B., K. Donohue, and P.A. Carter. 1996. Adaptation at specific loci. VI. Divergence vs. adaptation of polymorphic allozymes in molecular function and fitness-component effects among *Colias* species (Lepidoptera, Pieridae). *Mol. Biol. Evol.* 13:699–709.

Watterson, G.A. 1975. On the number of segregating sites in genetical models without recombination. *Theor. Pop. Biol.* 7:256–276.

Watterson, G.A. 1978a. An analysis of multi-allelic data. *Genetics* 88:171–179.

Watterson, G.A. 1978b. The homozygosity test of neutrality. *Genetics* 88:405–417.

Wayne, M.L. and K.L. Simonsen. 1998. Statistical tests of neutrality in the age of weak selection. *Trends Ecol. Evol.* 13:236–240.

Weaver, R.F. and P.W. Hedrick. 1992. *Genetics,* 2nd ed. Dubuque, IA: Wm. C. Brown Publishers.

Weaver, R.F. and P.W. Hedrick. 1997. *Genetics,* 3rd ed. Dubuque, IA: Wm. C. Brown.

Weber, J.L. and C. Wong. 1993. Mutation of human short tandem repeats. *Hum. Mol. Genet.* 2:1123–1128.

Wehausen, J.D. 1999. Rapid extinction of mountain sheep populations revisited. *Conserv. Biol.* 13:378–384.

Weibust, R.S. 1973. Inheritance of plasma cholesterol levels in mice. *Genetics* 73:303–312.

Weinberg, W. [1908] 1963. Ueber den nachweis der vererbung beim Menschen. *Jahreshefte des vereins fur vaterlandische naturkunde in Wurttemburg* 64:368–382. Translated and reprinted in S.H. Boyer, *Papers on Human Genetics.* Englewood Cliffs, NJ: Prentice Hall, 4–15.

Weir, B.S. 1996. *Genetic Data Analysis II.* Sunderland, MA: Sinauer Associates, Inc.

Weir, B.S. and C.C. Cockerham. 1984. Estimating *F*-statistics for the analysis of population structure. *Evolution* 38:1358–1370.

Weiss, K.M. and A.G. Clark. 2002. Linkage disequilibrium and the mapping of complex human traits. *Trends in Genet.* 18:19–24.

Weitcamp, L.R., T. Arends, M.L. Gallango, J.V. Neel, J. Schultz, and D.C. Shreffler. 1972. The genetic structure of a tribal population, the Yanomama Indians. III. Seven serum protein systems. *Annal. Human Genet.* 35:271–279.

Westemeier, R.L., J.D. Brawn, S.A. Simpson, T.L. Esker, R.W. Jansen, J.W. Walk, E.L. Kershner, J.L. Bouzat, and K.N. Paige. 1998. Tracking the long-term decline and recovery of an isolated population. *Science* 282:1695–1698.

Westneat, D. 1990. Genetic parentage in Indigo Buntings: a study using DNA fingerprinting. *Behav. Ecol. Sociobiol.* 27:67–76.

Whitlock, M.C. and D.E. McCauley. 1999. Indirect measures of gene flow and migration: $F_{ST} \neq 1/(4Nm + 1)$. *Heredity* 82:117–125.

Whittam, T.S., H. Ochman, and R.K. Selander. 1983. Geographic components of linkage disequilibrium in natural populations of *E. coli. Mol. Biol. Evol.* 1:67–83.

Willis, J.H., J.A. Coyne, and M. Kirkpatrick. 1991. Can one predict the evolution of quantitative characters without genetics? *Evolution* 45:441–444.

Wilson, D.S. and M. Turelli. 1986. Stable underdominance and the evolutionary invasion of empty niches. *Amer. Natur.* 127:835–850.

Wilson, E.O. 1992. *The diversity of life.* New York: WW Norton.

Wirtz, P. 1997. Sperm selection by females. *Trends Ecol. Evol.* 12:172–173.

Woese, C.R. 1996. Whither microbiology. *Curr. Biol.* 6: 1060–1063.

Wood, R.J. 1981. Insecticide resistance: Genes and mechanisms. In J.A. Bishop and L.M. Cook (eds.) *Genetic Consequences of Man Made Change.* New York and London: Academic Press, 53–96.

Wood, R.J. and J.A. Bishop. 1981. Insecticide resistance: Populations and evolution. In J.A. Bishop and L.M. Cook (eds.) *Genetic Consequences of Man Made Change.* New York and London: Academic Press, 97–127.

Workman, P.L. 1969. The analysis of simple genetic polymorphisms. *Human Biology* 41:97–114.

Wright, S. 1921a. Systems of mating. I. The biometric relations between parent and offspring. *Genetics* 6:111–123.

Wright, S. 1921b. Systems of mating. II. The effects of inbreeding on the genetic composition of a population. *Genetics* 6:124–143.

Wright, S. 1921c. Systems of mating. III. Assortative mating based on somatic resemblance. *Genetics* 6:144–161.

Wright, S. 1921d. Systems of mating. IV. The effects of selection. *Genetics* 6:162–166.

Wright, S. 1921e. Systems of mating. V. General considerations. *Genetics* 6:167–178.

Wright, S. 1922. Coefficients of inbreeding and relationship. *Amer. Natur.* 56:330–338.

Wright, S. 1931. Evolution in Mendelian populations. *Genetics* 16:97–159.

Wright, S. 1932. The roles of mutation, inbreeding, crossbreeding and selection in evolution. *Proc. 6th Intl. Congr. Genetics* 1:356–366.

Wright, S. 1934. Physiological and evolutionary theories of dominance. *Amer. Natur.* 63:24–53.

Wright, S. 1939. Statistical genetics in relation to evolution. Actualites scientifiques et industrielles 802: *Exposes de Biometrie et de la statistique biologique XIII.* Paris: Hermann et Cie, 5–64. Reprinted in Wright (1986).

Wright, S. 1940. Breeding structure of populations in relation to speciation. *Amer. Natur.* 74:232–248.

Wright, S. 1943. Isolation by distance. *Genetics* 28:114–138.

Wright, S. 1951. The genetical structure of populations. *Annals of Eugenics* 15:323–354.

Wright, S. 1968. *Evolution and the Genetics of Populations I. Genetic and Biometric Foundations.* Chicago: University of Chicago Press.

Wright, S. 1969. *Evolution and the Genetics of Populations. II The Theory of Gene Frequencies.* Chicago: University of Chicago Press.

Wright, S. 1977. *Evolution and the Genetics of Populations III. Experimental Results and Evolutionary Deductions.* Chicago: University of Chicago Press.

Wright, S. 1978. *Evolution and the Genetics of Populations IV: Variability Within and Among Natural Populations.* Chicago: University of Chicago Press.

Wright, S. 1986. *Evolution: Selected Papers.* Edited by W.B. Provine. Chicago: University of Chicago Press.

Wright, S. and Th. Dobzhansky. 1946. Genetics of natural populations. XII. Experimental reproduction of some of the changes cauese by natural selection in certain populations of *Drosophila pseudoobscura. Genetics* 31:125–156.

Wyttenbach, A., J. Goudet, J-M Cournuet, and J. Husser. 1999. Microsatellite variation reveals low genetic subdivision in a chromosome race of *Sorex araneus* (Mammalia, Insectivora). *J. Hered.* 90:323–327.

Yan, G., D.D. Chadee, and D.W. Severson. 1998. Evidence for genetic hitchhiking effect associated with insecticide resistance in *Aedes aegypti. Genetics* 148:793–800.

Yang, Z. and B. Rannala. 1997. Bayesian phylogenetic inference using DNA sequences: A Markov Chain Monte Carlo method. *Mol. Biol. Evol.* 14:717–724.

Yang, Z. and J.P. Bielawski. 2000. Statistical methods for detecting molecular adaptation. *Trends Ecol. Evol.* 15: 496–503.

Young, A.G., A.H.D. Brown, and F.A. Zich. 1999. Genetic structure of fragmented populations of the endangered daisy *Rutidosis leptorrhynchoides. Conserv. Biol.* 13:256–265.

Young, D.L., M.W. Allard, J.A. Moreno, M.M. Miyamoto, C.R. Ruiz, and R.A. Perez-Rivera. 1998. DNA fingerprint variation and reproductive fitness in the plain pigeon. *Conserv. Biol.* 12:225–227.

Yuhki, N. and S.J. O'Brien. 1990. DNA variation of the mammalian major histocompatibility complex reflects genomic diversity and population history. *Proc. Natl. Acad. Sci. USA* 87:836–840.

Yule, G.U. 1902. Mendel's laws and their probably relations to intra-racial heredity. *New Phytologist* 1:193–207, 222–238.

Yule, G.U. 1907. On the theory of inheritance of quantitative compound characters on the basis of Mendel's laws—A preliminary note. *Report of the Third International Conference on Genetics,* 140–142.

Zeng, L-W, J.M. Comeron, B. Chen, and M. Kreitman. 1998. The molecular clock revisited: the rate of synonymous vs. replacement change in *Drosophila. Genetica* 102/103:369–382.

Zeyl, C. and J.A.G.M. DeVisser. 2001. Estimates of the rate and distribution of fitness effects of spontaneous mutation in *Saccharomyces cerevisiae. Genetics* 157:53–61.

Zink, R.M. 1991. The geography of mitochondrial DNA variation in two sympatric sparrows. *Evolution* 45:329–339.

Zink, R.M. and D.L. Dittman. 1993. Gene flow, refugia, and evolution of geographic variation in the song sparrow (*Melispiza melodia*). *Evolution* 47:717–729.

Zouros, E. and W. Johnson. 1976. Linkage disequilibrium between functionally related enzyme loci of *Drosophila mojavensis. Canad. J. Genet. Cytol.* 18:245–254.

Zuckerkandl, E. and L. Pauling. 1962. Molecular disease, evolution, and genic heterogeneity. In M. Kasha and B. Pullman (eds.), *Horizons in Biochemistry.* New York: Academic Press, 189–223.

Zuckerkandl, E. and L. Pauling. 1965. Evolutionary divergence and convergence in proteins. In V. Bryson and H.J. Vogel (eds.), *Evolving Genes and Proteins.* New York: Academic Press, 97–166.

Index